Control of Cognitive Processes

Attention and Performance

Attention and Performance XIV: Synergies in Experimental Psychology, Artificial Intelligence, and Cognitive Neuroscience, edited by David E. Meyer and Sylvan Kornblum, 1992

Attention and Performance XV: Conscious and Nonconscious Information Processing, edited by Carlo Umiltà and Morris Moscovitch, 1994

Attention and Performance XVI: Information Integration in Perception and Communication, edited by Toshio Inui and James L. McClelland, 1996.

Attention and Performance XVII: Cognitive Regulation of Performance: Interaction of Theory and Application, edited by Daniel Gopher and Asher Koriat, 1998.

Control of Cognitive Processes: Attention and Performance XVIII, edited by Stephen Monsell and Jon Driver, 2000.

Control of Cognitive Processes

Attention and Performance XVIII

edited by Stephen Monsell and Jon Driver

This book is based on the papers presented at the Eighteenth International Symposium on Attention and Performance, held at Cumberland Lodge, The Great Park, Windsor, Berkshire, England, July 12–18, 1998.

A Bradford Book
The MIT Press
Cambridge, Massachusetts
London, England

This book was set in Palatino on 'Quark' by Asco Typesetters, Hong Kong and was printed and bound in the United States of America.

Library of Congress Cataloging-in-Publication Data

International Symposium on Attention and Performance (XVIII : 1998 : Berkshire, England)
Control of cognitive processes : attention and performance XVIII / edited by Stephen Monsell and Jon Driver.
p. cm
"Based on the papers presented at the Eighteenth International Symposium on Attention and Performance, held at Cumberland Lodge, The Great Park, Windsor, Berkshire, England, July 12–18, 1998."—CIP T.p.
ISBN 0-262-13367-9 (hc : alk. paper)
1. Cognition—Congresses. I. Monsell, Stephen. II. Driver, Jon. III. Title.
BF311.157 1998
153.7'33—dc21 00-030732

Contents

Acknowledgments

Financial support for the symposium was provided by the Wellcome Trust, the United States Air Force Office of Scientific Research, and the United States Office of Naval Research. We are grateful for their support.

The symposium took place at Cumberland Lodge, in the peaceful and beautiful surroundings of Windsor Great Park, Berkshire, England. We thank the Principal of Cumberland Lodge, Dr. John Cook, and his staff for looking after us so well. The administrative control functions required during the week of the symposium were largely delegated to an international triumvirate of graduate students, themselves experts on control: Sander Nieuwenhuis (Amsterdam), Glenn Wylie (Oxford), and Nicholas Yeung (Cambridge). There was warm praise for their efforts from the participants, to which we add our own appreciation. We are also grateful to Elaine Funnell of Royal Holloway College for local advice and help, and to Dennis and Jane Driver for child care throughout the meeting.

Preparation for the meeting and compilation of this volume were facilitated by the staff of the Department of Experimental Psychology at the University of Cambridge. Our special thanks to Maureen Staples, Louise White, Ian Cannell, Lynne Astell, Tracey Greaves, David Nicholas, and Robert Fishwick.

All chapters based on papers at the symposium were anonymously reviewed by two other participants and went through an extensive revision process. We are indebted to the reviewers. The Executive Committee and the Advisory Council of the International Association for Attention and Performance provided most useful advice on the organization of the meeting and suggestions of participants.

Anyone inspecting the acknowledgements in past volumes will detect an invariant: the expression of gratitude to Sylvan Kornblum, secretary and treasurer of the association. In due turn, we add our own heartfelt thanks for his help in organizing and raising funds for this meeting. On this occasion, the gratitude is tinged also with regret. Sylvan, who has been secretary and treasurer since the association's founding after the fourth symposium, has decided the time has come to lay down these offices. As the accompanying brief history explains, the Attention and

Performance symposium series has been both long-lived and of exceptional quality and influence. Sylvan's role in these achievements cannot be overstated. From meeting to meeting, he has deftly operated the intricate executive arrangements, rules, and principles devised by the association's founding fathers, among whom he numbered, to keep the symposium series continuously refreshed with new blood and ideas, while maintaining its quality and the continuity of themes. He has nurtured its slender financial base. He has guarded the flame of its spirit. He has been our Moses, and whether the community of information-processing researchers has reached the Promised Land or not, we now have to manage without him. The silver lining to this cloud is that Sylvan has retired just in time for the executive committee to elect him association lecturer (an honor that, though overdue, was previously prohibited by his office) at the nineteenth symposium, whose theme, "Perception and Action," happens to match his own research interests perfectly.

The Attention and Performance Symposia

Since the first was held in the Netherlands in 1966, the Attention and Performance symposia have been a major influence in experimental psychology and related disciplines. Meetings are now held every two years, in a different country. The International Assocation for the Study of Attention and Performance exists solely to run the meetings and publish the papers presented at them. An executive commitee with seven or eight members selects the organizers for each meeting, and develops the program in collaboration with them, with advice on potential participants from an advisory council of up to one hundred members. Participation is by invitation only, but the rules of the association are constructed to ensure participation from a wide range of countries, and a healthy proportion of young researchers, with a substantial number of new participants from meeting to meeting.

Held usually in a somewhat isolated locale, each meeting has four-and-a-half days of papers presented by a maximum of twenty-six speakers. There are up to forty other participants (including current members of the executive committee) who do not present formal papers, but contribute to informal discussions and sometimes present posters. A leading figure in the field is also invited to deliver the "association lecture." There are no parallel sessions, and participants commit themselves to attend all the sessions. Time is available for substantial papers and discussion periods, and of course discussion continues outside the formal sessions. The intensive workshop atmosphere has been one of the major strengths and attractions of these meetings.

Manuscript versions of the papers are reviewed anonymously by other participants and sometimes by external referees and, if accepted, are published in a volume edited by the organizers. The resulting series of volumes has attracted widespread praise: "unfailingly present[s] the best work in the field" (S. Kosslyn, Harvard); "most distinguished series in the field of cognitive psychology" (C. Bundesen, Copenhagen); "held in high esteem throughout the field because of its attention to rigor, quality and scope ... indispensable to anyone who is serious about understanding the current state of the science" (M. Jordan, MIT); "the books are an up-to-the-minute tutorial on topics fundamental to understanding mental processes" (M. Posner, Oregon).

In the early days of the series, when the scientific analysis of human information processing was in its infancy, thematic coherence could be generated merely by gathering together the most active researchers in the field. More recently, experimental psychology has ramified, and cognitive science and cognitive neuroscience have been born. Participation has therefore become interdisciplinary. Neuroscientists, neuropsychologists, and computational modelers have joined the experimental psychologists, and each meeting has focused on a restricted theme under the general heading of "Attention and Performance." Recent themes include the psychology of reading (U.K., 1986), motor representation and control (France, 1988), synergies in experimental psychology, artificial intelligence, and cognitive neuroscience (U.S., 1990), conscious and unconscious processes (Italy, 1992), integration of information (Japan, 1994), and cognitive regulation of performance: interaction of theory and application (Israel, 1996). The Association maintains a Web site at http://go.to/A&P.

Participants

Alan Allport
Department of Experimental Psychology
University of Oxford
alan.allport@psy.ox.ac.uk

Claus Bundesen
Psychological Laboratory
University of Copenhagen
bundesen@axp.psl.ku.dk

Paul Burgess
Department of Psychology
University College London
p.burgess@ucl.ac.uk

Richard Carlson
Department of Psychology
Pennsylvania State University
cvy@psu.edu

Jonathan Cohen
Department of Psychology
Princeton University
jdc@princeton.edu

Mark D'Esposito
Neuroscience Institute and Department of Psychology
University of California, Berkeley
despo@socrates.berkeley.edu

Ritske De Jong
Experimental and Occupational Psychology
University of Groningen
R.de.Jong@ppsw.rug.nl

Roberto Dell'Acqua
Dipartimento di Psicologia
Università di Padova,
dellacqu@ux1.unipd.it

Sergio Della Sala
Department of Psychology
University of Aberdeen
sergio@abdn.ac.uk

Guiseppe Di Pellegrino
Instituto di Psicologia
Università di Urbino
pellegri@b.b.uniurb.it

Jon Driver
Institute of Cognitive Neuroscience
University College London
j.driver@ucl.ac.uk

John Duncan
MRC Cognition and Brain Sciences Unit
Cambridge
john.duncan@mrc-cbu.cam.ac.uk

Martin Eimer
Department of Psychology
Birkbeck College London
m.eimer@bbk.ac.uk

Chris Frith
Institute of Neurology
University College London
cfrith@fil.ion.ucl.ac.uk

Luis Fuentes
Dipartimento Psicologia
Experimental y Psicobiologica
Universidad de Almería
lfuentes@ualm.es

Daniel Gopher
Faculty of Industrial Engineering
and Management
Technion, Haifa
ierbw05@ie.technion.ac.il

Thomas Goschke
Max-Planck-Institut für
Psychologische Forschung
München
goschke@mpipf-muenchen.mpg.de

Bernhard Hommel
Department of Experimental &
Theoretical Psychology
University of Leiden
hommel@fsw.leidenuniv.nl

George Houghton
School of Psychology
University of North Wales, Bangor
g.houghton@bangor.ac.uk

Glyn Humphreys
School of Psychology
University of Birmingham
g.w.humphreys@bham.ac.uk

Toshio Inui
Department of Psychology
University of Kyoto
inui@kupsy.kyoto-u.ac.jp

Richard Ivry
Department of Psychology
University of California, Berkeley
ivry@socrates.berkeley.edu

Pierre Jolicoeur
Psychology Department
University of Waterloo
pjolicoe@watcgl.uwaterloo.ca

Nancy Kanwisher
Department of Brain and
Cognitive Sciences
Massachusetts Institute of
Technology
ngk@psyche.mit.edu

Avi Karni
Department of Neurobiology
Weizmann Institute of Science
abraham.karni@.weizmann. ac.il

Steven Keele
Department of Psychology
University of Oregon
skeele@oregon.uoregon.edu

David Kieras
Department of Electrical
Engineering and Computer
Science
University of Michigan
kieras@eecs.umich.edu

Daniel Kimberg
Department of Neurology
University of Pennsylvania
Medical Center
kimberg@mail.med.upenn.edu

Raymond Klein
Department of Psychology
Dalhousie University
ray.klein@dal.ca

Sylvan Kornblum
Mental Health Research Institute
University of Michigan
kornblum@umich.edu

Arthur Kramer
Department of Psychology
University of Illinois
akramer@s.psych.uiuc.edu

Nilli Lavie
Department of Psychology
University College London
n.lavie@ucl.ac.uk

Gordon Logan
Department of Psychology
University of Illinois
glogan@s.psych.uiuc.edu

George Mangun
Center for Cognitive Neuroscience
Duke University
mangun@duke.edu

Ulrich Mayr
Institut für Psychologie
Universität Potsdam
mayr@rz.uni-potsdam.de

Nachshon Meiran
Department of Behavioral Sciences
Ben-Gurion University of the Negev
nmeiran@bgumail.bgu.ac.il

David Meyer
Department of Psychology
University of Michigan
demeyer@umich.edu

Earl Miller
Department of Brain and Cognitive Sciences
Massachusetts Institute of Technology
ekm@ai.mit.edu

David Milner
Department of Psychology
University of Durham
a.d.milner@durham.ac.uk

Stephen Monsell
School of Psychology
University of Exeter
s.monsell@exeter.ac.uk

Hermann Müller
Institut für Allgemeine Psychologie
Universität Leipzig
muellerh@rz.uni-leipzig.de

Roberto Nicoletti
Instituto di Psicologia
Università di Urbino
r.nicoletti@bib.uniurb.it

Adrian Owen
MRC Cognition and Brain Sciences Unit
Cambridge
adrian.owen@mrc-cbu.cam.ac.uk

Harold Pashler
Department of Psychology
University of California, San Diego
hpashler@ucsd.edu

Michael Petrides
Montreal Neurological Institute and Department of Psychology
McGill University
petrides@ego.psych.mcgill.ca

Molly Potter
Department of Brain and Cognitive Sciences
Massachusetts Institute of Technology
molly@psyche.mit.edu

Wolfgang Prinz
Max-Planck-Institut für Psychologische Forschung
München
kprinz@kf1.mpipf-muenchen.mpg.de

Robert Rafal
School of Psychology
University of North Wales, Bangor
r.rafal@bangor.ac.uk

Jane Riddoch
School of Psychology
University of Birmingham
m.j.riddoch@bham.ac.uk

Trevor Robbins
Department of Experimental Psychology
University of Cambridge
twr2@cus.cam.ac.uk

Ian Robertson
Department of Psychology
Trinity College, Dublin
ian.robertson@tcd.ie

Robert Rogers
Department of Psychiatry
University of Oxford
robert.rogers@psych.ox.ac.uk

David Rosenbaum
Department of Psychology
Pennsylvania State University
dar12@psu.edu

Yves Rossetti
INSERM 94
Bron
rossetti@lyon151.inserm.fr

Walter Schneider
Department of Psychology
University of Pittsburgh
waltsch@vms.cis.pitt.edu

Gregor Schöner
CNRS-CRNC
Marseilles
gregor@lnf.cnrs-mrs.fr

Tim Shallice
Institute of Cognitive Neuroscience
University College London
t.shallice@ucl.ac.uk

Gregory Stevens
Mental Health Research Institute
University of Michigan
gregs@umich.edu

Don Stuss
Rotman Research Institute
North York, Ontario
stuss@psych.utoronto.ca

Jan Theeuwes
Department of Cognitive Psychology
Vrije Universiteit, Amsterdam
j.theeuwes@psy.vu.nl

Steven Tipper
School of Psychology
University of North Wales, Bangor
s.tipper@bangor.ac.uk

Carlo Umiltà
Dipartimento di Psicologia Generale
Università di Padova
umilta@psico.unipd.it

Dirk Vorberg
Institut für Psychologie
Technische Universität Braunschweig
d.vorberg@tu-bs.de

Steven Yantis
Department of Psychology
Johns Hopkins University
yantis@jhu.edu

Key to Group Photograph

1 Ulrich Mayr
2 Martin Eimer
3 George Houghton
4 Sander Nieuwenhuis
5 Adrian Owen
6 Steven Tipper
7 Sergio Della Salla
8 Ian Robertson
9 Walter Schneider
10 Glenn Wylie
11 Bernhard Hommel
12 Dirk Vorberg
13 Hermann Müller
14 Robert Rogers
15 Gregor Schöner
16 Don Stuss
17 Ron Mangun
18 Glyn Humphreys
19 Wolfgang Prinz
20 Carlo Umiltà
21 David Kieras
22 Pierre Jolicoeur
23 Thomas Goschke
24 Roberto Dell'Acqua
25 Harold Pashler
26 Toshio Inui
27 Jan Theeuwes
28 Arthur Kramer
29 Yves Rossetti
30 Tim Shallice
31 Nachshon Meiran
32 Chris Frith
33 Stephen Monsell
34 David Meyer
35 Arvi Karni
36 Ritske De Jong
37 Michael Petrides
38 David Rosenbaum
39 Roberto Nicoletti
40 Trevor Robbins
41 Jane Riddoch
42 Ray Klein
43 Gordon Logan
44 Gregory Stevens
45 Molly Potter
46 Richard Ivry
47 Steven Yantis
48 Richard Carlson
49 Earl Miller
50 Mark D'Esposito
51 David Milner
52 Guiseppe Di Pellegrino
53 Jonathan Cohen
54 John Duncan
55 Paul Burgess
56 Robert Rafal
57 Daniel Gopher
58 Luis Fuentes
59 Sylvan Kornblum
60 Nancy Kanwisher
61 Alan Allport
62 Jon Driver
63 Nilli Lavie
64 Claus Bundesen
65 Steven Keele
66 Daniel Kimberg
67 Nick Yeung
68 Shoni Lavie-Driver

Introduction

1 Banishing the Control Homunculus

Stephen Monsell and Jon Driver

ABSTRACT We define the problem addressed at the eighteenth Attention and Performance symposium as that of explaining how voluntary control is exerted over the organization and activation of cognitive processes in accordance with current goals, without appealing to an all-powerful but ill-defined "executive" or controlling "homunculus." We provide background to the issues and approaches represented in the seven parts of the volume and review each chapter, mentioning also some other contributions made at the symposium. We identify themes and controversies that recur through the volume: the multiplicity of control functions that must be invoked to explain performance even of simple tasks, the limits of endogenous control in interaction with exogenous influences and habits, the emergence of control through top-down "sculpting" of reflexive procedures, the debate between structural and strategic accounts of capacity limits, the roles of inhibition and working memory, the fertile interactions between functional and neural levels of analysis. We identify important control issues omitted from the symposium. We argue that progress is at last being made in banishing—or fractionating—the control homunculus.

When we invited the psychologists and neuroscientists whose contributions constitute this volume to speak at the Eighteenth International Symposium on Attention and Performance, we declared the theme of the meeting to be "Control of Cognitive Processes: Banishing the Homunculus." We took the provocative subtitle from a contribution by the late Alan Newell to the eighth symposium:

> A major item on the agenda of cognitive psychology is to banish the homunculus (i.e., the assumption of an intelligent agent (little man) residing elsewhere in the system, usually off stage, who does all the marvelous things that need to be done actually to generate the total behavior of the subject). It is the homunculus that actually performs the control processes in Atkinson and Shiffrin's (1968) famous memory model, who still does all the controlled processing (including determining the strategies) in the more recent proposal of Shiffrin and Schneider (1977), who makes all the confidence judgments, who analyses all the payoff matrices and adjusts the behavior appropriately, who is renamed the "executive" in many models (clearly a promotion).... (Newell 1980, 715)

The eighteenth symposium, twenty years later, seemed a suitable occasion to take stock of progress on this agenda item. On the one hand, it is our impression that, far from leading the furtive life of a fugitive, the homunculus has continued to parade about in broad daylight, its powers largely intact and indeed dignified by even grander titles—not merely the "executive" but the "central executive," or the "supervisory attention system," or the "anterior attention system"—and flagrantly laying claim to prime real estate in the frontal lobes. On the other hand, there has been a substantial increase in research by neuroscientists, neuropsychologists, and experimental psychologists on "executive" functions, and on interactions between endogenous (voluntary) and exogenous (stimulus-driven) control of cognitive processes. We may now have a sufficient database for a serious attack on the problem to which the control homunculus has been the default solution.

1.1 THE PROBLEM OF VOLUNTARY CONTROL

Viewed from a subjective perspective, the problem of control is as old as philosophical speculation about the nature of our mental faculties. We feel able to exercise voluntary control over our thoughts and behavior, yet we also experience limitations to that control: we sometimes feel at the mercy of habits, impulses, compulsions, obsessions, or reflexes; we do things we apparently did not intend to do and leave undone things we intended to do. Can the seemingly voluntary aspects of our mental life be given the same sort of mechanistic causal explanation that we happily apply to the more reflexive aspects? If they can—if our exercise of voluntary control is no less determined than our reflexes—then in what sense do we have free will?

Posed in these subjective terms, the problem of control carries with it considerable philosophical, theological, and moral baggage. The computational theory of mind that now underlies cognitive psychology and neuroscience provides us with a more objective perspective from which to pose the problem. The mind/brain evidently contains many representations of information (perceptual, semantic, motoric, etc.), and procedures for translating between and transforming those representations. The performance of any one cognitive skill involves only a subset of these resources, which must be organized in a particular fashion for the task at hand, and defended against disruption by other influences. Although some processes (such as the earliest stages of sensory processing) may be triggered by appropriate input in an inflexible manner regardless of current goals, others may not. These other processes may have to be flexibly enabled or disabled, connected or disconnected, tuned, organized, directed, scheduled, and monitored (or some subset of these) to accomplish particular goals. The goals often change from moment to moment. The problem of voluntary control is thus: How are goal-relevant organizations of particular mental resources created or activated, and how are goal-appropriate processes triggered, when they are appropriate, and suppressed when they are inappropriate? And what constrains the flexibility of this deployment?

Although the problem of control is inherent in virtually every task that people can perform, it is often overlooked. As an example, consider the much studied task of reading a fixated word aloud. Typically, successive levels of abstraction from the retinal input are thought to result in a representation of letter identities and their sequence: an abstract orthographic representation. This is followed by translation into an abstract phonological sequence by several parallel pathways—perhaps two or three, depending on how one views proposed separations between assembling an abstract description of the pronunciation through letter-sound "rules," versus retrieving the learned pronunciation of a recognized letter string. Discrepancies between assembled and retrieved pronunciation are somehow reconciled, and the resulting abstract phonological description is then translated into articulatory form and executed, again via several substages. The literature contains many fine-grained analyses of this general sequence.

If you examine recent volumes of, say, the *Journal of Experimental Psychology* or the *Journal of Cognitive Neuroscience*, you cannot but be impressed by the wealth of data, the detailed models, and the critical debate about the overall architecture and substages involved in this single skill of word reading. But ask a slightly different, and rather simple, question, and you will scan the pages in vain. Why did the subjects in all these experiments speak each word aloud, as instructed? At a whim, they could instead have elected to perform innumerable other tasks in response to each word, using some parts of the same mental machinery but other resources as well. Given the same written input, they could readily have performed semantic categorization, letter counting, phoneme counting, free association, translation into French, and numerous other tasks (on many of which there is also an experimental literature). Given current mechanistic models of word reading, why is it that skilled readers do not *always* say a fixated word aloud, and how can they flexibly choose which task to perform on a given word? The same questions can be asked about all the other tasks psychologists have studied intensively. In each case, we may have detailed analyses of the components underlying performance of a given task, but little understanding of how that task rather than another comes to be performed.

Of course, there is an extensive literature on one aspect of the control of reading, concerning the difficulty of control in situations where the conversion from print to phonology seems to happen even though not required, as when naming the ink color of a printed color word (MacLeod 1991; Stroop 1935). That we can have difficulty *not* reading is, however, only part of the problem; that we can perform innumerable alternative tasks at will is just as important. Moreover, in comparison to the sophisticated analysis of, say, the translation from orthography to phonology, theoretical analyses of control seem quite crude, even for cases of control failure. Essentially, these boil down to the dichotomy between "controlled" and "automatic" processes, as in the influential Shiffrin and Schneider (1977) paper to which Newell referred, and its numerous precedents (e.g., James 1890). Like most dichotomies, this has been softened by use, so that "automaticity" may now be seen more as a matter of degree than as an all-or-none state. Nonetheless, the important question that the dichotomy begs, about exactly how any "controlling" is done, still tends to be neglected. Indeed, most theoretical claims in existing work have primarily concerned *what* is controlled (or cannot be controlled), rather than *how* that control is exercised.

1.2 HOMUNCULITIS

To the extent that control problems have been explicitly considered in psychology and neuroscience, until recently it seems to have been assumed that if control is exercised, then there must be a controller. Another common assumption is that the controller responsible for one "controlled" process (rehearsal, say) is likely to be the very same controller that controls another process (rotation of mental images, say, or direction of visual attention). Doubtless the readiness with which this assumption is made has deep roots in our culture, in the Cartesian doctrine of the soul as (singular) director of the material and mechanical brain, and its evolution into the "Will" of nineteenth-century philosophy and psychology (e.g., James 1890). The assumption may also reflect our familiarity with the pyramidal control structure of many social organizations, such as schools, armies, or governments. Even within modern information-processing approaches to the mind, the powerful metaphor of the "operating system" that directs—or at least schedules—program-specific processes in the standard computer (Johnson-Laird 1983) has proved very seductive. Yet the wide world also contains many examples of complex systems that are flexibly controlled without containing anything identifiable as a singular controller (e.g., termite communities). The notion of distributed control also has a long and respectable history in computer science, and some have already made the speculative extrapolation to the computational architecture of the human mind (e.g., Minsky 1985).

In defense of many theorists who talk of the "executive," it might be said that they are often not so much advocating a singular controller as declaring the problem of control to be extrinsic to their current concerns, which lie elsewhere, in the processes being controlled, or the limits of control. Talk of the "executive," then, is just a placeholder for mechanisms unknown. In one influential example, Baddeley and Hitch (1974) proposed a tripartite model of working memory, placing a "central executive" at center stage, flanked by two subsidiary systems, the "articulatory loop" and the "visuospatial sketchpad." Baddeley (1990, 117) later reflected on complaints that the central executive had remained underspecified: "Most of the research in the working memory tradition has tended to concentrate on the subsidiary systems, principally on the grounds that they offer more tractable problems than the central executive, which from time to time has tended to become something of a ragbag for consigning such important but difficult questions as how information from the various slave systems is combined, and how strategies are selected and operated." This may be a sensible strategy for dealing with complexity. We cannot understand every component of the system at once, and everyone is entitled to a ragbag. It is unfortunate, however, that language enforces a choice between singular and plural referential terms: a diagram with a big oval at the center labeled the "central executive" may seem more assertively homuncular than a cloud labeled "unknown executive functions."

Even when we address issues of control directly, to invoke homunculus-like entities may still be a productive strategy if used with sufficient caution. Consider Daniel Dennett's remarks (1978, 124) on the modeling of intentional systems ("intentional" in the philosophical sense of representing beliefs, goals, etc.) in artificial intelligence:

> Homunculus talk is ubiquitous in AI, and almost always illuminating. AI homunculi talk to each other, wrest control from each other, volunteer, subcontract, supervise, and even kill. Homunculi are bogeymen only if they duplicate entirely the talents they are rung in to explain.... If one can get a team or committee of relatively ignorant, narrow-minded, blind homunculi to produce the intelligent behavior of the whole, this is progress. A flow chart is typically the organizational chart of a committee of homunculi (investigators, librarians, accountants, executives); each box specifies a homunculus by prescribing a function *without saying how it is to be accomplished* (one says, in effect: put a little man in there to do the job). If we then look closer at the individual boxes we see that the function of each is accomplished by subdividing it via another flow chart into still smaller, more stupid homunculi. Eventually this nesting of boxes within boxes lands you with homunculi so stupid ... that they can be, as one says, "replaced by a machine." One *discharges* fancy homunculi from one's scheme by organizing armies of such idiots to do the work.

Although Dennett was addressing a somewhat different issue, this may prove a good blueprint for analyzing voluntary control over mental processes. Perhaps our slogan should be, not "Banish the homunculus!", but "Dissolve, deconstruct, or fractionate, the executive! Let a hundred idiots flourish!" Of course, there may still be those who will defend an indissoluble, unitary controller at the heart of the system, against the suggested "army" (or platoon) of "idiots." If however, their argument is based solely on a desire for parsimony in the number of agents proposed, then that parsimony must be fully costed against the concomitant extravagance of attributing multiple powers to a singular controller.

One common motive for proposing a central agency with plenipotentiary powers has been to provide a seat for "consciousness," in deference to the supposedly unitary nature of awareness (see, for example, Carlson and Sohn, chap. 19, this volume). Although processes associated with conscious awareness may well turn out to play an important functional role in cognitive control, we regard questions about the nature, unity, and substrate of consciousness as logically separable from those about the functional architecture and neural substrate of control, and have tried to keep this volume focused on the latter issues. The problem of control seems hard enough without confounding it with an even greater mystery. It is perhaps better first to model control functions from the "outside," and only then to worry about how they relate to what control or lack of control feels like from the "inside."

1.3 THE COMING OF AGE OF RESEARCH ON CONTROL

Theoretical developments often need decisive data, and vice versa. One reason this symposium seemed timely is that a sufficiently rich set of data on control functions is at last beginning to accumulate, across several different areas and disciplines. Research on normal human performance has increasingly concerned itself with issues of control, not only in familiar paradigms (such as the Stroop effect, visual search, and the psychological refractory period), but also in newly invented or rediscovered paradigms, such as task switching and the antisaccade task. As noted earlier, much psychological work has failed to address the control problem directly because it has been concerned primarily with just a single task, such as reading. Work on task switching specifically aims to determine how people reconfigure their cognitive resources, in accordance with arbitrary goals, to deal with stimuli that can afford several possible tasks in the experimental context. The antisaccade task provides an example of a paradigm that artificially and deliberately pits endogenous control against exogenous control to explore their behavioral and neural correlates.

Another powerful engine driving research on control is neuropsychological work on brain-damaged patients. As with the psychological analysis of individual tasks, many neuropsychological studies have overlooked control issues, focusing on how individual cognitive skills (e.g., reading, recognizing objects, reaching, etc.) are affected by brain injury. Recently, however, following observations of "dysexecutive" behavior after damage to the frontal lobes and associated structures (Luria 1966; Shallice 1988), impairments to cognitive control per se have become the focus of much neuropsychological work. Although the behavior of patients suffering such brain damage may be unimpaired on tests of specific perceptual, linguistic, spatial, or motoric functions, their behavior in daily life is often chaotically disorganized and often captured and diverted down task-irrelevant routes by a potent stimulus. Unlike other neuropsychological syndromes such as acquired dyslexia, these difficulties cannot be explained by damage to the standard components of models for particular tasks. Instead, they seem to suggest damage to mechanisms that coordinate these components, though not necessarily to a single central executive.

In the past, such patients have often been studied using batteries of complex clinical tests for "frontal impairment," such as Wisconsin card sorting, that involve many cognitive components, only some of which relate to control. Increasingly, however, paradigms adapted from the "normal" experimental laboratory, together with further custom-designed tests, are being used to isolate particular control demands. Moreover, these tests are being applied to patient groups with increasingly specific types and regions of frontal damage, and to those with lesions in other parts of interconnected neurotransmitter networks, leading to a neuropsychology of control with the potential to document the neural structures associated with particular control deficits.

In neuroscience more generally, there has been a substantial growth of interest in control processes, as part of a shift toward studying higher-level function. One impetus for this was the development of single-cell recording in awake rather than anesthetized animals, making it possible to study the effect of current goal state on neural activity. A further spur has been the increasing sophistication of tracing methods and pharmacological blockades for understanding interactions between "higher" and "lower" areas in network terms. But perhaps the biggest methodological advance has been the development of new technologies for measuring brain activity in humans. Functional neuroimaging can reveal neural activity as people perform any cognitive task, including tasks that exercise control functions. Unlike animals, people can be instructed to perform almost any arbitrary task "at will," with very little practice. By contrast, massive training is often required to get animals to perform tasks of the necessary complexity and contingency; in such cases, there is a danger of observations being restricted to overlearned skills, thus missing the heart of the control problem. Functional neuroimaging has already been used with humans in an effort to pin down specific functions for areas of prefrontal cortex and to characterize their interactions with other cortical and subcortical regions. These developments in neuroscience have led to increasing recognition of the plurality of control functions and the wide distribution of their neural substrate.

These trends in experimental psychology, neuropsychology and neuroscience are amply represented in this volume, which also demonstrates the considerable scope for mutual education on the control theme between these research traditions. As Newell (1973) complained in another celebrated paper, it is the besetting sin of experimental psychology, including the chronometric tradition represented in past Attention and Performance volumes, to become "phenomenon driven"—trapped in minute exploration of paradigm-specific effects. We firmly believe that the way to keep sight of the big picture and thus avoid the trap of paradigm-bound research is discourse and interaction with others using very different approaches to tackle related problems. We also take it as self-evident that experimental psychologists whose primary interest is at the functional level can learn much from appropriate study of the neural substrate. But the relationship between psychological and neuroscientific research must be reciprocal. As the focus in neuroscience shifts from cortical and subcortical regions close to the sensory and motor periphery, to brain activation in so-called association cortex during performance of complex tasks, neuroscience surely needs the sophistication in task analysis—specifying the functional components—that has been developed by several decades of human information-processing research. Where better to promote this two-way interaction than at an Attention and Performance symposium?

Data are not enough, of course, no matter how many different methods are used to collect them. What we would all like is a theory of control, or at least a theoretical framework, at a level above the specific behavioral paradigm or brain region. Although the problem of voluntary control has long been recognized, there have been few theories of control. The most influential of these, proposed by Norman and Shallice (1980, 1986) and further developed by Shallice (1988), was motivated largely by observations of action errors in everyday life (e.g., driving straight to work rather than taking the intended unusual detour for an errand), and of the more extreme but similar behavior seen in "dysexecutive" patients. Such errors seem to result from a stimulus "seizing" control of behavior, against current intention, by evoking a well-established habit or an action schema recently associated with the stimulus. To account for this, Norman and Shallice followed the theory-building strategy (recommended by Dennett) of hiving off from the control homunculus an additional (dumb) layer of control, conceived in production system terms, which they called "contention scheduling." The organization of components of a familiar task was attributed to stored "schemata" activated by appropriate input. Competition between different schemata that might simultaneously be activated by current input, and prohibition of mutually incompatible actions, were mediated by the dumb "contention-scheduling" level of control. Based on competition at this level alone, recently or frequently exercised schemata would tend to dominate (as in action errors or the dysexecutive syndrome) due to their greater competitive strength. For less well established or less recently used schemata to win the competition (as required in relatively novel situations and some experimental tasks), input from a superordinate layer of control—the "supervisory attention system" (SAS)—was assumed to modulate activation levels of schemata according to current goals.

Recognizing that "higher" control processes do not direct domain-specific resources in a hands-on, omniscient manner, but merely modulate or "sculpt" the activation of lower-level schemata organizing those resources, is an important step in our understanding of control mechanisms. It gives due weight to the role in our mental life of relatively automatic routines that can be exogenously triggered by stimuli. Indeed, the theme that controlled behavior may arise from subtle "sculpting" of more automatic response tendencies runs through this volume. Nevertheless, it must be acknowledged that the SAS as originally proposed was a homunculus only marginally reduced in powers. It retained sufficient omniscience to set activations so that lower-level contention scheduling would generally achieve the right outcome, and it was somehow clever enough to assemble and schedule the elements of a novel task (i.e., one for which lower level schemata do not yet exist), to troubleshoot when things went wrong, and to overcome temptation (Shallice 1988). Clearly, further deconstruction of the SAS and of the interaction of its parts with lower-level processes is required, ideally in explicitly computational terms that can be tested in simulations. This seems at last to be happening, and the volume includes illustrative contributions from both production system and connectionist traditions of computational modeling.

1.4 OVERVIEW OF VOLUME

Parts I–IV focus on specific forms of control in particular cognitive domains: control of visual attention (part I), translation between perception and action in the face of competing response tendencies (part II), coordination of simultaneous or closely successive performance of different tasks (part III), and management of successive elements in multistep tasks (part IV). Parts V–VII, although they speak to the particular forms of control described in parts I–IV, are organized around methodologies. Part V illustrates work on control functions using the techniques of neuroscience—anatomy, single-unit recording, lesions, and functional neuroimaging—and focusing on the functions of particular brain regions or circuits; part VI illustrates work on pathological control in neurological and developmental populations; and part VII, work on computational modeling of control functions, with the phenomena modeled ranging from reaction time data to the effects of neurotransmitters. Of course, the interplay among the various methodologies for studying control mechanisms is already sufficiently advanced that these divisions are somewhat arbitrary. For example, experiments with neuropsychological patients also appear in parts I–IV, and electrophysiological and functional neuroimaging research is discussed in part I.

Most contributors to this volume were invited to present their own recent research, and all were encouraged to consider "how" control operates, not merely "what" is controlled. A few agreed to contribute tutorial reviews rather than focus on their own research. Each group of papers presented at the meeting led to a discussion, initiated by a discussant. Several discussants agreed to contribute short commentaries on the field covered by the papers in their group. Some participants who did not present papers at the meeting presented posters or described new data in extended discussion sessions. Without trying to be exhaustive, we have included some mentions of these valuable contributions to the meeting.

At every Attention and Performance symposium it is customary to honor an eminent researcher's distinguished contribution by an invitation to give the Association Lecture. We were fortunate to have as association lecturer Alan Allport, who has both posed and challenged many of the critical questions about control and attention (e.g., Allport 1980, 1989, 1993; Allport, Styles, and Hsieh 1994). In chapter 2, he describes his recent research on task switching, which relates most closely to part III; we shall discuss it under that heading.

Although there are many points of contact among the chapters, some overarching themes are apparent, including

• *the "limits" theme*—the deliberate exploration of cases where our exercise of control is limited, typically through stimuli tending to drive processing irrespective of intentions, or in opposition to them;

• *the "sculpting" theme*—seeing control as top-down modulation of lower-level reflexlike circuitry and "reflexes" as potential building blocks rather than the enemy of control;

• *the "no simple dichotomy" theme*—general dissatisfaction with and superseding of the dichotomy between "controlled" and "automatic" processes;

• *the "multiple control functions" theme*—identifying and distinguishing between distinct control functions: some recruits to the "army of idiots";

• *the "working memory" theme*—recognizing that goal-appropriate processing requires short-term maintenance both of procedural "instructions" and of the information operated on; and

• *the "interdisciplinary convergence" theme*—recognizing that the functional and neural levels of the description of control functions should be complementary.

The papers, posters, and discussions at the meeting also highlighted running controversies about theory or methodology that cut across the topics:

• *Is inhibition necessary?* Do we need inhibitory processes to prevent undesired processes from occurring, or is it sufficient that the appropriate procedure or representation be the most activated, in a purely facilitatory manner?

• *Structural versus strategic bottlenecks.* Are apparent limits on information processing the result of immutable structural constraints on the architecture of the mind/brain, or of strategic choices about how best to deploy and coordinate available resources, or even of motivational limitations?

• *How apt is the operating system metaphor?* Control problems that are trivial for computers may be more challenging for brains, and vice versa.

• *Complex versus simple tasks.* Will we discover more about control by studying performance in complex situations that challenge many control functions or in simplified paradigms that seek to isolate specific control functions.

• *Is prefrontal cortex the control center?* To what extent should control functions be attributed to subcortical centers or regions of cortex other than prefrontal cortex? Indeed, is a search for discrete control "centers" misguided? Are the extensive network circuits that connect them a more appropriate level for analyzing the neural substrate?

• *Explicit versus emergent control?* Is it appropriate to see control systems, whether in prefrontal cortex or elsewhere, as "higher" mechanisms modulating dumber "lower" mechanisms, or is control better seen as an emergent property of interactions between equally dumb domain-specialist modules and organization-specialist modules?

We now provide a brief overview of each section, highlighting the overarching themes and controversies where space permits.

Part I: Control of Visual Attention

Visual attention seems a good model system for introducing our confrontation of control issues. Much is now known about "what" is controlled in this domain, but rather less about "how" such control is exerted. The early stages of vision are well characterized both in psychophysical and in neural terms, and there is good evidence that even these early stages of perception can be modulated to some extent by voluntary attention, in both people and animals. Overt eye movements can be dissociated from covert attention, although these are usually coordinated. The overarching themes and controversies of the volume are evident for both forms of attention. Limits in voluntary control are apparent: under some situations, and in some pathological states, salient stimuli attract attention, gaze, or both, regardless of intention. Inhibitory mechanisms of control have often been invoked to explain phenomena such as inhibition of return, antisaccades, or negative priming. Moreover, there is a long controversy over whether the limits of attentional capacity reflect an inflexible bottleneck or strategic filtering. Finally, the neurophysiology and anatomy of vision and eye movements are perhaps better understood than any other part of the system, and many attention researchers are already combining the research tools of human performance with those of neuroscience.

Yantis (chapter 3) presents a tutorial review of the limits on voluntary visual attention, describing how goal states interact with stimulus factors to determine what will be attended. He reviews the controversy over whether salient features that "pop out" when deliberately searched for likewise attract attention even when task irrelevant, or whether even this apparently early segregation of the visual field is subject to top-down modulation. Theeuwes, Atchley, and Kramer (chapter 4) take up this theme, with a fine-grained analysis of the time course of the interaction between endogenous and exogenous factors, suggesting that initial processing is driven solely by stimulus salience, with top-down modulation developing only later. A poster presented at the meeting by Kramer, Theeuwes, Hahn and Irwin provided further data on attentional capture by irrelevant but salient distractor stimuli: interestingly, while subjects often fixated the distractor and were sometimes aware of its presence, they were sure they never fixated it when attempting a deliberate saccade to the target. The role of strategy in visual search was also addressed in a poster by Müller, Krummenacher, and Heller, on situations where the target could be defined in predictable or unpredictable dimensions (e.g., color or orientation). Evidence for top-down dimension weighting was found, but also for limits in control in the form of a bias toward recently experienced target dimensions.

Rafal, Ro, Ingle, and Machado (chapter 6) focus on saccade preparation and the mechanisms that allow us to modulate the primitive fixation reflex to achieve voluntary control over our visual orienting by appropriate "sculpting" of reflexes. They discuss the neural substrates of these mechanisms and the deficits in eye movement control that can follow neurological damage. Klein and Shore's commentary (chapter 8) compares and contrasts exogenous and endogenous mechanisms for both overt and covert visual orienting in an integrative review.

Hopfinger, Jha, Handy, and Mangun (chapter 5) show that combining the temporal precision of event-related potentials (ERPs) with the spatial precision of functional imaging can reveal how early in the visual system top-down attentional modulation can penetrate: top-down gain modulation is found in extrastriate cortex. (In a poster, Worden and Schneider reported fMRI data suggesting attentional modulation even earlier, in striate cortex.) In addition to their detailed look at "what" is controlled, Hopfinger et al. also provide some preliminary data on the possible control structures. Finally, Lavie (chapter 7) proposes psychological boundary conditions for when such modulation of early sensory processing by top-down attention is possible. She provides a novel answer to the classic controversy of early versus late selection. In her view, perceptual categorization of irrelevant stimuli cannot be prevented unless the processing of relevant stimuli exhausts perceptual capacity. Early selection, as revealed by immunity to irrelevant distractors, is therefore apparent only under conditions of high perceptual load in relevant processing. Perceptual capacity may shrink with aging, with the paradoxical effect that, under some circumstances, the elderly can be less susceptible to distractor effects than the young. Lavie argues that perceptual load should not be equated with task difficulty: a task that loads working memory (hence control functions) rather than perceptual capacity may lead to more distraction rather than less.

Part II: Control of Perception-Action Coupling

One of the most familiar manifestations of "control difficulty" in the human performance laboratory arises in cases where the required response differs from the most natural response to the stimulus, as in the Stroop effect. In some cases, the more compatible, though currently undesired, response (e.g., reading the word) is in some sense just as arbitrary as the required response (color naming), but has been massively overlearned. In other cases, to respond according to the compatible mapping is not only well practiced, but is also assisted by phylogenetically ancient action systems (e.g., those guiding looking or reaching towards an object). In his commentary (chapter 9), Milner reviews evidence for the multiplicity of such systems (a veritable platoon of "idiots") that can transform visuospatial input directly into natural actions, bypassing pathways responsible for perception in the traditional sense. He considers the problems of coordination and integration posed by all these systems. In discussion, Rossetti supplemented the evidence mentioned by Milner, describing striking dissociations in both patients and normal subjects between immediate pointing (under "direct" control?) and somewhat delayed pointing (controlled by considered perception?) to tactile or visual targets.

The impact of direct affordances for action from visuospatial input are studied by Tipper, Howard, and Houghton (chapter 10), who describe findings from a paradigm in which subjects must move eye or hand to a visual target, while ignoring a concurrent visual distractor. Taking a strong position on the disputed need for inhibition in control, they argue that kinematic properties of the eye and hand trajectory reveal not only competition between representations of the actions directly evoked by the two stimuli, but also inhibition of the unwanted action.

Hommel (chapter 11) provides a tutorial review of results from choice reaction time situations in which interference is caused by irrelevant stimuli (as in the "flanker" effects) or stimulus properties (as in the Simon and Stroop effects) when associated with a competing response. His survey integrates these phenomena with stimulus-response compatibility effects. All have been interpreted as indicating difficulty in suppressing activation of an inappropriate response via a relatively direct and automatic pathway. (A different kind of theory was represented in a poster by Stevens and Kornblum, who presented their connectionist model, which locates the interference observed in the flanker paradigm at the stimulus identification level.) Hommel takes to task theorists who hold that these interference effects arise from competition between concurrent "intentional" and "automatic" translation processes. In a revival of Exner's late-nineteenth-century "prepared reflex" concept, he argues that the intentional and automatic components of processing operate at different points in time, with the intentional process (prior to the stimulus) setting the stage for automatic translation when the stimulus arrives—a clear example of the "sculpting" theme.

Part III: Task Switching and Multitask Performance

Dual-task performance has been a frequent theme at Attention and Performance meetings. In one of the most popular paradigms, subjects are required to perform two different reaction time tasks, with the stimuli so close in time that the second stimulus often occurs before a response to the first. The delay in response to the second stimulus when the interval between the stimuli is very short—the "psychological refractory period" (PRP) effect—has traditionally been attributed to a bottleneck in processing: the second task must wait until some critical processing stage of the first is completed (see Pashler 1993 for review). Meyer and Kieras (1997) have argued that the PRP effect may arise instead as the result of strategic control processes: a voluntary organization of processing priorities to ensure that the first stimulus is responded to first. In a poster presented at the symposium, Schumacher, Seymour, Glass, Lauber, and Meyer displayed their evidence that, when given appropriate instructions, subjects achieve almost perfect time-sharing (i.e., no PRP effect) with certain combinations of audiovocal and visuomanual tasks and a moderate amount of practice. But when these subjects are given different instructions about task priorities, a PRP effect appears. Thus the PRP paradigm has recently become particularly relevant to the controversy over strategic and structural bottlenecks.

At the same time, there has been a sudden flurry of research using variants of a "task-switching" paradigm in which subjects perform just one reaction time task at any time for each of a sequence of stimuli, but with the task frequently changing (either predictably or signaled by a cue). The focus of interest is the increased reaction time and error rate on the trial following a switch of task. This "switch cost" might seem to offer an index of the control processes involved in reconnecting and reconfiguring the various modules in our brains, so as to perform one task rather than another given the same input (e.g., naming an object aloud versus classifying or grasping it). It may thus provide a point of attack on the control problem traditionally referred to as "task set."

Both PRP and task-switching paradigms typically involve two choice reaction time tasks and thus require subjects to keep two task sets available. In the PRP case, the tasks may overlap in time, whereas in the task-switching paradigm the task sets must be enabled successively. In the belief that there may be at least some theoretical commonality between these apparently similar domains, we solicited several contributions under a common heading. In his tutorial review (chapter 12), Pashler takes on the difficult integrative task of surveying both paradigms and exploring possible commonalties between the processing limitations they reveal. He argues that the PRP effect cannot be attributed to strategic limitations and that there is a structural bottleneck associated with response selection, speculating also that this may be a special case of a more general principle: only one memory retrieval operation can be carried out at a time. He considers but rejects the notion that the same difficulty in maintaining more than one task-set (or stimulus-response mapping) in an active state is responsible for both the PRP effect and switch costs.

Jolicoeur, Dell'Acqua, and Crebolder (chapter 13) perform a detailed experimental comparison of the PRP effect and a seemingly similar phenomenon known as the "attentional blink" (AB): the decline in the ability to detect a second target in a very rapid stream of stimuli for half a second or so after a first target is detected. Here, too, there have been suggestions, especially by Potter (who presented a poster on the AB at the meeting) and her colleagues, that some instances of the limitation may be due to the need to change task sets. However, Jolicoeur, Dell'Acqua, and Crebolder argue that the PRP effect and the attentional blink reflect similar "bottlenecks" in processing, of structural rather than strategic origin. Ivry and Hazeltine (chapter 17) present experiments following up earlier work with a split-brain patient, which had suggested that despite still exhibiting a PRP effect, the divided brain is not subject to the same response selection bottleneck as an intact brain. Their new experiments, which combine the PRP and task-switching paradigms, suggest that, unlike normal subjects, the commisurotomy patient can maintain two S-R mappings for the same stimuli simultaneously, in separate hemispheres.

Pashler's review stresses one type of account of the switch cost—that it reflects the duration of control processes needed for reconfiguring task set, although some aspects of the reconfiguration may not be possible until after the stimulus. This latter idea provides an account of the "residual cost" (Rogers and Monsell 1995) observed even when subjects have ample time to prepare for a change of tasks. Alan Allport's Association Lecture (chapter 2), coauthored by Glenn Wylie, presents a development of Allport, Styles, and Hsieh's very different theory (1994). The residual cost is attributed to proactive interference with task-specific processing, a positive priming of the now-irrelevant task set through its recent association with the same stimulus or class of stimuli. For certain task pairs, it may also reflect carryover from an earlier trial of inhibition needed then to suppress the now-appropriate task set. Thus associations among stimuli, responses, and task can constrain the efficiency of task switching, even when ample preparation time for the switch is provided.

Other types of evidence that apparently inhibitory priming may contribute to residual task switch costs were presented at the symposium. For example, Goschke (chapter 14), combines the two prevailing views of switch cost, showing that switch costs arise in part from an active preparatory control process, which may be disrupted by certain concurrent tasks. But they may also arise in part when a stimulus affords competing responses, so that inhibition is applied to the irrelevant dimension or stimulus-response (S-R) mapping, and this carries over to the next trial. A poster by Mayr and Keele showed that when subjects must switch from judging dimension A to judging dimension B and then back to judging dimension A, performance is slowed relative to a C-B-A sequence, suggesting that inhibition is applied to a task set (e.g., "Attend to A") when it is abandoned and can persist for at least a few trials.

Another poster, by Monsell, Azuma, Eimer, Le Pelley, and Strafford, while acknowledging that priming from previous trials can slow performance on post-switch trials, showed some data comparing lateralized readiness potentials on switch and nonswitch trials, suggesting that, for at least one task pair, response selection was postponed, rather than merely prolonged, by the need to switch tasks. This implies that the residual cost was in part due to the insertion of an extra (control) process on switch trials. Meiran (chapter 16) partitions switch costs into component processes reflecting separate reconfiguration of a stimulus task set and a response task set.

Hence as data on task-switching costs accumulate, their causation begins to look far from simple. Some data suggest that the duration of active control processes forms one component of switch costs. Other data demonstrate the contribution of passive priming—both positive and negative priming—at the levels of both task sets and individual responses. By the standards of many reaction time (RT) "effects," the switch cost can be substantial (hundreds rather than tens of milliseconds). Thus we should not be surprised if this total is composed of several elements. Logically, too, most instances of task switching seem to require most of the following distinct functions: reorienting perceptual attention; resetting the criteria for classification; readying a response mode, a set of responses within it, or both; enabling or disabling S-R mappings; adjusting criteria for response initiation to balance speed and accuracy appropriately.

Most authors distinguish between a component of the cost of task switching that can be overcome by anticipatory preparation (if time and opportunity permit) and a component that cannot. This distinction is challenged, however, by De Jong (chapter 15), who presents evidence that RT distributions on switch and nonswitch trials can be fit by a model in which costs are attributed to a single "intention-activation" process, but that even with time to prepare, most subjects succeed in engaging this process before the stimulus only on a proportion of trials, due in part to the cognitive effort required. He discusses the necessary compromise between minimizing control effort and maximizing task performance, and shows that the balance can to some extent be manipulated experimentally. Results from the task-switching paradigm in neuropsychological patients are also reported in later parts, by Robbins and Rogers (chapter 21), and by Keele and Rafal (chapter 28).

Clearly, both the classic PRP paradigm and the newer task-switching paradigm are producing research that addresses many of the themes highlighted earlier: the limits to control, the role of inhibition, structural versus strategic bottlenecks, and the multiplicity of control functions. On the other hand, later in the volume, Burgess (chapter 20), and Kieras et al. (chapter 30) argue that the task-switching paradigm puts only a minor load on control processes compared to many multitasking situations in daily life, which require multiple goals to be fulfilled in tasks interleaved over a much longer time span. This may be so. The value of the task-switching paradigm, as for a number of the other paradigms surveyed in this volume, may precisely be that it offers the possibility of *isolating* for study a small subset of the controlling "army of idiots," such as those specifically responsible for reconfiguration of S-R mappings. Other paradigms are needed to tap planning, decision making, monitoring, troubleshooting, managing a goal-subgoal task structure, and a host of other potential control functions, some of which are considered in part IV.

Part IV: Control of Multistep Tasks

Much of the research under the previous three headings concerned simple tasks requiring discrete speeded responses to single events (e.g., classifying an object, or reaching for a target). More complex multistep tasks in daily life (such as cooking a meal or finding a route to a destination) may require additional layers of control. Subgoals need to be established and prioritized, triggers set in prospective memory to initiate subtasks when the conditions for them become ripe, transitions between subtasks managed to avoid capture of behavior by habitual transitions, and so on. The outcomes of each processing step may be have to be matched to intended outcomes, so that troubleshooting can be initiated if sufficient divergence from the goal or subgoal is detected or anticipated.

The natural history of "action slips" made by people in daily life (Norman 1981; Reason 1984) has suggested a number of different kinds of failure in multistep tasks, and the more frequent and pathological slips of "dysexecutive" patients have proved equally illuminating. Schwartz and colleagues (e.g., 1991) pioneered the detailed analysis of errors in familiar multistep tasks, such as making a cup of coffee, by patients with frontal brain damage. Humphreys, Forde, and Francis (chapter 18) describe neuropsychological research in this tradition, and extend it to the performance of normal subjects under dual-task conditions.

The performance of multistep tasks typically places considerable load on "working memory" to maintain representations both of the operations to be performed and of the information to be operated on, raising issues of how external instructions about the structure and content of complex tasks may most readily be assimilated. Carlson and Sohn (chapter 19) present research in which subjects perform multistep numerical and spatial tasks whose sequence is determined by the experimenter. Examining whether it is better for subjects to know the operator or operand in advance, they interpret their data within a "procedural frame" hypothesis derived from a more general theory of cognitive control.

In his commentary (chapter 20), Burgess points to the many control demands of real-life multitasking—the planning and interleaved execution of several multistep tasks—a demand familiar to the busy parent no less than to the fighter pilot or astronaut. Burgess argues that such complex situations may be more amenable to experimentation than is widely supposed, and may tax surprisingly specific brain areas. With Shallice and other colleagues, he has pioneered the study of frontal patients performing everyday tasks of real-world complexity, such as carrying out a series of errands in a busy shopping center. He has also developed simplified laboratory analogues that have considerable diagnostic utility. Burgess reports that if one studies a large range of control-dependent tasks in frontal patients, clusters of associated symptoms emerge, which suggest a particular fractionation of control functions that can be mapped to specific brain regions.

Part V: The Neural Substrate of Control

While focusing on brain mechanisms of control, especially in prefrontal cortex (PFC) and related areas, part V also emphasizes psychological function wherever possible. Robbins and Rogers (chapter 21) present a tutorial review of the anatomy, physiology, and function of "cortico-striatal loops" linking frontal cortex to the striatum and associated subcortical structures. They make it abundantly clear that, contrary to many textbook summaries, PFC cannot be considered in isolation with regard to executive function. They also present convergent evidence from lesion effects in humans and animals and from functional imaging on the role of various structures in the formation, maintenance, and shifting of cognitive set, in new paradigms that isolate specific components of the traditional Wisconsin card-sorting task, and in the task-switching paradigm.

Miller (chapter 22) describes research on single-unit activity in monkey PFC for tasks requiring control of visual attention and task set, analogous to some of the human tasks discussed in earlier parts. When the animals are cued to attend to an object or location in a subsequent display of several objects, prefrontal neurons show activity specific to anticipated objects and locations, maintaining this activity over the interval following the cue. Unlike activity in inferotemporal neurons, PFC activity is maintained in the face of distractors occurring during the interval. These PFC neurons appear to be functioning as (part of) a procedural working memory, maintaining a representation of where to attend, what to attend to, and what to do with the attended information. One intriguing question is how much of this PFC activity depends on extensive training of the animals, although considerable flexibility is nevertheless shown.

Chapters 23 and 24 identify specific control functions of lateral regions of PFC. Petrides (chapter 23) reviews his hypothesis, based on lesion effects in monkeys and humans, and on functional imaging of normal humans, that dorsal and ventral regions are specialized for different working memory functions. He sees dorsolateral PFC as responsible for "monitoring and manipulating" information in working memory, while ventrolateral PFC is specialized for active retrieval of information stored in posterior cortical association regions. In a poster, Owen described fMRI activation during forward and backward digit span tasks that supported a similar contrast between these two lateral frontal regions. Frith (chapter 24) attributes a somewhat different role to dorsolateral PFC. On the basis of functional imaging data indicating activation of this region during tasks requiring subjects to select from among response alternatives, he suggests that dorsolateral PFC selects responses, or response sets, in situations where these responses are otherwise underconstrained, by means of a top-down biasing of populations of cells in more posterior regions that represent particular responses. The recurring theme of control as a "sculpting" process is particularly explicit here. Frith attempts to reconcile his own perspective with that of Petrides, and both agree that many different control processes may be subsumed under the general heading of "working memory," a point to which we return below.

In their commentary (chapter 25), Duncan and Owen sound a caution on the inferences currently being drawn from functional imaging and from comparisons of lesion groups about specialization of function within PFC. They point out that the full double-dissociation design is rarely used, and that inferences in neuroimaging must guard against overinterpreting the locus of the "most active" voxel in particular tasks, when in fact very broad regions of lateral PFC and dorsal anterior cingulate are often activated by several types of increase in cognitive demand. They suggest that the present data justify only rather crude functional distinctions, for example, between the aforementioned regions, on the one hand, and medial and orbital frontal cortex, on the other, the latter being associated with affective and motivational processes.

Part VI: Disorders of Control

Although deficits in control following brain injury or disruption crop up throughout the volume, they form the central theme of part VI. D'Esposito and Postle (chapter 26) provide a further perspective on the role of PFC in working memory, reporting a meta-analysis of studies where patients with focal PFC lesions performed short-term memory tasks; a further behavioral study of patients with head injury and frontal involvement, or Parkinson's disease; and a pharmacological study of brain-injured and normal subjects. They argue for a functional and anatomical dissociation between tasks that require only passive maintenance of information in short-term memory and tasks that require rehearsal and other control processes, attributing the latter to PFC (cf. Petrides, chap. 23, this volume).

Riddoch, Humphreys, and Edwards (chapter 27) present data from patients who have difficulty in suppressing actions triggered via the "direct" pathways between perceptual affordances and motor control discussed in part II by Milner (chapter 9) and by Tipper, Howard, and Houghton (chapter 10). Such patients, who typically have frontal damage or disconnection, may exhibit behaviors such as "anarchic hand syndrome" (where one hand performs object-appropriate actions against the intention of the patient, who may use the other hand to try to suppress this action) or "utilization behavior" (where patients pick up and use the objects before them in schematic ways, such as lighting a match or cutting paper with scissors, even when such actions are quite inappropriate in the current context). Riddoch, Humphreys, and Edwards illustrate how such deficits in control of "afforded actions," which have hitherto been described mainly in informal clinical terms, can be studied experimentally.

Keele and Rafal (chapter 28) present data from patients with damage to left or right PFC in a task-switching paradigm similar to those discussed in part III. They find a deficit in patients with left frontal damage, but unlike Rogers et al. (1998), who found an exaggerated switch cost in patients with left frontal damage in a related but subtly different paradigm, they find that the abnormality remains apparent several trials after a switch. These patients seem to be showing abnormally large proactive interference effects of the type documented in normals by Allport and Wylie (chap. 2, this volume). Keele and Rafal speculate that this is due to deficient inhibition.

Whereas chapters 25–28 concern the effects of acquired lesions, Logan, Schachar, and Tannock (chapter 29) discuss research on a developmental disorder of control—attention deficit hyperactivity disorder. Although the impulsivity, hyperactivity, and inattentiveness of such children may be all too apparent in the classroom and at home, it has been hard to pinpoint the underlying functional deficits. Logan, Schachar, and Tannock describe the development and application to this group of a particular experimental test—the stop signal paradigm—which appears to provide a relatively pure measure of impulse control. Illustrating research on yet another patient group increasingly seen as manifesting control impairments, a poster by Fuentes described abnormalities in inhibition of return and negative priming in schizophrenic patients.

Leading the discussion on part VI, Stuss reviewed several examples of functional dissociations from his long-term study of patients with focal lesions of frontal lobe using variants of traditional clinical tests such as the fluency and Wisconsin card-sorting tests. For example, patients with right dorsolateral lesions were impaired in the fluency test, but those with inferior medial lesions were not. Inferior medial patients showed a tendency to lose set in a variant of the Wisconsin Card-Sorting Test (WCST), when told the relevant dimensions and that the rule would change, while superior medial patients did not. The latter patients, but not the former, showed marked perseveration on the classical version of the test, where they had to detect a change of rule for themselves.

Parts V and VI, together with a few of the earlier chapters, clearly illustrate the developing complementarity between behavioral and neuroscientific approaches to control, as well as revisiting many of the recurring themes and controversies. The presentations led to a lively discussion at the meeting of whether PFC plays the cardinal role in control. The emerging consensus was that although this large brain region clearly plays many vital roles, many other cortical and subcortical structures with which it interacts must also be considered.

Part VII: Computational Modeling of Control

As we noted earlier, a major need is for further development of a theoretical framework within which specific control functions can be modeled. The final part illustrates approaches to modeling control in explicit computational terms. General computational models of cognition have been developed within the production system tradition pioneered by Newell and colleagues, as developed in their SOAR project (Newell 1990; Newell, Rosenbloom, and Laird 1989) and by Anderson in the various generations of ACT* (Anderson 1983). Being global systems that pursue goals, these systems have of necessity to address important control problems, especially in problem-solving contexts—for example, how to escape from an impasse when the achievement of a particular subgoal is blocked. Such models, however, have generally not been aimed at fine-grained modeling of the temporal structure of human information processing studied in the Attention and Performance tradition (see Shallice 1994 for a further critique of SOAR as a model of control).

Kieras and Meyer have recently engaged in an ambitious project to develop a production system architecture they call "executive process interactive control" (EPIC). Its purpose is explicitly to model executive control processes, task-specific processes, and their interaction, and in so doing to account for the detailed chronometry of performance in paradigms like the PRP (Meyer and Kieras 1997) as well as more complex "real-life" multitasking situations such as those of the telephone operator or fighter pilot (Meyer and Kieras 1999). These theorists have taken a strong position on the structural versus strategic bottleneck, with strategic factors being to the fore in their interpretation of the PRP effect.

Kieras, Meyer, Ballas, and Lauber (chapter 30) illustrate EPIC modeling with applications to the task-switching paradigm, the PRP effect, and more complex combinations of two continuous tasks. Based on an analysis of general operating system principles from computer science, they also propose the next step in their project. Hitherto, the achievement has been to model control processes explicitly and to show that this can account for objective performance data, as in the combination of two particular tasks. But thus far, the control processes have been hand-crafted for each paradigm. Now the challenge is to model control processes that are more general in their application, so that they can coordinate and control a number of different task pairs.

Kieras et al. suggest that, as we learn to coordinate a particular pair of tasks, the improvement with practice reflects in part an evolution from control by general-purpose executive routines, to control by a learned set of executive procedures specialized for that particular coordination problem. In essence, they propose to model explicitly, within the EPIC framework, the contents of Norman and Shallice's SAS and schemata (1986), respectively. They argue convincingly that operating system principles from computer science can shed light on many psychological issues, although it remains unclear how literally the parallel should be taken. For example, it turns out that task switching is a relatively trivial operation for most computer operating systems, even though it produces very substantial costs in human performance, including proactive interference effects from previous tasks (cf. Allport and Wylie, chap. 2, and Keele and Rafal, chap. 28, this volume) that would never arise in standard computer architectures.

Braver and Cohen's contribution (chapter 31) comes from a connectionist tradition that seeks to make computational models more brain-like. Their approach has grown out of Cohen, Dunbar, and McClelland's model (1990) of the Stroop effect and Cohen and Servan-Schreiber's attempt (1992) to ground elements of the model in particular brain regions and neurotransmitter systems. In the model, activation by context of a representation of the current task biases processing, so as to achieve information transmission via the appropriate set of S-R associations. The problems addressed by Braver and Cohen are (1) how this task representation can be maintained in the face of other input to prevent irrelevant information from overwriting the short-term memory representation of the task context; and (2) how the system can learn what elements of the context to respond to as task cues. The computational solution is a gating mechanism they identify with interactions between prefrontal cortex neurons and the dopamine system. The recurring themes of control as a sculpting process, and of a critical role for working memory representations of the current task, resonate through this chapter.

Kimberg and Farah's commentary (chapter 32), which closes part VII, makes a single but crucial point on the controversial role of inhibition in modeling control functions. As we have seen in many of the previous chapters, there are numerous phenomena suggesting that neurological patients (as well as subjects with immature brains or developmental disorders, and normal subjects under load or distraction) may lack the ability to overcome the effects of a prepotent response tendency or procedure. The immediate temptation is to model this as impairment of an inhibitory mechanism, often thought to be located in PFC. But, such behavioral "disinhibition" can just as readily be modeled by loss of facilitatory activation of the "working memory" representation of the intended action as by loss of inhibition of the habitual action. Applying Occam's razor, we should deploy an inhibitory mechanism to explain behavioral disinhibition only when there is positive evidence for it.

1.5 SOME OMISSIONS

Although the range of research areas addressed within a symposium must necessarily be limited, we should acknowledge certain omissions. First, like most previous Attention and Performance symposia, ours focused on cognitive processes lasting between a fraction of a second and several seconds, in tasks that are speedily executed, rather than on tasks that fulfill goals over days or years. Only part IV explicitly considers multistep tasks. Moreover, although many of the authors refer to the role of memory for what to do, it is usually memory for what to do when the next stimulus of a particular kind appears within a few trials (i.e., procedural working memory), not what to do tomorrow, or by the end of next week. There is now a substantial body of research on "prospective memory" over these longer time spans (see Brandimonte, Einstein, and McDaniel 1996). The equally extended process of "automation" of a cognitive skill, or combinations of skills, through substantial practice likewise receives rather little analysis here (though see Allport and Wylie, chapter 2; and part VII, this volume)

We have also neglected some important control functions that operate at our chosen timescale. Although there were contributions on rehearsal or "monitoring" in working memory, the meeting did not address the important distinction between "automatic" and "intentional" components of retrieval from long-term memory (see Jacoby 1994), and strategies of retrieval (see Barnes et al. 1999). Another important set of control functions, as Newell (1980) put it in the quotation with which we open this chapter, "make all the confidence judgments, analyze all the payoff matrices and adjust the behavior appropriately." That is, there is the need, in addition to arranging cognitive resources suitably to accomplish a given task, to evaluate performance, detect errors, assess efficiency, and adjust decision and response criteria as appropriate. Relevant research includes that on reaction times following errors, and on the "blunder blip"—error-related negativity in the evoked potential (Gehring et al. 1993), and its possible localization in the anterior cingulate (Holroyd, Dien, and Coles 1998). As Robertson pointed out in discussion, the symposium addressed neither sustained attention nor the interactions between alerting and control (Robertson and Manly forthcoming). How is it, for example, that by "making an effort" we can prevent ourselves, for at least a while, from dropping asleep at the steering wheel when driving at night?

Another major research domain that clearly involves aspects of control, and on which we would have liked to include more is the planning and conduct of complex problem solving, a favorite domain for production system modeling (e.g., Newell 1980). There have been a number of neuropsychological (e.g., Shallice 1988) and neuroimaging (e.g., Baker et al. 1996) studies of problem-solving tasks, such as the "Tower of Hanoi" and the "Tower of London," as well as the beginnings of a mental chronometry of such tasks (Ward and Allport 1997)

Further disorders of control for which we had no space include delusions of control and auditory hallucinations in schizophrenia (analyzed by Frith 1996 as due to loss of the signal conveying intention to act or speak), intrusive thoughts in obsessive-compulsive disorders, and neurodevelopmental conditions such as Tourette's syndrome (see Georgiou, Bradshaw, and Chiu 1996). We largely neglected the effects of aging on cognitive control (see Kramer et al. 1999; Rabbitt 1997) and the normal development of frontal control mechanisms (see Diamond 1990). We also largely neglected individual differences in the ability of normal adults to maintain goals and coordinate multiple tasks, and the relation of these abilities to measures of intelligence (see Duncan, Emslie, and Williams 1996).

Perhaps our most fundamental omission is that while we have tried to focus on how the deployment of cognitive resources is controlled by "goals," little is said in this volume about the source of those goals in the interface between affective and cognitive systems (but see Robbins and Rogers, chap. 21, this volume). Typically, goals are simply provided by experimental instructions or training in laboratory studies, but they presumably derive from motivational states and reward values in the natural world. There has been some recent progress on this neglected topic, including neuropsychological work on the association between loss of affect and inappropriate decision-making in patients with orbitofrontal damage (e.g., Bechara et al. 1998; Damasio, 1996); comparative work on possible motivational bases for "perseverative" errors in different species of monkey (Hauser 1999); and research showing activation of orbitofrontal cortex in evaluative decision making (Rogers et al. forthcoming). Nevertheless, the interface between cognitive control and motivation remains a challenging issue for future research.

1.6 CONCLUSIONS

Would the present volume lead Alan Newell or a like-minded skeptic to think that some progress was at last being made in banishing the control homunculus? We think so. Although the contributions are varied in their mode and level of analysis, a number of encouraging general trends are apparent.

First, there is relatively little sign in these pages of any simple dichotomy between opposed "controlled" and "automatic" processes, save for some nailing down of its coffin lid. There is, instead, gratifying elaboration of the fundamental insight, captured in the Norman and Shallice (1980, 1986) model, of the complex and delicate interactions that are found between endogenous and exogenous control wherever we look, plus some explicit modeling of the functional and neural architecture of these interactions in specific domains, such as control of eye movements. In many cases, reflexes are no longer seen as the defining opposite of control, but as the fundamental building blocks from which controlled cognition can be built, given suitable top-down modulation.

Second, there is evident appreciation of the multiplicity of control functions. Even for a control problem considered relatively simple by some of our theorists (Burgess, Kieras et al.), namely, reconfiguring "task set" between two alternatives, we seem to need to invoke several subfunctions. Researchers are developing experimental paradigms that can dissect and isolate the contribution of these multiple control functions to performance. A similar growing sophistication is apparent in neuroscientific analyses of control, and the potential for combining psychological and neural analyses seems enormous. Although we have nothing yet as formal as a taxonomy of control processes and related neural substrates, it is beginning to seem possible that one could be compiled.

The multiplicity of control functions does not of itself entail a multiplicity of controlling mechanisms (after all, the single central processing unit of a standard computer has many functions). Nevertheless, the progressive fractionation and localization of control subfunctions, through the combination of chronometric performance analyses, neuropsychology, functional imaging, electrophysiology, and neuropharmacology, is surely making the traditional view of a singular controller at the apex of the system hard to sustain. It remains to be seen whether, in due course, the control homunuculus will turn out to have been merely fractionated or completely dissolved. That is, despite the progressive fractionation of executive *function*, it may still turn out to be appropriate to postulate an executive *system* (with interdependent and interacting parts) distinct from the domain-specific resources controlled. Alternatively, it may end up no more appropriate to ascribe functional coherence to all "control" functions (and their neural substrates) than to mechanisms as diverse as those that compute binocular stereopsis and segmentation of speech input into words. It is too early to tell. Either way, a basic lesson from much psychology and neuroscience is that our intuitive notion of a unitary self is largely illusory; we are composites of interacting subsystems, and this seems no less true for our experience of "free will" than for other aspects of mental life. Although the picture remains murky, each new result adds a little light, and we are beginning to discern the identities of some recruits to the army of control "idiots." At the same time, computational modelers writing explicit code to get control jobs done in their simulations are discovering what may be needed to do these jobs, and hence what we should look for in the emerging scene.

Although the picture has many complex details, some simple patterns and generalities are also apparent. Most basically, our capacity for voluntary control over mental processes is not absolute. In many cases, processes are driven in part (or, more rarely, entirely) by salient stimuli, past associations, or both, instead of by our intentions. Moreover, it is now self-evident that to overcome such exogenous triggering, cognitive control requires further "input" to be added endogenously to the computations, in the form of activating some representation of current task goals. This has become apparent for many different situations where a prepotent response tendency has to be overcome. In the chapters of this volume, these situations range from making antisaccades, dissociating covert attention from fixation, Stroop- and Simon-like interference effects, selective reaching, negative priming of concepts, responses, or S-R relationships, to explaining the anarchic hand and utilization behavior in frontal patients, the "A not B" error committed by babies toward the end of their first year, and stop signal failures in children with attention disorders. The evidence of dissociable deficits in these different situations and pathologies suggests that each may involve some unique neural structures at a fine-grained level of analysis. From a broader theoretical perspective, however, all these situations have in common the need to overcome prepotent response tendencies, and control for each may be implemented in computationally similar ways. Indeed, it is quite striking how many independent researchers in this volume propose that activating some form of "working memory" for current task goals may be the critical step.

Of course, numerous traps still lie on the path of progress. We must, for example, be wary of using "working memory" as an explanatory catchall. Clearly, many forms of control require short-term maintenance of procedural directives: where to orient, what the current contingencies are between cue and S-R mapping, what the current operators are, what the current goals and subgoals are in a multistep task, and so on. The PFC neurons studied in monkeys by Miller, and the regions of human PFC studied by Petrides or Frith with functional imaging, or by Rafal and colleagues and by D'Esposito and Postle in lesioned patients, are clearly doing something that might be broadly described under the banner headline of "working memory." However, note that this is "procedural" working memory (i.e., of what to do), rather than the more commonly studied "declarative" working memory for phonological sequences and spatial patterns, now thought to be held in posterior cortical regions. Note also that, just as declarative working memory has multiple levels and components even for language input and output (Monsell 1984), so procedural working memory may also comprise many components. Having chided Baddeley and colleagues for labeling the ragbag at the center of their working memory model *the* "central executive," we should not place a similarly singular rag bag at the center of a model of voluntary control, and label it "procedural working memory," as if that explained everything.

In a show-stopping dramatic monologue on the final evening of our meeting, Ian Robertson suggested that we could now declare the control homunculus extinct, with the few remaining examples of the species having been slain by the heroic efforts of those present. In reality, we suspect that the species will linger on in the pages of some learned journals and in the minds of their writers and readers, if only because its pelt provides such a convenient ragbag. Nevertheless, we hope readers of this volume will agree that the control homunculus is now an endangered species, and that a variegated genus of control "idiots" is beginning to colonize the vacated niches.

NOTE

We thank Tim Shallice for his comments on an earlier draft of this chapter.

REFERENCES

Allport, A. (1980). Attention and performance. In G. Claxton (Ed.), *Cognitive psychology: New directions*, pp. 26–64. London: Routledge and Kegan Paul.

Allport, D. A. (1989). Visual attention. In M. I. Posner (Ed.), *Foundations of cognitive science*, pp. 631–682. Cambridge, MA: MIT Press.

Allport, D. A. (1993). Attention and control: Have we been asking the wrong questions? A critical review of twenty-five years. In D. E. Meyer and S. Kornblum (Eds.), *Attention and Performance XIV: Synergies in experimental psychology, artificial intelligence and cognitive neuroscience*, pp. 183–218. Cambridge, MA: MIT Press.

Allport, D. A., Styles, E. A., and Hsieh, S. (1994). Shifting intentional set: Exploring the dynamic control of tasks. In C. Umiltà and M. Moscovitch (Eds.), *Attention and Performance XV: Conscious and nonconscious information processing*, pp. 421–452. Cambridge, MA: MIT Press.

Anderson, J. R. (1983). *The architecture of cognition*. Cambridge, MA: Harvard University Press.

Atkinson, R. C., and Shiffrin, R. M. (1968). Human memory: A proposed system and its control processes. In K. W. Spence and J. T. Spence (Eds.), *The psychology of learning and motivation*, vol. 2, pp. 89–195. London: Academic Press.

Baddeley, A. D. (1990). *Human memory: Theory and practice.* Hove, U.K.: Erlbaum.

Baddeley, A. D., and Hitch, G. (1974). Working memory. In G. A. Bower (Ed.), *Recent advances in learning and motivation,* vol. 8, pp. 47–90. New York: Academic Press.

Baker, S. C., Rogers, R. D., Owen, A. M., Frith, C. D., Dolan, R. J., Frackowiack, R. S. J., and Robbins, T. W. (1996). Neural systems engaged by planning: A PET study of the Tower of London task. *Neuropsychologia, 34,* 515–526.

Barnes, A. E., Nelson, T. O., Dunlosky, J., Mazzoni, G., and Narens, L. (1999). An integrative system of metamemory components involved in retrieval. In D. Gopher and A. Koriat (Eds.), *Attention and Performance XVII: Cognitive regulation of performance: Interaction of theory and application,* pp. 288–311. Cambridge, MA: MIT Press.

Bechara, A., Damasio, H., Tranel, D., and Anderson, S. W. (1998). Dissociation of working memory from decision making within human prefrontal cortex. *Journal of Neuroscience, 18,* 428–437.

Brandimonte, M., Einstein, G. O., and McDaniel, M., Eds. (1996). *Prospective memory: theory and applications.* Mahwah, NJ: Erlbaum.

Cohen, J. D., Dunbar, K., and McClelland, J. L. (1990). On the control of automatic processes: A parallel distributed processing account of the Stroop effect. *Psychological Review, 97,* 332–361.

Cohen, J. D., and Servan-Schreiber, D. (1992). Context, cortex and dopamine: A connectionist approach to behavior and biology in schizophrenia. *Psychological Review, 99,* 45–77.

Damasio, A. R. (1996). The somatic marker hypothesis and the possible functions of the prefrontal cortex. *Philosophical Transactions of the Royal Society of London, B351,* 1413–1420.

Dennett, D. (1978). *Brainstorms: Philosophical essays on mind and psychology.* Cambridge, MA: MIT Press.

Diamond, A. (1990). Developmental time course in human infants and infant monkeys, and the neural bases of, inhibitory control in reaching. *Annals of the New York Academy of Sciences, 608,* 637–676.

Duncan, J., Emslie, H., and Williams, P. (1996). Intelligence and the frontal lobe: Goal selection in the active control of behavior. *Cognitive Psychology, 30,* 257–303.

Frith, C. (1996). The role of the prefrontal cortex in self-consciousness: The case of auditory hallucinations. *Philosophical transactions of the Royal Society of London, B351,* 1505–1512.

Gehring, W. J., Goss, B., Coles, M. G. H., Meyer, D. E., and Donchin, E. (1993). A neural system for error detection and compensation. *Psychological Science, 4,* 385–390.

Georgiou, N., Bradshaw, J. L., and Chiu, E. (1996). The effect of Huntington's disease and Gilles de la Tourette's syndrome on the ability to hold and shift attention. *Neuropsychologia, 34,* 843–851.

Hauser, M. D. (1999). Perseveration, inhibition and the prefrontal cortex: A new look. *Current Opinion in Neurobiology, 9,* 214–222.

Holroyd, C. B., Dien, J., and Coles, M. G. H. (1998). Error-related scalp potentials elicited by hand and foot movements: Evidence for an output-independent error-processing system in humans. *Neuroscience Letters, 242,* 65–68.

Jacoby, L. (1994). Measuring recollection: Strategic versus automatic influences of associative context. In C. Umiltà & M. Moskovitch (Eds.), *Attention and Performance XV: Conscious and unconscious information processing,* pp. 661–679. Cambridge, MA: MIT Press.

James, W. (1890). *Principles of psychology.* New York: Holt.

Johnson-Laird, P. N. (1983). *Mental models.* Cambridge: Cambridge University Press.

Kramer, A. F., Larish, J. L., Weber, T. A., and Bardell, L. (1999). Training for executive control: Task coordination strategies and aging. In D. Gopher and A. Koriat (Eds.), *Attention and Performance XVII: Cognitive regulation of performance: Interaction of theory and application,* pp. 617–652. Cambridge, MA: MIT Press.

Luria, A. R. (1966). *Higher cortical functions in man.* New York: Basic Books.

MacLeod, C. M. (1991). Half a century of research on the Stroop effect: An integrative review. *Psychological Bulletin, 109,* 163–203.

Meyer, D. E., and Kieras, D. E. (1997). A computational theory of executive control proceses and human multiple task performance: 1. Basic mechanisms. *Psychological Review, 104,* 3–65.

Meyer, D. E., and Kieras, D. E. (1999). Précis to a practical unified theory of cognition and action: Some lessons from EPIC computational models of human multiple-task performance. In D. Gopher and A. Koriat (Eds.), *Attention and Performance XVII: Cognitive regulation of performance: Interaction of theory and application,* pp. 17–88. Cambridge, MA: MIT Press.

Minsky, M. (1985). *The society of mind.* New York: Simon and Schuster.

Monsell, S. (1984). Components of working memory underlying verbal skills: A "distributed capacities" view. In H. Bouma and D. G. Bouwhuis (Eds.), *Attention and Performance X: Control of language processes,* pp. 327–350. London: Erlbaum.

Newell, A. (1973). You can't play twenty questions with nature and win. In W. A. Chase (Ed.), *Visual information processing,* pp. 283–308. New York: Academic Press.

Newell, A. (1980). Reasoning, problem-solving and decision processes. In R. Nickerson (Ed.), *Attention and Performance VIII,* pp. 693–718. Hillsdale, NJ: Erlbaum.

Newell, A. (1990). *Unified theories of cognition.* Cambridge, MA: Harvard University Press.

Newell, A., Rosenbloom, P. S., and Laird, J. E. (1989). Symbolic architectures for cognition. In M. I. Posner (Ed.), *Foundations of cognitive science,* pp. 93–131. Cambridge, MA: MIT Press.

Norman, D. A. (1981). Categorization of action slips. *Psychological Review, 88,* 1–15.

Norman, D. A., and Shallice, T. (1980). *Attention to action: Willed and automatic control of behaviour.* Technical Report no. 99. San Diego, CA: University of San Diego, Centre for Human Information Processing.

Norman, D. A., and Shallice, T. (1986). Attention to action: Willed and automatic control of behaviour. In R. J. Davidson, G. E. Schwartz, and D. Shapiro (Eds.), *Consciousness and self-regulation,* vol 4, pp. 1–18. New York: Plenum Press.

Pashler, H. (1993). Dual-task interference and elementary mental mechanisms. In D. E. Meyer and S. Kornblum (Eds.), *Attention and performance XIV,* pp. 245–264. Cambridge, MA: MIT Press.

Rabbitt, P., Ed. (1997). *Methodology of frontal and executive function.* Hove, U.K.: Psychology Press.

Reason, J. T. (1984). Lapses of attention in everyday life. In R. Parasuraman and R. D. Davies (Eds.), *Varieties of attention,* pp. 515–549. Orlando, FL: Academic Press.

Robertson, I., and Manly, T. (Forthcoming). Sustaining attention in time and space. In G. W. Humphreys, J. Duncan, and A. M. Treisman (Eds.), *Attention, space and action: Studies in cognitive neuroscience,* pp. Oxford: Oxford University Press.

Rogers, R., and Monsell, S. (1995). The costs of a predictable switch between simple cognitive tasks. *Journal of Experimental Psychology: General, 124,* 207–231.

Rogers, R. D., Owen, A. M., Middleton, H. C., Williams, E. J., Pickard, B. J., Sahakian, T. W., and Robbins, T. W. (1999). Choosing between small, likely rewards and large, unlikely rewards activates inferior and orbital prefrontal cortex. *Journal of Neuroscience, 19,* 9029–9038.

Rogers, R. D., Sahakian, B. J., Hodges, J. R., Polkey, C. E., Kennard, C., and Robbins, T. W. (1998). Dissociating executive mechanisms of task control following frontal lobe damage and Parkinson's disease. *Brain, 121,* 815–842.

Schwartz, M. F., Reed, E. S., Montgomery, M., Palmer, C., and Mayer, N. H. (1991). The quantitative description of action disorganisation after brain damage: A case study. *Cognitive Neuropsychology, 8,* 381–414.

Shallice, T. (1988). *From neuropsychology to mental Structure.* Cambridge: Cambridge University Press.

Shallice, T. (1994). Multiple levels of control processes. In C. Umiltà and M. Moscovitch (Eds.), *Attention and Performance XV,* pp. 395–420. Cambridge, MA: MIT Press.

Shiffrin, R. M., and Schneider, W. (1977). Controlled and automatic human information processing: 2. Perceptual Learning, automatic attending, and a general theory. *Psychological Review, 84,* 127–190.

Stroop, J. R. (1935). Studies of interference in serial verbal reactions. *Journal of Experimental Psychology, 18,* 643–662.

Ward, G., and Allport, A. (1997). Planning and problem solving using the five-disc Tower of London task. *Quarterly Journal of Experimental Psychology, 50A,* 49–78.

Association Lecture

2 Task Switching, Stimulus-Response Bindings, and Negative Priming

Alan Allport and Glenn Wylie

ABSTRACT This chapter is about the effects of successively shifting between conflicting stimulus-response (S-R) mappings in speeded selective response tasks. Even after some time to prepare for a shift of task, there can still be a large reaction time (RT) cost on the first trial of the shifted task, generally referred to as a "residual switch cost." In five experiments, subjects performed Stroop color naming (in response to incongruent combinations of color and a distractor color word) and word reading. The word-reading task was in response to both "Stroop" and "neutral" word stimuli.

Our results show that at least a large component of the so-called switch costs results from a form of negative priming—or negative transfer of learning—arising from earlier performance of the competing selection task (Stroop color naming), interfering with the execution of the current task (word reading). The competing task need not have been performed on the immediately preceding trial to generate these effects. Hence these interference effects cannot be due to a time-consuming "switch of set" on the current trial.

The data also point to the special status of the first trial, in any run of speeded RT trials, even without any shift of task. In our experiments, the first trial of each block of speeded-response trials was consistently slower (and more accurate) than later trials. (We refer to this as the "restart" effect.) Following the Stroop color-naming task, however, word-reading RT was hugely increased, not only on the first trial of the next word-reading block (i.e., the "switch" trial), but also on the first trial of later (pure task) blocks of word reading *without any switch of task*. Some of the negative priming—or negative transfer—from the Stroop color-naming task to subsequent word reading turns out to be stimulus specific, depending on the occurrence of the same individual stimulus items (as distractors, in one task; as target stimuli in the other), rather than on competing, abstract "task sets." The results are interpreted in terms of a process of stimulus-response (S-R) binding in selection-for-action. Later S-R events can trigger retrieval of previously formed (conflicting or consistent) S-R bindings, resulting in positive or negative priming.

2.1 TASK SET AND TASK SWITCHING

The term *task switching* seems to suggest to many people the operation of some kind of a control switch, which shunts the processing system from one configuration of task readiness to another. These control operations presumably take time, and so should be detectable in performance data, in the form of reaction time (RT) "switching costs." Some part of the control operation may also require triggering by an imperative task stimulus (on a "switch" trial) for its completion. Several recent models of task switching appear to suggest a general view of this kind (e.g., De Jong 1996, chap. 15, this volume; Meiran 1996, chap. 16, this volume; Rogers and Monsell 1995; Rubinstein, Meyer, and Evans forthcoming). Although the models differ in certain respects, they share two fundamental assumptions: (1) "task set" corresponds to a certain configuration of the processing pathways: in effect, facilitation of some task-relevant processes and (at least partial) unenabling or "disengagement" of competing pathways—crucially, task set configuration directly determines the level of "task readiness"; and (2) the processing system is essentially a finite-state machine: once it has been "switched" into a given task configuration, it should *stay* that way until it is "switched" again. The processing system should thus remain in the same state of readiness for subsequent "nonswitch" trials at the same task.[1] From these assumptions it is inferred that the difference between switch and nonswitch RTs may be taken as a measure of (or at least include) the time needed to complete the relevant control operation. If these assumptions are correct, the measurement of behavioral switching costs should thus provide a valuable window into the control operations themselves.

This conception of task set and the associated metaphor of a control switch are attractive, not least because of their intuitive simplicity. However, these simple ideas are not easy to reconcile with the performance data, as we shall try to show. In fact, the data lead us to question both assumptions 1 and 2 above. Let us be clear. We do not doubt that there is endogenous control of task set, in the sense of controlling *which task* is performed. However, as we shall argue (following Fagot 1994), "task set" in this sense should not be simply equated with "task readiness"—where "readiness" is measured by speed of performance.

2.2 A LOOK AHEAD

In this chapter, we investigate speed of performance (task readiness) as a function of certain other tasks that subjects have previously had to perform. Our experiments focus on the origin of the performance (RT) costs—usually referred to as "switching costs"—when first one, then another, competing stimulus-response (S-R) mapping is executed, in response to the same "bivalent" stimuli (see Pashler, chap. 12, this volume). The stimuli we used include Stroop color-word combinations, for example a printed color name (e.g., "GREEN") presented in a different or "incongruent" color (e.g., blue); response may be based either on the color or the shape of the stimulus (Stroop 1935; MacLeod 1991).

Most current models of attention and control represent an essentially "memory-less" (finite-state) processing system; implicit learning (or "priming") effects from earlier S-R processing operations typically play little or no part in such models. To the contrary, we shall argue that the RT switching costs include a large component of (*long-term*) negative priming—or better, negative transfer—resulting from learning processes that occurred in the prior, competing task(s). These priming effects can be long-lasting; they appear to depend on the retrieval of conflicting S-R associations formed in earlier processing episodes, their retrieval being triggered by the same (bivalent) task stimuli.

To introduce our experiments, section 2.3 first outlines an earlier version of this idea, linking switching costs and S-R priming; section 2.4 then recaps some of the available evidence in its support; finally, section 2.5 briefly reviews current ideas on the mechanisms of both short- and long-term priming, with emphasis on memory-based retrieval accounts.

2.3 AN EARLIER MODEL OF TASK SWITCHING AND PRIMING: "TASK SET INERTIA"

Allport, Styles, and Hsieh (1994) postulated that the task set (or task readiness) for a given task is liable to persist, involuntarily, over successive trials, as a form of higher-order priming of competing S-R mappings (see also Allport and Wylie 1999; Meuter and Allport 1999). This priming, they supposed, took the form of persisting facilitation of the previously task-relevant S-R mappings or processing pathways, and persisting suppression of the previously competing (but now task-relevant) pathways. The result: negative priming of the current task, and "competitor priming" of the other (no longer intended) task. Allport and colleagues referred to this as "task set inertia" (TSI). Their conjecture was that positive and negative priming of this kind underlay the performance costs, in RT and errors, of switching between competing tasks, cued by the same, bivalent stimuli.

Note that the TSI hypothesis, as formulated by Allport, Styles, and Hsieh (1994), was consistent with—it certainly did not deny the existence of—some active or endogenous control operation (goal setting?), that determines *which task does in fact get performed*, and which may also be able to reduce the performance costs of task switching, at least in part, when the upcoming task is cued in advance.[2] It denied only that the time cost of task switching (i.e., the RT difference between switch and nonswitch trials) directly reflects the time needed to complete a shift of task readiness ("task set reconfiguration"), prior to executing the shifted task. If the latter were the case, they argued, a switch cost of, say, 200 msec (in terms of mean RT) should be eliminated simply by allowing an advance preparation interval of this order or longer. To the contrary, several studies (e.g., Allport, Styles, and Hsieh 1994; De Jong 1996, chap. 15, this volume; Fagot 1994; Goschke, chap. 14, this volume; Meiran 1996; Meiran et al. forthcoming; Rogers and Monsell 1995; Sudevan and Taylor 1987) have found that a preparation interval even of several seconds still left a large "residual" switch cost. Clearly, the performance costs of task switching can be at least partially offset by some process of task preparation, or goal setting, in advance of the imperative stimulus for the shifted task. This is not in dispute. In all of these cases, however, the reduction in switch cost (or the benefit of endogenous task preparation) is very much smaller (sometimes an order of magnitude smaller) than the length of the preparation interval needed. Thus the endogenous component of the RT switch cost, that is, the difference between the RT switch cost at zero and at long precue intervals (Meiran 1996), does not correspond, in any direct way, to the time needed for this preparation process (whatever it may be) *before* the task stimulus.

However, the nature of this preparation remains unclear. Fagot (1994) suggested the need to distinguish task "setting" and task "readiness."[3] According to Fagot, "setting" determines which task is in fact performed (the task goal) and can be executed during a preparation interval, whereas "readiness" determines the speed or efficiency with which the task is performed; according to him, it depends on the preceding trial and is unaffected by any intentional preparation. In Fagot's formulation, a subject can thus be "set" for one task but "ready" for another one. Clearly, this formulation is closely related to the TSI hypothesis, although TSI may have effects that last over many intervening trials (e.g., Allport, Styles, and Hsieh 1994, exp. 4). We return to this issue in sections 2.4 and 2.6.

2.4 LONG-TERM NEGATIVE PRIMING AND TASK SWITCHING

Allport, Styles, and Hsieh (1994) offered a number of empirical arguments for their interpretation of task-switching costs in terms of involuntary S-R priming (TSI) over an intended shift of tasks. We recapitulate two of these arguments here.

Earlier studies of task switching suggested that alternation between tasks resulted in substantial performance costs (relative to "pure" tasks) *only* if the alternation was between potentially competing or divergent S-R mappings, in response to the *same* stimulus set (Jersild 1927; Spector and Biederman 1976). However, Allport, Styles, and Hsieh (1994, exp. 4) showed that even tasks using dissimilar and entirely nonoverlapping stimuli and responses could exhibit large alternation costs, relative to pure task performance, if these task stimuli had previously—in an earlier experimental condition—been involved in *different* (competing) S-R mappings to those currently specified. These priming effects of the previous, competing S-R mappings (as they interpreted them) declined over successive runs at the new tasks, but were still detectable after more than 100 responses with the new S-R mappings. Clearly, the time course of TSI effects can be long-lasting, favoring stimulus-driven retrieval, rather than simply persisting facilitation or suppression of S-R pathways. We are not aware of any explanation of these results, to date, in terms of the time taken by a postulated switch operation. This experiment

(Allport, Styles, and Hsieh 1994, exp. 4) shows that task-switching costs can be the product of varied S-R mappings that occurred, not just on the preceding trial, but even in a prior experimental session. We present several further examples of this point in our experiments 2–5 (section 2.6).

Switching from Stroop color naming to word reading resulted in another, equally striking effect: word reading now showed large (~140 msec) "reverse Stroop" interference from the incongruent color, even after a preparation interval of over a second (Allport, Styles, and Hsieh 1994, exp. 5). As is well known, in the Stroop color word and picture word tasks, the interference is, typically, strongly asymmetrical (MacLeod 1991; Smith and Magee 1980). Thus, with an incongruent Stroop stimulus, color naming suffers interference from the word, but word reading normally shows no interference from the incongruent color (interference from color to word is known as the "reverse Stroop" effect). This asymmetry has been attributed to differences in the long-term, relative "strength" of the competing pathways (MacLeod and Dunbar 1988; Cohen, Dunbar, and McClelland 1990).

The reverse Stroop interference found by Allport, Styles, and Hsieh (1994) seems difficult to account for if there were some kind of control switch, before word processing (or indeed before response selection), capable of disengaging or unenabling the processing pathways for color naming, and selectively facilitating the processing pathways for word reading. On the contrary, the interference seems to provide rather direct evidence that the S-R bindings needed for (Stroop) color naming (and the suppression of word reading) either simply persist or, as we shall argue, are strongly reelicited, on a subsequent, intended switch to word reading, in response to the same type of (multivalent) task stimuli. In the new experiments to be described in section 2.6, we attempt to exploit these rather dramatic, reverse Stroop interference effects further, to explore the origins of switching costs, and the negative transfer between successive, competing tasks.

2.5 VARIETIES OF (LONG-TERM) PRIMING

As is well known, selective attention (and selective response) to a target stimulus can show persisting aftereffects in the form of item-specific, positive and negative priming. Thus attending to a particular stimulus attribute (in a "pop-out" search task) can facilitate a later selective response to the same target value, over a number of intervening trials (Maljkovic and Nakayama 1994; Nakayama and Joseph 1997). Moreover, a previously ignored distractor, now presented as a subsequent attentional target, can show a negative priming effect, over short lags, that has been attributed to persisting distractor inhibition (Houghton and Tipper 1994; Milliken and Tipper 1998). Longer-lasting negative priming, it is argued, reflects associative retrieval, based on prior, conflicting processing episodes, rather than persisting inhibition (e.g., Allport, Tipper, and Chmiel 1985; De Schepper and Treisman 1996; Lowe 1998; Neill et al. 1992; Park and Kanwisher 1994). There is an ongoing debate about the extent to which positive and negative priming effects (both short- and long-term) are due to the retrieval of associative bindings—formed in prior, congruent or conflicting processing episodes—rather than merely persisting activation or inhibition (see, for example, Becker et al. 1997; Fox 1995; Hommel 1998; Kane et al. 1997; Lowe 1998; Milliken and Tipper 1998; Neill 1997). Stimulus-driven retrieval of prior processing episodes (or rather, of composite "echoes" of those episodes) is the basis also of some models of long-term schema abstraction and automatization (Goldinger 1998; Hintzman 1986; Logan 1988; Logan and Etherton 1994).

Involuntary priming effects are not restricted to item-specific stimulus representations. Rabbitt and Vyas (1973, 1979) established the existence of RT facilitation effects when the same, abstract S-R mapping rule was repeated, independent of the repetition of individual stimulus or response items. Long-term semantic priming, where the stimuli are related only in terms of higher-order categories, has also been demonstrated (Joordens and Becker 1997; Becker et al. 1997). Each of these priming mechanisms—positive and negative, item-specific and higher-order effects; temporary activation or inhibition mechanisms; and long-term competitive retrieval processes—may, in principle, contribute to task set inertia effects following a switch of tasks. As we shall see, memory-based retrieval effects appear to play a major role.

2.6 STROOP COLOR NAMING AND WORD READING: EXPERIMENTS

In all of these experiments, we used the same pair of tasks: color naming and word reading. The stimuli for both color-naming and word-reading responses included "Stroop" (incongruently colored color words, that is, bivalent stimuli) and "neutral" (univalent) task stimuli, which afford only one or other of these tasks. We shall focus primarily on the effects of a shift from color naming to word reading. Word-reading latencies for a familiar word set have the advantage that they are exceptionally stable, with a very compact RT distribution and low error rate.

Experiments 1 and 2 used the "alternating-runs" method of Rogers and Monsell (1995). Experiments 3–5 used a different experimental paradigm, in which the probe task (word reading) was performed in "pure task" conditions, following a shift from color naming. In all five experiments, subjects responded by orally naming the target stimulus as fast as possible, and their RT was recorded by means of a voice key.

As a systematic constraint on stimulus sequences, the color or word identities (concepts) on trial *n* were not allowed to occur, either as color or word, on trial $n+1$; thus positive and negative item priming across immediately successive trials was excluded. This constraint applied to all the experiments reported in this chapter.

Finally, because we aimed to study "residual" switch costs, we provided relatively long (1.0 to 2.0 sec) preparation intervals before each switch of task (0.5 sec in experiment 2); the subjects were encouraged to do their best to prepare in advance, on each trial, for the upcoming task. (Experiments 1 and 2 are described in greater detail in Wylie and Allport forthcoming.)

Experiment 1: Task Alternation Costs on "Nonswitch" Trials and Effects of S-R Mappings in the "Other" Task

The goal of this experiment was to assess to what extent the cost of alternation between mutually competing tasks (color naming, word reading) depends on the specific S-R mappings in the prior, *competing* task, or on the control demands of the current task. Subjects switched between color naming and word reading, with three different pairings of Stroop and neutral stimuli. Our prediction was that switching performance would depend primarily on the status (Stroop versus neutral) of the "other" task, namely, the task switched *from*.

For this experiment, as also for experiment 2, we used the alternating-runs method introduced by Rogers and Monsell (1995). Subjects saw a large black cross, which divided the screen into four quadrants. On successive trials, the task stimulus was presented successively in adjacent quadrants, in continuous, clockwise rotation: top left, top right, bottom right, bottom left, top left, and so on. (In experiment 1 and all the following experiments, the stimulus remained on until the subjects' response.) Half the subjects were instructed to name the color of the stimuli in the top two quadrants and to name the words appearing in the bottom two quadrants; for the remaining subjects, this instruction was reversed. For all subjects, therefore, responses to stimuli appearing in the top left and bottom right quadrants represent switch trials, whereas responses to stimuli in the other two quadrants are nonswitch or "repeat" trials. Rogers and Monsell (1995) proposed that the time cost of task switching can be appropriately measured as the difference between switch and repeat trial RTs, in the alternating-runs paradigm. We follow this convention, initially.

The stimuli were the six color words, "red," "green," "blue," "yellow," "pink," and "brown," and the corresponding six colors. Words could appear in any one of the colors except the color named by the word; that is, these were incongruent Stroop stimuli. We also presented neutral stimuli, designed to afford the execution of only one of the two tasks (see details below). The intertrial interval was approximately 1.2 to 1.5 sec,

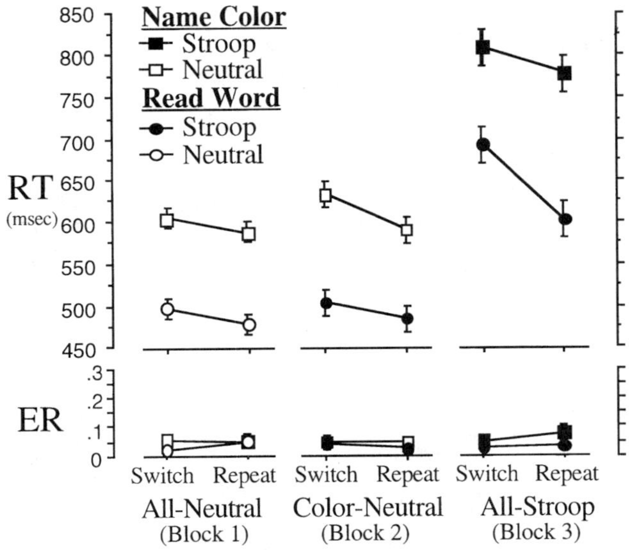

Figure 2.1 Reaction times (RT) and error rates (ER) for word reading and color naming in experiment 1. Error bars show 95% within-subject confidence intervals (Loftus and Masson 1994). Subjects alternated every second trial between naming colors and reading words, in three successive stimulus conditions.

varying randomly from trial to trial,[4] and subjects were encouraged to use this interval to prepare for the upcoming task. This was designed to permit asymptotic, "endogenous" task preparation between trials; hence there should be "residual" switch costs only (Meiran 1996; Meiran et al. forthcoming; Rogers and Monsell 1995).

The experiment was divided into three successive blocks (of 120 trials each), in a fixed order. In block 1, the "all-neutral" condition, the stimulus for color naming was a row of colored Xs; for word reading, it was a neutral (black) word. In block 2, the "color-neutral/word-Stroop" condition, the stimulus for color naming was a row of colored Xs, as before, whereas for word reading, it was an incongruent Stroop stimulus. Finally, in block 3, the "all-Stroop" condition, the stimuli for both tasks were incongruent Stroop stimuli. In block 1 (all-neutral), the respective stimulus types afforded only one of the two tasks, word reading or color naming, whereas in block 3 all stimuli were bivalent, affording *both* word-reading and color-naming responses. In block 2—the critical color-neutral/word-Stroop condition—the stimuli for the word-reading task were bivalent, as in block 3, but the stimuli for the color-naming task were univalent. Each experimental block of 120 trials was preceded by 30 trials of practice with the new stimulus conditions.

The results of experiment 1 are illustrated in figure 2.1. Consider first the results for blocks 1 and 3, all-neutral and all-Stroop. As described above, Allport, Styles, and Hsieh (1994, exp. 5) studied task switching between color naming and word reading with the same stimulus types as in blocks 1 and 3 (all-neutral, all-Stroop). The present results generally replicate their findings: in the all-neutral condition, switch costs (defined as the RT difference between switch and repeat trials) were small (about 20 msec) and symmetrical; mean color-naming RT was about 110 msec slower than word reading. In the all-Stroop condition, switch costs were larger, and markedly asymmetrical (about 30 msec for color naming and over three times this value for word reading). The same asymmetry in the switch costs between (all-Stroop) color naming and word reading is found in all five experiments reported here. (The theoretical interpretation of this—at first sight—counterintuitive result is discussed in detail, with reference to the TSI hypothesis, by Allport and Wylie (1999.)

The intended focus of the experiment, however, was on condition 2—color-neutral/word-Stroop—and the word-reading task in particular. We argued that, if alternation costs depended on the characteristics (e.g., bivalent task stimuli) of the task *to* which a switch is made, then the cost of shifting to the word-reading task should be about the same in this condition as in the all-Stroop condition because both conditions require responses to the same set of incongruent Stroop stimuli. In contrast, if switching costs depended on priming effects from the prior, competing task, as postulated by the TSI hypothesis, then the cost of shifting to the word-reading task in condition 2 should be about the same as in the all-neutral condition because both conditions have neutral color naming as the competing task. The results are extremely clear. As predicted by the TSI account, the cost of switching to the word-reading task was practically identical in conditions 1 and 2, and significantly larger ($p<0.0005$) in condition 3. That is, the switch cost here appears to be a function of the complementary task set (*from* which the switch is made, in this case), rather than depending on the intrinsic demands of the task *to* which the switch is made.

Comparison of the color-naming performance in conditions 2 and 3 is also relevant. Predictably, color-naming RTs to Stroop stimuli (condition 3) are much longer than to the neutral color stimuli in condition 2: a classic Stroop effect of about 180 msec. If switch costs reflected the control demands of the task set *to* which the switch is made, we should presumably expect a much larger switch cost for the color-naming task in condition 3, in response to Stroop stimuli, than in condition 2, with neutral color stimuli. However, this is clearly not the case. Switch costs for color naming did not differ significantly between the two conditions.

Returning to the word-reading task, we may also consider possible reverse Stroop interference effects. Taking condition 1 as the available baseline for reading neutral word stimuli, it is clear that condition 2, with Stroop word stimuli, shows essentially no such reverse Stroop effect. Word-reading performance in conditions 1 and 2 is practically identical. In contrast, comparing word-reading RTs in condition 3—also in response to Stroop word stimuli—against the neutral baseline of condition 1 appears to show a large (~200 msec) reverse Stroop effect on switch trials ($p < 0.0001$), and a still very substantial (~120 msec) performance cost on repeat trials ($p < 0.0001$). Given that, in most experimental conditions (that do not involve switching between color and word), word reading shows no interference from an incongruent stimulus color, the appearance of reverse Stroop effects in the all-Stroop condition is strong evidence that some components of the task set, and/or the specific S-R mappings, for color naming were still active (or were *reactivated*) during the word-reading trials. We note that these task interference effects occurred after a relatively long preparation interval (over 1 sec) between trials. Even more strikingly, a large performance cost for word reading was still present on *repeat* trials. In other words, readiness for the word-reading task (including effective disengagement from the complementary color-naming task) appears to be very far from complete on the repeat trials of condition 3. This observation undermines a widespread assumption of the alternating-runs method, namely, that task set reconfiguration can be assumed to be complete on nonswitch trials after a single switch trial, as several students of task switching have proposed (e.g., De Jong 1996, chap. 15, this volume; Rogers and Monsell 1995). Further discussion is deferred until after experiment 2.

Experiment 2: Time Course Effects of Priming between Competing S-R Mappings

Experiment 1 demonstrated large task interference effects from color naming to word reading, in the all-Stroop condition, not only on switch trials but also in the subsequent nonswitch or repeat trial RTs. Allport and Wylie (1999) interpreted these effects as a form of task priming (or task set inertia) resulting from the Stroop color-naming task. We may now ask: How long do these priming effects persist? This is clearly an important empirical question, both for the design of future studies of task switching and for the interpretation of existing data. For example, consider the word-reading performance in experiment 1, specifically, in conditions 2 and 3. (Recall that, in both conditions, word reading was in response to Stroop stimuli. The conditions differed only in the type of stimuli presented for color naming: Stroop stimuli in condition 3 and neutral stimuli in condition 2). Suppose that, after performing condition 3 for some time, the color task stimuli changed abruptly from Stroop to neutral, that is, to condition 2, while subjects continued to perform both word-reading and color-naming tasks in alternating runs. Prior to the stimulus change, performance in condition 3 might be expected to resemble that observed in experiment 1 for the same condition. How many trials (or how many iterated cycles of alternating runs) will it take, with neutral *color-naming* stimuli, before the Stroop word-reading performance approaches that observed in condition 2? This is the question that we attempted to answer in experiment 2.

For this purpose, subjects successively performed the all-Stroop and the color-neutral/word-Stroop condition (*"color-neutral,"* for short) of experiment 1, in continuously repeating "miniblocks" of 6 cycles in each condition (a "cycle" is four trials in the alternating-runs paradigm, with double alternation—two color-naming trials and two word-reading trials; a "miniblock" was 6 successive cycles). Stimuli appeared in successive screen locations, without a break, between successive all-Stroop and color-neutral miniblocks. (We had no way of knowing, in advance, how many cycles of the color-neutral/word-Stroop condition would be needed to track the decline of priming by the preceding all-Stroop color-naming task, on the word-reading RTs. Six cycles (24 trials) was arbitrarily chosen as long enough, we hoped, to show a substantial—and possibly complete—transition, after the change to color-neutral stimuli, to the no-interference pattern in word-reading RTs found in experiment 1.)

Two modifications to the neutral stimulus displays of experiment 1 were introduced in experiment 2. First, instead of being presented in solid print, as in experiment 1, the neutral word stimuli were presented in outline print. The letter outlines were in black, but they were not filled with any color, thus appearing "transparent" to the screen background. (This format was also used for the neutral word stimuli in all subsequent experiments.) For the incongruent Stroop stimuli, the same outline characters were filled in with the appropriate color. Second, instead of a string of colored Xs, in experiment 2 the neutral color stimuli were filled, colored rectangles of the relevant color, occupying approximately the area of a five-character word. The words (and colors) used in experiment 2 were "red," "green," "blue," "purple," "pink," and "orange." The response-stimulus interval (RSI) was fixed at 500 msec.

Subjects completed a total of 30 alternating, 24-trial cycles of all-Stroop and color-neutral stimulus conditions. They had 3 practice blocks of 30 trials each, with all-Stroop stimuli, immediately before the main experiment. This began with 6 cycles (24 trials) with all-Stroop stimuli, followed by 6 cycles (24 trials) of color-neutral stimuli, followed without a break by a further 6 cycles of all-Stroop stimuli, and so on. Subjects were allowed a rest pause after every 120 trials. Data from the first miniblock after a rest pause were excluded from analysis. A cycle always began with the two color-naming trials. The start of each cycle was also redundantly cued by a high (800 Hz) tone, for all-Stroop cycles, and a low (220 Hz) tone for color-neutral cycles, immediately before the first color trial of each cycle. Nine subjects from the Oxford University subject panel participated in the experiment, four men and five women, mean age 38 years. (For further experimental details, see Wylie and Allport forthcoming.)

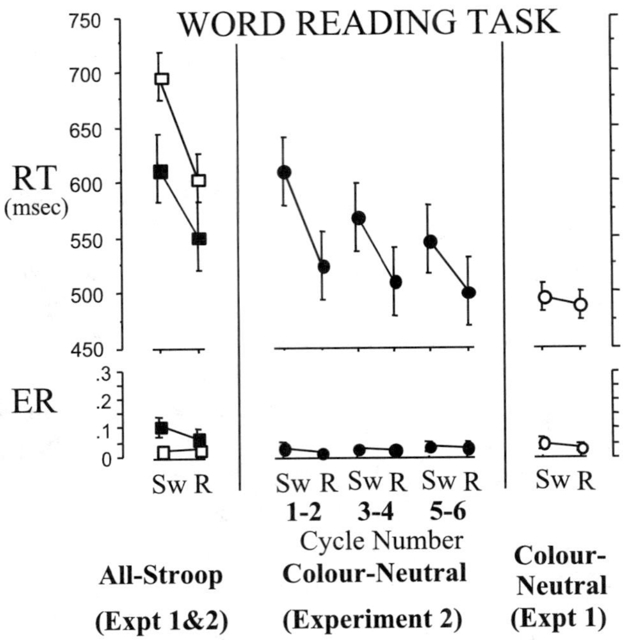

Figure 2.2 Reaction times (RT) and error rates (ER) for word reading in experiment 2 (filled symbols). Error bars show 95% within-subject confidence intervals. Subjects successively performed six cycles (24 trials) with "all-Stroop" stimuli, followed without a break by six cycles of "color-Neutral," and so on. The only difference between all-Stroop and color-Neutral conditions was in the stimuli presented for color naming. All word-reading RTs were in response to incongruent "Stroop" stimuli. Data from experiment 1 (open symbols) are shown for comparison.

The resulting mean RTs and error rates for the word-reading task are shown in figure 2.2 and for the color-naming task in figure 2.3. For comparison, we also include the results of the same two stimulus conditions from experiment 1. To track performance during the color-neutral miniblocks, the data were collapsed across cycles 1 and 2, 3 and 4, and 5 and 6, respectively, to give 24 observations per subject per cell.

The main focus of interest is the color-neutral condition. As expected, word-reading RTs (figure 2.2) showed a progressive reduction over successive cycles ($p < 0.005$), affecting both switch and repeat trials. The further away (either in time or number of trials) from the preceding all-Stroop miniblock, the smaller the task interference from the preceding Stroop color naming appears to be. Switch costs—defined as the difference between switch and repeat trial RTs—also diminished progressively over successive cycles ($p < 0.01$). However, as figure 2.2 shows, even after 6 cycles (24 trials) of the color-neutral condition, word-reading performance on switch trials had still not come down to the level of performance obtained in the color-neutral condition of experiment 1. Switch costs in cycles 5–6 were still larger ($p < 0.05$) than in the color-neutral condition of experiment 1, where subjects had, so far, done *no* Stroop color naming.

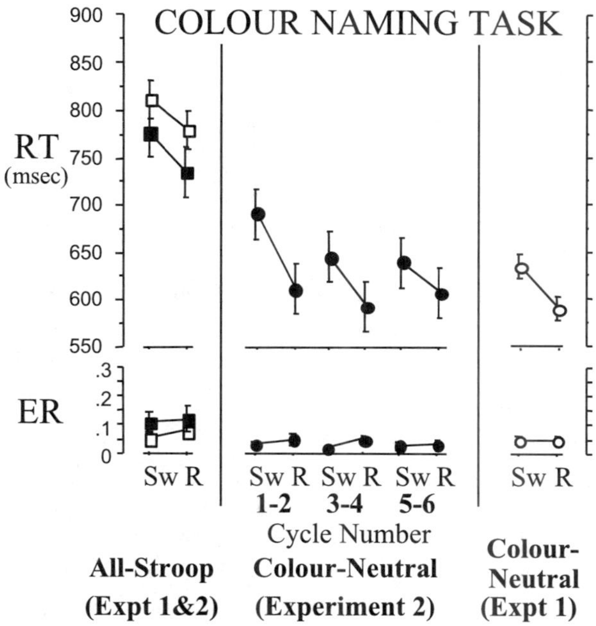

Figure 2.3 Reaction times (RT) and error rates (ER) for color naming in experiment 2 (filled symbols). Error bars show 95% within-subject confidence intervals. Data from experiment 1 (open symbols) are shown for comparison.

These data thus provide a clear, but incomplete answer to the question to which the experiment was addressed: How long do task-priming (i.e., interference) effects between color naming and word reading (generated in the all-Stroop conditions) persist, after the color-naming task shifts from Stroop to neutral stimuli? The incomplete answer is, evidently, longer than 24 trials, or 6 cycles.

RTs and error rates in the color-naming task are shown in figure 2.3. Data from the comparable conditions in experiment 1 are again included for comparison. As expected, there was a large difference in the speed of color naming in response to Stroop and neutral stimuli ($p < 0.0001$) and between switch and repeat trials ($p < 0.001$). Error rates for color naming were also significantly higher, as usual, in the all-Stroop condition. Color-naming RTs in the all-Stroop miniblock were similar to those in the corresponding all-Stroop condition of experiment 1. In the color-neutral condition, repeat trial performance was broadly similar to the equivalent (color-neutral) repeat trials in experiment 1; this was so already in the first two cycles, and showed no significant change on subsequent cycles. Switch trial RTs, on the contrary, decreased significantly ($p < 0.0005$) from cycles 1–2 to cycles 5–6. This combination resulted in a progressive reduction in the nominal switch costs (switch trial RT minus repeat trial RT) for color naming, as also in the word-reading task, over successive cycles of the color-neutral condition.

Discussion of Experiments 1 and 2

It may be useful to discuss experiments 1 and 2 together. Both experiments used the alternating-runs method (Rogers and Monsell 1995), with a fully predictable switch of task every second trial and a comparatively long intertrial interval. In these conditions, it has been argued, any anticipatory or endogenous task preparation is likely to be more or less asymptotic. In these conditions, as several authors have postulated, the residual switch costs (defined as the difference between switch and repeat RT at these longer intervals) are taken to reflect the time cost of a control operation (task set reconfiguration) executed during the course of the switch trial (e.g., Meiran 1996, chap. 16, this volume; Rogers and Monsell 1995; Rubinstein, Meyer, and Evans forthcoming).

Experiment 1 suggested that the switch cost, measured in this way, is a function primarily of the task requirements on the complementary, preceding task. However, the same experiment also demonstrated that the *repeat* (or nonswitch) trials, used as the baseline for this assessment of switch costs, by no means represent a fully or optimally prepared state of task readiness. Word reading on repeat trials, in the all-Stroop condition, still showed very large interference effects from the preceding color-naming task. This finding seems inconsistent with a simple model of an (exogenous) control switch that shunts the processing system from one discrete task configuration to another on a single switch trial, to leave the system fully prepared ("reconfigured") for the new task on subsequent trials.

Experiment 2 provided even more problematic results for such a conception. According to a simple executive switch model, in the color-neutral condition, task reconfiguration to word reading should be completed on the first cycle (indeed, on the first switch trial to word reading). It should then presumably *remain* in that state throughout the following five cycles because the alternating, complementary task was now cued by univalent stimuli (colored rectangles) that do not in any way afford word reading. Consequently, after the first trial of word reading, color-neutral performance should resemble that in the color-neutral condition of experiment 1, where switch costs for word reading amounted to no more than 20 msec. However, contrary to these expectations, in experiment 2 (color-neutral), we found switch costs for the word-reading task of between two and four times this size, decreasing slowly over successive cycles. The critical difference between the two experiments, we suggest, was that in experiment 2, but not in experiment 1, subjects had also recently been required to perform the Stroop color-naming task, in response to the same set of bivalent stimuli. It seems clear that any account of these results will need to refer to the priming effects of previous, competing tasks—up to at least some 24 trials earlier—cued by the same, bivalent stimuli.

Experiment 2 provided results that also seem inconsistent with Allport, Styles, and Hsieh's interpretation (1994) of task priming (task set inertia): that is, simply as the persisting facilitation or suppression of competing processing pathways. If the cost of performing a previous, divergent S-R mapping simply reflected persisting pathway activation or suppression, then, without further priming, such performance costs should presumably decrease monotonically over successive trials—they should certainly not *rebound* on the *next* switch trial. This, however, is precisely what we observed over successive cycles of the color-neutral condition in experiment 2, in both word-reading and color-naming RTs: in each case, a relatively fast repeat trial was followed by a slower switch trial on the next cycle or cycles (see figures 2.2–2.3). Task set inertia, interpreted simply as persisting pathway activation and inhibition, is not easily reconciled with this pattern of results.

On the other hand, this pattern of results could be consistent with a *retrieval* account of S-R priming by the prior, competing task. Suppose that a Stroop stimulus, previously associated with color naming, triggers the *re*activation of the same S-R associative links ("bindings"), previously associated with those same stimulus attributes. These S-R bindings might be postulated to include both "positive" links between the (previously) task-relevant stimulus attributes and their associated responses, and also "negative" links between (what were previously) distractor attributes and "do not respond" (or "nonresponse") action codes (cf. Allport, Tipper, and Chmiel 1985; Hommel 1998; Lowe 1998; Neill et al. 1992; Stoet and Hommel forthcoming). To account for the rebound effect on successive switch trials, however, the postulated retrieval of competing S-R bindings would have to be in some way more effective, or to trigger a greater interfering effect, at the start of each new run of trials, that is (in these experiments), on the *switch* trials.

As we shall demonstrate in the following experiments, the RT interference from color naming to word reading (and from word reading to color naming) is greatly enhanced on the first trial of each run of trials. Moreover, a similar, massive rebound of RT interference from a prior task occurs also on the first trial of a run, with *no explicit switch of task*. Indeed, it turns out that even a brief interruption (as brief as two seconds) in a regular series of speeded response trials, and subsequent restart of the *same* task, is liable to trigger renewed task interference from earlier, competing S-R mappings, executed in response to the same stimuli. The performance costs on a task switch trial may thus include (or be a special case of) a much more general phenomenon of competing, reevoked S-R mappings (both "positive" and "negative" associative bindings), triggered by the onset of a new run of trials.

Switching between "Pure" Tasks Experiments 1 and 2, using the alternating-runs paradigm, found very large task interference (reverse

Stroop) effects on word reading in the all-Stroop condition, including on the nonswitch or *repeat* trials. This interference was still detectable up to 6 cycles, or 24 trials, after the requirement to switch between competing S-R mappings—in response to the same, bivalent stimulus set—had been lifted. These results suggest that, as a measure of switch costs, the difference in RTs between switch and repeat trials in the alternating-runs paradigm may not represent a clean or appropriate contrast between an unprepared ("not-yet-reconfigured") and a completely prepared ("reconfigured") state.

This rather discouraging observation prompted us to search for other possible procedures for studying the costs of task switching. Rogers and Monsell (1995) argued forcefully that the procedure pioneered by Jersild (1927) of comparing performance in alternating and fixed (pure) tasks confounded the requirement to shift tasks and the requirement to "hold in mind" two tasks versus just one. This argument was of critical importance in motivating the measurement of switch and repeat trials within the same switching block (see also Meiran 1996). On the other hand, the results of experiments 1 and 2 raise serious doubts about this procedure, too, as a method for measuring straightforwardly interpretable switch costs. However, we might still escape the postulated confound between switching and task memory load if we could somehow probe the effects of task alternation within pure task blocks. An experiment reported in the much-cited landmark paper Stroop 1935 in fact suggests such a possible method.

Stroop (1935) reported three experiments. His experiment 3 described the following sequence of events, in three main stages. Subjects were first asked to read aloud lists of printed color names, both in neutral lists (words printed in black) and in lists of incongruently colored Stroop stimuli, 50 words to a sheet, to provide a baseline measure of word-reading performance. Their mean list completion times corresponded to an average time per item of 388 msec for Stroop stimuli and 382 msec for neutral word stimuli. Apparently, there was little or no reverse Stroop interference effect here. In stage 2 of the experiment, the subjects practiced *color-naming* similar lists of incongruent Stroop stimuli, again 50 items per list, 4 lists per session, for 8 successive days. Finally, in stage 3, after an intervening session of naming neutral color patches, they returned to their original task of word reading, though now only in response to incongruent Stroop stimuli. They again read aloud 50-word lists, 4 lists per session, on 2 successive days. Their list completion times in (postcolor) stage 3 corresponded to an average time per item of 696 msec for Stroop word reading on day 1, and 440 msec on day 2. Word-reading performance on postcolor day 1 thus revealed a mean cost of 308 msec per item, averaged over the first 200 trials of postcolor word reading. This was with Stroop stimuli. It is to be regretted that Stroop (1935) did not also include a condition of neutral postcolor word reading to assess the possible presence of reverse Stroop interference. Neither did he report list-reading times separately for successive lists in stage 3, to provide an indication of the possible decline in the word-reading performance costs over the session. Even on postcolor day 2, however, averaged over all 200 trials, there was evidently still some 50 msec per item performance cost on Stroop word reading.

These rather dramatic results are not often referred to (but see MacLeod 1991, 164–165). They appear to represent a particularly powerful and long-lasting demonstration of task set inertia. They might perhaps also be described as the "long-term costs of task alternation, observed in pure task conditions." It would be interesting to know whether similar, though perhaps more transitory, effects could be generated by a very much briefer induction phase than Stroop's eight days (1935) of color naming. It would be of interest also to track the time course of such effects, trial by trial, using discrete RTs. To what extent is the *first* trial of a run (as in a switch trial) differentially affected by long-term priming of a competing S-R mapping?

Experiment 3: The "Before and After" Paradigm

Our first explorations of these questions (described in Allport and Wylie 1999) used just 30 trials of Stroop color naming, sandwiched between an initial, baseline phase of both Stroop and neutral word reading, and a following postcolor phase of word reading, again in response to both Stroop and neutral stimuli. Allport and Wylie referred to this as the "before and after" paradigm. For half the subjects, stimuli for color naming appeared in the upper half of the screen, above a horizontal line, and stimuli for word reading appeared in the lower half of the screen. For the remaining subjects, this arrangement was reversed. All word-reading trials were performed under pure task conditions. Thus, in phase 2 (the color-naming phase), after 10 practice color trials, subjects were instructed that they would perform a further, single block of 20 color-naming trials; there would then be a 2 sec pause, with instructions on the monitor screen to return to the earlier word-reading task. Thereafter, they were assured, there would be no further color-naming trials.

The stimuli for word reading, in both the baseline (phase 1) and postcolor phase (phase 3) of the experiment, occurred in successive blocks of Stroop and neutral stimuli (10 trials per block). There was a 2 sec pause between blocks, during which the instruction "Read words" appeared on the screen. RSI within a block was fixed at 300 msec. In phase 1, all subjects performed ten 10-trial blocks of word reading, with alternate blocks of Stroop and neutral stimuli. The first 3 blocks of each type, in phase 1, were treated as practice. In phase 3, one group of subjects saw Stroop stimuli in postcolor block 1 and neutral stimuli in postcolor block 2 ("Stroop-first" subjects); the other experimental group saw neutral stimuli in postcolor block 1, and Stroop stimuli in postcolor block 2 ("Neutral-first" subjects). The same order of Stroop and neutral blocks was also used in phase 1. Experiment 3 also included a control group who completed the same phase 1 and phase 3 word-reading tasks (in neutral-first order), but simply rested during phase 2 (Stroop color naming). There were 10 subjects in each group.

The results are illustrated in figure 2.4. The initial interference effects on postcolor word reading, recorded in discrete reaction times, were even larger than the mean effects on overall list completion times reported by Stroop (1935), but lasted a very much shorter time. After a total of just 30 trials of Stroop color naming, the first trial of Stroop word reading (the nominal switch trial) showed an RT cost of over 450 msec, compared to the control group or to the experimental subjects' baseline (phase 1) first-trial performance. (In a partial replication experiment, we found an even larger cost, of approximately 600 msec; see Allport and Wylie 1999.) Errors on the first (postcolor) Stroop word-reading trial also increased sharply, to over 35%. Subsequent nonswitch Stroop trials, in postcolor block 1, also continued to show large, but rapidly diminishing, performance costs. Thus immediate postcolor trials 2–5 (all of them *nonswitch* trials) showed a mean Stroop word-reading cost relative to controls of over 200 msec. Postcolor interference was still present throughout the rest of this block (trials 6–10), with a mean RT cost of 135 msec. Subjects who read neutral words in postcolor block 1 ("neutral-first" subjects) also exhibited significant performance costs, though very much smaller than for Stroop word reading: well over 100 msec on trial 1, and around 20–30 ms over the remainder of the block. Comparison between postcolor Stroop and neutral word-reading performance indicates a reverse Stroop interference effect in immediate postcolor word reading on the order of 350 msec on trial 1, diminishing to around 180 and 100 msec over trials 2–5 and 6–10, respectively.

The most revealing feature of these results, however, was found in postcolor word-reading block 2. Between postcolor blocks 1 and 2 there was simply a 2 sec pause (a 1 sec screen prompt to continue to "Read words," followed by a 1 sec blank interval). Despite there being *no switch of task* from color naming to word reading between blocks 1 and 2, word-reading performance in postcolor block 2 again showed massive task interference effects on trial 1. That is—*over and above* the first-trial RT increment seen in the control subjects (who had not performed the previous color-naming task)—Stroop word reading on trial 1 of postcolor Block 2 showed an additional RT cost of over 300 msec, whereas neutral word reading showed an additional RT cost of around 150 msec (see figure 2.4). These large and highly significant task interference costs were only seen on trial 1 (the restart trial) of block 2, and not on any later trials in the block. Restart trials in later word-reading blocks also continued to show significant, but very much smaller reverse Stroop interference (i.e.,

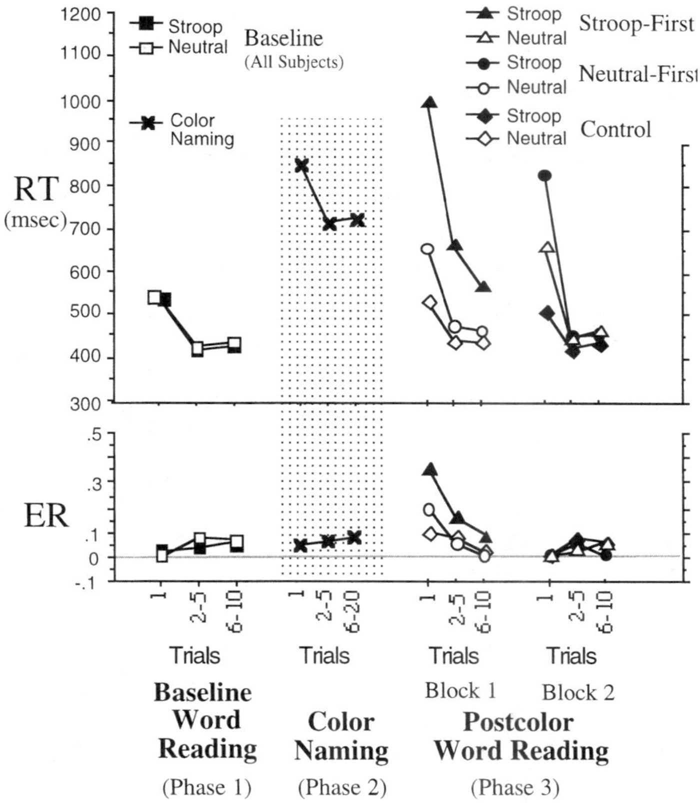

Figure 2.4 Reaction times (RT) and error rates (ER) in experiment 3, using the "before and after" paradigm. All subjects first performed a baseline condition of word reading (phase 1) and, later on, a further series of "pure task" word-reading blocks (phase 3). Between phases 1 and 3 the experimental groups performed a short period of Stroop color naming (phase 2). The control group performed phases 1 and 3 (in neutral-first order), but rested during phase 2.

compared to neutral word reading) over a number of subsequent blocks of word reading (not shown in figure 2.4).

However, first-trial or restart effects were not confined to the postcolor phase of word reading. As figure 2.4 shows, the baseline word-reading performance, before any mention to the subjects of a color-naming task, also showed a consistent ($p < 0.0001$) first-trial RT cost, on the order of 100 ms, accompanied by an equally consistent ($p < 0.0005$) reduction in errors. A similar pattern can also be seen in the color-naming task. (Note that trial 1 of this block of 20 color-naming trials was preceded by 10 previous color-naming trials, thus is also a restart trial, not a switch trial.) The control subjects, who simply rested during phase 2, showed a similar RT cost, on the first trial of each word-reading block, both in phase 3 and in their previous baseline data.

Very similar RT costs on the first trial of a run, also in a fixed-task condition, have been reported by De Jong et al. (forthcoming, exp. 3); the first-trial RT cost was very much larger in old than in young subjects. Error rate was not reported. A possibly related effect has been studied by Gopher and colleagues (Gopher, Greenshpan, and Armony 1996; Gopher, Armony, and Greenshpan forthcoming). In their experiments, a run of RT trials was briefly interrupted by an instruction cue, which requested the subject either to shift tasks ("switch") or to continue as before with the same task ("reconsider"). The first trial following both "switch" and "reconsider" instructions showed a large RT increment, the latter nearly as large as the former in some conditions.

It seems evident that the initial trial of a run of successive, speeded-response trials, even *without* any requirement to switch tasks, presents some additional processing demand, relative to all subsequent trials in the run. In the baseline performance (and in the control subjects throughout), the data clearly show a shift in speed-accuracy criterion toward greater caution, on the first trial of a run. Moreover, when the task stimulus on the first trial of a run is of a type that has been associated recently with a competing S-R mapping, the conflict latent in these divergent S-R mappings appears to be strongly reevoked, even though previous repeat trials in a preceding run may have exhibited apparently reduced conflict effects. The possibility arises, therefore, that RT switch costs, confined to the first trial of a run of alternating tasks, may reflect in large measure the same conjunction of effects. (Further discussion is deferred until after experiment 4.)

Experiment 4: "Restart" Costs and Repeated Task Switching

Experiment 4 represents a modified version of the "before and after" paradigm. There were several modifications. The principal difference was that, after the baseline word-reading phase (which was unchanged), the sequence of a short block of incongruent Stroop color-naming trials followed by two postcolor blocks of word reading was *iterated* in successive cycles throughout the experiment. (As in experiment 3, for half the subjects, stimuli for color naming appeared in the upper half of the screen, above a horizontal line, and stimuli for word reading appeared in the lower half of the screen. For the remaining subjects, this arrangement was reversed.) The control group, instead of performing the color-naming task, on each cycle performed what was intended to be (as far as possible) an unrelated RT task (size and luminance comparisons, with two-alternative keypress responses) followed by the two blocks of word reading.

Each Stroop color-naming block consisted of just 10 trials. At the end of this color-naming block (and at the end of the keypress block, for the control subjects), the instruction "Read words" appeared on the screen for 1 sec, followed by a horizontal line on a blank screen for 1 sec, followed by (postcolor) block 1 of word reading, consisting of 20 trials. The instruction

"Read words" then appeared on the screen again for 1 sec, followed by a blank screen (with the horizontal line) for 1 sec, immediately followed by (postcolor) block 2 of word reading, again consisting of 20 trials. (RSI within a block was fixed at 300 msec, as in Experiment 3. After a short rest pause, the sequence then recommenced with the next block of color naming (or keypress), then two blocks of word reading, and so on, throughout the remainder of the experiment. Subjects were encouraged to do whatever they could to prepare for the next word-reading block, during each 2 sec preparation interval.

Each block of 20 word-reading trials consisted of either 10 trials of Stroop stimuli followed (without a break) by 10 trials of neutral word stimuli, or the reverse sequence. Thus the color-naming (or keypress) block could be followed immediately by either Stroop or neutral word stimuli. Further, if postcolor block 1 consisted of 10 Stroop stimuli followed by 10 neutral stimuli, block 2 contained the reverse sequence. In this way, the break between postcolor blocks 1 and 2 never involved a change either of task or of stimulus type. There were thus three different types of transition to word-reading trials that might trigger a possible restart effect: (1) a 2 sec task interrupt with renewed instructions and also with a switch of tasks, at the start of postcolor block 1; (2) a 2 sec task interrupt with renewed instructions but *without* a switch of task or a change of stimulus type, at the start of postcolor block 2; and (3) a change of stimulus type, but without a task interrupt or a switch of task, at the transition from the first to the second 10 trials of each block.

The results are illustrated in figure 2.5. As in experiment 3, the first trial of each block, in each of the experimental conditions (baseline word reading; color naming; postcolor word reading, block 1; postcolor word reading, block 2), showed a highly consistent ($p < 0.0001$) increase in RT, relative to trials 2–10, and a reduction in errors ($p < 0.001$). In addition, on the first trial of postcolor block 1 (i.e., on the switch trial immediately following the color-naming block, for the experimental subjects) this restart RT effect appears massively enhanced, when compared either to the control group or to the precolor baseline; the effect (i.e., the RT difference betwen trial 1 and all subsequent trials in the block) was also significantly ($p < 0.0001$) larger for Stroop than for neutral trials. The control group showed *no* additional performance cost (switching cost) on shifting from the keypress task back to word reading, relative to their first-trial baseline performance where there was no shift of task.

Unlike experiment 3, however, the performance costs on postcolor word reading, for the experimental group, appear to be confined entirely to the first trial of the run. The large performance costs on later postcolor word-reading trials, found in experiment 3 and also (on a much longer timescale) in Stroop 1935, are absent on the later repeat trials of experiment 4. One factor that varies considerably between these different experiments is the ratio of (Stroop) color-naming to word-reading trials.

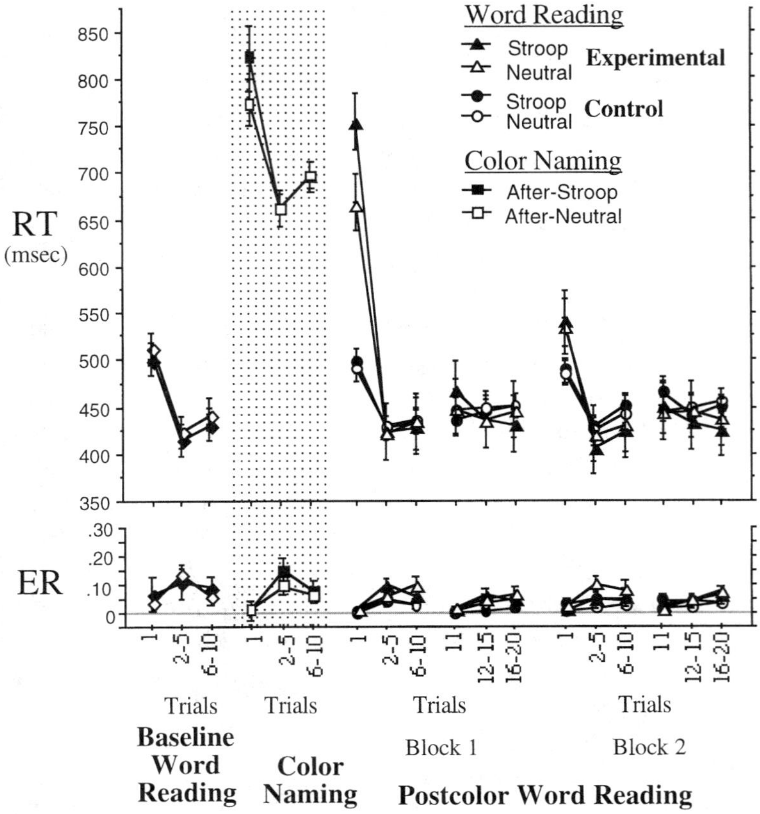

Figure 2.5 Reaction times (RT) and error rates (ER) in experiment 4. Error bars show 95% within-subject confidence intervals. All subjects first performed a baseline condition of word reading. They then performed repeated, successive blocks of either Stroop color naming (for the experimental groups) or an unrelated (keypress) RT task (for the control group), followed immediately by two blocks of (Stroop and neutral) word reading.

In Stroop 1935, subjects began the postcolor word-reading phase with a massive preponderance of color naming, in response to Stroop stimuli, in their recent experience. Experiment 4—the only experiment in this series where repeat trials showed no between-task interference—also had the lowest ratio (1:4) of color-naming to word-reading trials. Wylie and Allport (forthcoming) provide further evidence suggesting that the (recency-weighted) *ratio* of color-naming to word-reading trials, in response to the same set of bivalent Stroop stimuli, massively affects switch costs, as well as repeat trial RTs.

Color-naming RTs also showed a small effect of the immediately preceding word-reading condition. Trial 1 of color-naming was some 45 msec slower, on average, when the last ten trials of block 2, in the preceding word-reading cycle, consisted of ten Stroop, rather than ten neutral, word-reading trials.

However, experiment 4 was designed primarily to investigate the possible effects of three different types of restart trials on postcolor word-reading performance (the effects on the switch trials (i.e., block 1, trial 1) have already been discussed). The second type of possible restart trial was simply a change of stimulus type, with no temporal interrupt and no switch of task (i.e., trial 11 of postcolor blocks 1 and 2). As figure 2.5 shows, this manipulation had relatively little effect, besides a transient increase in accuracy, though there was a hint of an RT cost (trial 11 versus later trials in the block; $p = 0.077$) on the change from neutral to Stroop stimuli in postcolor block 1.

The third type of restart trial was at the start of postcolor block 2. The first trial of block 2 followed a 2 sec interrupt with renewed instructions but *without* a switch of task or a change of stimulus type. At this point, the experimental subjects had last engaged in Stroop color naming 20 trials before. Nevertheless, here again their RTs showed a significantly ($p < 0.025$) enhanced restart cost, relative to the control subjects' first-trial RT, analogous to the renewed task interference found previously in experiment 3 (in trial 1 of postcolor block 2). Unlike experiment 3, however, there was no sign of a differential cost for Stroop and neutral word stimuli. In experiment 3, subjects who read incongruent Stroop words in postcolor block 2 had only responded to neutral words in block 1 ("neutral-first" subjects). By contrast, in the present experiment all subjects had read both Stroop and neutral words in the preceding block 1. It seems plausible that this difference in prior exposure to bivalent stimuli, during postcolor word reading, may be responsible for this difference between experiments.

In summary experiment 4 confirms and extends three major findings from experiment 3:

1. The first trial of a run of speeded-response trials shows a substantial RT cost—the restart cost—generally (but not always) associated with a reduction in errors. (In experiment 3, there was a marked *increase* in errors on the first (postcolor) word trials.) This restart effect is found on the first trial of a run, without any switch of task (cf. also De Jong et al. forthcoming; Gopher, Greenshpan, and Armony 1996; Gopher, Armony, and Greenshpan forthcoming);

2. Prior performance of divergent S-R mappings (e.g., Stroop color naming) in response to the same (or related) stimuli as the current task (e.g., word reading), greatly amplifies or enhances the basic RT cost on restart trials (e.g., in the first postcolor block), relative to control subjects who have not been exposed to the competing, divergent task;

3. An enhanced RT cost (relative to the basic first-trial RT pattern seen in control subjects) occurs also on the first trial of *subsequent* trial blocks, many trials later (postcolor block 2). The effect looks like a rebound of the earlier—so-called—"switch cost", except that, in this case, there was *no* switch of task from the preceding trials.

Discussion of Experiments 3 and 4

Together, these findings raise a number of provocative and important questions. First, there is the restart RT cost itself. What is the causal relation (if any) between this effect and the RT switch cost, typically also found only on the first trial of a run (cf. Rogers and Monsell, 1995)? Second, what is the relation between either of these phenomena and finding 3 above, namely, the rebound of *enhanced* RT costs (over and above the basic restart effect seen in the control subjects' RT) on the first trial of later, *nonswitch* runs of word reading? Clearly, this is a rebound of task interference, resulting from earlier performance of the Stroop color-naming task because the enhanced RT cost is defined precisely by comparison with the first-trial RTs of the control subjects, who had not encountered the color-naming task. That such interference can be *re*elicited in later pure task blocks, with no intervening trials of the competing color-naming task—hence with *no intervening switch of set* from color naming to word reading—strongly favors some kind of learning or memory-based account, whereby the task stimulus (at the start of a new run of trials) triggers retrieval of the prior (conflicting) S-R bindings. The rebound phenomenon appears inconsistent with Allport, Styles, and Hsieh's interpretation (1994) of task priming (or task set inertia), purely in terms of the persisting activation or inhibition of task-relevant processing pathways. It would be consistent, however, with current models of long-term negative (and positive) priming, as the product of associative learning (S-R and S-S bindings), formed in the course of previous processing episodes (e.g., Becker et al. 1997; Lowe 1998; Neill 1997; cf. also Goldinger 1998). We suggest that the retrieval of conflicting S-R mappings further delays the system from settling to an internally consistent set of stimulus-to-task or stimulus-to-response bindings—consistent also with the currently activated task "goals." In terms of this speculative account of the results, the fact that *neutral* word stimuli also showed massive, first-trial rebound interference, as a result of prior Stroop color naming, implies that the S-R bindings formed during the course of the Stroop color task must have included associative bindings between the *distractor* (word) stimuli and some inhibitory ("do not respond") action codes, thus generating long-term negative priming—or negative transfer to the word-reading task (cf. Allport, Tipper, and Chmiel 1985; Lowe 1998; Neill et al. 1992).

These rebound interference effects at the start of later *nonswitch* trial blocks appear strikingly similar to the RT costs found on "true" switch trials, referred to generally, hitherto, as "residual switching costs." The apparent similarity of these two effects inevitably raises the question whether the same causal process may be responsible for both. (Two entirely separate mechanisms would seem uneconomical, to say the least.) Because the rebound cost occurs with no immediate switch of set between the two competing tasks, the resemblance of this phenomenon to the RT costs on immediate switch trials thus calls in question whether "residual switching costs" are appropriately so named. Of course, there may be some additional processing cost on immediate switch trials, not present in the rebound RT costs; if so, however, the results of experiment 3 suggest that this additional component may contribute only a small part of the residual switch cost, at least in some conditions; further discussion will be postponed until after experiment 5.

Experiment 5: Item-Specific Priming, S-R Bindings, and Task Switching

An important issue we have not yet addressed is the extent to which the priming of competing S-R mappings applies to *processing pathways* as a whole, namely, in these experiments, separable pathways for color naming or word reading *in general*, for example, the "grapheme-phoneme correspondence" (GPC) system (Coltheart 1985), and the extent to which these priming effects might be *item specific*, pertaining to individual S-R mappings. The distinction is fundamental (cf. Monsell, Taylor, and Murphy forthcoming). Positive or negative priming of a postulated processing pathway, as a whole, can be thought of as a possible mechanism of task readiness, or task set (e.g., Cohen, Dunbar, and McClelland 1990), whereas positive or negative priming of individual, item-specific S-R mappings cannot.

In our experiments 1–4, as in most other studies of task switching, we used a fixed set of stimuli and responses, each of which occurred many times in the course of the experiment. Moreover, in the "all-Stroop" conditions, which resulted in by far the largest RT interference costs, subjects encountered the identical, incongruent conjunctions of color and color word in both color-naming and word-reading tasks, with complete overlap of stimulus sets. What would happen if we reduced this stimulus overlap, even in part? To begin to address this question, we designed an experiment in which we probed subjects' postcolor word-reading performance (1) on words that had been presented as distractors in the Stroop color-naming task, as in previous experiments; and (2) on words that subjects had never encountered in the color-naming task.

The experimental rationale is as follows. Insofar as long-term negative priming, across a switch of task from Stroop color naming to word reading, is *item specific*, this effect should apply *only* to the particular subset of distractor words, word-color conjunctions, or both, encountered during Stroop color naming. On the other hand, insofar as the negative priming mechanism applies to the *word-processing pathway* as a whole, Stroop color naming should result in equal performance costs, for word reading, in response to all word stimuli, regardless of whether they had occurred as distractors during the prior Stroop color naming or not.

In experiment 5, subjects again alternated between short runs of color naming and word reading. All color naming was in response to incongruent Stroop stimuli; word reading was probed in response to both Stroop and neutral word stimuli. We used a set of eight possible colors (red, green, blue, purple, pink, orange, brown, and yellow) and the corresponding eight color words. The specific manipulation of stimulus overlap in experiment 5 was as follows. For the Stroop color-naming task, subjects saw, and named all eight colors; however, these were presented in conjunction with only *four* of the possible color words as distractors, resulting in just 28 (4×7) possible incongruent conjunctions of color and word, which occurred equiprobably. (Different subsets of four distractor words were presented to different subjects.) For the Stroop *word-reading* task, in contrast, all eight colors and color words occurred equiprobably, in each of the 56 possible incongruent conjunctions of color and word. Similarly, for neutral word reading, all eight color words were presented. (As in previous experiments, neutral words appeared in the form of outline letters, appearing "transparent" to the gray color of the screen background. Stroop stimuli used the same outline letters, but incongruently "colored in.")

Recall that, as in the previous four experiments, the color or word identities (concepts) presented on trial n could not occur, either as color or word, on trial $n+1$. Thus "negative priming" across *immediately successive* trials, either within or between tasks, was excluded.

Experiment 5 began with 30 trials of practice at the word-reading task, in response to both Stroop and neutral stimuli, followed by 30 test runs of word reading (3 trials per run, alternate runs of Stroop and neutral), to provide a baseline of pure task word performance. In the baseline condition, as in later parts of the experiment, each run of word reading was preceded by a 2 sec precue interval (see "Task Cuing" below). After the baseline word reading, for the remainder of the experiment, subjects alternately and repeatedly performed short runs of color naming followed by word reading (as in experiment 4), for a total of 60 cycles. Unlike experiment 4, however, in each cycle there were *seven* trials of the Stroop color-naming task followed by just *three* trials of word reading. (Thus, in the repeating cycles, the ratio of color naming to word reading trials was 7:3, in contrast to the 1:4 ratio in experiment 4.)

Task Cuing The monitor screen was bisected by a bold horizontal line. For half the subjects, stimuli for the color-naming task appeared in the top half of the screen, 2 cm above the horizontal line, and stimuli for word reading appeared 2 cm below the line; for the other subjects, this arrangement was reversed. (To ensure that subjects did not forget this rule, the word "WORD" remained present, as a reminder, at the top (or bottom) edge of the screen, respectively, and a bar of eight colors at the bottom (or top) edge, throughout the alternating runs.) The stimulus location (and

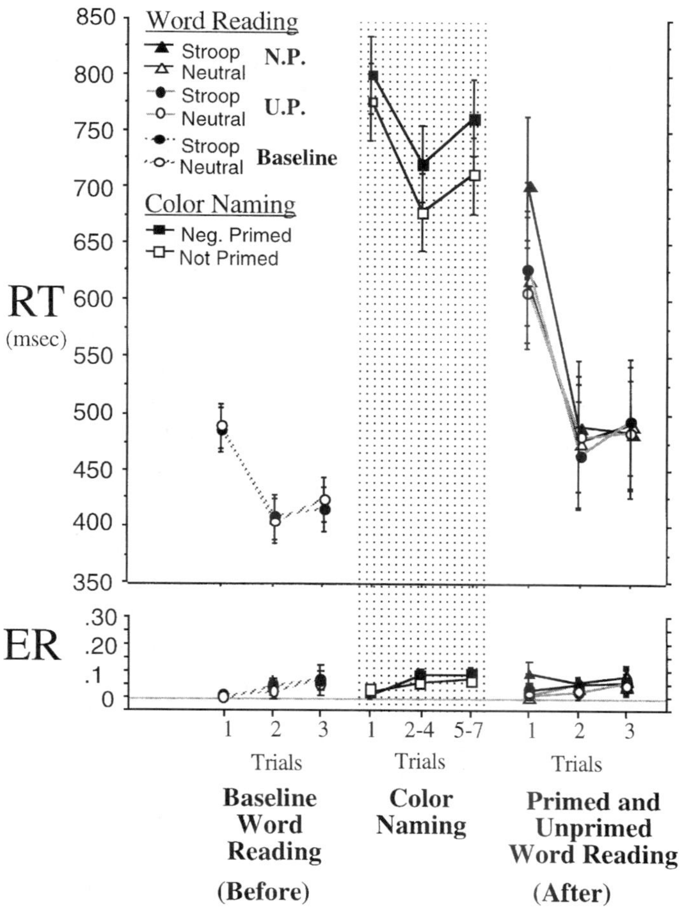

Figure 2.6 Reaction times (RT) and error rates (ER) in experiment 5. All subjects first performed a baseline condition of word reading. As in experiment 4, they then performed repeated, successive blocks of Stroop color naming and word reading. One set of words presented for word reading had appeared also as the distractors in the Stroop color-naming trials (NP or "negatively primed" items); another set of words appeared *only* in the word-reading task (UP or "unprimed" items).

hence the task) was precued by the appearance of a lighter gray rectangle, outlined in black, on the darker gray screen, in the location where the next color-naming or word-reading stimulus would appear. The light gray rectangle then remained present during the remaining trials in the run. Each cycle (starting with the seven color-naming trials) was initiated by the subject, by pressing a key when ready. The first color-naming stimulus then appeared after a delay of 600 msec, and remained on until the subject's response. Within a run, RSI between successive color-naming trials was fixed at 300 msec. After subjects had responded to the last color-naming trial, there was a blank interval of 800 msec; then the light gray rectangle reappeared, in the word location, surrounded by a bold black outline, for 600 msec; the black outline was then removed, leaving the light gray rectangle for 600 msec before the first word-reading stimulus appeared. The RSI between color-naming and word-reading runs was thus 2.0 sec. There were 8 subjects, of whom 6 were female (mean age 37 years).

The results are illustrated in figure 2.6. Baseline word reading again showed a highly reliable first-trial RT cost of about 80 msec (relative to trials 2 and 3), combined with a significant drop in the error rate. There was no reverse Stroop interference in the baseline condition. In postcolor word reading, the first-trial RT cost increased from 80 msec (baseline) to about 140 msec for unprimed stimuli, and to 220 msec for the negatively primed Stroop stimuli. The difference between the first trial RT to the negatively primed Stroop stimuli and the first trial RTs in the other three postcolor conditions was highly reliable ($p < 0.0001$ in each case). On the nonswitch trials 2 and 3, by contrast, the only reliable differences in word-reading RT were between the baseline and *all* other (postcolor) word-reading conditions (trial 2, $p < 0.0005$; Trial 3, $p < 0.0001$); the postcolor performance cost, relative to baseline, on these nonswitch trials was 50–70 msec in mean word-reading RT.

The color-naming task also showed a substantial first-trial effect both in RTs ($p < 0.01$) and errors ($p < 0.005$), though with a tendency for RTs to increase again later in the run. The color-naming task was included primarily to induce negative priming in word reading. However, the manipulation of presenting only half of the word set as distractors, in the color-naming task, means that long-term (*within*-task) negative priming can also be tested for in the color-naming RTs. Consider: four of the color name responses, in the color task, were also potentially elicited (on other color-naming trials) by the corresponding word distractors; hence (on most accounts of the Stroop color-naming task) these color names would have had to be actively suppressed, when they occurred as potential responses to the word distractors. There were four *other* color name responses, however, that were never evoked by their corresponding word distractors, because these distractor items were not presented in the color-naming task. The first set thus includes (long-term) distractor-to-target repetition; the second does not. Note that distractor-to-target repetition on immediately successive trials was excluded by the experimental design. Comparison of the color-naming RTs to these two stimulus subsets should thus provide an index of (long-term) negative priming within the color-naming task. This comparison resulted in a highly reliable negative priming effect ($p < 0.0005$) on *color-naming* RTs, which did not interact reliably with trial position.

Discussion of Experiment 5

The five principal results of this experiment can be summarized as follows:

1. All postcolor word reading showed a substantial performance cost, relative to the prior baseline performance. All stimulus types also showed a further, *enhanced* performance cost on postcolor trial 1, relative to trial 1 in the baseline (precolor) performance.

2. On trials 2 and 3, the postcolor performance cost was the same, regardless of whether the individual words had occurred as distractors in the color-naming task—and hence (nonresponse to) these stimulus items could have been, individually, negatively primed—or not. In other words, on trials 2 and 3, there was interference—long-term "negative priming"—affecting (some element of) the word-reading task or the word-processing pathway *as a whole*, independent of any item-specific priming. This assertion receives its most compelling support from the observation that postcolor word-reading RTs, in response to ("primed" or "unprimed") *neutral* words, were consistently slower than in the baseline condition (with the identical set of stimuli). Neutral word stimuli, as such, were of course never encountered in the color-naming task, although half of these word stimuli were presented, in the color-naming task, as distractors in a color-word conjunction; the other half were not. This latter manipulation had no effect whatever, either on trial 1 or on later trials. The question of whether these postcolor performance costs on nonswitch trials apply to *all* word reading, or specifically to the reading of English color names, or words in the same typeface, or words sharing other contextual features with the stimuli (or responses) in the color task, is beyond the scope of this experiment. Clearly, these are key questions to be resolved by future research.

3. In addition, however, postcolor word-reading RTs to negatively primed (NP) *Stroop* stimuli showed an enhanced performance cost, on trial 1 only, that appears to be strongly *item specific*. Interestingly, this is the only condition in which word-reading responses were made to the same conjunctions of color and word that had been presented previously in the color-naming task. This rather surprising pattern of results—a large first-trial cost for NP *Stroop* stimuli; no additional first-trial cost for NP *neutral* words—would thus be consistent with the possibility that S-R bindings or connection weights, formed in the Stroop color-naming task, might be specific to the individual conjunctions of task-relevant and -irrelevant stimulus attributes. An alternative possibility, also consistent with these results, would be that this component of the negative priming from the color-naming task was specific to individual words, as distractors, but encoded simply as being "colored in" (in any colour?). Although these intriguing conjectures remain to be established, the theoretically crucial point is that a substantial component of the first-trial switch cost, with repeated stimuli, is apparently *item specific*.

4. Unprimed (UP) word stimuli (i.e., items not presented as distractors in the color-naming task) do not appear to show any performance differ-

ences between Stroop and neutral words, even on postcolor trial 1.[5] This would suggest that the large, first-trial, reverse Stroop effects on postcolor word-reading RTs found in all our previous experiments may also reflect *item-specific* priming from the prior color-naming task. This question, also, clearly invites further experiment.

5. The color-naming task provided clear evidence of within-task negative priming (distractor-to-target repetition costs) across stimulus domain, that is, from the word distractors to later color-naming response; cf. Neill 1977; Neill and Westberry 1987; Tipper and Driver 1988). Recall, however, that distractor-to-target concept repetition over immediately successive trials was excluded in experiment 5, as in the previous experiments. Analyses of lag effects, between the occurrence of an individual distractor and its re-presentation as a target probe, are beyond the scope of the present chapter; suffice to say that the within-task negative priming in color naming is relatively long-lasting, consistent with our account of the implicit retrieval of earlier S-R bindings that link specific distractors and "do not respond" codes. Analyses of distance (lag) effects in the *between-task* negative priming from color naming to word reading on postcolor trial 1 failed to find any reliable effect of the number of intervening trials between the most recent occurrence of an item, in the color-naming task, and its being probed on a switch to Stroop word reading. Again, further experiments, specifically designed to examine these issues, are required.

2.7 CONCLUSIONS

What have we learned from all this, as regards selection-for-action (Allport 1980, 1987, 1989) in Stroop-like tasks, and the effects of task alternation? (We note that our conclusions may—or may not—turn out to be confined to task switching in response to incongruent Stroop stimuli. Only further research can tell.)

Priming versus "Switching" Costs

Negatively, the results of each of our five experiments challenge what has appeared to many people as the intuitively obvious interpretation for the residual switch costs (switch minus repeat RTs, at long RSIs), namely as the time cost of an interpolated control operation that shunts the processing system from one configuration of task readiness to another. These arguments have been presented at various points in the chapter, and will not be laboured further here.

More positively, we have shown that negative priming (or negative transfer) from prior, divergent S-R mappings can have massive and long-lasting interference effects on the speed and accuracy of response to the same or overlapping stimuli, following a shift (or reversal) of those mappings. This between-task proactive interference can be observed on both switch and nonswitch trials, and even in pure task conditions. The prior, competing task may have been last performed some considerable time before (cf. Stroop 1935, exp. 3; Allport, Styles, and Hsieh 1994, exp. 4; and experiments 2, 3, and 4 in this chapter). Between-experiment comparisons suggest that the relative frequency and recency of the competing S-R mappings strongly affect the size of the RT interference costs (see also Wylie and Allport forthcoming). Related to this, Lowe (1998) has reported evidence that *within*-task negative priming (over a 5 min delay) increased with the number of times that an item had been previously ignored.

Strikingly, however, the present between-task proactive interference—that is, long-term negative priming or negative transfer resulting from prior execution of competing S-R mappings—has by far its greatest effects on the *first trial* of a run of speeded RT trials. It seems clear, therefore, that such proactive interference forms a major component of what have hitherto been referred to as the "residual switch costs," obtained by subtracting nonswitch trial RTs from switch trial (i.e., first-trial) RTs, at long RSIs, following a shift of tasks. Critically, for the interpretation of switch costs, we have shown that there is no need for the source of this proactive interference to be the execution of the competing task on the *immediately preceding trial.* RT interference effects from competing S-R mappings can also be reevoked on the first trial of *later* trial blocks in pure task conditions. They can also be triggered by alternation between intrinsically noncompeting tasks, with no overlap of either stimuli or responses, hence no need, in principle, to disengage one task set and reengage another (Allport, Styles, and Hsieh 1994, exp. 4).

Moreover, even in the absence of any obvious source of negative priming by competing tasks (e.g., neutral word reading in the precolor, baseline conditions) the *first trial* of a run of speeded RT trials appears to be characteristically slow (and accurate): the restart effect (see also De Jong et al. forthcoming; Gopher, Armony, and Greenshpan forthcoming). The combination of the processes underlying these two effects, we are tempted to speculate, may be responsible in large part for the so-called "residual switch costs" in the cueing and alternating runs paradigms.

Finally, we have shown that the negative priming from prior, competing S-R mappings includes a substantial, *item-specific* component. Evidence for this component was confined to the switch trial itself in the present data (experiment 5). It seems plausible to infer that item-specific priming may also have contributed substantially to the observed differences between switch and repeat trial RTs (switch costs) in other experiments on task switching that similarly used the *same*, repeated set of stimulus items in the pre- and postswitch tasks. It is important to note that item-specific RT costs cannot be explained by (i.e., are logically beyond the scope of) models of task switching that postulate a discrete, stagelike control operation (task set reconfiguration) that precedes stimulus identification. Nor, for that matter, can they be explained by task set inertia—in which "task set" is conceived of as a control state that affects the efficiency of different tasks (or task processing pathways) *as a whole*.

A Tentative Model of Goal Setting and Selection for Action

As already suggested by several authors, it seems clear that alternation between competing S-R tasks (typically, tasks with divergent S-R mappings, in response to overlapping stimulus sets) involves a number of different processes and effects. We identify at least three. Following Fagot (1994), we earlier distinguished "goal setting" (or goal activation, including presumably the deactivation of other, competing goals) and "performance readiness" (i.e., the time needed for the system to "settle" to a unique response). Performance readiness, we tentatively propose, depends on at least three further factors: (1) the prior acquisition of both congruent and conflicting S-R bindings, learned in the course of earlier processing interactions, and giving rise to the negative—and positive—transfer (priming) effects we have attempted to illustrate here; (2) the cue-dependent activation of task-relevant (or -irrelevant) subsystems (e.g., subsystems involved in the coding of cue-related stimulus attributes, response attributes, or both); and (3) suppression or inhibition of subsystems that encode competing (distractor-related) attribute domains.

The process of goal activation—not directly studied in these experiments—can presumably be triggered in advance of an imperative task stimulus by appropriate externally or internally generated cues. Task precues (like task stimuli themselves) may also evoke activation or suppression of appropriate (or inappropriate) stimulus attribute domains, in advance of the task stimulus (cf. Chelazzi et al. 1993; Luck 1998; Miller 1999).[6] In contrast, neither of these processes (temporary goal activation; preactivation of domain-specific subsystems) should have any effect on the potentially conflicting, learned S-R connection weights, which may simply not be susceptible to direct modification by "control processes." On the other hand, our results lead us to believe, S-R connection weights are indeed subject to continuous (and very substantial) modification, through learning, in the course of trial-by-trial sensory-motor processing.

We find it helpful to think of attention and "control" issues in terms of the integrated competition (IC) hypothesis, as put forward by Duncan and colleagues (Duncan 1996; Duncan, Humphreys, and Ward 1997; see also Phaf, van der Heijden, and Hudson 1990). Ward (1999) has described a simple model of selection-for-action that illustrates some of the basic assumptions of IC, in the form of a multimodule, interactive activation and competition (IAC) network (McClelland and Rumelhart 1981), similar to that put forward by Phaf, van der Heijden, and Hudson (1990). In this model, if one "goal node" is strongly activated (i.e., clamped on), and other competing nodes inactivated, the IAC network can only settle to states consistent with the activated goal. In other words, goal activation determines (constrains) *which task* is performed.[7] The *processing time* (number of cycles) needed to settle to a unique response, on the other hand, will depend on the amount of conflict in the network. Associations (connection weights) formed in the execution of a prior, competing task, we suggest, can contribute massively to such conflict.

NOTES

Work on this chapter was funded by a studentship to Glenn Wylie from the McDonnell-Pew Foundation, through the Oxford Center for Cognitive Neuroscience. We gratefully acknowledge this support.

1. Of course, over a longer timescale there may be loss of arousal, or of "task activation," but over immediately successive trials, in motivated subjects, any such effects should not normally be expected to play a substantial role.

2. At the time that Allport, Styles, and Hsieh (1994) put forward the TSI hypothesis, there was little evidence available that the RT cost of a switch of tasks could be reduced by precuing or anticipatory preparation. Since then, a number of studies (e.g., De Jong, chap. 15, this volume; Meiran 1996; Rogers and Monsell 1995) have shown clear evidence of RT benefits of these manipulations.

3. A related distinction between "goal setting" and S-R "rule activation" is an important feature also of the model by Rubinstein, Meyer, and Evans (forthcoming).

4. The intertrial interval (ITI) varied in experiment 1 only, depending on the time needed by the experimenter to code the subject's response, via a keypress. In experiments 2–5, ITI was computer controlled.

5. Analysis by items in experiment 5 showed a small RT advantage for neutral words in postcolor word reading, but the contrast was not reliable.

6. After extended practice, more abstract task cues may come to do the same (see the striking data of Sudevan and Taylor 1987, fig. 3). Learned (and, in principle, arbitrary) cue-to-attribute-domain activation links may be an important component of the "endogenous" task preparation effects that many authors have reported.

7. Goal setting will be effective to the extent that the goal has been sufficiently strongly activated, that is, is effectively "clamped on," and competing goals inactivated. It is a matter of degree. In human subjects many factors (motivation, practice, etc.) are liable to affect this (Goschke and Kuhl 1993).

REFERENCES

Allport, A. (1980). Attention and performance. In G. Claxton (Ed.), *Cognitive psychology: New Directions*, pp. 112–153. London: Routledge and Kegan Paul.

Allport, A. (1987). Selection-for-action: Some behavioural and physiological considerations of attention and action. In H. Heuer and A. F. Sanders (Eds.), *Perspectives on perception and action*, pp. 395–419. Hillsdale, NJ: Erlbaum.

Allport, A. (1989). Visual attention. In M. I. Posner (Ed.), *Foundations of cognitive science*, pp. 631–682. Cambridge, MA: MIT Press.

Allport, A., Styles, E. A., and Hsieh, S. (1994). Shifting intentional set: Exploring the dynamic control of tasks. In C. Umiltà and M. Moscovitch (Eds.), *Attention and Performance XV*, pp. 421–452. Cambridge MA: MIT Press.

Allport, A., Tipper, S. P., and Chmiel, N. R. J. (1985). Perceptual integration and postcategorical filtering. In M. I. Posner and O. S. M. Marin (Eds.), *Attention and Performance, XI,* pp. 107–132. Hillsdale, NJ: Erlbaum.

Allport, A., and Wylie, G. (1999). Task-switching: Positive and negative priming of task-set. In G. W. Humphreys, J. Duncan, and A. M. Treisman (Eds.), *Attention, space and action: Studies in cognitive neuroscience,* pp. 273–296. Oxford: Oxford University Press.

Becker, S., Moscovitch, M., Behrmann, M., and Joordens, S. (1997). Long-term semantic priming: A computational account and empirical evidence. *Journal of Experimental Psychology: Learning Memory and Cognition, 23,* 1059–1082.

Chelazzi, L., Miller, E. K., Duncan, J., and Desimone, R. (1993). A neural basis for visual search in inferior temporal cortex. *Nature, 363,* 245–347.

Cohen, J. D., Dunbar, K., and McClelland, J. L. (1990). On the control of automatic processes: A parallel distributed processing account of the Stroop effect. *Psychological Review, 97,* 332–361.

Coltheart, M. (1985). Cognitive neuropsychology and the study of reading. In M. I. Posner and O. S. M. Marin (Eds.), *Attention and Performance, XI,* pp. 3–37. Hillsdale, NJ: Erlbaum.

De Jong, R. (1996). Cognitive and motivational determinants of switching costs in the task-switching paradigm. Paper presented at the Ninth ESCOP Conference, Würzburg, Germany, September.

De Jong, R., Emans, B., Eenshuistra, R., and Wagenmakers, E.-J. (Forthcoming). Strategies and intrinsic limitations in intentional task control.

De Schepper, B., and Treisman, A. (1996). Visual memory for novel shapes: Implicit coding without attention. *Journal of Experimental Psychology: Learning, Memory, and Cognition, 22,* 27–47.

Duncan, J. (1996). Cooperating brain systems in selective perception and action. In T. Inui and J. L. McClelland (Eds.), *Attention and Performance, XVI,* pp. 549–578. Cambridge, MA: MIT Press.

Duncan, J., Humphreys, G. W., and Ward, R. (1997). Competitive brain activity in visual attention. *Current Opinion in Neurobiology, 7,* 255–261.

Fagot, C. (1994). Chronometric investigations of task switching. Ph.D. diss., University of California, San Diego.

Fox, E. (1995). Negative priming from ignored distractors in visual selection: A review. *Psychonomic Bulletin and Review, 2,* 145–173.

Goldinger, S. D. (1998). Echoes of echoes? An episodic theory of lexical access. *Psychological Review, 105,* 251–279.

Gopher, G., Greenshpan, Y., and Armony, L. (1996). Switching attention between tasks: Exploration of the components of executive control and their development with training. In *Proceedings of the Annual Meeting of the Human Factors and Ergonomics Society.*

Gopher, G., Armony, L., and Greenshpan, Y. (Forthcoming). Switching tasks and attention policies and the ability to prepare for such shifts. *Journal of Experimental Psychology: General.*

Goschke, T., and Kuhl, J. (1993). Representation of intentions: Persisting activation in memory. *Journal of Experimental Psychology: Learning, Memory, and Cognition, 19,* 1211–1226.

Hintzman, D. L. (1986). "Schema abstraction" in a multiple-trace memory model. *Psychological Review, 93,* 411–428.

Hommel, B. (1998). Event files: Evidence for automatic integration of stimulus-response episodes. *Visual Cognition, 5,* 183–216.

Houghton, G., and Tipper, S. P. (1994). A model of inhibitory mechanisms in selective attention. In D. Dagenbach and T. Carr (Eds.), *Inhibitory processes in attention, memory and language,* pp. 53–112. San Diego, CA: Academic Press.

Jersild, A. T. (1927). Mental set and shift. *Archives of Psychology, no. 89.*

Joordens, S., and Becker, S. (1997). The long and the short of semantic priming effects in lexical decision. *Journal of Experimental Psychology: Learning, Memory, and Cognition, 23,* 1083–1105.

Kane. M. J., May. C. P., Hasher, L., Rahhal, T., and Stoltzfus, E. R. (1997). *Journal of Experimental Psychology: Human Perception and Performance, 23,* 632–650.

Loftus, G. R., and Masson, M. E. J. (1994). Using confidence intervals in within-subject designs. *Psychonomic Bulletin and Review, 1,* 476–490.

Logan, G. D. (1988). Toward an instance theory of automatization. *Psychological Review, 95,* 492–527.

Logan, G. D., and Etherton, J. L. (1994). What is learned during automatization? The role of attention in constructing an instance. *Journal of Experimental Psychology: Learning, Memory, and Cognition, 20,* 1022–1050.

Lowe, D. (1998). Long-term positive and negative identity priming: Evidence for episodic retrieval. *Memory and Cognition, 26,* 435–443.

Luck, S. J. (1998). Neurophysiology of selective attention. In H. Pashler (Ed.), *Attention,* pp. 257–295. Hove, U.K.: Psychology Press.

MacLeod, C. M. (1991). Half a century of research on the Stroop effect: An integrative review. *Psychological Bulletin, 109,* 163–203.

MacLeod, C. M., and Dunbar, K. (1988). Training and Stroop-like interference: Evidence for a continuum of automaticity. *Journal of Experimental Psychology: Learning, Memory, and Cognition, 14,* 126–135.

Maljkovic, V., and Nakayama, K. (1994). Priming of popout: 1. Role of features. *Memory and Cognition, 22,* 657–672.

McClelland, J. L., and Rumelhart, D. E. (1981). An interactive activation model of context effects in letter perception: 1. An account of basic findings. *Psychological Review, 88,* 375–407.

Meiran, N. (1996). Reconfiguration of processing mode prior to task performance. *Journal of Experimental Psychology: Learning, Memory, and Cognition, 22,* 1423–1442.

Meiran, N., Chorev, Z., and Sapir, A. (Forthcoming). Component processes in task-shifting. *Cognitive Psychology.*

Meuter, R. F. I., and Allport, A. (1999). Bilingual language switching and naming: Asymmetrical costs of language selection. *Journal of Memory and Language, 40,* 25–40.

Miller, E. K. (1999). Prefrontal cortex and the neural basis of executive functions. In G. W. Humphreys, J. Duncan, and A. M. Treisman (Eds.), *Attention, space and action: Studies in cognitive neuroscience,* pp. 251–272. Oxford: Oxford University Press.

Milliken, B., and Tipper, S. P. (1998). Attention and inhibition. In H. Pashler (Ed.), *Attention,* pp. 191–221. Hove, U.K.: Psychology Press.

Monsell, S., Taylor, T. J., and Murphy, K. (Forthcoming). Naming the colour of a word: Is it responses or task-sets that compete? *Memory and Cognition.*

Nakayama, K., and Joseph, J. S. (1997). Attention, pattern recognition, and popout in visual search. In R. Parasusaman (Ed.), *The attentive brain.* Cambridge, MA: MIT Press.

Neill, W. T. (1977). Inhibitory and facilitatory processes in attention. *Journal of Experimental Psychology: Human Perception and Performance, 3,* 444–450.

Neill, W. T. (1997). Episodic retrieval in negative priming and repetition priming. *Journal of Experimental Psychology: Learning, Memory, and Cognition, 23,* 1291–1305.

Neill, W. T., Valdes., Terry. K. M., and Gorfein, D. S. (1992). Persistence of negative priming: 2. Evidence for episodic trace retrieval. *Journal of Experimental Psychology: Learning, Memory, and Cognition, 18,* 993–1000.

Neill, W. T., and Westberry, R. L. (1987). Selective attention and the suppression of cognitive noise. *Journal of Experimental Psychology: Learning, Memory, and Cognition, 13,* 327–334.

Park, J., and Kanwisher, N. (1994). Negative priming for spatial locations: Identity mismatching, not distractor inhibition. *Journal of Experimental Psychology: Human Perception and Performance, 20,* 613–623.

Phaf, R. H., van der Heijden, A. H. C., and Hudson, P. T. W. (1990). SLAM: A connectionist model for attention in visual selection tasks. *Cognitive Psychology, 22,* 273–341.

Rabbitt, P. M. A., and Vyas, S. (1973). What is repeated in the "repetition effect"? In S. Kornblum (Ed.), *Attention and Performance IV,* pp. 327–342. London and New York: Academic Press.

Rabbitt, P. M. A., and Vyas, S. (1979). Memory and data-driven control of selective attention in continuous tasks. *Canadian Journal of Psychology, 33,* 71–87.

Rogers, R. D., and Monsell, S. (1995). The cost of a predictable switch between simple cognitive tasks. *Journal of Experimental Psychology: General, 124,* 207–231.

Rubinstein, J., Meyer, D. E., and Evans, J. E. (Forthcoming). Executive control of cognitive processes in task switching.

Smith, M. C., and Magee, L. E. (1980). Tracing the time course of picture-word processing. *Journal of Experimental Psychology: General, 109,* 373–392.

Spector, A., and Biederman, I. (1976). Mental set and shift revisited. *American Journal of Psychology, 89,* 669–679.

Stoet, G., and Hommel, B. (Forthcoming). Action planning and the temporal binding of response codes. *Journal of Experimental Psychology: Human Perception and Performance.*

Stroop, J. R. (1935). Studies of interference in serial verbal reactions. *Journal of Experimental Psychology, 18,* 643–662.

Sudevan, P., and Taylor, D. A. (1987). The cuing and priming of cognitive operations. *Journal of Experimental Psychology: Human Perception and Performance, 13,* 89–103.

Tipper, S. P., and Driver, J. (1988). Negative priming between pictures and words: Evidence for semantic analysis of ignored stimuli. *Memory and Cognition, 16,* 64–70.

Ward, R. (1999). Interactions between perception and action systems: A model for selective action. In G. W. Humphreys, J. Duncan, and A. M. Treisman (Eds.), *Attention, space and action: Studies in cognitive neuroscience,* pp. 311–332. Oxford: Oxford University Press.

Wylie, G., and Allport, A. (Forthcoming). Task switching and the measurement of "switch costs." *Psychological Research.*

I Control of Visual Attention

3 Goal-Directed and Stimulus-Driven Determinants of Attentional Control

Steven Yantis

ABSTRACT Selective visual attention to objects and locations depends both on deliberate behavioral goals that regulate even early visual representations (goal-directed influences) and on autonomous neural responses to sensory input (stimulus-driven influences). In this chapter, I argue that deliberate goal-directed attentional strategies are always constrained by involuntary, "hard-wired" computations, and that an appropriate research strategy is to delineate the nature of the interactions imposed by these constraints. To illustrate the interaction between goal-directed and stimulus-driven attentional control, four domains of visual selection are reviewed. First, selection by location is both spatially and temporally limited, reflecting in part early visual representations of the scene. Second, selection by feature is an available attentional strategy, but it appears to be mediated by location, and feature salience alone does not govern the deployment of attention. Third, early visual segmentation processes that parse a scene into perceptual object representations enable object-based selection, but they also enforce selection of entire objects, and not just isolated features. And fourth, the appearance of a new perceptual object captures attention in a stimulus-driven fashion, but even this is subject to some top-down attentional control. Possible mechanisms for the interaction between bottom-up and top-down control are discussed.

People are perceptually selective: they subjectively experience and respond to only a subset of the sensory signals evoked by objects and events in the local environment. The psychological and neural mechanisms that mediate perceptual selectivity are collectively termed *attention*. Although often used to refer to other psychological phenomena (e.g., the ability to perform two or more tasks at the same time, or the ability to remain alert for long periods or time), for the purposes of this chapter, "attention" shall refer exclusively to perceptual selectivity, and the examples will concern visual selection in particular.

Like any function of the brain, attention is adaptive: it supports behavior that achieves goals and ultimately promotes survival. Visual selection comprises an exquisite interaction between two mutually constraining factors. First, current behavioral goals can modulate processing of sensory input (top-down or goal-directed influences on selection). Second, properties of the stimulus and "hard-wired" architectural properties of the brain (i.e., properties that do not change with task set) together constrain the implementation of attentional goals (bottom-up or stimulus-driven influences on selection).

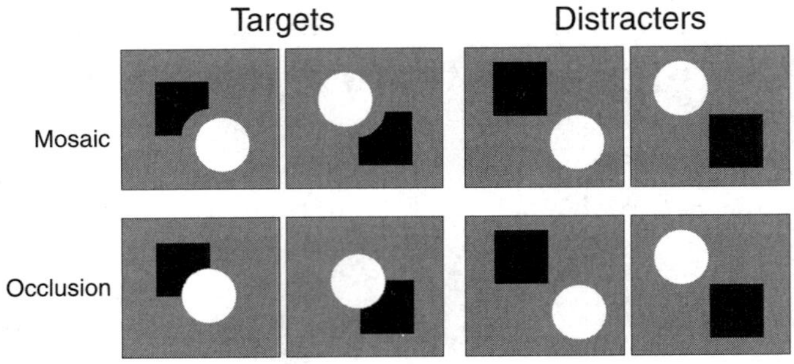

Figure 3.1 Displays from Rensink and Enns 1998. Viewing displays of zero or one target among varying numbers of distractors, subjects were to indicate whether the target was present by pressing one of two buttons as quickly as possible. *Top*. In the "mosaic" condition, targets consisted of fragmented squares and distractors were complete squares. Here, visual search for the target was highly efficient, with search slopes of 7 msec/element. *Bottom*. In the "occlusion" condition, targets now abutted the circles to give the impression of partly occluded squares. Performance was inefficient, with search slopes of 36 msec/element in the target-present case, and 66 msec/element in the target-absent case. Visual search apparently required a deliberate deployment of selective attention to one item at a time. This suggests that the completion of the partly occluded square occurred automatically before visual selection could operate.

Because "attention" is most often used in everyday language to refer to an intentional and deliberate mental process, the autonomous bottom-up influences on selection are sometimes overlooked. They arise as a result of brain mechanisms that perform certain types of computation efficiently and automatically; these are often referred to as "preattentive processes." For example, some forms of perceptual organization (e.g., figure-ground segregation and perceptual grouping) occur without a deliberate intent on the part of the perceiver, although these computations can sometimes be modulated to some extent by task goals. These autonomous computations presumably evolved to speed identification and ensure rapid responses to threatening events, and to free computational resources for higher tasks such as decision making and planning.

Thus when a perceiver with a particular goal encounters a scene, certain early, hard-wired visual computations will occur whether or not they are consistent with the goal. An example of this sort of conflict is reported in Rensink and Enns 1998. Observers were asked to search for a notched square in an array of multiple complete disks and squares (see figure 3.1, top). This was an easy task, and search was highly efficient, suggesting that the target's unique shape could be used to guide search. If, however, the target was placed so that it appeared to be partly occluded by a disk (figure 3.1, bottom), then the task was very difficult. Rensink and Enns concluded that the partly occluded square was perceptually completed by early vision without any deliberate intent to do so; additional scrutiny was required to recover the "real" (proximal) shape of each item in order to detect the target shape. Although this sort of perceptual completion is normally desirable, in this case, it interfered with the perceptual goal, and thereby revealed how an early automatic process can constrain top-down visual selection.

Although the goal-directed and stimulus-driven aspects of attentional control are typically treated as separate and distinct, with most empirical studies focusing on only one of the two factors, it has become increasingly clear that this distinction is untenable. Every episode of selection necessarily manifests both types of influence. The observer always occupies some sort of goal state, and of course the stimulus and its representation in the brain always exert an influence. The question thus becomes not whether or when attention is controlled in a bottom-up or top-down fashion, but how autonomous stimulus-driven influences constrain attentional goals in any given situation.

In this chapter, I review four domains of visual selection with an emphasis on how stimulus-driven factors constrain deliberate attentional deployment. These domains are neither mutually exclusive nor exhaustive; they merely provide a convenient framework for organizing the principles of stimulus-driven constraints on selection.

3.1 SELECTION BY LOCATION

Among the earliest ideas concerning the mechanisms of visual attention was that one can attend to a restricted region of space (e.g., Helmholtz 1866, 455). We have all had the experience of turning our heads and eyes when we are told, "Look over there!" Less obvious, however, is whether one can selectively attend to one spatial location in a scene containing many objects that are all equally visible (e.g., objects that are all equidistant from the center of gaze) without moving one's head or eyes (i.e., attend covertly). How rapidly can selection by location be accomplished, and how efficiently does one reject information to be ignored?

Among the earliest empirical demonstrations that covert selection by location is possible was Sperling's observation (1960) that observers can direct their attention to a specified region of a persisting visual memory of a display. An array of letters was briefly flashed on a screen, and very shortly after the array disappeared, a tone signaled the part of the display to be reported, and hence to be attended (e.g., "If the tone is high, report the items in the top row of the display"). Because the letters themselves were physically absent from the display by the time the tone sounded, overt eye movements to fixate the indicated row were not possible; attention was instead directed to a spatial location through covert movements of "the mind's eye."

A vast body of work carried out in the last four decades has revealed the spatial and temporal limitations of covert spatial selection. Eriksen and colleagues (e.g., Colegate, Hoffman, and Eriksen 1973; Erkisen and Hoffman 1972, 1973) used a cuing paradigm in which a circular array of letters (centered on fixation) was presented, and a small bar marker (the cue) appeared next to one of the letters. Subjects were to identify the cued target letter as rapidly as possible. Performance improved as the distance between the cued letter and its neighbors increased, suggesting a limitation in the spatial precision of attention. Similarly, performance improved as the duration between the onset of the cue and the onset of the letter array increased, suggesting that about 100–300 msec was required to focus attention at the cued location. Subjects were instructed not to move their eyes (in some cases, adherence to this instruction was verified by monitoring eye position) to ensure that covert attentional deployments, and not overt eye movements, were being measured.

Selective attention, as the name implies, entails selection of attended items and rejection of unattended ones; the efficiency of nontarget rejection was the focus of the work by Eriksen and colleagues. In addition, however, there is evidence that attention can speed detection and identification of single targets. Posner and his colleagues (e.g., Posner 1978, 1980; Posner, Snyder, and Davidson 1980) conducted a series of experiments varying the predictive validity of a spatial cue. For example, within a block of trials, the cue might indicate the target location on 80% of the trials (valid cues) and a nontarget location on 20% of the trials (invalid cues); participants were always informed of this contingency. These experiments revealed both benefits for valid cues and costs for invalid cues, relative to "neutral" cues that indicated no particular location.

Eriksen and colleagues originally estimated that nontarget rejection was efficient (i.e., that the identity of adjacent nontargets failed to affect response time and accuracy) as long as the stimuli were at least 1 degree of visual angle apart. Later, LaBerge and colleagues (LaBerge 1983; LaBerge and Brown 1986; LaBerge et al. 1991, 1997) and Downing and Pinker (1985) measured the spatial distribution of attention by cuing attention to a location likely to contain a target element, and then presenting a probe stimulus at other locations in space (see also Engel 1971; Hoffman and Nelson 1981). They generally reported a smooth gradient of selection surrounding the attended location for several degrees of visual angle, rather than a sharp boundary separating attended and unattended regions. More recently, other investigators have refined these techniques to explore the two-dimensional (e.g., Egly and Homa 1984; Eriksen and Yeh 1985; Henderson and Macquistan 1993; Kim and Cave 1995; Usai, Umiltà, and Nicoletti 1995) and three-dimensional (e.g., Atchley et al. 1997; Ghirardelli and Folk 1996; Iavecchia and Folk 1995) profiles of spatial selection. For example, Bahcall and Kowler (1998) have found that attended locations are surrounded by a local inhibitory region, analogous to a center-surround receptive field, which causes the attended target to stand out perceptually against its immediate background in crowded displays.

The intuitive conception of attention as a deliberate, strategic process led early investigators to consider spatial cues that were highly task relevant (e.g., they indicated the likely location of the upcoming target), and often indirect and symbolic (e.g., an arrow appearing in the center of the display pointing to a peripheral location, or a digit indicating a labeled location). In these cases, of course, the observer had an incentive to interpret the cue and actively use it to select the content of the cued location; the emphasis was on the efficiency of goal-directed, controlled deployments of attention.

Many studies have since investigated the extent to which certain stimulus events may be said to capture attention despite either contrary or "neutral" intentions. Jonides (1980, 1981) drew a distinction between peripheral, direct cues (i.e., cues near the impending stimulus locations), and central, symbolic cues (i.e., cues that indicated a location other than the one they occupied and therefore required some translation before the cued location could be decoded). He found that direct cues draw attention even when they are known to be uninformative and should be ignored. In contrast, symbolic cues affected performance only when task instructions required that they be used to direct attention. Evidently, the visual system is hard-wired to select peripheral abrupt onsets, with little need for top-down control (see section 3.4 for a more detailed discussion of this issue).

Investigations of the time course of selection produced by indirect central cues versus direct peripheral cues revealed distinct and characteristic patterns of performance for the two cases (e.g., Cheal and Lyon 1991; Koshino, Warner, and Juola 1992; Müller and Rabbitt 1989; Nakayama and Mackeben 1989). Symbolic cues (e.g., a central arrowhead that points to a likely target location) produce relatively sluggish and sustained attentional effects at the cued location, but only when the cue is task relevant, suggesting that voluntary control is necessary. Direct peripheral cues, in contrast, produce transient performance advantages for cued targets relative to uncued ones within as little as 100 msec after the cue, although these effects dissipate rapidly. Furthermore, the effects of peripheral cues appear subject to little voluntary control. For example, a peripheral cue that observers knew would never appear in the target location, and which should therefore be ignored, nevertheless slowed target identification by drawing attention automatically (Remington, Johnston, and Yantis 1992).

The overall picture that has emerged from these studies is that when attention is directed to a location in space, a spatial gradient is established around the attended location such that items near it are processed more efficiently than comparably visible items elsewhere. The time course of selection by location depends on whether the deployment of attention is deliberate and controlled or an "automatic" consequence of a peripheral visual onset. These spatiotemporal constraints on the deployment of selective attention are very likely imposed by hard-wired properties of the visual system such as the receptive field structure and the temporal precision of early vision.

3.2 SELECTION BY FEATURE

According to Marr (1980, p. 3), the purpose of vision is to "know what is where by looking." This might imply simply opening one's eyes to see what is present, but often it entails searching for a particular object (e.g., red berries). While selection by location (either by moving the eyes or through covert shifts of attention) is sometimes a reasonable strategy because one knows where to look, one may also seek objects with known visual properties (e.g., round and red) but unknown location, which suggests that selection by feature is possible.

Among the first to investigate this issue, von Wright (1970) asked whether selection in the partial report paradigm used by Sperling (1960) could be based on simple attributes other than location, such as color, or more abstract properties, such as meaning. von Wright found that attention could be guided efficiently by simple features (e.g., "Report the names of the red letters"), but not so efficiently by meaning (e.g., "Report the names of the vowels"). Corroborating evidence from studies of visual search by Neisser (1967) and by Egeth, Jonides, and Wall (1972) showed that simple shape differences (e.g., searching for a 4 among Cs) could be used to direct attention efficiently.

In their seminal paper on search for features or conjunctions of features, Treisman and Gelade (1980) found that "feature search," in which the target differs from nontargets in a single salient property (e.g., search for a red target among green nontargets) was much more efficient (as measured by visual search slopes) than was "conjunction search," in which the target was defined by the conjunction of two properties (e.g., search for a red vertical target—a conjunction of color and orientation—among red horizontal and green vertical nontargets). They were able to account for the efficiency of visual search in these tasks by proposing a framework called "feature integration theory." By offering a specific function for attention, the theory led to a surge in research on visual selection during search.

The central claims of feature integration theory were, first, that the visual system represents simple visual features like distinct colors and orientations in separate feature maps (roughly consistent with the neurophysiological results of Hubel and Wiesel 1968, and many others since); and, second, that the function of attention is to bind together the separately represented features belonging to a given object via their common spatial locations. According to the theory, feature search is efficient because one need only monitor, say, a "red map" and if any activation occurs there, a positive response can be made; attention need not be devoted to spatial locations one at a time. When, however, a target is defined by a conjunction of features (e.g., red vertical target among red horizontal and green vertical nontargets), search is inefficient because attention must be directed to one location in the scene at a time, binding the features at that location and allowing a decision about whether the representation so created is the target.

Feature integration theory, in its original form, held that visual selection was essentially an unguided spatial search, at least in conjunction search tasks. Egeth, Virzi, and Garbart (1984) showed, however, that even conjunction search could be guided to some extent. They asked observers to search for targets defined by a conjunction of features (e.g., a red *O* in a field of black *O*s and red *N*s) and found that search could be restricted to the red target among the *O*s, or the *O* target among the red items. Subsequent experiments verified and expanded on this finding (e.g., Driver, McLeod, and Dienes 1992; Nakayama and Silverman 1986; Wolfe, Cave, and Franzel 1989; see Wolfe 1998 for a comprehensive review).

Wolfe and colleagues (Wolfe 1994; Cave and Wolfe 1990, Wolfe, Cave, and Franzel 1989) proposed a theory of visual search called "guided search." Although strongly influenced by feature integration theory, guided search takes into account the guidance by feature values revealed in studies such as Egeth, Virzi, and Garbart 1984. An initial parallel stage represents items in features maps (as in feature integration theory), and then assigns priorities to items according to two criteria: items differing significantly from their neighbors in any given dimension (e.g., color or orientation) receive high bottom-up activation and items similar to the target in any given dimension receive high top-down activation. These two types of activation are combined in a priority map that determines the order of search. The second stage (again as in feature integration theory) involves selecting an item, binding its features into an object representation, and making a decision about whether it is the target. The order in which items are selected is determined by the priority map. Even though search is strictly serial, the guidance provided by the priority map yields efficient search under many circumstances where feature integration theory would have predicted inefficient search. For example, guided search provides a straightforward account of the results of Egeth, Virzi, and Garbart (1984).

Is Location Special?

The research reviewed thus far suggests that one can select an object by directing attention to a location (either randomly or according to an attentional priority schedule). In some sense, guided search holds that an item's features can guide search, but only indirectly through locations that have been assigned high attentional priority in the activation map.

These studies do not reveal whether an item can be selected directly by virtue of its having a particular feature value (e.g., the *red* object). Among the few theories of attention that offer a mechanism for purely feature-based selection is Bundesen's "theory of visual attention" (1990), which holds that only the discriminability of values within a feature dimension affects the efficiency of selection; all dimensions, including location, are assumed to be otherwise equivalent.

There is now substantial evidence, however, that feature-guided selection typically operates by directing attention to a spatial location containing the target-defining feature value (e.g., Tsal and Lavie 1993). In their investigation of this issue, Shih and Sperling (1996) asked whether selection by feature was possible without spatial mediation. On each trial of their experiment, several circular arrays of six letters were presented in rapid succession in the same location, each replacing the previous one. One array contained a single digit, and subjects were to report its identity, location, and color. In the alternating-feature condition, the letters in each array were of the same color, but the color of the arrays alternated (e.g., red, green, red, green); the target was known to be, say, red with high probability. In this condition, if feature-based selection was possible, an improvement in performance should be observed (relative to a baseline in which the target's color is unknown) because at least some of the green items should have been rejected. In the feature-defined location condition, an array consisted of five red items and one green item, alternating with arrays containing five green items and one red item. Again, the target was known to be, say, red with high probability. Shih and Sperling found that when spatial selection was impossible (in the alternating-feature condition), knowledge of the target's feature did not improve performance at all. In contrast, when attention could be directed to a location, as in the feature-defined location condition, performance improved dramatically. They concluded that feature-based selection is mediated by location. Several other reports corroborate this conclusion (e.g., Cepeda et al. 1998; Johnston and Pashler 1990; Moore and Egeth 1998; Tsal and Lavie 1993, 1988; but see van der Heijden et al. 1996). These studies suggest that location should not be viewed as just another feature dimension; instead, it is the medium in which all features are expressed and therefore enjoys a privileged status in visual selection (as Kubovy 1981 put it, location is an "indispensable attribute" in vision).

A potential exception to this claim is worth noting. O'Craven et al. (1997) reported evidence using functional magnetic resonance imaging that observers can selectively attend to stimuli exhibiting motion or stationarity, respectively. Their display consisted of a field of black and white dots on a gray background; the white dots were moving as a converging flow field toward fixation, while the black dots remained stationary. The observer was to shift attention every 20 seconds from the white dots to the black dots and vice versa. Activation in the cortical

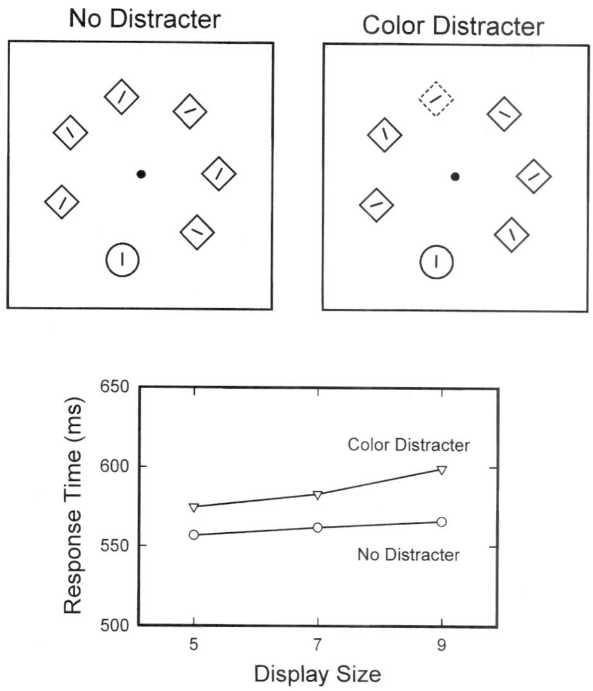

Figure 3.2 Displays and data from Theeuwes 1992. *Top.* Solid contours are green, dashed contours are red. In this example, subjects were to press a button corresponding to the orientation (vertical or horizontal) of the line segment contained within the shape singleton (i.e, the circle). In the "no distractor" condition, all the shapes were the same color (in this case, green). In the "color distractor" condition, one nontarget shape was a different color (here, red). Display size 7 is illustrated. *Bottom.* Response time was significantly slowed in the color distractor condition, suggesting that the color distractor captured attention despite its irrelevance to the task.

motion area MT was strongly modulated by the observer's attentive state. The authors concluded that selective attention to motion per se, not just to a particular spatial location, was possible.

Salience and Attentional Capture by Feature Singletons

An item unique in some feature dimension (e.g., a red item in a scene containing only blue items) is subjectively salient, and is sometimes said to "pop out" of the display. The possibility that such feature singletons may capture attention in a purely stimulus-driven fashion has proven to be a contentious issue. A series of studies by Theeuwes and colleagues starting in the early 1990s suggested that salient feature singletons indeed capture attention despite strategic efforts to the contrary. In particular, Theeuwes showed that when an observer searches for a singleton element, then singletons in a different feature dimension capture attention even though they are known to be irrelevant (see Theeuwes, Atchley, and Kramer, chap. 4, this volume). In Theeuwes 1992, subjects were to report the orientation of a target line segment (horizontal or vertical) that appeared within a green circle, presented together with 4, 6, or 8 green diamonds (figure 3.2). On half the trials, one of the green diamond distractors was replaced by a red diamond, and observers were told that this color singleton distractor could never contain the target and should be ignored. Theeuwes found that the presence of this distracting singleton slowed responses (figure 3.2, bottom), suggesting that even a singleton in an irrelevant visual dimension may capture attention. He concluded that, when searching for a target singleton—an important proviso—there is virtually no top-down control over attention: stimuli will be attended in order of their salience.

A similar result was reported earlier by Pashler (1988): the time required to find a target differing in orientation from the nontargets (i.e., an orientation singleton) was slowed by the presence of an irrelevant color singleton. These findings were interpreted by Bacon and Egeth (1994) as specifically reflecting the observer's intent to select items that are distinct from their neighbors in some feature dimension (see Nothdurft 1993). Calling this state of attentional readiness "singleton detection mode," they argued that, in such a state, an observer relies on a mechanism that computes the magnitude of local feature difference, but that does not supply the identity of the singleton's feature dimension or value (e.g., shape or circle). Search is somewhat unselective in this case: if one must rely on an item's status as a feature singleton, one cannot restrict search to the circle singleton or to the green singleton; instead, one selects the item that differs most from its neighbors, and this may not be the target of search (as in figure 3.2). Singleton detection mode is thus only effective under circumstances in which the target happens to be the most salient element in the display.

Bacon and Egeth (1994) supported this idea by showing that when subjects are prevented from using singleton detection mode in tasks such as those in Theeuwes 1992—for example, by using displays in which there were multiple instances of the target so that the target was no longer a singleton—irrelevant singletons no longer produce the distraction effect observed by Theeuwes. Thus the apparent capture of attention by feature singletons appears to be the result of a deliberate search strategy adopted by observers and that sometimes yields inefficient search. This is a clear example of how a top-down selective strategy is modulated by an early, autonomous visual process such as the computation of local feature contrast.

In a refinement of this conclusion, Folk and Remington (1998) suggest that there are at least two possible causes for the sort of slowing observed by Theeuwes (1992, 1994, 1996). The first is the account offered by Theeuwes himself: when searching for a shape singleton, the local salience of each element in the display is computed, and attention is directed spatially to the most salient element without regard to its featural identity. If this happens to be a color singleton, then responses are slowed because additional time is required to redirect attention to the next most salient element, and so forth. Folk and Remington offer a second possible mechanism for the interference observed by Theeuwes: the presence of a distracting singleton may slow the deployment of attention to the target item by requiring an effortful and time-consuming operation to filter the distractor, but this may not entail a shift of spatial attention to the distractor's location.

In order to determine which of these two possible sources of the interference effect is operative in Theeuwes's experiments, Folk and Remington (1998) employed a paradigm used successfully by Folk and colleagues (e.g., Folk, Remington, and Johnston 1992) to study the interaction between stimulus-driven and goal-directed attentional control (discussed in greater detail below). Their approach was to ask observers to search for a target that differed from nontargets in a single dimension (e.g., a red target among white nontargets, thus inducing an attentional set for red elements). A distracting display was briefly presented before the target display; this display could contain a red singleton or a singleton in some other dimension, and that singleton could appear at the same location as the target or not. Folk and colleagues had already demonstrated that a distracting singleton that matched the target-defining feature (in this case, a red distractor) slowed performance when it did not appear in the target location and speeded performance when it appeared in the upcoming target location (relative to a nondistractor baseline condition); this pattern is taken as showing that the distractor captured visual attention.

In the present case, Folk and Remington (1998) observed that a nonmatching distractor (e.g., green distractor when searching for a red target) failed to show position effects (that is, response time was the same whether the distractor appeared in the target location or elsewhere) and yet it did produce an overall slowing relative to a no-distractor baseline. Thus distracting feature singletons were shown to have two dissociable effects: if they matched the target's defining feature, then they showed location-specific effects, suggesting that they influenced the deployment of spatial attention; if they did not match the target's defining feature, then they produced a filtering cost that was not spatially specific. Theeuwes, Atchley, and Kramer (chap. 4, this volume) have argued that the paradigm used by Folk and Remington (1998) probed the deployment of attention too late to reveal an early spatial capture of attention that could be overridden by top-down attentional control. Nevertheless, attentional capture by an irrelevant but salient feature singleton in the experiments described thus far has been observed only when the target of search is itself a feature singleton. This sort of attentional capture must therefore be viewed as a stimulus-driven modulation of a top-down selection strategy (i.e., singleton detection mode).

In a study that does not appear to involve a strategic singleton detection mode, Joseph and Optican (1996) used a paradigm similar to the one invented by Folk, Remington and Johnston (1992). Observers were required to search for a target *L* shape in a dense array of *T* shapes—a difficult search task that would not be expected to evoke singleton detection mode because the target consists of a particular arrangement of oriented line segments (Beck and Ambler 1972). The target array was flashed briefly and then masked, and the task was to report in which quadrant of the display the *L* appeared. A distractor array preceding the target display consisted of a single vertical (or horizontal) bar embedded in an array of horizontal (or vertical) bars. This orientation singleton appeared at one of the possible target locations, but was unpredictive of the target's location. Subjects were told to ignore the distractor because it contained no relevant information. Localization accuracy was substantially greater when the target appeared in the location previously occupied by the distractor singleton than otherwise; Joseph and Optican concluded that attention was involuntarily drawn to the location of the singleton, which suggests that, at least under some conditions, irrelevant feature singletons may capture attention. On the other hand, Hendel and Egeth (1998) found that even the difficult search for an *L* in an array of *T*s may cause observers to adopt a search strategy in which oriented bars are task relevant: when the target was a color singleton, an orientation singleton distractor failed to capture attention. Thus here, too, feature singletons apparently capture attention only when they are part of the subject's search strategy.

Folk and colleagues (Folk, Remington, and Johnston 1992; Folk and Remington 1998; Folk, Remington, and Wright 1994) have suggested that all deployments of attention, including those that may appear to be purely stimulus driven, are necessarily implementations of a top-down attentional control setting. The idea is that all organisms are at all times perceptually set for some input, and this perceptual set biases the visual system to give higher priority to sensory representations satisfying the contents of the current attentional set. As suggested by Hendel and Egeth (1998), the findings of Joseph and Optican (1996) can be viewed as arising from an intent to attend to a particular combination of vertical and horizontal bars, and because the singleton location alone contained, say, a vertical bar, the attentional set caused a seemingly "involuntary" shift of attention to it.

In all of their experiments on this topic, Folk and colleagues adopted the following experimental approach (see figure 3.3). Subjects were explicitly instructed to search a multielement array for an object defined by one or more features (e.g., a red element among white elements), and to report some other property of that object (e.g., its shape). Instructions are presumed to establish a known and well-defined attentional control setting in the observer, a current top-down state of attentional readiness that will influence the observers' perception in the upcoming events. Each trial then consisted of two parts: a to-be-ignored distractor display and a closely following target display. In Folk, Remington, and Johnston

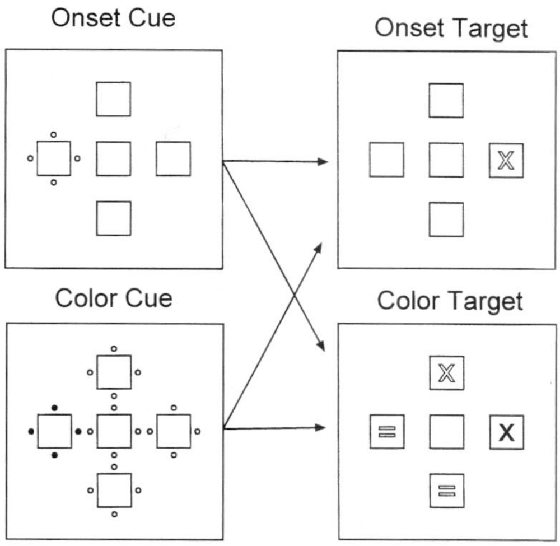

Figure 3.3 Procedure used by Folk, Remington, and Johnston (1992). On each trial, subjects were to press one key if the target was an "=" and another key if it was an "X." The target was defined as the only stimulus in the display (onset target) or as the uniquely colored element in the display (color target). Each target display was preceded by either an onset or a color cue.

1992, exp. 1, the target display consisted of two "X"s and two " = "s; the primary task was to report whether the element with the target-defining feature (say, red) was an "X" or an " = ." The distractor display consisted of elements clearly distinguishable from the targets: a cluster of four small dots surrounding one or more of the potential target locations. One of the distractor clusters could either match the target-defining feature (e.g., it, too, could be a red element) or have another irrelevant feature (e.g., abrupt onset). The distractor display appeared briefly (e.g., for 50 msec), followed after 100 msec by the target display. Subjects were told to ignore the distractor because typically it would not occur in the target location and was therefore irrelevant to the task. Folk and colleagues found that response time was longer when the distractor and target locations were different than when they were the same; this indicated that the distractor drew attention even though subjects were instructed to ignore it. However, this pattern only obtained when the distractor matched the target-defining feature; that is, a color distractor failed to draw attention when the target was defined by abrupt onset. Subsequent experiments verified and extended this observation. Folk and colleagues concluded that attentional capture is often, perhaps always, a manifestation of some top-down attentional set.

Although the claims of Folk and colleagues seem to directly contradict those of Theeuwes (e.g., 1994), who has asserted that there is *no* top-down control of attention when subjects are engaged in "preattentive visual search," the conflict may be more apparent than real. Theeuwes has shown quite clearly that if one is to search for a target that differs from its neighbors in some dimension, the feature difference computation only represents the magnitude of the difference, and not the identity of the dimension exhibiting that difference. In this sense, there is limited top-down control. The adoption of singleton detection mode is a strategic choice, however, and therefore represents a clear case of an interaction between top-down strategic control and modulating stimulus factors.

Several studies have shown that when a feature singleton is completely task irrelevant (both in its identity and, critically, in that subjects need not enter singleton detection mode to find the target), then the presence of a salient singleton distractor has virtually no effect on performance, suggesting that feature singletons do not autonomously capture attention in a purely stimulus-driven fashion. For example, Yantis and Egeth 1999 asked subjects to search for a target that was difficult to discriminate from nontargets (in this case, a vertical bar among bars tilted slightly to the left and right), so that singleton detection mode was not a viable strategy. In a control condition, the tilted target was always colored red and the nontargets blue, and this yielded highly efficient search, verifying that the color difference was sufficient to be labeled "salient." In the experimental condition, one item was always red, but because it was only rarely the target, there was no incentive for subjects to use it to guide attention. Response times to color singleton targets were no faster than to nonsingleton targets, suggesting that the singleton failed to draw attention. Other examples of this result have been reported (Folk and Annett 1994; Gibson and Jiang 1998; Hillstrom and Yantis 1994; Jonides and Yantis 1988; and Theeuwes 1990; for counterexamples, see Todd and Kramer 1994; Theeuwes and Burger 1998). In other words, the salience of feature singletons apparently does not control the deployment of attention unless it is licensed to do so by the adoption of singleton detection mode.

3.3 SELECTION BY SEGMENTED OBJECT

Kahneman and Henik (1981, 183) asked the following prescient question: "If attention selects a stimulus, what is the stimulus that it selects?" The standard answer at the time would have been that attention selects the contents of a spatial location; an attentional "spotlight" illuminates a convex region of space. Our everyday commerce with the world, however, involves interactions with segmented objects, not with empty locations or with free-floating features. Although objects occupy spatial locations, we are sometimes faced with a scene in which two objects spatially overlap one another (e.g., a cat that is partly occluded by foliage); in such cases, visual selection of an object via its location is not straightforward. It seems possible to select one object and ignore another even when the two objects occupy a common two-dimensional spatial location. This is the advantage offered by early scene segmentation.

Object-based selection also imposes a stimulus-driven constraint on the implementation of a goal-directed selection strategy. As I stated at the beginning of this chapter, among the earliest hard-wired computations carried out by the visual system is the segmentation of a scene into its constituent objects and the separation of figure from background: the perceptual organization of the spatiotemporally fragmented retinal image (e.g., parts of a cat intermingled with parts of occluding foliage) into a collection of coherent object representations (a single cat whose head, legs, and tail are linked by their common motion, color, texture, depth, and collinear contour). The principles of perceptual organization articulated by the Gestalt psychologists in the early part of this century (e.g., proximity, similarity, common fate) describe how image features guide grouping and segmentation (see Nakayama, He, and Shimojo 1996 for a recent review) . For example, edges that are collinear will tend to be perceived as bounding a common object even if they are partly occluded by an intervening surface; image regions with the same color, texture, and motion will tend to be perceived as part of a common surface; and so forth. These grouping mechanisms are autonomous and indeed may require cognitive effort to override, as suggested by Rensink and Enns 1998, discussed earlier (figure 3.1).

Among the first to demonstrate the constraints on selection imposed by scene segmentation, Duncan 1984 clearly articulated the distinction between space-based and object-based theories of visual selection. In Duncan's experiments, a display containing two superimposed objects (an outline square and a tilted line) was flashed briefly and followed by a mask. Each object had two attributes (e.g., line tilt and texture) with two possible values per attribute (e.g., tilt left or right). Subjects were to report one or two attributes, and in the latter case, the two attributes could belong to the same object or come from two different objects. Duncan found that whereas there was little cost in accuracy for reporting two attributes from the same object, compared to reporting just a single attribute, there was a larger cost when the two attributes came from different objects. He suggested that when an object is selected, all of its attributes automatically become available for report. When attributes from two different objects are to be reported, there is a time cost associated with selecting the second object.

The effects of object segmentation appear to occur even when the task does not require it. For example, in Egly, Driver, and Rafal 1994, a display containing two parallel outline rectangles was presented vertically to the right and left of fixation or horizontally above and below fixation (figure 3.4). Attention was cued to one end of one rectangle by brightening the contours of the rectangle in that region. Shortly thereafter, one end of one rectangle was filled in, and this target event had to be detected as rapidly as possible. The cue was valid on most trials, and response time (RT) was shortest when the target appeared in the cued location. On

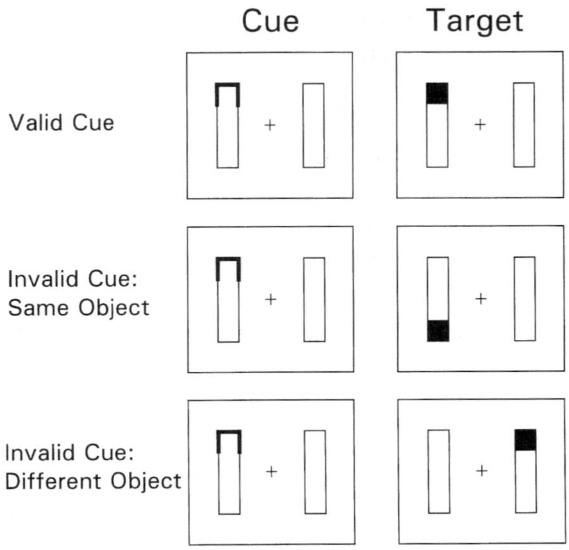

Figure 3.4 Sample displays from Egly, Driver, and Rafal 1994. Each trial began with a 100 msec cue, indicating one end of one of the rectangles, followed after 200 msec by a target, the filling in of one end of a rectangle. As illustrated in the top row, the cue was valid on 75% of the trials. On the 25% of the trials where the cue was invalid, the target appeared within the cued object on half the trials (same-object condition) and within the uncued object on the other half (different-object condition). The distances between the cued location and each of the two uncued locations were the same.

trials where the cue was invalid, however, RT was shorter when the target was in the uncued end of the cued object (the same-object condition) than when it was in the uncued object (the different-object condition). This object-specific advantage occurred even though both locations were equidistant from the cued location and equally likely to contain a target, and even though there was no need to respect object boundaries in this task. Moore, Yantis, and Vaughan (1998) observed a similar object-specific benefit for targets appearing in uncued regions of a cued object even when the object was partly occluded. Behrmann, Zemel, and Mozer (1998), Lavie and Driver (1996), and Vecera and Farah (1994) have reported related corroborating evidence. Thus the process of perceptual organization operating on visual scenes proceeds automatically and influences the attentional priorities within the scene even when segmentation into distinct objects is not part of the current attentional set.

Several other studies (e.g., Bacon and Egeth 1991; Baylis and Driver 1993; Grossberg, Mingolla, and Ross 1994; Humphreys and Müller 1993; Treisman 1982), have shown that the Gestalt principles governing the construction of perceptual object representations systematically influence visual selection. Such influences are manifestly object-based and not space-based ones, and they provide a further instance of stimulus-driven constraints modulating top-down control settings. Duncan and Humphreys's attentional engagement theory (1989, 1992) emphasizes the role of perceptual grouping in visual search. According to the theory, the similarities among targets and nontargets determine the efficiency of selection in two ways. First, when target items are similar to nontarget items, search will be inefficient because the detectability of the targets will be low in a signal detection–theoretic sense. In other words, the targets will tend to be grouped with the nontargets, making selection difficult. Second, when nontarget items are similar to one another, search will be relatively efficient because similar items are grouped into structures whose constituents are treated similarly. If one is to reject (or suppress the representation of) an item because it contains features known to be task irrelevant, then all other items grouped with that item are also going to be suppressed (what Duncan and Humphreys call "spreading suppression"). This promotes efficient search when the nontargets are all similar because they can be grouped and rejected all at once.

Much empirical support exists for the role of perceptual grouping in visual selection. Baylis and Driver 1992 showed that the identification of a central red target letter was influenced more by the (conflicting or congruent) identity of distant red distractor letters than by adjacent green distractor letters. Here grouping by color similarity caused the red letters to be perceived "together," even though color similarity was not relevant to the task. Other examples of this sort include Driver and Baylis 1989 and Kramer and Jacobson 1991.

Thus we see that although perceptual objects can be selected according to a top-down selection criterion, object-based selection seems to require that the object or perceptual group be selected or rejected as a whole, bottom-up, even when only a single part or attribute is desired.

3.4 STIMULUS-DRIVEN ATTENTIONAL CAPTURE

As we have shown, certain highly efficient forms of visual search (e.g., search for a red object in an array of green objects) sometimes produce the subjective impression that the target item effortlessly "pops out" of the display. In these cases, however, the feature singleton is the target of search, or the subject has entered singleton detection mode, which amounts to much the same thing. The observer is thus deliberately searching for that stimulus, and there is almost certainly a goal-directed component to the search strategy. In other words, such searches cannot be characterized as purely stimulus driven. The question then remains whether any search is purely stimulus driven.

My colleagues and I (Remington, Johnston, and Yantis 1992; Yantis and Hillstrom 1994; Yantis and Johnson 1990; Yantis and Jonides 1984, 1990; see also Oonk and Abrams 1998) have argued that an abrupt visual onset enjoys high priority in vision and often captures attention in the absence of a specific attentional set for abrupt onset. Our studies were designed

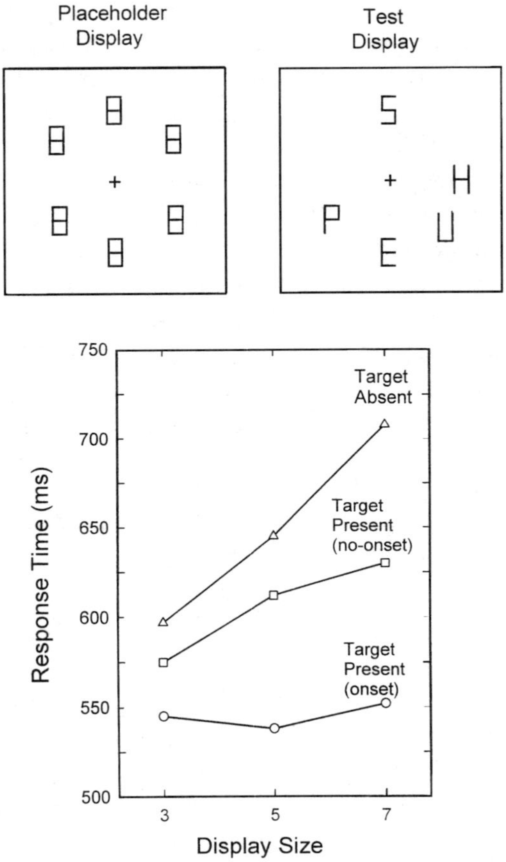

Figure 3.5 Displays and data from Jonides and Yantis 1988. *Top.* Each trial began with the presentation of a target letter for that trial (not shown), followed by a set of six figure-eight placeholders presented for 1 second. At the end of this interval, a subset of the line segments in some of the figure eights disappeared to reveal letters. In conditions with display size 3 and 5, some of the figure-eight placeholders disappeared altogether. The test display contained one abrupt onset letter and 2, 4, or 6 no-onset letters (display size 5 is illustrated). Subjects were to press one of two buttons to indicate whether the specified target was present or absent. The target was the onset item on $1/n$ of the trails, where n is display size. *Bottom.* Response time for trials in which the target was the onset item did not increase with display size, suggesting that the onset item captured attention despite its irrelevance to the task.

specifically to ask whether a visual event would capture attention when it was explicitly not part of the observers' state of attentional readiness. In the experiments of Yantis and Jonides (1984, 1990; Jonides and Yantis, 1988; see figure 3.5), for example, the task was to search for a prespecified target letter in a multielement array containing one element that appeared abruptly in a previously blank location (the onset element) and several other elements that were present but camouflaged before the appearance of the search array (the no-onset elements). The display items were easily confusable, which typically would require an inefficient serial search for the specified letter. Because the target happened to have an abrupt onset only rarely, there was no incentive for observers to adopt an attentional set that conferred high priority on such elements. Nevertheless, we found that when the target was the onset element, RT was short and did not depend on the number of elements in the display, whereas when one of the no-onset elements was the target, RT increased almost linearly with the number of elements in the display (figure 3.5, bottom). This pattern strongly suggests that the abrupt onset element captured attention in a purely stimulus-driven fashion. This distinguishes abrupt onset from other salient features, such as color or brightness singletons, that do not capture attention (e.g., Yantis and Egeth 1999; see section 3.2).

The capture produced by abrupt onsets is not absolute, however. Yantis and Jonides (1990) presented a central arrow at various moments in time before a search display was to appear. The target of search was likely (in some experiments, certain) to appear in the location indicated by the arrow (eye position was monitored to ensure that fixation was maintained). An abrupt onset always appeared at the same time as the target, sometimes in the expected (cued) location, and sometimes elsewhere. If an abrupt visual onset captures attention regardless of the observers' attentive state (in this case, their spatial focus of attention), then we would expect performance to be disrupted (in this case, slowed) when the onset appeared at an uncued location, reflecting the involuntary capture of attention by the onset, followed by the effortful redeployment of attention to the target location. Instead, Yantis and Jonides (1990) found that when sufficient time was provided to shift attention in advance to the cued location, and when the predictive validity of the cue was high enough, capture by an abrupt onset was averted. That the cue could override attentional capture by the abrupt onset only when the cue was predictive is crucial evidence of goal-directed attentional control, rather than of competition between two abrupt onsets (i.e., the cue and the onset letter). Several other studies (e.g., Juola, Koshino, and Warner 1995; Koshino, Warner, and Juola 1992; Müller and Rabbitt 1989; and Theeuwes 1991) have corroborated the conclusion that deliberate deployments of attention can prevent capture by abrupt onset. It remains an open question whether the low-level, reflexive neural responses to abrupt onsets such as those discussed by Rafal et al. (chap. 6, this volume) still occur but are dominated by the top-down attentional set, or are suppressed entirely by top-down deployments of attention.

Jonides and Yantis 1988, Hillstrom and Yantis 1994, and Yantis and Egeth 1999 have demonstrated that the uniqueness of the onset element per se cannot be the crucial factor in the observed attentional capture; they showed that highly salient but uninformative feature singletons in dimensions other than onset (e.g., color) do not draw attention (see section 3.2), whereas an abrupt onset does. In an effort to determine the mechanism for attentional capture by abrupt onsets, Yantis and Hillstrom (1994) considered two possibilities. First, attentional capture might be mediated by the abrupt luminance change associated with the onset letter, which would implicate a low-level visual mechanism sensitive to the spatiotemporal profile of the stimulus. Alternatively, the appearance of a new perceptual object, independent of the luminance change, might draw attention automatically.

Yantis and Hillstrom found that the appearance of a target letter defined by equiluminant discontinuities in texture, motion, or depth nevertheless captures attention, showing that a new object is sufficient to capture attention without a luminance increment (see also Oonk and Abrams 1998; Gellatly, Cole, and Blurton 1999). According to recent studies in our lab, new objects defined by equiluminant discontinuities in color, using the flicker photometry method, also capture attention.[1] These studies show that luminance change is not necessary to produce attentional capture by new perceptual objects, but rather that the appearance of new objects alone can capture attention.

Moreover, Enns, Yantis, and Di Lollo (1998) have shown that a luminance change is not sufficient to capture attention. Subjects were asked to search for a target letter (*E* or *H*) in an array of black and white letters on a gray background. In an initial control experiment, we verified that when one of the letters appeared in a previously blank location among no-onset letters, it captured attention, even though the new object was not predictive of the target location. In other words, the heterogeneity of the letter colors (some black and some white) did not affect the standard result that new objects capture attention. A second experiment then examined search performance when all of the stimuli were no-onset letters. At the moment the camouflage was removed from the figure-eight placeholders to reveal the letter forms, one of the objects exhibited a polarity reversal (e.g., black to white or white to black). Although the luminance change in this case was at least as much as that exhibited by the onset letter in the control condition, the element undergoing the luminance change nevertheless failed to capture attention. Thus luminance change is not sufficient to produce attentional capture by abrupt onsets. Together, these two lines of evidence suggest that the appearance of a new object captures attention, not by virtue of the luminance change that typically accompanies it, but because the visual system is predisposed to attend to the creation of a new perceptual representation in a purely stimulus-driven fashion.

This claim has not gone unchallenged. Miller (1989) and Martin-Emerson and Kramer (1997) have reported that contour offsets can compete to some extent for attention with abrupt onsets. Folk, Remington, and Johnston (1992) found that when searching for a color singleton target, a preceding peripheral onset cue failed to capture attention; this may well be an instance of top-down control over attentional capture by abrupt onset analogous to the findings of Yantis and Jonides (1990). Recent experiments by Gellatly, Cole, and Blurton (1999) have shown that a new object defined by equiluminant discontinuities in motion failed to capture attention, suggesting that new objects do not always capture attention, although there is a question about the strength of the object representation in the equiluminant case. Generally speaking, items defined by equiluminant discontinuities in dimensions other than luminance are difficult to see, as evidenced by the relatively slow response times observed in these tasks. It would not be surprising if a near-threshold object failed to capture attention when it appeared.

Folk, Remington, and Johnston (1992, 1993; see also Yantis 1993) have argued that attentional capture by new objects is a result of an implicit attentional control setting for abrupt onset, based on the assumption that there is a subtle contingency in the experimental procedure that encourages subjects to selectively attend to luminance change (each trial begins with a luminance change, for example), and that capture by abrupt onset is therefore a side effect of top-down attentional control. This assumption is undermined, however, by the experiments of Enns, Yantis, and Di Lollo (1998), which showed no attentional advantage for items reversing their polarity at the beginning of each trial, even though the procedures were precisely analogous to the onset case.

Folk, Remington, and Johnston (1992) have also argued that there may be a "default" attentional control setting for new perceptual objects that operates in the absence of any specific feature-based attentional set. This suggests, of course, that the visual system is predisposed, perhaps even hard-wired, to treat new perceptual objects with higher priority than other attributes. That the visual system should have this bias for new objects is hardly surprising: new perceptual objects are of obvious behavioral significance and such an "early warning system" (Breitmeyer and Ganz 1976, 31) for new objects would be expected to increase reproductive fitness. As we have seen, however, even this form of stimulus-driven attentional capture is subject to some degree of top-down modulation (Yantis and Jonides 1990).

3.5 INTERACTIVE CONTROL OF VISUAL ATTENTION

The experiments reviewed in this chapter reveal that most instances of visual selection involve an interaction between top-down attentional control and autonomous neural responses to visual stimuli. For example, when directing attention to locations in space, we can readily observe constraints on the spatial and temporal precision of selection imposed by neuroanatomical and neurophysiological properties of the brain. In the case of search for a feature singleton, the adoption of singleton detection mode is a deliberate strategy that has implications for the efficiency with which a target can be found, and for whether items to be ignored will draw attention. Autonomous scene segmentation and perceptual grouping mechanisms cause all the features of an object to be selected as a unit, whether that is part of the attentional goal or not. And while the appearance of new perceptual objects can capture attention in the absence of a specific intent to attend to such changes, top-down control can be exerted to avert such capture. Overall, the evidence suggests that purely stimulus-driven attentional capture is rare; instead, interactions between top-down attentional control settings and stimulus-driven factors that modulate deliberate control are the rule.

How might current behavioral goals interact with early visual modules to yield the observed influences on selection? The goals that drive top-down attentional deployment are presumably contained in working memory representations of the observer's current task. In most cases reviewed here, the current task is stipulated by the instructions conveyed to the participant in an experimental psychology laboratory. These memory representations generally contain the target-defining features (e.g., the *red* item), the reported attribute of the target (e.g., its presence or its name), the manual or vocal responses associated with the various possible outcomes of search, the contingencies in the experimental design (e.g., the relative probabilities with which various objects will appear), along with expectations about the properties of the nontargets, the display in general, and other aspects of the testing session. All of these are aspects of the observer's explicit state of attentional readiness. Other factors that may influence the implementation of attentional goals may include long-term or implicit memory representations, such as the participant's memory of previous similar experiences, together with autonomous early perceptual mechanisms, such as perceptual organization and object segmentation or pure attentional capture by new perceptual objects.

The stored memory representation of the task at hand comprises an attentional set, which gates the neural representation of the sensory input. For example, if there is a positional expectancy (e.g., "Name the letter appearing four degrees to the right of the present point of fixation"), then the neurons with receptive fields in that location may be primed to receive input, those with receptive fields elsewhere may be suppressed if a target appears in the expected location, or both (Moran and Desimone 1985). Duncan and colleagues (Desimone and Duncan 1995; Duncan, Humphreys, and Ward 1997) have articulated an approach to this problem in which multiple brain systems exhibit competitive responses to inputs (i.e., a given brain region will tend to represent one object at a time and suppress representations of other objects), whereas cooperative integration across brain regions will tend to yield concurrent activation of the same object. According to this idea, there is no one place where attention originates; instead, it is an emergent property of the competition and cooperation among multiple brain regions. Recent neurobiological and computational models of attention (e.g., Mozer and Sitton 1998; Niebur, Koch, and Rosin 1993; Olshausen, Anderson, and Van Essen

1993; Tsotsos 1995; Usher and Niebur 1996) suggest how this neural gating might be implemented.

Although it is well known that there are massive feedback pathways in the brain from higher centers to early visual areas (e.g., Van Essen and DeYoe 1995), details of the mechanism by which memory representations of the current behavioral goal modulate sensory responses are not well understood. Rather than discuss any specific proposal, I will simply outline the properties any candidate mechanism must have. First, there must be a working memory representation that specifies task requirements and relevant stimulus attributes (much recent evidence points to prefrontal regions as being particularly involved in such representation; e.g., Courtney et al. 1998; Goldman-Rakic 1995). Miller (chap. 22, this volume) provides neurophysiological evidence that prefrontal neural representations are modulated by task demands. Second, there must be direct or indirect feedback connections between the neural representation for the current attentional set (including the properties of the desired object, its probable location, or both) and the early visual areas whose responses are subject to attentional modulation, including V1 and V2 (Gandhi, Heeger, and Boynton 1999; Motter 1993), extrastriate areas including V4 (e.g., Connor et al. 1997; Hopfinger et al., chap. 5, this volume; Motter 1994; Moran and Desimone 1985), IT (e.g., Miller, Li, and Desimone 1991), MT (e.g., Beauchamp, Cox, and DeYoe 1997; O'Craven et al. 1997; Treue and Maunsell 1996), and LIP (e.g., Gottlieb, Kusuoki, and Goldberg 1998). Candidate areas that appear to have the requisite feedback connections and that are active during attentive tasks include the posterior parietal cortex (e.g., Bushnell, Goldberg, and Robinson 1981; Corbetta et al. 1993; Mountcastle, Anderson, and Motter 1981), parts of the thalamus, including the pulvinar and the reticular formation (Crick 1984; LaBerge 1995, Olshausen, Andersen, and Van Essen 1993), and some prefrontal areas (Miller, chap. 22, this volume).

Visual pathways specialized to represent rapidly changing input (the M-pathway) may well mediate stimulus-driven attentional capture by new perceptual objects, which can in turn produce efficient control over eye movements (Rafal et al., chap. 6, this volume). Early scene segmentation mechanisms, presumably operating primarily in occipital and occipital-temporal areas, further constrain the implementation of selection strategies (Driver 1995).

Despite recent efforts to characterize attentional control in terms of neural systems, the problem of translating task requirements into specific attentional goals has not yet been solved. The working memory representations that specify attentional set remain as givens, outside the scope of most models. We are usually left with a "central executive" to sort out multiple competing goals. Although tackling this problem is among the most compelling challenges we face in this area, any attempt to explain the efficiency of visual selection must confront the constraints imposed by bottom-up factors on the successful implementation of behavioral goals.

NOTES

Preparation of this chapter was supported by National Institute of Mental Health grant R01-MH43924. I thank Howard Egeth for many valuable discussions, and Jon Driver, Nilli Lavie, Stephen Monsell, and Jan Theeuwes for valuable comments on an earlier version.

1. Theeuwes (1995) has reported that the appearance of a new object equiluminant with its background failed to capture attention. For technical reasons, however, the duration of the "old" items in the display was only 50 msec (as compared to 1,000 msec in many previous studies). Thus the new object may not have been perceived as "new" relative to these fairly new "old" objects, thereby preventing it from capturing attention.

REFERENCES

Atchley, P., Kramer, A. F., Andersen, G. J., and Theeuwes, J. (1997). Spatial cuing in a stereoscopic display: Evidence for a "depth-aware" attentional spotlight. *Psychonomic Bulletin and Review, 4,* 524–529.

Bacon, W. F., and Egeth, H. E. (1991). Local processes in preattentive feature detection. *Journal of Experimental Psychology: Human Perception and Performance, 17,* 77–90.

Bacon, W. F., and Egeth, H. E. (1994). Overriding stimulus-driven attentional capture. *Perception and Psychophysics, 55,* 485–496.

Bahcall, D. O., and Kowler, E. (1998). Attentional interference at small spatial separations. *Vision Research, 39,* 71–86.

Baylis, G. C., and Driver, J. S. (1992). Visual parsing and response competition: The effect of grouping factors. *Perception and Psychophysics, 51,* 145–162.

Baylis, G. C., and Driver, J. S. (1993). Visual attention and objects: evidence for hierarchical coding of locations. *Journal of Experimental Psychology: Human Perception and Performance, 19,* 451–470.

Beauchamp, M. S., Cox, R. W., and DeYoe, E. A. (1997). Graded effects of spatial and featural attention on human area MT and associated motion processing areas. *Journal of Neurophysiology, 78,* 516–520.

Beck, J., and Ambler, B. (1972). Discriminability of differences in line slope and in line arrangement as a function of mask delay. *Perception and Psychophysics, 12,* 201–204.

Behrmann, M., Zemel, R., and Mozer, M. (1998). Object-based attention and occlusion: Evidence from normal subjects and a computation model. *Journal of Experimental Psychology: Human Perception and Performance, 24,* 1011–1036.

Breitmeyer, B. G., and Ganz, L. (1976). Implications of sustained and transient channels for theories of visual pattern masking, saccadic suppression, and information processing. *Psychological Review, 83,* 1–36.

Bundesen, C. (1990). A theory of visual attention. *Psychological Review, 97,* 523–547.

Bushnell, M. C., Goldberg, M. E., and Robinson, D. L. (1981). Behavioral enhancement of visual responses in monkey cerebral cortex: 1. Modulation in posterior parietal cortex related to selective visual attention. *Journal of Neurophysiology, 46,* 755–772.

Cave, K. R., and Wolfe, J. M. (1990). Modeling the role of parallel processing in visual search. *Cognitive Psychology, 22,* 225–271.

Cepeda, N. J., Cave, K. R., Bichot, N. P., and Kim, M. S. (1998). Spatial selection via feature-driven inhibition of distractor locations. *Perception and Psychophysics, 60,* 727–746.

Cheal, M. L., and Lyon, D. R. (1991). Central and peripheral precuing of forced-choice discrimination. *Quarterly Journal of Experimental Psychology, 43A,* 859–880.

Colegate, R. L., Hoffman, J. E., and Eriksen, C. W. (1973). Selective encoding from multielement visual displays. *Perception and Psychophysics, 14,* 217–224.

Connor, C. E., Preddie, D. C., Gallant, J. L., and Van Essen, D. C. (1997). Spatial attention effects in macaque area V4. *Journal of Neuroscience, 17,* 3201–3214.

Corbetta, M., Miezin, F. M., Shulman, G. L., and Petersen, S. E. (1993). A PET study of visuospatial attention. *Journal of Neuroscience, 13,* 1202–1226.

Courtney, S. M., Petit, L., Maisog, J. M., Ungerleider, L. G., and Haxby, J. V. (1998). An area specialized for spatial working memory in human frontal cortex. *Science, 279,* 1347–1351.

Crick, F. (1984). The function of the thalamic reticular complex: The search-light hypothesis. *Proceedings of the National Academy of Sciences, U.S.A., 81,* 4586–4590.

Desimone, R., and Duncan, J. (1995). Neural mechanisms of selective visual attention. *Annual Review of Neuroscience, 18,* 193–222.

Downing, C. J., and Pinker, S. (1985). The spatial structure of visual attention. In M. Posner and O. Martin (Eds.), *Attention and Performance XI,* pp. 171–187. Hillsdale, NJ: Erlbaum.

Driver, J. (1995). Object segmentation and visual neglect. *Behavioural Brain Research 71,* 135–146.

Driver, J. S., and Baylis, G. C. (1989). Movement and visual attention: The spotlight metaphor breaks down. *Journal of Experimental Psychology: Human Perception and Performance, 15,* 448–456.

Driver, J., McLeod, P., and Dienes, Z. (1992). Motion coherence and conjunction search: Implications for guided search theory. *Perception and Psychophysics, 51,* 79–85.

Duncan, J. (1984). Selective attention and the organization of visual information. *Journal of Experimental Psychology: General, 113,* 501–517.

Duncan, J., and Humphreys, G. W. (1989). Visual search and stimulus similarity. *Psychological Review, 96,* 433–458.

Duncan, J., and Humphreys, G. W. (1992). Beyond the search surface: Visual search and attentional engagement. *Journal of Experimental Psychology: Human Perception and Performance, 18,* 578–588.

Duncan, J., Humphreys, G. W., and Ward, R. (1997). Competitive brain activity in visual attention. *Current Opinion in Neurobiology, 7,* 255–261.

Egeth, H., Jonides, J., and Wall, S. (1972). Parallel processing of multielement displays. *Cognitive Psychology, 3,* 674–698.

Egeth, H. E., Virzi, R. A., and Garbart, H. (1984). Searching for conjunctively defined targets. *Journal of Experimental Psychology: Human Perception and Performance, 10,* 32–39.

Egly, R., Driver, J., and Rafal, R. D. (1994). Shifting visual attention between objects and locations: Evidence from normal and parietal lesion subjects. *Journal of Experimental Psychology: General, 123,* 161–177.

Egly, R., and Homa, D. (1984). Sensitization of the visual field. *Journal of Experimental Psychology: Human Perception and Performance, 10,* 778–793.

Engel, F. L. (1971). Visual conspicuity, directed attention and retinal locus. *Vision Research, 11,* 563–576.

Enns, J. T., Yantis, S., and Di Lollo, V. (1998). Luminance transients capture attention, but only in "new" display locations. Paper presented at the Thirty-eighth Annual Meeting of the Association for Research in Vision and Ophthalmology, Fort Lauderdale, FL, May.

Eriksen, C. W., and Hoffman, J. E. (1972). Temporal and spatial characteristics of selective encoding from visual displays. *Perception and Psychophysics, 12,* 201–204.

Eriksen, C. W., and Hoffman, J. E. (1973). The extent of processing noise elements during selective encoding from visual displays. *Perception and Psychophysics, 14,* 155–160.

Eriksen, C. W., and Yeh, Y. (1985). Allocation of attention in the visual field. *Journal of Experimental Psychology: Human Perception and Performance, 11,* 583–597.

Folk, C. L., and Annett, S. (1994). Do locally defined feature discontinuities capture attention? *Perception and Psychophysics, 56,* 277–287.

Folk, C. L., and Remington, R. W. (1998). Selectivity in distraction by irrelevant featural singletons: Evidence for two forms of attentional capture. *Journal of Experimental Psychology: Human Perception and Performance, 24,* 847–858.

Folk, C. L., Remington, R. W., and Johnston, J. C. (1992). Involuntary covert orienting is contingent on attentional control settings. *Journal of Experimental Psychology: Human Perception and Performance, 18,* 1030–1044.

Folk, C. L., Remington, R. W., and Johnston, J. C. (1993). Contingent attentional capture: A reply to Yantis (1993). *Journal of Experimental Psychology: Human Perception and Performance, 19,* 682–685.

Folk, C. L., Remington, R. W., and Wright, J. H. (1994). The structure of attentional control: Contingent attentional capture by apparent motion, abrupt onset, and color. *Journal of Experimental Psychology: Human Perception and Performance, 20,* 317–329.

Gandhi, S. P., Heeger, D. J., and Boynton, G. M. (1999). Spatial attention affects brain activity in human primary visual cortex. *Proceedings of the National Academy of Sciences USA, 96,* 3314–3319.

Gellatly, A., Cole, G., and Blurton, A. (1999). Do equiluminant object onsets capture visual attention? *Journal of Experimental Psychology: Human Perception and Performance, 25,* 1609–1624.

Gibson, B. S., and Jiang, Y. (1998). Surprise! An unexpected color singleton does not capture attention in visual search. *Psychological Science, 9,* 176–182.

Ghirardelli, T. G., and Folk, C. L. (1996). Spatial cuing in a stereoscopic display: Evidence for a "depth-blind" attentional spotlight. *Psychonomic Bulletin and Review, 3,* 81–86.

Goldman-Rakic, P. S. (1995). Architecture of the prefrontal cortex and the central executive. In J. Grafman, K. J. Holyoak, and F. Boller (Eds.), *Structure and functions of the human prefrontal cortex,* pp. 71–83. New York: New York Academy of Sciences.

Gottlieb, J. P., Kusunoki, M., and Goldberg, M. E. (1998). The representation of visual salience in monkey parietal cortex. *Nature, 391,* 481–484.

Grossberg, S., Mingolla, E., and Ross, W. D. (1994). A neural theory of attentive visual search: Interactions of boundary, surface, spatial, and object representations. *Psychological Review, 101,* 470–489.

Helmholtz, H. von (1866). *Treatise on physiological optics.* 3d ed. Vol. 3, ed. and trans. J. P. C. Southhall. Reprint, Washington, DC: Optical Society of America, 1925.

Hendel, S. K., and Egeth, H. E. (1998). Orientation singletons do not automatically capture attention. Poster presented at the thirty-ninth Annual Meeting of the Psychonomic Society, Dallas, TX.

Henderson, J. M., and Macquistan, A. D. (1993). The spatial distribution of attention following an exogenous cue. *Perception and Psychophysics, 53,* 221–230.

Hillstrom, A. P., and Yantis, S. (1994). Visual motion and attentional capture. *Perception and Psychophysics, 55,* 399–411.

Hoffman, J. E. (1979). A two-stage model of visual search. *Perception and Psychophysics, 25,* 319–327.

Hoffman, J. E., and Nelson, B. (1981). Spatial selectivity in visual search. *Perception and Psychophysics, 30,* 283–290.

Hubel, D. H., and Wiesel, T. N. (1968). Receptive fields and functional architecture of monkey striate cortex. *Journal of Physiology* (London), *195,* 215–243.

Humphreys, G. W., and Müller, H. J. (1993). Search via recursive rejection (SERR): A connectionist model of visual search. *Cognitive Psychology, 25,* 43–110.

Iavecchia, H. P., and Folk, C. L. (1995). Shifting visual attention in stereographic displays: A timecourse analysis. *Human Factors, 36,* 606–618.

Johnson, D. N., and Yantis, S. (1995). Allocating visual attention: Tests of a two-process model. *Journal of Experimental Psychology: Human Perception and Performance, 21,* 1376–1390.

Johnston, J. C., and Pashler, H. (1990). Close binding of identity and location in visual feature perception. *Journal of Experimental Psychology: Human Perception and Performance, 16,* 843–856.

Jonides, J. (1980). Towards a model of the mind's eye's movement. *Canadian Journal of Psychology, 34,* 103–112.

Jonides, J. (1981). Voluntary versus automatic control over the mind's eye's movement. In J. B. Long and A. D. Baddeley (Eds.), *Attention and Performance IX,* pp. 187–203. Hillsdale, NJ: Erlbaum.

Jonides, J. (1983). Further toward a model of the mind's eye's movement. *Canadian Journal of Psychology, 34,* 103–112.

Jonides, J., and Yantis, S. (1988). Uniqueness of abrupt visual onset in capturing attention. *Perception and Psychophysics, 43,* 346–354.

Joseph, J. S., and Optican, L. M. (1996). Involuntary attentional shifts due to orientation differences. *Perception and Psychophysics, 58,* 651–665.

Juola, J. F., Koshino, H., and Warner, C. B. (1995). Tradeoffs between attentional effects of spatial cues and abrupt onsets. *Perception and Psychophysics, 57,* 333–342.

Kahneman, D., Henik, A. (1981). Perceptual organization and attention. In M. Kubovy and J. R. Pomerantz (Eds.), *Perceptual organization,* pp. 181–211. Hillsdale, NJ: Erlbaum.

Kim, M. S., and Cave, K. R. (1995). Spatial attention in visual search for features and feature conjunctions. *Psychological Science, 6,* 376–380.

Koshino, H., Warner, C. B., and Juola, J. F. (1992). Relative effectiveness of central, peripheral, and abrupt-onset cues in visual search. *Quarterly Journal of Experimental Psychology, 45A,* 609–631.

Kramer, A. F., and Jacobson, A. (1991). Perceptual organization and focused attention: The role of objects and proximity in visual processing. *Perception and Psychophysics, 50,* 267–284.

Kubovy, M. (1981). Concurrent-pitch segregation and the theory of indispensable attributes. In M. Kubovy and J. R. Pomerantz (Eds.), *Perceptual organization,* pp. 55–98. Hillsdale, NJ: Erlbaum.

LaBerge, D. (1983). The spatial extent of attention to letters and words. *Journal of Experimental Psychology: Human Perception and Performance, 9,* 371–379.

LaBerge, D. (1995). *Attentional processing.* Cambridge, MA: Harvard University Press.

LaBerge, D., and Brown, V. (1986). Variations in size of the visual field in which targets are presented: An attentional range effect. *Perception and Psychophysics, 40,* 188–200.

LaBerge, D., Brown, V. Carter, M., Bash, D., and Hartley, A. (1991). Reducing the effects of adjacent distrators by narrowing attention. *Journal of Experimental Psychology: Human Perception and Performance, 17,* 90–95.

LaBerge, D., Carlson, R. L., Williams, J. K., and Bunney, B. G. (1997). Shifting attention in visual space: Tests of moving-spotlight models versus an activity distribution model. *Journal of Experimental Psychology: Human Perception and Performance, 23,* 1380–1392.

Lavie, N., and Driver, J. (1996). On the spatial extent of attention in object-based visual selection. *Perception and Psychophysics, 58,* 1238–1251.

Marr, D. (1980). *Vision.* San Francisco: W. H. Freeman and Co.

Martin-Emerson, R., and Kramer, A. F. (1997). Offset transients modulate attentional capture by sudden onsets. *Perception and Psychophysics, 59,* 739–751.

Miller, E. K., Li, L., and Desimone, R. (1991). A neural mechanism for working and recognition memory in inferior temporal cortex. *Science, 254,* 1377–1379.

Miller, J. (1989). The control of attention by abrupt visual onsets and offsets. *Perception and Psychophysics, 45,* 567–571.

Moore, C. M., and Egeth, H. (1998). How does feature-based attention affect visual processing? *Journal of Experimental Psychology: Human Perception and Performance, 24,* 1296–1310.

Moore, C. M., Yantis, S., and Vaughan, B. (1998). Object-based visual selection: Evidence from perceptual completion. *Psychological Science, 9,* 104–110.

Moran, J., and Desimone, R. (1985). Selective attention gates visual processing in the extrastriate cortex. *Science, 229,* 782–784.

Motter, B. C. (1993). Focal attention produces spatially selective processing in visual cortical areas V1, V2, and V4 in the presence of competing stimuli. *Journal of Neurophysiology, 70,* 909–919.

Motter, B. C. (1994). Neural correlates of attentive selection for color or luminance in extrastriate area V4. *Journal of Neuroscience, 14,* 2178–2189.

Mountcastle, V., Anderson, R., and Motter, B. (1981). The influence of attentive fixation upon the excitability of the light-sensitive neurons of the posterior parietal cortex. *Journal of Neuroscience, 1,* 1218–1232.

Mozer, M. C., and Sitton, M. (1998). Computational modeling of spatial attention. In H. Pashler (Ed.), *Attention,* pp. 341–393. London: Psychology Press.

Müller, H. J., and Rabbitt, P. M. A. (1989). Reflexive and voluntary orienting of visual attention: Time course of activation and resistance to interruption. *Journal of Experimental Psychology: Human Perception and Performance, 15,* 315–330.

Nakayama, K., He, Z. J., and Shimojo (1996). Visual surface representation: A critical link between lower-level and higher-level vision. In S. M. Kosslyn and D. N. Osherson (Eds.), *An invitation to cognitive science: Visual cognition,* pp. 1–70. Cambridge, MA: MIT Press.

Nakayama, K., and Silverman, G. H. (1986). Serial and parallel processing of visual feature conjunctions. *Nature, 320,* 264–265.

Nakayama, K., and Mackeben, M. (1989). Sustained and transient components of focal visual attention. *Vision Research, 29,* 1631–1647.

Neisser, U. (1967). *Cognitive psychology.* New York: Appleton-Century-Crofts.

Niebur, E., Koch, C., and Rosin, C. (1993). An oscillation-based model for the neuronal basis of attention. *Vision Research, 33,* 2789–2802.

Nothdurft, H. C. (1993). Saliency effects across dimensions in visual search. *Vision Research, 33,* 839–844.

O'Craven, K. M., Rosen, B. R., Kwong, K. K., Treisman, A., and Savoy, R. L. (1997). Voluntary attention modulates fMRI activity in human MT-MST. *Neuron, 18,* 591–598.

Olshausen, B. A., Anderson, C. H., and Van Essen, D. C. (1993). A neurobiological model of visual attention and invariant pattern recognition based on dynamic routing of information. *Journal of Neuroscience, 13,* 4700–4719.

Oonk, H. M., and Abrams, R. A. (1998). New perceptual objects that capture attention produce inhibition of return. *Psychonomic Bulletin and Review, 5,* 510–515.

Pashler, H. (1988). Cross-dimensional interaction and texture segregation. *Perception and Psychophysics, 43,* 307–318.

Posner, M. I. (1978). *Chronometric explorations of mind.* Englewood Cliffs, NJ: Erlbaum.

Posner, M. I. (1980). Orienting of attention. *Quarterly Journal of Experimental Psychology, 32,* 3–25.

Posner, M. I., Snyder, C. R. R., and Davidson, B. J. (1980). Attention and the detection of signals. *Journal of Experimental Psychology: General, 109,* 160–174.

Remington, R. W., Johnston, J. C., and Yantis, S. (1992). Involuntary attentional capture by abrupt onsets. *Perception and Psychophysics, 51,* 279–290.

Rensink, R. A., and Enns, J. T. (1998). Early completion of occluded objects. *Vision Research, 38,* 2489–2505.

Shaw, M. L. (1978). A capacity allocation model for reaction time. *Journal of Experimental Psychology: Human Perception and Performance, 4,* 586–598.

Shih, S. I., and Sperling, G. (1996). Is there feature-based attentional selection in visual search? *Journal of Experimental Psychology: Human Perception and Performance, 22,* 758–779.

Sperling, G., (1960). The information available in brief visual presentations. *Psychological Monographs, 74,* 1–29.

Theeuwes, J. (1990). Perceptual selectivity is task-dependent: Evidence from selective search. *Acta Psychologica, 74,* 81–99.

Theeuwes, J. (1991). Exogenous and endogenous control of attention: The effect of visual onsets and offsets. *Perception and Psychophysics, 49,* 83–90.

Theeuwes, J. (1992). Perceptual selectivity for color and form. *Perception and Psychophysics, 51,* 599–606.

Theeuwes, J. (1994). Endogenous and exogenous control of visual selection. *Perception, 23,* 429–440.

Theeuwes, J. (1995). Abrupt luminance change pops out; abrupt color change does not. *Perception and Psychophysics, 57,* 637–644.

Theeuwes, J. (1996). Perceptual selectivity for color and form: On the nature of the interference effect. In A. F. Kramer, M. Coles, and G. Logan (Eds.), *Converging operations in the study of visual selective attention*, pp. 297–314. Washington, DC: American Psychological Association.

Theeuwes, J., and Burger, R. (1998). Attentional control during visual search: The effect of irrelevant singletons. *Journal of Experimental Psychology: Human Perception and Performance, 24,* 1342–1353.

Todd, S., and Kramer, A. F. (1994). Attentional misguidance in visual search. *Perception and Psychophysics, 56,* 198–210.

Treisman, A. (1982). Perceptual grouping and attention in visual search for features and for objects. *Journal of Experimental Psychology: Human Perception and Performance, 8,* 194–214.

Treisman, A., and Gelade, G. (1980). A feature-integration theory of attention. *Cognitive Psychology, 12,* 97–136.

Treue, S., and Maunsell, J. H. R. (1996). Attentional modulation of visual motion processing in cortical areas MT and MST. *Nature, 382,* 539–541.

Tsal, Y., and Lavie, N. (1988). Attending to color and shape: The special role of location in selective visual processing. *Perception and Psychophysics, 44,* 15–21.

Tsal, Y., and Lavie, N. (1993). Location dominance in attending to color and shape. *Journal of Experimental Psychology: Human Perception and Performance, 19,* 131–139.

Tsotsos, J. K. (1995). Toward a computational model of visual attention. In T. V. Papathomas, C. Chubb, A. Gorea, and E. Kowler (Eds.), *Early vision and beyond,* pp. 207–218. Cambridge, MA: MIT Press.

Usai, M. C., Umiltà, C., and Nicoletti, R. (1995). Limits in controlling the focus of attention. *European Journal of Cognitive Psychology, 7,* 411–439.

Usher, M., and Niebur, E. (1996). Modeling the temporal dynamics of IT neurons in visual search: A mechanism for top-down selective attention. *Journal of Cognitive Neuroscience, 8,* 311–327.

van der Heijden, A. H. C., Kurvink, A. G., de Lange, L., de Leeuw, F., and van der Geest, J. N. (1996). Attending to color with proper fixation. *Perception and Psychophysics, 58,* 1224–1237.

Van Essen, D. C., and DeYoe, E. A. (1995). Concurrent processing in the primate visual cortex. In M. S. Gazzaniga (Ed.), *The cognitive neurosciences,* pp. 383–400. Cambridge, MA: MIT Press.

Vecera, S. P., and Farah, M. J. (1994). Does visual attention select objects or locations? *Journal of Experimental Psychology: General, 123,* 146–160.

von Wright, J. M. (1970). On selection in immediate visual memory. In A. F. Sanders (Ed.), *Attention and Performance III,* pp. 280–292. Amsterdam: North-Holland.

Warner, C. B., Juola, J. F., and Koshino, H. (1990). Voluntary allocation versus automatic capture of visual attention. *Perception and Psychophysics, 48,* 243–251.

Wolfe, J. M. (1994). Guided search 2.0: A revised model of visual search. *Psychonomic Bulletin and Review, 1,* 202–238.

Wolfe, J. M. (1998). Visual search. In H. Pashler (Ed.), *Attention,* pp. 13–73. London: Psychology Press.

Wolfe, J. M., Cave, K. R., and Franzel, S. L. (1989). Guided search: An alternative to the feature integration model for visual search. *Journal of Experimental Psychology: Human Perception and Performance, 15,* 419–433.

Yantis, S. (1992). Multielement visual tracking: Attention and perceptual organization. *Cognitive Psychology, 24,* 295–340.

Yantis, S. (1993). Stimulus-driven attentional capture and attentional control settings. *Journal of Experimental Psychology: Human Perception and Performance, 19,* 676–681.

Yantis, S., and Egeth, H. E. (1999). On the distinction between visual salience and stimulus-driven attentional capture. *Journal of Experimental Psychology: Human Perception and Performance, 25,* 661–676.

Yantis, S., and Hillstrom, A. P. (1994). Stimulus-driven attentional capture: Evidence from equiluminant visual objects. *Journal of Experimental Psychology: Human Perception and Performance, 20,* 95–107.

Yantis, S., and Johnson, D. N. (1990). Mechanisms of attentional priority. *Journal of Experimental Psychology: Human Perception and Performance, 16,* 812–825.

Yantis, S., and Jonides, J. (1984). Abrupt visual onsets and selective attention: Evidence from visual search. *Journal of Experimental Psychology: Human Perception and Performance, 10,* 601–621.

Yantis, S., and Jonides, J. (1990). Abrupt visual onsets and selective attention: Voluntary versus automatic allocation. *Journal of Experimental Psychology: Human Perception and Performance, 16,* 121–134.

4 On the Time Course of Top-Down and Bottom-Up Control of Visual Attention

Jan Theeuwes, Paul Atchley, and Arthur F. Kramer

ABSTRACT Previous research showed that a salient feature singleton captured attention bottom-up (Theeuwes 1991a, 1992, 1994a). A salient color singleton interfered with search for a less salient shape singleton, which suggested that early processing was driven by bottom-up saliency factors. The present experiments examined how bottom-up and top-down processing develops over time. Subjects searched for a shape singleton target and had to ignore a color singleton distractor presented at different stimulus onset asynchronies prior to the search display. The results indicate that when the target and distractor were presented simultaneously, the salient singleton distractor captured attention, whereas when the distractor singleton was presented about 150 msec before the target singleton, the distractor did not disrupt performance. The findings suggest a stimulus-driven model of selection in which early processing is solely driven by bottom-up saliency factors. In later processing, the early bottom-up activation of the distractor can be overridden by top-down attentional control.

One of the most basic questions in the study of attention is the extent to which top-down attentional control can prevent distraction from irrelevant stimuli. Visual selective attention is thought to be (1) *goal directed* when attentional priority is given to only those objects and events that are in line with the current goals of the observer: and (2) *stimulus driven* when, irrespective of the intentions or goals of the observer, objects and events involuntarily receive attentional priority—a phenomenon referred to as "attentional capture" (for recent reviews, see Egeth and Yantis 1997; Theeuwes 1993, 1994b; Yantis, 1993, 1996). These two mechanisms of selection have been referred to as "top-down" and "bottom-up" attentional control, respectively, (e.g., Eriksen and Hoffman 1972; Posner 1980; Theeuwes 1991b; Yantis and Jonides 1984).

Many models of visual search assume that visual selection is the result of an interaction between goal-directed and stimulus-driven factors (e.g., Cave and Wolfe 1990; Treisman and Sato 1990). Typically, it is assumed that bottom-up activation occurs during early preattentive processing in which the visual field is segmented into functional perceptual units. Bottom-up activation is a measure of how salient an item is in its context. An item that is locally unique in some basic visual dimension—usually referred to as *a* "feature singleton" or simply a "singleton"—will generate a large bottom-up activation (e.g., a red poppy in a field of green grass). Top-down activation may also operate during attentional processing. Various studies have demonstrated that in more complex search tasks, knowledge of the specific task demands may guide attention to only those locations that match the target-relevant feature. For example, Kaptein, Theeuwes, and van der Heijden (1995) showed that when searching for a red vertical line segment between red tilted and green vertical line segments, subjects searched serially among the red items while ignoring the green line segments (see also Egeth, Virzi, and Garbart 1984). Top-down guidance is typically assumed to proceed either by *activation* of features that match those of the target (e.g., Wolfe 1994) or by *inhibition* of features that do not (Treisman and Sato 1990).

In a series of experiments, Theeuwes (1991a, 1992, 1996) showed that a salient feature singleton captured attention bottom-up. Even though subjects had a clear top-down attentional set to search for a particular singleton, performance was disrupted by a distractor with a salient, unique feature in a task-irrelevant dimension. Top-down control of attention could not entirely override bottom-up interference from a singleton distractor known to be irrelevant. For example, Theeuwes (1992) presented subjects with displays consisting of colored circles or diamonds appearing on the circumference of an imaginary circle. Line segments of different orientations appeared in the circles and diamonds. Subjects had to determine the orientation of the line segment appearing in the target shape. Subjects searched for a shape singleton, a single green diamond among green circles. Time to find the shape singleton increased when an irrelevant color singleton was present (i.e., one of the circles was red). Even though subjects had a clear top-down set to search for the shape singleton (i.e., the single green diamond), the presence of an irrelevant singleton (i.e., the single red circle) caused interference. It was shown that selectivity depended on the relative salience of the stimulus attributes: when the color singleton was made less salient than the shape singleton (by reducing the color difference between the target and the nontarget elements), the shape singleton interfered with search for the color singleton, whereas the color singleton no longer interfered with the search for the shape singleton.

Based on these experiments, Theeuwes (1991a, 1992, 1994a, 1996) concluded that early preattentive processing is driven by bottom-up factors such as salience. Attention is captured by the most salient singleton in the display, regardless of whether the property defining that singleton is relevant for the task or not (for more recent evidence, see Bacon and Egeth 1994, exp. 1; Caputo and Guerra 1998; Joseph and Optican 1996; Kawahara and Toshima 1996; Kim and Cave 1999; Kumada 1999; Todd and Kramer 1994). When engaged in parallel search for a particular feature singleton (e.g., a diamond among circles), the extent to which singletons capture attention is determined by the relative salience of the singletons present in the visual field. It was suggested that, irrespective of any top-down control, spatial attention is automatically and involuntarily captured by the most salient singleton. The shift of spatial attention to the location of the singleton implies that the singleton is selected for further processing. If this singleton is the target, a response is made. If it is not the target, attention is directed to the next most salient singleton. The initial shift of attention to the most salient singleton is thought to be the result of relatively inflexible, "hard-wired" mechanisms, triggered by the presence of these difference signal interrupts. Consistent with proposals by Sagi and Julesz (1985) and Koch and Ullman (1985), it is assumed that the parallel process can only perform *local mismatch* detection (i.e., indicating the presence of a discontinuity, but not its nature) followed by a serial stage directed to areas of the visual field with the largest magnitude mismatches.

Contrary to these findings, a group of other researchers have claimed that the ability of a singleton to capture attention is contingent on whether an attention-capturing stimulus is consistent with top-down settings established "off-line" on the basis of current attentional goals (Folk, Remington, and Johnston 1992; Folk and Remington 1998). According to this "contingent capture" model, only stimuli that match the top-down control settings will capture attention; stimuli that do not match the top-down settings will be ignored. Top-down control is thus possible even when target and distractor are both singletons. Along these lines, it was argued that in Theeuwes's experiments the irrelevant singleton captured attention because subjects were set to find a singleton (e.g., a local mismatch) rather than a particular feature, such as a red circle (see Bacon and Egeth 1994). It was claimed that irrespective of the bottom-up saliency, the singleton that matched the top-down setting would capture attention. These claims are based on evidence from experiments in which subjects had to ignore a cue that appeared 150 msec before the presentation of the target display (Folk, Remington, and Johnston 1992). Subjects responded to a character shape (*X* versus =) that, in different conditions, had either a unique color or a unique abrupt onset. When the search display was preceded by a to be ignored featural singleton (the cue) that matched the singleton for which they were searching, the cue captured attention as evidenced by a prolonged reaction time to identify the target (i.e., when the cue and target appeared in different spatial locations). On the other hand, if the to be ignored featural singleton cue did not match the singleton for which they were searching, its appearance apparently did not capture attention. This "contingent" capture of attention occurred for both color and onset conditions, and is considered evidence that involuntary capture is contingent on the adoption of some attentional set.

The critical finding in these studies is that a cue that does not match the top-down search goal (i.e., the defining property of the target) does not affect response time (RT), whereas a cue that matches the search goal does. In other words, if subjects were searching for a red plus sign, they were more likely to be distracted by a red cue than by an abrupt onset cue, and vice versa. Folk, Remington, and Johnston (1992) have suggested that the absence of an effect on RT for a cue that does not match the target indicates that the cue did not capture attention. On the other hand, the irrelevant cue may indeed have captured attention, but because the cue display came on 150 msec before the search display, subjects may have been able to overcome the attentional capture by the time the search display was presented (see also Theeuwes 1994a,b). Disengagement of attention from the cue may have been relatively fast when the cue and target did not share the same defining properties (e.g., the cue is red and the target is an onset), whereas disengagement from the cue may have been relatively slow in the case where the cue and target share the same defining properties (e.g., both were red). Such a mechanism could explain why there are RT costs when the cue and target have the same defining characteristics—and no costs when cue and target are different. This does not imply, however, that there is no capture of attention by the irrelevant cue singleton; it simply indicates that, after a certain time, subjects are able to exert top-down control over the erroneous capture of attention by the irrelevant singleton, to overcome its effects.

This account holds that early preattentive processing is driven by solely bottom-up feature salience factors, generating an activation pattern on which later attentive processing may then exert control to give priority to elements that match the top-down attentional set. It thus remains consistent with the claim of Theeuwes (1991a, 1992, 1994a, 1996) that during early *preattentive* processing, top-down control is not possible. It is also in line with models of visual search suggesting that during *attentive* processing either top-down inhibition may be applied to features that match the distractors (Treisman and Sato 1990) or top-down activation, to features that match the target (Wolfe 1994).

If the model presented above is correct, it should be possible to reveal how bottom-up and top-down processing develop over time. As in previous studies (e.g., Theeuwes 1992), subjects searched multielement displays for a shape singleton and reported the letter located inside the shape singleton. On some trials, an irrelevant salient color singleton was presented along with a premask display at different stimulus onset asynchronies (SOAs) before the onset of the search display. When the target and distractor singleton are presented close in time, and attention is captured by the distractor, search for the target singleton should be slowed. If, however, the singleton distractor is presented well in advance of the search display, subjects may be able to exert top-down control over the irrelevant singleton, ensuring that, by the time of the arrival of the search display, attention is directed to the target singleton. In these latter conditions, there should be no effect of the singleton distractor on search time.

4.1 EXPERIMENT 1

A visual search task similar to that in Theeuwes 1992 was employed, where subjects had to search for a feature singleton, and where this singleton is typically detected by means of preattentive parallel search. Subjects searched for a shape singleton (a single gray diamond among eight gray circles) and had to determine the orientation of the letter *C* (*C* or reversed *C*) appearing in the diamond. Determining the orientation of the letter *C* requires the allocation of focal attention to the location of the shape singleton. In the distractor condition, one of the circles was red. Because previous studies (see Theeuwes 1991a, 1992) have shown that such a color singleton is more salient than a shape singleton, it was expected that, in line with previous studies, the color singleton (i.e., the distractor) would interfere with the search for the shape singleton (i.e., the target). To determine the time course of bottom-up and top-down activation, the singleton distractor (the red circle) appeared at different SOAs prior to the presentation of the target display.

Subjects

Twelve subjects, ranging in age from 18 to 30, participated as paid volunteers. All had self-reported normal or corrected-to-normal vision and reported having no color vision defects.

Apparatus

A 486 computer with an SVGA color monitor controlled the timing of the events, generated stimuli and recorded reaction times. The "/"-key and the "z"-key of the computer keyboard were used as response buttons. All subjects were tested in a sound-attenuated, dimly lit room, with their head resting on a chinrest. The monitor was located at eye level, 60 cm from the chinrest.

Stimuli

Subjects performed a visual search task in which they searched for a uniquely shaped element (a diamond located between circles) and responded to the letter located inside this uniquely shaped singleton. The display consisted of nine elements equally spaced around the fixation point on an imaginary circle whose radius was 3.4 degrees. In the control condition each display contained one gray outline diamond (1.4 degrees on a side) surrounded by eight gray outline circles (1.4 degrees in diameter). In the distractor condition, one of the gray outline circles was replaced by a red circle producing a condition identical to that of Theeuwes 1992, in which the target had a unique shape (shape singleton)

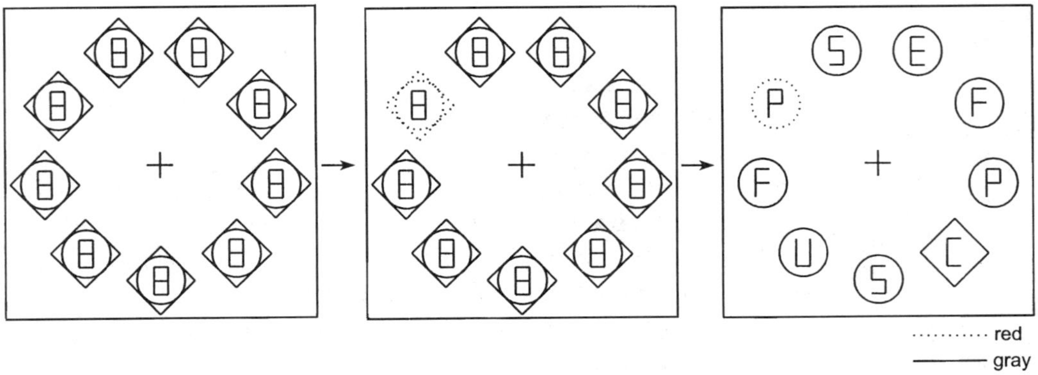

Figure 4.1 Trial events in experiment 1. The premask display (left panel) was presented for 700 msec. At stimulus onset asynchronies of 50, 100, 150, 200, 250, or 300 msec before the presentation of the search display, the color of one of the elements of the premask changed from gray into equiluminant red (middle panel). The search display (right side) contained both the color singleton distractor (the red element) and a shape target singleton (the diamond).

while the distractor had a unique color (color singleton). To ensure that distractor effects were not due to attention encompassing both the target singleton and the neighboring color singleton, the color singleton distractor was never placed adjacent to the target (i.e, there was always one gray element between the target and color distractor).

Each display element contained a letter (0.4×0.8 degrees). The uniquely shaped outline diamond (i.e., the target) contained either a *C* or a reversed *C*, the orientation of the letter determining the response (subjects pressed the "z"-key for a *C* and the "/"-key for a reversed *C*). The letters inside the other eight circles were randomly sampled from the set *E*, *P*, *F*, *U* and *S*. The letters were presented in white (11.0 cd/m^2) and the circle and diamond were presented in gray (6.4 cd/m^2). The color singleton distractor was presented in red (6.3 cd/m^2).

Design and Procedure

The sequence of events was as follows: Initially, a fixation dot was presented for 1,000 msec. Then the premask display came on consisting of nine premask elements, each composed of a single outline circle and diamond, and each containing a figure-eight premask letter (see figure 4.1). The premask display was presented for 700 msec. The color of one of the elements of the premasks changed from gray to equiluminant red with SOAs of 50, 100, 150, 200, 250, or 300 msec before the presentation of the search display. The search display was revealed by removing particular diamonds or circles of the premask display resulting in a search display consisting of eight circles and one diamond. Simultaneously with the removal of the premask, the letters inside the outline elements were

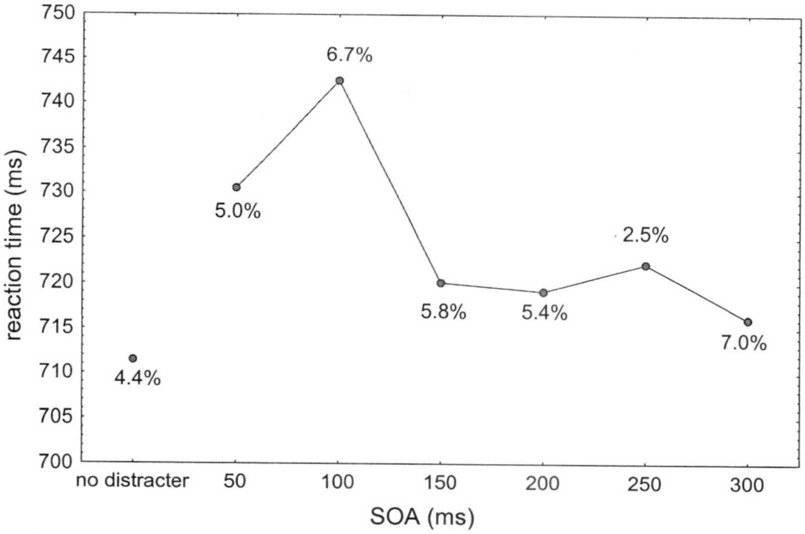

Figure 4.2 Experiment 1: Mean RTs and error percentages as a function of stimulus onset asynchrony (SOA) for the distractor and no-distractor conditions.

displayed by removing line elements from the figure eights. The search display remained present for a maximum of 2 sec until a response was emitted.

Each subject performed 240 trials, 120 no-distractor and 120 distractor trials which were presented randomly within blocks of trials. SOA between premask and search display was varied randomly between trials as well. Subjects were told to keep their eyes fixated at the fixation dot. Subjects received 240 practice trials prior to the experimental trials, as well as feedback about their performance in terms of RT and error rates after each block of 60 trials. Prior to the start of the experiment, subjects were instructed to search for the diamond and respond to the orientation of the letter inside the diamond by pressing the appropriate response key. They were told to ignore the uniquely colored red singleton.

Results

Response times longer than 1,200 msec were counted as errors, which led to a loss of less than 1% of the trials. A one-way analysis of variance (ANOVA) with no distractor or SOAs of 50, 100, 150, 200, 250, and 300 msec as levels showed a significant main effect: $F(6, 66) = 2.3$; $p < 0.05$. Additional planned comparisons showed that the RT at SOAs of 50 msec (731 msec) and 100 msec (742 msec) were significantly slower ($p < 0.05$) than the RT in the no-distractor control condition (711 msec), indicating that at the early SOA the singleton distractor interfered with search for the target singleton. However, the RTs of the later SOAs (150, 200, 250, and 300 msec) were not significantly different from the no-distractor condition suggesting that in these conditions search for the target singleton was not affected by the presence of the singleton distractor (see figure 4.2). Note that a distractor, when presented close in time to the target, slows down search by about 25 msec, an effect size very similar to that reported in Theeuwes 1992. The error rates were low (about 4.9%) and did not vary systematically with any of the conditions.

Discussion

The present results confirm earlier findings (e.g., Theeuwes 1991a, 1992) that the presence of a irrelevant salient distractor interferes with search for a relevant target singleton. The analysis of SOA suggests that there is a reliable effect of the distractor at the early SOAs (50 and 100 msec) but not at the later SOAs (150, 200, 250, 300 msec).

The results regarding SOA are in line with our predictions: at the early SOAs when distractor and target are presented in close succession, there is a clear interference effect of the distractor. It was argued that in these conditions, when target and distractor were presented in close temporal proximity, there was not enough time to exert top-down control that could have overcome attentional capture by the salient distractor. When, however, the singleton distractor was presented a considerable time (SOAs of 150 to 300 msec) before the presentation of the target singleton, sufficient top-down control could be exerted that there was no sign of attentional capture by the distractor. Indeed, response times at SOAs of 150 to 300 msec did not differ significantly from that in the no-distractor condition.

4.2 EXPERIMENT 2

Experiment 1 suggests that early in processing attention is captured by the salient distractor and that, later, attentional capture is overcome by top-down attentional control. To determine whether spatial attention was indeed captured by the distractor, we used the response congruency paradigm (Eriksen and Eriksen 1974; Eriksen and Hoffman 1972), in which subjects have to ignore a stimulus that is either congruent or incongruent with the response to the target. In previous studies (Theeuwes 1996; Theeuwes and Burger 1998; Theeuwes et al. 1999) investigating whether subjects could intentionally ignore salient but irrelevant singleton elements, the element to ignore was either identical to or different from the target element they were looking for. The results showed that the identity of the element to be ignored had an effect on response time suggesting that indeed spatial attention was directed at the location of the distractor element. Subjects were faster when the distractor element was identical to the target (congruent with the response) than when the distractor element was different from the target (incongruent with the response).

To determine whether spatial attention was shifted to the location of the color singleton distractor, we also presented a *C* or reversed *C* inside the color singleton distractor at the various SOAs used here. This letter was either identical with the letter inside the target shape singleton (and therefore congruent with the response) or different from the letter inside the target shape singleton (and therefore incongruent with the response). If attention is indeed captured by the color singleton distractor, then the identity of the letter inside the colored singleton distractor should have an effect on responding, that is, a letter congruent with the response should produce faster RTs than a letter incongruent with the response. If attention is not captured by the colored singleton, then there should be no congruency effect on RT.

Subjects

Fifteen subjects, ranging in age from 18 to 30, participated as paid volunteers.

Stimuli

The stimuli were identical to those in experiment 1. The letter located inside the irrelevant color singleton distractor was either a *C* or a reversed *C*, and this could be congruent or incongruent with the target letter inside the relevant shape singleton.

Design and Procedure

Only SOAs of 50, 100, 200, and 400 msec were used. In the current experiment, there was always a red singleton distractor present in each display. A congruent or incongruent letter was presented inside the distractor singleton.[1] Note that the letter inside the singleton distractor was revealed simultaneously with the red singleton distractor element. In other words, the letter (which could be congruent or incongruent with the response) was presented simultaneously with the singleton distractor and therefore this letter was presented 50, 100, 200, or 400 msec before the presentation of the other letters of the search display (including the target letter). SOA was varied randomly within blocks of trials. Subjects received 240 practice trials and 240 experimental trials.

Results

Response times longer than 1,300 msec were counted as errors, which led to a loss of 0.9% of the trials. An ANOVA with SOA (50, 100, 200, 400 msec) and congruency (congruent versus incongruent) as orthogonal within subject factors showed an effect of SOA: $F(3, 42) = 7.8$; $p < 0.001$;

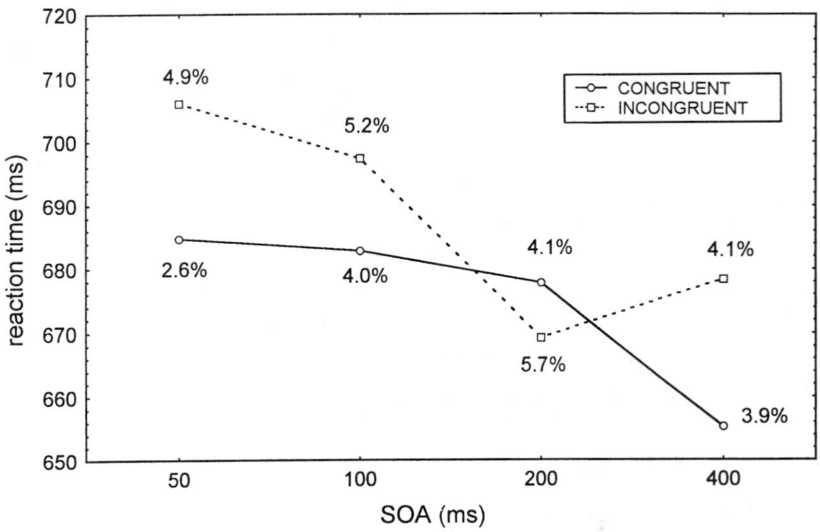

Figure 4.3 Experiment 2: Mean RTs and error percentages as a function of stimulus onset asynchrony (SOA) for the congruent and incongruent conditions.

and of congruency: $F(1, 14) = 10.3$; $p < 0.001$. The interaction between SOA and congruency was also reliable: $F(3, 42) = 3.37$; $p < 0.05$. As is clear from figure 4.3, response times become faster with increasing SOA, suggesting that (in line with experiment 1) the effect of the singleton distractor diminishes with increasing SOA. Additional planned comparisons show that a reliable congruency effect at SOA, 50 and 400 msec ($p < 0.05$) and a marginally significant congruency effect at SOA 100 msec ($p = 0.07$). At SOA 200 msec, congruency failed to reach significance ($p = 0.29$; 678 msec versus 670 msec). Also, as is clear from figure 4.3, when the letter inside the singleton distractor was congruent with the response to the letter inside the target singleton response times were faster than when it was incongruent. The finding that the letter inside the singleton distractor did affect responding to the target singleton can only be explained by assuming that at some point attention resided at the location of the singleton distractor (but see Folk and Remington 1998). The error rates were low (4.4%) and did not vary systematically with any of the conditions.

Discussion

Experiment 2 shows that response latencies become shorter with increasing SOA, suggesting again that presenting the distractor in advance of the target overcomes attentional capture by the distractor. As in experiment 1, the distractor seems to slow search by about 25 msec at the two short SOAs (50 and 100 msec).

The overall congruency effect indicates that the identity of the letter inside the singleton distractor had an effect on the response to the letter appearing inside the target singleton. When the letter inside the distractor was identical to the letter inside the target singleton, and therefore congruent with the response, response times were faster than when the letter inside the distractor was incongruent with the response to the letter inside the target singleton, a result identical to that in Theeuwes 1996. These findings are consistent with attention being *captured* by the irrelevant singleton. Because capturing attention implies that focal attention was directed to the irrelevant singleton, the identity of the letter became available, thereby affecting the speed of responding to the target.

Folk and Remington (1998) have suggested an alternative explanation for such findings. Instead of assuming that attention was captured by the irrelevant singleton, they suggested that the congruency effect as observed in Theeuwes 1996 and in Theeuwes and Burger 1998 was the result of processing the target and distractor letter in parallel. Such an explanation, though possible, is unlikely: at the eccentricities used in the current experiments, letters cannot be processed efficiently in parallel for discriminations such as *C* versus reversed *C* (see, for example, Theeuwes 1991c; Wolfe 1994). Usually, when subjects search for a target letter among nontarget letters, search time increases linearly with the number of nontarget letters in the display, a result typically seen as evidence for spatially serial search. Given these considerations, the most likely explanation is that the identity of the letter in the irrelevant singleton affected responding because attention was directed at the location of the singleton distractor before a response was made. In addition, the control experiment (see note 1), in which a congruent or incongruent letter was placed in a nonsingleton item, showed no effect of congruency ($F = 1$), providing evidence that the congruency effect only shows when attention is attracted to the location of the colored singleton. This finding suggests that parallel processing of *all* letters (including the congruent or incongruent letter placed in the nonsingleton) is highly unlikely.

It is important to note that there is a clear congruency effect at SOA 400 msec ($p = 0.0065$). This finding is important because it implies that even when the singleton distractor (with the congruent or incongruent letter inside) is presented 400 msec before the presentation of the search display, attention was captured by the singleton. In other words (as demonstrated in experiment 1), at SOAs of 200 msec, subjects had enough time between the presentation of distractor and target, not to prevent attentional capture, but to gain attentional control after their attention had been erroneously captured by the salient distractor.

Another interesting finding is that at SOA 200 msec, the congruency effect is absent (if anything, the effect is reversed). This suggests that to gain attentional control subjects may have inhibited the singleton distractor location and thereby reduced the influence of the letter inside the singleton distractor. Because of this inhibition, the letter inside the distractor no longer influences responding to the target letter. The fact that the congruency effect is absent at SOA 200 msec but not at SOA 400 msec suggests that the inhibition may be transient.

4.3 EXPERIMENT 3

The goal of experiment 3 was to investigate the possible role of inhibition of the distractor color over trials. Experiment 3 was identical to experiment 2 except that the color of the singleton distractor could be either red or green and changed randomly from trial to trial. The results of experiment 2 suggesting inhibition of the distractor at SOA 200 msec and not at SOA 400 msec implies that the inhibition may be relatively short-lived. If inhibition is relatively brief, then changing the color of the distractor from trial to trial should produce the same pattern of effects as that observed in experiment 2.

If, however, attentional set (e.g., in the sense of inhibiting a specific color) is carried over from one trial to the next, then response latencies should be faster when the singleton distractor has the same color as on the previous trial than when it does not. Such a result would be consistent with Maljkovic and Nakayama 1994, which showed that visual search responses were faster when the color of the target singleton was repeated from the previous trial than when it was changed. Subjects were considered to be relatively fast on same-color trials because they could retrieve an attentional set identical to the one used in the previous trial. Although it is not clear whether such a repetition effect also occurs when distractor rather than target colors are changed, if repeating the same attentional set produced a general effect, then a repetition effect should also be observed for the distractors in the present studies. Note that, as in Maljkovic and Nakayama 1994, any repetition effect in the current experiment cannot be a response-based effect because subjects did not respond to the color, but to the letters inside the elements.

Subjects

Twelve subjects, ranging in age from 18 to 30, participated as paid volunteers.

Stimuli

The stimuli were identical to those in experiment 2. The singleton distractor was either red or green and changed color randomly from trial to trial.

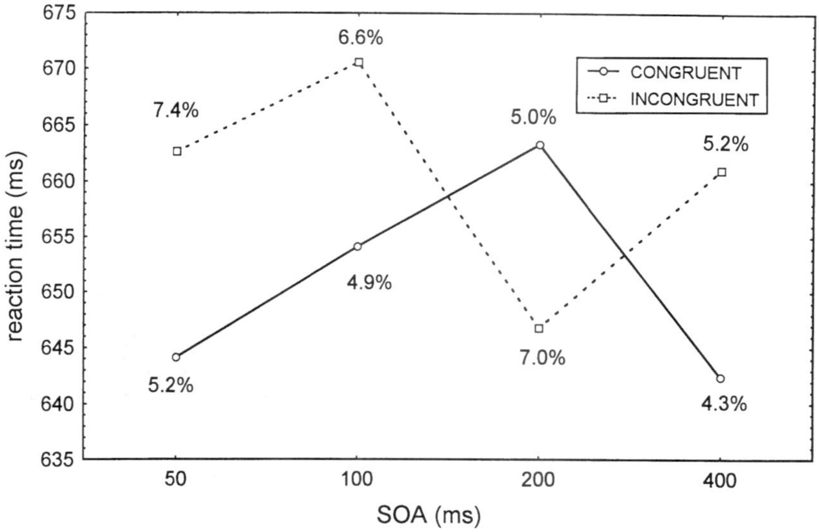

Figure 4.4 Experiment 3: Mean RTs and error percentages as a function of SOA for the congruent and incongruent conditions.

Design and Procedure

Subjects received 512 practice and 512 experimental trials.

Results

Response times longer than 1,300 msec were counted as errors, which led to a loss of 1.1% of the trials. An ANOVA with SOA (50, 100, 200, 400 msec) and congruency (congruent versus incongruent) as factors showed an effect of congruency: $F(1, 11) = 4.90$; $p < 0.05$; and of congruency × SOA: $F(3, 33) = 6.54$; $p < 0.001$. Planned comparisons indicate that, for all SOAs, the difference between the congruent and incongruent conditions is reliable (all $p < 0.05$). Note, however, that at SOA 200 msec this effect is reversed ($p = 0.02$), that is, incongruent responses are *faster* than congruent responses (see figure 4.4).

An additional analysis was carried out to determine whether changing the color of the singleton distractor over trials had an effect on response latencies. An ANOVA showed no effect of color change: $F(1, 11) = 3.0$; $p = 0.11$; nor did distractor color change interact with any of the other variables—color change × congruency: $F(1, 11) = 0.07$; color change × SOA: $F(3, 33) = 0.86$. This suggests subjects were not able to carry over the attentional set (including the color to inhibit) from the previous trial in order to speed up responding. The error rates were low (5.5%) and did not vary systematically with any of the variables.

Discussion

The finding that RTs in trials in which the distractor color switched were the same as when the color remained the same suggests that attentional set in the sense of which color to inhibit does not carry over from one trial to the next. Unlike the findings in Maljkovic and Nakayama 1994, which showed a repetition effect for the *target* color, the current findings indicate that this does not hold for the *distractor* color. The results suggest that the color to inhibit may not be part of the attentional set that transfers from one trial to the next. Note, however, that in experiment 3 the target remained fixed over trials. If specifying the target is the most important feature of the attentional set, then one may argue that repetition effects of the distractor color were not observed in experiment 3 because the target remained the same. Future studies may address whether switching the color of the distractor produces a repetition effect when the color of the target also changes from trial to trial. Overall, consistent with the findings of experiment 2 that overcoming of the distractor effect was relatively short-lived, the current findings suggest that rejection of the relevant color singleton does not transfer from one trial to the next.

The congruency effects are similar to those of experiment 2. For SOAs 50, 100, and 400 msec, there is a clear congruency effect in the sense that congruent responses are faster than incongruent responses. Yet, consistent with a trend in experiment 2, at an SOA of 200 msec, the congruency effect is reversed, that is, congruent responses are faster than incongruent. The findings suggest that in order to redirect attention away from the singleton distractor location, subjects may have inhibited the location of the distractor, and thereby inhibited the letter inside the singleton distractor. When the inhibited letter is identical to the target letter (i.e., the congruent condition), subjects are relatively slow. On the other hand, when the inhibited letter is different from the target letter (the incongruent condition), the letter is not inhibited and subjects are relatively fast.

Distractor inhibition also appears in many experiments demonstrating negative priming, in which the response to a stimulus is slowed when the previously inhibited stimulus becomes relevant for responding (e.g., Neill and Valdes 1996). For example, Tipper and Cranston (1985) showed that when subjects ignored a letter on trial n, the response to a letter with the same identity on trial $n + 1$ was impaired, a condition comparable to the congruency manipulation in experiments 2 and 3. It is hypothesized that actively inhibiting the potentially competing response from the letter in the singleton to be ignored, may cause a reversal of the congruency effect; that is, a response congruent with the letter inside the distractor is slower than a response that is incongruent. Note that this reversal only occurs when the distractor is presented 200 msec before the presentation of the target, suggesting it takes time for inhibition to accrue. A similar pattern of facilitation and inhibition appears in experiments addressing "inhibition of return": targets appearing on the cued side show an RT advantage for the first 150 msec, which is replaced by an inhibition after 250 to 300 msec (Posner and Cohen 1984).

The observation in experiment 3 that the congruency effect returns after a time interval of 400 msec is not in line with findings from either the "negative priming" nor the "inhibition of return" literature. At the early SOAs, in which distractor and target are presented within 100 msec, active top-down inhibition at the location of the distractor may start to build up, yet, before it is complete, the appearance of the target singleton causes attention to be automatically captured by the location of the target singleton. In other words, there may not be enough time to allow active top-down inhibition at the early SOA, resulting in a "typical" congruency effect, as observed in previous studies, where target and singleton distractor were presented simultaneously (see Theeuwes 1996).

At the later SOA of 200 msec, as evidenced by the absence of an interference effect of the distractor, subjects may have enough time to exert top-down control. Top-down control results in active inhibition of the singleton distractor, including the letter located inside the distractor. Inhibition is important because the distractor and target are presented in relatively close succession (i.e., within a 200 msec time frame), and will compete for attention. At this point, it is not clear why the congruency effect returns at SOA 400 msec. Perhaps it is impossible to maintain this type of inhibition over a longer time period.

4.4 GENERAL DISCUSSION

The current experiments were designed to examine the time course of bottom-up and top-down processing in visual search. The results indicate that a salient singleton distractor presented close in time to the target singleton causes interference, as demonstrated by response times that are significantly longer than those in the no-distractor condition. The finding that the letter inside the singleton distractor had an effect on responding to the target (i.e., the congruency effect) also suggests that spatial attention was drawn to the location of the distractor providing evidence that the increase in RT is indeed due to attentional capture.

When a singleton distractor is presented at least 150 msec in advance of the target, the interference effect is no longer observed, although the finding that the letter inside the singleton distractor has an effect on RT at still longer SOAs indicates that attention was captured by the singleton distractor. Yet, with an interval of 150–200 msec between the presentation of distractor and target, there was sufficient time to reorient spatial attention from the location of the distractor. When, at that point, the target singleton is presented, attention is immediately directed to the target singleton resulting in response times equivalent to those in the no-distractor condition.

When a singleton distractor is presented 150–200 msec in advance of the target, it is assumed that top-down control can reduce or eliminate the effect of the distractor. Note, however, that the presence of a congruency effect at the longer SOAs indicates that top-down attentional control cannot *prevent* attention from being captured by the singleton distractor, but rather it allows a fast and efficiently redirection of attention from the distractor to the target location.

The present findings are consistent with those in Kim and Cave 1999, which investigated the temporal interaction between top-down and bottom-up control of attention by means of probe RTs. Kim and Cave also used a task similar to that in Theeuwes 1992, where subjects searched for a shape singleton (a circle among diamonds) while an irrelevant color singleton distractor (a red element among green elements) was present. Either 60 or 150 msec after the presentation of the search display containing the target and singleton distractors, probes could appear at any of the locations. It was hypothesized that if the early preattentive processing is solely driven by bottom-up salience, as suggested by Theeuwes (1991, 1992), then the location of the salient singleton distractor should be attended first, and thus the probe RT at the distractor location should be faster than at any of the other locations in the short-SOA condition regardless of whether the unique feature is relevant. On the other hand, if top-down control is possible somewhat later in time, as the current experiments suggest, then in the late-SOA condition, attention should no longer be at the distractor location but instead at the location of the target singleton. For conditions in which target and distractor were locally unique (and therefore salient enough) Kim and Cave (1999) did indeed find these results. At an SOA of 60 msec, the probe RT at the location of the singleton distractor was about 20 msec faster than at the target singleton location. At an SOA of 150 msec, however, this pattern was reversed: the probe RT at the target location was about 15 msec faster than at the distractor location.

The current findings fit very well with those reported in Kim and Cave 1999, namely, that after 150 msec, attention is no longer at the location of the distractor but instead at the location of the target. In our experiment 1, we show that, at an SOA of at least 150 msec, the singleton distractor no longer interferes with search for the target singleten: there is no difference in RTs between the long-SOA conditions and the no-distractor condition. These findings both suggest that it takes somewhere between 100 and 150 msec to disengage attention from the location of the distractor and redirect it to the location of the target singleton.

The current results shed some new light on the findings obtained with the spatial cuing paradigm of Folk and colleagues (Folk, Remington, and Johnston 1992; Folk and Remington 1998) in which subjects have to ignore a cue that appears 150 msec before the search display. The critical finding is that a cue that does not match the top-down search goal (e.g., as in our experiments, the search goal is a shape singleton; the cue is a color singleton) does not affect RT, whereas a cue that does match the search goal slows search. The current findings and those in Kim and Cave 1999 show why with an SOA of 150 msec, a cue that does not match the search goal has no effect on RT: by the time the search display is presented, subjects are able to exert enough top-down control to allow a redirection of attention from the location of the distractor to the location of the target. The finding that there is an effect on RT in Folk and colleagues' experiments when the cue and target share the same defining property (e.g., the cue is red and the target is red) is not surprising because it is likely that disengagement and redirection of attention from the distractor location will take much longer when the distractor and target have the same defining property. It will be clear that this explanation of Folk and colleagues' data does not suggest anything like a "contingent capture" hypothesis, but merely confirms Theeuwes's stimulus-driven model of selection (1992), in which early processing is driven by bottom-up saliency factors. Note that our current findings and those of Kim and Cave (1999) also disconfirm a more recent notion put forward by Folk and Remington (1998), which suggests that irrelevant singletons do not capture spatial attention but merely cause a "filtering" cost. Both the effect of congruency of the letter inside the distractor, as found in our experiments, and spatial RT probe effect, as found in Kim and Cave 1999, clearly indicate that *spatial* attention was in fact captured by the location of the distractor.

Even though we suggested that the effect of the distractor at the later SOAs was reduced because of top-down control, the time course of the distractor effect could also be explained in a purely bottom-up fashion. Along these lines, it is assumed that attention is captured bottom-up by the most salient singleton and, after being disengaged from the most salient singleton, automatically reoriented to the next most salient singleton. If it takes about 150 msec to disengage and reorient attention, then it is not surprising that, at an SOA of 150 msec the interference effect was reduced. Note, however, that we assume that top-down control (i.e., knowing that one is looking for a diamond shape) does facilitate the disengagement of attention from the colored distractor singleton. After selecting the colored distractor, knowing that one is looking for a diamond and not for a red circle will most likely speed up the disengagement of attention and facilitate reorienting (see Theeuwes 1994b for a similar account).

We interpreted the current results in a strictly serial fashion, assuming that attention is first shifted to the most salient singleton and then to the next. Parallel processing models could also explain the current findings, assuming that on some trials the distractor finishes processing first, while on others, the target singleton finishes first. A purely parallel model, in which not only the two singletons are processed in parallel but all items are processed in parallel is somewhat less likely, given the findings of the control experiment (see note 1), which showed no congruency effect when a congruent or incongruent letter was placed in a nonsingleton. If all letters were processed in parallel, there should have been a clear congruency effect in the control study because the response-related letter inside a nonsingleton would have been processed at least as fast as any of the other letters in the display (possibly faster because of a top-down setting to look for this letter).

The current study indicates that during early preattentive processing, selection is driven bottom-up, that is, attention is captured by the most salient singleton present in the visual field. After attention is captured by the location of the singleton distractor, "attentive" processing exerts top-down activation that allows attention to be shifted elsewhere. The current model assumes that visual selection is the result of an interaction between goal-directed and stimulus-driven factors, consistent with models of visual search (e.g., Cave and Wolfe 1990; Wolfe 1994; Treisman and Sato 1990). Yet, unlike other models, the current model assumes that early preattentive parallel processing (assumed to calculate differences among stimulus features) is not accessible to top-down control. Only after an item has been selected does top-down processing help to speed up the disengagement of attention, allowing attention to be shifted to the next location.

NOTES

This research was supported by cooperative research agreement DAAAL01-96-2-0003 with the U.S. Army Research Laboratory. We thank Angela Glass and Meredith Minear for their assistance in conducting the studies.

1. To ensure that any congruency effect does indeed depend on attention being attracted to the colored singleton distractor, we ran a control study in which twelve subjects searched for a shape singleton while a congruent or incongruent letter was placed in one of the nonsingletons, instead of being placed in the colored distractor. There were no reliable effect of congruency on RT: $F(1,11) = 1.06$; $p = 0.32$; nor on error rate: $F(1,11) = 1.32$; $p = 0.27$.

REFERENCES

Bacon, W. F., and Egeth, H. E. (1994). Overriding stimulus-driven attention capture. *Perception and Psychophysics, 55,* 485–496.

Cave, K. R., and Wolfe, J. M. (1990). Modeling the role of parallel processing in visual search. *Cognitive Psychology, 22,* 225–271.

Caputo, G., and Guerra, S. (1998). Attentional selection by distractor suppression. *Vision Research, 38,* 669–689.

Egeth, H. E., Virzi, R. A., and Garbart, H. (1984). Searching for conjunctively defined targets. *Journal of Experimental Psychology: Human Perception and Performance, 10,* 32–39.

Egeth, H. E., and Yantis, S. (1997). Visual attention: control, representation and time course. *Annual Review of Psychology, 48,* 269–297.

Eriksen, B. A., and Eriksen, C. W. (1974). Effects of noise letters upon the identification of a target letter nonsearch task. *Perception and Psychophysics, 16,* 143–149

Eriksen, C. W., and Hoffman, J. E. (1972). Temporal and spatial characteristics of selective encoding from visual displays. *Perception and Psychophysics, 12,* 201–204.

Folk, C. M., and Remington, R. W. (1998). Selectivity in distraction by irrelevant featural singletons: Evidence for two forms of attentional capture. *Journal of Experimental Psychology: Human Perception and Performance, 24,* 847–858.

Folk, C. L., Remington, R. W., and Johnston, J. C. (1992). Involuntary covert orienting is contingent on attentional control settings. *Journal of Experimental Psychology: Human Perception and Performance, 18,* 1030–1044.

Joseph, J. S., and Optican, L. M. (1996). Involuntary attentional shifts due to orientation differences. *Perception and Psychophysics, 12,* 201–204.

Kaptein, N. A., Theeuwes, J., and van der Heijden, A. H. C. (1995). Search for a conjunctively defined target can be selectively limited to a color-defined subset of elements. *Journal of Experimental Psychology: Human Perception and Performance, 21,* 1053–1069.

Kawahara, J., and Toshima, T. (1996). Stimulus-driven control of attention: Evidence from visual search for moving target among static nontargets. *Japanese Journal of Psychonomic Science, 15,* 77–87.

Kim, M. S., and Cave, K. R. (1999). Top-down and bottom-up attentional control: On the nature of interference from a salient distractor. *Perception and Psychophysics, 61,* 1009–1023.

Koch, C., and Ullman, S. (1985). Shifts in selective visual attention: Towards the underlying neural circuitry. *Human Neurobiology, 4,* 219–227.

Kumada, T. (1999). Limitations in attending to a feature value for overriding stimulus-driven interference. *Perception and Psychophysics, 61,* 61–79.

Maljkovic, V., and Nakayama, K. (1994). Priming of pop-out: 1. Role of features. *Memory and Cognition, 22,* 657–672.

Neil, W. T., and Valdes, L. A. (1996). Facilitatory and inhibitory aspects of attention. In A. F. Kramer, M. G. H. Coles and G. D. Logan (Eds.), *Converging operations in the study of visual attention,* pp. 77–106. Washington DC: American Psychological Association.

Posner, M. I. (1980). Orienting of attention. *Quarterly Journal of Experimental Psychology, 32,* 3–25.

Posner, M. I., and Cohen, Y. (1984). Components of visual orienting. In H. Bouma and D. Bouwhuis (Eds.), *Attention and Performance X,* pp. 531–556. Hillsdale, NJ: Erlbaum.

Sagi, D., and Julesz, B. (1985). Detection versus discrimination of visual orientation. *Perception, 14,* 619–628.

Theeuwes, J. (1991a). Cross-dimensional perceptual selectivity. *Perception and Psychophysics, 50,* 184–193.

Theeuwes, J. (1991b). Exogenous and endogenous control of attention: The effect of visual onsets and offsets. *Perception and Psychophysics, 49,* 83–90.

Theeuwes, J. (1991c). Categorization and identification of simultaneous targets. *Acta Psychologica, 76,* 73–86.

Theeuwes, J. (1992). Perceptual selectivity for color and form. *Perception and Psychophysics, 51,* 599–606.

Theeuwes, J. (1993). Visual selective attention: A theoretical analysis. *Acta Psychologica, 83,* 93–154.

Theeuwes, J. (1994a). Stimulus-driven capture and attentional set: Selective search for color and visual abrupt onsets. *Journal of Experimental Psychology: Human Perception and Performance, 20,* 799–806.

Theeuwes, J. (1994b). Endogenous and exogenous control of visual selection. *Perception, 23,* 429–440.

Theeuwes, J. (1996). Perceptual selectivity for color and form: On the nature of the interference effect. In A. F. Kramer, M. G. H. Coles and G. D. Logan (Eds.), *Converging operations in the study of visual attention,* pp. 297–314. Washington DC: American Psychological Association.

Theeuwes, J., and Burger, R. (1998). Attentional control during visual search: The effect of irrelevant singletons. *Journal of Experimental Psychology: Human Perception and Performance, 24,* 1342–1353.

Theeuwes, J., Kramer, A. F., Hahn, S., Irwin, D. E., and Zelinsky, G. J. (1999). Influence of attentional capture on oculomotor control. *Journal of Experimental Psychology: Human Perception and Performance, 25,* 1595–1608.

Todd, S., and Kramer, A. F. (1994). Attentional misguidance in visual search. *Perception and Psychophysics, 56,* 198–210.

Tipper, S. P., and Cranston, M. (1985). Selective attention and priming: Inhibitory and facilitory effects of ignored primes. *Quarterly Journal of Experimental Psychology, 37A,* 591–611.

Treisman, A. M., and Sato, S. (1990). Conjunction search revisited. *Journal of Experimental Psychology: Human Perception and Performance, 16,* 451–478.

Wolfe, J. M. (1994). Guided search 2.0: A revised model of visual search. *Psychonomic Bulletin and Review, 1,* 202–238.

Yantis, S. (1993). Stimulus-driven attentional capture. *Current Directions in Psychological Science, 2,* 156–161.

Yantis, S. (1996). Attentional capture in vision. In A. F. Kramer, M. G. H. Coles, and G. D. Logan (Eds.), *Converging operations in the study of visual attention,* pp. 45–76. Washington DC: American Psychological Association.

Yantis, S., and Jonides, J. (1984). Abrupt visual onsets and selective attention: Evidence from selective search. *Journal of Experimental Psychology: Human Perception and Performance, 10,* 601–621.

5 Electrophysiological and Neuroimaging Studies of Voluntary and Reflexive Attention

Joseph B. Hopfinger, Amishi P. Jha, Jens-Max Hopf, Massimo Girelli, and George R. Mangun

ABSTRACT Powerful brain systems specialized for voluntary and reflexive attentional control influence visual information processing. Studies of voluntary selective attention have shown that the amplitudes of visual event-related potentials (ERPs) are greater for events occurring at attended locations. Using ERPs, we recently investigated the neural correlates of reflexive attention and found that early visual processing in the cortex is also modulated by reflexive orienting. By integrating functional imaging with ERP recording, we related ERP signs of voluntary attention to underlying neural mechanisms. We used event-related functional magnetic resonance imaging (fMRI) and trial-by-trial spatial cuing to investigate the time course and functional anatomy of these attentional control systems. Attentional mechanisms in frontal, parietal, and temporal cortex were found to produce changes in visual cortical processing at multiple loci in the visual hierarchy, facilitating or attenuating information from competing loci to reduce interference from irrelevant events during perception and performance.

Over the past three decades, psychophysical studies of the effects of selective attention on perception and performance in humans have established that voluntarily directing covert attention to selected locations or events facilitates perception and performance. For example, observers are typically faster and more accurate in responding to stimuli at attended than at unattended locations (see Yantis, chap. 3, this volume). In order to elucidate the neurobiological underpinnings of these attentional phenomena, physiological approaches have been employed in both humans and animals. Much of this work initially investigated where in the ascending processing stream top-down attentional control could influence stimulus analysis. Studies in both humans and animals have now clearly demonstrated that sensory-perceptual processes are modulated by top-down spatial attention (see Mangun, Hillyard, and Luck 1993; Desimone and Duncan 1995).

Other research has aimed to understand the control systems themselves (e.g., Corbetta et al. 1993; Harter et al. 1989; Posner et al. 1984). Widespread regions of the brain, including frontal and parietal cortex, subcortical structures such as the basal ganglia, portions of the thalamus, and brain stem structures such as the superior colliculus, have been implicated in attentional control (see Posner and Petersen 1990).

After contrasting the effects of top-down and bottom-up control over sensory processing in visual cortex, we review our recent efforts to integrate electrophysiological recordings with functional neuroimaging measures to provide detailed anatomical information about where in the human visual cortex top-down control acts to influence sensory processing. We conclude by examining the control circuitry itself.

The emerging picture is that visual analyses in multiple areas of human extrastriate cortex are modulated under the control of both top-down and bottom-up mechanisms during spatial attention. These modulations include changes in the gain of sensory inputs for attended and ignored locations. As a result, the signal-to-noise ratio is improved for relevant versus irrelevant inputs across visual space. At the earliest levels of visual cortical analysis subject to attentional control, these effects of attention are distributed in a gradient across the visual field, and are specific for selection based on location (see, for example, Mangun 1995).

5.1 VOLUNTARY ATTENTIONAL CONTROL OF VISUAL PROCESSING VIA SPATIAL SELECTION

In the late 1960s, Eason, Harter, and White (1969) used scalp-recorded ERPs in studies of spatially selective visual attention. They instructed their subjects to attend and respond manually to stimuli flashed to one visual field, and to ignore those flashed to the opposite field. Comparing attended versus passively viewed ERPs to the same physical stimuli, the authors observed changes in occipital sensory-evoked responses at latencies between 150 and 200 msec after stimulus onset. Subsequently, Van Voorhis and Hillyard (1977) replicated and extended these findings in experiments designed to control for nonselective effects such as behavioral arousal. They did so by comparing one attention condition directly to another roughly equivalent condition (e.g., attend right versus attend left location), rather than to passive conditions. Amplitude modulations of sensory-evoked components as a function of spatial selective attention began as early as 70–80 msec after the onset of the visual stimulus (see also Eason 1981). The earliest effect of spatial attention was on the amplitude of the so-called P1 component, a positive polarity wave recorded over the lateral occipital scalp at 70–130 msec after the onset of the stimulus and believed to be a reflection of activity in striate cortex, or even in subcortical pathways. We now know, however, that the P1 attention effect arises at later stages of visual cortical processing. Numerous studies have demonstrated the reliability of these spatial attention effects in cortical sensory-evoked ERPs and have significantly clarified their properties (e.g., Eimer 1994; Harter, Aine, and Schroeder 1982; Hillyard and Münte 1984; Mangun and Hillyard 1991; Mangun, Hillyard, and Luck 1993). One key finding is that the P1 component is affected only by spatial attention, and not by attention selectively directed to nonspatial stimulus features

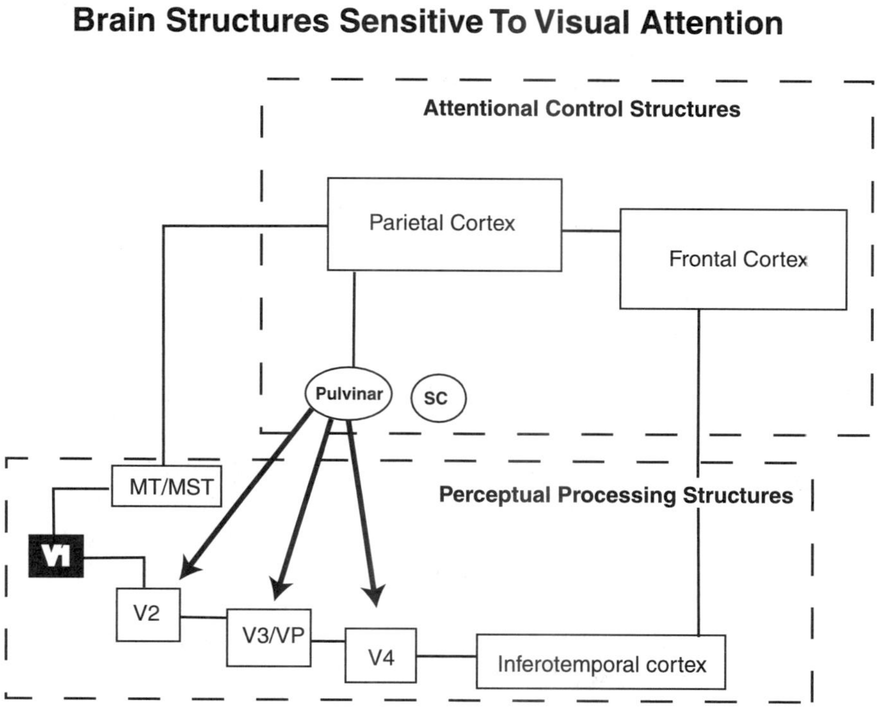

Figure 5.1 Schematic diagram of brain structures sensitive to visual attention. Various brain regions have been implicated in attentional processing based on single-neuron recordings and pharmacological manipulations in monkeys, plus neuropsychological, event-related potential (ERP), and functional imaging studies in humans. These structures can be segregated into those hypothesized to be involved in attentional control, such as the frontal and parietal cortex and subcortical structures like the pulvinar nuclei of the thalamus and superior colliculus (SC), and those involved in perceptual analyses influenced by attentional control systems.

such as color (e.g., Anllo-Vento, Luck, and Hillyard 1998; Harter and Aine 1984; Hillyard and Münte 1984).

Stages of Information Processing Influenced by Top-Down Control

Although much evidence indicated that sensory ERPs were affected by attention, until very recently, there was little direct evidence about which anatomically defined brain areas were being affected. Precisely where in the complex visual system of the primate brain are top-down attentional control processes able to influence incoming information?

Recent evidence indicates that spatial selective attention exerts its greatest control over visual input processing at the cortical rather than subcortical levels of the ascending pathways (figure 5.1), and it is now well established that the P1 attention effect reflects modulation in the extrastriate cortex (e.g., Heinze et al. 1994; Mangun et al. 1997; Woldorff et al. 1997). Consistent with evidence from single-cell studies of spatial attention in nonhuman primates (e.g., Luck et al. 1997; Moran and Desimone 1985), ERP activity in the P1 latency range is thought to arise from visual areas V2, V3/VP, and V4, although it remains unclear whether the P1 attention effect is generated in one or several of these areas. Some evidence suggests that, under certain stimulus and task conditions, incoming sensory signals can be weakly modulated earlier in primary visual cortex (V1). We will return to this issue later. First, however, we turn to our studies of *reflexive* attention, in which we ask whether visual cortical processing is influenced by bottom-up control as well as by the more well-established top-down mechanisms discussed so far.

5.2 REFLEXIVE ATTENTIONAL CONTROL OF VISUAL PROCESSING

Visual attention can be oriented reflexively (automatically) as well as voluntarily when sensory events trigger (cue) attention to their locations in the visual field (for a review, see Yantis, chap. 3, this volume). Both reflexive and voluntary attention produce facilitation in reaction times (RTs) to target stimuli occurring at attended or cued locations, but the time courses of these effects differ. Reflexive attention is more rapidly engaged and more transient than voluntary attention (Posner, Snyder, and Davidson 1980; Jonides 1981). In addition, reflexive attention includes inhibitory processes that lead to slowed RTs for cued-location events as time between the reflexive cue and target increases (e.g., Posner and Cohen 1984). Known as "inhibition of return" (IOR), this may lead the reflexive system to favor novel locations, promoting effective search of the scene. The neural correlates of reflexive attention are less well understood than those for voluntary attention. Does, for example, the fast RT facilitation observed with reflexively attended stimuli involve changes in visual input processing, or does it reflect later changes in decision criteria or motor activation for cued-location events?

Event-Related Potential Evidence for Reflexive Attentional Control over Visual Cortical Processes

To test the effects of reflexive attention on cortical processing, we (Hopfinger and Mangun 1998) presented spatially nonpredictive cues (a brief offset-onset of white dots) in left or right hemifields, and followed these with task-relevant targets in either the same or opposite field. The targets were either tall or short bars (0.5 probability) and required a discriminative button press. The interstimulus interval (ISI) between reflexive cues and targets was either short (34–234 msec) or long (566–766 msec). Targets that followed on the same side as the reflexive cues at the short ISI elicited occipital ERPs of enhanced amplitudes, in addition to faster RTs. Furthermore, these ERP enhancements appeared to be at the same neural locus as the earliest enhancements typically produced by voluntary spatial attention, in the occipital P1 component (90–140 msec latency). In contrast, at the longer ISI, the facilitation in ERPs was replaced by a reversal in the ERP pattern. That is, the P1 tended to be of smaller amplitude to cued location targets, a pattern reminiscent of IOR. We interpreted these data as evidence that reflexive attention produces a short-lived, spatially restricted facilitation in visual processing in extrastriate visual cortex. Interestingly, this appears to occur at the same stage of visual processing influenced by voluntary attention.

A limitation in the foregoing ERP study of reflexive attention was that the subjects' task involved a discrimination of target features (tall versus short vertical bars). Although the subjects were informed that the cues were completely uninformative about where the target would occur, because the analysis of target features required focal allocation of attention for task performance, the discrimination task might have introduced a voluntary component (e.g., Egly et al. forthcoming; Treisman 1988); under these stimulus and task conditions, enhancements of the P1 component might not reflect activity of a purely reflexive mechanism. This seems unlikely given that the reflexive P1 attention effect was rapidly engaged and transient, as is typically seen with RT facilitation for reflexive cues at short ISIs. To eliminate this possibility, however, and to further investigate the relationship between IOR and processing in visual cortex, we conducted a study that utilized the same stimulus parameters as Hopfinger and Mangun (1998) but required only a simple, speeded RT response to the suprathreshold target bars.

Methods Four small white dots were continuously displayed in the left and right visual hemifields, forming an imaginary rectangle 1.0 by 1.4 degrees in size (the center of the imaginary rectangle was 1.5 degrees above fixation and 6.4 degrees lateral to fixation). Each trial began when the four dots on one side of fixation (equally probable on the left or right of fixation) blinked off for 34 msec before reappearing (reflexive cue). Subjects were explicitly told that the blinking of the dots was nonpredictive of the location of the subsequent target bar, and were told not to attend to the dots because this would be an unproductive strategy in responding rapidly to the targets.

As in our prior study, the targets followed the reflexive cues by either 34–234 or 566–766 msec, in a random fashion. The intertrial interval varied randomly between 1,500 and 2,000 msec. The target remained on the screen for 50 msec and was randomly either 1.8 or 2.3 degrees in height by 0.60 degree in width, but the height was irrelevant for the present study. Subjects pressed a button with their index finger as soon as the bar was detected (response hand was counterbalanced). Trials were presented in 80 total blocks over two separate days of testing for each subject; 20% were catch trials, in which no target followed the reflexive cue—and to which subjects virtually never responded.

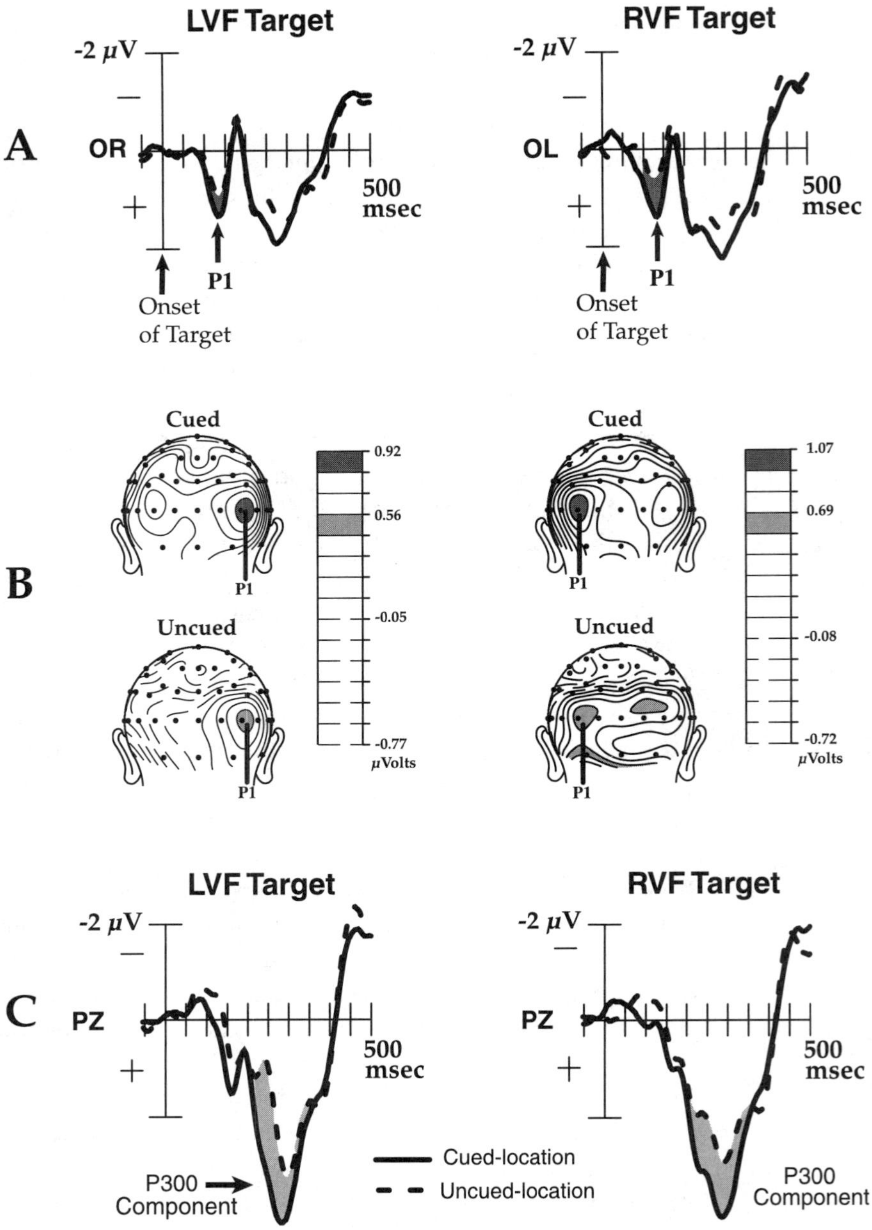

Figure 5.2 A. Event-related potential (ERP) waveforms and topographic maps in reflexive cuing study. ERPs at lateral occipital scalp sites to left (LVF) and right (RVF) target bars are shown at the top. Tick marks are 50 msec, target onset is indicated by the arrow and upright calibration bar. Positive voltages are plotted downward. Overlaid are the responses to targets when preceded by a reflexive cue in the same location (cued location) and when preceded by a cue in the opposite visual hemifield (uncued location) after correction with Adjar filtering. Differences in the amplitude of the occipital P1 component are shaded. B. Topographic voltage maps of the responses to left and right targets during the time range of the P1 effect (100–150 msec). Each line on the head represents an isovoltage contour. The scalp topographic maxima of the P1 for cued and uncued targets is shaded and labeled. C. ERPs from midline parietal electrode site Pz are shown for left and right visual field target bars. Cued-location targets (solid line) elicited larger P300 components than did uncued-location targets (dashed line), and the difference is shaded in the figure.

The ERPs were collected from 64 tin electrodes placed on the scalp, but only selected sites are shown in the figures. Eye position was monitored with an infrared video camera system and by recording the electrooculogram from electrodes placed around each eye. Trials with eye movements or blinks were rejected. The adjacent response filter (Adjar) method (Woldorff 1993) was employed to separate the brain responses to the cues from those to the targets, something that is critical at short ISIs. The details of the recording and analysis were identical to those in Hopfinger and Mangun 1998.

Results and Conclusions The subjects were significantly faster in responding to targets at the cued location than at the uncued location (cued = 282 msec versus uncued = 290 msec; $p < 0.05$) at short cue-to-target ISIs, although this pattern changed at the longer ISI, where a typical IOR pattern was observed (cued = 290 msec versus uncued = 277 msec; $p < 0.05$). The data presented in figure 5.2A (top) are the mean (i.e., group-averaged) ERP responses over 8 right-handed subjects. In line with our prior report (Hopfinger and Mangun 1998), cued location targets in the short-ISI range elicited significantly enhanced P1 components in comparison to targets at the uncued location (cued = 0.79 μV versus uncuedd = 0.31 μV; $p < 0.05$). At the longer ISIs, this pattern was no longer present, and the P1 tended to be smaller at the cued location, although this difference was not statistically reliable (cued = 0.79 μV versus uncued = 0.92 μV, $p > 0.05$; not shown in figure 5.2). This reflexive effect appears at the same latency (P1 latency range) as has been observed for the effects of voluntary attention, suggesting that a similar processing stage is being modulated by reflexive and voluntary attentional control, although clearly the control mechanisms may not be identical.

The topographic maps of figure 5.2B (middle) show the scalp maxima of the P1 components for cued and uncued targets. The scalp distribution of these effects is quite similar to that observed for the P1 in studies of voluntary attention, being maximal over contralateral occipital scalp locations. These topographic distributions are consistent with activity in the ventral extrastriate cortex (e.g., Heinze et al. 1994).

To assess whether the targets at cued and uncued locations are treated differently at later stages of analysis, we also examined activity in the P300 latency range (200–400 msec) elicited to the targets. The P300 is a cognitive ERP elicited by stimuli that have higher perceived relevance or require contextual updating (e.g., Donchin and Coles 1988). At short cue-to-target ISIs, when the P1 to the target was enhanced by reflexive attention, the P300 to that target was also larger (cued = 2.41 μV versus uncued = 1.69 μV; $p < 0.001$). At longer cue-to-target ISIs, however, there was no difference in the amplitude of the P300 component (3.02 μV versus 2.76 μV; $p > 0.05$; figure 5.2C, bottom).

The data from the present study demonstrate that reflexive attention triggered by sensory events leads to a brief facilitation of target processing for subsequent stimuli. Because they were obtained for simple as well as more difficult detection tasks (Hopfinger and Mangun 1998), these data strengthen our proposal that the effects of reflexive attention on the P1 component are automatic. An unexpected result of this work has been to demonstrate that both voluntary and reflexive attention manifest their effects on sensory signals at similar stages of cortical analysis, the stages reflected in the P1 attention effect.

5.3 THE FUNCTIONAL ANATOMY OF EARLY SPATIAL ATTENTION

An important next step is to determine where in the visual system the modulation of the P1 component of the ERP is generated. Studies of the intracranial generators of scalp-recorded ERPs all suffer from the same general limitation—the recordings are made relatively far from the site of generation, making accurate localization difficult. Neuroelectric modeling can be used to infer the intracranial locus of scalp-recorded activity, but the well-known "inverse problem" limits this approach. Although a given distribution of charges inside the head will specify a unique pattern on the scalp (the so-called forward solution), the inverse is not true (e.g., Dale and Sereno 1993). Thus no unique solution can be obtained when going in the inverse direction from scalp recordings to neural generators. Many studies have used inverse modeling to investigate the neural generators of scalp-recorded activity, but for the reasons noted above, it is difficult to accept or reject any particular model.

Nonetheless, inverse modeling with computer algorithms can be employed to test possible models, especially when combined with additional information. For example, Dale and Sereno (1993) outlined the use of anatomical information obtained from anatomical MRI scans to constrain the locations of possible neuroelectric sources to regions of the cortex, thereby eliminating many areas of the head from consideration as possible sites of generation of scalp-recorded ERPs. Similarly, we used functional neuroimaging to identify active brain regions during a spatial selective attention task that could serve to constrain source localization models of ERPs (Heinze et al. 1994; see Mangun, Hopfinger, and Heinze 1998 for a review).

Integrating Event-Related Potentials and Neuroimaging in Studies of Attention

In our first study integrating electrophysiology and functional imaging methods, we combined ERP recording and positron-emission tomography (PET; Heinze et al. 1994). The design was similar to those in several of our ERP experiments (e.g., Heinze and Mangun 1995). Subjects were presented with bilateral stimulus arrays containing two nonsense symbols within each lateral hemifield at a rate of about 3 per second. The task was to fixate a central point and, by attending covertly to the symbol pair in one hemifield, to determine whether the two symbols on that side were identical. Matching symbol targets required a rapid button press. The symbols in the opposite field were ignored during that block. In different blocks, subjects were instructed to attend to the left or right field stimuli.

PET activations showed that spatial selective attention activated extrastriate visual cortex (posterior fusiform gyrus) in the hemisphere contralateral to the attended stimuli. This PET information was used to constrain modeling of ERP sources. We modeled neuroelectric sources at the anatomical loci identified using PET, and calculated the patterns of electrical activity that sources at these sites would produce on the scalp model. Because we placed (or seeded) these model neural sources within the PET-defined brain loci in the computer simulation, we referred to them as "seeded forward solutions." We found that dipoles located within the PET-defined brain loci yielded highly accurate accounts of the scalp-recorded ERP attention data, but only in the time range corresponding to the P1 component (80–130 msec latency). This suggests that changes in input processing in extrastriate visual cortex, in the region of the posterior fusiform gyrus, were generating the P1 attention effect in the ERPs. We were able to localize the site of top-down attentional control over ascending visual sensory processing in both time (80–130 msec poststimulus) and space (posterior fusiform gyrus). An important methodological feature of this experiment was that we compared identical experimental conditions, in the same volunteers, to isolate the same attention effects in the functional imaging and ERP data.

Covariations in Event-Related Potential and Functional Imaging Measures If the P1 attention effect in the ERPs really is related to the attentional modulation revealed by changes in regional cerebral blood flow (rCBF) in the posterior fusiform gyrus, then these measures should covary with one another as a function of experimental manipulations. In Mangun et al. 1997, we tested this directly by manipulating the perceptual load of the task (see Lavie, chap. 7, this volume) to determine whether the P1 attention effect and the posterior fusiform activations would be similarly affected. As before, subjects viewed bilateral arrays, and in separate blocks attended to either the right or left of the arrays. Two different tasks were now compared. One task was identical to that in Heinze et al. 1994, with subjects having to respond to matching symbols at the attended location (high-load task). In the other, only a simple luminance detection was required (low-load task); subjects were required to respond to a small dot appearing on one side within the confines of the bilaterally flashed symbol arrays. ERPs and PET measures were obtained in separate sessions for each subject.

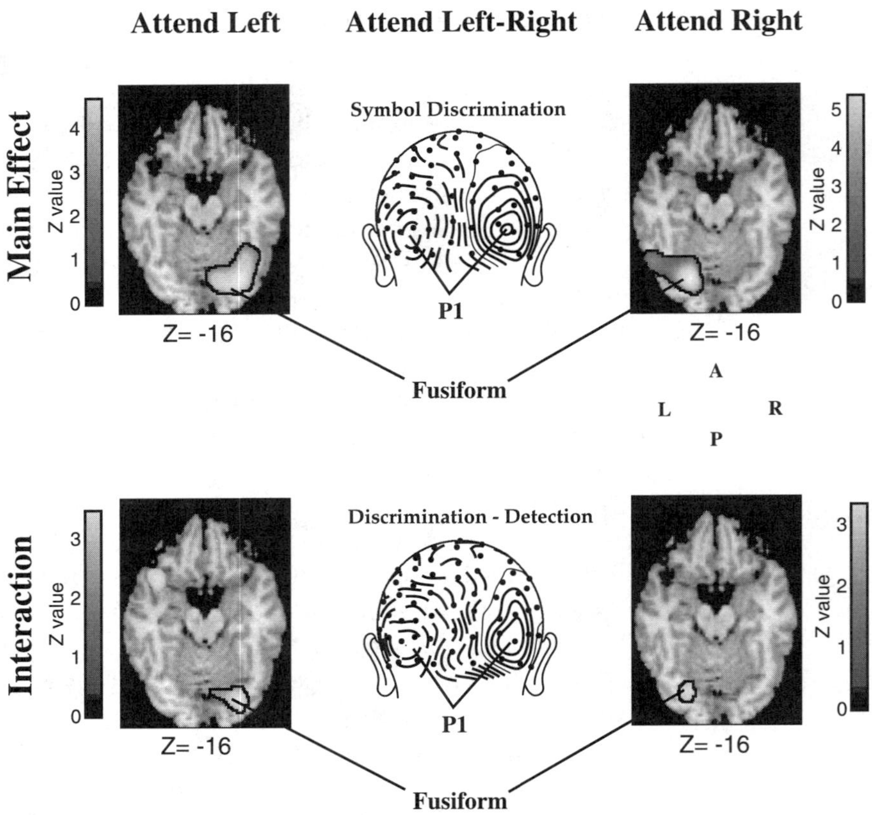

Figure 5.3 Positron emission tomography (PET) activations and event-related potential (ERP) topographic maps during voluntary spatial attention. *Top*. Main effects of attending left versus right are shown when the subjects performed the symbol discrimination task. The PET activations (outlined in black lines) are overlaid onto a horizontal section from MRI scans. The Z-value scale next to each MRI scan refers only to the activated regions outlined with black lines. The topographic voltage map shown at the top is the attend-left minus attend-right difference map in the P1 latency range. The contour lines on the topographic maps indicate polarity and voltage (thick solid = positive; dashed = negative). Because, however, the polarity is an artifact of the direction of subtraction (left minus right), the P1 over the left hemisphere has a negative polarity in the maps, but is actually a positive enhancement. *Bottom*. Plots of PET activation are statistical interaction maps of regions where the amplitude of the attention effect was different for symbol discrimination versus luminance detection (high versus low perceptual load). The topographic difference map was derived by subtracting the attend-left minus attend-right attention map for the luminance detection task from that in the symbol discrimination task. A = anterior, P = posterior, L = left, and R = right.

There was a complete replication of our earlier study with respect to the ERP and PET effects in the fusiform gyrus during discrimination. When subjects attended to one visual hemifield, there was a significant increase in the P1 component over contralateral scalp sites, and a corresponding increase in rCBF in the contralateral posterior fusiform gyrus (figure 5.3, top). Additional activations were also found in the contralateral middle occipital gyrus, probably due to the use of more sensitive PET methods (i.e., 3-D imaging).

Importantly, the amplitude of the attention effects (attend left versus attend right) in both the ERP (P1 component) and PET (posterior fusiform activation) measures were found to covary with perceptual load. This was observed as significant interactions between attention (attend left versus attend right) and task (symbol discrimination versus dot detection) for both the P1 component of the ERP and the activations in the posterior fusiform gyrus (figure 5.3, bottom). The attention effects were larger when perceptual load was higher. This covariation between the P1 effect and the fusiform gyrus PET effect supports the idea that the stage of visual processing indexed by the P1 component occurs in extrastriate cortex in the posterior fusiform gyrus. Although the PET activity in the medial occipital gyrus showed a tendency in the same direction as the fusiform activity, this was not reliable (no statistical interaction).

The increased attention effects with higher perceptual load can be interpreted as the result of more attentional resources being dedicated to the attended location, so that the differences between attended versus unattended locations are enlarged. These data provide physiological support for the proposal of Lavie and colleagues (Lavie, chap. 7, this volume; Lavie and Tsal 1994) that perceptual load of target discrimination influences early selection processes (see also Handy and Mangun 2000).

Attentional Modulations in Functionally Defined Visual Areas

Having demonstrated that modulations of incoming sensory signals occur as a function of spatial attention within visual cortex, we must now identify whether these mechanisms are occurring within a single visual cortical area or in multiple visual areas. The presence of multiple areas in visual cortex is now well established in nonhuman primates based on single-cell studies (e.g., Van Essen and DeYoe 1995). Homologous visual cortical areas can now be mapped in humans using functional neuroimaging (e.g., Engel et al. 1994; Sereno et al. 1995). Such mapping allows one to refine the localization of visuo–spatial attention effects by relating them to visual areas (e.g., V1, V2, V3/VP, and V4), not merely to anatomical structures (e.g., lingual, fusiform, and middle occipital gyri), as we (Jha et al. 1997) have done.

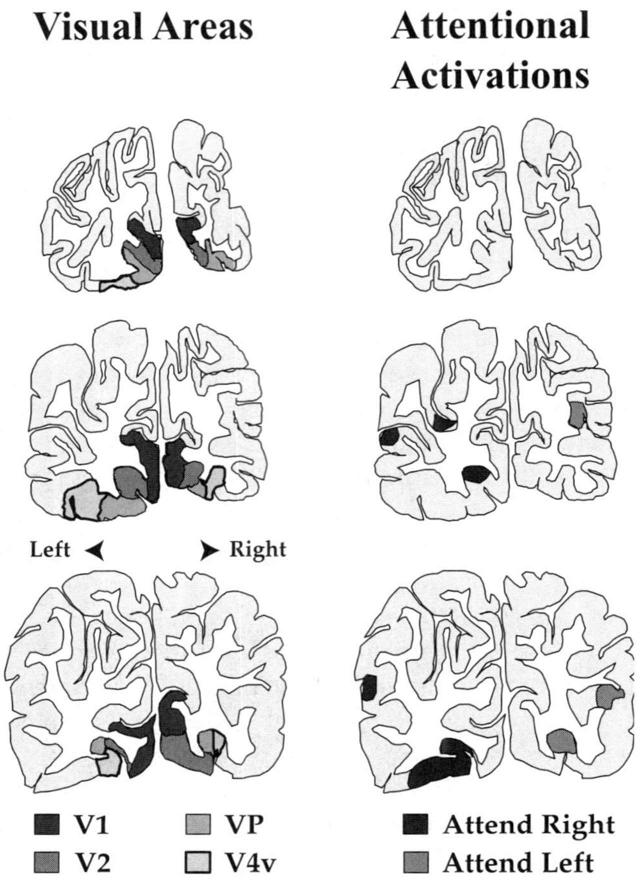

Figure 5.4 Derivations of visual areas and activations during spatial selective attention from fMRI. Data from one representative subject (the first of six to be analyzed). Based on the activations to meridia stimuli, the extent of visual areas V1 through V4v is shown for the upper visual hemifield field representation on the ventral surface of the brain. Traced sections are sequential coronal slices beginning near the occipital pole (*top*) and continuing anteriorly. The attentional activations from the same subject (*right*) are shown for the attend-right (dark) and attend-left (lighter) conditions. By comparing these to the derived boundaries of the visual areas (*left*), one can observe that attention effects in the posterior fusiform gyrus/lingual gyrus include activity in visual areas V2, VP, and V4v, as well as in other regions that may be homologous to area TEO in monkeys (see Kastner et al. 1998).

Methods and Results We used fMRI to functionally define the borders of the early visual areas in each of six subjects. The methods, though similar to those of Engel et al. (1994) and Sereno et al. (1995), stimulated only the meridia of the visual field (Kastner et al. 1998; Tootell et al. 1995). Under passive viewing conditions, the upper and lower vertical meridia, and left and right horizontal meridia were separately stimulated by pattern-reversing checkerboard stimuli. Because the visual borders between V1 and V2, between V2 and VP/V3, and between VP/V3 and V4 occur at the meridia of the visual field, we were able to determine the extent of the first few visual areas, whose derivation from the fMRI data for one subject is shown in the left column of figure 5.4.

The subjects also performed a visual attention task that required matching symbols at the attended location, every 16 sec a central arrow cue instructed the subjects where to attend (see Mangun et al. 1998). It was then possible to determine which early visual areas were modulated during the spatial attention task by comparing the attention-related activations (attend left versus attend right) to the functionally defined visual areas for each subject (compare left versus right columns of figure 5.4). Attention-related activations were found in multiple visual areas, including V2, VP, and V4.

Conclusions and Discussion Prior studies in humans using ERPs or functional imaging have been unable to identify the precise areas of visual cortex displaying attentional modulations. In this study, we used fMRI to define the borders of cortical visual areas V1–V4, and were thus able to demonstrate that spatial attention modulates neuronal processing in multiple visual areas (V2–V4), but not in V1. Knowing that activations previously viewed as singular sources of activity in extrastriate cortex (as in our earlier PET studies) actually reflect activities in adjacent visual cortical maps should allow more complex neuroelectric models to be developed and tested. These will prove crucial in helping to define the role that attention plays within different regions of visual cortex.

For example, modeling of ERP activity constrained by functional activations in adjacent, functionally defined visual areas might help resolve how the primary visual cortex (V1 or striate cortex) is involved in visual spatial selective attention. Many studies have failed to find any evidence that the striate cortex could be modulated by spatial *selective* attention either in animals (Luck et al. 1997; Moran and Desimone 1985) or in humans, using ERPs (e.g., Clark et al. 1996; Mangun, Hillyard, and Luck 1993) or functional neuroimaging (e.g., Heinze et al. 1994; Kastner et al. 1998; Mangun et al. 1997, 1998). On the other hand, single-neuron recording in monkeys (Motter 1993; Vidyasagar 1998) and fMRI in humans (Somers et al. 1999; Worden, Schneider, and Wellington 1996) have occasionally detected modulations of striate cortex during spatial selective attention as well as during nonselective attention, where the nonspecific effects of arousal are not well controlled (e.g., Watanabe et al. 1997). These findings raise the possibility that, under certain conditions, incoming sensory signals can be influenced by top-down attentional processes as early as striate cortex (see Posner and Gilbert 1999 for review).

With the exception of Motter 1993, most studies showing attention effects in V1 have measured regional cerebral blood flow in humans in ways that could not specify the time course of the effects, or have measured effects at very long latencies not consistent with input gating (Roelfsema, Lamme and Spekreijse 1998). To interpret fMRI attention effects in V1, it is essential that the time course of the activations be established. Martinez and colleagues 1999 combined ERPs and fMRI to do precisely this. Their subjects selectively attended to stimuli in the left or right visual field (ignoring the opposite hemifield). In separate sessions, fMRI and ERP measures of attention were obtained (attend left versus attend right). Mapping their effects onto functionally defined visual cortical areas, the authors found that attention-related fMRI activations occurred in visual areas V1–V4, but that short-latency ERPs generated in V1 were not affected by attention. Rather, attentional modulations in the ERPs, occurred later, at latencies consistent with activity in extrastriate cortex. The Martinez et al. study suggests that increased rCBF in V1 during spatial selective attention does not reflect an early gain control process over incoming signals in striate cortex. Instead, V1 modulation may be a reflection of reafferent activation of V1 from later stages in the visual hierarchy, a view consistent with observations of long-latency attention effects in V1 from single-neuron recordings in monkeys (Roelfsema, Lamme, and Spekreijse 1998).

5.4 ATTENTIONAL CONTROL CIRCUITRY

Thus far we have considered the effect of attentional control on incoming sensory signals. In the remainder of this chapter, we turn to consideration of the control systems responsible for top-down effects of attention. The issue of which brain systems participate in attentional control is somewhat less well understood than where attention influences sensory inputs. Research in neurological patients, animals, and also in healthy observers using neuroimaging suggests that the control of visuospatial attention involves a complex network of widely distributed neuronal populations, including those in dorsolateral-prefrontal, anterior cingulate, posterior parietal cortex, and thalamic and midbrain structures (e.g., Bushnell, Goldberg, and Robinson 1981; Corbetta 1998; Goldberg and Bruce 1985; Heilman, Watson, and Valenstein 1994; Mesulam 1981; LaBerge 1997; Posner and Petersen 1990; Posner and Driver 1992). The specific functions of these structures in attentional control are only partially understood, however, perhaps in part because the time course of their relative activations during attentional orienting has not yet been clarified. ERP and functional imaging studies incorporating new analytical approaches can be used to address the time course and functional anatomy of attentional control systems, just as they have been used to investigate their modulatory effects on perceptual processes.

Electrophysiological Studies of Attentional Orienting and Control

Although most ERP studies of attention focused on how attention affects sensory processing, some have also investigated neuroelectric correlates of attentional preparation prior to the arrival of the target stimulus. For example, in voluntary, trial-by-trial spatial cuing paradigms, ERPs can be recorded in response to an attention-directing cue, and brain activity can be monitored in the period after the instruction about where to attend, but before the target is delivered (Harter et al. 1989; Mangun 1994; Yamaguchi, Tsuchiya, and Kobayashi 1994).

Harter and coworkers (e.g., 1989) first studied the ERP correlates of shifts of visuospatial attention. In their studies, a small (~0.5 degree) central arrow cue (located at fixation) pointed either to the right or left visual field. The cues defined the relevant side for that trial. Targets appearing on the cued side were responded to as fast as possible, while targets appearing on the uncued side were ignored. The subtraction of ERP responses triggered by left-pointing cues from that of right-pointing cues, revealed two principal attention shift-related ERP effects. The first, denoted "early directing attention negativity" (EDAN), was a negative polarity deflection over the parietal scalp contralateral to the direction indicated by the attention-directing cue, starting 200 msec after cue onset and lasting until 400 msec past cue onset. Presumably the EDAN is related to attentional control processes that establish selective spatial attention. Later (500–700 msec after the cue), at occipital electrode sites in the hemisphere contralateral to the arrow direction, the ERP was more positive (in comparison to the ipsilateral hemisphere). Referred to as "late directing attention positivity" (LDAP), this second effect was proposed to reflect the modulation of cortical excitability in regions of the brain corresponding to where attention has been directed in space. The LDAP has been difficult to observe in adults, however, and has been identified only in studies of children. Because no differences were observed in the early (< 200 msec) sensory-evoked responses to the left versus right cues, it is unlikely that any of the foregoing effects resulted from simple physical differences between the left and right cue stimuli.

The attention-orienting ERP effects described above have been replicated and extended in several reports. In Mangun 1994, we reported an EDAN-like effect in adult subjects over parietal-temporal scalp sites contralateral to the direction of the cue between 250 and 350 msec after an endogenous central arrow cue. We also reported a longer-latency (300–500 msec) right-hemisphere negative wave over frontal scalp sites. Replicating the EDAN effect described by Harter et al. (1989), Yamaguchi, Tsuchiya, and Kobayashi (1994) reported that it first occurred at posterior temporal and parietal sites, but then appeared to spread over central and frontal sites after about 380 msec. Independent of cue direction, a right posterior temporal negativity was found starting 500 msec after cue onset and lasting until the target onset. None of these studies observed the LDAP of Harter and colleagues when recording in adults.

Although the foregoing studies correlated ERPs with an instruction to shift attention from one location to another, we still do not know how these potentials relate to underlying brain structures implicated in attentional processes. Evidence from single-neuron recordings in monkeys (e.g., Bushnell, Goldberg, and Robinson 1981; Colby, Duhamel, and Goldberg 1993; Steinmetz et al. 1994), from studies of patients with focal cortical lesions (e.g., Posner et al. 1984), and from human functional neuroimaging (e.g., Corbetta et al. 1993; Corbetta 1998) indicates that the parietal cortex is involved in visuospatial attention. This has led to various models for the role of parietal cortex in attention, most of which emphasize attentional control processes such as shifting attention, disengaging attention from a current locus to enable shifting, or mapping locations to be attended so that visual processing can be influenced in a spatially defined manner (e.g., Posner and Petersen 1990; LaBerge 1997).

If the EDAN component of Harter et al. 1989 reflects neural processes involved in the control of visual attention by parietal cortex, then one might expect it to have a narrow scalp maximum over parietal cortical regions. Moreover, this topography should be distinct from that of the later LDAP, which, if related to effects in visual cortex, should have a distribution similar to that for attention-related enhancements of target processing (e.g., P1 attention effect).

In a recent study, we sought to clarify the scalp distributions of cue-related ERPs using detailed topographical analysis of the ERPs in the cue-target interval in a voluntary, trial-by-trial spatial cuing paradigm (Hopf and Mangun, forthcoming). High-resolution mappings of the ERP components were obtained.

Methods ERPs from 92 scalp sites were recorded from 14 healthy, right-handed student volunteers. Subjects fixated a point (0.19 degree diameter) in the center of a computer monitor. Two white outline boxes (3.3 degrees wide by 5.5 degrees tall) were continuously present (10.7 degrees lateral to fixation, measured to center) in the upper visual field to demarcate possible target locations. Each trial began with an arrow cue (100 msec duration) flashed to fixation. It randomly pointed to the left or right, and was followed 900 msec later by a pair of symbols flashed (35 msec duration) to one of the two lateral locations.

While maintaining fixation, subjects had to attend covertly to the cued box and discriminate whether the symbols presented there were identical. Approximately 16% of cued-location stimuli were matching targets, and required a button press response (response hand was counterbalanced between experimental blocks within and across subjects). The cue was not predictive of target location, but instructive, indicating to the subjects that the cued box was relevant for that trial. They were to ignore the uncued location, and no response was required on trials where the targets appeared in the uncued location.

The perceptual load of the target discrimination was manipulated in separate experimental blocks. In the low-load condition, the symbols dif-

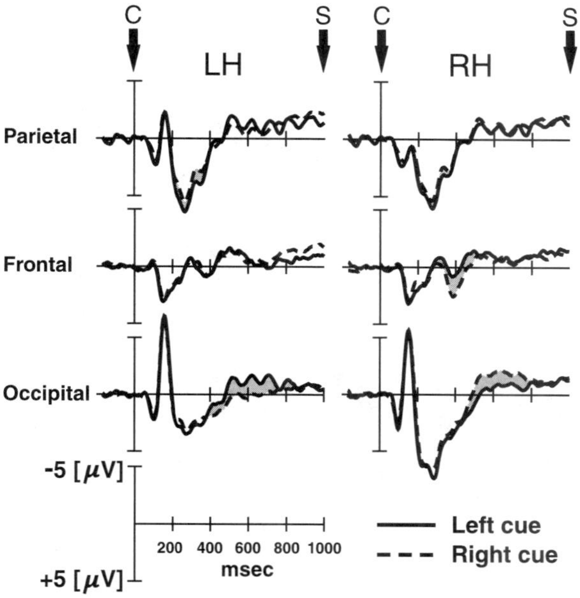

Figure 5.5 Event-related potentials (ERPs) to attention-directing cues. Grand average ERPs over 14 subjects are shown in response to an attention-directing arrow cue located near fixation. ERPs to left cues are shown in solid lines; those to right cues in dashed lines, with differences shaded. The onsets of the cues (C) and subsequent target stimuli (S) are indicated above the ERPs. Left-hemisphere (LH, left column) and right-hemisphere (RH, right column) electrode sites were from parietal (top row), frontal (middle row), and occipital (bottom row) scalp regions.

fered in more than one detail (mean of 2–3). In the high-load condition, the symbols differed in only one detail, a small horizontal bar of about 1.1 degrees.

Analysis of variance (ANOVA) was used to assess significant differences between conditions for the target-evoked ERPs (e.g., P1 and N1) where a priori predictions guided statistical tests at specific electrode sites. For the cue-related ERPs whose topographies were under investigation, however, statistical evaluation was performed within the context of topographic mapping using a procedure developed to assess the scalp maps. An *F*-test was performed for each electrode location separately, and a modified Bonferroni correction applied. The corrected critical *F*-value in the present experiment is $F(1, 13) = 10.58$. The *F*-value for the single-electrode test where each effect discussed below was maximal will be reported in the text.

Results and Discussion Figure 5.5 shows superimposed grand-averaged waveforms (over the 14 subjects) for left- and right-pointing cues in the time interval between the onset of the cue and the onset of the target symbols. The ERPs are from parietal, frontal, and occipital electrodes. Starting approximately 200 msec after cue onset, a statistically significant differ-

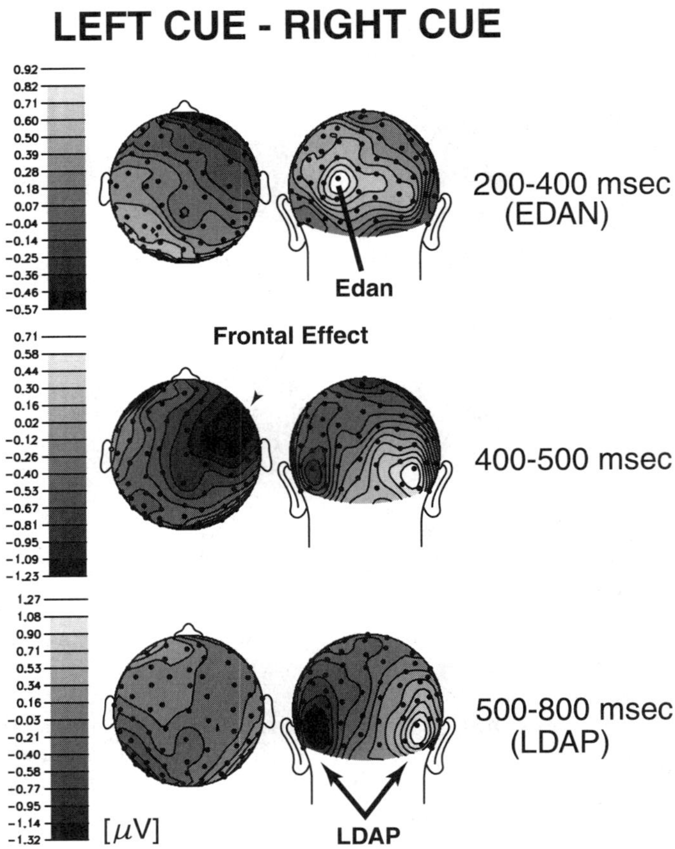

Figure 5.6 Topographic voltage maps of the differences obtained by subtracting right from left cue-related activity (grand averages over the 14 subjects). The topographic maps were computed over the indicated time ranges, and thus correspond to the EDAN, frontal effect, and LDAP components. Darker shades indicate negative voltages. Because, however, polarities are dependent on the direction of subtraction, the EDAN has a positive polarity in these maps, and the LDAP over the two hemispheres appear to be of opposite polarity, but each is actually a contralateral positive-going deflection in the ERPs.

ence for right- versus left-pointing cues can be observed over the posterior parietal scalp of the left hemisphere (figure 5.5, top, shaded areas): $F(1, 13) = 15.90$. The scalp voltage topographies of this posterior left hemisphere effect are shown in the difference maps (left-right cue, low-load condition only) in figure 5.6 between 200 and 400 msec from the cue onset. (Note that as a result of the direction of the subtraction, this effect is seen as a positive focus in the topographic maps of figure 5.6, but can be interpreted as greater left hemisphere negativity for right-pointing versus left-pointing cues.) This effect may be related to the EDAN component of Harter and colleagues (1989; Harter and Anllo-Vento 1991).

By about 300 msec after cue onset, there was a statistically significant effect over right frontal scalp regions. The shaded area in figure 5.5 (middle) illustrates this effect in the average waveforms for lateral frontal

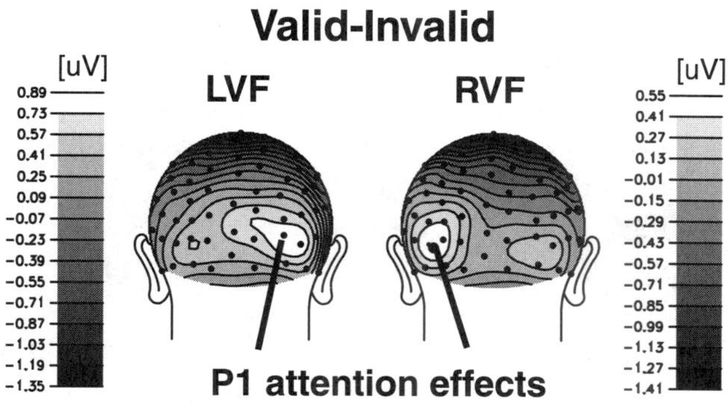

Figure 5.7 Scalp topographic maps of the attention difference effects (attended minus unattended) for left (LVF) and right (RVF) target stimuli in the P1 time range. Attentional modulations of the P1 evoked by left-field stimuli showed a scalp maximum over right lateral occipital regions, whereas attention effects on for right-field stimuli it showed a scalp maximum over left lateral occipital regions.

electrodes. This cue-related effect lasted until 500 msec after cue onset and was reliably larger at right than at left frontal sites (figure 5.6, middle row, left map). Statistically, the right frontal effect was most robust between 300 and 400 msec latency: $F(1, 13) = 40.49$, consistent with the right frontal component described in Mangun 1994.

A longer-latency effect of cue direction began at about 400 msec after arrow onset (figure 5.5, bottom) at occipital sites over both hemispheres. This took the form of a statistically significant focal scalp positivity contralateral to the direction of the attention-directing arrow cue: $F(1, 13) = 14.6$ at electrode T5 in the left hemisphere; $F(1, 13) = 25.73$ at electrode T6 in the right hemisphere. The topographic maps of figure 5.6 show these effects as a posterior right-hemisphere positive focus, and a mirror image left-hemisphere negative focus. (Note again that the opposite polarities over the two hemispheres were caused by the direction of the subtraction of the ERPs in the topographic maps.) These contralateral positivities lasted until 850 msec after cue onset, but terminated approximately 150 msec before the onset of the symbols. This occipital effect may be related to the LDAP effect previously observed in children (Harter and Anllo-Vento 1991). Finally, at around 600 msec after the cue onset, a left anterior scalp difference occurred as a function of the direction of the attention-directing cues (figure 5.5, middle, and figure 5.6, bottom left map), and this frontal effect lasted until the end of the cue-target interval: $F(1, 13) = 15.82$ from 700–900 msec latency.

Bearing close similarity to prior studies, a sequence of ERPs related to the direction of attention-directing cues were obtained in the period following the cue, but prior to the onset of the target stimuli. Did perceptual load influence these cue-related ERPs? Although there were significant main effects of perceptual load in the ERPs to the cues, these effects did not interact with the responses to the attention-directing cues, with the possible exception of a marginally significant effect between 300 and 400 msec latency over the left frontal scalp: $F(1, 13) = 5.8$. Thus, to our surprise, the load manipulation did not influence the ERP signs of attentional control to the cues. Likewise, the effects of the load manipulation only weakly influenced the attention effects on the subsequent target-evoked ERPs. Although cued symbols showed a significantly larger occipital positivity between 80 and 120 msec (the P1 attention effect; see topographic maps in figure 5.7) and an enhanced negativity between 130 and 200 msec (the N1 attention effect—not shown in figures), the magnitude of the attention effects were not different for low versus high load for the P1 or N1 (c.f., Handy and Mangun 2000).

Conclusions The present results provide some insights into the nature of attentional control processes during visual spatial attention. First, when attention is directed in space by an endogenous cue, a series of ERP components is generated, providing additional tools for investigating attentional controls processes in the absence of overt behavioral responses. Second, these cue-related effects have distinct scalp topographies and time courses.

The EDAN had a focus over parietal scalp, but contrary to prediction, it was only significant over left parietal scalp regions. Although the left lateralization of the EDAN effect might signal a special role of the left hemisphere in spatial attention, such an interpretation would be inconsistent with other evidence. For example, neuropsychological evidence has demonstrated that hemispatial neglect is worse following right versus left parietal lesions (e.g., Heilman, Watson, and Valenstein 1994). Neuroimaging suggests that both left and right parietal lobes are involved in spatial attention (Corbetta et al. 1993), but that the right parietal lobe may play a greater role.

In light of the foregoing evidence for a right-hemisphere role in spatial processing and attentional orienting in space, an alternative interpretation for the present left-lateralized effects is that they may be an artifact of our comparison, which subtracted right- from left-pointing cues. That is, if the right hemisphere were equally active for both left- and right-pointing cues, then the subtraction would yield no difference over the right parietal scalp. If the left hemisphere were differentially active for leftward versus rightward orienting, then the subtraction would yield a left-lateralized effect such as we have observed here. Such a model closely fits our behavioral studies in split-brain patients (Mangun et al. 1994), where we found evidence that the right hemisphere was affected by both right and left attention-directing cues, whereas the left hemisphere was not. We return to this question later in this chapter.

Interestingly, the earliest ERP signs of cue-related activity were over the parietal scalp, with the frontal effects occurring later. Although many models of attention posit the frontal cortex as the seat of initiation of attentional control (e.g., LaBerge 1997; Posner and Petersen 1990), the time course information of the present study is not consistent with such a proposal, at least not for activity differential for right- and left-pointing cues.

The LDAP showed topographic foci over the lateral occipital scalp. The maxima of the LDAP on the scalp closely matched those for the attentional enhancements of the P1 component to subsequent target stimuli. Thus the occipital activity in response to the cues and that in response to the subsequent early attention effects had similar topographies, as predicted by the hypothesis that the LDAP is a sign of increased neural excitability in the neurons coding the regions of space to be attended. A key finding, however, was that the LDAP effect terminated prior to the onset of the target stimuli, even though the targets showed significant modulation by attention. This raises significant questions about the function of the neural processes involved in the LDAP effect. For example, perhaps the LDAP is a sign of control signals to neurons coding the region of visual space to be attended, rather than evidence of the resultant enhanced excitability of those visual neurons.

ERP data as used here can provide only indirect clues about the underlying neural structures involved in top-down control of spatial attention. We must turn to functional imaging to help fill in the details about functional anatomy.

Event-Related Functional Magnetic Resonance Imaging Studies of the Control of Spatial Attention

Until about 1995, neuroimaging studies of attentional control systems suffered from two serious methodological limitations: (1) the inability to provide temporal information; and (2) the use of "blocked" designs, precluding separation of individual trials within blocks, as commonly done with behavioral and electrophysiological data (see Rugg 1998). By contrast, event-related fMRI now permits the brain activity related to different intermingled trials to be decomposed (e.g., Buckner 1998; McCarthy et al. 1997), an approach conceptually similar to that used in ERP research. For example, in the cuing paradigms described in the last section, we separately derived ERPs to the cues and those to the subsequent targets, and a similar approach can now be used with fMRI. We employed this method to investigate control of spatial attention in a trial-by-trial cuing task (Hopfinger, Buonocore, and Mangun 2000).

Methods Three healthy adult subjects participated in the data reported here. All were right-handed and had normal vision. Each trial began with a voluntary cue at fixation (500 msec duration) that randomly pointed either to the left or right visual hemifield. The direction of the arrow was an instructive cue, telling the observer to attend selectively to the cued hemifield. To eliminate any differences in sensory activations to the cue between conditions, the cue consisted of two overlapping arrows, one blue and one yellow, pointing in opposite directions. Some subjects were told to use the blue cue to direct attention, while the others were told to use the yellow.

Following the cue by a random interval of 1,000 msec (17% of trials) or 8,160 msec (83% of trials), bilateral black-and-white checkerboard targets were presented to the upper visual field (4 Hz reversal rate; 750 msec duration). This permitted brain responses to be modeled when the cues and targets were separated by several seconds (long-ISI trials), but required the subjects to prepare for the possibility that a target would appear at shorter ISIs. The task was to maintain fixation on central cross-hairs while covertly attending to the cued side to discriminate whether elements of the checkerboard on that side were missing (on 50% of trials, some checks were gray). From 3 to 9 checks were missing for infrequent target checkerboards. Subjects pressed one button for targets and another for nontargets. Approximately 8 sec separated trials (from target off to onset of next cue).

The hemodynamic responses to the cues and targets were separately modeled with event-related fMRI methods using SPM97 for the trials where the cue and target were separated by 8,160 msec. The hemodynamic response was modeled as the sum of two gamma functions and its temporal derivative. A statistical significance level of $p < 0.001$ was set. The activated regions were overlaid on the canonical T1-weighted MRI scans of SPM97 for the images shown here.

Results Figure 5.8 shows the average activations over three subjects to the onset of the arrow-directing cues (left panels) and the subsequent target stimuli (right panels). For the present discussion, we will consider only a single posterior parieto-occipital slice in which cue-related activations and target-related activations were both visible, and we will not consider activations present in other brain regions (e.g., frontal, temporal, or subcortical regions).

Let us turn first to the hemodynamic responses to the cues and targets collapsed over the direction of cuing (figure 5.8, top row). The cue resulted in bilateral inferior parietal cortex activations, whereas the targets produced no such inferior parietal activity, being restricted to the visual cortex. Thus parietal and occipital cortical regions were differentially activated by attention directing cues and subsequent targets.

Were these activations different as a function of the direction in which attention was cued? The activations in the inferior parietal region to the cues were not significantly different for left versus right cues (figure 5.8, left panels, middle and bottom rows). In contrast to the findings for the parietal cortex, the visual cortex activations in both the right and left hemispheres were significantly modulated by the direction of attention in

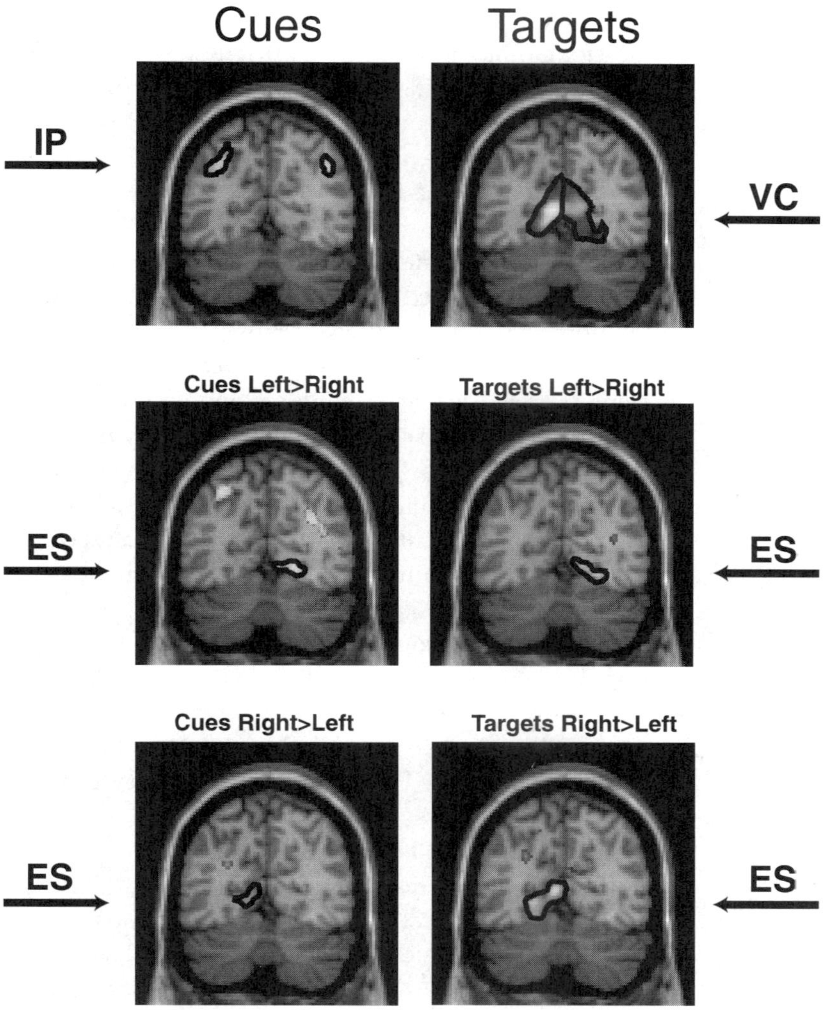

Figure 5.8 Event-related fMRI during spatial cuing. Activations in response to cues (*left*) and targets (*right*) in the spatial cuing paradigm described in the text. Bilateral inferior parietal (IP) activations were obtained in responses to cues (left, collapsed over cue direction), and visual cortical (VC) activation in response to the bilateral target stimuli (right, also collapsed over cue direction). There were contralateral activations in the visual cortex to cues. For the targets, attention to the left and right (right, middle, and bottom) halves of the bilateral target stimulus showed contralateral medioventral occipital activations, consistent with activity in extrastriate cortex (ES).

response to both cues and targets. In responses to the cues, there were differential activations of the contralateral medioventral occipital cortex.

These responses to the cues in occipital cortex may represent either attentional control signals acting in visual cortex to alter neuronal excitability or regional cerebral blood flow related to the increased neuronal excitability itself. They do not, however, represent simple sensory activations by the cues, which were localized more posterior in the brain, near the foveal representation on the occipital pole as determined by control sessions (not shown in figures). The ERP data described in the last section suggest that, in response to cues, the occipital effects followed the parietal activations in time. That is, the occipital activations may be related to the LDAP component in the ERPs. This must remain speculative until combined ERP and fMRI studies are performed with this paradigm.

In response to the targets, activations contralateral to the direction of attention were observed in visual cortex (figure 5.8, right panels, middle and bottom rows). These effects on target processing are in line with our prior PET (Heinze et al. 1994; Mangun et al. 1997) and fMRI (Mangun et al. 1998) studies. Increased rCBF was observed in lingual and fusiform gyri in the hemisphere contralateral to the attended visual hemifield.

Conclusions These findings while preliminary, demonstrate differential activation of parietal and occipital cortical areas during distinct time periods of a spatial cuing task. They support the thesis that the parietal cortex is engaged during attentional orienting, presumably as part of a cortical network for top-down attentional control (e.g., LaBerge 1997). The result of these orienting processes is the selective activation of extrastriate cortex to filter inputs from relevant versus irrelevant locations. Although the complete circuit remains to be elucidated, these and our ERP findings for related tasks paint similar pictures.

5.5 GENERAL CONCLUSIONS

While the physiological mechanisms of visuospatial selective attention are certainly not completely understood, there has been significant progress in that direction. The findings we have presented demonstrate that not only is extrastriate cortex modulated by top-down attentional mechanisms, as we showed previously (Heinze et al. 1994; Mangun et al. 1997), but this occurs in multiple visual areas, including V2, V3/VP, and V4. The related ERP recordings indicate that these effects represent modulations of initial input processing, as opposed to reafferent activations of these areas by later stages in the visual hierarchy. Our ERP studies also suggest that visual cortical processing is significantly modulated by reflexive attention, at sites similar to those for voluntary attention, although fine-grained functional anatomical studies remain to be done. Finally, the control systems involved in top-down voluntary spatial attention can be studied by monitoring brain activity in response to attention-directing cues. Although spatially distinct brain regions, including frontal, parietal, and occipital brain areas, are engaged by instructions to shift attention, these activities have very different time courses, as indexed by ERP recordings, with activity over parietal scalp appearing earliest in response to an informative cue. Event-related fMRI studies show that cue-related hemodynamic responses in inferior parietal cortex occur, and that this activity can be modeled separately from that in the visual cortex in response to an attention-directing cue or subsequent target stimulus. Together, these results support a model in which parietal brain regions are involved in the initial control of attention to affect changes in stimulus input processing by visual cortex.

NOTE

This research was supported by grants to Joseph B. Hopfinger from the National Science Foundation, to Amishi P. Jha from the National Institute of Mental Health, to Jens-Max Hopf from the German Academic Exchange Program (DAAD) and to George R. Mangun from the National Institute of Mental Health, the National Institute of Neurological Disorders and Stroke, and the Human Frontier Science Program. We are grateful to Maryam Soltani and Michael H. Buonocore for their contributions, and to Karl Friston and Christian Buchel for their assistance.

REFERENCES

Anllo-Vento, L., Luck, S. J., and Hillyard, S. A. (1998). Spatio-temporal dynamics of attention to color: Evidence from human electrophysiology. *Human Brain Mapping, 6,* 216–238.

Buckner, R. (1998). Event-related fMRI and the hemodynamic responses. *Human Brain Mapping, 6,* 394–398.

Bushnell, M. C., Goldberg, M. E., and Robinson, D. L. (1981). Behavioral enhancement of visual responses in monkey cerebral cortex: 1. Modulation in posterior parietal cortex related to selective visual attention. *Journal of Neurophysiology, 46,* 755–772.

Clark, V. P., and Hillyard, S. A. (1996). Spatial selective attention affects extrastriate but not striate components of the visual evoked potential. *Journal of Cognitive Neuroscience, 8,* 387–402.

Colby C., Duhamel, J., and Goldberg, M. (1993). Ventral intraparietal area of the macaque: Anatomic location and visual response properties. *Journal of Neurophysiology, 69,* 902–914.

Corbetta, M. (1998). Frontoparietal cortical networks for directing attention and the eye to visual locations: Identical, independent, or overlapping neural systems? *Proceedings of the National Academy of Sciences, U.S.A., 95,* 831–838.

Corbetta, M., Miezin, F. Shulman, G., and Petersen, S. (1993). A PET study of visuospatial attention. *Journal of Neuroscience, 13,* 1202–1226.

Dale, A. M., and Sereno, M. I. (1993). Improved localization of cortical activity by combining EEG and MEG with MRI cortical surface reconstruction: A linear approach. *Journal of Cognitive Neuroscience, 5,* 162–176.

Desimone, R., and Duncan, J. (1995). Neural mechanisms of selective visual attention. *Annual Review of Neuroscience, 18,* 193–222.

Donchin, E., and Coles, M. G. H. (1988). Is the P300 component a manifestation of context updating? *Behavioral and Brain Sciences, 11,* 357–427.

Eason, R. G. (1981). Visual evoked potential correlates of early neural filtering during selective attention. *Bulletin of the Psychonomic Society, 18,* 203–206.

Eason, R., Harter, M., and White, C. (1969). Effects of attention and arousal on visually evoked cortical potentials and reaction time in man. *Physiology and Behavior, 4,* 283–289.

Egly, R., Rafal, R., Henik, A., and Berger, A. (Forthcoming). Reflexive and voluntary covert orienting in detection and discrimination tasks. *Journal of Experimental Psychology: Human Perception and Performance.*

Eimer, M. (1994). Sensory gating as a mechanism for visuospatial orienting: Electrophysiological evidence from trial-by-trial cuing experiments. *Perception and Psychophysics, 55,* 667–675.

Engel, S. A., Rumelhart, D., Wandell, B., Lee, A., Glover, G., Chichilinsky, E., and Shadlen, M. (1994). FMRI of human visual cortex. *Nature, 370,* 106.

Goldberg, M. E., and Bruce, C. J. (1985). Cerebral cortical activity associated with the orientation of visual attention in the rhesus monkey. *Vision Research, 25,* 471–481.

Handy, T. C., and Mangun, G. R. (2000). Attention and spatial selection: Electrophysiological evidence for modulation by perceptual load. *Perception and Psychophysics, 62,* 175–186.

Harter, M. R., and Aine, C. J. (1984). Brain mechanisms of visual selective attention. In R. Parasuraman and D. R. Davies (Eds.), *Varieties of attention,* pp. 293–321. New York: Academic Press.

Harter, M. R., Aine, C., and Schroeder, C. (1982). Hemispheric differences in the neural processing of stimulus location and type: Effects of selective attention on visual evoked potentials. *Neuropsychologia, 20,* 421–438.

Harter, M. R., and Anllo-Vento, L. (1991). Visual-spatial attention: preparation and selection in children and adults. In C. H. M. Brunia and M. N. Verbaten (Eds.), *Event-related brain research,* vol. EEG suppl. 42, pp. 183–194. Amsterdam: Elsevier.

Harter, M. R., Miller, S. L., Price, N. J., LaLonde, M. E., and Keyes, A. L. (1989). Neural processes involved in directing attention. *Journal of Cognitive Neuroscience, 1,* 221–237.

Heilman, K. H., Watson, R. T., and Valenstein, E. (1994). Localization of lesions in neglect and related disorders. In A. Kertesz (Ed.), *Localization and neuroimaging in neuropsychology,* pp. 495–524. San Diego, CA: Academic Press.

Heinze, H. J., and Mangun, G. R. (1995). Electrophysiological signs of sustained and transient attention to spatial locations. *Neuropsychologia, 33,* 889–908.

Heinze, H. J., Mangun, G. R., Burchert, W., Hinrichs, H., Scholz, M., Münte, T. F., Gös, A., Johannes, S., Scherg, M., Hundeshagen, H., Gazzaniga, M. S., and Hillyard, S. A. (1994). Combined spatial and temporal imaging of spatial selective attention in humans. *Nature, 392,* 543–546.

Hillyard, S. A., and Münte, T. F. (1984). Selective attention to color and location: An analysis with event-related brain potentials. *Perception and Psychophysics, 36,* 185–198.

Hopf, J.-M., and Mangun, G. R. (Forthcoming). Shifting visual attention in space: An electrophysiological analysis using high spatial resolution mapping. *Clinical neurophysiology.*

Hopfinger, J., and Mangun, G. R. (1998). Reflexive attention modulates processing of visual stimuli in human extrastriate cortex. *Psychological Science, 9,* 441–447.

Hopfinger, J. B., Buonocore, M. H., and Mangun, G. R. (2000). The neural mechanisms of top-down attentional control. *Nature Neuroscience, 3,* 284–291.

Jha, A. P., Buonocore, M., Girelli, M., and Mangun, G. R. (1997). FMRI and ERP studies of the organization of spatial selective attention in human extrastriate visual cortex. *Society for Neuroscience Abstracts, 23,* 301.

Jonides, J. (1981). Voluntary versus automatic control over the mind's eye movement. In J. B. Long and A. D. Baddeley (Eds.), *Attention and Performance IX*, pp. 187–203. Hillsdale, NJ: Erlbaum.

Kastner, S., DeWeerd, P., Desimone, R., and Ungerleider, L. C. (1998). Mechanisms of directed attention in the human extrastriate cortex as revealed by functional MRI. *Science, 282*, 108–111.

LaBerge, D. (1997). Attention, awareness, and the triangular circuit. *Consciousness and Cognition, 6*, 149–181.

Lavie, N., and Tsal, Y. (1994). Perceptual load as a major determinant of the locus of selection in visual attention. *Perception and Psychophysics, 56*, 183–197.

Luck, S. J., Chelazzi, L. Hillyard, S. A. and Desimone, R. (1997). Neural mechanisms of spatial selective attention in areas V1, V2 and V4 of macaque visual cortex. *Journal of Neurophysiology, 77*, 24–42.

Mangun, G. R. (1994). Orienting attention in the visual fields: An electrophysiological analysis. In H.-J. Heinze, T. F. Münte, and G. R. Mangun (Eds.), *Cognitive electrophysiology*, pp. 81–101. Boston: Birkhäuser.

Mangun, G. R. (1995). Neural mechanisms of visual selective attention in humans. *Psychophysiology, 32*, 4–18.

Mangun, G. R., Buonocore, M., Girelli, M., and Jha, A. (1998). ERP and fMRI measures of visual spatial selective attention. *Human Brain Mapping, 6*, 383–389.

Mangun, G. R., and Hillyard, S. A. (1991). Modulation of sensory-evoked brain potentials provide evidence for changes in perceptual processing during visual-spatial priming. *Journal of Experimental Psychology: Human Perception and Performance, 17*, 1057–1074.

Mangun, G. R., Hillyard, S. A., and Luck, S. J. (1993). Electrocortical substrates of visual selective attention. In D. Meyer and S. Kornblum (Eds.), *Attention and Performance XIV*, pp. 219–243. Cambridge, MA: MIT Press.

Mangun, G. R., Hopfinger, J., Kussmaul, C., Fletcher, E., and Heinze, H. J. (1997). Covariations in ERP and PET measures of spatial selective attention in human extrastriate cortex. *Human Brain Mapping, 5*, 273–279.

Mangun, G. R., Hopfinger, J., and Heinze, H. J. (1998). Integrating electrophysiology and neuroimaging in the study of human cognition. *Behavior Research Methods, Instruments and Computers, 30*, 118–130.

Mangun, G. R., Luck, S. J., Plager, R., Loftus, W., Hillyard, S. A., Clark, V., Handy, T., and Gazzaniga, M. S. (1994). Monitoring the visual world: Hemispheric asymmetries and subcortical processes in attention. *Journal of Cognitive Neuroscience, 6*, 265–273.

Martinez, A., Anllo-Vento, L., Sereno, M. I., Frank, L. R., Buxton, R. B., Dubowitz, D. J., Wong, E. C., Heinze, H. J., and Hillyard, S. A. (1999). Involvement of striate and extrastriate visual cortical areas in spatial selective attention. *Nature Neuroscience, 2*, 364–369.

McCarthy, G., Luby, M., Gore, J., and Goldman-Rakic, P. (1997). Infrequent events transiently activate human prefrontal and parietal cortex as measured by functional MRI. *Journal of Neurophysiology, 77*, 1630–1634.

Mesulam, M.-M. (1981). A cortical network for directed attention and unilateral neglect. *Annals of Neurology, 10*, 309–325.

Moran, J., and Desimone, R. (1985). Selective attention gates visual processing in the extrastriate cortex. *Science, 229*, 782–784.

Motter, B. C. (1993). Focal attention produces spatially selective processing in visual cortical areas V1, V2 and V4 in the presence of competing stimuli. *Journal of Neurophysiology, 70,* 909–919.

Posner, M. I., and Cohen, Y. (1984). Components of visual attention. In H. Bouma and D. Bowhuis (Eds.), *Attention and Performance X,* pp. 531–556. Hillsdale, NJ: Erlbaum.

Posner, M. I., and Driver, J. (1992). The neurobiology of selective attention. *Current Opinion in Neurobiology, 2,* 165–169.

Posner, M. I., and Gilbert, C. D. (1999). Attention and primary visual cortex. *Proceedings of the National Academy of Science, U.S.A., 96,* 2585–2587.

Posner, M. I., and Petersen, S. E. (1990). The attention system of the human brain. *Annual Review of Neuroscience, 13,* 25–42.

Posner, M. I., Snyder, C. R. R., and Davidson, B. J. (1980). Attention and the detection of signals. *Journal of Experimental Psychology: General, 109,* 160–174.

Posner, M. I., Walker, J. A., Friedrich, F. A., and Rafal, R. D. (1984). Effects of parietal injury on covert orienting of attention. *Journal of Neuroscience, 4,* 1863–1874.

Roelfsema, P. R., Lamme, V. A. F., and Spekreijse, H. (1998). Object-based attention in the primary visual cortex of the macaque monkey. *Nature, 395,* 376–381.

Rugg, M. D. (1998). Convergent approaches to electrophysiological and hemodynamic investigations of memory. *Human Brain Mapping, 6,* 394–398.

Sereno, M., Dale, A., Reppas, J., Kwong, K., Belliveau, J., Brady, J., Rosen, B., and Tootell, R. (1995). Borders of multiple visual areas in humans revealed by functional magnetic resonance imaging. *Science, 268,* 889–893.

Somers, D. C., Dale, A. M., Seiffert, A. E., and Tootell, R. B. H. (1999). Functional MRI reveals spatially specific attentional modulation in human primary visual cortex. *Proceedings of the National Academy of Science, U.S.A., 96,* 1663–1668.

Steinmetz, M. A., Connor, C. E., Constantinidis, C., and McLaughlin, J. R. (1994). Covert attention suppresses neuronal responses in area 7a of the posterior parietal cortex. *Journal of Neurophysiology, 72,* 1020–1023.

Treisman, A. M. (1988). Features and objects: The fourteenth Bartlett memorial lecture. *Quarterly Journal of Experimental Psychology, 40A,* 201–237.

Tootell, R., Reppas, J., Kwong, K., Malach, R., Born, R., Brady, T., Rosen, B., and Belliveau, J. W. (1995). Functional analysis of human MT and related visual cortical areas using magnetic resonance imaging. *Journal of Neuroscience, 15,* 3215–3230.

Van Essen, D., and DeYoe, E. (1995). Concurrent processing in the primate visual cortex. In M. S. Gazzaniga (Ed.), *The cognitive neurosciences,* pp. 383–400. Cambridge, MA: MIT Press.

Van Voorhis, S. T., and Hillyard, S. A. (1977). Visual evoked potentials and selective attention to points in space. *Perception and Psychophysics, 22,* 54–62.

Vidyasagar, T. R. (1998). Gating of neuronal responses in macaque primary visual cortex by an attentional spotlight. *NeuroReport, 9,* 1947–1952.

Watanabe, T., Sasaki, Y., Miyauchi, S., Putz, B., Fujimaki, N., Nielsen, M., Takino, R., and Miyakawa, S. (1997). Attention-regulated activity in human primary visual cortex. *Journal of Neurophysiology, 79,* 2218–2221.

Woldorff, M. G. (1993). Distortion of ERP averages due to overlap from temporally adjacent ERPs: Analysis and correction. *Psychophysiology, 30,* 98–119.

Woldorff, M. G., Fox, P., Matzke, M., Lancaster, J., Veeraswamy, S., Zamarripa, F., Seabolt, M., Glass, T., Gao, J., Martin, C., and Jerabeck, P. (1997). Retinotopic organization of the early visual-spatial attention effects as revealed by PET and ERPs. *Human Brain Mapping, 5,* 280–286.

Worden, M., Schneider, W., and Wellington, R. (1996). Determining the locus of attentional selectivity with functional magnetic resonance imaging. *Cognitive Neuroscience Society Abstracts, 3,* 101.

Yamaguchi, S., Tsuchiya, H., and Kobayashi, S. (1994). Electroencephalographic activity associated with shifts of visuospatial attention. *Brain, 117,* 553–562.

6 Looking Forward to Looking: Saccade Preparation and Control of the Visual Grasp Reflex

Robert Rafal, Liana Machado, Tony Ro, and Harris Ingle

ABSTRACT Studying eye movements is a useful avenue to understanding the neural basis of automaticity and control from an evolutionary perspective. Two experiments investigated control of the midbrain circuits responsible for the primitive visual grasp reflex. Experiment 1 showed that saccade preparation by normal participants reduced the fixation offset effect—the benefit in saccade latency afforded by elimination of a fixation stimulus. This implies strategic control over fixation neurons in the rostral pole of the superior colliculus, suggesting some autonomy from reflexive activation by external signals at fixation. Experiment 2 showed that in patients with chronic, unilateral lesions involving the frontal eye fields the visual grasp reflex is selectively disinhibited to signals in the contralesional field. These findings are consistent with the view of the frontal cortex as a tool maker that manipulates phylogenetically older neural circuits, evolved to provide for reflexive responses to the environment, putting them to new uses in the service of coherent, goal-directed behavior and creative problem solving.

Our neural machinery for visual orienting, like the rest of us, is the product of a long evolutionary history (Ingle 1973). All vertebrates have midbrain circuits for reflexively orienting the eyes toward salient events occurring in the visual periphery—the visual grasp reflex (VGR). In foveate mammals, including humans, these archetypal pathways function to align high-acuity regions of the retina to the location of a sudden change in the visual periphery; they must also be integrated with cortical mechanisms involved in strategic search under endogenous control. In everyday life, the outside world and internally generated goals place constantly competing demands on visual orienting systems. Coherent and adaptive behavior requires control mechanisms to arbitrate between these competing demands and to coordinate responding. Much is known about the neural basis of visual orienting; the coordination of its reflexive and voluntary deployment provides a useful framework for understanding the psychobiology of automaticity and control from an evolutionary perspective.

In this chapter, we consider strategic control over the midbrain circuits responsible for the primitive VGR. We proceed from a perspective that the nervous system routinely goes about its business through an orchestration of reflexes by endogenous processes that can activate or inhibit them (Easton 1973). The first experiment examines the effects of voluntary saccade preparation on the fixation offset effect (FOE; Klein and Kingstone 1993). We use the FOE—the reduction in saccade latency afforded by the offset of a fixation point—as a marker for the status of the collicular circuits with respect to the generation and inhibition of the VGR. This experiment shows that normal individuals are endogenously able to control the midbrain fixation reflex that inhibits the VGR. The second experiment examines the effects of chronic unilateral lesions, restricted to either dorsolateral prefrontal cortex (DLPFC) or parietal cortex, on errors in an antisaccade task. The antisaccade task requires both the inhibition of the VGR toward the peripheral stimulus and the endogenous generation of a voluntary saccade in the other direction. It allows examination of direct competition between reflexive and voluntary neural systems for visual orienting.

6.1 EXPERIMENT 1: THE EFFECTS OF VOLUNTARY SACCADE PREPARATION ON THE FIXATION OFFSET EFFECT

The visual grasp reflex is controlled by an opponent process mechanism within the superior colliculus (SC) of the midbrain. Its activation—or inhibition—is determined by competition between collicular neurons for maintaining fixation and others for generating a saccadic eye movement to a new location. Both in the superficial sensory layers and in the deeper layers in which movement cells generate saccadic eye movements, the SC has a topographic map of the visual field (Wurtz and Albano 1980). The colliculus uses a place code for eye movements such that the vector of a saccade is contingent on the location activated within the colliculus. Cells in the rostral pole help to hold the eyes anchored at fixation (Munoz and Wurtz 1992). As one moves caudally in the SC, neurons code saccades of increasing amplitude into the contralateral field. Like all other collicular neurons, fixation neurons in the rostral pole have mutually inhibitory connections with movement neurons throughout the colliculus (Munoz and Istvan 1998), and pharmacological inactivation of these neurons is associated with disinhibition of reflexive saccades (Munoz and Wurtz 1993a,b). Thus the potential for an eccentric visual stimulus to trigger a VGR—pulling the eyes to the stimulus—is determined by the relative activity of neurons in the caudal colliculus, with a movement field toward the stimulus, and of rostral pole neurons that inhibit movement cells and maintain fixation.

Fixation neurons are active during fixation, even in darkness, and increase their activity when a fixation signal is present. Hence the presence of a fixation point inhibits the VGR, increasing saccade latencies. The reduction in saccade latency when a fixation point offsets—the fixation offset effect—thus provides a measure of the degree to which fixation neurons are being driven by a fixation stimulus. Saslow (1967) first observed that the offset of a fixation point decreases the latency of saccades to visual targets in an experiment where there was a temporal gap between fixation offset and target onset. This facilitation has been called the "gap effect." A gap of approximately 200 msec is optimal for reducing saccade latencies. Under some circumstances, a robust gap effect may generate a bimodal distribution of saccade latencies, with a separate peak of very fast "express saccades" having latencies of less than 100 msec (Fischer and Ramsperger 1984). On the other hand, Kingstone and Klein (1993a) and Reuter-Lorenz, Hughes, and Fendrich (1991) emphasize that a temporal gap between fixation offset and the target provides a component of general alerting, in addition to the specific effects on saccades afforded by fixation offset. This component, specifically due to oculomotor disengagement and not to general alerting, represents the FOE and can be obtained even when fixation offset is simultaneous with target onset (Fendrich, Demirel, and Danziger 1999). The decrease in activity of the fixation neurons with fixation offset represents a neural correlate of express saccades (Dorris and Munoz 1995).

The fixation reflex, in which a visual signal at the point of fixation reflexively activates fixation neurons that inhibit the VGR to eccentric events, is especially strong in early infancy (Johnson 1990). At about 2 months of age, the colliculi come under the unopposed inhibitory influence of the basal ganglia (substantia nigra pars reticulata). Infants may become distressed because they are unable to break the lock of a visual stimulus in order to move their eyes. The FOE decreases during infant development (Hood, Atkinson, and Braddick 1997; Johnson and Gilmore 1997), marking maturation of frontobasal ganglia-colliculus circuits that brings the fixation reflex under voluntary control, permitting efficient visual search with alternating saccades and fixations.

Moreover, normal adults can learn to make express saccades with practice, even while the fixation point remains visible (Fischer and Breitmeyer 1987). The implication here is that normal participants may be able to voluntarily inhibit collicular pole fixation neurons even in the presence of a visual fixation stimulus. With fixation cell activity being more under strategic control than under the exogenous influence of a fixation point, the FOE should be reduced because fixation cell activity is less influenced by the presence or absence of a fixation point. In experiment 1, we examined whether normal adults can modulate the FOE when they voluntarily prepare an eye movement.

Participants

Twenty-six undergraduates participated for course credit: 13 in experiment 1a and 13 in experiment 1b.

Apparatus, Stimuli, and Procedure

The apparatus for stimulus display and eye movement recording is detailed in Ro, et al. 1997. Saccade latency was defined as the time when eye velocity exceeded 60 degrees/second. Participants were tested sitting in a quiet, dimly lit room facing a video display monitor 54 cm in front of them. The display consisted of white stimuli on a black background. After an intertrial interval of 1 sec, each trial began with presentation of a 0.35-degree fixation dot flanked by 2-degree unfilled squares, 8-degrees to the left and right. After 500 msec, the fixation dot was replaced by a 1-degree central precue. On one half of the trials, the precue was an arrowhead that pointed (with 100% probability, except on catch trials—(16.7% of total trials) to the location of the forthcoming saccade target, thus permitting saccade preparation prior to target appearance. On the other random half of trials, the precue was a double-headed arrow that was uninformative about the location of the forthcoming target, so that participants could not fully program a saccade until the target appeared. Fixation stimulus (the precue) offset versus overlap with target onset was also manipulated independent of precue validity. On half the trials, the precue offset simultaneously with target onset (fixation offset condition); on the other half, the precue remained visible until response to the target (fixation overlap condition). The target was a 1.5-degree asterisk that appeared with equal frequency in the center of either of the two peripheral boxes and remained visible until either a response was recorded or 3 sec lapsed. In experiment 1a, the target appeared either 200 or 700 msec after precue onset; in experiment 1b, the cue-target intervals were 75 and 700 msec.

Participants were instructed to prepare a saccade if a single arrowhead was presented, but to maintain fixation until a target appeared, at which time they were to make an eye movement to the target as quickly as possible. Catch trials were used to test for compliance with these instructions, and participants practiced until it was clear that they understood the task. Eye movements initiated during the interval between fixation onset and target onset terminated the trial and triggered the computer to buzz. In both experiment 1a and 1b, participants completed a single session of test trials (384 and 192 trials, respectively).

Results

In both experiments, trials were excluded from analysis if saccade latencies were less than 75 msec (2.6%) or more than 1,000 msec (4.5%). Responses on catch trials (8.9%, not including responses that appeared to be blinks) exceeded 10% for 6 individuals in experiment 1a and for 4 in experiment 1b. The analyses presented below included all participants, although analyses that excluded individuals with more than 10% catch trial responses gave comparable results. Saccade latencies for each par-

Table 6.1 Mean Saccade Latencies in Milliseconds (Standard Deviation in Parentheses) for Experiments 1a and 1b

	Experiment 1a 200 msec precue-target interval			700 msec precue-target interval		
Condition	Fixation overlap	Fixation offset	Fixation offset effect	Fixation overlap	Fixation offset	Fixation offset effect
Uninformative	330	271	**59**	322	255	**67**
Precue	(34)	(40)		(44)	(31)	
Informative	292	253	**39**	323	274	**54**
Precue	(33)	(36)		(32)	(36)	
	Experiment 1b 75 msec precue-target interval			**700 msec precue-target interval**		
Condition	Fixation overlap	Fixation offset	Fixation offset effect	Fixation overlap	Fixation offset	Fixation offset effect
Uninformative	347	285	**62**	334	280	**54**
Precue	(40)	(36)		(43)	(39)	
Informative	298	268	**30**	324	284	**40**
Precue	(40)	(49)		(44)	(36)	

ticipant were subjected to an analysis of variance (ANOVA). Within-subject factors included precue (informative or uninformative), fixation (offset or overlap), and cue-target interval (stimulus onset asynchrony of 200 or 700 msec) in experiment 1a, and of 75 or 700 msec in experiment 1b).

Both experiment 1a and experiment 1b showed effects of precue: $F(1, 12) = 4.6$, $p = 0.05$; $F(1, 12) = 7.4$, $p < 0.02$, respectively; and fixation offset: $F(1, 12) = 96.0$, $p < 0.001$; $F(1, 12) = 95.7$, $p < 0.001$, respectively. Both showed an interaction between precue and cue-target interval: $F(1, 12) = 31.6$, $p < 0.001$; $F(1, 12) = 14.3$, $p < 0.005$, respectively. As shown in table 6.1, there was a benefit of an informative precue on saccade latency at the short intervals. Because saccade latency increased in the informative precue condition between the short and long intervals, however, no benefit of an informative precue was present at the 700 msec interval in either experiment 1a or 1b. Both showed a reduction of the FOE in the saccade preparation (informative precue) condition: ($F(1, 12) = 14.4$, $p < 0.005$; $F(1, 12) = 10.9$, $p < 0.01$), respectively. The effect of precue on the FOE was present at both cue-target intervals and did not interact with their SOA.

Discussion

The major result of this experiment, with regard to strategic control of the VGR, is that saccade preparation reduced the FOE. This finding suggests that the midbrain fixation reflex—the otherwise automatic tendency of a fixation stimulus to hold the eyes—can be voluntarily controlled. It appears that collicular fixation neurons can be inhibited voluntarily when a saccade is prepared, even when a fixation point is visible. Because the frontal eye fields are considered responsible for initiating voluntary saccades, one possibility is that their projections can inhibit the rostral pole fixation neurons in the ipsilateral colliculus, either directly or through the basal ganglia, even in the presence of a fixation stimulus that would otherwise activate those neurons.

It seems likely that the reduction of the FOE afforded by an informative precue is attributable to voluntary preparation of a saccade, not simply from covert orienting to the cued field. Walker, Kentridge, and Findlay (1995) showed that precues that enabled covert orienting of attention, but not saccade preparation, did not influence the gap effect. In their experiment, saccade targets could appear at either of two possible locations (near or far) in either the left or right field. Informative precues enabled participants to orient their attention covertly toward the cued field. Because, however, the cue did not indicate which location the target would appear at in the cued field, near or far, participants could not fully program a saccade until the target appeared.

An intriguing dissociation in this experiment was observed between the effects of an informative precue over the fore period on mean saccade latency and on the FOE. The benefit of an informative precue on saccade latency at the short precue-target interval was not sustained at the long interval, whereas the reduction in the FOE afforded by informative precues was maintained throughout the fore period. Our previous work (Rafal et al. 1989) has shown that preparation, and then cancellation, of an endogenous saccade activates an inhibitory tag called "inhibition of return" (IOR) at the location toward which the saccade had been prepared. One possibility suggested by the current results is that IOR is generated even when an endogenously prepared eye movement is not canceled, resulting in the loss of informative precue benefit over time. However, while the sustained effect of an informative precue on the FOE indicates that some preparatory state was sustained, it is not clear what type of preparation was being maintained: the orienting of covert attention, the preparation of a saccade program, or some other aspect of preparation. We are conducting further experiments to test whether the loss of the informative precue benefit is due to generation of IOR by saccade preparation; and hence whether IOR can influence target detection even at actively attended locations.

6.2 EXPERIMENT 2: EFFECTS OF CORTICAL LESIONS ON THE VISUAL GRASP REFLEX IN AN ANTISACCADE TASK

The antisaccade task, in which a saccade must be made *away* from a peripheral target, demands both that the visual grasp reflex be inhibited and that a voluntary saccade be generated toward the opposite field. One mechanism for preventing reflexive glances toward a peripheral visual stimulus could be to increase the level of activity of rostral pole fixation neurons that inhibit the VGR. The fixation offset effect has been shown to be much reduced in an antisaccade situation (Forbes and Klein 1996; Reuter-Lorenz, Hughes, and Fendrich 1991; Reuter-Lorenz et al. 1995), possibly because frontal eye field (FEF) projections increase activity in fixation neurons. In this situation, strategic control over fixation neurons could allow individuals to maintain a high level of fixation cell activity even in the absence of an external fixation point, reducing reflexive eye movement errors.

Thus, both in antisaccade tasks and when prosaccades are prepared, the FOE is reduced because fixation neurons are less influenced by the presence or absence of the external fixation point. In the case of antisaccades, a high rate of fixation cell activity is maintained even in the absence of a fixation point, whereas in experiment 1, saccade preparation may have decreased the rate of fixation cell activity, even in the presence of a fixation point. Thus the FOE is reduced in both cases, but for different reasons in response to opposite strategic requirements.

Neuropsychological evidence for a role of dorsolateral prefrontal cortex in the control of the VGR in the antisaccade task was first reported by Guitton, Buchtel, and Douglas (1985), who showed that unilateral damage caused an increase in reflexive glances, that is, prosaccade errors, in the antisaccade task. One possible mechanism for this deficit postulates that the frontal lobes have a critical role in inhibition, and that lesions of oculomotor cortex result in collicular disinhibition. On the other hand, it has also been shown that the demands of working memory are critical determinants of errors in the antisaccade task (Roberts, Hager, and Heron 1994). Because lesions of DLPFC cause impairments of working memory (e.g., Funahashi, Bruce, and Goldman-Rakic 1991), the reflexive glances made by patients with frontal lobe damage may not necessarily reflect loss of inhibitory control, but instead may result from reduced working-memory capacity (Kimberg and Farah 1993; see also Kimberg and Farah, chap. 32, this volume). A reduction in working memory that prevents patients from maintaining task instructions should cause patients to make errors to both the ipsilesional and contralesional field. In contrast, if regions of oculomotor cortex are normally involved in inhibiting the ipsilateral colliculus, then a unilateral lesion in that cortex might result in an asymmetry where more reflexive glances are made to targets appearing in the contralesional visual field. Thus lesions involving the FEF might be expected to result in disinhibition of the ipsilesional colliculus, and an asymmetric pattern of reflexive glances. Some patients with extensive DLPFC lesions may have bilateral increases in reflexive glances due to impaired working memory, but if the lesion also involves the FEF, the deficit may be asymmetric, with more contralesional than ipsilesional errors.

Table 6.2 Clinical Information for Patients with Frontal Lesions

Patient	Lesion side	FEF damage	Age/ Sex	Volume (cc)	Etiology	Chronicity (years)
A.A.	Left	Yes	30F	59	Stroke	3
W.A.	Left	Yes	74F	26	Stroke	10
O.A.	Left	Yes	64M	18	Stroke	12
J.C.	Left	Yes	71M	106	Stroke	9
M.K.	Right	Yes	64M	200	Aneurysm	17
K.K.	Left	Yes	65M	14	Stroke	13
A.L.	Left	Yes	68F	56	Stroke	16
R.M.	Left	Yes	71M	14	Stroke	8
J.M.	Left	Yes	71M	15	Shrapnel	3
L.S.	Left	Yes	67F	28	Resection*	15
R.T.	Left	Yes	80M	46	Stroke	11
E.B.	Right	No	79F	17	Stroke	14
W.T.	Right	No	52M	9	Resection*	8

* L.S. and W.T.'s brain lesions resulted from surgical resection of a meningioma and colloid cyst, respectively.

Experiment 2 studied antisaccade performance in patients with chronic, unilateral lesions of the FEF. Its goal was to determine whether chronic lesions of frontal cortex cause an asymmetric impairment in inhibiting the VGR attributable to frontocollicular disinhibition, or bilateral deficits attributable to reductions in working memory. One further goal of this research was to examine the effect of cortical lesions on the ability to use precues to modulate the FOE strategically in both prosaccade and antisaccade tasks. Here, however, we report only the effect of these cortical lesions on the incidence of reflexive glances made toward the ipsilesional and contralesional field, that is, errors due to disinhibition of the VGR.

Participants

Eleven patients with chronic (tested at least two years after brain injury), unilateral lesions of oculomotor cortex in the superior dorsolateral prefrontal region were studied (FEF group). Control subjects included 24 normal elderly individuals (mean age: 72 years; standard deviations: 6), and neurological patients with unilateral lesions sparing the FEF: 2 with DLPFC lesions and 9 with lesions of parietal cortex (PAC; clinical details of individual patients are provided in tables 6.2 and 6.3). The region of lesion overlap in the FEF group is depicted in figure 6.1. All of the patients with FEF damage except one had left-hemisphere lesions. Both patients with DLPFC lesions sparing the FEF had right-hemisphere lesions. Of the 9 PAC-lesioned patients, 4 had left-hemisphere lesions and five had right-hemisphere lesions.

Table 6.3 Clinical Information for Patients with Parietal Lesions

Patient	Lesion side	TPJ damage?	Age/ Sex	Volume (cc)	Etiology	Chronicity (years)
R.A.	Left	Yes	65M	71	Stroke	7.5
L.L.	Left	Yes	73M	40	Stroke	2.5
J.S.	Right	Yes	38M	73	Shrapnel	10
R.S.	Right	Yes	51M	80	Stroke	6.5
M.K.	Left	No	51M	33	Shrapnel	29
T.E.	Left	No	46F	27	Stroke	2
K.T.	Right	No	48F	46	Resection*	22
L.P.	Right	No	72M	6	Stroke	>6
R.R.	Right	No	69M	34	Stroke	11

* K.T.'s brain lesion resulted from surgical resection of a glioma.

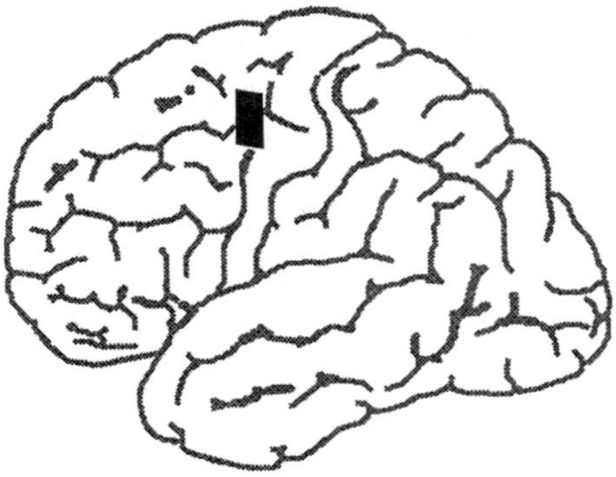

Figure 6.1 Composite reconstruction of neuroimages showing the common area of lesion at the intersection of the superior frontal sulcus and the precentral sulcus in all patients in the FEF group.

All of the patients had normal oculomotor function on standard clinical testing of saccades, pursuit eye movements, and optokinetic nystagmus. None had any clinical signs of hemispatial neglect, a scotoma in the region of the stimuli, or a coexisting neurological disorder (e.g., Parkinson's disease). Patients were not included in the study if review of their neuroimages showed the lesion to approach or to undercut the margins of the FEF such that they could not be clearly classified as having or not having FEF involvement. PET activation studies (reviewed in Paus 1996) and a study in our laboratory using transcranial magnetic stimulation (Ro et al. 1999) have localized the brain region responsible for generating voluntary, contraversive saccades to cortex centered at the intersection of the superior frontal sulcus and the precentral sulcus (Brodmann's area 6), approximately 2 cm rostral to the motor hand area. Our selection criterion for patients assigned to the FEF group was guided by these findings, and also by stimulation studies in humans that have

Table 6.4 Mean Percentage of Reflexive Eye Movement Errors (Standard Deviation in Parentheses) Made by Each Group of Subjects

	Uninformative precue		Informative precue		
	Overlap	Offset	Overlap	Offset	Total
Field: Normal elderly (n = 24)					
Left	14	14	8	10	**12**
	(11)	(10)	(7)	(9)	
Right	17	19	11	11	**14**
	(15)	(16)	(12)	(10)	
PAC-lesioned patients (n = 9)					
Ipsilesional	16	17	12	12	**14**
	(12)	(12)	(10)	(10)	
Contralesional	11	9	6	10	**9**
	(7)	(9)	(6)	(13)	
DLPFC-lesioned patients with FEF involvement (n = 11)					
Ipsilesional	11	13	9	13	**11**
	(10)	(11)	(7)	(9)	
Contralesional	22	27	23	21	**23**
	(10)	(18)	(10)	(16)	
DLPFC-lesioned patients without FEF involvement (n = 2)					
Ipsilesional	16	20	11	18	**16**
	(5)	(4)	(12)	(25)	
Contralesional	13	11	9	15	**12**
	(18)	(9)	(9)	(21)	

demonstrated that saccades may be elicited from regions somewhat more rostral, including Brodmann's area 8 (Penfield and Rasmussen 1950).

Apparatus, Stimuli, and Procedure

The same apparatus was used as in experiment 1. We measured the number of reflexive glances toward peripheral targets during an antisaccade task using the same stimuli and procedure as in experiment 1 except for the following: (1) participants were instructed to make an eye movement to the box *opposite* the target as soon as the target appeared; (2) the unfilled squares that marked the two possible target locations, and hence the targets, appeared 6-degrees to the left and right of fixation; (3) only the 200 msec precue-target interval was used; (4) the target was a 0.35-degree white dot; (5) the intertrial interval was 1,500 msec; and (6) there was a total of 384 test trials, 320 target-present trials plus 64 catch trials.

Results

All reflexive glances (i.e., errors in which saccades were made toward, rather than away from, the peripheral target) with saccadic latencies less

Table 6.5 Percentage of Reflexive Glances to Ipsilesional and Contralesional Fields for Each Patient

	Frontal lesion			Parietal lesion	
Patient	Ipsilesional	Contralesional	Patient	Ipsilesional	Contralesional
	FEF lesion			*TPJ lesion*	
O.A.	2	12	R.S.	14	1
J.C.	19	28	L.L.	22	9
J.M.	4	11	R.A.	14	9
R.M.	4	12	J.S.	11	5
M.K.	13	19		*No TPJ lesion*	
A.A.	10	11	L.P.	28	21
K.K.	18	28	K.T.	5	3
L.S.	8	43	M.K.	1	4
A.L.	19	40	R.R.	25	23
W.A.	25	36	T.E.	9	3
R.T.	3	17			
	No FEF lesion				
E.B.	24	22			
W.T.	8	2			

than 80 msec or greater than 500 msec were excluded from analysis. We conducted an ANOVA of the percentage of remaining reflexive eye movements made by the FEF-lesioned, PAC-lesioned, and normal elderly groups, *collapsed across field of target*, with precue (informative or uninformative) and fixation (offset or overlap) as within-subject factors. There was a main effect of precue condition, showing that informative precues reduced the number of reflexive glances to targets: $F(1,41) = 16.896$, $p < 0.001$. No other main effects or interactions were statistically reliable ($p > 0.1$). The FEF and PAC groups were each separately compared to the normal elderly group, confirming that neither of these patient groups differed from the normal elderly group in its overall percentage of reflexive eye movements ($p > 0.1$ for each comparison).

Table 6.4 shows the percentage of erroneous reflexive eye movements made toward each visual field for each group. Individual ANOVAs of the normal elderly, FEF, and PAC group data, with precue type, fixation condition, and target side as within-subject variables, were conducted to test for any visual field asymmetries. The results showed that the normal elderly group did not show a significant asymmetry ($p > 0.2$) between the left and right fields. FEF patients made more reflexive glances toward contralesional targets: $F(1,10) = 18.018$, $p = 0.002$; however, the two patients with DLPFC lesions that spared the FEF did not make more reflexive glances toward the contralesional field than the ipsilesional field. PAC patients made significantly *fewer* reflexive glances toward contralesional targets: $F(1, 8) = 10.840$, $p = 0.011$. (See table 6.5 for individual patient data.)

Because difficulties in disengaging attention in PAC patients have been associated with lesions of the temporal-parietal junction (TPJ; Friedrich et al. 1998), five of whom also participated in the current study, we did a lesion subregion analysis of the PAC group, comparing the patients whose lesions involved the TPJ to those whose lesions did not, based on whether neuroimaging showed the lesion to involve the posterior part of the superior temporal gyrus (area 22). The results showed an interaction between lesion site and target side: $F(1,7) = 5.555$, $p < 0.05$. Separate analysis of the PAC patients with TPJ involvement ($n = 4$) and those without TPJ involvement ($n = 5$) revealed that only the TPJ group made significantly fewer reflexive glances toward the contralesional field: $F(1,3) = 19.667$, $p = 0.02$. The group whose lesions spared the TPJ did not show a significant effect of target side ($p > 0.2$). Because the average lesion volume was greater in TPJ than in non-TPJ-lesioned patients (66.0 versus 29.4 cc), the data were examined to determine whether the apparent differences between the two groups could be attributed to differences in lesion volumes. Inspection of tables 6.3 and 6.5 shows no correlation between the degree of field asymmetry and lesion volume.

Discussion

The current observations extend our understanding of corticocollicular interactions as a model system for the control of automaticity. FEF lesions resulted in a disinhibition of the VGR specifically to contralesional signals. This is consistent with evidence that chronic FEF lesions result in hyperactivity of the ipsilesional colliculus (Henik, Rafal, and Rhodes 1994). Note, however, that in our study there were only two DLPFC control patients without FEF involvement, and both of them had right-hemisphere lesions. By contrast, 10 of the 11 patients with FEF lesions had left-hemisphere involvement. Unlike the FEF patients, patients with lesions of the TPJ showed a decrease in reflexive glances toward the contralesional field. This is consistent with evidence that parietal lesions result in hypoactivity of the ipsilesional colliculus (Sprague 1966).

The field-specific effect shown by the FEF patients in our study suggests that the FEF may normally suppress unwanted reflexive eye movements by exciting fixation cells in the ipsilateral SC. Fixation cells in turn project to the brain stem omnipause neurons (Paré & Guitton, 1994) which, like fixation cells, show a high firing rate when the eyes are stationary, but which pause during saccades and inhibit brain stem premotor neurons that innervate the oculomotor muscles (reviewed in Büttner-Ennever and Horn 1997; Everling et al. 1998). Alternatively, the FEFs may inhibit reflexive glances through their direct projection to the omnipause neurons (reviewed in Moschovakis, Scudder, and Highstein 1996).

To date, reports of antisaccade performance in groups of patients with unilateral frontal damage have been inconsistent. Several investigators reported bilateral disinhibition of the VGR (Fukushima et al. 1994; Guitton, Buchtel, and Douglas 1985; Pierrot-Deseilligny et al. 1991), while some found either ipsilateral disinhibition (Fukushima et al. 1994) or no disinhibition (Rivaud et al. 1994).

Discrepancies in whether unilateral disruption of FEF activity results in contralesional or ipsilesional deficits are also present in the monkey literature. Burman and Bruce (1997) found that electrically stimulating some cells in the monkey FEF, but not anterior or posterior to the FEF, inhibited both memory and visually guided saccades, especially those directed toward the contralesional field. Accordingly, permanent lesions of the monkey FEF led to frequent premature saccades toward contralesional targets during a memory-guided saccade task (Deng et al. 1986). On the other hand, acute chemical inactivation of the monkey FEF led to premature saccades primarily toward ipsilateral targets during the delay of a memory-guided saccade task (Dias, Kiesau, and Segraves 1995; Dias and Segraves 1997; Sommer and Tehovnik 1997).

The apparent inconsistencies in the effects of FEF damage may relate to the heterogeneity of the lesions, both in terms of (1) anatomical extent—not all of the frontal patients in previous investigations had FEF damage; and (2) chronicity—most previous investigations examined patients in the acute stage of illness, when diaschesis (the remote effects of an acute lesion on neural structures with which damaged tissue has been interconnected) may have contributed to collicular dysfunction on the side of the frontal lesion. After acute lesions of the FEF in monkeys, there is hypometabolism of the SC on the side of the lesion (Deuel and Collins 1984), and clinical hemispatial neglect during this acute stage is common because both cortical and subcortical orienting systems on the side of the lesion are dysfunctional. After the acute phase of diaschesis resolves, however, neglect recovers, and monkeys are able, within weeks, to make saccades to contralesional targets with normal latencies (Schiller, Sandell, and Maunsell 1994).

Humans with chronic lesions of the FEF show evidence that the ipsilesional colliculus becomes hyperactive and the contralesional colliculus becomes hypoactive. Henik, Rafal, and Rhodes (1994) demonstrated that patients with chronic, unilateral FEF lesions (four of whom also participated in the current study) have *shorter* latencies to initiate saccades to contralesional targets. The patients in the current study were also tested in a prosaccade task using the same display as used for the antisaccade task. The results, to be reported in more detail elsewhere, replicated Henik et al.: latencies of saccades toward contralesional targets were shorter.

Two single-case studies of patients with unilateral FEF damage demonstrated the effects of the chronicity of the lesion on changes in collicular activity for behavior in an antisaccade task. In a series of sessions beginning 5 days after a stroke and ending 170 days after the stroke, Butter et al. (1988) reported that a patient with a right frontal infarct including the FEF initially exhibited contralesional sensory neglect. Although the percentage of saccades erroneously made toward ipsilesional targets did not change over the testing sessions, the percentage of reflexive saccades made toward contralesional targets increased dramatically as the patient recovered from neglect. Kwon and Heilman (1991) later replicated this effect of neglect on disinhibition of the VGR, and extended it to the limb system using a line bisection task, in a patient with right frontal damage, including Brodman's areas 6 and 8.

In addition to the evolving effects of FEF lesions on collicular circuitry, some patients with frontal lobe lesions also have reduced working memory that may disrupt antisaccade performance bilaterally (Walker et al. 1998). Some models providing a unified account of the deficits associated with frontal lobe damage suggest that the ability to inhibit automatic responses depends on working memory (e.g., Kimberg and Farah 1993; see also Kimberg and Farah, chap. 32, this volume). Asking college students to perform an antisaccade task simultaneously with a variety of secondary tasks that placed different demands on working memory, Roberts, Hager, and Heron (1994) found that performing a concurrent task with high working memory demands caused normal participants to make twice as many erroneous reflexive glances. These authors suggested that the difficulty in inhibiting reflexive glances shown by frontal patients, and by normal participants when working memory is taxed, stems from not "maintaining a high enough level of activation of the relevant self-instructions." Guitton, Buchtel, and Douglas (1985) reported that simplifying their antisaccade task in patients with frontal lesions, by eliminating an identification response that was required after each eye movement, diminished the frequency of reflexive glances. This reduction in reflexive errors supports the possibility that working memory plays a role in inhibiting reflexive glances and that this effect could be independent of the unilateral effect that results from disruption of a corticocollicular pathway.

The unique contribution of the current investigation among both animal and patient studies of antisaccade performance is the specific examination of the effects of *chronic* cortical lesions. The results help to reconcile apparent inconsistencies in the neurological literature. The cortico-subcortical circuits for controlling eye movements are a dynamic system, and the effects of FEF lesions on collicular function may evolve over time. Initially, frontal lesions cause hypometabolism in the ipsilesional colliculus (Deuel and Collins 1984). In the acute stage, patients often have transient hemispatial neglect because both the cortical and subcortical components of the orienting system are dysfunctional on the side of the lesion. This causes hyperorienting toward the side of the lesion, and a disinhibited ipsilesional VGR. By contrast, the current findings and our earlier findings in a prosaccade task (Henik, Rafal, and Rhodes 1994) suggest that, in the chronic stage, the ipsilesional colliculus is hyperactive, causing a disinhibited contralesional VGR, whereas the colliculus contralateral to the FEF lesion is relatively hypoactive. The result is an asymmetric deficit, with a disinhibited VGR only toward contralesional targets. The effects of frontal lesions on antisaccade performance are thus determined by (1) the chronicity of the lesion; (2) the state of dynamic interaction between cortex and subcortex; (3) the specific regions of DLPFC involved; and (4) the degree of working-memory impairment.

6.3 CONCLUSIONS

This research was motivated by the hypothesis that fixation neurons in the rostral pole of the superior colliculus, which are activated by stimuli at fixation to hold the eyes in place, may also be under endogenous control via frontocollicular pathways. These pathways enable us to free the oculomotor system from reflexive responses to the environment, and permit coherent behavior based on strategic goals. The results of the current experiments are consistent with this hypothesis. Saccade preparation in response to informative precues reduced the fixation offset effect in normal individuals, and damage to the frontal eye field released the visual grasp reflex toward signals appearing in the contralesional field.

The current evidence for endogenous control over fixation converges with other evidence for cortical control of collicular circuitry. Express saccades are classically associated with conditions in which there is a temporal gap between fixation offset and target onset. Nevertheless, with practice, some individuals can be trained to make express saccades even when a fixation point is present (Fischer and Breitmeyer 1987). Machado and Rafal (in press) have recently examined another manipulation of strategic set on the FOE—the proportion of catch trials (in which no saccade target appeared and fixation had to be maintained). In these experiments, prosaccades were made, and there were no informative precues; rather, the proportion of catch trials was manipulated systematically. When catch trials were less frequent, the FOE was smaller. Kingstone and Klein (1993b) have shown that decreasing the proportion of catch trials decreases saccade latency in a gap condition. A reduction in the FOE due to fewer catch trials, as found by Machado and Rafal, would require a greater decrease in saccade latencies in the fixation overlap condition than in the offset condition. We can infer, then, that the decrease in the FOE with fewer catch trials may result from cortical inhibition of fixation neurons even when a fixation point is present—the same mechanism we are proposing for the reduction of the FOE in experiment 1.

The reduction of the FOE in an antisaccade task may also be due to strategic control of the VGR (Forbes and Klein 1996). In this circumstance, however, the FOE is presumably decreased because cortical control can maintain a high level of activity of fixation neurons, even when there is no external stimulus at fixation. In both cases, prosaccade and antisaccade, the FOE is reduced because the oculomotor set makes fixation neurons more autonomous from external stimuli and more under endogenous control.

Experiment 2 showed that FEF lesions caused disinhibition of the VGR specifically to *contralesional* targets. It demonstrated that the same region of the superior dorsolateral prefrontal cortex, the FEF, that is responsible for generating contralateral voluntary saccades is also responsible for inhibiting the VGR. The critical role of the FEF in voluntary saccade generation in humans has been established by converging evidence. It is activated during voluntary saccades (Paus 1996); and lesions in it (Henik, Rafal, and Rhodes 1994) or its transient inactivation by transcranial magnetic stimulation (TMS; Ro et al. 1999, 1997) increases the latency of contralateral, voluntary saccades. An important mission of the FEF in controlling voluntary saccades is the ability to inhibit reflexive eye movements, when necessary. We suggest that this mission is accomplished by the same kind of voluntary regulation of the opponent neural circuitry of the colliculus that normal participants exhibited in experiment 1.

Although the antisaccade task is an artificial situation contrived in the laboratory, humans (as well as prey and other social animals) frequently inhibit reflexive eye movements in natural conditions. The ability to inhibit eye movements may have evolved in conjunction with the ability to covertly orient visual attention to meet specific adaptive requirements, for example, the need of a prey animal to attentively track a predator without establishing eye contact that could attract attention to itself; the need of a juvenile primate to keep track of the doings of the alpha male while avoiding confrontation; or the need of a human to pay attention to the hands of an approaching stranger while maintaining eye contact.

One way of thinking about the frontal cortex is as the brain's tool user and tool maker. Its great expansion in humans is paralleled by increasing flexibility in goal-directed behavior, and creativity in problem solving. The frontal cortex orchestrates the novel use of phylogenetically older neural circuits, which may have evolved to meet the needs of entirely different environmental pressures. In this sense, the circuitry of reflexes may be thought of as tools used by frontal cortex to make new mental tools to solve new problems. In the mission of tool maker, the frontal lobes require competence in a number of functions, to include holding the computations of several operations of a complex task on line in working memory; sequencing their implementation in time with frontostriatal switching circuitry; and, in circumstances like the antisaccade task, inhibiting the primitive functions for which the component circuits initially evolved.

In summary, we have focused on a simple primitive midbrain visuomotor reflex as a model system for understanding the neural basis of automaticity and control. The strategic control of the VGR provides for an efficient coordination of two opponent oculomotor systems for moving and fixing the eyes. Visual search involves a sequence of saccades and intervening fixations during which attention dwells on the attended item to extract visual information. The oculomotor system implements this through anatomically distinct pathways with mutually inhibitory interactions for fixation and saccades. This implementation occurs at the level of an opponent physiological process within the SC itself, and is under control of the FEF.

NOTE

This research was supported by U.S. Public Health Service grants MH41544, MH54100, and MH19930.

REFERENCES

Burman, D. D., and Bruce, C. J. (1997). Suppression of task-related saccades by electrical stimulation in the primate's frontal eye field. *Journal of Neurophysiology, 77,* 2252–2267.

Butter, C. M., Rapcsak, S. Z., Watson, R. T., and Heilman, K. M. (1988). Changes in sensory inattention, directional akinesia, and release of fixation reflex following a unilateral frontal lesion: A case report. *Neuropsychologia, 26,* 533–545.

Büttner-Ennever, J. A., and Horn, A. K. E. (1997). Anatomical substrates of oculomotor control. *Current Opinion in Neurobiology, 7,* 872–879.

Deng, S.-Y., Goldberg, M. E., Segraves, M. A., Ungerleider, L. G., and Mishkin, M. (1986). The effect of unilateral ablation of the frontal eye fields on saccadic performance in the monkey. In E. L. Keller and D. S. Zee (Ed.), *Adaptive processes in visual and oculomotor systems,* pp. 201–208. Oxford: Pergamon Press.

Deuel, R. K., and Collins, R. C. (1984). The functional anatomy of frontal lobe neglect in monkeys: Behavioral and 2-deoxyglucose studies. *Annals of Neurology, 15,* 521–529.

Dias, E. C., Kiesau, M., and Segraves, M. A. (1995). Acute activation and inactivation of macaque frontal eye field with GABA-related drugs. *Journal of Neurophysiology, 74,* 2744–2748.

Dias, E. C., and Segraves, M. A. (1997). The primate frontal eye field and the generation of saccadic eye movements: Comparison of lesion and acute inactivation/activation studies. *Revista Brasileria de Biologia, 56,* 239–255.

Dorris, M. C., and Munoz, D. P. (1995). A neural correlate for the gap effect on saccadic reaction times in monkey. *Journal of Neurophysiology, 73,* 2558–2562.

Easton, T. A. (1973). On the normal use of reflexes. *American Scientist, 60,* 591–599.

Everling, S., Paré, M., Dorris, M. C., and Munoz, D. P. (1998). Comparison of the discharge characteristics of brainstem omnipause neurons and superior colliculus fixation neurons in monkey: Implications for control of fixation and saccade behavior. *Journal of Neurophysiology, 79,* 511–528.

Fendrich, R., Demirel, S., and Danziger, S. (1999). The oculomotor gap effect without a foveal fixation point. *Vision Research, 39,* 833–841.

Fischer, B., and Breitmeyer, B. (1987). Mechanisms of visual attention revealed by saccadic eye movements. *Neuropsychologia, 25,* 73–84.

Fischer, B., and Ramsperger, E. (1984). Human express saccades: Extremely short reaction times of goal-directed eye movements. *Experimental Brain Research, 57,* 191–195.

Forbes, K., and Klein, R. M. (1996). The magnitude of the fixation offset effect with endogenously and exogenously controlled saccades. *Journal of Cognitive Neuroscience, 8,* 344–352.

Friedrich, F. J., Egly, R., Rafal, R., and Beck, D. (1998). Spatial attention deficits in humans: A comparison of superior parietal and temporal parietal junction lesions. *Neuropsychology, 12,* 193–207.

Fukushima, J., Fukushima, K., Miyasaka, K., and Yamashita, I. (1994). Voluntary control of saccadic eye movement in patients with frontal cortical lesions and parkinsonian patients in comparison with that in schizophrenics. *Biological Psychiatry, 36,* 21–30.

Funahashi, S., Bruce, C. J., and Goldman-Rakic, R. P. (1991). Neuronal activity related to saccadic eye movements in the monkey's dorsolateral prefrontal cortex. *Journal of Neurophysiology, 65,* 1464–1483.

Guitton, D., Buchtel, H. A., and Douglas, R. M. (1985). Frontal lobe lesions in man cause difficulties in suppressing reflexive glances and in generating goal-directed saccades. *Experimental Brain Research, 58,* 455–472.

Henik, A., Rafal, R., and Rhodes, D. (1994). Endogenously generated and visually guided saccades after lesions of the human frontal eye fields. *Journal of Cognitive Neuroscience, 6,* 400–411.

Hood, B. M., Atkinson, J., and Braddick, O. J. (1997). Selection-for-action and the development of visual selective attention. In J. E. Richards (Ed.), *Cognitive neuroscience of attention: A developmental perspective,* pp. 219–250. Hillsdale, NJ: Erlbaum.

Ingle, D. (1973). Two visual systems in the frog. *Science, 181,* 1053–1055.

Johnson, M. H. (1990). Cortical maturation and the development of visual attention in early infancy. *Journal of Cognitive Neuroscience, 2,* 81–95.

Johnson, M. H., Gilmore, R. O., and Lsibra, G. (1997). Toward a computational model of the development of saccade planning. In J. E. Richards (Ed.), *Cognitive neuroscience of attention: A developmental perspective,* pp. 103–130. Hillsdale, NJ: Erlbaum.

Kimberg, D. Y., and Farah, M. J. (1993). A unified account of cognitive impairment following frontal lobe damage: The role of working memory in complex, organized behavior. *Journal of Experimental Psychology: General, 122,* 411–428.

Kingstone, A., and Klein, R. M. (1993a). Visual offsets facilitate saccadic latency: Does predisengagement of visuospatial attention mediate this gap effect? *Journal of Experimental Psychology: Human Perception and Performance, 19,* 1251–1265.

Kingstone, A., and Klein, R. M. (1993b). What are express saccades? *Perception and Psychophysics, 54,* 260–273.

Klein, R., and Kingstone, A. (1993). Why do visual offsets reduce saccadic latencies? *Behavioral and Brain Sciences, 16,* 583–584.

Kwon, S. E., and Heilman, K. M. (1991). Ipsilateral neglect in a patient following a unilateral frontal lesion. *Neurology, 41,* 2001–2004.

Machado, L., and Rafal, R. (Forthcoming). Strategic control over the visual grasp reflex: Studies of the fixation offset effect. *Perception and Psychophysics.*

Moschovakis, A. K., Scudder, C. A., and Highstein, S. M. (1996). The microscopic anatomy and physiology of the mammalian saccadic system. *Progress in Neurobiology, 50,* 133–254.

Munoz, D. P., and Istvan, P. J. (1998). Lateral inhibitory interactions in the intermediate layers of the monkey superior colliculus. *Journal of Neurophysiology, 79,* 1193–1209.

Munoz, D. P., and Wurtz, R. H. (1992). Role of the rostral superior colliculus in active visual fixation and execution of express saccades. *Visual Neuroscience, 9*, 409–414.

Munoz, D. P., and Wurtz, R. H. (1993a). Fixation cells in monkey superior colliculus: 1. Characteristics of cell discharge. *Journal of Neurophysiology, 70*, 559–575.

Munoz, D. P., and Wurtz, R. H. (1993b). Fixation cells in monkey superior colliculus: 2. Reversible activation and deactivation. *Journal of Neurophysiology, 70*, 576–589.

Paré, M., and Guitton, D. (1994). The fixation area of the cat superior colliculus: Effects of electrical stimulation and direct connection with brainstem omnipause neurons. *Experimental Brain Research, 101*, 109–122.

Paus, T. (1996). Location and function of the human frontal eye-field: A selective review. *Neuropsychologia, 34*, 475–484.

Penfield, W., and Rasmussen, T. (1950). *The cerebral cortex of man: A clinical study of localization of function*. New York: Macmillan.

Pierrot-Deseilligny, C., Rivaud, S., Gaymard, B., and Agid, Y. (1991). Cortical control of reflexive visually-guided saccades. *Brain, 114*, 1473–1485.

Rafal, R., Calabresi, P., Brennan, C., and Sciolto, T. (1989). Saccade preparation inhibits reorienting to recently attended locations. *Journal of Experimental Psychology: Human Perception and Performance, 15*, 673–685.

Reuter-Lorenz, P. A., Hughes, H. C., and Fendrich, R. (1991). The reduction of saccadic latency by prior offset of the fixation point: An analysis of the gap effect. *Perception and Psychophysics, 49*, 167–175.

Reuter-Lorenz, P. A., Oonk, H. M., Barnes, L. L., and Hughes, H. C. (1995). Effects of warning signals and fixation point offsets on the latencies of pro- versus antisaccades: Implications for an interpretation of the gap effect. *Experimental Brain Research, 103*, 287–293.

Rivaud, S., Muri, R. M., Gaymard, B., Vermersch, A. I., and Pierrot-Deseilligny, D. C. (1994). Eye movement disorders after frontal eye field lesions in humans. *Experimental Brain Research, 102*, 110–120.

Ro, T., Cheifet, S., Ingle, H., Shoup, R., and Rafal, R. (1999). Localization of the human frontal eye fields and motor hand area with transcranial magnetic stimulation and magnetic resonance imaging. *Neuropsychologia, 37*, 225–231.

Ro, T., Henik, A., Machado, L., and Rafal, R. D. (1997). Transcranial magnetic stimulation of the prefrontal cortex delays contralateral endogenous saccades. *Journal of Cognitive Neuroscience, 9*, 433–440.

Roberts, R. J., Hager, L. D., and Heron, C. (1994). Prefrontal cognitive processes: Working memory and inhibition in the antisaccade task. *Journal of Experimental Psychology: General, 123*, 374–393.

Saslow, M. G. (1967). Effects of components of displacement-step stimuli upon latency for saccadic eye movements. *Journal of the Optical Society of America, 57*, 1024–1029.

Schiller, P. H., Sandell, J. H., and Maunsell, J. H. (1994). The effect of frontal eye field and superior colliculus lesions on saccadic latencies in the rhesus monkey. *Experimental Brain Research, 98*, 179–190.

Sommer, M. A., and Tehovnik, E. J. (1997). Reversible inactivation of macaque frontal eye field. *Experimental Brain Research, 116*, 229–249.

Sprague, J. M. (1966). Interaction of cortex and superior colliculus in mediation of peripherally summoned behavior in the cat. *Science, 153*, 1544–1547.

Walker, R., Husain, M., Hodgson, T. L., Harrison, J., and Kennard, C. (1998). Saccadic eye movements and working memory deficits following damage to human prefrontal cortex. *Neuropsychologia, 36,* 1141–1159.

Walker, R., Kentridge, R. W., and Findlay, J. M. (1995). Independent contributions of the orienting of attention, fixation offset and bilateral stimulation on human saccadic latencies. *Experimental Brain Research, 103,* 294–310.

Wurtz, R. H., and Albano, J. (1980). Visual-motor function of the primate superior colliculi. *Annual Review of Neuroscience, 3,* 189–226.

7 Selective Attention and Cognitive Control: Dissociating Attentional Functions through Different Types of Load

Nilli Lavie

ABSTRACT Selective attention acts to ensure that behavior is controlled by goal-relevant rather than goal-irrelevant stimuli. Here I propose two very different mechanisms for attentional control. The first, a passive mechanism, excludes irrelevant stimuli from perception, but can only operate in situations of high perceptual load which exhaust perceptual capacity with processing of relevant material. In situations of low perceptual load, a second active mechanism comes into play to suppress responses to irrelevant distractors that cannot be excluded from perception. Such suppression requires higher-level control functions (e.g., working memory). High cognitive load (e.g., in working memory) engages these functions, and therefore leads to inefficient rejection of distractors. Evidence for the distinction between passive and active control mechanisms is obtained from the contrasting effects of different types of load on distractor processing in behavioral studies. The implications of this distinction for neural activity, as well as for aging of attentional functions, are also considered.

Selective attention usually allows efficient and focused processing of goal-relevant stimuli, with minimal intrusions from goal-irrelevant stimuli. In this chapter, I argue for two forms of such attentional selection. The first is a relatively passive form of control, whereby interference from irrelevant distractors is prevented simply because they are not perceived. Efficient exclusion of irrelevant stimuli from perception is not always possible, however.[1] A second, more active control function therefore comes into play in situations where irrelevant as well as relevant stimuli are perceived. In such cases, active control is crucial for suppressing response tendencies toward the irrelevant, yet processed distractors.

Specifically, I argue that high load in the perception of relevant stimuli results in reduced perception of distractor stimuli because there is insufficient capacity to process them all. Low perceptual load in the relevant task, on the other hand, results in the processing of irrelevant as well as relevant information, and therefore requires some active means of rejecting distractors for maintaining appropriate control of behavior. These active control processes rely on higher mental functions, such as working memory (WM), which are required to maintain current priorities and thus ensure that low-priority items can be suppressed. Contrary to the predicted effect for perceptual load, however, increasing the load on these higher mental functions will drain the capacity available for active control and result in more, rather than fewer, intrusions from irrelevant distractors.

The two proposed mechanisms for selective attention, active and passive, should thus be dissociable from one another through the opposite effects that different types of load are predicted to have on the efficiency of selective attention. In particular, the extent to which intrusions from distractors can be avoided should depend on the level and type of load in relevant processing. In the following sections, the two mechanisms are described in greater detail and evidence is discussed for the role of different processing loads in determining whether the passive or active mechanism of control comes into play. Some implications of this distinction for the normal aging of selective attention are also discussed.

7.1 PASSIVE CONTROL OF SELECTIVE ATTENTION

In this section, I consider the role of perceptual load in selective attention, placing the perceptual load model in the context of previous research on selective attention, then describing the model in greater detail and reporting the empirical evidence available thus far to support it.

The Perceptual Load Theory: Background and Model

The extent to which selective attention can prevent perception of irrelevant distractors has been debated for the last four decades, within the tradition of debate on whether attentional selection has an early or a late locus in processing (e.g., Kahneman and Treisman 1984). On the one hand, numerous reports of distractors excluded from perception (e.g., Treisman 1969; Yantis and Johnston 1990) seemed to support the early-selection view of attention. On the other, reports of failures to exclude distractors from perception (e.g., Shiffrin and Schneider 1977; Driver and Tipper 1989) seemed to support the rival, late-selection view, according to which attention can affect only processes of response selection that occur *after* distractors are perceived. These conflicting reports led to a theoretical impasse for many years.

A resolution to the early- versus late-selection debate may be possible, however, if we consider a hybrid model for attention, which combines aspects of both views (Lavie 1995). According to this model, perceptual processing has capacity limits (as in the early-selection approach) but operates automatically on all stimuli (as in the late-selection approach) accommodated within those limits. Thus voluntary control is restricted to determining priorities between relevant and irrelevant stimuli, although any remaining capacity from processing relevant stimuli will "spill over" to the processing of irrelevant distractors.

The extent to which the perception of *irrelevant* distractors can be prevented should thus depend on the perceptual load imposed by the processing of *relevant* information. Situations of low perceptual load will inevitably result in perception of irrelevant stimuli, despite the assignment of low priority to their processing. By contrast, when relevant perceptual processing imposes a high load, this will exhaust perceptual capacity leaving none for distractors, so that perception of distractors is prevented by passive exclusion: there is simply insufficient capacity to process them.[2] The more active process of maintaining current priorities in WM is important for ensuring the correct distinction between relevant and irrelevant information, in situations of both high and low perceptual load. But this is in itself insufficient to cause selective perception, as long as perceptual load in relevant processing is low, and thus leaves spare perceptual capacity. Irrelevant information that is involuntarily perceived must therefore be actively suppressed at a later stage.

A review of past studies in the early- versus late-selection debate (Lavie and Tsal 1994) has lent support to this hybrid model. Results indicating selective perception have typically been obtained under high perceptual load in the relevant task; those indicating unselective perception, under low perceptual load.

The Role of Perceptual Load in Distractor Interference: Behavioral Studies

A series of new experiments using various manipulations of perceptual load (Lavie 1995; Lavie and Cox 1997) has provided further evidence for the model, showing that perceptual processing of distractors is reduced only by a high perceptual load in the relevant task. A new method was used to manipulate the perceptual load imposed by processing of relevant stimuli and to assess its effect on processing irrelevant stimuli. Stimuli for the relevant task were presented in the display center, while an irrelevant distractor stimulus was presented in a peripheral position. Subjects were told to ignore the peripheral distractor and to focus on the central task, and the extent to which distractors were nonetheless processed was assessed by measuring response competition effects (cf. Eriksen and Eriksen 1974) from distractor stimuli that were either response incongruent, response congruent, or neutral with respect to the target response. Perceptual load was manipulated in the central task by varying either the number of stimuli relevant for processing (i.e., set size of the relevant items in the display center) or the processing requirements for constant items (e.g., requiring feature versus conjunction processing for the same displays). For example, Lavie and Cox (1997) presented subjects with a visual search task at the center of the display, asking them to ignore an irrelevant peripheral distractor (presented outside the relevant search area) while searching for one of two target letters (e.g., *X* or *N*) among other nontarget letters in the central array.[3] Note that, unlike the peripheral distractor, these nontarget stimuli were always neutral with respect to target response, and only served to load target perception by forcing subjects to search for the target among them. Search load was manipulated in one experiment by varying the similarity between targets and nontargets, (e.g., the *X* or *N* targets were presented either among *O*s in the low-load condition, or among angular letters in the high-load condition). In another experiment, search load was manipulated by varying the set size of similar targets and nontarget letters.

We found that *efficient* searches, involving target pop-out, led to *inefficient* rejection of the peripheral distractor because the search load of the relevant task was low. By contrast, *inefficient* searches, with a steep search slope indicating that each potentially relevant item imposed an additional demand on attention, led to *efficient* rejection of irrelevant peripheral distractors, as long as more than four items were involved in the relevant search to exhaust capacity (see also Fisher 1982; Yantis and Jones 1991; Kahneman, Treisman, and Gibbs 1993; Pylyshyn et al. 1994 for similar reports of capacity limits).

Because these manipulations of search load involved either varying the nontarget letters (e.g., curved versus angular letters) that appeared in the relevant central area or varying their number, the appearance of the display in this study differed between the load conditions. Another study (Lavie 1995) demonstrated that the load imposed by the processing of relevant items determines irrelevant distractor processing even for cases that did not involve any variation in the stimulus displays with load, which was now manipulated via different processing requirements for the same displays. In one experiment for example, subjects made speeded choices discriminating the identity of a central target letter, while attempting to ignore an irrelevant peripheral distractor. Whether subjects should respond to the central target, however, was conditional on another shape adjacent to the target. In the low-load condition, the mere presence of this additional shape was sufficient to license a response (which was to be withheld if that shape was absent). In the high-load condition, subjects had to identify the combination of shape, exact size, and position of the adjacent shape to decide whether to respond to the target letter. Distractor interference was observed when mere detection of the shape was required (low load), but was significantly reduced when it had to be identified (high load) for the very same displays.

The consistent decrease in interference from incongruent distractors with higher perceptual loads in all these previous experiments was taken as supporting our hypothesis that perceptual load imposed by relevant stimuli reduces the perceptual processing of irrelevant distractors, thus also supporting early selection.

Recent late-selection views (e.g., Tipper and Milliken 1996) offer an alternative account for our results, however, one that stresses the role of inhibitory mechanisms in distractor exclusion, as revealed by negative priming effects (i.e., the slowing down of subsequent responses to items that served as the irrelevant distractor on the preceding trial). If inhibition

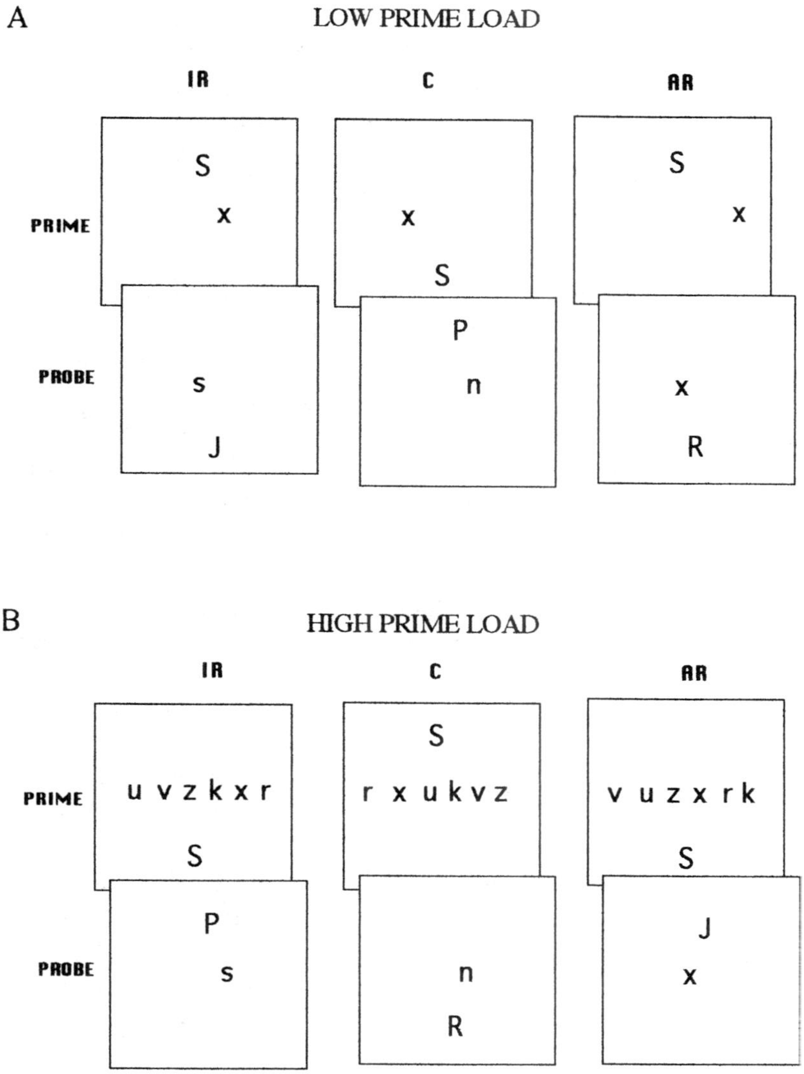

Figure 7.1 Example displays from the low-load (panel A) and high-load (panel B) conditions in Experiment 1 of Lavie and Fox 2000. A prime display and the immediately following probe display are shown for each condition. IR = ignored repetition; C = control; AR = attended repetition condition.

is the primary means for selective processing (e.g., Driver and Tipper 1989), then the reduced interference we found in situations of high load may not necessarily reflect reduced distractor processing, but rather increased inhibition of processed distractors. On the other hand, if perceptual load determines distractor processing, as we claim, inhibition will only be required when distractors are perceived, that is, only under low-load conditions.

A study of negative priming (NP) by the author and Elaine Fox (Lavie and Fox 2000) provided support for the perceptual load hypothesis in this respect. Presenting subjects with pairs of prime and probe displays, we assessed NP effects from prime distractors as a function of perceptual load in the processing of prime targets. Figure 7.1 displays the task we used in this study. Subjects searched for a target letter among a varying number of nontarget letters in the center of the prime display and ignored an irrelevant peripheral distractor, which was always incongruent with the prime target. NP from this distractor was found to depend on the relevant search set size, decreasing as this set size was increased.

Several experiments allowed us to rule out alternative accounts for this effect of load on NP. For example, the effect of load in our first experiment might be attributed to the greater similarity between prime and probe displays in the low-load versus high-load conditions (as they both involved the same number of items, with a relevant set size 1; see figure 7.1). In experiment 2, however, prime and probe similarity was greater in the high-load than in the low-load prime conditions: all the probes in experiment 2 involved a relevant set size of 6. The same result was found in both experiments, namely, less negative priming for conditions with a high perceptual load in the prime display. The two experiments taken together thus rule out any account of the results in terms of retrieval of episodic memory for the distractor, which can depend on the similarity between the prime and probe displays (see Fox and DeFockert 1998; Neil, 1997).

In addition, these experiments demonstrated that NP crucially depends on the level of perceptual load in the relevant processing for the *prime* displays, rather than on general task difficulty, as the same effect of prime load on NP was obtained regardless of the level of load in the probes. Finally, NP did not depend on reaction times overall (RTs; and their associated variability): it was obtained in all the conditions of low prime load even when their overall probe RTs were just as slow as those for our high prime loads (as was the case for experiment 2, which had high probe loads).

We conclude that high perceptual load in the relevant task reduces perceptual processing of distractors, hence protects the postperceptual processing of relevant stimuli from distractor intrusions. Moreover, because high perceptual load reduces response competition effects from distractors on concurrent targets (Lavie and Fox 2000, exp. 4; Lavie and Cox 1997)—as well as any NP in responses to subsequent targets—we think that distractor interference is reduced by high perceptual load in a rather passive manner, without requiring any active inhibition mechanisms such as those indicated by NP.

We conclude that with high perceptual loads, the reason distractors do not interfere is simply that they are not identified (i.e., early selection). On the other hand, a more active means of suppressing distractor responses (as indexed by NP effects; see Tipper and Milliken 1996) may become crucial in situations of low perceptual load, when distractors are processed

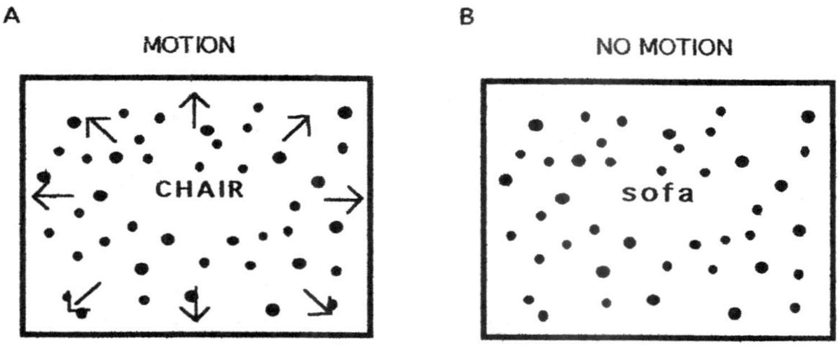

Figure 7.2 Example displays of the moving (panel A) or static (panel B) dot distractors from the functional imaging study by Rees et al. (1997). The stimuli were the same for low- and high-load conditions; only the task performed on the central stream of words differed. Comparing the fMRI response for moving versus static dot distractors allowed a measure of processing for irrelevant background motion.

and may thus compete to control behavior (i.e., late selection). The nature of these active control mechanisms will be discussed in section 7.2.

The Role of Perceptual Load in Determining Neural Activity for Distractors

Our perceptual load hypothesis raises some interesting predictions for the brain activity that should be produced by distractors. If a high load in relevant processing actually reduces irrelevant distractor *perception*, as we claim, then neural responses in sensory cortices associated with distractor perception should depend on the load imposed by the relevant task, even if that task is quite unrelated to the distractors in question. Specifically, we claim that brain activity for entirely irrelevant distractors should be found despite subjects' attempts to ignore them, provided the relevant task load is low. Activity to irrelevant distractors should only be reduced by higher load in the relevant task.

We recently tested these predictions using fMRI to assess the neural responses to moving distractors (Rees, Frith, and Lavie 1997; see figure 7.2). A stream of words was presented at fixation at a rate of one word per second. A full field of dots was presented in the periphery. These dots were either static (figure 7.1B) or moving (figure 7.1A), and subjects were requested to focus on the words and ignore the dots under two task conditions. In the low-load condition, subjects discriminated between lower- and uppercase letters in the word stream, and in the high-load condition, they discriminated between bisyllabic and mono- or trisyllabic words for the same streams.

The results confirmed our predictions exactly: motion related activity in cortical area V5/MT varied as a function of the word task. Activity in V5/MT for moving versus static dots was apparent in the case discrimination conditions (low load), but was eliminated in the syllable discrimination conditions (high load). This interaction between load and the neural responses to background motion was also found in other areas likely to be involved in motion perception, such as the V1/V2 border, and the superior colliculus (SC; Shipp and Zeki 1985; Ungerleider et al. 1984).[4] In sum, we found that a whole network of sensorimotor areas that are likely to be involved in motion perception were active in the presence of irrelevant motion distractors, provided that the relevant task involved only low load; but that this distractor-induced activity was then significantly reduced as load in relevant processing was increased.

In a further psychophysical experiment, we used the same task and displays as those used in our scanning experiment, while assessing the processing of irrelevant motion via the duration of the motion aftereffect it induced (see Chaudhuri 1991). We found that the duration of the motion aftereffect induced by the irrelevant moving distractors was significantly reduced in the syllable discrimination condition. These two experiments thus provide evidence for our claim that the relevant task load can decrease perception of irrelevant moving distractors. Additional evidence that attentional modulation of neural activity in early visual cortices is most apparent under high perceptual load comes also from some recent fMRI and single-cell studies (see Motter 1994; Kastner et al. 1998). It is important to note, however, that our current conclusion about the role of load in distractor processing, as determined by fMRI, is confined to the particular manipulation of load we used, and the specific type of moving distractors presented. As with our behavioral studies, additional experiments with different load manipulations as well as various types of distractors need to be run to allow us to reach a more definitive conclusion

The Role of Perceptual Load in the Normal Aging of Selective Attention

Elizabeth Maylor and I tested some implications of the perceptual load theory for the normal aging of attention (Maylor and Lavie 1998). It is often claimed (e.g., Ball et al. 1988) that aging can lead to a greater restriction in perceptual processing capacity. Because, according to our model, distractor processing depends on the amount of available processing capacity, we predicted that older adults should benefit more than younger adults from *smaller* increases in relevant perceptual load with respect to susceptibility to distractors. Because smaller increases in load should be needed to exhaust capacity for the elderly group.

To test this prediction, we compared the effect of a graded increase in perceptual load on distractor processing for 16 younger (aged 19–30) and 16 older (aged 65–79) adults. Subjects were presented with a relevant set of letters in the center of the display, and had to make speeded choices indicating which of two target letters was present among these relevant

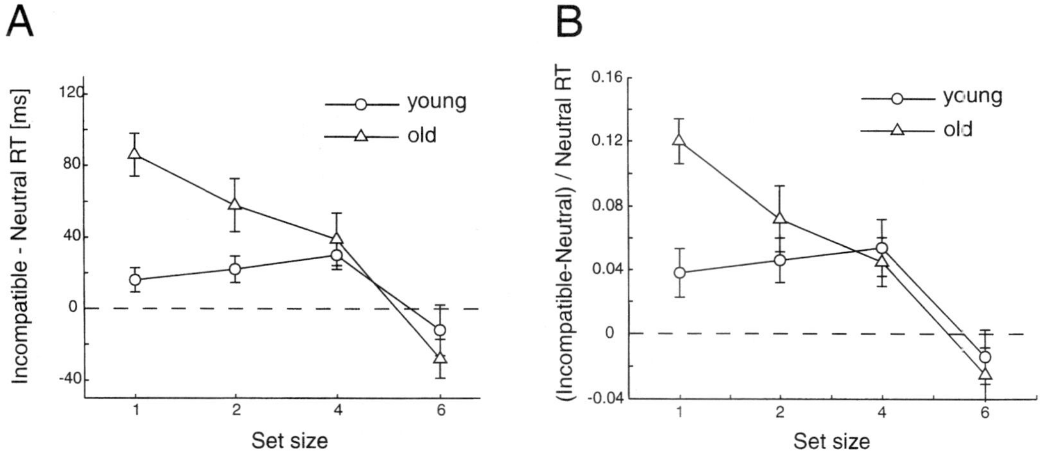

Figure 7.3 Distractor effects as a function of relevant set size and age group in experiment 1 of Maylor and Lavie 1998. A. Mean differences in reaction time between incongruent and neutral conditions, with standard error bars. B. Proportional differences in reaction time for these conditions.

letters, while attempting to ignore an irrelevant distractor in the periphery. This irrelevant distractor was either response incongruent or neutral with respect to the current target letter, to provide a response competition measure for distractor processing. Perceptual load in the relevant target processing was manipulated by varying the set size of the central array (i.e., by adding neutral nontarget letters).

The results (presented in figure 7.3) support our prediction. Although elderly subjects suffered from greater distraction in situations of very low load (i.e., with just 1 target and 1 distractor), very small increases in load (e.g., to just 2 relevant items, rather than only 1) were indeed sufficient for reducing distractor effects in the old but not in the young subjects. Thus older adults seem capable of benefiting from their greater restriction in the available capacity for perception to reduce irrelevant distractor processing at intermediate perceptual loads.

The finding of greater distraction for older versus younger adults at very low levels of load (i.e., relevant set size 1; see figure 7.3) cannot, however, be explained by such capacity limits in *perception* for the elderly because response compatibility effects indicate identification of the distractor and its associated response for both groups (as we might expect under situations of low perceptual load). Also, this larger distractor effect in the elderly cannot be explained by general slowing with age because the distractor effect at low load in the elderly was significantly larger than that for the young even when the effect was calculated as a proportion of the overall RTs for each population (figure 7.3).[5] Finally differential eye movements toward the distractor (e.g., Olincy et al. 1997) cannot account for this aging effect because the display durations used (100 msec) were too brief to allow eye movements. This effect seems therefore to reflect an additional age-related deficit in the ability to suppress irrelevant response tendencies to distractors when these do get processed (as at very low perceptual loads). The hypothesis that aging involves a specific decline in inhibitory control mechanisms (see Hasher and Zacks 1988) has received support from a number of previous studies. For example, it is often found that NP is reduced with age (e.g., Hasher et al. 1991; Kane 1994; McDowd and Oseas-Kreger 1991; but see Kramer et al. 1994; Sullivan and Faust 1993 for evidence of some age-related equivalence in negative priming). Thus we conclude that the normal aging of attention seems to involve (at least) two components. First, there is a decreased capacity for perception, which can actually lead to some improvement in *passive* selectivity: reduced processing of distractors as a natural consequence of perceptual capacity being more readily exhausted by relevant processing. Second, there is an additional age-related decline in the ability to *actively* reject distractors that do get processed, in situations of very low load. Thus aging also seems to involve a deficit in the mechanisms of active control.

7.2 ACTIVE CONTROL OF SELECTIVE ATTENTION

Our previous perceptual load studies support a simple account of an early selective attention mechanism that can prevent distractors from being perceived. We have provided substantial evidence for our claim that distractors are excluded from perception as a matter of course in situations of high perceptual load, which exhaust perceptual capacity in the relevant processing. This is a somewhat passive form of early selection: distractors do not interfere simply because they are not processed. A complete account of selective attention, however, also requires consideration of a more active form of selection, one that allows appropriate selective behavior even in situations of low perceptual load. In such situations, our results show that irrelevant as well as relevant stimuli are perceived, and thus can compete to guide behavior. Some late-selection mechanism is then needed to actively suppress responses to processed distractors, and thus ensure that behavior is appropriately controlled by relevant rather than irrelevant stimuli.

The importance of such active mechanisms of attentional control can be seen from the various "slips of action" that can occur if irrelevant response tendencies are not suppressed. While such failures of attention are relatively infrequent in young healthy adults, they become more pronounced in older adults (see Maylor and Lavie 1998; see also Hasher and Zacks 1988). Moreover, they can arise in extreme form for patients suffering from frontal lobe damage (see, for example, Shallice and Burgess 1991). Indeed, the greater distraction we found at *low* perceptual loads in older versus younger subjects might be explained by deterioration of the frontal lobes. Although aging involves a loss of cells in both posterior and anterior cortices, the greatest proportion of cell loss is frontal (e.g., Kramer et al. 1994). Moreover, frontal areas are known to be involved in various high-level cognitive processes, such as working memory (Baddeley 1986; D'Esposito, and Postle, chap. 26, this volume; Goldman-Rakic and Friedman 1991; Petrides, chap. 23, this volume), multiple task coordination (e.g., Burgess, chap. 20, this volume; Della Sala et al. 1995; Shallice and Burgess 1996), and inhibition of irrelevant responses (e.g., Foster, Eskes and Stuss 1994; Posner and DiGirolamo 1998; Tipper, Howard, and Houghton, chap. 10, this volume; but see Kimberg and Farah, chap. 32, this volume), all of which seem crucial for maintaining priorities between relevant and irrelevant stimuli, to guide behavior in accordance with current goals. Thus our functional distinction between early-selection and late-selection mechanisms of attentional control seems likely to map onto an anatomical distinction that has been made between posterior and anterior attention systems in the brain (e.g., Posner and Petersen 1990).

In drawing an analogy with Posner's general distinction between anterior and posterior mechanisms, however, I do not wish to imply that there is only one mechanism of frontal control. Evidence from imaging and lesion studies in monkeys and humans suggests several distinct control functions, with particular frontal areas being differentially involved (see D'Esposito and Postle, chap. 26, Keele and Rafal, chap. 28, Petrides, chap. 23, and Robbins and Rogers, chap. 21, this volume). My point is that selective attention, and in particular the ability to reject irrelevant distractors, might depend on all of these various control functions being intact (i.e., not lesioned, or not loaded).

Crucially, I propose that these two major psychological functions of attention can be distinguished by contrasting the effects of different types of load on selective processing. As described above, the exclusion of distractors improves with high perceptual load in the relevant task. However, a high load on the "frontal" processes important for cognitive control (e.g., working memory, task coordination) should lead to a deterioration of selective attention—an effect functionally similar to the effect of a frontal lobe lesion. This is because increased load on those anterior processes involved in cognitive control should leave less capacity for the active suppression of intrusions from irrelevant but perceived distractors into behavior. Thus the two major control functions of attention, namely, selective perception and control of response selection, should be distinguishable from one another by contrasting the effects of different types of load on distractibility. Increases in *perceptual* load should *decrease* distraction, by engaging perceptual capacity in the relevant processing. By contrast, increases in higher-level *cognitive control* load (at low perceptual load) should *increase* distraction, by engaging cognitive control mechanisms so that they become less able to block responses to perceived irrelevant distractors.

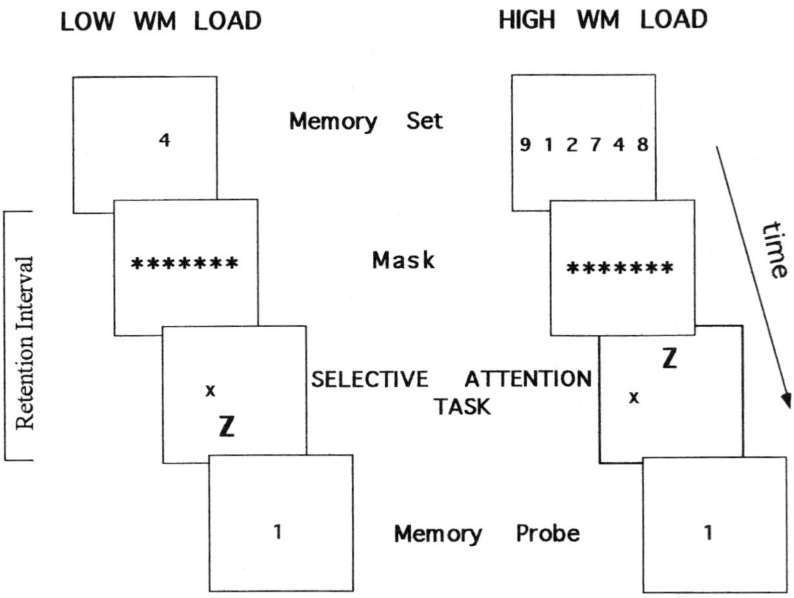

Figure 7.4 Example display sequences from single trials for the low-working-memory-load (panel A) and high-working-memory-load (panel B) conditions of experiment 1 in Hirst and Lavie 1998.

I now review evidence for these contrasting effects of different types of load on selective attention from a series of new studies conducted with my graduate student Sandra Hirst (Lavie et al. in preparation).

The Role of Working-Memory Load in Distractor Rejection for Selective Attention Tasks

Directing attention appropriately requires the active maintenance of goals and task priorities in working memory (WM), specifying which stimulus types are currently relevant, and which irrelevant. Accordingly, we reasoned that loading WM in a situation of low perceptual load should lead to reduced differentiation between high- and low-priority items (i.e., between targets versus distractors), and hence result in more intrusions from items that should have been given low priority. To manipulate WM load during a selective attention task, we developed the following new paradigm (figure 7.4). A selective attention task was interleaved with a WM task. Each trial began with a memory set (e.g., several visual digits) that subjects had to maintain in WM. The identity of the characters in the memory set differed on each trial, to ensure that active memorizing of items was required, so that any process of recency detection (e.g., Monsell 1978) that might be involved in recognition would still require active maintenance through rehearsal.

During the retention interval (which typically lasted for about 1.6 sec) a display for the selective attention task appeared (e.g., a central target

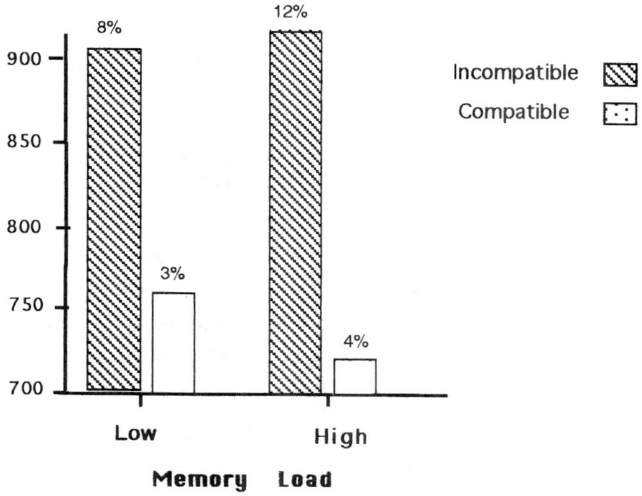

Figure 7.5 Mean correct reaction time and percentage error (in parentheses) for performance in the selective attention task, plotted as a function of the compatibility between concurrent target and distractor in that task, and also as a function of the working memory load in the interleaved task

letter for speeded discrimination, together with a flanking distractor letter). After a speeded-choice response was made in this selective attention task, a single memory probe then appeared, and subjects had to indicate whether it had been present in the memory set that began the trial. WM load was manipulated by the size of the memory set. In the low-WM-load condition, only one digit was present in the memory set for each trial. In the high-WM-load condition, six digits were presented in this set (see figure 7.4). Our prediction was that increasing WM load in this way should lead to greater distractibility in the unrelated selective attention task, by drawing on resources otherwise used to control selection in situations of low perceptual loads. Recall that, according to our model, distractors are always perceived in such situations, so that active control is required to prevent a response to them.

The results supported our prediction. As can be seen in figure 7.5, a greater distractor effect was found in the selective attention task with a high WM load (mean interference of 193 msec), than with a low WM load (mean of 140 msec).[6] Hence an increase in WM load can lead to increased distractibility, supporting our hypothesis that loading WM engages active mechanisms of attentional control, and therefore leads to a reduced ability to reject perceived distractors under low perceptual load. In further experiments, we replicated this effect of WM load for additional memory tasks (e.g., implementing recall procedures within our interleaved paradigm).

Note that, as predicted, WM load led to a result *opposite* to that typically found for perceptual load. As shown repeatedly earlier (see section 7.1), higher perceptual load *reduces* distractor interference, while here we found that higher WM load *increases* distractor interference. This contrast seems to confirm our distinction between the two control mechanisms of selective attention. To corroborate this, we checked that the usual perceptual load effect could still be found when the selective attention task was interleaved with a WM task, as in the new paradigm described above. Perceptual load was manipulated by varying the set size or the relevant central letters presented in each display for the selective attention task, while again interleaving this task with the WM task. WM load was now held constant (and low), with a memory set of just one item on each trial. As predicted, higher perceptual load again resulted in a decreased distractor effect (a distractor effect of 128 msec was obtained with low perceptual load, but one of only 11 msec with high perceptual load). These two experiments, with interleaved tasks, thus confirm that effects of perceptual and WM load on selective attention can indeed be dissociated within the same paradigm. Whereas perceptual load decreases the effects from irrelevant distractors, WM load increases these effects.

Note also that these experiments provide an entirely new form of evidence for the importance of active control mechanisms in attention. Many previous studies have shown that increasing the load on cognitive control functions can lead to a performance cost (e.g., to a greater dual-task decrement with a greater load in WM; Baddeley 1986). However, such an overall drop in performance is the typical result of *any* increase in load; for instance, higher perceptual load in a relevant task will also produce an overall decline in performance. Our approach differs from previous work on the loading of control processes because we specifically measure processing of *irrelevant distractors*, rather than merely overall performance in the *relevant* task. Thus our approach allows us to tie control processes more closely to specific functions of selective attention (i.e., to the rejection of perceived distractors in particular).[7]

While these experiments clearly showed greater distractor effects under high WM load, even under low WM load, levels of distractor effects were fairly high (provided that perceptual load was low). For example, compare the distractor effects produced in the conditions of low perceptual and low WM load of our experiment 1 (mean interference of 140 msec) and of our second study (128 msec) in this series, against the typical range of distractor effects found in all the studies from section 7.1, or in traditional studies of response competition effects from flanking distractor letters (where distractor interference effects typically range from 20 to 50 msec; see Lavie and Tsal 1994 for a comprehensive review). Why should the overall level of distractor interference be so much increased in our WM studies, even when WM was low? One likely reason is that our new paradigm required subjects to switch back and forth between the WM task and the selective attention task. As noted earlier, the coordination of multiple tasks has long been associated with frontal control processes (e.g., Della Sala et al. 1995; Shallice and Burgess 1996; see also Keele and Rafal, chap. 28, this volume; Robbins and Rogers, chap. 21, this volume). Accordingly, we hypothesized that the requirement to coordinate task switching between our interleaved WM and selective attention tasks may have loaded another anterior component of cognitive control, which would again impair the ability to reject perceived distractors in both low and high memory loads. The following subsection directly considers whether a task-switching requirement can impair the active rejection of perceived distractors.

Effects of Task Coordination on Distractor Rejection in Selective Attention Tasks

Much previous work has established the importance of anterior control functions in the coordination of multiple tasks. Patients suffering from frontal lesions are impaired at such coordination (e.g., Baddeley 1986; Shallice and Burgess 1996). Recent functional imaging studies also demonstrate the involvement of frontal areas in dual-task coordination within the normal brain (see Keele and Rafal, chap. 28, this volume). Finally, behavioral studies of normals have also highlighted the special demand that is posed by the requirement to coordinate two tasks rather than one. For example, Della Sala et al. (1995) and others have reported that the cost involved in coordinating two WM tasks versus performing one of them far exceeds the performance decrement caused by increasing the load in either one of the tasks alone. In this subsection, we test whether imposing a greater demand on task coordination can lead to greater failures of selective attention, by exhausting subjects' control capacity, and thus leaving them less able to reject perceived distractors.

Although the procedure described in the previous subsection involved an aspect of dual-task coordination (because a selective attention task was interspersed with a WM task), this was held constant across the experimental conditions, with only WM load or only perceptual load being varied within an otherwise constant setting of two interleaved tasks. We (Lavie et al. in preparation) now manipulated the requirement for dual-task coordination directly, while keeping memory set and perceptual load constant. Distractor processing in a selective attention task was measured as before, but was now compared between single- and dual-task situations.

We compared distractor effects between single- and dual-task conditions, in a similar paradigm to the one used in our previous WM experiments, except for two major changes. First, we now presented the memory probe *before* the display for the selective attention task, so that the entire WM task and the entire selective attention task now alternated, rather than the WM task spanning the interval during which the selective

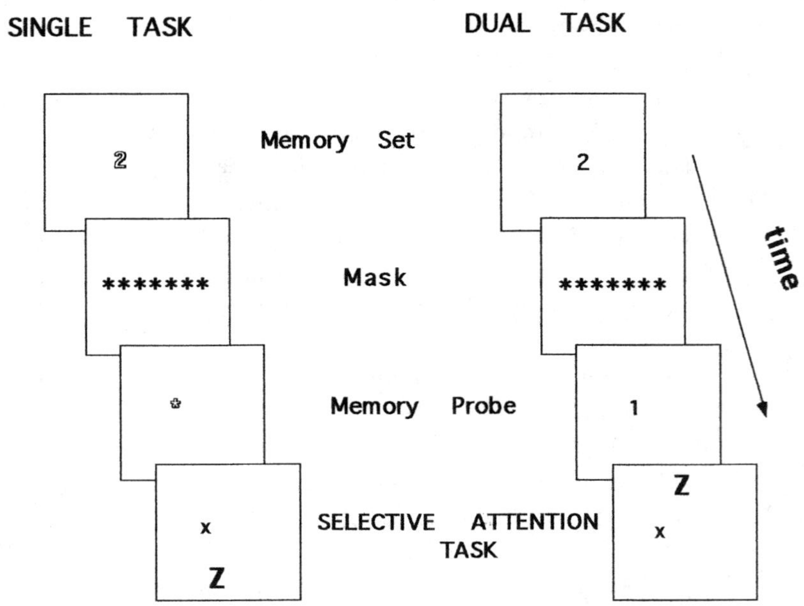

Figure 7.6 Example display sequences for individual trials from the single- and dual-task conditions of the experiment on task coordination. Note that task conditions were blocked and the display sequences were the same in all respects between the blocks except that the "memory probe" was always an asterisk in the single-task condition, and thus required no response.

attention task was performed. Second, we kept the memory set constant (with a set size of one item), and varied only whether subjects had to perform the attentional or both tasks.

Figure 7.6 presents the sequence of events and the experimental conditions used. In the dual-task condition, subjects were presented with a memory set, followed by a retention interval, and then a memory probe to which they had to respond. After the memory response had been made, and a further 1 sec had elapsed (blank except for a fixation dot), the display for the selective attention task appeared. This again required subjects to make a speeded-choice response to a central letter, while any compatibility effects were measured from an irrelevant flanking letter to provide a measure of distractor interference. In the single-task condition, a similar sequence of events was used, except subjects did not have to make any response to the "memory" probe (which was now simply an asterisk on every trial). As the two conditions were presented in separate blocks, the subjects presumably made no attempt to memorize the digit in the single-task blocks.

We predicted more distractor interference for the selective attention task in the dual-task condition. Even though the memory set no longer had to be maintained in WM while performing the selective attention task, we expected that the requirement to alternate between the WM and selective attention tasks would load the anterior control processes that coordinate task switching, and thus disrupt the ability to actively suppress perceived distractors. This prediction was confirmed. The distractor effect of 63 msec (M = 704 for incongruent RTs; M = 641 for congruent RTs) in the single-task condition was significantly increased to 90 msec in the dual-task condition (M = 781 for incongruent RTs; M = 691 for congruent RTs); and error rates were increased from 3% to 7% in the single- versus dual-task conditions. This experiment confirms that task coordination is another important component in the active control of selective attention. Note that, once again, the loading of anterior control processes is found to have the opposite effect to increases in perceptual load, leading to greater rather than less distractibility.

7.3 CONCLUSIONS

In summary, the work presented here establishes a distinction between active late-selection mechanisms, and passive early-selection mechanisms for the control of selective attention, and demonstrates the important role of relevant processing load in determining the extent of distraction by irrelevant information. In our work on late-selection mechanisms of active attentional control, we have started to lay out in greater detail the involvement of specific anterior control functions in determining distractibility for selective attention tasks. This work already indicates the importance of control functions loaded by working memory in selective attention, and also provides some preliminary evidence that control functions involved in coordinating multiple tasks may also play a crucial role. Future work should further specify the nature of these control functions, and determine whether other components of anterior control are similarly involved in distractor rejection for selective attention tasks. Finally, working out whether the distinction between passive and active mechanisms of selective attention can be related, respectively, to posterior and anterior attentional networks in the brain, should provide further insights into the influences of control processes on selective attention.

NOTES

This work was supported by Medical Research Council (U.K.) grant G9805400. I thank Jon Driver, Art Kramer, Stephen Monsell, and Steven Yantis for their helpful comments.

1. My usage of the term *perception* throughout this chapter follows the conventional usage in the early- versus late-selection debate, namely, referring to processes that lead to stimulus identification. From this perspective, elaborative semantic activation, memory, response selection, and response execution are conceived as postperceptual processes. See Pashler 1989 and Pashler and Johnston 1989 for discussion of distinctions between perception, in this sense, and later processes.

2. Situations of high perceptual load will result in selective perception even if the correct set cannot be actively maintained. Selection in such cases may not be the correct one, however,

which is to say, it may not follow the appropriate attentional set (e.g., some irrelevant stimuli may be perceived *instead* of some relevant stimuli).

3. The term *nontargets* is used to refer to stimuli presented for the relevant task in central positions that could contain the target.

4. Although SC has been implicated in oculomotor control, eye movements do not provide a plausible explanation for these findings: no significant eye movements were found during the performance in the experimental conditions (when measured outside the scanner). For a full discussion of this, see Rees, Frith, and Lavie (1997).

5. It should be noted that our analysis of proportional RTs can only discount linear general slowing models. For a more detailed treatment of general slowing accounts, see Maylor and Lavie (1998).

6. Although it might appear from figure 7.4 that the increased distractor effect with high memory load was due to reduced RTs in the congruent condition, this simple effect was not statistically significant. Moreover, our additional WM and selective attention experiments typically showed a WM load effect on both incongruent and congruent RTs.

7. The opposite effects that different types of load have on distractibility allow us to rule out alternative accounts of the effect on distractibility from each type of load alone. For example, at the meeting Daniel Gopher suggested that the result of better selectivity obtained with high perceptual load might be due to subjects increasing their motivation for selective processing when anticipating a difficult trial. If this were the case, we should presumably have also found better selectivity with higher WM load because this also led to a substantial increase in task difficulty. Any account of load effects in terms of general task difficulty thus seems insufficient.

REFERENCES

Baddeley, A. D. (1986). *Working memory*, Oxford: Oxford University Press.

Baddeley, A. D., Bressi, S. Della Sala, S., Logie, R., and Spinnler, H. (1991). The decline of working memory in Alzheimer's disease: A longitudinal study. *Brain, 114,* 2521–2542.

Ball, K. K., Beard, B. L., Roenker, D. L., Miller, R. L., and Griggs, D. S. (1988). Age and visual search: Expanding the useful field of view. *Journal of the Optical Society of America A5,* 2210–2219.

Chaudhuri, A. (1991). Modulation of the motion aftereffect by selective attention. *Nature, 344,* 60–62.

Della Sala, S., Baddeley, A., Papagno, C., and Spinnler, H. (1995). Dual-task paradigm: A means to examine the central executive, structure and functions of the human prefrontal cortex. *Annals of the New York Academy of Sciences, 769,* 161–171.

Driver, J., and Tipper, S. P. (1989). On the nonselectivity of "selective" seeing: Contrasts between interference and priming in selective attention. *Journal of Experimental Psychology: Human Perception and Performance, 15,* 304–314.

Eriksen, B. A., and Eriksen, C. W. (1974). Effects of noise letters upon the identification of a target letter in a non search task. *Perception and Psychophysics, 16,* 143–149.

Fisher, D. L. (1982). Limited-channel models of automatic detection: Capacity and scanning in visual search. *Psychological Review, 89,* 662–692.

Foster, J. K., Eskes, G. A., and Stuss, D. T. (1994). The cognitive neuropsychology of attention: A frontal lobe perspective. *Cognitive Neuropsychology, 11,* 133–147.

Fox, E., and De Fockert, J. W. (1998). Negative priming depends on prime-probe similarity: Evidence for episodic retrieval. *Psychonomic Bulletin and Review, 5,* 107–113.

Goldman-Rakic, P. S., and Friedman, H. R. (1991). The circuitry of working memory revealed by anatomy and metabolic imaging. In H. S. Levin, H. M. Eisenberg, and A. L. Benton, (Eds.), *Frontal lobe function and dysfunction,* pp. 72–91. New York: Oxford University Press.

Hasher, L., Stoltzfus, E. R., Zacks, R. T., and Rypma, B. (1991). Age and inhibition. *Journal of Experimental Psychology: Learning, Memory and Cognition, 17,* 163–169.

Hasher, L., and Zacks, R. T. (1988). Working memory, comprehension, and aging: A review and a new view. In G. H. Bower (Ed.), *The psychology of learning and motivation,* vol. 22, pp. 193–225. New York: Academic Press.

Kane, M. J., Hasher, L., Stoltzfus, E. R., Zacks, R. T., and Connelly, S. L. (1994). Inhibitory attentional mechanisms and aging. *Psychology and Aging, 9,* 103–112.

Kahneman, D., and Treisman, A. (1984). Changing views of attention and automaticity. In R. Parasuraman and D. R. Davies (Eds.), *Varieties of attention,* pp. 29–61. New York: Academic Press.

Kahneman, D., Treisman, A., and Gibbs, B. (1992). The reviewing of object files: Object-specific integration of information. *Cognitive Psychology, 24,* 175–219.

Kastner, S., De Weerd, P., Desimone. R., and Ungerleider, L. G. (1998). Mechanisms of directed attention in the human extrastriate cortex as revealed by functional MRI. *Science, 282,* 108–111.

Kramer, A. F., Humphrey, D., Larish, J. F., Logan, G. D., and Strayer, D. L. (1994). Aging and inhibition: Beyond a unitary view of inhibitory processing in attention. *Psychology and Aging, 9,* 491–512.

Lavie, N. (1995). Perceptual load as a necessary condition for selective attention. *Journal of Experimental Psychology: Human Perception and Performance, 21,* 451–468.

Lavie, N., and Cox, S. (1997). On the efficiency of attentional selection: Efficient visual search results in inefficient rejection of distraction. *Psychological Science, 8,* 395–398.

Lavie, N., and Fox, E. (2000). The role of perceptual load in negative priming. *Journal of Experimental Psychology: Human Perception and Performance, 26.*

Lavie, N., Hirst, S., and Colledge, E. (In preparation). The role of working memory load in control of selective visual attention.

Lavie, N., and Tsal, Y. (1994). Perceptual load as a major determinant of the locus of selection in visual attention. *Perception and Psychophysics, 56,* 183–197.

Maylor, E., and Lavie N. (1998). The influence of perceptual load on age differences in selective attention. *Psychology and Aging, 13,* 563–573.

McDowd, J. M., and Oseas-Kreger, D. M. (1991). Aging, inhibitory processes, and negative priming. *Journal of Gerontology: Psychological Sciences, 46,* 340–345.

Monsell, S. (1978). Recency, immediate recognition memory and reaction time. *Cognitive Psychology, 10,* 465–501.

Motter, B. C. (1994). Neural correlates of attentive selection for color or luminance in extrastriate area V4. *Journal of Neuroscience, 14,* 2178–2189.

Neill, W. T. (1997). Episodic retrieval in negative priming and repetition priming. *Journal of Experimental Psychology: Learning, Memory, and Cognition, 18,* 565–576.

Olincy, A., Ross, R. G., Youngd, D. A., and Freedman, R. (1997). Age diminishes performance on an antisaccade eye movement task. *Neurobiology of Aging, 18,* 483–489.

Pashler, H. (1989). Dissociations and dependencies between speed and accuracy: Evidence for a two-component theory of divided attention in simple tasks. *Cognitive Psychology, 21,* 469–514.

Pashler, H., and Johnston, J. C. (1989). Chronometric evidence for central postponement in temporally overlapping tasks. *Quarterly Journal of Experimental Psychology, 41A,* 19–45.

Posner, M. I., and DiGirolamo, G. J. (1998). Conflict, target detection and cognitive control. In R. Parasuraman (Ed.), *The attentive brain.* Cambridge, MA: MIT Press.

Posner, M. I., and Petersen, S. E. (1990). The attention system of the human brain. *Annual Review of Neuroscience, 13,* 25–42.

Pylyshyn, Z., Burkell, J., Fisher, B., and Sears, C. (1994). Multiple parallel access in visual attention. *Canadian Journal of Experimental Psychology, 48,* 260–283.

Rees, G., Frith, C., and Lavie, N. (1997). Modulating irrelevant motion perception by varying attentional load in an unrelated task. *Science, 278,* 1616–1619.

Rogers, R., and Monsell, S. (1995). Costs of a predictable switch between simple cognitive tasks. *Journal of Experimental Psychology: General, 124,* 207–231.

Salthouse, T. A. (1992). *Mechanisms of age-cognition relations in adulthood.* Hillsdale, NJ: Erlbaum.

Shallice, T., and Burgess, P. (1991). Deficits in strategy application following frontal lobe damage in man. *Brain, 114,* 727–741.

Shallice, T., and Burgess, P. (1996). The domain of supervisory processes and temporal organization of behaviour. *Philosophical Transactions of the Royal Society of London, 351,* 1405–1412.

Shiffrin, R. M., and Schneider, W. (1977). Controlled and automatic human information processing: 2. Perceptual learning, automatic attending and a general learning theory. *Psychological Review, 84,* 127–190.

Shipp, S., and S. Zeki, (1985). Segregation of pathways leading from area V2 to areas V4 and V5 of macaque monkey visual cortex. *Nature, 315,* 322–324.

Sullivan, M. P., and Faust, M. E. (1993). Evidence for identity inhibition during selective attention in old adults. *Psychology and Aging, 8,* 589–598.

Tipper, S. P., and Milliken, B. (1996). Distinguishing between inhibition-based and episodic retrieval-based accounts of negative priming. In A. F. Kramer, M. G. H. Coles, and G. D. Logan (Eds), *Converging operations in the study of visual selective attention,* pp. 77–106. Washington, DC: American Psychological Association.

Treisman, A. M. (1969). Strategies and models of selective attention. *Psychological Review, 76,* 282–299.

Ungerleider, R., Desimone, T. W. Galkin, and M. Mishkin, (1984). Subcortical projections of area MT in the macaque. *Journal of Comparative Neurology, 223,* 368.

Yantis, S., and Johnston, J. C. (1990). On the locus of visual selection: Evidence from focused attention tasks. *Journal of Experimental Psychology: Human Perception and Performance, 16,* 135–149.

Yantis, S., and Jones, E. (1991). Mechanisms of attentional selection: Temporally modulated priority tags. *Perception and Psychophysics, 50,* 166–178.

8 Relations among Modes of Visual Orienting

Raymond M. Klein and David I. Shore

ABSTRACT Selective processing of the visual world is accomplished through overt shifts of gaze direction and covert shifts of attention. Such visual orienting, whether overt or covert, can be controlled exogenously by environmental stimulation, or endogenously by the observer's intentions. The modes of orienting implied by these distinctions may interact cooperatively or competitively. We illustrate, with examples often drawn from the chapters in this section, (1) the interactions between endogenous and exogenous control over overt and covert orienting, and (2) the relations between overt and covert orienting and between endogenous and exogenous control.

8.1 MODES OF VISUAL ORIENTING

Visual orienting is a set of processes used to give a region of space and the objects in it preferential access to the visual and cognitive routines that control behavior. The need for eye movements (overt orienting) is apparent when one considers that primate vision is characterized by a relatively small central area of fine resolution that must be aligned with potentially important targets. In contrast, a shift of attention (covert orienting) involves an internal selection, accomplished without an overt shift of gaze, whereby some regions or objects in space are given processing preference over others. Overt and covert orienting can be directed by environmentally generated inputs (spatially asymmetric stimulation to the visual, auditory or tactile modalities) or by observer-generated signals (based on momentary intentions or enduring dispositions). Following Posner (1980; see also Klein, Kingstone, and Pontefract 1994), we will refer to these sources of control as "exogenous" (coming from outside the organism, this source is also referred to as "bottom-up" and characterized as "reflexive" or "automatic") and "endogenous" (coming from within the organism, this source is also referred to as "top-down" and characterized as "voluntary" or "strategic"). Working within the 2 × 2 matrix implied by these two distinctions (figure 8.1), this chapter will explore several issues dealing with the control of orienting.

A high degree of coordination between these modes of orienting characterizes everyday behavior. A compelling demonstration is provided by Yarbus (1967) who, in one study, presented observers the same stimulus

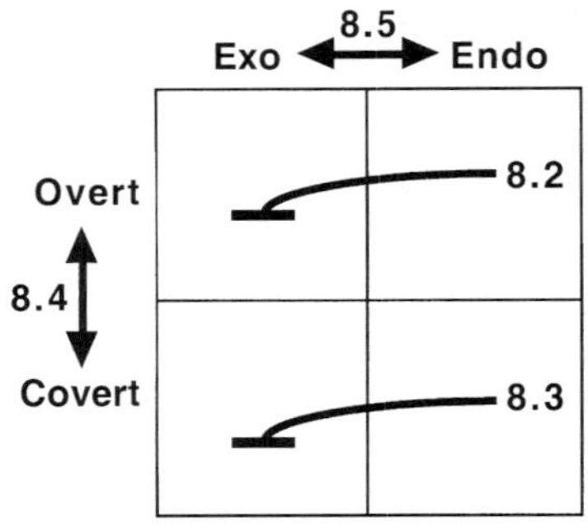

Figure 8.1 Modes of orienting and the relations among them (↔) and competitive interactions between them (⊣—) discussed in the chapter sections (indicated here by number).

(the painting *The Unexpected Visitor*, by I. E. Repin) repeatedly and each time asked them a different question. Their oculomotor scanning behavior was dramatically affected by the question they were trying to answer. Thus, while the exogenous input remained the same, knowledge-driven endogenous control of overt orienting was used to select those regions where the answers might be found (faces for emotional expression; objects and clothes for material circumstances; etc.). Although Yarbus did not use an independent method to verify which regions were being attended, because the regions fixated were those containing the most useful information for answering the questions, it seems reasonable to assume that the path of attention and the path of fixations would have been highly correlated. Whereas the modes of orienting appear cooperative in Yarbus's study, dissociations between and competitive interactions among them have been well demonstrated, as will be illustrated in the sections that follow.

8.2 ENDOGENOUS VERSUS EXOGENOUS OVERT ORIENTING

Within the oculomotor machinery there is a network of gating mechanisms in the brain stem and superior colliculus whose activation can prevent overt orienting (Everling et al. 1998). Removal of a fixated stimulus provides an exogenous signal to deactivate this inhibitory gating mechanism, and when such removal precedes the appearance of a target with a delay sufficient to allow endogenous preparatory processes to become active, extremely rapid ("express") saccades can be initiated (Fischer and Ramsperger, 1984; Kingstone and Klein 1993). Because natural scene components rarely disappear, to permit overt orienting in the presence of a fixated stimulus, an internally generated disinhibitory signal must be hypothesized. Evidence for such an "endogenous oculomotor disengagement" from fixation has been reported by Taylor, Kingstone, and Klein (1998). Similarly, Rafal et al. (chap. 6, this volume) found that when a stimulus driven saccade is highly probable, the retarding effect on saccadic latencies of the fixation stimulus is reduced (their experiment 1). This suggests that endogenous disengagement from the fixation stimulus is part of the preparation for overt orienting.

A compelling interaction between exogenous and endogenous control of overt orienting was reported by Theeuwes et al. (1998). Initially, 6 gray objects were placed around an imaginary circle; when five of the objects changed to red, subjects were to foveate the remaining gray object in order to identify a target contained within it. Although the singleton nature of the target object made selection unambiguous, the exogenous system might not be optimally engaged because there was no change in that item. On one-half of the trials, an irrelevant, red item was added to the array at the same time the target location was revealed. Overall, responses were slowed by the appearance of this new object; more important, subjects executed a large proportion of eye movements toward this irrelevant distractor despite their intention to move to the target. These errors were often rapidly redirected toward the target, and the observers were usually not aware of their overt orienting errors. The authors propose that two eye movements were programmed in parallel—one, endogenously, to move to the location of the target; the other, exogenously elicited by the abrupt onset—and that when the incorrect response toward the distractor was launched by the exogenous system it was quickly inhibited and overwritten by endogenous control. Left with the intended result of the final fixation, subjects were unaware of the exogenously generated behavior. This reinforces prior evidence (Kaufman and Richards 1969) that we are often not aware of where our eyes have been.

Although these results are reminiscent of the antisaccade task (Everling and Fischer 1998; Forbes and Klein 1996), where a stimulus presented in one location instructs observers to move their eyes in the opposite direction, one important difference is that in the antisaccade task the exogenous signal is task relevant until its location has been encoded and inverted. Although one might expect that this would give the onset stimulus greater salience and that errors would thus be higher in the antisaccade task than in the distractor paradigm, this is not the case. We think this is due in part to the timing of oculomotor disengagement. In the antisaccade task, endogenous release from fixation is likely delayed until the exogenous signal has been converted to an endogenous command. In the distractor paradigm, the color change in the displayed items, revealing the singleton target, is likely used to initiate disengagement from fixation, which would thus occur in close temporal proximity to the onset distractor. To the reflexive machinery in the superior colliculus, the distractor is much more salient than the unchanging target, and therefore it frequently captures control of overt orienting. Exploring the performance of patients with damage to different cortical systems on the antisaccade task, Rafal et al. (chap. 6, this volume) noted the importance of the frontal eye fields and posterior parietal cortex (particularly the parietotemporal junction) in successful performance in this task (their experiment 2). We think that exciting new data will be generated by examining the performance of patients, such as those used by Rafal et al., on the distractor paradigm of Theeuwes et al. (1998).

8.3 ENDOGENOUS VERSUS EXOGENOUS COVERT ORIENTING

The interaction described above for overt orienting is paralleled in studies of covert orienting and highlighted in the debate concerning the extent to which salient exogenous signals (e.g., abrupt onsets or singletons) capture attention or can be ignored when they are task irrelevant (see Yantis, chap. 3, this volume, and Theeuwes, Atchley and Kramer, chap. 4, this volume, for reviews). Whereas early studies (Jonides and Yantis 1988; Yantis and Jonides 1984, 1990) suggested that abrupt onsets capture attention, this proposal has been modified in three distinct ways. First, Theeuwes and colleagues (see also Joseph and Optican 1996) have maintained that attention is captured by any salient stimuli (at least initially). In support of this conclusion, they have repeatedly shown that when searching for a singleton (color or shape) the presence of a second unique singleton (defined on a different dimension) retards search performance. Second, others (Folk, Remington, and Johnston 1992; Folk and Remington 1998) have claimed that exogenous attention can be prevented from orienting to an irrelevant distractor if an appropriate attentional control setting (ACS) is instantiated. ACSs are conceptualized as endogenously generated rules that determine which exogenous signals will result in orienting. Hence, if the unique singleton is task irrelevant, orienting can be avoided. The findings that led to the ACS proposal provide a powerful demonstration of the ability of endogenous control to modulate, even countermand, exogenous orienting. Third, Yantis (chap. 3, this volume) has also considered attentional control settings to be a primary determinant of exogenous features that might capture attention, with one proviso—new objects have a special status in their ability to capture attention whether or not they appear as abrupt onsets (Yantis and Hilstrom 1994).

Yantis responds to the finding that singleton distractors attract attention even when they are in a dimension (e.g., color) different from that of the target (e.g., form) by noting that in these tasks the observers may be adopting a "singleton detection mode" (cf. Bacon and Egeth 1994). While this may be true, we feel that an important question remains unanswered: What is the default control setting? Schmidt (1994; McColl and Schmidt 1995) approached this question by avoiding a search task altogether. He asked whether a singleton in an otherwise homogeneous visual array would support the phenomenon of illusory line motion (ILM). If a line, presented all at once, is preceded by a cue adjacent to one end of the line, the observer perceives the line drawn on the screen away from the cue (Hikosaka, Mayauchi, and Shimojo 1993). When Schmidt presented an array of stimuli around fixation, ILM was observed away from an orientation singleton in the array. Because there was no task to perform with the array, an interpretation in terms of a "singleton detection mode" seems unlikely. Instead, it seems that the singularity in the visual array creates an "attractor" region. All other things being equal (the default setting), in paradigms where attention is labile, this region is more likely than any other region to attract attention.

The empirical conflict between studies supporting salience-driven orienting (exogenous control) to irrelevant distractors and successful filtering via ACS (endogenous control) might be resolved by considering methodological differences. Theeuwes, Atchley, and Kramer (chap. 4, this volume; Theeuwes et al. 1998) show that the identity contained within an irrelevant singleton is processed automatically when it appears at the same time as the relevant singleton containing the target; whereas the irrelevant identity does not affect performance when it precedes the target by 200 msec. Because 200 msec is close to the cue-target interval (150 msec) used in the studies by Folk and colleagues, those studies do not directly challenge the salience-driven attention hypothesis.[1] Neurophysiological data (Chelazzi et al. 1993; see also Desimone and Duncan 1995) show that the initial response of neurons in inferotemporal cortex to a stimulus is uninfluenced by task-relevance, whereas the cell's subsequent (100–200 msec after target onset) response rate is much higher to the task-relevant stimulus. This converges with Theeuwes's proposal that some early exogenously driven processes can be immune to endogenous control. Behaviorally speaking, there are two time course issues that warrant investigation: How much forewarning does a subject require to establish an ACS in order to avoid distraction from an irrelevant singleton? And how soon after the target does the distractor have to be presented to interfere with ongoing processing?

It is interesting to consider how the ACS concept might be related to concepts developed within the task-switching literature (see Allport, Styles, and Hsieh 1994; Rogers and Monsell 1995; see also Allport and Wylie, chap. 2, this volume and Pashler, chap. 12, this volume). When an observer is searching an array for a target item, and the array is preceded by an uninformative cue, it seems reasonable to assume that the task will be accomplished by instantiating rules for finding the target and for ignoring the irrelevant cue. Nonconflicting rules (e.g., ignore color, attend onset; ignore onset, attend color) can be maintained in parallel, hence one should be able to avoid reflexive orienting. Because conflicting rules (ignore onset, attend onset), cannot be maintained in parallel, however, switching between them may take considerable time. Hence, with conflicting rules and short cue-target stimulus onset asynchronies (SOAs) the observer may merely maintain the rule necessary to find the target, in which case the irrelevant cue necessarily attracts attention.

The control exerted in order to implement ACS (whereby orienting toward task-irrelevant singletons is inhibited) may be similar to the control hypothesized by guided search (Wolfe, Cave, and Franzel 1989; Treisman and Sato 1990) to be exerted against distractors sharing a nontarget attribute in some conjunction search tasks. It would be interesting to use a neuropsychological or individual differences approach to obtain evidence for or against such an association. For example, the work of Kingstone et al. (1995), showing that the left, but not the right, hemisphere can implement guided search, might be extended by looking at ACS in the left and right hemisphere of the split-brain subject. Or it might be found that aging disrupts both ACS and guided search while leaving other attention functions relatively unaffected.

8.4 RELATIONS BETWEEN COVERT AND OVERT ORIENTING

One question that follows from our ability to attend where we are not looking was posed during the discussion by David Meyer: What is the relationship between overt and covert orienting? With exogenous orienting, there is a consensus that overt and covert orienting are strongly linked, perhaps because the same kinds of stimuli that tend to attract attention also activate the oculomotor system. In contrast, the literature on overt and covert relations with endogenous orienting is characterized by contradictory claims as to whether covert orienting is prepared, but unexecuted, overt orienting.

Following a long tradition in psychology suggesting that motor plans play an important role in perception, Klein (1980) proposed that endogenous shifts of attention are accomplished by oculomotor preparation to fixate the location to be attended—an idea captured in Rizzolatti et al.'s "premotor" theory (1987). When a saccade is executed, whether under endogenous or exogenous control, the gaze shift is preceded by a shift of attention toward the location to be fixated (cf. Posner 1980; Hoffman and Subramaniam 1995; Shepard, Findlay, and Hockey 1986). Although often taken as evidence for the idea that oculomotor readiness mediates endogenous covert orienting, this finding is actually not pertinent. Covert orienting is a shift of attention without a shift in gaze. By definition, a mechanism that could only shift attention when the eyes moved could not be responsible for covert orienting. If the oculomotor readiness proposal were true, then when a saccade is prepared, but *not* executed, there should be a corresponding attention shift, and conversely, when there is an endogenously generated attention shift, a corresponding saccade should be prepared. Direct tests of these predictions (Klein 1980; Klein and Pontefract 1994; Ennis and Kingstone 1998) seemed to disconfirm the oculomotor readiness proposal for endogenous covert visual orienting (and the similar premotor theory).[2]

Sheliga et al. (1994) found that endogenously generated probe saccades are biased away from a covertly attended location, whereas Kustov and Robinson (1996) seem to have found the opposite with electrically elicited saccades (for a review, see Klein, in preparation). Both teams claimed that their biased saccades provide evidence that covert endogenous orienting was accomplished by the endogenous preparation of overt orienting. The Sheliga et al. (1994) pattern can be explained, however, by assuming that there is a natural tendency to look where one is attending and that the instruction to attend covertly (without making an eye movement) causes the tendency to be inhibited, thus deflecting probe saccades away from the attended location. In this account, neither the natural tendency nor the inhibition would be responsible for causing the covert shift of attention. Evolution of the ability to endogenously attend without overtly looking would have required inhibitory control over the natural tendency to gaze at the attended object, as Rafal et al. (chap. 6, this volume) and Klein (in preparation) have argued. Kustov and Robinson's (1996) study, which comes the closest to providing evidence in favor of Klein's original proposal (1980), contains a serious confound: cues to attend spatially also indicated whether a right- or a left-limb response was likely to be required, which creates two ambiguities. Because the cuing effect could be due to motor preparation rather than to visual orienting, we cannot be sure that the endogenous cue elicited a shift of attention. Even if there were a shift of attention, it could not be confidently determined whether the effect on saccades was due to this shift or to the preparation of the likely manual response. The confound precludes firm conclusions about the oculomotor readiness hypothesis. This clever experimental test should certainly be repeated with this confound removed. If it is found that electrically elicited saccades are biased in the direction of attention, the oculomotor readiness proposal will be strongly supported.[3]

8.5 RELATIONS BETWEEN ENDOGENOUS AND EXOGENOUS ORIENTING

With respect to overt orienting, endogenous and exogenous signals control the same thing: where the fovea is directed. With respect to covert orienting, it is often assumed that they control the same attentional system, however, paraphrasing a remark by Nancy Kanwisher during the discussion, we might note that calling two processes "attention" does not make them the same and ask, Is there evidence linking or dissociating the attention oriented via endogenous and exogenous means?

Using central and peripheral cues to direct attention, Jonides (1976, 1981) provided the earliest evidence of differences between endogenous and exogenous control of covert orienting. He showed that covert orienting was faster under exogenous control, and that endogenous, but not

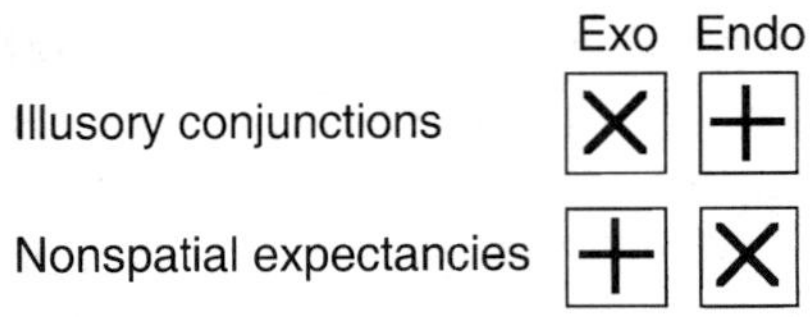

Figure 8.2 Whether covert orienting adds (+) or interacts (×) with opportunities for illusory conjunctions and with nonspatial expectancies depends on the type of control (exogenous or endogenous).

exogenous, control was sensitive to cognitive load and to the relative probabilities of the two types of cues. These important differences are consistent with a reflexive versus voluntary distinction (Müller and Rabbitt 1989). Most investigators have assumed that the attentional mechanisms brought by endogenous or exogenous control to a region of or object within space are the same; all that differs is how attention is "transported" to its spatial destination. In contrast, a behavioral double dissociation (see figure 8.2) we will briefly describe (see also Klein, Kingstone, and Pontefract 1992) suggests that these two types of "attention" might be fundamentally different.

According to Treisman's "feature integration theory" (FIT; Treisman and Gelade 1980), attention is required to correctly "glue" together the features present in a region that belong to an object. Briand and Klein (1987) tested whether feature integration was among the functions performed by the attention system recruited by an informative precue. They combined the Posner cuing paradigm with tasks where the target (the letter **R**) could be discriminated from the distractors (**P**, **B**; feature task) by a single feature (slanted line) or where the correct conjoining of features was required because of the possibility of an illusory conjunction from the distractors (**P**, **Q**; conjunction task). With exogenous orienting in response to an informative peripheral cue, there was a larger cuing effect for the conjunction task than for the feature task (see also Prinzmetal, Presti, and Posner 1986; and Treisman 1985), whereas with endogenous orienting in response to a similarly informative central cue, both tasks showed similar cuing effects. Briand (1998) recently replicated and extended this pattern, notably by using uninformative peripheral cues, features from different dimensions (form and color), and a range or cue-target SOAs. Thus the answer to the question posed by Briand and Klein's title (1987) "Is Posner's Beam the Same as Treisman's 'Glue'?" is yes, for the exogenous beam, and no, for the endogenous beam. This was the first dissociation between these modes of control (see figure 8.2) that suggested something more fundamental than how attention gets to its destination might differentiate exogenous from endogenous covert orienting.

Suppose there are several possible targets that might appear, and that one target type (form, color, orientation) is much more likely to occur than the other, causing the observer to generate a nonspatial expectancy. When a location is now cued, will covert orienting interact with the nonspatial expectancy? The findings obtained by Klein (1994; Klein and Hansen 1990), which have been replicated and extended by Kingstone and Egly (in preparation), reveal additivity with exogenous orienting and an interaction with endogenous orienting (see figure 8.2). This suggests that nonspatial expectancies and endogenously controlled orienting (spatial expectancies) involve overlapping mechanisms or stages of processing. In contrast, the effects on processing that follow exogenous orienting elicited by a peripheral cue are independent of those associated with the nonspatial expectancy.

Although it is possible to infer from this double dissociation that different types of "attention" are being oriented by endogenous and exogenous means (see Klein 1994; Briand 1998), one need not go so far. It is possible to assume that a common attention system is oriented in response to endogenous and exogenous signals, so long as one assumes that in addition unique stages of processing are affected by exogenous and endogenous orienting. Both systems operate relatively early on feature encoding or extraction stages. The evidence reviewed by Hopfinger et al. (chap. 5, this volume) showing amplification of the event-related potential within about 100 msec of stimulus onset, and Hopfinger et al.'s isolation of this modulation to sources in extrastriate cortex strongly suggest that endogenous orienting can involve early amplification of the sensory signals that might give rise to exogenous orienting. Interactions between endogenous and exogenous control (as implied by ACS, and discussed above) may also arise at this stage of operation. Exogenous control, perhaps because it typically entails visual information for peripheral pattern-recognizing routines to analyze, interacts with opportunities for illusory conjunction, suggesting that it plays a role in feature binding.[4] In contrast, endogenous control involves pigeonholing operations at the decision stage (cf. Broadbent 1971), which would interact with other, nonspatial expectancies that may be similarly implemented.

There are several examples of covert orienting, two of which we will mention here, that seem to have features associated with both exogenous and endogenous control. First, as noted earlier, when an eye movement is made under endogenous control, attention is drawn to the location to be foveated before the eyes get there. This shift of attention appears obligatory because the uniform distribution of probe stimuli (which were used to determine the locus of attention) would warrant a uniform distribution of attention. Second, recent studies (Langton and Bruce 1999; Driver et al. 1999; Friesen and Kingstone 1998) have shown that attention is shifted rapidly and automatically in the direction that a foveally presented representation of a conspecific (whether person or cartoon drawing) is looking. We believe that progress in understanding the nature of these hybrid forms of covert orienting will be advanced by exploring how they behave in relation to nonspatial expectancies and opportunities for illusory conjunctions.

8.6 SUMMARY

We have highlighted a subset of the relations among the different modes of orienting shown in figure 8.1. Both overt and covert orienting involve an interplay between control by endogenous signals and control by exogenous signals. This interplay can be cooperative or competitive. The concept of attentional control settings provides a powerful tool for understanding a wide range of phenomena, including the competitive interactions for which it was generated. The utility of the overall framework (figure 8.1) is accentuated when considering two questions raised during the meeting: What is the relation between overt and covert orienting? And is the same form of attention shifted by exogenous and endogenous signals? We answered the first question by noting the strong linkage when orienting is controlled exogenously and the implicit competitive interaction when attention is endogenously shifted in space while gaze direction is maintained. In answer to the second question, a double dissociation was briefly described (under exogenous versus endogenous control, attention behaves differently with respect to feature integration and nonspatial expectancies; see figure 8.2), which strongly suggests that when attention is endogenously or exogenously elicited, subtly different selective mechanisms are engaged. Phenomena such as inhibition of return, illusory line motion, meridian effects, and the disengage deficit seen with neglect patients present dissociations consistent with this conclusion. Finally, we propose that double dissociation be used to explore the nature of attention elicited in ambiguous cases.

NOTES

This chapter was made possible by a Collaborative Projects Grant to Raymond M. Klein from the Natural Sciences and Engineering Research Council of Canada and a Killam Fellowship to David I. Shore. We are grateful for the comments of Stephen Monsell and Jon Driver on an earlier version, and for the advice of Susan Boehnke, Jeff Hancock, Amelia Hunt, Jason Ivanoff, Bill Matheson, and William Schmidt.

1. Other differences in the methodology used by Theeuwes, Atchley, and Kramer and by Folk and colleagues may also be responsible for the conflicting results that have been observed. For example, Theeuwes, Atchley, and Kramer typically use a searchlike display, where the features that define the target and distractor are likely to be grouped with the elements of the array, whereas Folk and colleagues use a cuing-type display, where the distracting cue is unlikely to be grouped with the items that must be reported.

2. That inhibition of return (see Taylor and Klein 1998 for a review) is generated following endogenous motor preparation but not following endogenous covert orienting provides converging evidence for this disconfirmation.

3. In this case, the apparent conflict with Sheliga et al. (1994) may be resolved by considering whether the signal generating the probe saccade is imposed exogenously with no

opportunity to modify the state of the oculomotor system (Kustov and Robinson 1996), or whether it is generated endogenously and therefore might be accompanied by a voluntary cancellation of the prepared saccade (Sheliga et al. 1994).

4. Whether exogenous covert orienting interacts with feature integration because it entails peripheral visual stimulation that is an input to pattern recognition routines could be tested by exploring whether exogenous visual orienting in response to a localizable auditory event yields the same interaction with illusory conjunctions as does a peripheral visual stimulus.

REFERENCES

Allport, A., Styles, E., and Hseieh, S. (1994). Shifting intentional set: Exploring the dynamic control of tasks. In C. Umiltà and M. Miscovitch (Eds.), *Attention and Performance XV*, pp. 421–452. Cambridge, MA: MIT press.

Bacon. W. F., and Egeth, H. E. (1994). Overriding stimulus-driven attention capture. *Perception and Psychophysics, 55,* 485–496.

Briand, K. A. (1998). Feature integration and spatial attention: More evidence of a dissociation between endogenous and exogenous orienting. *Journal of Experimental Psychology: Human Perception and Performance, 24,* 1243–1256.

Briand, K. A., and Klein, R. M. (1987). Is Posner's "beam" the same as Treisman's "glue"? On the relation between visual orienting and feature integration theory. *Journal of Experimental Psychology: Human Perception and Performance, 13,* 228–241.

Broadbent, D. E. (1971). *Decision and stress.* London: Academic Press.

Chelazzi, L., Miller, E. K., Duncan, J., and Desimone, R. (1993). A neural basis for visual search in inferior temporal cortex. *Nature, 363,* 345–347.

Desimone, R., and Duncan, J. (1995). Neural mechanisms of selective viual attention. *Annual Review of Neuroscience, 18,* 193–222.

Driver, J., Davis, G., Ricciardelli, P., Kidd, P., Maxwell, E., and Baron-Cohen, S. (1999). Gaze perception triggers reflexive visuospatial orienting. *Visual Cognition, 6,* 509–540.

Ennis, T., and Kingtone, A. (1998). The oculomotor readiness hypothesis: Revisited (Again!). Paper presented at the Canadian Society for Brain, Behavior and Cognitive Science, Ottawa, June.

Everling, S., and Fischer, B. (1998). The antisaccade: A review of basic research and clinical studies. *Neuropsychologia, 36,* 885–899.

Everling, S., Paré, M., Dorris, M. C., and Munoz, D. P. (1998). Comparison of the discharge characteristics of brain stem omnipause neurons and superior colliculus fixation neurons in monkey: Implications for control of fixation and saccade behavior. *Journal of Neurophysiology, 79,* 511–528.

Fischer, B., and Ramsperger, E. (1984). Human express saccades: Extremely short reaction times of goal-directed eye movements. *Experimental Brain Research, 57,* 191–195.

Folk, C. L., and Remington, R. W. (1998). Selectivity in distraction by irrelevant featural singletons: Evidence for two forms of attentional capture. *Journal of Experimental Psychology: Human Perception and Performance, 24,* 847–858.

Folk, C. L., Remington, R. W., and Johnston, J. C. (1992). Involuntary covert orienting is contingent on attentional control settings. *Journal of Experimental Psychology: Human Perception and Performance, 18,* 1030–1044.

Forbes, K., and Klein, R. M. (1996). The magnitude of the fixation offset effect with endogenously and exogenously controlled saccades. *Journal of Cognitive Neuroscience, 8,* 344–352.

Friesen, C. K., and Kingstone, A. (1998). The eyes have it! Reflexive orienting is triggered by nonpredictive gaze. *Psychonomics Bulletin and Review, 5,* 490–495.

Hikosaka, O., Miyauchi, S., and Shimojo, S. (1993). Focal visual attention produces illusory temporal order and motion sensation. *Vision Research, 33,* 1219–1240.

Hoffman, J. E., and Subramaniam, B. (1995). The role of visual attention in saccadic eye movements. *Perception and Psychophysics, 57,* 787–795.

Jonides, J. (1976). Voluntary versus reflexive control of the mind's eye's movement. Paper presented at the annual meeting of the Psychonomics Society, St. Louis, November.

Jonides, J. (1981). Voluntary versus automatic control over the mind's eye's movement. In J. B. Long and A. D. Baddeley (Eds.), *Attention and Performance IX,* pp. 187–203. Hillsdale, NJ: Erlbaum.

Jonides, J., and Yantis, S. (1988). Uniqueness of abrupt visual onset in capturing attention. *Perception and Psychophysics, 43,* 346–354.

Joseph, J. S., and Optican, L. M. (1996). Involuntary attentional shifts due to orientation differences. *Perception and Psychophysics, 58,* 651–665.

Kaufman, L., and Richards, W. (1969). Spontaneous fixation tendencies for visual forms. *Perception and Psychophysics, 5,* 85–88.

Kingstone, A., and Egly, R. (In preparation). Endogenous and exogenous orienting during the maintenance of a form expectancy.

Kingstone, A., Grabowecky, M., Mangun, G. R., Valsangkar, M. A., and Gazzinaga, M. S. (1995). Paying attentionn to the brain: The study of selective visual attention in cognitive neuroscience. In J. Burak and J. Enns (Eds.), *Attention, development and psychopathology,* pp. 263–287. New York: Guilford.

Kingstone, A., and Klein, R. M. (1993). What are human express saccades? *Perception and Psychophysics, 54,* 260–273.

Klein, R. M. (1980). Does oculomotor readiness mediate cognitive control of visual attention? In R. Nickerson (Ed.), *Attention and Performance VIII,* pp. 259–275. New York: Academic Press.

Klein, R. M. (1994). Perceptual-motor expectancies interact with covert visual orienting under endogenous but not exogenous control. *Canadian Journal of Experimental Psychology, 48,* 151–166.

Klein, R. M. (In preparation). Shared neural control of attention and eye movements: Intuitively appealing but not quite established. *Journal of Experimental Psychology: Human Perception and Performance.*

Klein, R. M., and Hansen, E. (1990). Chronometric analysis of spotlight failure in endogenous visual orienting. *Journal of Experimental Psychology: Human Perception and Performance, 16,* 790–801.

Klein, R. M., Kingstone, A., and Pontefract, A. (1992). Orienting of visual attention. In K. Rayner (Ed.), *Eye movements and visual cognition: Scene perception and reading,* pp. 46–65. New York: Springer-Verlag.

Klein, R. M., and Pontefract, A. (1994). Does oculomotor readiness mediate cognitive control of visual attention? Revisited! In R. Nickerson (Ed.), *Attention and Performance XV: Conscious and nonconscious information processing,* pp. 333–350. Hillsdale, NJ: Erlbaum.

Klein, R. M., Schmidt, W. C., and Muller, H. J. (1998). Disinhibition of return: Unnecessary and unlikely. *Perception and Psychophysics, 60,* 862–872.

Klein, R. M., and Taylor, T. L. (1994). Categories of cognitive inhibition, with reference to attention. In D. Dagenbach and T. H. Carr (Eds.), *Inhibitory processes in attention, memory, and language,* pp. 113–150. San Diego, CA: Academic Press.

Kustov, A. A., and Robinson, D. L. (1996). Shared neural control of attention shifts and eye movements. *Nature, 384,* 74–77.

Langton, S. R. H., and Bruce, V. (1999). Reflexive visual orienting in response to the social attention of others. *Visual Cognition, 6,* 541–567.

Maylor, E. (1985). Facilitatory and inhibitory components of orienting in visual space. In M. I. Posner and O. S. M. Marin (Eds.), *Attention and Performance XI,* pp. 189–203. Hillsdale, NJ: Erlbaum.

McColl, S. L., and Schmidt, W. C. (1995). Orientation singletons evoke facilitation as measured using illusory line motion. *Investigative Opthalmology and Visual Science, 36,* S373.

McCormick, P. A., and Klein, R. (1990). The spatial distributon of attention during covert visual orienting. *Acta Psychologica, 75,* 225–242.

Müller, H. J., and Rabbitt, P. M. A. (1989). Reflexive and voluntary orienting of visual attention: Time course of activation and resistance to interruption. *Journal of Experimental Psychology: Human Perception and Performance, 15,* 315–330.

Posner, M. I. (1980). Orienting of attention. *Quarterly Journal of Experimental Psychology, 32,* 3–25.

Posner, M. I., Snyder, C. R., and Davidson, B. J. (1980). Attention and the detection of signals. *Journal of Experimental Psychology: General, 109,* 160–174.

Prinzmetal, W., Presti, D., and Posner, M. I. (1986). Does attention affect feature integration? *Journal of Experimental Psychology: Human Perception and Performance, 12,* 361–369.

Rizzolatti, G., Riggio, L., Dascola, I., and Umiltà, C. (1987). Reorienting attention across the horizontal and vertical meridians: Evidence in favor of a premotor theory of attention. *Neuropsychologia, 25,* 31–40.

Rogers, R. D., and Monsell, S. (1995). Costs of a predictable switch between simple cognitive tasks. *Journal of Experimental Psychology: General, 124,* 207–231.

Schmidt, W. C. (1994). Stimulus-driven attentional capture: Evidence from illusory line motion. Master's thesis, University of Western Ontario, London.

Sheliga, B. M., Riggio, L., and Rizzolatti, G. (1994). Orienting of attention and eye movements. *Experimental Brain Research, 98,* 507–522.

Shepherd, M., Findlay, J. M., and Hockey, R. J. (1986). The relationship between eye movements and spatial attention. *Quarterly Journal of Experimental Psychology, 38A,* 475–491.

Subramaniam, B., and Hoffman, J. E. (1991). Saccadic eye movement and visual selective attention. Paper presented before the Psychonomics Society, San Francisco, November.

Taylor, T., Kingstone, A. F., and Klein, R. M. (1998). Visual offsets and oculomotor disinhibition: Endogenous and exogenous contributions to the gap effect. *Canadian Journal of Experimental Psychology, 52,* 192–200.

Taylor, T. L., and Klein, R. M. (1998). On the causes and effects of inhibition of return. *Psychonomic Bulletin and Review, 5,* 625–643.

Theeuwes, J., Kramer, A. F., Han, S., and Irwin, D. E. (1998). Our eyes do not always go where want them to go: Capture of the eyes by new objects. *Psychological Science, 9,* 379–385.

Treisman, A. (1985). Preattentive processing in vision. *Computer Vision, Graphics and Image Processing, 31,* 156–177.

Treisman, A., and Gelade, G. (1980). A feature integration theory of attention. *Cognitive Psychology, 12,* 97–125.

Treisman, A., and Sato, S. (1990). Conjunction search revisited. *Journal of Experimental Psychology: Human Perception and Performance, 16,* 459–478.

Wolfe, J. M., Cave, K. R., and Franzel, S. L. (1989). Guided search: An alternative to the feature integration model for visual search. *Journal of Experimental Psychology: Human Perception and Performance, 15,* 419–433.

Yantis, S., and Hillstrom, A. P. (1994). Stimul us-driven attentional capture: Evidence from equiluminant visual objects. *Journal of Experimental Psychology: Human Perception and Performance, 20,* 95–107.

Yantis, S., and Jonides, J. (1984). Abrupt visual onsets and selective attention: Evidence from visual search. *Journal of Experimental Psychology: Human Perception and Performance, 10,* 601–621.

Yantis, S., and Jonides, J. (1990). Abrupt visual onsets and selective attention: Voluntary versus automatic allocation. *Journal of Experimental Psychology: Human Perception and Performance, 16,* 121–134.

Yarbus, A. L. (1967). In L. A. Riggs (Ed.), *Eye movements and vision.* New York: Plenum Press.

II Control of Perception-Action Coupling

9 The Control of Visuomotor Control

A. David Milner

ABSTRACT There is growing evidence for a number of parallel systems transforming visual inputs into action coordinates, as well as several transforming inputs into percepts. Moreover, the perceptual system is itself implicated in several aspects of visuomotor guidance. This multiplicity of routes participating in visuomotor control raises various questions of integration and coordination.

9.1 SEPARATION OF PERCEPTION FROM VISUOMOTOR CONTROL

Until recently, there has been a near-universal tendency in psychology and neuroscience to think of visual processing as a means of constructing a single, all-purpose perceptual representation, one that can serve us in all of our dealings with the world, whether motor, mnemonic, aesthetic, or social. Although it seems introspectively obvious, and almost absurd to deny, that "what we see is what we get," this view of vision has become increasingly untenable, and we are currently in the midst of a radical change in the way that visual scientists conceptualize their field of study (Georgeson 1997; Milner and Goodale 1995). The change in perspective toward seeing vision as having multiple endpoints has sprung largely from research in visual neuroscience—human neuropsychology, primate electrophysiology, and neurobehavioral studies of animals (Milner and Goodale 1995). Moreover, there are many observations in normal human subjects of dissociations between different visual processing systems (brought about, for example, through perceptual illusions) that fit well into the same framework (Goodale and Haffenden 1998).

"Human perception," as Von Hofsten (1987, 34) observed prophetically,

> may be specialized enough to make it appropriate to speak of a number of perception-action systems instead of regarding perception as a unitary process separate from action .... To believe that the study of, for instance, arbitrary finger movements as response to displayed letters or numbers will reveal anything essential about the coordination between perception and action is doubtful. It may actually be as erroneous as the belief that verbal memory could be studied through the use of nonsense syllable lists.

Current approaches have now led to a general recognition that there are quasi-independent visual subsystems in the primate brain, each serving a different motor domain—saccadic eye movements, ocular pursuit, locomotion, reaching with the arm, grasping with the hand. The neuronal machinery underlying such "pragmatic" coding of visual information (Jeannerod and Rossetti 1993) is vested principally in the occipitoparietal complex of visual areas known as the "dorsal stream" and its associated structures in the brain stem, thalamus, and frontal lobes. A quite different complex of visual areas, the occipitotemporal "ventral" stream, carries out the processing that provides the furniture for our perceptual experience and the raw materials for storing visual memories. This perceptual system, itself not unitary, appears in large part to operate independently of the dedicated visuomotor systems. While the efficient operation of both streams must depend on selective spatial gating, the visuomotor control systems appear to do this without the benefit of visual consciousness.

The separate functioning of these two broad systems of visual processing can be seen most dramatically in neurological "experiments of nature" in which one of the systems has been damaged, leaving the other to work largely in isolation. That is, patients may suffer from parietal lobe damage that impairs visuomotor control but largely spares visual perception, a condition known as "optic ataxia" (Perenin and Vighetto 1988); or more rarely, they may suffer damage that *preserves* visuomotor control despite a severe loss of form perception, a condition known as "visual form agnosia" (Milner et al. 1991).

The existence of multiple visual pathways in the brain raises important and puzzling questions as to their interrelationships: How do they interact in the guidance of behavior? Are they subject to overall orchestration by other systems, and if so, how? These questions, though scarcely addressed in empirical investigations to date, are central to the theme of this volume. I restrict my present comments to two broad issues:

1. How concurrently activated visuomotor systems or processes are coordinated;

2. When and how perceptual processing is co-opted in the service of visuomotor control.

9.2 COORDINATION OF VISUOMOTOR CONTROL

The first and most experimentally tractable question is how a movement through space, such as a manual reach or a saccadic eye movement, can be influenced by competing stimuli in other parts of space. Pioneering studies to address this question were carried out by Tipper, Lortie, and Baylis (1992) and Sheliga, Riggio and Rizzolatti (1994), discussed, along with more recent work, by Tipper, Howard, and Houghton (chap. 10, this volume). Depending on the distractor's proximity and salience, it has been found that the eye or hand trajectory might veer toward or away from the distractor, and Tipper and colleagues present a model that can encompass both of these types of effect.

Of course, given that studies of this kind reveal the operation of competing motor programs, it may be expected that such competition will occur not only between competing transport tendencies such as reaching, but also between nontransport movements such as the in-flight preformation of hand grip as a function of the size of target and nontarget objects.

Initial studies of this possibility (Chieffi et al. 1993; Jackson, Jackson, and Rosicky 1995) found little evidence for anticipatory grip size to be influenced by the size of a distractor during normal grasping movements. More recently, however, Gangitano, Daprati, and Gentilucci (1998) and Castiello (1998) have reported subtle interference effects under certain conditions. For example, Castiello (1998) has found changes in the rate of hand opening when three-dimensional distractors, but not two-dimensional ones, are used. This may be because a 2-D shape only partially activates the visuomotor module that governs the grasp component of a prehension movement.

A complementary question that arises is how we can make on-line changes during a reaching movement in response to a change in the visual array that actually is relevant to our overall goals and has to be responded to. This may require a change in the nature of the action itself, or a change in some parameter of the action, such as the direction of a reach. A number of experiments have addressed these issues. In an interesting example of the first type of study, Pisella, Arzi, and Rossetti (1998) have used a task in which the subject is required to either redirect or interrupt a reaching movement in response to particular changes in the stimulus. More commonly, however, researchers have used tasks in which a target change during the course of a movement requires only that the metrics, rather than the nature, of the movement be modified to deal with the change. Interestingly, if a visual target is abruptly moved by only a few degrees during a saccade, not only are subjects unaware of any shift (Bridgeman, Hendry, and Stark 1975), but they will make a perfectly normal recalibrated hand movement toward it, unwittingly incorporating an appropriate terminal correction (Goodale, Pélisson, and Prablanc 1986).

If the target is shifted well away from the "ballpark" of the original location, however, the kinematic characteristics of a reach do change. Even though the subject becomes subjectively aware of such major perturbations, however, this awareness may occur much later than the motor adjustments made. For example, Castiello, Paulignan, and Jeannerod

(1991), asked subjects to indicate (using a vocal response) when they perceived a sudden displacement in the position of an object to which they were directing a grasp. On trials in which the object was displaced at the onset of the reach, the vocal response was emitted 420 msec after the onset of the movement. In contrast, adjustments to the trajectory of the grasping movement could be seen as early as 100 msec after the perturbation, that is, more than 300 msec earlier than the vocal response.

Other studies have examined the effects of on-line perturbations of the intrinsic stimulus properties of a target object, such as its size or orientation, on wrist and finger movements preparatory to grasping the object (e.g., Paulignan et al. 1991). These investigations show that our visuomotor apparatus is flexible enough to respond adaptively to target changes, and indeed that some adjustments will be absorbed into the movement almost as if nothing had happened, offering a speed advantage that allows us, for example, to pursue a fly in midflight. But while "natural" changes (such as small displacements) are handled as a matter of course by the system, without any need for external modulation, other changes—particularly ones that would rarely or never occur in everyday life—may require a reprogramming dependent on a conscious perception of the change, resulting in a slowing of the action.

Another well-studied question is how we coordinate different visuomotor subsystems to work together smoothly during the execution of everyday actions. The unfolding of a prehension movement, in which the transport and grasp components of the arm and hand along with the requisite eye movements, are integrated in exquisite temporal and spatial harmony (Jeannerod 1988), provides a prime example. Jeannerod and his colleagues make a good case that this coordination can be achieved by means of direct interactions between the visuomotor subsystems involved. Yet such interactions would have to be able to cope with the fact that spatial location for action is visually coded in different ways in different subsystems within the parietal cortex (Colby and Duhamel 1997; Snyder et al. 1998).

If different spatial "languages" are used to perform these visuomotor transformations, how are the subsystems able to talk to each other? Current evidence in the monkey suggests a solution: in both the reaching and saccadic control areas of the monkey's posterior parietal cortex, many visuomotor neurons retain a retinotopic coding of target location, although their responses may be modulated by eye and sometimes head or limb position signals (Andersen 1997; Snyder et al. 1998). While this "gain field" modulation provides a mechanism for ensembles of neurons to code location with respect to the head or body, it does so without discarding the retinal information (Andersen 1997). This preserved retinal information could provide the single common language needed for the different systems to work together (Goodale 1998).

9.3 COORDINATION OF PERCEPTION AND ACTION

But if there is a binding problem between different visuomotor systems, there must be a greater one between the perceptual and visuomotor systems (first noted by Peter Milner in 1974). We have argued (Milner and Goodale 1995; Milner 1997) that the ventral stream, through its close connections with areas such as the perirhinal cortex (Parker and Gaffan 1998), can both inform and be informed by systems for storing the enduring visual characteristics of objects. Once the perceptual system has consulted its "semantic" knowledge base and identified a visual target as deserving of further action, it is presumed that suitable motor instructions can be issued and the "pragmatic" dorsal system be brought into play to guide the animal's actions.

This proposed relationship finds a nice analogy in the use of "teleassistance" in robotic control (Goodale 1998). In this case, an intelligent system in the form of a human operator may be able to identify an item of interest on, say, the surface of the moon by means of a video signal, and can send an instruction to a semiautonomous robot to carry out coordinated actions upon that item using its own sensing and output devices. This metaphor underlines the important point that in the proposed mode of communication between the two visual systems, the perceptual processing is not providing the visuomotor control but *preparing the way* for it.

Neuropsychological studies make this point graphically. For example, D.F., a patient with visual form agnosia (Milner et al. 1991), is well able to perform a number of reaching and grasping tasks with normal levels of skill even though she is unable to process the objects of her actions perceptually. Thus she is unable to judge the width or orientation of a rectangular block, and yet reaches out to pick it up with the same degree of visually based wrist and grip calibration as a normal subject. This means that providing she has some way of "tagging" the target object (e.g., through its color), she does not need to *perceive* its contours in order to grasp it successfully.

Indeed, visuomotor control can proceed with modest success without the intervention of any visual perception at all, provided the subject has some means of localizing the target stimulus. It has been known for many years that some patients with complete hemianopia caused by damage to the primary visual cortex may still be able to direct the eye or hand toward stimuli in the "blind" field despite having no conscious visual experience of those stimuli (Weiskrantz et al. 1974). More recently, it has been found that similar patients may show significant visual calibration of the wrist and fingers when reaching for objects in their blind field (Perenin and Rossetti 1996; Rossetti 1998; Marcel 1998). Evidently, an object need not be present in awareness for the brain to be able to "tag" it spatially and to process its characteristics to guide action.

The problem in understanding how the real brain can use "teleassistance" arises because of a profound difference in the way visual space is encoded in the perceptual and visuomotor systems (Bridgeman et al. 1979; Wong and Mack 1981; Goodale, Pélisson, and Prablanc 1986; Paillard 1987). The perceptual (sometimes called "cognitive" or "representational") system relates stimulus location to a contextual framework, and is consequently subject to various visual illusions, whereas the visuomotor system relates stimulus location directly to the observer, and is therefore much less prone to systematic error. How, then, can an item localized within a *relative* visual coordinate system be tagged in a way that can be accessed by a visuomotor system that operates in *egocentric* coordinates?

The answer may be that the perceptual system is itself able to bring about movements, albeit less directly than the dedicated system (Bridgeman et al. 1979; Wong and Mack 1981). This use of perceptual representations to drive action is exemplified by our ability to direct the eyes or hand to a target no longer physically present, a feat the dedicated system is not equipped to perform. This limitation is apparent in patients who cannot use their perceptual system, whether through visual form agnosia (Goodale, Jakobson, and Keillor 1994; Milner, Dijkerman, and Carey 1999) or hemianopia (Rossetti 1998). They are unable to guide their reaching and grasping on the basis of visual information presented just a few seconds earlier.

We may assume then that a normally functioning perceptual system must be able to guide our body and eyes towards the location of a relevant stimulus with respect to other items in the world. Once the appropriate action has been selected (probably through frontal systems; Riddoch, Humphreys, and Edwards, chap. 27, this volume), our dorsal visuomotor systems would then provide the precise guidance of limbs and other effectors needed for performing the action.

A converse form of mutual assistance will of course arise regularly whenever the visuomotor system initiates an orienting movement toward a novel or salient visual stimulus, bringing it onto the fovea for perceptual analysis. The ventral processing stream, which is specialized for analyzing the central region of the visual field, can then do its work. But how could a stimulus located through the dorsal stream that does *not* result in an overt orienting response still receive detailed processing by the perceptual system? We have suggested previously that a selected location might be "broadcast" from parietal areas to other visual areas (Milner and Goodale 1995), presumably again in retinal code. Some such mechanism could underlie instances of "cross-priming," whereby stimuli selected as the targets of saccadic eye movements (Deubel and Schneider 1996) or manual reaching movements (Deubel, Schneider, and Paprotta 1998) gain higher perceptual discriminability. Thus it may be that "the (dorsally based) preparation of a goal-directed motor response ... binds the (perceptual) processing capacities of the ventral stream to the same object" (Deubel, Schneider, and Paprotta 1998, 100).

So far we have considered relatively indirect forms of interaction between the visual streams. But there is little hard evidence that the dorsal stream can offer a detailed visual analysis of objects, beyond their axial orientation, size, and spatiotemporal disposition. Might not the brain need to supplement this in order to guide everyday actions adequately? If the perceptual system does need to be recruited in this rather more direct way, then the lack of perceptual ability in an agnosic patient such as D.F. would be expected to impose sharp limits on her visuomotor skills. There are in fact several examples of such limits. Thus, although she can post a flat object through an oriented slot, she makes many 90-degree errors when asked to post a T-shape into a T-shaped aperture (Goodale et al. 1994). Similarly, she can grasp an elongated block at any orientation with normal accuracy, yet fails to vary her hand orientation when reaching to grasp a cross-shaped object set at different orientations (Carey, Harvey, and Milner 1996). And although she can point accurately to single points in space, she cannot open her thumb and forefinger to match the separation of two holes set in a disk she is asked to grasp (Dijkerman, Milner, and Carey 1998).

This need for perceptual information also becomes apparent in grasping tasks that depend on 3-D information, whether for calibrating the amplitude of a reach (Marotta, Behrmann, and Goodale 1997) or for adjusting the wrist when grasping an object tilted in depth (Dijkerman et al. 1996). Whenever the use of binocular vision is prevented, the normal visual system can fall back on using "pictorial" cues provided by perspective and figural context. Because these are cues that D.F. cannot use, however, her performance is impaired under stationary monocular viewing conditions, although her monocular accuracy is restored to near normal when she moves her head sideways to provide herself with motion parallax cues (Dijkerman, Milner, and Carey 1999).

All of these limitations on D.F.'s visuomotor ability, then, illustrate the intact brain's ability to benefit from perceptual processing in its execution of visuomotor acts. Our prehension and other motor skills in the real world of complex objects cannot depend entirely on the basic, if quick and reliable, guidance that the dedicated visuomotor system provides.

But the ventral stream can offer other benefits to motor guidance beyond simply more elaborate bottom-up analysis. It can also, through its links with memory stores, access top-down information about the nature of the objects themselves, for example, their fragility and weight—information not given directly to the retina. That is, the perceptual system can modulate aspects of our actions beyond those that can be computed from the geometry of the target array. A good example of this is the visual calibration of grip force, which is measurable on initial contact with an object, prior to any proprioceptive feedback (Johansson and Cole 1992).

This modulation must be based on stored size-weight correlations. The assumption that this force calibration is achieved via the perceptual rather than the visuomotor system is consistent with recent studies of geometric visual illusions. Although such illusions fail to influence anticipatory grip aperture during our reaches for an object whose size we misperceive, they do influence the force of our grip in the grasping act itself (Brenner and Smeets 1996; Jackson and Shaw 2000).

Jeannerod, Decety, and Michel (1994) found what may be a clue to how ventral processing is able to influence action in these direct ways. Although their optic ataxic patient A.T. showed severe visuomotor problems in her attempts to grasp rectangular blocks of different widths, she became much better able to calibrate her grip when faced with familiar objects (e.g., a lipstick). Presumably, outputs from her functioning recognition system were able to bypass the badly damaged control networks in her parietal lobes and gain direct access to motor systems. An important question for the future will be to delineate which outputs of the perceptual system can follow such an independent route, without needing to implicate the dorsal stream.

9.4 CONCLUSIONS

In this commentary, I have drawn attention to a number of questions that arise from current conceptualizations of perception and action, but have offered only a few tentative answers. Perhaps the next Attention and Performance symposium, "Common Processes in Perception and Action," will provide a clearer picture.

NOTE

I am grateful to Mel Goodale for his comments on the draft manuscript, to Stephen Monsell and Jon Driver for their helpful remarks, and to the Wellcome Trust for their financial support.

REFERENCES

Andersen, R. A. (1997). Multimodal integration for the representation of space in the posterior parietal cortex. *Philosophical Transactions of the Royal Society of London, B352,* 1421–1428.

Brenner, E., and Smeets, J. B. J. (1996). Size illusion influences how we lift but not how we grasp an object. *Experimental Brain Research, 111,* 473–476.

Bridgeman, B., Hendry, D., and Stark, L. (1975). Failure to detect displacement of the visual world during saccadic eye movements. *Vision Research, 15,* 719–722.

Bridgeman, B., Lewis, S., Heit, G., and Nagle, M. (1979). Relation between cognitive and motor-oriented systems of visual position perception. *Journal of Experimental Psychology: Human Perception and Performance, 5,* 692–700.

Carey, D. P., Harvey, M., and Milner, A. D. (1996). Visuomotor sensitivity for shape and orientation in a patient with visual form agnosia. *Neuropsychologia, 34,* 329–338.

Castiello, U. (1998). Attentional coding for three-dimensional objects and two-dimensional shapes: Differential interference effects. *Experimental Brain Research, 123,* 289–297.

Castiello, U., Paulignan, Y., and Jeannerod, M. (1991). Temporal dissociation of motor responses and subjective awareness. *Brain, 114,* 2639–2655.

Chieffi, S., Gentilucci, M., Allport, A., Sasso, E., and Rizzolatti, G. (1993). Study of selective reaching and grasping in a patient with unilateral parietal lesion. *Brain, 116,* 1119–1137.

Colby, C. L., and Duhamel, J. R. (1997). Spatial representations for action in parietal cortex. *Cognitive Brain Research, 5,* 105–115.

Deubel, H., and Schneider, W. X. (1996). Saccade target selection and object recognition: Evidence for a common attentional mechanism. *Vision Research, 36,* 1827–1837.

Deubel, H., Schneider, W. X., and Paprotta, I. (1998). Selective dorsal and ventral processing: evidence for a common attentional mechanism in reaching and perception. *Visual Cognition, 5,* 81–107.

Dijkerman, H. C., Milner, A. D., and Carey, D. P. (1996). The perception and prehension of objects oriented in the depth plane: 1. Effects of visual form agnosia. *Experimental Brain Research, 112,* 442–451.

Dijkerman, H. C., Milner, A. D., and Carey, D. P. (1998). Grasping spatial relationships: failure to demonstrate allocentric visual coding in a patient with visual form agnosia. *Consciousness and Cognition, 7,* 424–437.

Dijkerman, H. C., Milner, A. D., and Carey, D. P. (1999). Prehension of objects oriented in depth: motion parallax restores performance of a visual form agnosic when binocular vision is unavailable. *Neuropsychologia, 37,* 1505–1510.

Gangitano, M., Daprati, E., and Gentilucci, M. (1998). Visual distractors differentially interfere with the reaching and the grasping components of prehension movements. *Experimental Brain Research, 122,* 441–452.

Georgeson, M. (1997). Vision and action: You ain't seen nothin' yet. *Perception, 26,* 1–6.

Goodale, M. A. (1998). Visuomotor control: Where does vision end and action begin? *Current Biology, 8,* R489–R491.

Goodale, M. A., and Haffenden, A. M. (1998). Frames of reference for perception and action in the human visual system. *Neuroscience and Biobehavioral Reviews, 22,* 161–172.

Goodale, M. A., Pélisson, D., and Prablanc, C. (1986). Large adjustments in visually guided reaching do not depend on vision of the hand or perception of target displacement. *Nature, 320,* 748–750.

Goodale, M. A., Jakobson, L. S., and Keillor, J. M. (1994). Differences in the visual control of pantomimed and natural grasping movements. *Neuropsychologia, 32,* 1159–1178.

Goodale, M. A., Jakobson, L. S., Milner, A. D., Perrett, D. I., Benson, P. J., and Hietanen, J. K. (1994). The nature and limits of orientation and pattern processing supporting visuomotor control in a visual form agnosic. *Journal of Cognitive Neuroscience, 6,* 46–56.

Jackson, S. R., and Shaw, A. (2000). The Ponzo illusion affects grip force but not grip aperture scaling during prehension movements. *Journal of Experimental Psychology: Human Perception and Performance, 26,* 1–6.

Jackson, S. R., Jackson, G. M., and Rosicky, J. (1995). Are non-relevant objects represented in working memory? The effects of non-target objects on reach and grasp kinematics. *Experimental Brain Research, 102,* 519–530.

Jeannerod, M. (1988). *The neural and behavioural organization of goal-directed movements.* Oxford: Oxford University Press.

Jeannerod, M., Decety, J., and Michel, F. (1994). Impairment of grasping movements following bilateral posterior parietal lesion. *Neuropsychologia, 32,* 369–380.

Jeannerod, M., and Rossetti, Y. (1993). Visuomotor coordination as a dissociable visual function: experimental and clinical evidence. In C. Kennard (Ed.), *Visual perceptual defects, Bailliere's clinical neurology, vol. 2, no. 2,* pp. 439–460. London: Bailliere Tindall.

Johansson, R. S., and Cole, K. J. (1992). Sensory-motor coordination during grasping and manipulative actions. *Current Opinion in Neurobiology, 2,* 815–823.

Marcel, A. J. (1998). Blindsight and shape perception: Deficit of visual consciousness or of visual function? *Brain, 121,* 1565–1588.

Marotta, J. J., Behrmann, M., and Goodale, M. A. (1997). The removal of binocular cues disrupts the calibration of grasping in patients with visual form agnosia. *Experimental Brain Research, 116,* 113–121.

Milner, A. D. (1997). Vision without knowledge. *Philosophical Transactions of the Royal Society of London, B352,* 1249–1256.

Milner, A. D., and Goodale, M. A. (1995). *The visual brain in action.* Oxford: Oxford University Press.

Milner, A. D., Dijkerman, H. C., and Carey, D. P. (1999). Visuospatial processing in a pure case of visual-form agnosia. In N. Burgess, K. Jeffery and J. O'Keefe (Eds.), *The hippocampal and parietal foundations of spatial cognition,* pp. 443–466. Oxford: Oxford University Press.

Milner, A. D., Perrett, D. I., Johnston, R. S., Benson, P. J., Jordan, T. R., Heeley, D. W., Bettucci, D., Mortara, F., Mutani, R., Terazzi, E., and Davidson, D. L. W. (1991). Perception and action in "visual form agnosia." *Brain, 114,* 405–428.

Milner, P. M. (1974). A model for visual shape recognition. *Psychological Review, 81,* 521–535.

Paillard, J. (1987). Cognitive versus sensorimotor encoding of spatial information. In P. Ellen and C. Thinus-Blanc (Eds.), *Cognitive processes and spatial orientation in animal and man.* Vol. 2, *Neurophysiology and developmental aspects,* pp. 43–77. Dordrecht: Nijhoff.

Parker, A., and Gaffan, D. (1998). Memory systems in primates: episodic, semantic, and perceptual learning. In A. D. Milner (Ed.), *Comparative neuropsychology,* pp. 109–126. Oxford: Oxford University Press.

Paulignan, Y., Jeannerod, M., MacKenzie, C., and Marteniuk, R. (1991). Selective perturbation of visual input during prehension movements: 2. The effects of changing object size. *Experimental Brain Research, 87,* 407–420.

Perenin, M. T., and Rossetti, Y. (1996). Grasping without form discrimination in a hemianopic field. *NeuroReport, 7,* 793–797.

Perenin, M.-T., and Vighetto, A. (1988). Optic ataxia: A specific disruption in visuomotor mechanisms: 1. Different aspects of the deficit in reaching for objects. *Brain, 111,* 643–674.

Pisella, L., Arzi, M., and Rossetti, Y. (1998). The timing of color and location processing. *Experimental Brain Research, 121,* 270–276.

Rossetti, Y. (1998). Implicit perception in action: Short-lived motor representations of space. *Consciousness and Cognition, 7,* 520–558.

Sheliga, B. M., Riggio, L., and Rizzolatti, G. (1994). Orienting of attention and eye movements. *Experimental Brain Research, 98,* 507–522.

Snyder, L. H., Grieve, K. L., Brotchie, P., and Andersen, R. A. (1998). Separate body- and world-referenced representations of visual space in parietal cortex. *Nature, 394,* 887–891.

Tipper, S. P., Lortie, C., and Baylis, G. C. (1992). Selective reaching: Evidence for action-centered attention. *Journal of Experimental Psychology: Human Perception and Performance, 18,* 891–905.

Von Hofsten, C. (1987). Catching. In H. Heuer and A. F. Sanders (Eds.), *Perspectives on perception and action,* pp. 33–46. Hillsdale, NJ: Erlbaum.

Weiskrantz, L., Warrington, E. K., Sanders, M. D., and Marshall, J. (1974). Visual capacity in the hemianopic field following a restricted occipital ablation. *Brain, 97,* 709–728.

Wong, E., and Mack, A. (1981). Saccadic programming and perceived location. *Acta Psychologica, 48,* 123–131.

10 Behavioral Consequences of Selection from Neural Population Codes

Steven P. Tipper, Louise A. Howard, and George Houghton

ABSTRACT The perceptual elicitation of actions takes place even for visual inputs that are not the intended target of subsequent overt behavior. According to our proposed model, this automatic analysis can activate competing population codes that represent different actions. Because these codes can overlap, inhibitory mechanisms are necessary to select one population to guide overt behavior. Selection should result in changes to the population vector that can produce deviations of the movement trajectory either toward or away from the stimulus to be ignored. Such trajectory deviations are observed in both manual reaches and saccades. The polarity and extent of the deviation are determined by the potency of the distractor.

The process of perception-action coupling can be so fluent that inhibitory mechanisms are sometimes required to overcome inappropriate but dominant responses to achieve behavioral goals. An interesting example of the costs of failure to inhibit a dominant response is provided by Stins (1998) from the work of Boysen (1993). It is relatively simple to train a chimpanzee to point to numerals to receive a reward. For example, it will quickly learn to point to the numeral 4 rather than 2 when it receives four candies in the former case and only two candies in the latter case. It is also easy to train the chimpanzee to point to the numeral 2 rather than 4 when another chimpanzee will receive the two candies while it receives the four candies. When, however, the actual candies are presented rather than numerals, performance changes dramatically. The chimpanzee cannot point to two candies to receive four. Rather, it always reaches to the location containing the four candies, even though these always go to the other chimpanzee. It seems clear that the stimuli are evoking action automatically, and the level of activation is determined by the reward value of the stimulus. The inhibitory mechanisms necessary to overcome the dominant response, and to maximize reward, are not available to the chimpanzee and hence it continually loses out to its companion.

As this example indicates, and as Diamond's analysis (1990) of the "A not B" error in babies also makes clear, to achieve free choice and control, it is essential that organisms develop the capacity to resist the strongest response of the moment. The ability to selectively direct action to achieve our goals is one of the most distinctive components of human behavior.

In sharp contrast to the chimpanzee, humans have evolved remarkably efficient inhibitory control mechanisms. Only in the earliest stages of development (e.g., Diamond 1990), or after brain damage, is action consistently captured by irrelevant objects (e.g., Lhermitte 1983; Riddoch et al. 1998; Riddoch, Humphreys, and Edwards, chap. 27, this volume).

Because visual information can automatically stimulate action, the critical mechanism for behavioral control is not necessarily some means for evoking a desired action, but rather, mechanisms for preventing undesired action (see Tipper, Howard, and Houghton 1998). The visuomotor system might function like a car with an automatic transmission: when the engine is running and the transmission is engaged in drive, the car is always attempting to act; indeed, one needs to depress the brake pedal (constant tonic inhibition) to prevent action from being automatically initiated. Hommel (chap. 11, this volume) reviews the evidence for the automatic initiation of actions and the complex interplay between control processes and intentions.

10.1 SELECTION-FOR-ACTION MECHANISMS

In this chapter, we focus on two aspects of the visual control of action: mechanisms of inhibition and the use of multiple frames of reference. We discuss each in turn.

There is evidence for parallel activation both of manual actions (e.g., Coles et al. 1985) and of saccades (e.g., Henderson and Ferreira 1990; Reichle et al. 1998; Theeuwes et al. 1998). The model we propose assumes that sometimes both objects relevant for action (targets) and objects irrelevant for action (distractors) are processed in parallel to the level of action planning, and compete for the control of effectors. Parallel encoding of actions requires selection mechanisms to be engaged. We suggest that the representations of the response activated by a distractor are inhibited to facilitate responding to the target (Houghton and Tipper 1994). Evidence for inhibition of irrelevant information has been found in studies of a number of cognitive processes, including working memory (Hasher and Zacks 1988), episodic retrieval (Anderson and Bjork 1994), language production (e.g., Dell and O'Seaghdha 1994), language comprehension (e.g., Gernsbacher and Faust 1995), serial order (Houghton and Tipper 1996), and selective attention (e.g., Tipper 1985; Tipper, Brehaut, and Driver 1990). The most direct evidence for inhibition comes from single-cell recordings in the monkey brain (e.g., Moran and Desimone 1985). For example, Schall and Hanes (1993) found that when monkeys were required to direct a saccade toward a target among distractors, a distractor stimulus initially evoked a competing saccade, which was encoded and subsequently inhibited. The role of inhibition in selective attention has been described formally, and simulated in a computational model (Houghton and Tipper 1994).

Our model also assumes that visual inputs can be represented in different reference frames. There are multiple spatial frames of reference, such that information can be encoded in retinotopic (e.g., Abrams and Pratt forthcoming), environment-based (e.g., Hinton and Parsons 1988), head-centered (e.g., Andersen and Zipser 1988), shoulder-centered (e.g., Soechting and Flanders 1989), or hand-centered (e.g., Graziano and Gross 1996; Tipper, Lortie, and Baylis 1992) frames.

In tasks demanding that the hand be moved to make direct contact with a target, as in the experiments to be described here, the information must be encoded in an oculomotor frame (to enable saccades) and a hand-centered frame, in which the distance and direction of the reach is represented by activation in motor networks (see also de Graaf, Sittig, and Denier van der Gon 1994; Ghez, Hening, and Gordon 1991). This hand-centered coding requires that proprioceptive information concerning hand location is integrated with visual input. For example, a visual receptive field may surround the hand, and move as the hand moves (Graziano and Gross 1996).

In our own work, we have demonstrated that target and distractor are encoded in parallel into such hand-centered frames of reference (e.g., Howard and Tipper 1997; Meegan and Tipper 1998; Tipper, Howard, and Jackson 1997; Tipper, Lortie, and Baylis 1992). From the pattern of interference effects produced by the distractor, we can infer the frame of reference into which visual inputs are analyzed. The amount of interference produced by a distractor could be explained only by assuming that the stimulus was represented in terms of the reaching action it evoked, and not by any other form of internal representation. As the hand started its reach from different positions, the pattern of distractor interference changed, even though visual information, and other body-centered frames (e.g., head and shoulder) remained static. Thus distractors close to the hand produced much more interference than those far from the hand. The results are consistent with the view that multiple objects evoke action in parallel, and that there is competition between these simultaneously active responses that is resolved by inhibition mechanisms (see also Meegan and Tipper 1998).

10.2 POPULATION CODING AND SELECTION

Most of the work studying selection for action has relied on temporal measures such as total time from stimulus onset to response completion, or reaction time (RT) to begin, and movement time to complete, the action. Consideration of the physiology mediating such behaviors suggests, however, that hand trajectory may yield further insights into these visuomotor processes.

A number of studies suggest that action parameters can be encoded in populations of neurons. Investigating the neural basis of primate reaching behavior in area 5 of the parietal cortex and in premotor cortex, Georgopoulos (1990a,b), Kalaska (1988), and Kalaska, Caminiti, and Georgopoulos (1983) observed distributed neural activity in which the direction of a particular reach is represented by the activity of a population of cells. Each individual neuron's level of activity is broadly tuned around a preferred direction, at which greatest activity is evoked. Accordingly, a given cell will contribute, to varying degrees, to reaching movements in different directions. The actual direction of the reach is determined by the sum of the single-cell contributions to the population vector. Importantly for our current concerns, Georgopoulos (1990a) has also argued that the specification of movement direction involves similar codes in both arm and eye movement systems. Thus information concerning oculomotor (e.g., Sparks, Holland, and Guthrie 1976) and manual behavior is distributed within neural ensembles in which direction of movement is uniquely specified only at the population level.

This form of coding has important implications for models of selection. Because each neuron's activity is broadly tuned, each cell will contribute to a variety of reaches. Thus, when two objects are present that both evoke reaches, the cell activities coding their directions can overlap, that is, some cells will be activated by both reaches. Inhibitory selection of one reach over the other may shift the population distribution in such a way that it affects the final reach to a target. In the model we have been developing, we have found that the form of the inhibition acting to control distractor activation can have differing effects on the reach path to targets (Houghton and Tipper forthcoming).

10.3 EXPERIMENTAL APPROACH AND MODEL PREDICTIONS

The experimental procedure used in these studies is developed from the work of Sheliga and colleagues (Sheliga et al. 1995, 1997; Sheliga, Riggio, and Rizzolatti 1994, 1995). Consider figure 10.1. While fixating the central cross and depressing the start key at the bottom of the board with the right hand, subjects were required to attend to one of the four light-emitting diode (LED) cues placed at equal distances around fixation. If the LED flashed green, subjects were required to reach to and depress the target key at the top of the board as fast as possible. In contrast, if the LED flashed red, no response was to be emitted. Subjects were precued before each trial as to which LED would contain the color cue, and hence attention was endogenously oriented to the LED until color onset. At color onset, attention was presumably withdrawn from the LED as action was directed toward the target. At no time in the first experiment was action ever directed toward the LED cue. Therefore we refer to the LED as the "distractor."

Following the premotor theory of Rizzolatti et al. (e.g., 1987), we assume that directing covert attention to a location will also activate

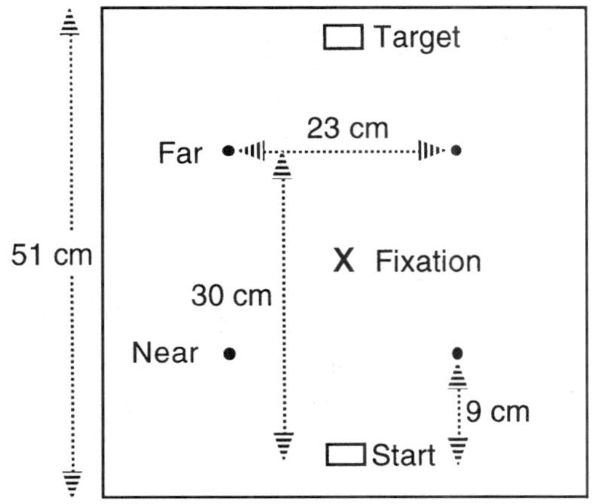

Figure 10.1 Stimulus board used in the present study.

motor systems (see also Deubel and Schneider 1996; Morrison 1984; and Hoffman 1998 for a recent review). Rizzolatti and colleagues have argued that saccades are automatically activated when attention is directed to a location in space, and that these activated saccades are suppressed under task instructions not to look at the LED. We employ Sheliga and colleagues' procedure to examine distractor effects in both oculomotor and manual frames of reference, and to test predictions derived from computer simulations of selection mechanisms acting on overlapping population codes.

In simulation work, we have investigated two ways in which distractor activation may be controlled, based on current neural network models of selective attention (Houghton and Tipper forthcoming). The first mechanism uses lateral inhibition between cells (units) coding direction. Units are organized in topographic fashion so that units coding similar directions are side by side. Each unit has excitatory connections to nearby units (those representing similar directions) and inhibitory connections to more distant units. This on-center, off-surround (oCoS) organization among directionally sensitive neurons has physiological support (Georgopoulos 1995). If target objects and their associated direction achieve enhanced activation due to attention (Houghton and Tipper 1994), then, provided distractor activation is not too strong, the oCoS interactions among the direction units can resolve the conflict. The neural activity caused by the distractor is largely suppressed, with all activity clustering around the target direction. A residual asymmetry in the distribution may persist, however, resulting in a shift of the population vector slightly toward the distractor.

Figure 10.2 illustrates this situation. Initially, subjects attend to a stimulus (an LED), which simply provides a cue as to whether to respond to a target; no action to this stimulus is required. Nevertheless, neural activity encoding the direction toward this stimulus is produced (figure

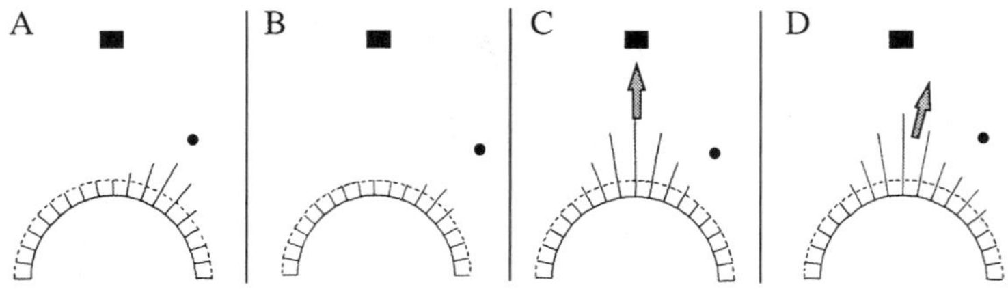

Figure 10.2 Example of a weakly evoked movement, such as a reach to an LED cue. In each panel the directional preference of each cell is represented by a line whose height represents the activity of the cell. The dotted line shows baseline activity. The small rectangle is the target key, and the small circle is an LED cue on the right. The arrows in panels C and D show the direction of movement resulting from the summation of the population. A. Low-level activity produced by the LED cue light. B. Reduced activity to the LED cue following lateral inhibition. C. High-level activity for the reach to the target. D. Summation of the activity in panels B and C. The resultant movement to the target deviates to the right (toward the cue; compare with panel C).

10.2A). Shortly afterward, a reach to a target is produced, represented in figure 10.2C. At this time, the activity associated with the LED distractor, though small, is still present (figure 10.2B). The resulting population code is shown in figure 10.2D, where panels B and C are summed. Because there are cells in common to both the reach evoked by the LED and the target, the population is shifted slightly toward the LED.

In other circumstances in which a stimulus to be ignored evokes a very powerful response, such lateral inhibition mechanisms are not sufficient to resolve response conflict. That is, action can be captured by the wrong stimulus. To resolve this level of competition, a further reactive inhibition mechanism is required, one that specifically acts on the activation caused by the distractor (this mechanism is described in detail in Houghton and Tipper 1994). In our model, inhibition feeds back onto the distractor, and the level of inhibition is related to the activation state of the distractor: more potent distractors produce greater levels of self-inhibition. Importantly, this form of inhibition has effects on population distributions distinct from that caused by the oCoS mechanism. In particular, it can lead to trajectories that veer away from distractors (see Houghton and Tipper forthcoming).

This situation, and its effect on trajectories, is shown in figure 10.3. Again, action (a saccade) is evoked by the LED stimulus. In contrast to the previous example, much greater activity is represented by higher neural activity in figure 10.3A than in figure 10.2A. To subsequently select against this stimulus and respond to the target (figure 10.3C), self-inhibition feeds back onto the population of cells encoding action toward the LED, and the effect of this reactive inhibition is shown in figure 10.3B. The summation of neural activity (combining figures 10.3B and 10.3C) is shown in figure 10.3D, in which it can be seen that trajectories veer away from the LED to be ignored.

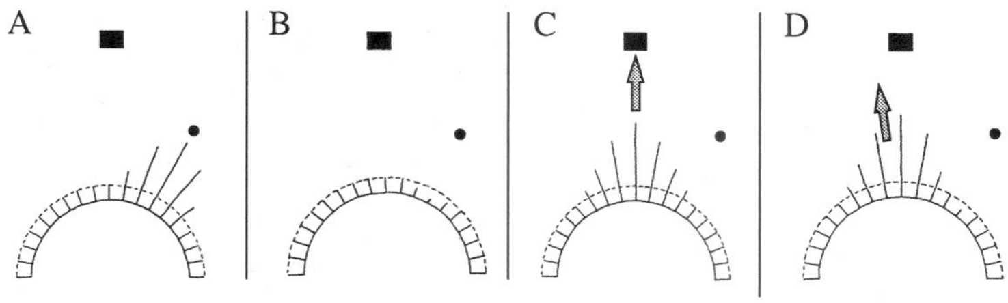

Figure 10.3 Example of a strongly evoked movement, such as a saccade to an LED cue. Conventions are as for figure 10.2. A. Cells respond strongly to the LED cue. B. Reactive inhibition reduces cell activity below baseline. C. Activity to the target. D. Summation of activity in panels B and C results in the saccade deviating away from the cue (compare with panel C).

To investigate the relative strength of the actions evoked by the LED, we examined both eye and hand trajectories. Recall that subjects initially attend to the LED to discriminate its color, while maintaining fixation at the center of the display. Subjectively, the urge to fixate the LED is extremely powerful. Clearly, orienting the cone-rich fovea would greatly facilitate the color discrimination task. We argue that reactive inhibition is necessary to prevent such saccades to the LED, and that saccades should therefore deviate away from the attended LED (figure 10.3).

In sharp contrast, there is no conscious urge to reach to the LED, suggesting only weak activation of reaches toward the LED. We argue that reaching actions are nevertheless automatically activated while attending to the LED (cf. Hommel, chap. 11, this volume; Simon 1969). As noted above, we have shown computationally that this conflicting activation can be largely resolved by oCoS interactions, resulting in a fairly straight movement path toward the target, though with a residual tendency in the direction of the distractor. Hence we predict that hand movements will exhibit small deviations toward the LED, as shown in figure 10.2.

Our other concern is frames of reference. Recall that other studies of selective reaching have provided evidence for hand-centered frames (e.g., Meegan and Tipper 1998; Tipper, Lortie, and Baylis 1992; Tipper, Howard, and Jackson 1997), because distractors close to the hand produce greater levels of interference than those far from the hand. In the current model, we therefore predict that LEDs close to the reaching hand will evoke more powerful reaches, and hence larger deviation effects, than LEDs far from the hand.

In contrast, our model predicts the opposite result for saccades. That is, saccades will deviate away from LEDs far from the hand more than they will from LEDs near the hand. This emerges from the amount of neural overlap between saccades evoked by the LED and subsequent target. For example, in figure 10.1, the right-side LED, which is far from the hand, has a saccade direction closer to that of the target saccade (i.e., they are both in the upper hemifield) than does the right-side LED, which is closer to the hand. Hence the populations of cells encoding the two saccades activated will overlap substantially more in the former case, and suppression of one group will have more of an effect on the other. Indeed, precisely this result, in which saccade deviation is greater when attending to an LED in the same hemifield as the target saccade (LED far from hand) than when attending an LED in the hemifield opposite the saccade direction (LED near hand) has been observed by Sheliga, Riggio, and Rizzolatti (1994).

In summary, we are attempting to demonstrate (1) that similar selection processes take place in both the saccade generating systems and in the manual reaching systems; (2) that two mechanisms (on-center, off-surround; and reactive feedback) enable selection between competing populations of cells; and (3) whether the second mechanism is engaged determines the direction of changed trajectory, and depends on task demands. Actions with a very low level of activity (e.g., manual reaches to the present color cues) will have little or no reactive inhibitory feedback, resulting in deviations toward the distractor. In contrast, actions that are powerfully evoked (e.g., saccades to color cues) will require substantial reactive inhibitory feedback to suppress them, resulting in deviations away from the distractor.

10.4 EXPERIMENT 1

Experiment 1A

Subjects In return for course credit, 11 right-handed subjects (all females) were recruited from our student subject pool, ranging in age from 18 to 45 (mean age: 23.3). One subject had poor stereopsis, but her performance on the task did not differ from that of the other subjects. Visual acuity was normal or corrected to normal in all subjects. All reported normal hearing.

Apparatus The experiment was programmed in LabView (version 4.0.1), running on an Apple Macintosh PowerPC 8100/100. A National Instruments NB-DIO-24 I/O card was used to send and receive digital signals. Hand movements were recorded using a MacReflex system with 2 infrared cameras recording at 50 Hz, plus video processors running on an Apple Macintosh Quadra 630. Subjects wore a reflective marker, approximately 9 mm diameter, on an elastic band on their wrist over the ulnar notch of the radius. The stimulus board, with start and target keys and 4 two-colored LEDs, arranged as in figure 10.1, was oriented with the start key closest to the subject.

Design A within-subject design was used. Cues were presented on the left or right by lighting up one of the LEDs. The near versus far cue locations were run in separate testing sessions separated by a short interval, with order counterbalanced between subjects. The experiment consisted of 8 "go" (green LED) and 2 "no-go" (red LED) trials per block, with an equal number of each color on the left and right. Each block of 10 trials was presented in a new random order, and there were 8 blocks per session.

Procedure The experiment's two sessions together lasted approximately 50 minutes, and took place in a dimly lit room. Subjects' vision was tested first, then the experimenter demonstrated the task, and then subjects were given two practice blocks of trials before beginning the experimental trials. When after eight blocks, the cue location changed from front to back or vice versa, subjects were given a further practice block. The start key lit up yellow when a trial was ready to begin. Subjects were told to fixate the blue dot in the center of the display and to press the start key and hold it down to begin a trial, at which point the light went out. If they released the key at this point, an error beep would sound and the start key would illuminate again. After a variable interval (range 510–1485 msec) a tone would sound for 250 msec. If it was a high-pitched tone (800 Hz), then subjects were to orient their attention to the LED on the left; if it was a low-pitched tone (300 Hz), they were to orient their attention to the right (the validity of the tone cue was 100%, and only one LED flashed on each trial). Fixation remained at the blue dot, and this was monitored by the experimenter (see Tipper, Brehaut, and Driver 1990 for similar procedure and reliability). There was then a 1,500 msec interval before the LED cue was presented for 100 msec, at which point the cameras were triggered to start recording. If the cue was green, subjects were to release the start key and press the target key as fast as they could. The start key would light up again to signal the start of the next trial 1,500 msec after depression of the target. If, however, the LED had flashed red, they were to keep holding the start key down (otherwise an error beep would sound) until the start key flashed to signal the start of the next trial.

Data Collection and Analyses Reaction time (RT) for the manual reach was the interval between onset of the cue and the time at which the wrist velocity achieved 25 mm/sec. Because no mean RT contrasts between attending to left/right or near/far LEDs were significant, these will not be discussed further (means were 312, 301, 309, and 304 msec for near-right, near-left, far-right and far-left cues, respectively).

Wrist trajectory was constructed by standardizing each reach spatially (see Tipper, Howard, and Jackson 1997). Location of reach onset was defined as for the RT, and the end of the reach was the greatest extent in the *Y*-dimension (sagittal plane) achieved by the wrist marker on each trial. In previous research using more than one target location, we have

Table 10.1 Mean and Standard Error (SE) Trajectory Deviation (in Millimeters in the X Dimension) at 25, 50, and 75% Stages through the Path (Y Dimension) of Hand and Eye Movements to the Target Following Cues at the Near-right, Near-left, Far-right, and Far-left Locations in Experiments 1 and 2

Experiment	n	Stage (%)	*Near right* M	*Near right* SE	*Near left* M	*Near left* SE	*Far right* M	*Far right* SE	*Far left* M	*Far left* SE
1A (hand)	11	25	17.92	3.15	14.54	3.10	15.97	2.54	15.11	2.70
		50	11.39	3.37	6.72	3.34	9.02	2.39	7.87	2.70
		75	1.26	3.43	−3.24	3.49	−0.50	2.26	−1.55	2.55
1B (eye)	7	25	0.00	2.13	6.40	3.69	−1.16	1.75	4.07	2.13
		50	−1.75	6.01	9.89	9.12	−1.75	6.21	6.40	4.65
		75	1.16	10.28	13.77	14.16	3.10	10.47	10.47	7.95
2 (hand)	21	25	9.72	2.50	8.40	2.77	13.21	2.47	9.62	2.78
		50	−4.20	3.49	−6.89	3.82	−0.70	3.19	−6.07	3.90
		75	−20.40	4.04	−23.80	4.28	−18.00	3.70	−23.44	4.35
2 (eye)	6	25	−8.34	3.30	11.06	5.04	−7.18	4.46	8.73	3.30
		50	−12.90	5.43	17.84	9.12	−12.20	5.82	6.40	4.65
		75	−12.20	6.98	18.81	11.06	−11.60	6.79	10.47	7.95

excluded outlying reaches by the extent to which they deviated from the norm. In the present series of experiments, which had only one target location, but which also had no-go trials, trials were excluded if they did not show a smooth trajectory, defined as any decrease in wrist velocity after onset but before peak velocity was achieved. On this basis, 19% of trials were excluded (but inclusion of these trials does not change the pattern of data). A further 7% of trials were excluded because of recording or subject errors.

The dependent variable was the amount of deviation of the reach path to left or right of the origin a quarter, half, and three-quarters of the way through the reach. Data were analyzed using analysis of variance (ANOVA), with cue distance (near and far), cue side (left and right), and stage through the reach (25%, 50%, and 75%, respectively) included as repeating factors.[1]

Results and Discussion

Hand Trajectory The main effect of the cue side just missed significance, with reaches deviating toward the cue: $F(1, 10) = 4.03$, $p = 0.07$. However, the interaction between cue distance and side was highly significant: $F(1, 10) = 10.64$, $p < 0.01$. Post hoc ANOVAs conducted on the data obtained with near and far cues separately indicated that the side effect was significant for the near cues: $F(1, 10) = 7.60$, $p < 0.05$; but not the far cues: $F(1, 10) = 0.68$, n.s. These effects are illustrated in figure 10.4A; mean scores are shown in table 10.1.

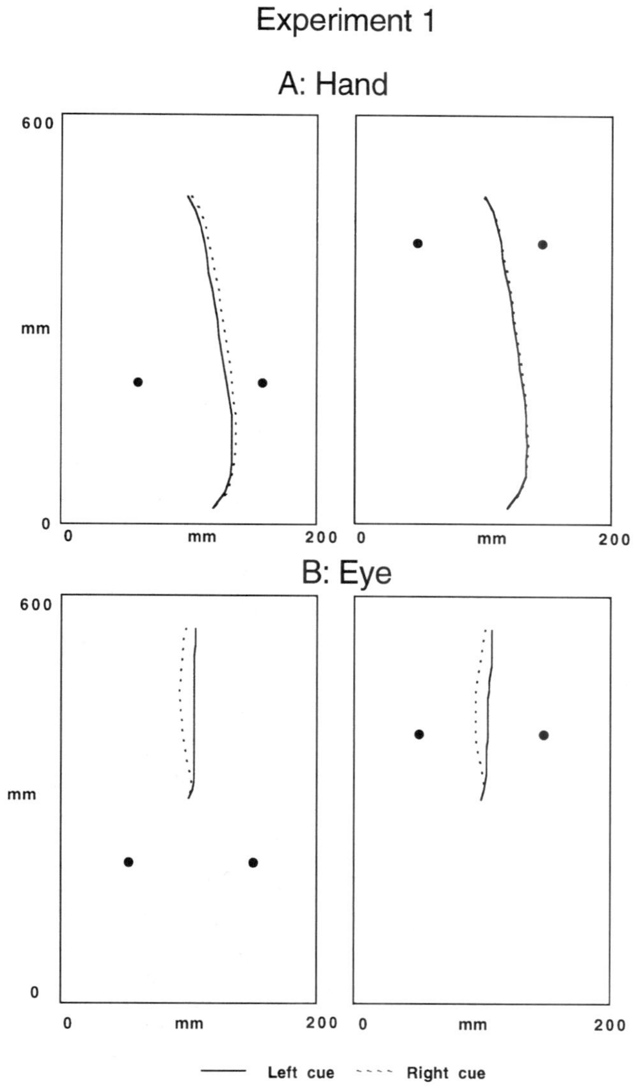

Figure 10.4 Mean trajectories from experiment 1 (go/no-go task). Hand trajectories (experiment 1A, panel A) and eye trajectories (experiment 1B, panel B). The approximate location of the near cues (left panels) and far cues (right panels) are shown.

Experiment 1A therefore provides some evidence that when subjects covertly attend to a location to analyze the color of a briefly presented cue, manual action to that location is evoked. Furthermore, in line with the theory of hand-centered frames, larger effects were produced by the LEDs closer to the hand's starting location. The effect observed was a small deviation of hand trajectory toward the attended LED cue. This result is consistent with the notion that reaching response activation is very weak in this procedure, so that little or no reactive inhibition feeds back onto this representation. Neural activity representing a reach to the distractor therefore remains slightly above baseline when the population encoding the reach to the target is activated. As a result, the latter population is slightly skewed such that trajectories veer toward the attended LED.

Experiment 1B

Task design and procedure for experiment 1B was as described in experiment 1A in that subjects reached for the target when the LED cue went green, except that we now recorded eye movements. Seven new subjects with normal or corrected-to-normal vision and a mean age of 23.6 (age range: 19 to 31) were recruited from our subject panel. Horizontal and vertical eye movements were recorded as electrooculograms (EOGs) on two separate amplifiers operating at 200 Hz, and a Biopac Systems MP100 processor.

Data Collection and Analyses Trials were excluded from both fixation and saccade analyses if fixation was not maintained during the fixation period according to the following criteria. Maximum, minimum, and mean voltages in the horizontal dimension for the 1 sec period before the LED cue was illuminated were collated for each of the four LED locations. Any trial in which the maximum or minimum value fell outside 2 standard deviations (SDs) of the mean of the maximum or minimum values was excluded. Difference scores were then calculated (maximum-minimum) of the remaining trials, and again, those in which the difference exceeded 2 SDs of the mean were excluded. To examine presaccadic drift, analyses were conducted on the mean of the means of the remaining trials. Trials in which fixation was maintained were also used to analyze subsequent eye movements.

Saccade RTs were defined as follows. The onset of saccadic eye movements was defined as the time at which either the vertical (Y) voltage stopped decreasing, or the velocity of the Y became less than the maximum velocity of the saccade divided by 150 (both measured backward in time from the end of the saccade). When saccade RTs were analyzed, the main effect of cue distance was significant: $F(1,6) = 5.65$, $p = 0.05$ (means were 347 and 334 msec for the near-right and near-left cues, 370 and 367 msec for the far-right and far-left cues). The end of the saccade was the maximum voltage in the Y-dimension. X- and Y-coordinates were standardized in the same manner as the hand movements, and means were obtained for the different cue location conditions (25%, 50%, and 75% through the eye movement). Repeated measures ANOVAs used the same factors as those in experiment 1A. Voltages were converted into approximate mm values for figure 10.4B, and table 10.1.

Table 10.2 Mean and Standard Error (SE) Values (in Volts in the X Dimension) while Attending to Each of Four Locations (Near-right, Near-left, Far-right, and Far-left) during the Fixation Period and Before Onset of the Cue in Experiments 1 and 2

		Near right		*Near left*		*Far right*		*Far left*	
Experiment	*n*	M	SE	M	SE	M	SE	M	SE
1B	7	0.15	0.02	0.12	0.06	0.25	0.03	0.24	0.02
2	6	0.27	0.05	0.24	0.04	0.28	0.05	0.23	0.03

Results and Discussion

There was a significant main effect of cue side, $F(1, 6) = 28.85$, $p < 0.005$, indicating saccadic deviations away from the cued side; and a significant interaction between stage and side, $F(2, 12) = 21.50$, $p < 0.0001$, indicating that the difference between right and left cues increased as the saccade progressed (figure 10.4). No other effects reached significance.

We also analyzed mean fixation in the 1 sec period before LED cue onset to ensure that presaccadic drift could not account for the effects obtained, with distance and side as repeated factors (see table 10.2). The effect of cue distance was significant: $F(1,6) = 18.72$, $p < 0.005$ (mean far = 0.245; mean near = 0.131 volts), indicating that in testing sessions with far cues, fixation was slightly to the right of that in sessions with near cues. Because, however, there was no reliable effect of cue side, presaccadic drift is very unlikely to account for the significant deviations in saccades that were found.

Experiment 1B clearly replicates Sheliga and colleagues' finding (Sheliga, Riggio, and Rizzolatti 1994, 1995; Sheliga et al. 1995, 1997) of a deviation of the saccade away from the location covertly attended. Furthermore, as Sheliga and colleagues have confirmed in numerous studies, this saccade deviation cannot be explained by eye drift to the attended side of space.

In line with our predictions, the hand (experiment 1A) and eye movements (experiment 1B) show quite different trajectory deviations (compare figures 10.4A and 10.4B). The eye clearly deviates away from the attended LED cue, whereas the hand deviates slightly toward the cue. Essentially similar principles are involved in the mechanisms governing representations of eye and hand movements. We propose that the only difference between these action systems is that the powerful saccades evoked when attending to the LED require reactive inhibition; this mechanism is not required when reaching, selection being achieved by lateral inhibition alone. On the other hand, although our model predicted larger deviations when neural populations overlap substantially than when they are more separate, and although Sheliga, Riggio and Rizzolatti (1994) have in fact observed this result, our data show a trend for the opposite result. Saccade deviations were slightly larger when attending to LEDs in the hemifield opposite to that of the subsequent saccade (near to hand) in which population overlap is relatively small, than when attending to LEDs in the same hemifield. This will be discussed further at the end of section 10.5.

10.5 EXPERIMENT 2

In experiment 1, the LED cue was never the target for overt behavior. Thus, even though subjects had to orient covert endogenous attention to the LED to analyze the color of the brief stimulus, they knew in advance that eye and hand movement to the keypress target was the only response required; we assumed that the oculomotor and manual responses to the target were prepared in advance. It is therefore surprising that reaching and saccade actions to the LED cue were nevertheless still covertly evoked merely by attending to the cue. Such data provide support for Rizzolatti's premotor theory of attention.

In experiment 2, we examined the effect on performance of actually making the LED relevant to behavior. That is, rather than the cue being a go (green) or no-go (red) signal, such that action was only ever directed to the keypress target, here action toward the LED cue was required on some trials. The green color cue now signaled a rapid response to the target key (66% of trials), whereas the red color cue signaled a visually guided reach toward the illuminated LED (33% of trials).

In this new procedure, such explicit coding of action should produce high levels of activity in neural populations, and hence the effects should be more pronounced. It is easy to predict the effect of this manipulation on saccades. The increased activity of the saccade to the potential LED target will produce greater reactive inhibition to prevent a saccade to the LED than observed in experiment 1. Therefore our model predicts that saccade deviations away from the LED cue will increase.

Unfortunately, it is not possible to predict the effect of this new procedure on reaching trajectory. We have argued that the weakly activated reaching response does not trigger reactive inhibition. Rather, selection can be resolved via lateral inhibition between cells in the activated populations. We simply did not know whether the increased salience of the reaching response to the LED would be sufficient to trigger reactive inhibition, and thus reduce deviations toward, or even cause deviations away from, the LED.

Experiment 2

Subjects Twenty-one new members of our subject panel took part (14 females; 7 males), all with normal vision. The mean age was 22.1 (age range: 17 to 32). The design and procedure were as for experiment 1A,

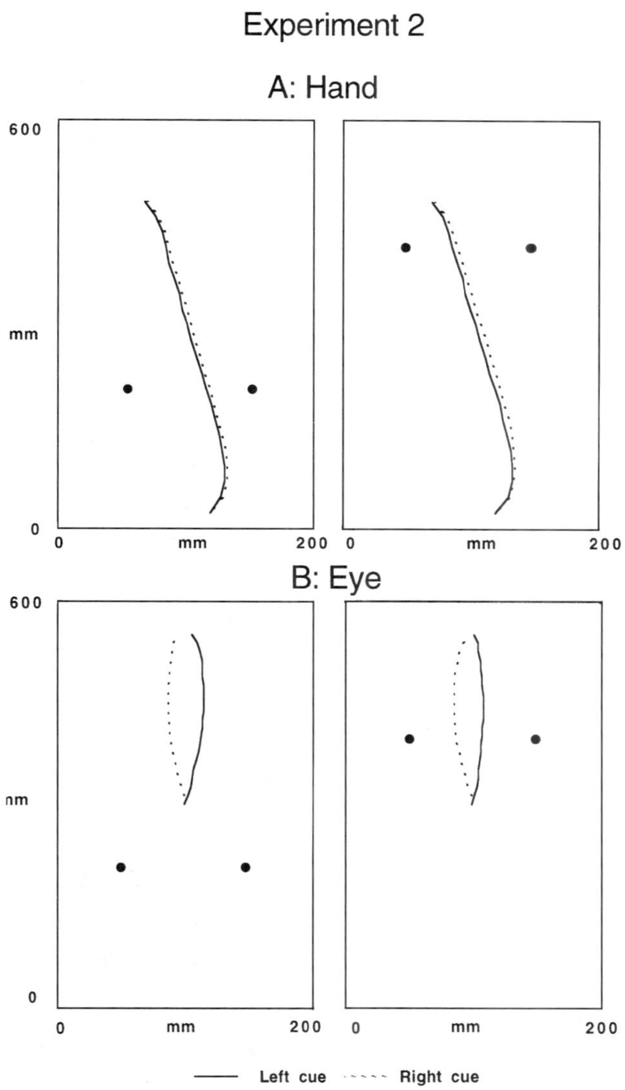

Figure 10.5 Mean trajectories from experiment 2 (LED cues are potential targets). Hand trajectories (panel A) and eye trajectories (panel B). The approximate location of the near cues (left panels) and far cues (right panels) are shown.

except that subjects were instructed to reach out and touch the surface of the LED if it flashed red, as happened on one in three trials (randomized in blocks of 12 trials). When the LED flashed green, they were to reach to the central target, as before. The apparatus was as for experiment 1B.

Data Collection and Analyses Only data for those trials on which subjects reached for the target are reported. Hand movement data were available for all 21 subjects, and eye movement data were available for 6 of these. Of the hand movement data, 3% of trials were excluded due to recording or subject errors, and a further 20% were excluded because the trajectories did not show a consistently increasing velocity, as described in experiment 1A. (The same effects were produced even when these trials were included.) Of the eye movement data, 10% of trials were excluded because fixation had not been maintained or because the saccade could not be determined.

Results and Discussion

Hand Trajectory Mean and standard error trajectory scores are shown in table 10.1. Averaged trajectories are shown in figure 10.5. There was a main effect of cue distance on hand trajectory, indicating that reaches were more to the right when the cues were in the far locations: $F(1, 20) = 8.75$, $p < 0.01$. There was also a significant stage effect, with reaches moving more to the left as they progressed: $F(2, 40) = 246.96$, $p < 0.0001$. Of most importance, there was also a significant side effect, with the hand deviating toward the side on which the cue appeared: $F(1, 20) = 4.44$, $p < 0.05$. The interaction of side and stage was significant: $F(2, 40) = 3.65$, $p < 0.05$. Post hoc ANOVAs showed that the side effect was greater in the middle and end of the reach than in the beginning: $p < 0.005$ versus $p = 0.01$ (figure 10.5A).

As in experiment 1A, we found that hand trajectory veers toward the LED cue. The main difference is that we no longer observed the asymmetry observed in experiment 1A. In experiment 1A, the reaching hand veered toward the cue only when the cue was relatively near the hand, in conformity with our view that visual stimuli are encoded in hand-centered frames for reaching. In experiment 2, this contrast was no longer observed. The amount of veering toward the attended LED was, if anything, greater in the far than in the near conditions. This difference is confirmed by a significant three-way interaction between side of cue (left or right), distance from hand (near or far) and experiment (1A or 2): $F(1, 30) = 5.33$, $p < 0.05$.

Although, as discussed, a priori predictions of hand trajectory were not possible, we can provide the following post hoc suggestions. In experiment 1A, the hand deviated toward the near LEDs. In experiment 2, the reach evoked by the LED was assumed to be more potent because it was now a potential target on some trials. If no reactive inhibition was triggered, then deviations toward the LED should have been greater. Clearly, this was not the case. We therefore suggest that reactive inhibition was evoked when attending to near LEDs. Indeed, there is a small trend for the deviations to be smaller in experiment 2 than in experiment 1A. In contrast, LEDs far from the hand had no effect in experiment 1A. Increasing the salience of the LEDs in experiment 2 may thus have increased their activation state, but not passed the threshold for triggering reactive inhibition. This implies that hand trajectory deviations should be greater in experiment 2 than in experiment 1A, and, indeed, that is what we observed. Clearly, these are only speculations, and more research is necessary to confirm the properties of our model.

Eye Trajectory Mean fixation scores were analyzed as in experiment 1B (table 10.2). There was a nonsignificant effect of attended side on fixation drift: $F(1, 5) = 3.48$ (left = 0.238; right = 0.273 volts), n.s. When saccades were analyzed, there was a significant main effect of cue side on eye trajectory, with the eyes deviating away from the side of the cues: $F(1, 5) = 10.30$, $p < 0.05$. This effect was greater at the middle and end of the eye movement than at the start, as evidenced by the interaction of stage with side: $F(1, 5) = 6.06$, $p < 0.05$ (figure 10.5B).

We predicted that when overt behavior toward the cue was required on some trials, subjects would explicitly prepare a saccade to that location. This explicit level of internal representation is assumed to correspond to greater activation levels than when action is never overtly directed toward the stimulus. Furthermore, the subsequent inhibition of the saccade toward the LED, when the saccade has to be directed to the keypress target, was predicted to be larger. These predictions were supported. Comparing figure 10.5B with figure 10.4B, it can clearly be seen that the saccade deviations away from LED cues were approximately twice as large in experiment 2, where saccades were sometimes directed to the LED, as in experiment 1B, where saccades were never directed to the LED. The two-way interaction between experiment (1B or 2) and side of LED (left or right) was significant: $F(1, 11) = 4.72$, $p = 0.05$.

As can be seen in figure 10.5, there is a trend for saccade deviations to be larger when attending to LEDs near the hand, which are in the opposite hemifield to the saccades directed to the target key. Recall that this trend was also observed in experiment 1B. As discussed previously, this is opposite to the predictions of our model, and to findings by Sheliga, Riggio, and Rizzolatti (1994).

We suggest that eye and hand movement systems have to make quite different computations based on different frames of reference. Separate neural systems will therefore be necessary, and this is supported by neurophysiological evidence: lateral intraparietal (LIP) to frontal eye fields (FEF) and ventral intraparietal (VIP) to dorsolateral prefrontal cortex (F4) control eye and hand, respectively. However, eye and hand have to also be closely coordinated in real-world interactions, such as when rapidly and accurately grasping an object (e.g., Abrams, Meyer, and Kornblum 1990; Jeannerod 1988). Various brain structures encode both eye and hand, and hence there are multiple neural sites for such interactions, such as the cerebellum (e.g., Brown et al. 1993), superior colliculus (e.g., Bekkering, Pratt, and Abrams 1996; Fries 1984, 1985; Werner 1993), and supplementary eye fields (SEFs; e.g., Mushiake, Fuji, and Tanji 1996). For example, a population of cells in an SEF responds only when saccade and reach are directed to the same object.

If the hand- and eye-centered frames are closely integrated, then there may be crosstalk between them. The main goal of the subjects in the present experiments was to reach to and depress the target key as fast as possible. Hand-centered frames are critical for achieving this behavioral goal. As we saw in experiment 1A, effects on hand trajectory were greater when attending to the nearer LEDs. It is therefore a reasonable assumption that movements to the nearer LEDs are more actively represented. If eye and hand movements are closely related in some neural systems, then there could be spreading activation from the highly active reach representations of locations near the hand, to the representations of the same locations in the eye movement system. The result of this would be to inflate the effects of the LEDs near the hand when making a saccade to a target in the upper hemifield.

Although the idea of crosstalk between different action systems is speculative, recent pilot data indicate that it is worth pursuing further. Thus when subjects undertake a task very similar to that of experiments 1A and 1B, but only make saccades to the target (never reaching), the asymmetries predicted by our model are obtained. That is, large saccade deviations are produced when attending to LEDs far from the hand, in the same (upper) hemifield as the saccade; whereas very small saccade deviations are observed when attending to LEDs near the hand in the opposite (lower) hemifield to the target saccade. Thus when reaching is removed from the experimental task, the pattern of saccade deviations changes completely.

10.6 GENERAL DISCUSSION

Parietal-frontal circuits do not encode space in some general form, but in ways relevant to the control of action. Neurophysiological studies demonstrate separate circuits for hand and eye movements, rather than a unified map of space. A single visual object can be represented in multiple ways in different brain areas depending on the actions that may be evoked, such as saccading toward (Goldberg and Colby 1989) or reaching to and grasping (Fogassi et al. 1992; Graziano and Gross 1993, 1996; Rizzolatti et al. 1981). Rizzolatti has argued that orienting attention will activate these motor circuits. Because of the fundamental role of eye movements in visual perception, orienting the fovea to the object of interest is automatically evoked. Hence orienting attention triggers saccades to the attended location.

Although it is less obvious that reaching to an attended location should also be evoked when only eye movements are required, when the goal is to reach rapidly and accurately, hand movement circuits become activated and merely attending to a location can begin to evoke a reach. Thus covertly attending to a position can activate separate visuomotor networks in parallel.

Action parameters can be represented by population codes, and models for the resolution of competition within such populations lead to predictions regarding the dynamics of effector movement. Clearly, this is our preferred interpretation of the trajectory deviation effects. However, an alternative account is offered by Duhamel, Colby, and Goldberg (1992), who have shown that the intention to make a saccade can change the location to which a neuron responds. Typically, the receptive field moves to the location it would occupy following the saccade. Sheliga and colleagues (Sheliga, Riggio, and Rizzolatti 1994, 1995; Sheliga et al. 1995, 1997) have used this finding to explain eye trajectory deviations away from cues by assuming that changes in the receptive field of the eye will also affect eye position information. That is, even though the eye stays at the fixation point, eye position information shifts to the attended side. When a vertical saccade has to be made, the eye trajectory will veer away from the side of attention.

On the other hand, Colby (1996) compared intention to act with attention in monkeys and reported that the remapping of saccades takes place *only* when the monkey intends to saccade. When the animal covertly attends to a location without making a saccade to it, no remapping is observed. The latter procedure is the same as that used in experiment 1, in which subjects never saccaded to the covertly attended LED. As Colby points out, remapping would be counterproductive if no saccade were made because it would introduce a mismatch between the external world and the internal parietal image of it. (Indeed, it is possible that the inhibition preventing the execution of the saccade also inhibits the remapping process.) If remapping does not occur, then it cannot explain the saccade deviations reported in this chapter and in the studies by Sheliga and colleagues (Sheliga, Riggio, and Rizzolatti 1994, 1995; Sheliga et al. 1995, 1997).

In conclusion, we have confirmed that two visuomotor circuits can be activated by covert attention and that these two systems (oculomotor and manual) function in essentially the same way. The basic properties of these systems are (1) action is represented in distributed neural populations; (2) both relevant and irrelevant action can be represented in the same networks; (3) inhibitory control mechanisms (lateral and reactive inhibition) are necessary for selective behavior; and (4) the potency of the stimulus determines whether reactive inhibition is triggered and hence the nature of the movement trajectory deviation.

NOTES

This work was supported by grant S07727 from Biotechnology and Biological Sciences Research Council. We would like to thank Darren Whitehouse and Daniel Fraser for their assistance with data collection and Matthew Paul and Beccy Pleasant for their assistance with data processing.

1. Greenhouse-Geisser and Huhn-Feldt corrections for sphericity appropriate to nonindependent repeating measures were used.

REFERENCES

Abrams, R. A., Meyer, D. E., and Kornblum, S. (1990). Eye-hand co-ordination: Oculomotor control in rapid aimed limb movements. *Journal of Experimental Psychology: Human Perception and Performance, 16,* 248–267.

Abrams, R. A., and Pratt, J. (Forthcoming). Retinal coding of inhibited eye movements to recently attended locations.

Andersen, J. R., and Zipser, D. (1988). The role of the posterior parietal cortex in co-ordinate transformations for visual-motor integration. *Canadian Journal of Physiological Pharmacology, 66,* 488–501.

Anderson, M. D., and Bjork, R. A. (1994). Mechanisms of inhibition in long-term memory: A new taxonomy. In D. Dagenbach and T. H. Carr (Eds), *Inhibitory processes in attention, memory, and language,* pp. 265–326. New York: Academic Press.

Bekkering, H., Pratt, J., and Abrams, R. A. (1996). The gap effect for eye and hand movements. *Perception and Psychophysics, 58,* 628–635.

Boysen, S. T. (1993). Counting in chimpanzees: Nonhuman principles and emergent properties of number. In S. T. Boysen and E. J. Capaldi (Eds), *The development of numerical competence: Animal and human models,* pp. 39–59. Hillsdale, NJ: Erlbaum.

Brown, S. H., Kessler, K. R., Hefter, H., Cooke, J. D., and Freund, H. J. (1993). Role of the cerebellum in visuomotor coordination: 1. Delayed eye and arm initiation in patients with mild cerebellar ataxia. *Experimental Brain Research, 94,* 478–488.

Colby, C. L. (1996). A neurophysiological distinction between attention and intention. In T. Inui and J. L. McClelland (Eds), *Attention and Performance XVI,* pp. 157–178. Cambridge, MA: MIT Press.

Coles, M. G., Gratton, G., Bashore, T. R., Eriksen, C. W., and Donchin, E. (1985). A psychophysical investigation of the continuous flow model of human information processing. *Journal of Experimental Psychology: Human Perception and Performance, 11,* 529–553.

de Graaf, J. B., Sittig, A. C., and Denier van der Gon, J. J. (1994). Misdirections in slow, goal-directed arm movements are not primarily visually based. *Experimental Brain Research, 99,* 464–472.

Dell, G. S., and O'Seaghdha, P. G. (1994). Inhibition in interactive activation models of linguistic selection and sequencing. In D. Dagenbach and T. H. Carr (Eds.), *Inhibitory processes in attention, memory, and language,* pp. 409–454. New York: Academic Press.

Deubel, H., and Schneider, W. X. (1996). Saccade target selection and object recognition: Evidence for a common attentional mechanism. *Vision Research, 6,* 1827–1837.

Diamond, A. (1990). Developmental time course in human infants and infant monkeys, and the neural bases of inhibitory control in reaching. *Annals of the New York Academy of Sciences, 608,* 637–676.

Duhamel, J.-R., Colby, C. L., and Goldberg, M. E. (1992). The updating of the representation of visual space in parietal cortex by intended eye movements. *Science, 255,* 90–92.

Fogassi, L., Gallese, V., Di Pellegrino, G., Fadiga, L., Gentilucci, M., Luppino, G., Matelli, M., Pedotti, A., and Rizzolatti, G. (1992). Space coding by premotor cortex. *Experimental Brain Research, 89,* 686–690.

Fries, W. (1984). Cortical projections to the superior colliculus in the macaque monkey: A retrograde study using peroxidase. *Journal of Computative Neurology, 230,* 55–76.

Fries, W. (1985). Inputs from the motor and premotor cortex to the superior colliculus in the rhesus monkey. *Behavioural Brain Research, 18,* 95–105.

Georgopoulos, A. P. (1990a). Neural coding of the direction of reaching and a comparison with saccadic eye movements. *Cold Spring Harbor Symposia on Quantitative Biology, 55,* 849–859.

Georgopoulos, A. P. (1990b). Neurophysiology of reaching. In M. Jeannerod (Ed.), *Attention and Performance XIII,* pp. 227–263. Hillsdale, NJ: Erlbaum.

Georgopoulos, A. P. (1995). Current issues in directional motor control. *Trends in Neurosciences, 18,* 506–510.

Gernsbacher, M. A., and Faust, M. E. (1995). Skilled suppression. In F. N. Dempster and C. J. Brainerd (Eds.), *Interference and inhibition in cognition,* pp. 296–327. San Diego, CA: Academic Press.

Ghez, C., Hening, W., and Gordon, J. (1991). Organisation of voluntary movement. *Current Opinion in Neurobiology, 1,* 664–671.

Goldberg, M. E., and Colby, C. L. (1989). The neurophysiology of spatial vision. In F. Boller and J. Grafman (Eds.), *Handbook of neurophysiology,* vol. 2, pp. 301–315. Amsterdam: Elsevier.

Graziano, M. S. A., and Gross, C. G. (1993). A bimodal map of space: Somatosensory receptive fields in the macaque putamen with corresponding visual receptive fields. *Experimental Brain Research, 97,* 96–109.

Graziano, M. S. A., and Gross, C. G. (1996). Multiple pathways for processing visual space. In T. Inui and J. L. McClelland (Eds.), *Attention and Performance XVI,* pp. 181–208. Cambridge, MA: MIT Press.

Hasher, L., and Zacks, R. T. (1988). Working memory, comprehension and aging: A review and new view. *Psychology of Learning and Motivation, 22,* 193–225.

Henderson, J. M., and Ferreira, F. (1990). Effects of foveal processing difficulty on the perceptual span in reading: Implications for attention and eye movement control. *Journal of Experimental Psychology: Human Perception and Performance, 16,* 417–429.

Hinton, G. E., and Parsons, L. M. (1988). Scene-based and viewer-centered representations for comparing shapes. *Cognition, 30,* 1–35.

Hoffman, J. E. (1998). Visual attention and eye movements. In H. Pashler (Ed.), *Attention,* pp. 119–153. Hove, East Sussex: Psychology Press.

Houghton, G., and Tipper, S. P. (1994). A model of inhibitory mechanisms in selective attention. In D. Dagenbach and T. Carr (Eds), *Inhibitory mechanisms in attention, memory and language,* pp. 53–112. Orlando, FL: Academic Press.

Houghton, G., and Tipper, S. P. (1996). Inhibitory mechanisms of neural and cognitive control: Application to selective attention and sequential action. *Brain and Cognition, 30,* 20–43.

Houghton, G., and Tipper, S. P. (Forthcoming). Attention and the control of action: An investigation of the effects of selection on population coding of hand and eye movement. In D. Heinke, G. W. Humphreys, and A. Olson (Eds.), *Proceedings of the 5th neural computational and psychological workshop*. Springer-Verlag.

Howard, L. A., and Tipper, S. P. (1997). Hand deviations away from visual cues: Indirect evidence for inhibition. *Experimental Brain Research, 113,* 144–152.

Jeannerod, M. (1988). *The neural and behavioural organisation of goal-directed movements.* Oxford: Clarendon Press.

Kalaska, J. F. (1988). The representation of arm movements in postcentral and parietal cortex. *Canadian Journal of Physiological Pharmacology, 66,* 455–463.

Kalaska, J. F., Caminiti, R., and Georgopoulos, A. P. (1983). Cortical mechanisms related to the direction of two-dimensional arm movements: Relations in parietal area 5 and comparison with motor cortex. *Experimental Brain Research, 51,* 247–260.

Kugler, P. N., and Turvey, M. T. (1987). *Information, natural law, and the self-assembly of rhythmic movements.* Hillsdale, NJ: Erlbaum.

Lhermitte, F. (1983). "Utilization behaviour" and its relation to lesions of the frontal lobes. *Brain, 106,* 237–255.

Meegan, D., and Tipper, S. P. (1998). Reaching into cluttered visual environments: Spatial and temporal influences of distracting objects. *Quarterly Journal of Experimental Psychology, 51A,* 225–249.

Moran, J., and Desimone, R. (1985). Selective attention gates visual processing in the extrastriate cortex. *Science, 229,* 782–784.

Morrison, R. E. (1984). Manipulation of stimulus onset delay in reading: Evidence for parallel programming of saccades. *Journal of Experimental Psychology: Human Perception and Performance, 10,* 667–682.

Mushiake, H., Fuji, N., and Tanji, J. (1996). Visually guided saccades versus eye-hand reach: Contrasting neuronal activity in the cortical supplementary and frontal eye fields. *Journal of Neurophysiology, 75,* 2181–2191.

Reichle, E. D., Pollatsek, A., Fisher, D. L., and Rayner, K. (1998). Toward a theory of eye movement control in reading. *Psychological Review, 105,* 125–157.

Riddoch, M. J., Edwards, M. G., Humphreys, G. W., West, R., and Heafield, T. (1998). Visual affordances direct action: Neuropsychological evidence from manual interference. *Cognitive Neuropsychology, 15,* 645–683.

Rizzolatti, G., Riggio, L., Dascola, I., and Umiltà, C. (1987). Reorienting attention across the horizontal and vertical meridians: Evidence in favor of a premotor theory of attention. *Neuropsychologia, 25,* 31–40.

Rizzolatti, G., Scandolara, C., Matelli, M., and Gentilucci, M. (1981). Afferent properties of periarcuate neurons in macaque monkey: 2. Visual responses. *Behavioural Brain Research, 2,* 147–163.

Schall, J. D., and Hanes, D. P. (1993). Neural basis of saccade target selection in frontal eye fields during visual search. *Nature, 366,* 467–469.

Sheliga, B. M., Craighero, L., Riggio, L., and Rizzolatti, G. (1997). Effects of spatial attention on directional manual and ocular responses. *Experimental Brain Research, 114,* 339–351.

Sheliga, B. M., Riggio, L., Craighero, L., and Rizzolatti, G. (1995). Spatial attention-determined modifications in saccade trajectories. *Neuroreport: Cognitive Neuroscience and Neuropsychology, 6,* 585–588.

Sheliga, B. M., Riggio, L., and Rizzolatti, G. (1994). Orienting of attention and eye movements. *Experimental Brain Research, 98,* 507–522.

Sheliga, B. M., Riggio, L., and Rizzolatti, G. (1995). Spatial attention and eye movements. *Experimental Brain Research, 105,* 261–275.

Simon, J. (1969). Reactions toward the source of stimulation. *Journal of Experimental Psychology, 78,* 344–346.

Soechting, J. F., and Flanders, M. (1989). Sensorimotor representations for pointing to targets in three-dimensional space. *Journal of Neurophysiology, 62,* 582–594.

Sparks, D. L., Holland, R., and Guthrie, B. L. (1976). Size and distribution of movement fields in the monkey superior colliculus. *Brain Research, 113,* 21–26.

Stins, J. F. (1998). Information-action compatibility. Ph.D. diss., Vrije University, Amsterdam.

Theeuwes, J., Kramer, A. F., Hahn, S., and Irwin, D. E. (1998). Our eyes do not always go where we want them to go: Capture of the eyes by new objects. *Psychological Science, 9,* 379–385.

Tipper, S. P. (1985). The negative priming effect: Inhibitory priming by ignored objects. *Quarterly Journal of Experimental Psychology, 37A,* 571–590.

Tipper, S. P., Brehaut, J. C., and Driver, J. (1990). Selection of moving and static objects for the control of spatially directed action. *Journal of Experimental Psychology: Human Perception and Performance, 16,* 492–504.

Tipper, S. P., Howard, L. A., and Houghton, G. (1998). Action-based mechanisms of attention. *Proceedings of the Royal Society of London, Series B, 353,* 1385–1393.

Tipper, S. P., Howard, L. A., and Jackson, S. R. (1997). Selective reaching to grasp: Evidence for distractor interference effects. *Visual Cognition, 4,* 1–32.

Tipper, S. P., Lortie, C., and Baylis, G. C. (1992). Selective reaching: Evidence for action-centered attention. *Journal of Experimental Psychology: Human Perception and Performance, 18,* 891–905.

Tipper, S. P., Weaver, B., and Houghton, G. (1994). Behavioural goals determine inhibitory mechanisms of selective attention. *Quarterly Journal of Experimental Psychology, 47A,* 809–840.

Werner, W. (1993). Neurons in the primate superior colliculus are active before and during arm movements to visual targets. *European Journal of Neuroscience, 5,* 335–340.

11 The Prepared Reflex: Automaticity and Control in Stimulus-Response Translation

Bernhard Hommel

ABSTRACT This chapter reviews a number of empirical and theoretical approaches to the translation of stimulus information into action in choice reaction tasks. Abundant evidence shows that stimulus-response (S-R) translation does not always conform to people's intentions, which rules out the notion that it is a highly selective control (or intentionally controlled) operation. This has led to the conception of dual-route models, which view action control as the outcome of a competition between intentional and automatic S-R translation processes. Although these conceptions have many advantages, they also have their limitations. In particular, there is evidence for more than two routes from perception to action; intention-related S-R translation can shown to be triggered automatically; and effects attributed to "automatic translation" often depend on the actor's intentions. An alternative view conceives of intentional and automatic processes, not as being different in kind, but rather as taking place at different points in time, with intentional processes setting the stage for automatic S-R translation.

Higher organisms exhibit an enormous flexibility in responding and adapting to immediate changes in environmental conditions. Their behavior is not only controlled by direct and persistent input-output connections but mediated by internal states and modified through experience. A wealth of cognitive processes is involved in transforming sensory inputs into observable muscle contractions. This chapter will focus on a central stage in the transformation process—the interface between perceptual processing and action selection—emphasizing the role intentional and automatic processes play in translating stimulus information into response activation.

Theories of human information processing commonly deal with this interface under the heading of "stimulus-response translation" (or "S-R translation"), "response determination," "response identification," or "response selection." Although most models include a box carrying one of these labels surprisingly little is known about how stimulus information is actually translated into action plans. However, to speak of S-R translation at least two requirements need to be met.

First, there has to be some indication, whatever the level of analysis—that response-related functional codes or brain structures are activated, at least to some degree. These indications may be relatively direct, such as the increase in activation of some part of the motor cortex in a brain-imaging study; or indirect, such as a reaction time pattern revealing competition between alternative responses. The consequences of S-R translation differ widely between situations, ranging from the unobservable activation of a mild response tendency, overcome within a few milliseconds, to the actual execution of the activated response; clearly, these differences are of great theoretical and practical moment. Yet, in this chapter, all that counts is whether there is any indication of response activation under particular stimulus conditions and task instructions.

The second requirement is that the measured arousal of response tendencies, be systematically related to the present stimulus conditions. Obviously, merely observing that some situations induce response tendencies or increase the likelihood of responding does not yet allow one to assume that some kind of stimulus information was translated into a corresponding response. To be sure that S-R translation actually took place requires one to predict which response tendency was aroused as a function of which stimulus information. Its logical dependency on available stimulus information already puts some constraints on the temporal and functional locus of S-R translation in the sequence of stages in human information processing. Indeed, most authors (e.g., De Jong 1993; Frith and Done 1986; Kornblum, Hasbroucq, and Osman 1990; Meyer and Kieras 1997; Pashler 1994) locate the S-R translation or response selection stage in between what is commonly called the "perceptual" or "stimulus identification" stage and those stages having to do with "response initiation" and "response execution." Although some stimulus processing is required before the processed information can be translated into a response, this does not mean that S-R translation has to await full processing or identification of a stimulus. For instance, Miller (1988) and others have argued that perceptual stages may pass partial output to response stages before stimulus identification is complete. For our present purposes, any specific, stimulus-related activation of response-related codes or structures will count as evidence that S-R translation has taken place, irrespective of the type of the corresponding stimulus information and the degree to which it is processed.

Authors have characterized intentional and automatic processes in many different ways (for overviews, see Neumann 1984; Schweickert and Boggs 1984): intentional translation processes have been characterized as controlled (by whatever state or mechanism), working serially (implying only one translation at a time), capacity limited, effortful, conditional (on intentions), and conscious, whereas automatic processes have been characterized as uncontrolled, working in parallel (implying more than one translation at a time), capacity unlimited, effortless, unconditional, and unconscious. However, most of the data to be discussed here speak only to the issue of whether, or how much, translation processes depend on the perceiver's or actor's intentions, apart from some preliminary hints about whether these processes work serially or in parallel (thus being capacity limited or unlimited).[1]

From a phenomenological perspective, it may seem odd to ask whether S-R translation depends on intentions. We commonly feel that we perceive an environmental event, think about it, and then deliberately select an appropriate action without further ado. This view, which has so obviously motivated many stage models of information processing, strongly suggests that S-R translation is a more or less direct reflection of the perceiver's or actor's intentions. There is increasing empirical evidence, however, for stimulus-induced and unwanted response activation, which challenges the idea of S-R translation being under direct, immediate intentional control.

11.1 THE DEMONSTRATION OF AUTOMATIC STIMULUS-RESPONSE TRANSLATION

Under normal circumstances, we do not have the slightest doubt that the actions we perform originate within ourselves, that we are the causal agents in the process of transforming mere willing into actual moving. Accordingly, many early psychological approaches to action control, especially those based on the theorist's introspection, assumed that human action was guided and controlled by human will.

A well-known proponent of such an intentional view was Donders (1868), who attributed the responsibility of translating perceptual information into movement to an "organ of will" (*wilsorgaan*). To measure how long this organ would need to make a decision, Donders manipulated S-R uncertainty in a number of ways. In one experiment, subjects responded to the electrical stimulation of their left or right foot by moving their left or right hand, respectively. It turned out that subjects were faster to respond correctly if they knew in advance which stimulus would occur than when they did not, and Donders took this difference in reaction time as an estimate for the combination of stimulus discrimination and "determination of the will." To further disentangle these two processes, Donders employed a go/no-go task that required a selective response to a specified subset of the stimulus set, pairing stimulus uncertainty with response certainty. He reasoned that such a task would not require any further will determination processes (assuming that the response could be selected in advance), so that their duration could then be estimated by subtracting the go/no-go reaction time from that obtained in conditions requiring a response decision. He calculated will determination to take 36 msec.

The outdated expression "will determination" easily translates into the more fashionable "S-R translation" or "response selection" (Gottsdanker and Shragg 1985). Indeed, despite marked changes in terminology, some information-processing models (e.g., Hasbroucq, Guiard, and Ottomani 1990; Pashler 1994; Sanders 1980; Teichner and Krebs 1974; Welford 1968) are still based on the (sometimes implicit) idea of S-R translation as a process that exclusively serves to realize the actor's intention. Conceived this way, S-R translation represents a control operation by means of which the "will," or some functional equivalent, decides what to do by selecting one stimulus and activating the corresponding response. Fitting well into this picture are claims (e.g., Pashler 1994; Welford 1952) that S-R translation draws heavily on mental resources and thus constitutes a rather fixed, structural bottleneck in the flow of information through the cognitive system. On the other hand, a number of robust empirical findings cast doubt on whether an account of S-R translation as purely intentional is tenable. These findings fall into four categories, each suggesting a different type of nonintentional and sometimes even counter-intentional S-R translation.

Compatibility: Effects of Stimulus-Response Similarity

Since the classical work of fitts and Seeger (1953), it is known that the speed of S-R translation depends not only on the stimulus or the response but also on the relationship or mapping between stimuli and responses.[2] If stimuli and responses vary on the same dimension, such as with left- and right-hand responses to left- and right-side stimuli, then responses to stimuli having the same value on the respective dimension (e.g., left response to left stimulus) can be initiated faster than responses that do not (e.g., left response to right stimulus).

Of greater interest for our purposes is that feature overlap between stimulus and response affects performance even if this overlap is irrelevant to the task, as convincingly demonstrated by the Simon effect (Simon and Small 1969; for an overview, see Lu and Proctor 1995). This is observed when people make a spatial response, such as a left versus a right keypress, to a nonspatial stimulus attribute, such as color. If the location of the stimulus varies randomly, and if it does so on the same spatial dimension as the response, performance is better if the stimulus spatially corresponds to the response than if it does not. Importantly, this is true not only for absolute spatial S-R correspondence, but also when left and right stimuli appear within the same visual hemifield (Nicoletti and Umiltà 1989; Umiltà and Liotti 1987) or when subjects respond with two fingers of the same hand (Arend and Wandmacher 1987; Heister, Ehrenstein, and Schroeder-Heister 1987). That is, anatomical linkage between hemifield and hand is insufficient to account for the Simon effect.

If S-R translation exclusively reflected the instructed S-R mapping rules, stimulus location would have no effect. The location of the stimulus is obviously processed, however, which leads to at least partial activation of the spatially corresponding response. Presenting a left or right stimulus can be shown to activate the corresponding response even when the relevant stimulus feature calls for the alternate response—whether response activation is assessed by means of lateralized readiness potentials (De Jong, Liang, and Lauber 1994; Sommer, Leuthold, and Hermanutz 1993), electromyographical recordings (Zachay 1991), or registration of subthreshold movements (Zachay 1991). Even symbolic stimuli with a spatial meaning, such as left- or right-pointing arrows, can under certain conditions automatically activate the corresponding response (Eimer 1995). Clearly, these findings provide strong evidence against S-R translation being purely under the control of intentions, all the more so because the critical spatial stimulus feature is evidently not relevant for the task at hand.

One might argue that, for some reason, the wrong stimulus feature is "intentionally" translated into response activation, perhaps because the (nonspatial) relevant stimulus dimension is sometimes confused with the (spatial) relevant response dimension. Or S-R translation might always need to take into account all the features of a relevant stimulus, so that irrelevant features cannot be excluded. However, these attempts to save the intentional translation notion are inconsistent with the observation of Simon-type effects in tasks that, on a given trial, do not require any translation between attributes of the critical stimulus and the appropriate response. For instance, if subjects are signaled to prepare a left- or right-hand keypress in advance of a go/no-go signal—so that all relevant S-R translations can be completed before that signal appears—performance is still better with spatial correspondence between go signal and response (Hommel 1995a, exp. 1). Correspondence effects show up even with 100% go- signal probability, that is, in simple reactions, and even when responses are blocked over 80 consecutive trials (Hommel 1996).

Altogether, these findings clearly undermine the idea that the translation of stimulus location into response activation is wholly under the control of intentional processes. There is more evidence from nonspatial tasks. The best known example is the Stroop effect (Stroop 1935; for an overview, see MacLeod 1991), which occurs when people verbally name the color of ink in which color words are written. Performance is better if the color word denotes the color of ink to be named (e.g., "RED" written in red ink) than if it refers to another color (e.g., "GREEN" written in red ink). On the one hand, the occurrence of the Stroop effect can be taken to show that the meaning of the stimulus word cannot be ignored but is automatically translated into a (congruent or incongruent) response.[3] On the other, requiring subjects to name or respond to the color of the word clearly introduces color as a task-relevant dimension, and it may be exactly this task relevance that makes the word so difficult to ignore.

Habits: Effects of Overlearned Stimulus-Response Associations

From everyday life, we know how difficult it is to escape bad habits, that is, to change or inhibit overlearned responses to particular stimuli (Ouellette and Wood 1998). In what appears to be the first empirical study of the interplay between will and habits, Ach (1910) argued that human will can be studied best when opposed by overlearned habits that need to be overcome. In his "combined method," he first had subjects acquire particular S-R associations by asking them, for instance, to produce a rhyme to a stimulus syllable (e.g., "zup" → "tup"). After extensive practice, he presented the same stimuli but asked for another response, such as reading the syllable backward (e.g., "zup" → "puz"; cf. Hommel 2000). According to Ach, practice leads to direct associations between stimuli and responses, so that presenting a stimulus later on will automatically activate the corresponding response. If this response is not the correct one, it is up to the will to counteract the now dysfunctional habit and to make sure that the intended response is made. This extra demand should show up in two measures: (1) increased reaction time to stimuli previously associated with a different response; and (2) increased occurrence of what Ach called "intended errors," that is, production of the previously associated but now incorrect response.

Although the methodological standards of experiments in these early days certainly do not meet today's expectations—especially the lack of inferential statistics and the small number of subjects per study—both increased reaction times and increased frequency of "intended errors" after the task switch were replicated many times by Ach and several of his students (summarized in Ach 1935). According to Ach, these findings indicate that a stimulus event not only provokes an intentional translation into an appropriate response; it may also, and at the same time, automatically retrieve a previously acquired S-R association, thereby activating the previously associated response.

In a better-controlled study, MacLeod and Dunbar (1988) followed the same logic as Ach in trying to manipulate the relative strength of S-R associations through differential practice (cf., Stroop 1935 for a very similar approach). They first trained their subjects to give verbal color word responses to the shapes of polygons. Then colored polygons were presented, and subjects either named the color (color → color word, shape being irrelevant) or responded to the shape (shape → color word, color being irrelevant). In congruent conditions, stimulus color and shape called for the same response, and in incongruent conditions the implied responses were different. As it turned out, testing after only a little practice produced substantial effects of congruence on shape naming but not on color naming, suggesting that the associations between stimulus colors and color word responses were stronger than those between the shapes and the just acquired color word responses. However, after more practice, congruence also affected color naming; after even more practice, incongruent shapes had a stronger effect on color naming than incongruent colors had on shape naming. Obviously, then, the relative impact of irrelevant stimuli on response selection varies with the relative strength of S-R associations, which suggests that the speed or likelihood of automatic S-R translation, or both, can be affected by learning.

A similar conclusion might be drawn from the findings of Proctor and Lu (1999). Their subjects practiced a spatial compatibility task for three sessions with either a compatible S-R mapping (left stimulus → left response; right stimulus → right response) or an incompatible mapping (left stimulus → right response; right stimulus → left response) before performing a standard Simon task requiring left-right responses to letter stimuli. After compatible mapping practice, a Simon effect of normal size was obtained, but an inverted effect was observed after incompatible mapping practice. Possibly, learning an incompatible mapping leads to the formation of S-R associations that are then automatically activated in the Simon task, too, and thus cancel out, and even overwrite the usual benefits of spatial correspondence.

Rules: Effects of Involuntary Application of the Mapping-Rule

Thus far, the evidence for automatic S-R translation discussed has been restricted to S-R pairs that were either compatible or highly overlearned. However, indications of automatic translation have also been observed in single-session experiments (with no opportunity for extensive S-R learning) using arbitrary S-R mappings. In none of these studies were the translation-inducing stimulus attributes really task irrelevant, nor was the translation completely unrelated to the task or the subject's intentions. Nevertheless, the translation indicated by the results was involuntary and inappropriate, either translating the wrong stimulus or occurring at the wrong time—the right rules used in a wrong way.

If people make a discriminative response to a visual target, their reaction time is strongly affected by irrelevant stimuli surrounding the target. For instance, if a left versus right keypress is made to the centrally presented letters *H* and *S*, which are flanked by other letters, performance is better if target and flankers look the same (e.g., *H* flanked by *H*s) than if the flankers resemble the other, alternative target (e.g., *H* flanked by *S*s; Eriksen and Eriksen 1974). This is not just an effect of visual similarity or distraction. If two dissimilar letters are assigned to each response, flankers assigned to the same response as (but different from) the present target produce better performance than flankers assigned to the alternate response (Miller 1991). Such an effect suggests that both flankers and target are processed and activate their corresponding responses. Indeed, incongruent flankers activate their assigned (incorrect) response to the extent that the activation can be observed in the lateralized readiness potential (LRP; Coles et al. 1985), or in electrophysiological activity (Eriksen et al. 1985) and overt, subthreshold movements of the wrong hand (St. James 1990).

The flanker effect demonstrates that stimuli are not translated into response activation only in strict conformity with the actor's intention and thus indicates some kind of automatic processing.[4] It is also true, however, that the incorrectly selected and translated flanker stimuli in a flanker task are not completely irrelevant; after all, they are valid targets that merely appear at a wrong location. On the one hand, S-R translation in a flanker task is intentional in the sense that it realizes the intention to respond to stimuli in a particular way. On the other, it seems that not every aspect of the resulting translation can be controlled, so that, somewhat paradoxically, intended S-R rules are automatically applied.

A very similar picture emerges from studies on task-switching performance: moving from one task to another does not switch off the previously used S-R mapping rules completely. Consider, for instance, Sudevan and Taylor 1987, whose subjects responded to single digits ranging from 2 to 9 by pressing a left or right key. There were two different S-R mapping rules, varying randomly from trial to trial, that were signaled by a letter cue preceding the stimulus. According to one rule, odd digits were assigned to one response key and even digits to the other, while the alternate rule assigned low digits (2–5) to one key and high digits (6–9) to the other. Obviously, such mappings introduce conditions of rule or intertask S-R congruence and incongruence, inasmuch as some stimuli require the same response under either S-R assignment (e.g., "3" if "odd" and "low" stimuli were assigned the same key), while other stimuli imply different responses (e.g., "2"). In fact, intertask congruence had a strong impact on performance, with response-congruent stimuli (i.e., stimuli that in the alternate task would require the same response) speeding up performance even if the mapping rule was precued as early as four seconds before the stimulus set in. Similar effects have been observed in Rogers and Monsell 1995, in Meiran 1996, and in several experiments in our lab, suggesting that cross talk between tasks is a reliable phenomenon (cf., Allport and Wylie, chap. 2, this volume). As observed by Otten et al. (1996), this cross talk can have far-reaching consequences, with stimuli belonging to the currently invalid task triggering their associated response up to a level of response-related LRPs.

Note that cross talk between different tasks can occur only if the mapping rules of these tasks are concurrently applied to translate the stimulus into response activation. In fact, participants in task-switching studies seem not so much confused about what to do as uncertain about which (of the simultaneously applied) rules to follow. For instance, Meiran and Daichman (forthcoming) had people switch between tasks under high time pressure, so that many errors were made. Analyses and simulations showed that the types of errors made were not random but rather reflected the correct use of the incorrect S-R mapping rule, which fits well with the (commonly less pronounced) error patterns observed in other task-switching studies. Obviously, then, we have here the same kind of interplay between intentional and automatic processes as seen before. On the one hand, we find evidence of S-R translation that is neither needed nor helpful, which indicates a high degree of automaticity even in the absence of extensive practice and S-R similarity. On the other, the outcomes of these automatic processes do not seem erratic or habitlike, but rather are strongly related to the actor's intentions.

The same conclusion can be drawn from Hommel 1998a. Subjects performed two tasks in a row (response order was strictly controlled), a manual left-right keypressing response (R_1) to the color (S_1) of a stimulus, followed by a verbal color name response (R_2) to the form (S_2) of the same stimulus. As often found in such double tasks, the second response was delayed relative to the first by a half second or more, hence there was a "psychological refractory period" (PRP) effect (Telford 1931). However, the type of R_2 strongly affected reaction time in the primary manual task. If the meaning of R_2 corresponded to the color of S_1 (e.g., S_1 = red; R_2 = "red") the response to S_1 was much faster than if R_2 and S_1 did not match (e.g., S_1 = red; R_2 = "green"). This could only happen if R_2 was activated before the primary task was completed, which again implies that (at least some) S_2-R_2 translation must have taken place with or even before the processing of S_1 and R_1. Obviously, then, S-R translation is unlikely to be the "structural bottleneck" that is widely believed to be responsible for dual-task costs and PRP effects (e.g., Pashler 1994, chap. 12, this volume; Welford 1952). Whatever or wherever this bottleneck may be, it does not seem to prevent different stimuli from being translated into response activation at about the same time. Indeed, the application of arbitrary S-R translation rules seems to be so automatic that it occurs even if it or its timing produces unintended and inappropriate results.

Integration: Aftereffects of Stimulus-Response Binding

The previous examples show that extensive learning may promote, but is not always necessary to bring about, automatic S-R translation. Even single-trial learning can produce stimulus-triggered response activation. Hommel (1998b) used a task that required two responses (R_1 and R_2) to two stimuli (S_1 and S_2) on each trial. Participants were presented with a response cue that signaled the identity of R_1 (e.g., left versus right keypress). R_1 was prepared but not performed until S_1 was presented. Although S_1 varied randomly in shape, color, and location (e.g., green versus red; *X* versus *O*; top versus bottom position), R_1 did not depend on or covary with any of the features of S_1. About 1 sec later, S_2 appeared; it varied on the same dimensions as S_1, with one feature (shape, say) signaling R_2. That is, the already prepared, simple R_1 was made to the mere onset of S_1, and the binary forced-choice discrimination R_2 was made to the relevant feature of S_2. For example, a left-pointing arrow might cue a left-hand response, which is then prepared and performed at S_1 onset, independent S_1 being, say, a red *X* in the top position. After 1 sec, S_2 would appear (e.g., a green *X* at the bottom position), with its shape signaling a left-hand response. (Note that this example implies repetition of stimulus shape and response, and alternation of stimulus color and location.)

One might expect several kinds of repetition effects with a task like this, such as better performance if a stimulus feature or the response is repeated. Indeed, repetition effects were obtained, although not very reliably so and only in task versions with very short intervals between S_1 and S_2 (Hommel forthcoming-a). Much more interesting, however, is the consistent observation that stimulus- and response-related repetition effects interacted. In particular, repeating stimulus shape or location was beneficial *only* if the response was also repeated; if not, shape or location repetition yielded interference instead (Hommel 1998b). Apparently, a single co-occurrence of S_1 and R_1 resulted in an association or binding of stimulus (features) and response (features). As a consequence, presenting the same stimulus (feature) reactivated the associated response, which caused a problem if this response was not the correct one, that is, if the repeated stimulus required a new response. That automatic response activation is indeed involved is also suggested by experiments in which the forced-choice R_2 was replaced by a free-choice response to S_2. Even if urged to avoid any strategy and produce random behavior, participants tend to repeat R_1 if S_1 is also repeated (Hommel forthcoming-b). Being unintended, unwanted, and unhelpful, these S-R binding effects fulfill the most common criteria for automaticity and hence represent a case of automatic S-R translation. Interestingly, however, they clearly do not result from practice or S-R compatibility, or from applying S-R rules in an inappropriate way.

Automaticity: Types versus Degrees

The foregoing examples making the case for automatic translation stem from a broad range of tasks and paradigms and may therefore seem to indicate very different kinds of automaticity. However, it is tempting to try ordering them on a common dimension, such as the length of the learning history involved. Binding effects, which result from experiencing a single S-R co-occurrence, clearly have the shortest history, followed by effects indicating inappropriate rule use, which can be measured after only 50 trials or less. Then we have effects of S-R associations that seem to take several sessions of practice to emerge and, finally, effects of S-R compatibility, which are sometimes attributed to extreme overlearning of S-R relationships (cf. Umiltà and Zorzi 1997). Indeed, the available demonstrations of automatic S-R translation may differ only with respect to the strength of the underlying S-R associations and thus indicate merely different degrees or states, not different types of automaticity.

Although such an account is attractively parsimonious, it is not supported by the (still few) findings that speak to this issue. First, there is no evidence available as to whether binding effects increase with the number of consistent S-R occurrences, so that it is not clear whether binding is the first stage of S-R associative learning or only a temporary phenomenon. Second, there is no indication that effects of inappropriate rule use would increase over practice. On the contrary, whereas Hommel (1998a) found no systematic relationship between effects of automatic, inappropriate rule use and practice within a single session, Sudevan and Taylor (1987) observed a general decrease of such effects over 20 sessions of task-switching practice. Third, whereas there is strong evidence for the impact of task-irrelevant S-R associations on performance increasing with practice (MacLeod and Dunbar 1988), the studies on automatic rule use (Hommel 1998a; Sudevan and Taylor 1987) have found no evidence of such a relationship, suggesting that the two kinds of effect are of different origin.

Fourth, up to now there is no convincing evidence that S-R compatibility effects are due to S-R learning. Of course, testing this assumption is difficult—if we are talking about lifelong experience (e.g., responding with the right hand to objects on the right side or verbally responding to objects with their name), it would be unethical to prevent subjects from having this experience and impractical to experimentally induce an equivalent number of (counter-) practice trials. Nevertheless, several studies have investigated whether S-R compatibility effects could be eliminated through extensive practice. For instance, Fitts and Seeger (1953) found better performance with spatially compatible than with incompatible S-R mappings even after 32 sessions of practice. Later studies all showed the same pattern of results. During the very first trials, subjects have much more difficulty getting into the task with an incompatible than with a compatible mapping, but then the difference between compatible and incompatible conditions stabilizes very quickly and is more or less unaffected by further practice (e.g., Brebner, Shephard, and Cairney 1972; Dutta and Proctor 1992; Morin and Grant 1955). A similar pattern has been observed in Simon tasks. Although Simon, Craft, and Webster (1973) did find a reduction during 5 sessions of 216 trials each, a pronounced Simon effect was still observed in the final session (see also Proctor and Lu 1999). Even 30 sessions of 210 trials each do not suffice to eliminate the effect, as demonstrated by the performance of a single, heroic subject in Hommel 1995b.

To sum up, the available findings do not support the assumption of a single dimension of automaticity or associative S-R strength on which the observed phenomena could be easily ordered. However, given that some relationships between significant phenomena have not yet been investigated and that some of these relationships are difficult to investigate in any meaningful way, it would be premature to draw definitive conclusions.

11.2 MULTIPLE ROUTES FROM STIMULUS TO RESPONSE

We have seen substantial evidence against the intuitive, but perhaps naive idea that S-R translation is a control operation that realizes the intentions of a perceiver or actor, thereby shielding the action system against unwelcome stimulus-induced action tendencies. The insight that S-R translation is only partially under intentional control has led to the formulation of several models that assume both an intentional and an automatic route from perception to action. I shall review some of the most influential dual-route models, considered state-of-the art in many domains of information-processing psychology, pointing out limitations that need to be overcome if we are to achieve a comprehensive model of S-R translation.

Dual-Route Models

Part of the reasoning behind today's dual-route models can already be found in Ach 1910, which distinguished between will, a capacity-limited mechanism in charge of S-R translation and action control, and habits, S-R associations that result from and become stronger with S-R learning. Habits are assumed to lead to fully automatic S-R translation, that is, to the activation of the response most often associated with the given stimulus in the past. If the outcome of this translation is in agreement with (i.e., functional for reaching) the intended action goal, only minimal effort (or will power) needs to be applied, and execution is facilitated. If an existing habit activates a counterproductive tendency, however, this needs to be overcome by an increase in effort deployed.

Although current dual-route models are often more specific as to the processes involved and the conditions that need to be fulfilled, the general idea that habit and will compete for action control is still alive—even if habitual S-R translation is now called "automatic" or "unconditional" and willed translation referred to as "intentional" or "conditional." A model that has much in common with Ach's has been suggested by Logan (1988), who assumes that each experience of a S-R episode leaves a memory trace of an "instance." Attended stimulus events necessarily retrieve the instances associated with them and, through that retrieval, activate the associated response. The more S-R co-occurrences experienced in the past, the more instances retrieved; the more instances retrieved, the more likely it is that the corresponding response will be activated, which then will compete with intentional, rule-governed S-R translation processes for action control. Although the two models implement habits in different ways—through the strengthening of single S-R associations (Ach) and through a separate trace for each experience (Logan)—the general way they characterize the relationship between intentional and automatic S-R translation is very similar.

Perhaps the most general of dual-route models, Kornblum, Hasbroucq, and Osman's "dimensional overlap model" (1990) attributes S-R compatibility effects to a competition between automatic response activation and voluntary S-R translation. If, and only if, a stimulus event shares features with a response, such as with spatial S-R correspondence in a Simon task, the stimulus activates the corresponding response automatically and in parallel to the controlled translation of the relevant stimulus feature into the correct response. If the automatically activated response happens to be appropriate, response execution is faster and performance better. If not, the system must suppress the misleading response tendency before the correct response can be issued—a time-consuming process. This basic architecture is shared by other, less general models of S-R compatibility (e.g., De Jong, Liang, and Lauber 1994; Hommel 1993a; Lu 1997; Virzi and Egeth 1985).

In the last decade, computational parallel distributed processing (PDP) or neural network models of S-R compatibility have spelled out the dual routes in increasing detail, often implementing intentional and automatic routes in very similar ways. Typically, stimulus feature codes are assumed to be permanently connected to codes of responses they share features with, such as a left stimulus code and a left response code (e.g., Barber and O'Leary 1997; Kornblum et al. 1999; Zorzi and Umiltà 1995). Consequently, registering and coding a stimulus leads to a spreading of activation to the feature-overlapping response, hence to automatic S-R translation. In contrast, intentional translation is modeled by introducing temporary, short-term associations connecting codes of the relevant stimulus feature or features and the respective response. These associations are task specific and intention dependent and may be taken to represent something like S-R rules temporarily stored in working memory.

The notion of dual routes from perception to action has advanced our basic understanding of S-R compatibility and motivated a wealth of empirical investigations. It has played a crucial role in explaining, among other things, the consistent observation that the Simon effect decreases with increasing task difficulty (De Jong, Liang, and Lauber 1994; Hommel 1993a) and the dependence of spatial compatibility effects on task preparation (De Jong 1997; Shaffer 1965). There are several reasons, however, why the basic idea and architecture of dual-route models may fail to fully capture the essence and diversity of S-R translation. I shall discuss three.

Multiple Routes to Action

Constructed to serve rather specialized purposes, such as accounting for practice effects or effects of S-R compatibility, existing dual-route models emphasize one particular type of automaticity and neglect others. Inasmuch as there is more than one type or cause of automatic S-R translation, however, none of the available models seems sufficiently developed to serve as a comprehensive model of S-R translation. Such a model would need more than two routes or pathways from perception to action. To model such multiple pathways, we need to understand the relationships between the various phenomena indicative of automatic translation.

First, we need to know whether S-R binding is only a process for short-term temporary integration or whether it also represents the mechanism that forms long-term S-R associations—what Logan (1988) has called "instances." Second, we need to know when, how, and why S-R rules, stored in working memory to guide current behavior, can be accessed and used by other, inappropriate or irrelevant stimuli to activate the corresponding responses, and what roles short-term binding and long-term learning play in this context. Third, we need to know more clearly what the relationship is between habits or overlearned S-R associations and S-R compatibility. Take, for instance, MacLeod and Dunbar's finding (1988) that practicing at naming shapes with color words results in Stroop-like interference with naming colors. If this effect indicates some kind of acquired compatibility between irrelevant stimulus shape and response (which are defined on nonoverlapping dimensions), this would seem to argue against, say, the dimensional overlap model of Kornblum, Hasbroucq, and Osman (1990). Alternatively, if the effect is assumed to be mediated by different mechanisms and simply to mimic compatibility effects, we need to specify these mechanisms and how they differ from those mediating compatibility effects. This in turn requires compatibility models to be specific as to *why* similarity between stimulus and response sets lead to automatic S-R translation—an issue commonly neglected in dual-route models (but see Eimer, Hommel, and Prinz 1995; Hommel 1997).

Automaticity of Intentional Translation

Obviously, people can respond to the same stimulus in many different ways, depending on the task or context and, most important, depending on their intentions and strategies. To account for this enormous degree of flexibility in S-R translation, dual-route models have been equipped with "intentional" or "controlled" pathways, that is, with perception-action links that are under full control of the perceiver's or actor's intentional states. On the other hand, we have already seen that intentional or controlled translation is not always as intended and controlled as it should be: irrelevant flankers activate arbitrarily assigned responses, and task-specific S-R rules are inappropriately applied while performing another task. This means that stimuli can activate responses automatically not only via the automatic pathway proposed by dual-route models but also by the intentional route. If so, it cannot be the *process* of S-R translation that is under intentional control, but rather the *implementation* of the underlying S-R rules. That is, although intentional states may determine which rules are selected, formed, and implemented, once they are established, stimuli seem to have direct and uncontrolled access to these rules, leading to automatic translation via intentional routes.

This conclusion has important theoretical implications. First, as far as S-R translation is concerned, it shifts the time point of intentional control from the interval between stimulus perception and response selection to the beginning of a task. In a sense, such a view stands in contrast to Donders's idea (1868) that "will determination" follows perception—an idea that has made its way into many modern information-processing models. In fact, if the preconditions for S-R translation are already set before a stimulus comes up, at least part of the will has already been determined in advance, a consideration I will develop in the section 11.3.

Second, if intentional S-R translation is really as automatic as the available findings suggest, it is unlikely to represent the processing bottleneck that has always been associated with it by single-channel models of dual-task performance since Welford 1952. Obviously, if more than one stimulus at a time can be translated into a response, there is no reason why costs observed in dual-task performance should have something to do with S-R translation. Rather, it may be the automaticity of intentional translation, not the lack of it, that causes the trouble. If more than one stimulus at a time is translated into its response, the system may need to find out which response belongs to which stimulus, and in what order the responses are to be carried out. This may be called a problem of "response selection," but not one of S-R translation (Hommel 1998a).

Intentionality of Automatic Translation

Although exact criteria for automaticity are still under debate (e.g., Bargh 1989; Hasher and Zacks 1979; Neumann 1984), dual-route models explicitly or implicitly share the definition of Kornblum, Hasbroucq, and Osman (1990, 261) that the automatic route can "under some conditions be attenuated or enhanced" but "under no conditions ... ignored or bypassed," and that, accordingly, people "whether instructed to use or to suppress an automatized process would therefore produce evidence of its operation in their performance." There are reasons to believe, however, that automatic S-R translation is not independent of the task at hand and the instructions given to acting participants. In particular, it has been shown that the occurrence of effects attributed to automatic translation depends on attention (i.e., the way stimuli are selected and coded), intention (i.e., the way responses are prepared and coded), and on task-specific strategies.

Attention and Stimulus Coding A first demonstration of the impact of instructions on "automatic" S-R translation comes from the observation that the Simon effect occurs not only with unilateral, but also with symmetrical, bilateral stimulation. That is, even when people are presented with a left and a right stimulus at the same time, with the relevant one defined by its form (Grice, Boroughs, and Canham 1984), color (Hommel 1993b; Proctor and Lu 1994), or meaning (O'Leary and Barber 1993), they are faster if the relevant stimulus comes up on the same side as the required response. Thus it is not the spatial correspondence between any stimulus and the response that matters for the Simon effect, but the spatial relationship between the *attended* stimulus of a display and the response (Stoffer and Umiltà 1997). Given that the task instruction specifies which stimulus to attend to, this implies that there is no Simon effect without specific task instructions, at least when more than one stimulus is presented at a time. Inasmuch as the Simon effect is attributed to automatic S-R translation, this kind of translation cannot be completely independent from the task.

There are more challenging findings. Consider, for example, Eimer's observation (1995) that response-irrelevant arrows automatically activate corresponding responses, a finding consistent with dual-route models of S-R compatibility. In a recent lateralized readiness potential study, Eimer and Schlaghecken (1998) showed that even subliminal (i.e., not consciously perceivable) arrowheads preceding a target arrow activated the corresponding response. However, as soon as the relevant arrow stimuli were replaced by letters without any spatial meaning, arrow primes no longer produced "automatic activation." Obviously, the translation of stimulus information into the activation of spatially congruent responses can depend critically on what relevant information a perceiver or actor intends to translate—hence automatic translation depends on intentions.

A very similar conclusion is suggested by the findings of Cohen and Shoup (1997), who modified the standard flanker task by manipulating targets and distractors on two dimensions: color and orientation. For instance, one response key could be assigned to a red vertical line and a blue right diagonal line and the other key to a green vertical line and a blue left diagonal line. If target and flankers were defined on the same dimension (e.g., red vertical line flanked by red vertical lines versus green vertical lines), the standard flanker effect was obtained, that is, congruent flankers produced better performance than incongruent flankers. If, however, target and flankers were defined on different dimensions (e.g., red vertical line flanked by blue right diagonal lines versus blue left diagonal lines), there was no congruence effect—an observation also made by Fournier, Eriksen, and Bowd (1998) in a speeded feature judgment task. It seems that, although incongruent flankers are unintentionally translated into corresponding response activation, the probability of this translation is strongly determined by what is defined and identified as target, that is, by task-specific, attentional and intentional processes. This fits nicely with the results of Bauer and Besner (1997), who showed that Stroop words affect keypressing responses only if participants classify the ink of the words, but not if they judge whether a given color is present or absent (even if RT levels are comparable). Obviously, automatic processes are (or at least can be) task dependent.

Intention and Response Coding Evidence for a role of response sets in S-R translation comes from Hommel's 1996, study on spatial S-R compatibility in simple, prepared responses. One major outcome was that effects of S-R compatibility are not restricted to situations involving response uncertainty, as commonly believed (e.g., Berlucchi et al. 1977), but also occur if a completely prepared response is made to a spatially compatible or incompatible go stimulus. It also turned out that the size of the compatibility effect depended strongly on the task relevance of the responses. For instance, if the same (left- or right-hand) response was used throughout a long block of trials, the effect of spatial correspondence between response and go signal was very small and often insignificant. Interestingly, though, much larger and more reliable correspondence effects showed up when another spatial (i.e., right- or left-hand) response was used in a secondary task performed in between the trials of the compatibility task. Apparently, the overlap of stimulus and response features is not a sufficient predictor of automatic S-R translation. Whether a particular response possesses a particular feature and whether this feature overlaps with those of the stimulus are of little consequence if the task at hand does not require use of the response feature to discriminate the response from another one. In other words, similarity between a stimulus and a response produces "automatic" S-R translation only (or at least mainly) if the respective feature dimension is important to the given task context.

If this is so, one should be able to manipulate the kind of "automatic" S-R translation by asking the participant to attend more to some response features than to others. This is what Hommel (1993c) did in a version of the Simon task, where people responded to the pitch of a tone heard randomly on the left or right side by pressing a left- or right-hand key. Pressing a particular key flashed a light on the opposite side, so that each response had two spatial features: the location of the finger or key and the location of the action-contingent light. When subjects were instructed, as in a standard Simon task, to "press the left/right key in response to the low/high pitch," left-hand keypresses were faster to left-side tones and right-hand keypresses were faster to right-side tones—a standard Simon effect. When, however, subjects were instructed to "flash the right/left light in response to the low/high pitch," left-hand keypresses were faster to right-side tones and right-hand keypresses were faster to left-side tones. Obviously, the instruction not only had a strong impact on automatic S-R translation; it actually determined its outcome. Merely describing the task in terms of keypressing led the participants to code their responses with respect to the locations of the response keys, whereas describing the very same task in terms of light flashing persuaded them to code their responses with respect to the locations of the lights. If we attribute the Simon effect to automatic S-R translation, this is further evidence that automatic translation is not independent of how participants interpret the task and how they intend to solve it.

Strategies and Implementation of Stimulus-Response Rules Apart from stimulus- and response-related factors, automatic translation can also be affected by task-specific strategies and expectations. Evidence for this comes from variations of the relative frequency or likelihood of stimulus-stimulus-congruent or stimulus-response-compatible trials in Stroop tasks (Logan 1980; Logan and Zbrodoff 1979), flanker tasks (Gratton, Coles, and Donchin 1992), and Simon tasks (Hommel 1994; Toth et al. 1995), that is, from manipulations of the utility of irrelevant, but response-related information. Whatever the task, increasing the frequency of congruent or compatible trials increased, and decreasing the frequency decreased or even eliminated, the effect. In the same vein, Proctor, Lu, and Van Zandt (1992) found that the Simon effect gets larger if the likely response is precued and can be prepared in advance. Clearly, these observations suggest that the degree and outcome of automatic translation is modified by, and sometimes even depends on, task-specific strategies and preparatory processes.

Further evidence for a role of task preparation has been reported by Valle-Inclán and Redondo (1998), who measured response activation in a Simon task by means of LRPs. The relevant S-R mapping was not fixed in this study, but varied randomly from trial to trial, as did the temporal order in which mapping and stimulus were presented. When the mapping was presented before the stimulus, the stimulus immediately activated the spatially corresponding response, independently of which response was correct. That is, there was evidence of automatic S-R translation. On the other hand, when the stimulus appeared before the S-R mapping, automatic response activation was no longer observed. Apparently, although automatic S-R translation did not follow the relevant S-R rules, it required their implementation or at least, as Valle-Inclán and Redondo suggest, some degree of readiness to react. Whatever the correct answer may be, it seems clear that automatic routes proposed by dual-route models can be "ignored or bypassed," which stands in contradiction to how these routes are typically defined and characterized.

11.3 PROSPECTS: STIMULUS-RESPONSE TRANSLATION AS PREPARED REFLEX

The abundant evidence for several kinds of automatic access of stimuli to action control calls for a translation model with more than just one, highly controlled pathway from perception to action. As a consequence, several dual-route models have been developed to account for different aspects of the available evidence, and these models are quite successful in their respective empirical domains. On the other hand, if we want a comprehensive S-R translation model not restricted to particular experimental effects, we still have some way to go.

I have sketched three major theoretical problems that need to be solved. First, a comprehensive model is likely to comprise more than two routes. There is evidence of at least four kinds of automatic S-R translation, and the ways they differ do not suggest that they originate in the same type of process. It thus seems insufficient to distinguish just one intentional and one automatic route. We need more complex, multiroute models. Second, observations of inappropriate rule use suggest that the intentional route from perception to action is not very tightly controlled, but can be automatically accessed by task-related stimuli. This raises doubts about the usefulness of distinguishing between controlled and uncontrolled routes, or at least requires that we specify exactly when and how control is exerted. Third, phenomena that current dual-route models attribute to automatic S-R translation strongly depend on attentional set and action intentions, suggesting that the supposedly automatic route is not uncontrollable. Thus, all in all, there are reasons to doubt that the roles of, and the interplay between, control and automaticity in S-R translation are best captured by the distinction between intentional and automatic routes.

A more suitable approach to the control-automaticity relationship might be derived from consideration of Exner 1879. On the basis of his introspections in "reaction time" experiments (a term he had introduced to psychology six years earlier), Exner explicitly rejected the notion that intentional control (or the will) intervenes between stimulus and response—a notion that seemed quite natural to Donders and that still does to his followers. Exner argued that preparing for a task is accomplished by setting oneself, long before the first stimulus comes up, into a state that ensures that responses are carried out efficiently and as intended. Although evoking that state is a voluntary act requiring attention, once the state is created, the response is actually involuntary, that is, no further effort of will is needed to translate the upcoming stimulus into the response. In fact, stimuli trigger their respective response unless the mediating state is actively deactivated or inhibited. According to this conception, intentional processes do not actually carry out S-R translation, but only configure the cognitive system to do so automatically, once the defined target stimulus arrives—that is, as a "prepared reflex"

(Woodworth 1938). Interestingly, the old idea of theoretically distinguishing between intentional set implementation and set-dependent, but automatic S-R translation is currently experiencing a revival (see the overview by Monsell 1996), and recent models such as those of Cohen and Huston (1994) or Meyer and Kieras (1997; Kieras et al., chap. 30, this volume) can be viewed as first, systematic attempts to implement the major aspects of this distinction into a computational framework.

From a prepared reflex perspective, it is not so surprising to find evidence of both automaticity of intended S-R translation and intentional control of automatic routes. Obviously, a prepared cognitive reflex is neither exclusively automatic nor exclusively voluntary. On the one hand, it is implemented as a consequence of, and does express a voluntary decision to perform an action under particular circumstances in a particular way and thus necessarily depends on task and intention. If so, the resulting task set is likely to reflect the way the task is understood and interpreted by the perceiver or actor, and hence determines how stimuli are coded (e.g., which stimulus features are attended and linked to response features), how responses are coded (e.g., which response features are attended and linked to stimulus features), when stimulus information is expected, and when actions are prepared and issued. As we have seen, all these decisions have a strong impact on the occurrence of automatic processes, and therefore can be regarded as both implementing arbitrary, transient S-R connections (the intentional route) and directly or indirectly enabling learning- or compatibility-related S-R associations (the automatic route).

Once a task set is implemented (and automatic routes enabled), however, the whole system is prepared to act in an automatic fashion—and this may sometimes produce undesirable side effects. It is certainly an advantage that the cognitive system is able to automatize itself, so to speak, so that the onset of a stimulus immediately triggers the corresponding prepared action without (much) further ado. On the other hand, the price to pay for this economical solution is that unwanted information will sometimes lead to troublesome consequences, especially if an irrelevant stimulus fits the internal description of the triggering stimulus, such as in flanker or Stroop tasks, or in task-switching experiments. Nevertheless, even unhelpful and misleading S-R translations of this sort strictly depend on, and thus in some sense represent, the actor's intention.

Such a prepared reflex view may be helpful in developing a comprehensive theory of S-R translation. Indeed, it complements and extends recent attempts at computational modeling of S-R translation processes in compatibility and related tasks. Take, for instance, the models of Barber and O'Leary (1997) and of Zorzi and Umiltà (1995), which distinguish between transient S-R associations reflecting the instructed S-R mapping and permanent links that can be hard-wired or acquired through learning. Although this distinction maps onto that of intentional and automatic routes, once the transient links are implemented, they work in a purely stimulus-triggered fashion like their permanent counterparts. That is, the two types of pathway differ only in history and durability, not in automaticity. The same can be said of the model proposed by Cohen, Dunbar, and McClelland (1990) and Cohen and Huston (1994), who went one step further in attempting to deal with the process of route implementation itself (also treated in Meyer and Kieras 1997). To do so, task demand representations are postulated, the activation of which (e.g., through presenting task instructions) can directly modify the flow of information from stimulus to response codes. In this case, S-R links differ neither in permanence nor automaticity, but in task-specific strength only. Although it is clear that more work needs to be done to understand and model in greater detail how S-R associations are acquired in the first place, how stimulus and response coding can affect the implementation or use of S-R links, and how the preparation to act influences the likelihood of automatic S-R translation, current modeling attempts are very much in line with the idea of S-R translation as a prepared cognitive reflex.

To summarize, we have seen that S-R translation is not just a direct expression of human will, nor is it satisfactorily sketched as a competition between fully automatic, stimulus-triggered processes and autonomous control operations representing an on-line realization of task intentions. S-R translation is almost always modulated by the intentions of the perceiving or acting person. Rather than directly intervening between stimulus perception and response selection, and thus actually performing the translation, intentional processes seem merely to set the stage for later S-R translation and to leave the rest to the dynamic interplay between intentionally implemented and nonintentionally enabled automatic processes. Even though this kind of interplay may sometimes produce unwanted side effects, we must not forget that intentions usually refer to behavioral outcomes, not to processes realizing them. Therefore, the functionality of our intentionally controlled automatic processes should be judged in terms not of reaction times but of behavioral outcome. Given that, with sufficient time, no subject in a Stroop task would ever name the color word, this surely provides a much brighter perspective on our capacity for self-control.

NOTES

1. Some evidence pertaining to the relationship between conscious awareness and the control of manual pointing and grasping is reviewed by Milner (chap. 9, this volume), although the distinction made there between processing streams for conscious perception and for visuomotor control does not easily map onto the distinction between intentional and automatic S-R translation discussed here.

2. In this chapter, the terms *compatible* and *incompatible* refer to the relationship or mapping between stimuli and responses, whereas the terms *congruence* and *incongruence* refer to the relationship between stimuli or between responses.

3. The Stroop effect has also been observed with manual keypressing responses (e.g., in the absence of S-R feature overlap; Keele 1972), which might be taken to suggest a contribution of stimulus-stimulus (in)congruence to the overall Stroop effect (e.g., Kornblum 1994). Even if this were so, however, the robust finding that switching from manual to verbal responses substantially increases the effect (e.g., Redding and Gerjets 1977) shows that S-R compatibility makes an important contribution of its own.

4. Note that this conclusion in no way depends on the actual cause of the flanker effect. Whether the effect is due to a conflict between target- and flanker-activated responses (Eriksen and Schultz 1979) or to interactions between target- and flanker-coding processes (Kornblum et al. 1999)—implying that response activation only reflects, but does not produce, the flanker effect—it is clear that (1) flanker information is translated into response activation and (2) this particular translation is not intended.

REFERENCES

Ach, N. (1910). *Über den Willensakt und das Temperament.* Leipzig: Quelle and Meyer.

Ach, N. (1935). *Analyse des Willens.* In E. Abderhalden (Ed.), *Handbuch der biologischen Arbeitsmethoden.* Vol. 4, Berlin: Urban and Schwarzenberg.

Arend, U., and Wandmacher, J. (1987). On the generality of logical recoding in spatial interference tasks. *Acta Psychologica, 65,* 193–210.

Barber, P. J., and O'Leary, M. J. (1997). The relevance of salience: Towards an activational account of irrelevant stimulus-response compatibility effects. In B. Hommel and W. Prinz (Eds.), *Theoretical issues in stimulus-response compatibility,* pp. 135–172. Amsterdam: Elsevier.

Bargh, J. A. (1989). Conditional automaticity: Varieties of automatic influence in social perception and cognition. In J. S. Uleman, and J. A. Bargh (Eds.), *Unintended thought,* pp. 3–51. London: Guilford Press.

Bauer, B., and Besner, D. (1997). Processing in the Stroop task: Mental set as a determinant of performance. *Canadian Journal of Experimental Psychology, 51,* 61–68.

Berlucchi, G., Crea, F., Di Stefano, M., and Tassinari, G. (1977). Influence of spatial stimulus-response compatibility on reaction time of ipsilateral and contralateral hand to lateralized light stimuli. *Journal of Experimental Psychology: Human Perception and Performance, 3,* 505–517.

Brebner, J., Shephard, M., and Cairney, P. (1972). Spatial relationships and S-R compatibility. *Acta Psychologica, 37,* 93–106.

Cohen, A., and Shoup, R. (1997). Perceptual dimensional constraints in response selection processes. *Cognitive Psychology, 32,* 128–181.

Cohen, J. D., Dunbar, K., and McClelland, J. L. (1990). On the control of automatic processes: A parallel distributed processing account of the Stroop effect. *Psychological Review, 97,* 332–361.

Cohen, J. D., and Huston, T. A. (1994). Progress in the use of interactive models for understanding attention and performance. In C. Umiltà, and M. Moscovitch (Eds.), *Attention and Performance XV: Conscious and nonconscious information processing,* pp. 453–476. Cambridge, MA: MIT Press.

Coles, M. G. H., Gratton, G., Bashore, T. R., Eriksen, C. W., and Donchin, E. (1985). A psychophysiological investigation of the continuous flow model of human information processing. *Journal of Experimental Psychology: Human Perception and Performance, 11,* 529–553.

De Jong, R. (1993). Multiple bottlenecks in overlapping task performance. *Journal of Experimental Psychology: Human Perception and Performance, 19,* 965–980.

De Jong, R. (1997). Compatibility effects on performance and executive control in dynamic task settings. In B. Hommel and W. Prinz (Eds.), *Theoretical issues in stimulus-response compatibility,* pp. 223–239. Amsterdam: Elsevier.

De Jong, R., Liang, C.-C., and Lauber, E. (1994). Conditional and unconditional automaticity: A dual-process model of effects of spatial stimulus-response correspondence. *Journal of Experimental Psychology: Human Perception and Performance, 20,* 731–750.

Donders, F. C. (1868). Over de snelheid van psychische processen. *Onderzoekingen, gedann in het physiologisch laboratorium der Utrechtsche hoogeschool,* series 2, vol. 2, pp. 92–120.

Dutta, A., and Proctor, R. W. (1992). Persistence of stimulus-response compatibility effects with extended practice. *Journal of Experimental Psychology: Learning, Memory, and Cognition, 18,* 801–809.

Eimer, M. (1995). Stimulus-response compatibility and automatic response activation: Evidence from psychophysiological studies. *Journal of Experimental Psychology: Human Perception and Performance, 21,* 837–854.

Eimer, M., Hommel, B., and Prinz, W. (1995). S-R compatibility and response selection. *Acta Psychologica, 90,* 301–313.

Eimer, M., and Schlaghecken, F. (1998). Effects of masked stimuli on motor activation: Behavioral and electrophysiological evidence. *Journal of Experimental Psychology: Human Perception and Performance, 24,* 1737–1747.

Eriksen, B. A., and Eriksen, C. W. (1974). Effects of noise letters upon the identification of a target letter in a nonsearch task. *Perception and Psychophysics, 16,* 143–149.

Eriksen, C. W., Coles, M. G. H., Morris, L. R., and O'Hara, W. P. (1985). An electromyographic examination of response competition. *Bulletin of the Psychonomic Society, 23,* 165–168.

Eriksen, C. W., and Schultz, D. W. (1979). Information processing in visual search: A continuous flow conception and experimental results. *Perception and Psychophysics, 25,* 249–263.

Exner, S. (1879). Physiologie der Grosshirnrinde. In L. Hermann (Ed.), *Handbuch der Physiologie,* vol. 2, part 2, pp. 189–350. Leipzig: Vogel.

Fitts, P. M., and Seeger, C. M. (1953). S-R compatibility: Spatial characteristics of stimulus and response codes. *Journal of Experimental Psychology, 46,* 199–210.

Fournier, L. R., Eriksen, C. W., and Bowd, C. (1998). Multiple-feature discrimination faster than single-feature discrimination within the same object? *Perception and Psychophysics, 60,* 1384–1405.

Frith, C. D., and Done, D. J. (1986). Routes to action in reaction time tasks. *Psychological Research, 48,* 169–177.

Gottsdanker, R., and Shragg, G. P. (1985). Verification of Donders' subtraction method. *Journal of Experimental Psychology: Human Perception and Performance, 11,* 765–776.

Gratton, G., Coles, M. G. H., and Donchin, E. (1992). Optimizing the use of information: Strategic control of activation of responses. *Journal of Experimental Psychology: General, 121,* 480–506.

Grice, G. R., Boroughs, J. M., and Canham, L. (1984). Temporal dynamics of associative interference and facilitation produced by visual context. *Perception and Psychophysics, 36,* 499–507.

Hasbroucq, T., Guiard, Y., and Ottomani, L. (1990). Principles of response determination: The list-rule model of SR compatibility. *Bulletin of the Psychonomic Society, 28,* 327–330.

Hasher, L., and Zacks, R. T. (1979). Automatic and effortful processes in memory. *Journal of Experimental Psychology: General, 108,* 356–388.

Heister, G., Ehrenstein, W. H., and Schroeder-Heister, P. (1987). Spatial S-R compatibility with unimanual two-finger choice reactions: Effects of irrelevant stimulus location. *Perception and Psychophysics, 42,* 195–201.

Hommel, B. (1993a). The relationship between stimulus processing and response selection in the Simon task: Evidence for a temporal overlap. *Psychological Research, 55,* 280–290.

Hommel, B. (1993b). The role of attention for the Simon effect. *Psychological Research, 55,* 208–222.

Hommel, B. (1993c). Inverting the Simon effect by intention. *Psychological Research, 55,* 270–279.

Hommel, B. (1994). Spontaneous decay of response code activation. *Psychological Research, 56,* 261–268.

Hommel, B. (1995a). Stimulus-response compatibility and the Simon effect: Toward an empirical clarification. *Journal of Experimental Psychology: Human Perception and Performance, 21,* 764–775.

Hommel, B. (1995b). Unpublished study reported in W. Prinz, G. Aschersleben, B. Hommel, and S. Vogt, Handlungen als Ereignisse. In D. Dörner and E. van der Meer (Eds.), *Das Gedächtnis: Trends, Probleme, Perspektiven,* pp. 129–168. Göttingen: Hogrefe, 1995.

Hommel, B. (1996). S-R compatibility effects without response uncertainty. *Quarterly Journal of Experimental Psychology, 49A,* 546–571.

Hommel, B. (1997). Toward an action-concept model of stimulus-response compatibility. In B. Hommel and W. Prinz (Eds.), *Theoretical issues in stimulus-response compatibility,* pp. 281–320. Amsterdam: Elsevier.

Hommel, B. (1998a). Automatic stimulus-response translation in dual-task performance. *Journal of Experimental Psychology: Human Perception and Performance, 24,* 1368–1384.

Hommel, B. (1998b). Event files: Evidence for automatic integration of stimulus-response episodes. *Visual Cognition, 5,* 183–216.

Hommel, B. (2000). Intentional control of automatic stimulus-response translation. In Y. Rossetti and A. Revonsuo (Eds.), *Interaction between dissociable conscious and nonconscious processes,* pp. 223–244. Amsterdam: John Benjamins Publishing Company.

Hommel, B. (Forthcoming-a). Time course of feature binding.

Hommel, B. (Forthcoming-b). Single stimulus-response co-occurrences affect subsequent free-choice behavior.

Keele, S. W. (1972). Attention demands of memory retrieval. *Journal of Experimental Psychology, 93,* 245–248.

Kornblum, S. (1994). The way irrelevant dimensions are processed depends on what they overlap with: The case of Stroop- and Simon-like stimuli. *Psychological Research, 56,* 130–135.

Kornblum, S., Hasbroucq, T., and Osman, A. (1990). Dimensional overlap: Cognitive basis for stimulus-response compatibility—a model and taxonomy. *Psychological Review, 97,* 253–270.

Kornblum, S., Stevens, G, Whipple, A., and Requin, J. (1999). The effects of irrelevant stimuli: 1. The time-course of stimulus-stimulus and stimulus-response consistency effects with Stroop-like stimuli, Simon-like tasks, and their factorial combinations. *Journal of Experimental Psychology: Human Perception and Performance, 25,* 688–714.

Logan, G. D. (1980). Attention and automaticity in Stroop and priming tasks: Theory and data. *Cognitive Psychology, 12,* 523–553.

Logan, G. D. (1988). Toward an instance theory of automatization. *Psychological Review, 95,* 492–527.

Logan, G. D., and Zbrodoff, N. J. (1979). When it helps to be misled: Facilitative effects of increasing the frequence of conflicting stimuli in a Stroop-like task. *Memory and Cognition, 7,* 166–174.

Lu, C.-H. (1997). Correspondence effects for irrelevant information in choice-reaction tasks: Characterizing the stimulus-response relations and the processing dynamics. In B. Hommel and W. Prinz (Eds.), *Theoretical issues in stimulus-response compatibility,* pp. 85–117. Amsterdam: Elsevier.

Lu, C.-H., and Proctor, R. W. (1995). The influence of irrelevant location information on performance: A review of the Simon and spatial Stroop effects. *Psychonomic Bulletin and Review, 2,* 174–207.

MacLeod, C. M. (1991). Half a century of research on the Stroop effect: An integrative review. *Psychological Bulletin, 109,* 163–203.

MacLeod, C. M., and Dunbar, K. (1988). Training and Stroop-like interference: Evidence for a continuum of automaticity. *Journal of Experimental Psychology: Learning, Memory, and Cognition, 14,* 126–135.

Meiran, N. (1996). Reconfiguration of processing mode prior to task performance. *Journal of Experimental Psychology: Learning, Memory, and Cognition, 22,* 1423–1442.

Meiran, N., and Daichman, A. (Forthcoming). Parallel processing in rapid switching between tasks.

Meyer, D. E., and Kieras, E. D. (1997). A computational theory of executive cognitive processes and multiple task performance: 1. Basic mechanisms. *Psychological Review, 104,* 3–75.

Miller, J. (1988). Discrete and continuous models of human information processing: Theoretical distinctions and empirical results. *Acta Psychologica, 67,* 191–257.

Miller, J. (1991). The flanker compatibility effect as a function of visual angle, attentional focus, visual transients, and perceptual load: A search for boundary conditions. *Perception and Psychophysics, 49,* 270–288.

Monsell, S. (1996). Control of mental processes. In V. Bruce (Ed.), *Unsolved mysteries of the mind,* pp. 93–148. Hove, U.K.: Erlbaum.

Morin, R. E., and Grant, D. A. (1955). Learning and performance on a key-pressing task as function of the degree of spatial stimulus-response correspondence. *Journal of Experimental Psychology, 49,* 39–47.

Neumann, O. (1984). Automatic processing: A review of recent findings and a plea for an old theory. In W. Prinz, and A. F. Sanders (Eds.), *Cognition and motor processes,* pp. 255–293. Berlin: Springer.

Nicoletti, R., and Umiltà, C. (1989). Splitting visual space with attention. *Journal of Experimental Psychology: Human Perception and Performance, 15,* 164–169.

O'Leary, M. J., and Barber, P. J. (1993). Interference effects in the Stroop and Simon paradigms. *Journal of Experimental Psychology: Human Perception and Performance, 19,* 830–844.

Otten, L. J., Sudevan, P., Logan, G. D., and Coles, M. G. H. (1996). Magnitude versus parity in numerical judgements: Event-related brain potentials implicate response conflict as the source of interference. *Acta Psychologica, 94,* 21–40.

Ouellette, J. A., and Wood, W. (1998). Habit and intention in everyday life: The multiple processes by which past behavior predicts future behavior. *Psychological Bulletin, 124,* 54–74.

Pashler, H. (1994). Dual-task interference in simple tasks: Data and theory. *Psychological Bulletin, 116,* 220–244.

Proctor, R. W., and Lu, C.-H. (1994). Referential coding and attention shifting accounts of the Simon effect. *Psychological Research, 56,* 185–195.

Proctor, R. W., and Lu, C.-H. (1999). Processing irrelevant location information: Practice and transfer effects in choice-reaction tasks. *Memory and Cognition, 27,* 63–77.

Proctor, R. W., Lu, C.-H., and Van Zandt, T. (1992). Enhancement of the Simon effect by response precuing. *Acta Psychologica, 81,* 53–74.

Redding, G. M., and Gerjets, D. A. (1977). Stroop effect: Interference and facilitation with verbal and manual responses. *Perceptual and Motor Skills, 45,* 11–17.

Rogers, R. D., and Monsell, S. (1995). Costs of a predictable switch between simple cognitive tasks. *Journal of Experimental Psychology: General, 124,* 207–231.

Sanders, A. F. (1980). Stage analysis of reaction processes. In G. E. Stelmach, and J. Requin (Eds.), *Tutorials in motor behavior,* pp. 331–354. Amsterdam: Elsevier.

Schweickert, R., and Boggs, G. J. (1984). Models of central capacity and concurrency. *Journal of Mathematical Psychology, 28,* 223–281.

Shaffer, L. H. (1965). Choice reaction with variable S-R mapping. *Journal of Experimental Psychology, 70,* 284–288.

Simon, J. R., Craft, J. L., and Webster, J. B. (1973). Reactions toward the stimulus source: Analysis of correct responses and errors over a five-day period. *Journal of Experimental Psychology, 101,* 175–178.

Simon, J. R., and Small, A. M. (1969). Processing auditory information: Interference from an irrelevant cue. *Journal of Applied Psychology, 53,* 433–435.

Sommer, W., Leuthold, H., and Hermanutz, M. (1993). Covert effects of alcohol revealed by event-related potentials. *Perception and Psychophysics, 54,* 127–135.

St. James, J. D. (1990). Observations on the microstructure of response conflict. *Perception and Psychophysics, 48,* 517–524.

Stoffer, T. H., and Umiltà, C. (1997). Spatial stimulus coding and the focus of attention in S-R compatibility and the Simon effect. In B. Hommel and W. Prinz (Eds.), *Theoretical issues in stimulus-response compatibility,* pp. 181–208. Amsterdam: Elsevier.

Stroop, J. R. (1935). Studies of interference in serial verbal reactions. *Journal of Experimental Psychology, 28,* 643–662.

Sudevan, P., and Taylor, D. A. (1987). The cuing and priming of cognitive operations. *Journal of Experimental Psychology: Human Perception and Performance, 13,* 89–103.

Teichner, W. H., and Krebs, M. J. (1974). Laws of visual choice reaction time. *Psychological Review, 81,* 75–98.

Telford, C. W. (1931). The refractory phase of voluntary and associative responses. *Journal of Experimental Psychology, 14,* 1–36.

Toth, J. P., Levine, B., Stuss, D. T., Oh, A., Winocur, G., and Meiran, N. (1995). Dissociation of processes underlying spatial S-R compatibility: Evidence for the independent influence of what and where. *Consciousness and Cognition, 4,* 483–501.

Umiltà, C., and Liotti, M. (1987). Egocentric and relative spatial codes in S-R compatibility. *Psychological Research, 49,* 81–90.

Umiltà, C., and Zorzi, M. (1997). Commentary on Barber and O'Leary: Learning and attention in S-R compatibility. In B. Hommel and W. Prinz (Eds.), *Theoretical issues in stimulus-response compatibility,* pp. 173–178. Amsterdam: Elsevier.

Valle-Inclán, F., and Redondo, M. (1998). On the automaticity of ipsilateral response activation in the Simon effect. *Psychophysiology, 35,* 366–371.

Virzi, R. A., and Egeth, H. E. (1985). Toward a translational model of Stroop interference. *Memory and Cognition, 13,* 304–319.

Welford, A. T. (1952). The "psychological refractory period" and the timing of high-speed performance: A review and a theory. *British Journal of Psychology, 43,* 2–19.

Welford, A. T. (1968). *Fundamentals of skill.* London: Methuen.

Woodworth, R. S. (1938). *Experimental psychology.* New York: Holt, Rinehart and Winston.

Zachay, A. (1991). Diskrete und kontinuierliche Informationsverarbeitungsmodelle zur Erklärung von Reiz-Reaktions-Inkompatibilitäten: Evidenz für einen Antwortkonflikt beim Simon-Effekt. Master's thesis, University of Tübingen.

Zorzi, M., and Umiltà, C. (1995). A computational model of the Simon effect. *Psychological Research, 58,* 193–205.

III Task Switching and Multitask Performance

12 Task Switching and Multitask Performance

Harold Pashler

ABSTRACT Research on task switching and dual-task performance has spawned two literatures that have, to a surprising extent, developed independently. This tutorial reviews the principal findings of each tradition and considers how these phenomena may be related. Beginning with Jersild 1927, task-switching studies reveal that when people perform two tasks in succession, with each task requiring different responses to the same set of stimuli, substantial slowing occurs. Recent research suggests that while this slowing can be partially ameliorated by allowing sufficient time between tasks, advance reconfiguration is almost always incomplete. In studies of dual-task performance, stimuli are presented very close together in time, and subjects attempt concurrently to perform two wholly distinct tasks. A substantial slowing of one or both tasks is usually observed. The most stubborn source of this slowing appears to be queuing of central processing stages, sometimes supplemented by other kinds of interference. This queuing occurs even when the tasks are highly dissimilar and is unlikely to reflect voluntary strategies. A number of possibilities for how task switching and dual-task queuing might be related are discussed critically, including the possibility that queuing might stem from an inability to maintain two distinct task sets at the same time.

What happens when people try to switch rapidly between one task and another? What happens when they try to do more than one task at the same time? The first of these two fundamental questions is chiefly discussed in a modest-sized literature under the label "task switching" or "mental set"; the second, in a much larger literature under the label "divided attention" or "dual-task performance." The present chapter reviews main phenomena and theoretical issues in both areas and tries to draw some substantive connections between them.

12.1 TASK SWITCHING

In 1927, well before the modern era of information-processing psychology, an educational psychologist named Arthur T. Jersild published a pioneering study of people's ability to alternate between different tasks. Jersild measured the total time it took a person to work through a printed list of stimuli, making a response of some kind to each individual item in turn. In pure task blocks, subjects performed the same task on each item (for example, subtracting three from each number on a list). In alternating-task blocks, subjects performed one task on all the odd-numbered stimuli, and another task on the even-numbered stimuli. In some of the experiments, every stimulus was a potential input for either task (following Fagot 1994, I will refer to this arrangement of tasks and stimuli as a "bivalent" list or mapping). One of Jersild's bivalent alternating-task lists contained two-digit numbers; subjects were instructed to subtract three from the first number, add six to the second number, subtract three from the third, and so forth. They were substantially slower (more than 0.5 sec per item) in bivalent alternating lists than in pure lists, sometimes by more than 1 sec per item. This difference between pure and alternating bivalent lists will be referred to as the "alternation cost."

Jersild also examined the case of task alternation, where each stimulus was a potential input only for the appropriate task (henceforth referred to as a "univalent" list or mapping). For example, one univalent alternating list contained two-digit numbers and words, numbers alternating with words; subjects were instructed to subtract three from each number and to say aloud the antonym of each word. Remarkably, Jersild found that there was no alternation cost at all with these univalent lists; indeed, subjects were actually slightly *faster* in responding to alternating lists than to pure lists.

Some fifty years later, Spector and Biederman (1976) confirmed Jersild's basic results, finding a modest-sized benefit of alternation with univalent lists. This occurred, however, only when the items were printed as in Jersild's studies, and subjects were allowed to preview items ahead of the ones they were responding to.[1] When the items were placed on cards, so that subjects could not see the next stimulus until they turned a card over, there was actually a small alternation cost; the same was true when the experiment was run with a discrete-trials procedure. With the alternating bivalent lists (adding three, then subtracting three, etc.), Spector and Biederman found a large alternation cost (402 msec/item). This was cut about in half, to 188 msec/item, when a visual task cue ("+3" or "−3") was placed next to each item.

Several rather trivial potential explanations for the basic alternation cost need to be considered.

One might propose that the alternation cost merely reflects a tendency of subjects occasionally to forget what task they should perform next. If this is correct, the slowing should largely be confined to a few, very slow trials. This does not seem to be the case, however. Fagot (1994) had subjects make button-push responses to either the color or the identity of a letter (an *A*, *B*, or *C* in red, green, or blue). Figure 12.1 shows the Vincentized reaction time (RT) distributions for a zero response-stimulus interval (RSI) condition where the two tasks were performed in alternation.[2] The slowing is by no means confined to the slowest responses. Evidently, then, among the sources of the alternation cost are events that occur on at least a significant number of trials.

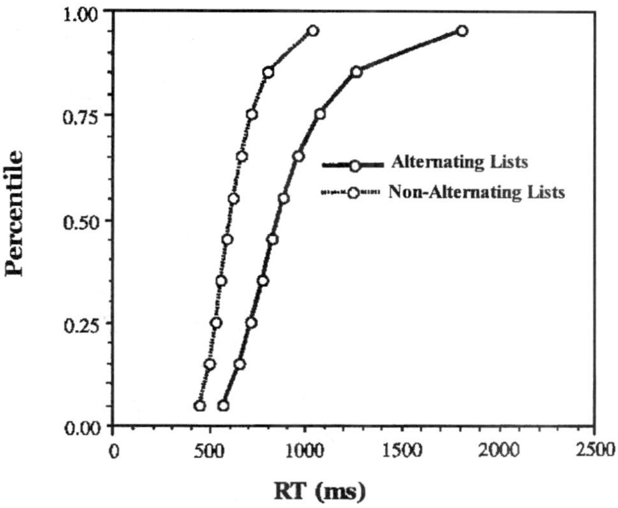

Figure 12.1 Vincentized reaction time (RT) distributions for a bivalent list alternating-task design. Alternation cost appears even among the fastest responses. From Fagot 1994; reprinted with permission.

One might also propose that the faster responses to pure as opposed to alternating lists arise because alternating lists do not include any stimulus repetitions. In any speeded-choice task, people respond much faster to stimuli that match whatever was presented on the preceding trial (Kornblum 1973; Pashler and Baylis 1991). This potential confound does not explain the effect, however. In the experiment by Fagot (1994) shown in figure 12.1, lists were selected with the constraint that there be no item repetitions, but the alternation cost was still found; the same was probably done informally in some of the earlier studies.[3]

What, then, accounts for the alternation cost with bivalent lists, and why is this cost sometimes virtually absent with univalent lists? Perhaps the most obvious interpretation is that depicted in figure 12.2. According to this "task set reconfiguration" (TSR) view (Monsell 1996), preparing to perform a task involves linking and/or configuring different processing modules. Different modules are assumed to be responsible for different aspects or stages of the task (e.g., perception, response selection, etc.). With bivalent lists, task alternation requires changing the links, settings, or both between when the central processing of one stimulus is completed and when the central processing of the next begins. In some cases, changes in the configuration of perceptual modules may also be involved. Given the conflicting response selection rules in the case of bivalent stimuli, the module responsible for response selection cannot be set the same way throughout the block of trials. At first blush in this account, one would assume that the alternation cost simply reflects the time needed to complete the switch. As for univalent alternating tasks, it should be possible for the two task mappings to coexist more or less

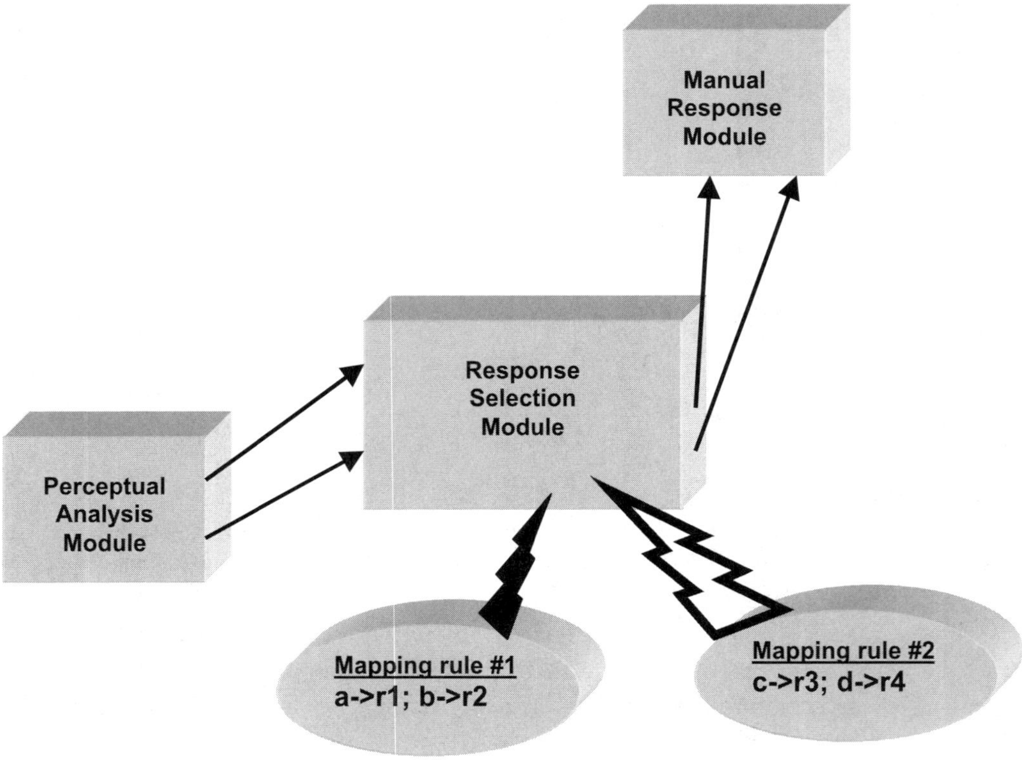

Figure 12.2 Discrete conception of task set switching. The response selection machinery is prepared at any one time to perform either of the two incompatible mappings, but not both. In the alternating-task blocks, one mapping is switched out and the other inserted, somewhat as a crystal in early radio sets.

happily, so that the union of the two mappings could simply be loaded into the response selection module. This may explain why there should be minimal cost in that situation, although of course by itself it does not explain why there should ever be a benefit.

If this account is correct, allowing extra time between the response to stimulus n and the presentation of stimulus $n + 1$ (RSI) might allow subjects to complete the switch in advance, thereby reducing or eliminating the alternation cost. Many recent studies of task alternation have found some reduction. A notable example is Rogers and Monsell 1995, which found an approximately 50% reduction as RSI was lengthened from 150 msec to 1,200 msec, so long as subjects could rely on having the long RSI. In Fagot 1994, conducted in my own laboratory, subjects were instructed to respond to colored letters by pushing a button to indicate either the color of the letter or its identity. In alternating-task blocks, RSI varied from 0 to 1.5 sec; the alternation cost fell from 314 msec at the zero RSI to 201 msec at the 1.5 sec RSI, with most of the reduction occurring over the range of RSIs between zero and 400 msec; this pattern was confirmed in several other experiments within that series.

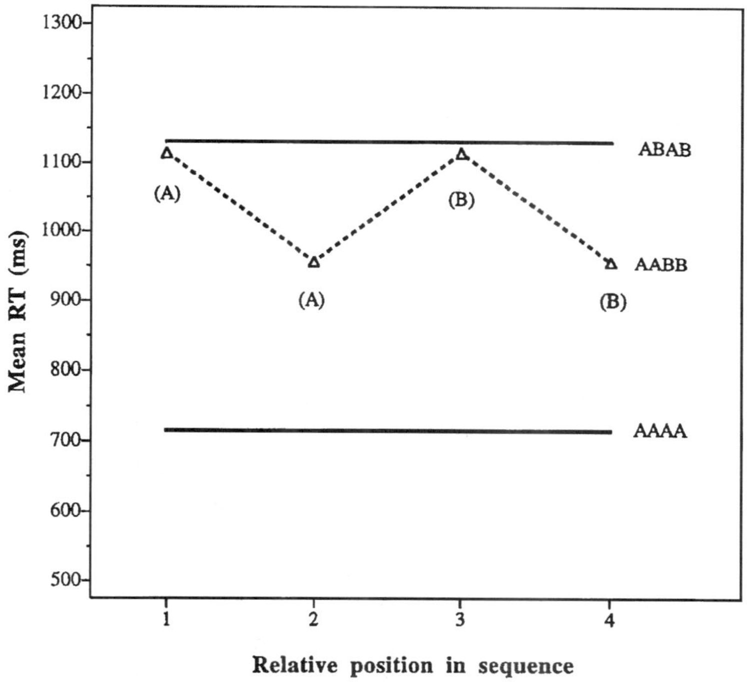

Figure 12.3 Reaction time (RT) results from an AABB task (Fagot 1994), redrawn in Pashler 1997. Subjects are faster on second performance of a given task, but still not so fast as in a pure block of trials.

As Rogers and Monsell (1995) point out, the alternation cost (difference between pure and alternating-task blocks) is likely to include several factors in addition to reconfiguration time. For example, there might be slowing due to the processing "overhead" needed to maintain and implement the intention to alternate. Because concurrent memory loads generally slow performance in reaction time tasks (Logan 1978), it seems reasonable to expect that holding onto a plan for alternating would impose a memory load of its own. In addition, differences in effort or arousal cannot be ruled out.

To help tease apart these factors, Rogers and Monsell used an "alternating-runs" procedure, wherein subjects performed first one task a number of times, then the other, and so forth. A pair of characters was presented on each trial, one a letter and the other a digit. Subjects either classified the letter as a vowel versus consonant, or the digit as odd versus even. Sometimes each task was performed twice in succession (AABB). The first response within a run of a given task (AABB) was substantially slower than the second (AABB), even at the long RSI. This was later confirmed by Fagot (1994) using the color and letter tasks described above. In AABB lists, subjects were required to perform the color task twice, the letter task twice, and so forth. Fagot also included pure blocks of trials and alternating (ABAB) blocks for comparison. As seen in figure

12.3, the first performance of a given task (AABB) was close to the ABAB blocks, but slower than the second performance (AABB), as in Rogers and Monsell's data. The second performance was still quite a bit slower than the pure block (AAAA), however, suggesting that the overhead cost is nontrivial.

In a further example of the stubbornness of the residual task switch cost at long RSIs, Goschke (chap. 14, this volume) allowed subjects 1.5 sec between two colored letters, each of which was to be classified by color or shape, and found responses were substantially slower when a different task had to be performed on the second letter.

Thus it appears, as Rogers and Monsell argued, that several factors play a role in the basic Jersild alternation effect. From the standpoint of conventional thinking in information-processing psychology, probably the most surprising of these factors is the switch cost, which persists even after ample time has been provided for reconfiguration. Some clues about the nature of this residual switch cost come from an additional experiment by Rogers and Monsell (1995), in which subjects performed a task four times in succession, then switched and performed the other task four times, and so forth. Performing a given task initially produced a substantial speedup for the second response, but over the next two responses, no additional improvement was detected (see figure 12.4). The authors concluded that the gain observed from performing the task once could not be attributed to "micropractice"—a small dose of the same optimization process that, over many trials, yields the familiar practice effect. After all, they reasoned, such a process could hardly reach an abrupt and final asymptote after one trial, as these data seem to show. The empirical basis for this conclusion has recently been challenged, however, by Salthouse et al. (1998), who had subjects switch tasks and then perform fairly long runs of a different task. They found RTs for the second trial within a run had still not reached a baseline in their data, and argued that Rogers and Monsell may have had insufficient experimental power to detect this continuing decline.

Cuing the Task Set

Control over task set is also illuminated by experimental designs in which the subject cannot tell which task to perform until a task cue is provided. Following in the footsteps of Shaffer (1965), Sudevan and Taylor (1987) had subjects perform one of two different tasks involving a digit. One task required classifying the digit as odd or even, while the other required classifying it as less than six or greater than five (bivalent mapping). The cue preceded the digit by an interval ranging between 400 msec and 4 sec. Responses became faster and more accurate as the interval was lengthened to about 2 or 3 sec. In his color/identity design, Fagot (1994) examined cue-target intervals ranging from zero to 4 seconds, and

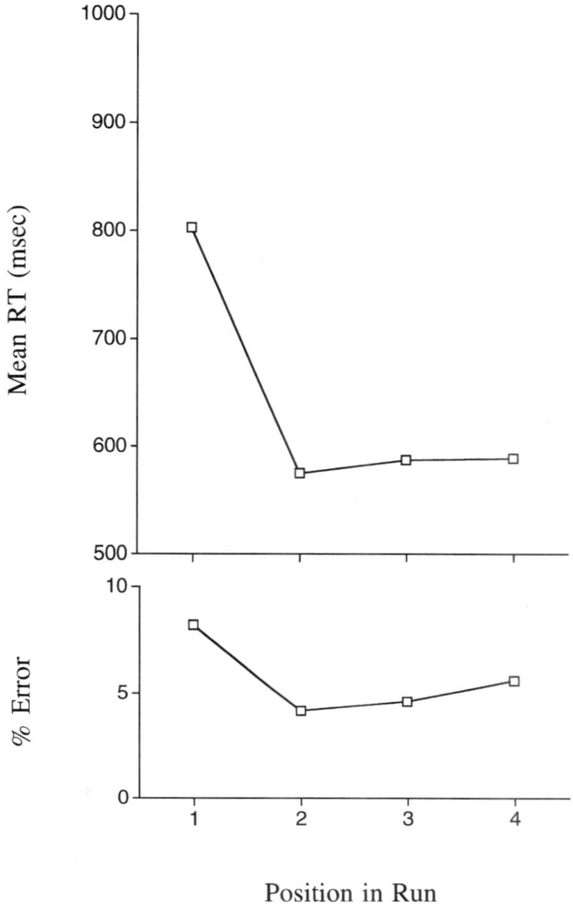

Figure 12.4 Reaction times (RT) and error rates in Rogers and Monsell 1995, experiment 6, as a function of position in a run of four trials (subjects performed one task four times, then switched to the other task; redrawn from Rogers and Monsell 1995, fig. 5).

found that the bulk of the benefit (over 200 msec) occurred over the range from 0 to 500 msec, with some further improvement out to about 1 sec; thereafter, performance was little changed. Other studies (e.g., Logan and Zbrodoff 1982) have also found a similar time course using cues that are helpful but not strictly necessary in performing the task. As Rogers and Monsell (1995) point out, one cannot directly derive an estimate of the time needed for reconfiguration based on these kinds of experiments, because at the shortest cue-target interval, the time needed to read and interpret the cue is presumably slowing responses, along with the requirement to reconfigure.

Recall that in the alternating-task procedure described earlier, the first response within a run of two successive instances of the same task is slower than the second response, even with an ample RSI. Based on that result, we would naturally expect that in the cuing procedure, no matter how long the cue-target interval, responses would be slower when the previous trial involved the other task. This is indeed the case. For example, Fagot (1994) presented task cues in blocks with a random or a fixed task sequence (either alternating or nonalternating). Even when subjects had four seconds to use the task cue, there was still a benefit of having performed the same task on the preceding trial; as expected, there was an additional benefit of having a fixed sequence.

In a clever recent study, Meiran (1996) cued subjects to respond to the vertical or horizontal position of a disk; the task varied from trial to trial within a block. Task cues (arrows pointing either up and down or left and right) appeared about 200 or 1,400 msec prior to onset of the imperative stimulus. Subjects were slower when they had to perform a different task from one trial to the next. This difference was substantially greater at the short cue-stimulus interval than at the long interval, but did not disappear at the longer interval. Meiran argued that the reduced task alternation effect produced by increasing the cue-target interval did not occur merely because lengthening this interval made the previous task more distant in time, reducing its impact by passive decay. When the interval between the previous response and the cue was decreased to make up for the increase in the cue-target interval, thereby holding the RSI constant, the longer cue-target interval still reduced the effect of a task switch. This strongly suggests that some, albeit incomplete, advance reconfiguration is indeed taking place.

Incompleteness of Reconfiguration

We have seen that in both the alternating-task procedure and the task-cuing procedure, providing subjects plenty of time to prepare reduces the cost of having to perform a task different from the one they just performed (in the bivalent situation), but it does not allow them to respond as quickly as if no switch of task had been required. This residual difference cannot be attributed to overhead cost because it appears also with the alternating-runs procedure (e.g., Rogers and Monsell 1995) as well as with the task-cuing procedure (e.g., Meiran 1996).

Why should there be a residual switch cost? Why is reconfiguration incomplete? De Jong (chap. 15, this volume) asked whether the residual switch cost stems from a constant slowing that appears on all trials or from a slowing that arises on only a fraction of the trials. Using the alternating-runs procedure of Rogers and Monsell, he had subjects classify colored letters according to either color or identity (consonant versus vowel). He found little evidence for a constant slowing component at the long RSI, and argued that incompleteness of reconfiguration is at least avoidable under certain conditions. His results may not rule out the possibility, however, that residual cost is always present, but imposes a delay whose magnitude varies from trial to trial.

Meiran (chap. 16, this volume) proposes that residual shift depends on a feature of certain switching designs not discussed thus far, namely, "ambiguity of responses," the use of an overlapping set of responses in the two tasks. As in his earlier experiments described above, Meiran used a design in which subjects respond to either the vertical or the horizontal position of a disk placed in one of four quadrants of the display. When the two tasks used the same two response keys (ambiguous responses), there was a positive residual task switch cost; when the responses were separate, the residual cost was reduced or absent.

According to Meiran, task preparation may involve not only the selective amplification or enabling of particular stimulus-response links, as depicted in figure 12.2, but also the selection of a response set, which can only be achieved by actually performing the task. This proposal is intriguing, and receives support from the data reported in this volume, although there are cases in the literature where residual task-switching costs have been observed even when two tasks did not involve "ambiguous" responses. For example, Fagot (1994) observed residual switch costs for mappings both with the same keys and with corresponding keys of different hands.

An alternative view of the residual cost of a task switch is that it results, not from the need to perform a time-consuming control process on the switch trial (as the authors described above have assumed), but from a prolongation on switch trials of the response selection process that happens on all trials. This prolongation, is caused by competition due to positive or negative priming of task sets or of S-R associations from previous trials on which the other task was performed. Such a view was first proposed by Allport, Styles, and Hsieh (1994), and a new version of it is presented by Allport and Wylie's chapter (chap. 2, this volume), to which the reader is referred for arguments and evidence. It seems clear from Allport and Wylie's work that there are carryover effects from recently performing the alternative task in response to the same stimulus or class of stimulus. What is not clear, however, is whether these carryover effects are sufficient to account for the dramatic drop in RT from the first to the second trial after a task swtich. Further, the notions of priming effects and control processes are by no means mutually exclusive.

Task Congruity Effects

The incompleteness of reconfiguration is revealed, not only by residual switch costs that persist despite long RSIs, but also by persisting effects of the purportedly disengaged mapping. Recall that Rogers and Monsell (1995) had subjects respond to either the letter or the digit in a letter-digit pair, using an alternating-runs procedure. The authors examined reaction times as a function of whether the irrelevant item in the pair would, according to the irrelevant (supposedly inactive) task mapping, yield the same response as that required on the current trial. The trial was called "congruent" when it did, and "incongruent" when it did not. There was a modest but significant tendency for slower responses on incongruent trials than on congruent trials, although responses trials with neutral stimuli were faster still. Similarly, in Fagot's color/letter design, where subjects responded either to color or to identity, responses were about 90 msec slower when the other feature was associated with a response inappropriate on the current trial. These congruency effects imply that the "competing task set is not entirely disabled" (Rogers and Monsell 1995, 216).

There is some controversy about whether the competing task set can be disabled when a sufficiently long RSI is provided. In their experiment 3, Rogers and Monsell found no significant reduction in the congruence effect (measured, as usual, in RTs) when they increased the R-S interval, although there was a marginally significant interaction in the error rates. Similarly, in three experiments, Fagot (1994) found only a weak reduction in congruency effects. By contrast, Meiran (1996), using his location button tasks, found a strong interaction, with congruence effects reduced but not eliminated. Finally, Sudevan and Taylor (1987) reported that congruence effects with their digit task disappeared at long cue-target intervals, while Goschke (chap. 14, this volume) reports having nearly eliminated the effect of task congruence with a long, unfilled RSI and after practice. Unhappily, then, the results run the full gamut from complete persistence of the congruence effect at a long RSI all the way to virtual disappearance. This issue remains to be sorted out.

Conclusions

Evidently, when subjects anticipate the need to perform a task incompatible with the one they just performed (as in the case of a bivalent list), whether this anticipation is based on the requirement to alternate (as in the Jersild paradigm and its spin-offs), or on the perception of a cue telling them to perform a task different from the one they just performed, some advance reconfiguration can occur, as depicted in figure 12.2. With the sorts of simple but arbitrary tasks studied in this literature, this reconfiguration usually seems to take under 0.5 sec when subjects have no other intervening task to perform. Reconfiguration may be accompanied by verbalization, usually covert, of the instructions for the upcoming trial.

The notion of advance reconfiguration illustrated in figure 12.2 seems to have some validity, but it misses important aspects of task switching. First, advance reconfiguration usually fails to eliminate the costs of having just performed a different task. Even with ample RSIs or cue-target intervals, subjects are still typically slower when they must perform a task different from the one they just performed (although Meiran's design reveals at least some exceptions). Actually performing a task once allows a significant amount of additional reconfiguration or tuning to take place. Rogers and Monsell refer to the tuning before first performing a task as the "endogenous component" of task preparation, and to that after performing the task as the "exogenous component." Although their data had suggested that exogenous reconfiguration is complete after one trial, subsequent data (Salthouse et al. 1998) suggest it may not be entirely complete until two trials.

Perhaps the most intriguing aspect of task switching is the lingering effect of the irrelevant mapping—the "task congruity effect." Not only advance reconfiguration, but indeed *all* reconfiguration accomplished up to the point of selecting a response in the new task seems incapable of wholly disabling the old mapping. While task congruity effects have on some occasions been observed to disappear with adequate preparation time, as noted above, more commonly they seem to persist, at least to some extent (an issue discussed in detail by Allport and Wylie, chap. 2, this volume).

12.2 DUAL-TASK PERFORMANCE

We turn now to the limitations that arise when people attempt to perform two different tasks at the same time. Our focus here will be on discrete tasks; with more continuous tasks, interference and switching are easily disguised for reasons that will emerge clearly below. Not surprisingly, limitations on simultaneous mental operations evidently arise at various different functional loci. Perceptual analysis of multiple stimuli often takes place in parallel, with capacity limitations sometimes becoming evident when perceptual demands exceed a certain threshold (Pashler 1997) although nonperceptual factors (such as statistical noise in search designs) often masquerade as capacity limitations (Palmer, 1995). These limitations appear largely, but probably not entirely, modality specific (Treisman and Davies 1973; Duncan, Mertens, and Ward 1997). Similarly, response conflicts arise when responses must be produced close together in time. These perceptual limitations are often most acute when similar or linked effectors are used, such as the two hands (Heuer 1985).

The most intriguing, and for the present topic most relevant, limitations arise in central stages of decision, memory retrieval, and response selection. Intuitively, most laymen assume that the cognitive aspects of two tasks can be performed simultaneously unless one or both are intellectually demanding. That this is not the case, however, is most clearly seen when people try to carry out two speeded but relatively simple tasks, each requiring a response to a separate individual stimulus. As Telford (1931) first observed, people almost invariably respond more slowly to the second stimulus when the interval between the two stimuli is reduced. Telford called this the "psychological refractory period" (PRP)

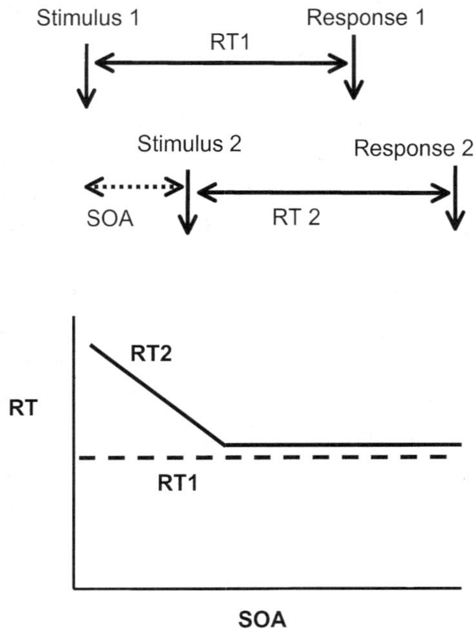

Figure 12.5 Schematic diagram of the psychological refractory period (PRP) design and idealized pattern of data (hypothetical numbers).

effect, by analogy to the refractory period of neurons. Though the analogy is probably not very apt, the label has stuck. In the PRP design, two stimuli (S1 and S2) are presented, their onsets separated by some stimulus onset asynchrony (SOA). The person makes a separate response to each stimulus (R1 and R2, respectively). Figure 12.5 (idealized data) shows the type of result usually obtained; the reaction time between S2 and R2 (RT2) grows as the SOA is shortened. Meanwhile, the reaction time between S1 and R1 (RT1) is usually relatively constant, although this depends on the instructions (see below). In some cases, the slope relating RT2 to SOA is as extreme as −1, which means that any reduction in SOA beyond a certain point merely increases RT2 by the same amount. To put it differently, presenting S1 and S2 closer together in time (once the interval reaches some minimum value) often fails to result in R2 being produced any earlier. Another important observation is that while processing required by the two tasks resists being "compressed" beyond a certain point, at short SOAs, the total time required to carry out both tasks (the interval between S1 and R2) is often substantially less than the sum of the times required to complete each task separately. In short, there is a saving in the total time for completing the two tasks, suggesting overlap in some aspects of processing.

The PRP effect has been observed in many different tasks, including simple reaction time (as in Telford's studies) and choice reaction time tasks (starting with Creamer 1963). Although early PRP experiments mostly used pairs of manual responses, sometimes made with the same finger, the effect can also be found when the two tasks involve completely different effectors. For example, PRP effects have been found with tasks combining manual and oculomotor responses (Pashler, Carrier, and Hoffman 1993), manual and vocal responses (Pashler 1990), manual and foot responses (Osman and Moore 1993), and vocal and foot responses (Pashler and Christian 1994). Thus it is not necessary for two tasks to use a common motor control system in order for a PRP effect to be observed.

The PRP effect is also found when the two stimuli involve different sensory modalities. For example, Borger (1963) and Creamer (1963) found PRP effects with visual and auditory stimuli, as have many more recent researchers. It is not clear whether the PRP effect is greater when S1 and S2 are presented in the same modality; this is hard to determine because changes in input modality are typically confounded with differences in the compatibility of the task mapping.

Limits of the Psychological Refractory Period Effect

The PRP effect is very robust, but over the past twenty-five years or so, a number of exceptions have emerged. Greenwald and Shulman (1973; Greenwald 1972) found that the effect of SOA on second-task RTs was virtually eliminated when one task involved repeating a spoken word (shadowing) and the other involved a highly compatible visuomanual task. They hypothesized that "ideomotor compatibility," the fact that the stimulus mimics the feedback produced by the response, might be critical. Although McLeod and Posner (1984) demonstrated noninterference with combinations of shadowing and other tasks in ways that seemed consistent with this proposal, other research suggests ideomotor compatibility is probably not sufficient to eliminate interference. For example, Brebner (1977) devised a novel ideomotor-compatible task, requiring subjects to press a button in response to upward pressure from a solenoid located under the corresponding finger. When task 1 involved left-hand stimulation and task 2 involved right-hand stimulation, a clear-cut PRP effect was observed. Tasks requiring a saccadic eye movement toward a single spot, or even the generation of an eye movement in response to a single stimulus based on its color, seem not to generate PRP effects (Pashler, Carrier, and Hoffman 1993). Visuomanual tasks with very high spatial stimulus-response compatibility may also be free of central interference (Koch 1994). At present, then, the conditions under which the PRP effect disappears are not well characterized. Indeed, it seems that dual-task interference in pairs of punctate tasks can be eliminated only with tasks that are, intuitively speaking, extremely natural and easy. Whether the critical factor is the existence of prewired neural circuits that bypass central machinery, a high degree of practice, or some combination of these factors remains unknown. Perhaps the more significant point is

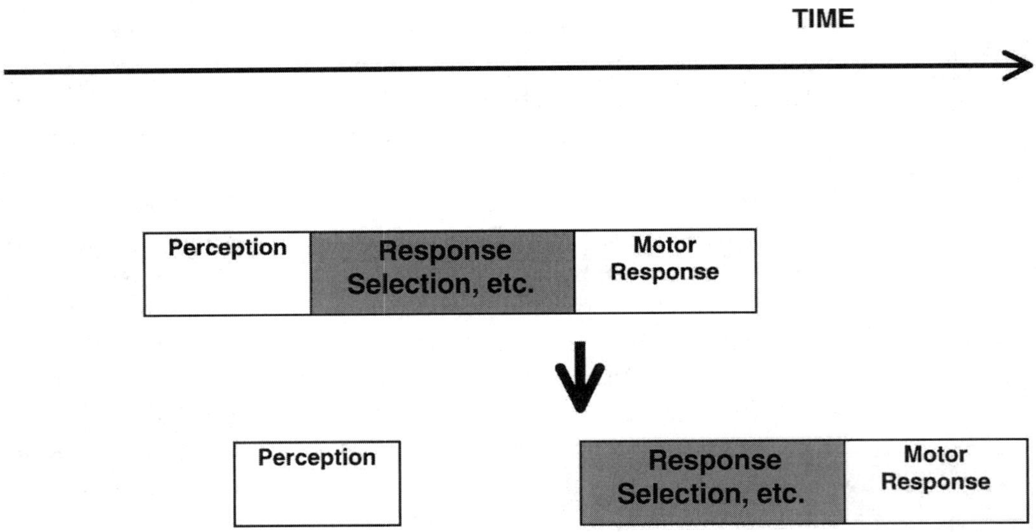

Figure 12.6 Central bottleneck account of the psychological refractory period (PRP) effect.

that it is easy to find tasks with minimal cognitive demands that produce robust PRP effects.

Sources of Dual-Task Slowing

Based largely on observations of PRP interference even where there is no overlap in stimulus or response modality, Welford (1952, 1980) proposed that dual-task slowing arises from a bottleneck in what he called "stimulus-response translation"—in more modern parlance, the stage of "response selection." The basic idea is illustrated in figure 12.6. According to this hypothesis, each task is composed of three broad stages (perception, response selection, and response execution); any stage of task 1 can overlap any stage of task 2, except for the shaded stage of response selection: while one response is being selected, selection of the other response must wait. As formulated, however, the hypothesis does not say what should happen in tasks more complicated than choice RT, where one often encounters mental operations that do not obviously fall into any one of the three categories.

From this account, one can derive very specific predictions for the results of dual-task experiments in which different stages of task 1 or task 2 are selectively prolonged. Increases in the duration of stages of task 1 up to and including the shaded stage should, at short SOAs, propagate and slow task 2 as well as task 1. Increasing the duration of the post-bottleneck stages of task 1, on the other hand, should slow only task 1, regardless of the SOA. Increasing the duration of stages in task 2 prior to the bottleneck should correspondingly slow the second response at long SOAs. At short SOAs, on the other hand, there is "slack" because the response selection in task 2 is not waiting for completion of perception in task 2, but rather for the completion of response selection in task 1. The result of the slack is that, at short (but not at long) SOAs, the perceptual slowing should cease to affect RT2. The prediction, then, is that manipulations of the prebottleneck processing stages in task 2 should exhibit underadditive interaction with SOA (see Jolicoeur, Dell'Acqua, and Crebolder, chap. 13, this volume, for further details and examples). Lengthening the duration of stages at or after the shaded portion of task 2, on the other hand, should always slow R2 to the same extent, regardless of SOA.

These predictions have been confirmed in many experiments involving fairly elementary choice RT tasks (for a review, see Pashler 1997). The predictions are distinctive in the sense that they not only favor the central bottleneck, but also rule out accounts that would place the bottleneck earlier or later in the sequence of processing stages. Several of the results also seem unfavorable to graded capacity-sharing models, especially the fact that increases in first-task response selection difficulty have at least as large an effect on RT2 as it has on RT1 (e.g., Broadbent and Gregory 1967). If task 1 were being performed with depleted capacity, and the manipulations increased the capacity required to carry out the stage in question, one would expect to see a greater effect on RT1 than on RT2 (see Pashler and Johnston 1998 for discussion).

Much recent work within the bottleneck framework has focused on the question of exactly which processes are subject to this limitation, and which are not. Manipulations of the duration of sensory processing in task 2 (e.g., contrast) show the underadditive pattern indicating that the stages affected are not subject to the bottleneck (Pashler 1984; De Jong 1993). Johnston and McCann (forthcoming) degraded letters by making them very squat or very narrow without altering stroke widths and contrast. In another experiment, they altered the tilt of strokes composing the letters (for instance, rotating the diagonal segments in the letter *A* inward so that the character looked something like a teepee). At long SOAs, these distortions slowed RT2 by about 30 msec. At short SOAs, however, RTs for distorted and undistorted were indistinguishable, suggesting absorption into slack. It seems likely, therefore, that letter identification, not merely visual feature extraction, can occur on task 2 while critical stages of task 1 are under way. On the other hand, when perceptual processing demands on task 2 include not just identifying stimuli, but performing additional manipulations such as mental rotation or comparisons, these operations are usually subject to the central bottleneck (Ruthruff, Miller, and Lachman 1995).

Recent evidence suggests that, not merely the planning of actions based on task-mapping instructions or difficult perceptual manipulations, but memory retrieval overall is subject to queuing. Carrier and Pashler (1996) combined a manual response to a tone (task 1) with paired associate retrieval cued by a visually presented word (task 2) in a PRP design. The duration of the memory retrieval was manipulated by varying the amount of practice subjects had carrying out any particular retrieval. Second-task RTs were, not surprisingly, faster for better-learned pairs. In the dual-task situation, this difference appeared additive with SOA (Carrier and Pashler 1996). Following the logic described above, this implies that memory retrieval was postponed by first-task processing and refutes the claim that only the execution of the motor response is delayed. The latter point seems especially clear because of the greater difficulty of task 2 compared to task 1. In the short (50 msec) SOA condition, subjects responded to the tone about 600 msec after it was presented; the paired-associate task was far more challenging, however: on average, the paired-associate response did not occur until about 1,100 msec later. If all interference were response related, it is hard to see what could be postponing a second response so temporally remote from the first. The results are to be expected, however, if one assumes that the central bottleneck encompasses both response selection in task 1 and memory retrieval in task 2 (and perhaps response selection as well, if that is a separate stage in this sort of task).

It seems to me a reasonable conjecture that the inability to select two responses at the same time, which is apparent in choice RT tasks (Welford's response selection bottleneck), may be just a special case of a broader constraint, namely, that two retrievals cannot be carried out at the same time. Within the confines of the choice RT experiment, it is an action plan that is to be retrieved, whereas in other situations, it may, for example, be a word or concept or episode. While the proposed constraint can be expressed very simply, it stands in great need of explication. For example, what is meant by "two retrievals"? If two stimuli are presented, each associated with the same single response, does the lookup of that single response based on the two stimuli constitute two retrievals or one? In choice RT tasks, two redundant stimuli produce what Miller (1982) calls "coactivation," a particularly strong form of parallel processing. The same is almost surely true of more time-consuming memory retrieval operations. What about one stimulus associated with two responses? Timothy Rickard and I (Rickard and Pashler 1998) trained subjects in one phase of training to associate each item on a list of ten words with a corresponding verbal paired associate, and then, in a second phase of training, to associate each item on the same list with a manual response.[4] In a final test phase, subjects were sometimes instructed to carry out both retrievals at once. Whichever response was produced second had on average a latency that was about twice as long as the single-task control. Other aspects of the data also argued that the retrievals were carried out sequentially. Thus, for the purpose of the proposed constraint, it is the number of outputs, not the number of inputs, that determines whether a single retrieval or multiple retrievals are required.

The term *retrieval* also needs clarification. A priori, one might have described letter identification, for example, as involving the retrieval of the letter identity corresponding to a visually presented character. Yet I have argued that object identification is not subject to the bottleneck. What differentiates retrieval from classification or identification? At this point, the answer must be vague: it seems that the operations subject to queuing involve retrieving some mental contents that are distinct from the input in that they are not an internal description of the input but some separate contents. Sharpening up this description will require at the very least testing a broader range of different types of retrievals in different dual-task contexts; conceivably, it will also require a better understanding of the neural substrates of these processes.

Strategic Interpretations

The apparent inability to execute the central stages of even fairly easy tasks concurrently is surprising from both an intuitive and a computational standpoint. It has recently been argued that postponement of central processing in the PRP design stems not from a fundamental inability to carry out the two tasks at the same time, but rather represents a strategic response to the explicit or implicit demands of the experiment. This idea has been developed in detail by Meyer and Kieras (1997), who proposed an ambitious theory of human performance ("executive process interactive control" or EPIC), discussed in detail by Kieras et al. (chap. 30, this volume). According to EPIC, there are no intrinsic limitations whatever in the ability to select responses or carry out memory retrievals concurrently. There are, however, structural limitations in the initiation and execution of responses. In addition, postponement of central processing (i.e., queuing of processing stages) may occur whenever subjects perceive this to be advantageous.

Why would subjects adopt a queuing strategy in a dual-task design when doing so means responding more slowly in one or both tasks? As Meyer and Kieras note, in many PRP experiments, subjects have been told to produce R1 as fast as possible (and even, in a few cases, to produce R1 before R2). Primarily, this has been done in order to avert the "grouping" strategy that people naturally fall into, whereby R1 is buffered and then emitted shortly before R2 (Borger 1963; Pashler and Johnston 1989, exp. 2). Given a strong emphasis on first-task speed, subjects might choose not to select the two responses in parallel because doing so might result in responding to task 2 before task 1.

One obvious question, then, is what happens when there is no emphasis on the speed of the first task and subjects try to respond to both tasks as quickly as possible. A number of studies that did not emphasize first-task speed have nevertheless shown evidence of central postponement. For example, in Carrier and Pashler 1996, even though subjects were not told to emphasize the speed of the first response, both slowing of R2 and postponement of central processing were observed. Similarly, in one of their experiments, Ruthruff, Miller, and Lachmann (1995) did not emphasize first-task speed but nonetheless found evidence of a central bottleneck.

There are also some other, rarely cited studies in which investigators looked at performance of two serial choice RT tasks, where subjects are instructed to respond to a train of signals in each task, rather than to two discrete signals, as in the PRP design. Here the order of responses is entirely up to the subjects, who simply attempt to achieve as much "throughput" as possible in each task. Gladstones, Regan, and Lee (1989), for example, had subjects perform serial tasks paced by the experimenter (e.g., pressing a key in response to the position of a light and pronouncing a letter in response to the color of a light). In some conditions, subjects performed just one such task, whereas in others, they performed two concurrently. The total rate at which information was processed summed over the two tasks (which corresponds roughly to the total number of responses in either task per unit time) was the same whether one task was performed or two. This was true even after considerable practice, and regardless of whether the tasks used the same or different input and output modalities. Similar findings were reported by Fisher (1975a,b) and Schouten, Kalsbeek, and Leopold (1960). Although, following Meyer and Kieras (1997), some interference might be expected due to conflicts in the initiation of responses, a bottleneck confined to response-related processing should allow a dramatic increase in total throughput rate to be achieved when two tasks are performed, instead of one.

My colleagues and I recently carried out other kinds of studies using discrete tasks to examine whether central queuing is strategic in origin. In one study, Eric Ruthruff, Alwin Klaassen, and I instructed subjects to perform two tasks and group the responses close together in time, a requirement subjects find quite natural. One task required judging whether a figure was a normal or a mirror image letter and making a corresponding keypress response. The other task, which could be performed more quickly, involved discriminating between a single 17 msec tone and a rapid-fire sequence of two 17 msec tones separated by 50 msec, with a vocal response (saying "one" or "two"). The first tone and the letter began simultaneously.

The instruction to group the two responses obviously does not provide any incentive to perform one task before the other. If there is no interference between the decision or response selection phases of the two tasks, the response should almost always be selected more quickly in the easier task, normally the tone judgment. Thus the grouped response should only be a bit slower than the response for the letter task alone, due to occasional trials in which the letter task happens to take longer than the tone task, plus any cost associated with grouping. In fact, there was very substantial slowing of mean RTs (1,475 msec for the dual-task grouped response, compared to 917 msec for the letter task alone). Monte Carlo simulations disclosed that this slowing could not be accounted for by the fact that the tone task was occasionally slower than the letter task. It is also not likely to reflect extra time taken to produce a grouped response; costs of producing grouped responses can be assessed directly, and prove negligible (e.g., Pashler and Johnston 1989, exp. 2).

As a further test, the difficulty of response selection in the easier task was varied: in compatible blocks, subjects responded by saying "one" to the single tone pulse, and "two" to the two pulses; in incompatible blocks, the mapping was reversed, producing about 200 msec of slowing. If central processing on the easier task were carried out in parallel with central processing on the harder task, much of the slowing of the tone task should be absorbed in "slack," and thus have minimal effect on the time to produce the grouped responses. In reality, compatibility had at least as large an effect on the grouped response in the dual-task context as it had on performance of the tone task by itself. Thus the whole pattern of results in this experiment favors the idea that central queuing was occurring in a situation where parallel processing would clearly have been advantageous.

In another recent study, Levy and I required subjects to make a three-alternative button-push response to the color of a large disk presented on a monitor screen, and to make a vocal response to its position (saying "one", "two" or "three" for left, middle, or central position). Here, rather than using grouping, we provided explicit payoffs designed to promote parallel processing and to place equal emphasis on the speed of each task. On blocks where both stimuli were presented, average reaction times for both tasks exhibited substantial slowing. Again it appears that encouragement to prioritize one task more than the other is by no means a necessary condition for dual-task interference to occur.

12.3 RELATING DUAL-TASK INTERFERENCE AND TASK SET

Having very briefly and selectively reviewed some of the main phenomena in the area of task switching and central limitations in simple dual-task performance, let us consider possible relations between the two topics. The research on dual-task interference bears on the issue of task set and task switching in several interesting respects. Two of these will be discussed here. The first is a very broad question of cognitive architecture: Do the phenomena of task set reconfiguration and dual-task interference (and specifically the sort of central queuing argued for in section 12.2) singly or jointly imply the existence of a "central executive" or "supervisory attention system"? The second question is narrower: Does the bottleneck itself reflect a limitation in task set, and perhaps the same limitation as is responsible for task-shifting costs, in which case the phenomena of dual-task queuing and task switching might really be one and the same?

Many writers have assumed that cognitive control requires the existence of a specific controlling mechanism whose function is to program (other) cognitive machinery. As discussed in several chapters in this volume, this controlling function is often associated with the frontal lobes or specific parts thereof. Several well-known theoretical frameworks in cognitive psychology, such as Baddeley's dissection (1986) of working memory and Norman and Shallice's theory (1986) of attention and control, famously invoked the idea of a "central executive." For present purposes, we can put aside the common criticism that invoking an executive as an account of mental control creates a sort of infinite regress (does the executive contain its own executive?). Rather, let us simply ask whether the phenomena of set and dual-task interference provide any sort of evidence for such a conception.

As several authors have pointed out (e.g., Allport 1987; Monsell 1996), the alternative is a scheme in which executive control emerges from the interaction of the very same machinery that ordinarily carries out the mental processes being controlled. The brute phenomena of executive control (e.g., that we can decide to perform one task or another; that verbal instructions can, if their recipient chooses to comply, completely determine which stimuli evoke which responses) emphatically do not require the existence of machinery dedicated for the purpose of control. Mutual competition between distributed mechanisms for the control of thought and action may well account for task set–switching phenomena. Indeed, work on "multiagent planning" in artificial intelligence suggests such a mechanism is capable of much more than that (e.g., Suarez, Winstanley, and Griffiths 1998). Furthermore, some of the phenomena of task set described above, such as the need to perform at least one trial of a new task in order to fully reconfigure processing machinery for that task, seem slightly more congenial to a distributed control architecture than to the notion of a distinct executive mechanism.

It is also commonly suggested that the idea of an all-or-none processing bottleneck (particularly a single bottleneck that spans diverse cognitive contents, as argued for above) naturally implies or at least suggests the existence of a single mechanism that carries out whatever cognitive operations are subject to queuing. Noting this, some writers (e.g., Kinsbourne 1981) have pointed out that the notion of a single-channel bottleneck seems hard to reconcile with the highly distributed processing that characterizes the human cerebral cortex.

It is certainly true that one very natural explanation for obligatory queuing of any given operation is the possibility that there is only a single device capable of carrying out the operation. That may not be the only explanation, however, let alone the correct one. Consider, for example, recent studies of processing bottlenecks in commisurotomy ("split-brain") patients. If the central bottleneck described above has a defined cortical locus, split-brain patients should show no PRP effect whenever each task is confined to a separate hemisphere (assuming they are capable of performing the tasks under such conditions). However, using lateralized stimuli and responses, Pashler et al. (1994) observed relatively normal performance and a relatively normal PRP effect in four split-brain patients. We concluded that the queuing underlying the PRP bottleneck must have a subcortical source because connections at these brain levels remain intact in split-brain patients (but see Ivry and Haseltine, chap. 17, this volume, for another view based on later studies conducted with one of these patients). It seems very unlikely that a brain stem mechanism would be responsible for actually carrying out memory retrieval and response selection. The natural alternative, then, is that the operations subject to queuing are themselves distributed and subcortical mechanisms trigger or control the queuing.

Is Queuing a Consequence of Task Set Limitations?

Is it possible that difficulties in selecting two responses at the same time (resulting in the PRP effect) stem from an inability to simultaneously maintain the task set for the two separate tasks? Although this idea has been suggested from time to time (e.g., Gottsdanker 1980), such a reduction seems hard to reconcile with the task-switching phenomena described earlier in this chapter.[5] Recall that in the Jersild paradigm, people usually incur only a fairly modest cost (and sometimes none at all) in shifting from one task to another so long as the mapping is univalent (i.e., where no stimulus is ever mapped onto different responses in the two tasks). Because, in the typical PRP task, the stimulus sets for the two tasks are nonoverlapping, the problem of concurrent task set maintenance should be comparable to that found with the univalent Jersild task, not with the bivalent task. Based on the results described earlier, one would therefore expect to find only a fairly modest slowing, presumably because both tasks sets can simultaneously coexist. Because the PRP effect often reaches several hundred milliseconds, presumably this concurrent maintenance problem cannot be the whole source of it.

On the other hand, one need not rely on indirect inferences; the concurrent maintenance contribution to PRP slowing can be assessed fairly directly, with a control seldom used until recently, by introducing to the PRP experiment blocks in which subjects prepare for both tasks, but are presented only one stimulus and are unable to predict which one this will be. In one unpublished study, Eric Ruthruff and I had subjects make a verbal response to a color patch, a manual response to a tone, or both. In the "or" task, subjects performed one task or the other, but not both (only one stimulus was presented). The "and" task was basically a PRP task with a zero SOA. There was some slowing in the "or" task compared to pure task blocks, but much more slowing on top of that in the "and" task.

The preparatory limitation responsible for the slowing in the "or" task as compared to a pure single task is likely to be responsible for slowing found in various single-task designs, as Gottsdanker (1980) pointed out. In a choice RT task, a greater number of stimulus-response (S-R) pairs is associated with longer RTs (Hick 1952)—an effect that depends chiefly on the number of alternatives subjects must prepare for, rather than the number of different alternatives they were exposed to during the current block of trials (Dixon 1981). Presumably, the need to prepare more S-R "links" means that each link cannot be prepared as fully, causing performance to be slowed (Gottsdanker 1980; Logan 1978). It is not merely the number of links that matters, however; the more conceptually cohesive the set of stimuli mapped onto any single response, the faster the task can be performed (Greenwald, McGhee, and Schwartz 1998; Seymour 1973). What is not clear is how preparatory costs should be understood. For example, does poorer preparation for larger or more heterogeneous mappings reflect more time having elapsed since a given link was prepared, or is "preparatory capacity" subject to continuous sharing, as proposed by Gottsdanker (1980)?

A Modified Reduction Hypothesis

Even though dual-task slowing is not reducible to the preparatory limitation for the reasons just discussed, one could still try to explain the PRP effect in terms of a limitation in task set. Consider the following hypothesis. In the "or" task experiments just described, the response selection module might not be preset at all, or it might be set in a "neutral" fashion. The shift from this unprogrammed state to the appropriate task set might occur very quickly, producing only a minor cost. Suppose, counter to what we have been assuming throughout this chapter, that, in the PRP design, despite a univalent mapping, the first task set must be disengaged and the second task set loaded before the second task can be processed. To explain why the dual-task case ("and" task) produces more slowing than the unknown single-task case ("or" task), one merely has to suppose that the response selection machinery cannot be reprogrammed while it is being used. This does not seem like an unreasonable supposition. The only problem is that because this account presumes that task set reconfiguration is necessary even with univalent mappings, it fails to explain why bivalent lists exhibit so much more alternation cost than univalent lists, although, with some ingenuity, it could probably be made to explain this as well.

Fortunately, however, we do not need to rely on such arguments. What would provide a critical test of the hypothesis that the bottleneck reflects a limitation in maintaining the set for each task? If the bottleneck reflects an inability to prepare the two task mappings simultaneously, then it should disappear when two or more tasks use the same mapping. That is, if the stimulus-response mapping rule remains fixed, and several stimuli must be processed, parallel central processing should be possible, unlike in the normal PRP case. One possible test of this claim would use a PRP task in which two distinct stimuli are presented and the response rule is the same.[6] Another method in which the mapping remains constant but subjects attempt to perform more than one task at the same time is the serial RT task, where subjects respond to a whole string of stimuli. In a recent study, we had subjects carry out a self-paced serial task, with and without preview (Pashler 1994). Letters unfolded from left to right, and subjects made a button-push response to the identity of each letter (four possible keys and four possible letters); ten letters unfolded, so that at the completion of the trial, there were ten letters on the screen and subjects had made ten responses. In the no-preview condition, the experiment began with the presentation of a single letter; stimulus $n+1$ was presented as soon as subjects responded to stimulus n. In the preview condition, the experiment began with two letters on the screen; stimulus $n+1$ was presented on the response to stimulus $n-1$. Due to the preview, subjects could potentially begin processing stimulus $n+1$ while still processing stimulus n. Is this logical possibility also a psychological possibility?

The rate of responding in the preview condition was greater than in the no-preview condition. First noted by Cattell (1886) and confirmed by Leonard (1953), this finding strongly suggests that *some* overlap of processing stages does indeed occur in the preview condition (as it does in the conventional PRP situation, too; see figure 12.7). The key question was whether the response selection stages associated with successive stimuli could overlap. To answer this question, several different task difficulty manipulations were used: targeting perception, response selection, and response production. When the mapping was made less natural, thereby increasing response selection duration (the manipulation was applied for the whole list of ten stimuli), the time between each response in the run was increased. The slowing was the same with or without preview. On the other hand, when perceptual processing was made more difficult, the time between the first stimulus and the first response lengthened, but the rate of responding thereafter was virtually unaffected. The results can be summarized by saying that response selection (but not perception or response production) seems to be rate limiting for serial performance even when stimuli are presented well before they are needed. Evidently, only one response can be selected at a time even if the rule for selecting responses does not change.

If the need to select new responses without any need to change task set is sufficient to produce response selection queuing, it seems gratuitous (or at least unparsimonious) to attribute the bottleneck in selecting com-

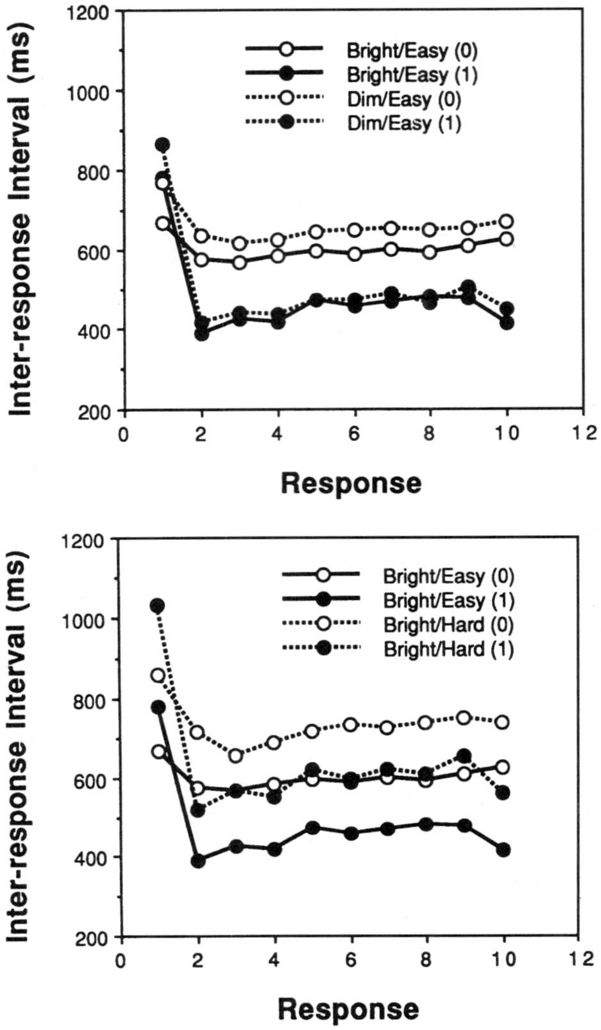

Figure 12.7 Effects of preview, stimulus quality and S-R compatibility on serial reaction time task. From Pashler 1994.

pletely distinct responses to an inability to maintain nonoverlapping (univalent) mappings simultaneously prepared. In view of this finding, plus the minimal cost of shifting in univalent lists (Jersild and others), it seems likely the limitation on carrying out two response selections at once cannot be reduced to a limitation on maintaining the two task sets at once. Presumably, because the mappings are univalent, the response selection module is loaded with both mappings (although not without cost, and not necessarily to the same degree at all times throughout the trial). That would suggest that the order of task performance in the PRP situation is probably not preplanned, a view that has been challenged by De Jong (1995). Logically speaking, there is no contradiction between saying that the two task mappings are simultaneously loaded and saying that the order of processing is planned or anticipated, although De Jong's evidence for preplanning of order involved tasks with two manual responses, and may therefore represent a rather special form of response selection.[7]

Alternative Explanations for Bottlenecks

We have considered two possible reasons for why a bottleneck might arise in the process of action planning (and, it was suggested above, memory retrieval as well). One explanation suggested that the bottleneck reflects strategic choices in scheduling mental operations, rather than a structural limitation: the other, that it reflects a limitation in simultaneously maintaining the two mappings in an active state. The evidence described above, although not fully conclusive, suggests that neither of these explanations is likely to be correct. If so, how else might one account for this puzzling limitation?

One intuitively very appealing idea, proposed by Allport (1987, 1993) and endorsed by De Jong (chap. 15, this volume), is that a bottleneck in planning might serve a positive function of preventing incompatible actions, thus maintaining the overall coherence of our behavior. The PRP effect, which appears as an obstacle to optimal performance within the contrived constraints of the dual-task experiment, might therefore be adaptive—in computer parlance, a "feature, not a bug." This proposal does not explain, however, why even time-consuming memory retrievals should be subject to queuing, as argued above. Nor, as formulated, does it specify exactly what sort of incoherence is meant to be prevented by queuing. One idea might be that preventing unrelated actions from being selected simultaneously would prevent the simultaneous execution of motor responses created by different action plans. This, it might be argued, would help maintain the coherence of behavior because a single planning operation will seldom (one might assume) generate behaviors that are mutually disruptive. The problem with this idea is that we are actually quite capable of simultaneously executing responses reflecting two or more independent planning operations. Casual observation of ordinary human activities reveals many examples. In a café, for example, a patron will lower a coffee cup while simultaneously beginning to speak; in a store, a clerk greets a customer while simultaneously putting the previous customer's groceries in a bag. It seems far-fetched to suppose that the speech and the hand movement, or the greeting and the hand movement, result from a single plan. These informal observations are confirmed by objective data. Van Galen and ten Hoopen (1976), for example, had people pronounce multisyllabic words in response to a letter and then make a button-push response to a second letter that followed soon after. The button-push response often occurred while the vocal response was still in progress; when this happened, there was no detectable interference.

One might suggest that what the brain is engineered to avoid is not the overlapping execution of independently selected responses, but rather the planning of an action that would terminate or disrupt a previously selected action. Such a constraint might, in de Jong's words (chap. 15, this volume), "protect task performance in progress from interference." Here again, there is little reason to believe that the constraint envisioned really exists. People can cancel actions that have just been launched, even when these are highly practiced. For example, Logan and Burkell (1983) showed that skilled typists could rapidly stop typing when an auditory stop signal was presented. In simple terms, action planning and the earliest stages of execution are not "ballistic." If they were, it might lend a certain form of coherence to our behavior, but probably a sort of coherence we should be glad not to possess.

The obvious alternative to accounts that view queuing as a positive benefit are accounts that claim the computational requirements of parallel retrieval would exceed available resources. This is somewhat puzzling, though, in view of the rather elementary kinds of task mappings that elicit queuing. The possibility of cross talk between tasks may help explain the ubiquity of queuing, if not quite as directly as some writers have supposed. Because similarity of tasks seems *not* to be a necessary condition for dual-task interference or queuing, attributing dual-task interference to content-specific cross talk within a given task combination seems rather unpromising (Pashler 1997). It is possible, however, that the system is wired up to require queuing as a general policy (conceivably one that can be overcome with sufficient practice) to prevent cross talk from unpredictably degrading performance in certain cases. Such an account seems consistent with several findings described earlier, including the proposed unity of limitations in action selection and memory retrieval, and the evidence from split-brain patients that anatomically distributed processing can be subject to queuing.

Open Questions

The study of task set is in its relative infancy, and the suggestions offered here about how we might relate task set to dual-task limitations are modest and preliminary. Many very basic questions remain to be addressed. One obvious question is whether the process of task reconfiguration itself can be carried out in parallel with another task. Goschke (chap. 14, this volume) finds that people are able to achieve the usual (partial) degree of reconfiguration if required concurrently to verbalize a description of the task they are about to perform. On the other hand, producing an irrelevant verbalization interfered with reconfiguration. What is not clear is whether carrying out an unrelated nonverbal task would interfere. This issue seems quite amenable to chronometric study.

Another open question is how the concepts useful in thinking about arbitrary choice reaction time tasks that have been the focus of the research described here might generalize to the more ordinary activities of everyday life. In activities like driving and conversing, one may speak of "task schemata" or "goals," but the notion of "mapping" seems inapt or contrived. Unfortunately, the implications of many of the concepts described here for such tasks remain to be clarified. This statement is not intended as a criticism of researchers who have, reasonably enough, started by studying relatively tractable cases. One area where some steps have been taken toward greater "ecological validity" is bilingual lexical production. Several investigators have given bilingual subjects cues telling them to name stimuli such as numbers in one language or another, and examined the effects of RSI and related variables. Thus far, the results with this task seem encouragingly similar to those found with non-linguistic laboratory tasks described above (MacNamara, Krauthammer, and Bolgar 1968; Meuter and Allport 1999). It is to be hoped that further efforts to examine tasks of this sort, as well as classic laboratory tasks, may shed greater light on the issues of task control and dual-task performance.

NOTES

This work was supported by National Institute of Mental Health grant 1-R01-MH45584 and by National Science Foundation grant SBR9729778.

1. Why preview should produce a switch benefit remains an open question. Conceivably, people can overlap more of the processing of each successive task when the mapping is changing.

2. In Vicentized distributions, the values for different percentiles are determined separately for each subject, then averaged across subjects; the results represent the typical shape of individuals' distributions, even if their speed of responding differs greatly.

3. In some cases (e.g., Rogers and Monsell 1995, exp. 4), a significant switch cost has been found with univalent lists that use compound stimuli, where the irrelevant stimulus was neutral (i.e., associated with no response).

4. The stimulus terms were color names and the verbal response terms were digits. During testing, single- and dual-task blocks were interspersed.

5. Note that the issue here is not whether the PRP effect arises merely as a consequence of temporal uncertainty about when S2 will arrive. This idea is clearly refuted by the finding that when the temporal parameters are unchanged, but subjects need not respond to S1, no PRP slowing occurs (e.g., Pashler and Johnston 1989).

6. One would naturally assume that sensory- or effector-specific interference would potentially contaminate such a study. If, however, the duration of central processing substantially exceeded that of more peripheral processing, reuse of the same sense and effector mechanisms should make very little difference; this deserves testing.

7. Manual response selection may ordinarily choose a spatial location, rather than a finger. If both a left-hand response and a right-hand response must be selected, the potential set of spatial locations may be unwieldy. As a strategy, the response-selection machinery might therefore choose within-hand spatial locations for each hand in turn, requiring a planned order (Pashler 1990). If this explains De Jong's findings, evidence for preplanning ought to disappear when one task is manual and the other vocal.

REFERENCES

Allport, A. (1987). Selection for action: Some behavioral and neurophysiological considerations of attention and action. In H. Heuer and A. F. Sanders (Eds.), *Perspectives on perception and action,* pp. 395–419. Hillsdale, NJ: Erlbaum.

Allport, A. (1993). Attention and control: Have we been asking the wrong questions? A critical review of twenty-five years. In D. Meyer and S. Kornblum (Eds.), *Attention and Performance XIV: Synergies in experimental psychology, artificial intelligence, and cognitive neuroscience,* pp. 183–218. Cambridge, MA: MIT Press.

Allport, A., Styles, E. A., and Hsieh, S. (1994). Shifting intentional set: Exploring the dynamic control of tasks. In C. Umiltà and M. Moscovitch (Eds.), *Attention and Performance XV: Conscious and nonconscious information processing,* pp. 421–452. Cambridge, MA: MIT Press.

Baddeley, A. D. (1986). *Working memory.* Oxford: Oxford University Press.

Biederman, I. (1973). Mental set and mental arithmetic. *Memory and Cognition, 1,* 383–386.

Borger, R. (1963). The refractory period and serial choice reactions. *Quarterly Journal of Experimental Psychology, 15,* 1–12.

Brebner, J. (1968). The search for exceptions to the psychological refractory period. In S. Dornic (Ed.), Attention and Performance V1, pp. 63–78. Hillsdale, NJ: Erlbaum.

Broadbent, D. E., and Gregory, M. (1967). Psychological refractory period and the length of time required to make a decision. *Proceedings of the Royal Society of London B168,* 181–193.

Carrier, M., and Pashler, H. (1996). The attention demands of memory retrieval. *Journal of Experimental Psychology: Learning, Memory, and Cognition, 21,* 1339–1348.

Cattell, J. M. (1886). The time it takes to see and name objects. *Mind, 11,* 63–65.

Creamer, L. R. (1963). Event uncertainty, psychological refractory period, and human data processing. *Journal of Experimental Psychology, 66,* 187–194.

De Jong, R. (1993). Multiple bottlenecks in overlapping task performance. *Journal of Experimental Psychology: Human Perception and Performance, 19,* 965–980.

De Jong, R. (1995). The role of preparation in overlapping-task performance. *Quarterly Journal of Experimental Psychology: Human Experimental Psychology, 48A,* 2–25.

Dixon, P. (1981). Algorithms and selective attention. *Memory and Cognition, 9,* 177–184.

Duncan, J., Martens, S., and Ward, R. (1997). Restricted attentional capacity within but not between sensory modalities. *Nature, 387,* 808–810.

Fagot, C. (1994). Chronometric investigations of task switching. Ph.D. diss., University of California, San Diego.

Fisher, S. (1975a). The microstructure of dual-task interaction: 1. The patterning of main-task responses within secondary-task intervals. *Perception, 4,* 267–290.

Fisher, S. (1975b). The microstructure of dual task interaction: 2. The effect of task instructions on attentional allocation and a model of attention-switching. *Perception, 4,* 459–474.

Gladstones, W. H., Regan, M. A., and Lee, R. B. (1989). Division of attention: The single-channel hypothesis revisited. *Quarterly Journal of Experimental Psychology: Human Experimental Psychology, 41A,* 1–17.

Gottsdanker, R. (1980). The ubiquitous role of preparation. In G. E. Stelmach and J. Requin (Eds.), *Tutorials in motor behavior,* pp. 355–371. Amsterdam: Elsevier.

Greenwald, A. G. (1972). On doing two things at once: I. Time-sharing as a function of ideomotor compatibility. *Journal of Experimental Psychology, 100,* 52–57.

Greenwald, A., and Shulman, H. (1973). On doing two things at once: II. Elimination of the psychological refractory period. *Journal of Experimental Psychology, 101,* 70–76.

Greenwald, A. G., McGhee, D. E., and Schwartz, J. L. K. (1998). Measuring individual differences in implicit cognition: The implicit association test. *Journal of Personality and Social Psychology, 74,* 1464–1480.

Heuer, H. (1985). Intermanual interactions during simultaneous execution and programming of finger movements. *Journal of Motor Behavior, 17,* 335–354.

Hick, W. E. (1952). On the rate of gain of information. *Quarterly Journal of Experimental Psychology, 4,* 11–26.

Jersild, A. T. (1927). Mental set and shift. *Archives of Psychology,* no. 89.

Johnston, J. C., and McCann, R. S. (forthcoming). On the locus of dual-task interference: Is there a bottleneck at the stimulus classification stage?

Kinsbourne, M. (1981). Single channel theory. In D. Holding (Ed.), *Human skills,* pp. 65–89. Chichester, U.K.: Wiley.

Kornblum, S. (1973). Sequential effects in choice reaction time: A tutorial review. In S. Kornblum (Ed.), *Attention and Performance IV,* pp. 259–288. New York: Academic Press.

Leonard, J. A. (1953). Advance information in sensorimotor skills. *Quarterly Journal of Experimental Psychology, 5,* 141–149.

Logan, G. D. (1978). Attention in character classification tasks: Evidence for the automaticity of component stages. *Journal of Experimental Psychology: General, 107,* 32–63.

Logan, G. D., and Burkell, J. (1986). Dependence and independence in responding to double stimulation: A comparison of stop, change and dual-task paradigms. *Journal of Experimental Psychology: Human Perception and Performance, 12,* 549–563.

Logan, G. D., and Zbrodoff, N. J. (1982). Constraints on strategy construction in a speeded discrimination task. *Journal of Experimental Psychology: Human Perception and Performance, 8,* 502–520.

MacNamara, J., Krauthammer, M., and Bolgar, M. (1968). Language switching in bilinguals as a function of stimulus and response uncertainty. *Journal of Experimental Psychology, 78,* 208–213.

McLeod, P., and Posner, M. I. (1984). Privileged loops from percept to act. In H. Bouma and D. G. Bouwhuis, (Eds.), *Attention and Performance X,* pp. 55–66. London: Erlbaum.

Meiran, N. (1996). Reconfiguration of processing mode prior to task performance. *Journal of Experimental Psychology: Learning, Memory, and Cognition, 22,* 1423–1442.

Meuter, R. F. I., and Allport, A. (1999). Bilingual language switching in naming: Asymmetrical costs of language selection. *Journal of Memory and Language, 40,* 25–40.

Meyer, D. E., and Kieras, D. E. (1997). A computational theory of executive cognitive processes and multiple-task performance: 2. Accounts of psychological refractory-period phenomena. *Psychological Review, 104,* 749–791.

Miller, J. (1982). Divided attention: Evidence for coactivation with redundant signals. *Cognitive Psychology, 14,* 247–279.

Monsell, S. (1996). Control of mental processes. In V. Bruce (Ed.), *Unsolved mysteries of the mind: Tutorial essays in cognition,* pp. 93–148. Hove, U.K.: Erlbaum and Taylor and Francis.

Norman, D. A., and Shallice, T. (1986). Attention to action: Willed and automatic control of behavior. In R. J. Davidson, G. E. Schwartz, and D. Shapiro (Eds.), *Consciousness and self-regulation*, vol 4, pp. 1–18. New York: Plenum Press.

Osman, A., and Moore, C. (1993). The locus of dual-task interference: Psychological refractory effects on movement-related brain potentials. *Journal of Experimental Psychology: Human Perception and Performance, 19*, 1292–1312.

Palmer, J. (1995). Attention in visual search: Distinguishing four causes of a set-size effect. *Current Directions in Psychological Science, 4*, 118–123.

Pashler, H. (1984). Processing stages in overlapping tasks: Evidence for a central bottleneck. *Journal of Experimental Psychology: Human Perception and Performance, 10*, 358–377.

Pashler, H. (1990). Do response modality effects support multiprocessor models of divided attention? *Journal of Experimental Psychology: Human Perception and Performance, 16*, 826–840.

Pashler, H. (1994). Overlapping mental operations in serial performance with preview. *Quarterly Journal of Experimental Psychology, 47*, 161–191.

Pashler, H. (1997). *The psychology of attention*. Cambridge, MA: MIT Press.

Pashler, H., and Baylis, G. C. (1991). Procedural learning: 2. Intertrial repetition effects in speeded-choice tasks. *Journal of Experimental Psychology: Learning, Memory, and Cognition, 17*, 33–48.

Pashler, H., Carrier, M., and Hoffman, J. (1993). Saccadic eye movements and dual-task interference. *Quarterly Journal of Experimental Psychology, 46A*, 51–82.

Pashler, H., and Christian, C. (n.d.). Dual-Task interference and motor response production.

Pashler, H., and Johnston, J. C. (1989). Interference between temporally overlapping tasks: Chronometric evidence for central postponement with or without response grouping. *Quarterly Journal of Experimental Psychology, 41A*, 19–45.

Pashler, H., and Johnston, J. C. (1998). Attentional limitations in dual-task performance. In H. Pashler (Ed.), *Attention*, pp. 155–189. Hove, U.K.: Psychology Press, Erlbaum, and Taylor and Francis.

Pashler, H., Luck, S. J., Hillyard, S. A., Mangun, G. R., and Gazzaniga, M. (1994). Sequential operation of disconnected cerebral hemispheres in split-brain patients. *Neuroreport, 5*, 2381–2384.

Rickard, T., and Pashler, H. (1998). A bottleneck in retrieval from a single cue. Talk presented at Thirty-ninth Annual Meeting of the Society for Psychonomics, Dallas, November.

Rogers, R. D., and Monsell, S. (1995). Costs of a predictable switch between simple cognitive tasks. *Journal of Experimental Psychology: General, 124*, 207–231.

Ruthruff, E., Miller, J., and Lachmann, T. (1995). Does mental rotation require central mechanisms? *Journal of Experimental Psychology: Human Perception and Performance, 21*, 552–570.

Salthouse, T. A., Fristoe, N., McGuthry, K. E., and Hambrick, D. Z. (1998). Relation of task switching to speed, age, and fluid intelligence. *Psychology and Aging, 13*, 445–461.

Schvaneveldt, R. W., and Staudenmayer, H. (1970). Mental arithmetic and the uncertainty effect in choice reaction time. *Journal of Experimental Psychology, 85*, 111–117.

Schouten, J. F., Kalsbeek, J. W. H., and Leopold, F. F. (1960). On the evaluation of perceptual and mental load. *Ergonomics, 5*, 251–260.

Seymour, P. H. (1973). Rule identity classification of name and shape stimuli. *Acta Psychologica, 37*, 131–138.

Shaffer, L. H. (1965). Choice reaction with variable S-R mapping. *Journal of Experimental Psychology, 70,* 284–288.

Simon, J. R., Webster, J. B., and Craft, J. L. (1981). Effects of changing compatibility relationships on reaction time to the first stimulus in a double-stimulation task. *Journal of Motor Behavior, 13,* 301–312.

Spector, A., and Biederman, I. (1976). Mental set and mental shift revisited. *American Journal of Psychology, 89,* 669–679.

Suarez, J. N., Winstanley, G. and Griffiths, R. N. (1998). A reluctance-based cost distribution strategy for multiagent planning. *Applied Intelligence, 9,* 39–55.

Sudevan, P., and Taylor, D. A. (1987). The cueing and priming of cognitive operations. *Journal of Experimental Psychology: Human Perception and Performance, 13,* 89–103.

Telford, C. W. (1931). The refractory phase of voluntary and associative responses. *Journal of Experimental Psychology, 14,* 1–36.

Treisman, A., and Davies, A. (1973). Dividing attention to ear and eye. In S. Kornblum (Ed.), *Attention and Performance IV,* pp. 101–117. New York: Academic Press.

Van Galen, G. P., and ten Hoopen, G. (1976). Speech control and single channelness. *Acta Psychologica, 40,* 245–255.

Welford, A. T. (1952). The "psychological refractory period" and the timing of high speed performance: A review and a theory. *British Journal of Psychology, 43,* 2–19.

Welford, A. T. (1967). Single-channel operation in the brain. *Acta Psychologica, 27,* 5–22.

Welford, A. T. (1980). The single-channel hypothesis. In A. T. Welford (Ed.), *Reaction time,* pp. 215–252. New York: Academic Press.

13 Multitasking Performance Deficits: Forging Links between the Attentional Blink and the Psychological Refractory Period

Pierre Jolicœur, Roberto Dell'Acqua, and Jacquelyn Crebolder

ABSTRACT This chapter presents new empirical work that bears on the issue of whether multitasking performance deficits are a by-product of strategic control as opposed to structural capacity limitations. Accuracy in reporting the identity of a masked visual target was measured at various delays following an auditory stimulus that required an immediate speeded response. A larger attentional blink (AB) effect was found when the auditory task had four possible stimuli and response alternatives than when it had only two. In a psychological refractory period (PRP) experiment, two speeded responses were required to stimuli presented in rapid succession. The auditory task used as the first task of the AB experiment served as the second task in the PRP experiment. Effects of number of response alternatives in the second task were additive with stimulus onset asynchrony (SOA), suggesting that the manipulation had an effect at or after the locus of PRP interference, and that a locus of AB interference existed at or after a locus of interference causing the PRP effect. Implications for the role of control versus structural limitations are discussed.

This chapter explores the relationship between two popular paradigms that require different responses to two stimuli presented in rapid succession: the "psychological refractory period" (PRP) paradigm (see Pashler 1994 for a review) and the "attentional blink" (AB) paradigm (see Shapiro and Raymond 1994 for a review). In the PRP paradigm, two speeded responses are made in rapid succession, usually to two distinct and unmasked stimuli. The PRP effect is a slowing of the second response as the stimulus onset asynchrony (SOA) between the two stimuli is reduced. In the AB paradigm, two stimuli are also shown in rapid succession, and usually both are masked. The AB phenomenon is a decrease in report accuracy of the second stimulus at short SOAs.

With some notable exceptions, most researchers have concluded that performance deficits in these paradigms arise because of capacity limitations in central processing mechanisms. These mechanisms are assumed to perform such operations as memory encoding and retrieval, response selection, and other complex cognitive operations that occur after perceptual encoding but before motor output.

As noted by Monsell (1996), discussion of these paradigms has focused largely on the locus of interference effects or on the specific combinations of tasks that result in performance limitations. Recently, however, the issue of control has been brought to the fore by theoretical arguments that relate major findings in both these paradigms to causal influences of strategic control. We consider two examples.

First, using their "executive process interactive control" (EPIC) model, Meyer and Kieras (1997a,b; see also Kieras and Meyer, chap. 30, this volume) propose that dual-task slowing in the PRP paradigm results from adaptive executive control processes designed to guarantee that the response to the first stimulus occurs before the response to the second stimulus (subjects are often instructed to perform the responses in this order). And second, finding evidence of a cross-modal AB deficit only when a switch in task (Rogers and Monsell 1995) associated with the first and second targets to be reported was required, Potter et al. (1998) propose that the AB deficit may have a "strategic" source, such as preparation (see De Jong, chap. 15, this volume; De Jong and Sweet 1994), different from the structural source of AB interference when the stimuli are presented in the same modality.

In the AB paradigm, the response to the second target stimulus is not speeded. The control-based account of Meyer and Kieras (1997a,b; Kieras and Meyer, chap. 30, this volume) for dual-task slowing in the PRP paradigm would therefore not appear to apply to the AB paradigm, regardless of whether stimulation is within modality or between modalities. Indeed, given their account, it would seem most natural to think of PRP and AB effects as two rather different manifestations of multitasking performance deficits. While the PRP effect would reflect central control required to sequence rapid responses, AB effects would presumably be taken to reflect some other form of system overload that occurs at a more peripheral level (e.g., perceptual, motor, or both) because, according to the EPIC model, the central operations required to perform the AB task are not likely to induce central interference. However, this hypothesis does not agree well with recent results suggesting that the AB effect may be caused by central interference (see "Locus of Factor Effects in Psychological Refractory Period and Attentional Blink Paradigms" in section 13.1).

The empirical work presented below explores the relationship between AB and PRP paradigms, and asks whether the sources of these multitasking deficits may share some fundamental functional similarity, which would in turn have implications for the role of control processes in causing these effects. We first consider the interpretation of factor effects in these paradigms.

13.1 THE ATTENTIONAL BLINK PHENOMENON

The AB phenomenon is a decrease in the accuracy of report of a second target (T_2), when that target follows rapidly after a first target (T_1) that must also be reported. The paradigm most commonly used to study the phenomenon embeds the two targets within a stream of other stimuli presented using rapid serial visual presentation (RSVP; e.g., Broadbent and Broadbent 1987; Chun and Potter 1995; Raymond, Shapiro, and Arnell 1992, 1995; Jolicœur 1998; but see Duncan, Ward, and Shapiro 1994).

For example, Jolicœur (1998, exp. 1) presented a red first target (T_1, *H* or *S*) embedded in an RSVP stream of white letters. The second target (T_2) was an *X* or a *Y*. T_2 occurred on every trial, but T_1 was presented on only half of the trials. The most interesting results concern the accuracy of report of the second target. In control trials (T_1 absent), the mean accuracy was about 85%. In the experimental condition (T_1 present), accuracy was at about 73% when T_2 followed T_1 immediately (lag 1); about 64% at lags 2–3; and about 71% at lag 4, with a continued recovery to near-baseline levels as lag was increased further. Raymond, Shapiro, and Arnell (1992) labeled the loss of accuracy of report for T_2, as a function of the lag between T_1 and T_2, the "attentional blink" (AB), and this label is now widely used.

Locus of Factor Effects in Psychological Refractory Period and Attentional Blink Paradigms

Much of the work on the AB phenomenon has focused on two issues:

- Where is the interference between task 1 and task 2 taking place?
- What is the nature of this interference?

This chapter primarily addresses the first issue. There is growing agreement that the locus of the interference is relatively late in processing, probably after stimuli have activated semantic-level representations (Shapiro et al. 1997; Luck, Vogel, and Shapiro 1996; Jolicœur 1998b, 1999c; Duncan, Ward, and Shapiro 1994; Chun and Potter 1995; see also Jolicœur and Dell'Acqua 1998, 1999, forthcoming; Dell'Acqua and Jolicœur 1998).

The simplest model of dual-task interactions assumes that some mechanisms cannot be shared across two tasks (Welford 1952). When two tasks both need the same mechanism, interference results—the mechanism constitutes a processing bottleneck. The top stage diagram in figure 13.1A represents the processing required to perform the first of two tasks in a PRP paradigm. The presentation of the first target (T_1) triggers the stages labeled A_1, which represent all stages before the bottleneck. Processing stages that require the bottleneck are labeled B_1. Finally, stages after the bottleneck are labeled C_1. The sum of prebottleneck, bottleneck, and post-bottleneck stage durations equals the response time in task 1, or RT_1.

When the SOA between T_1 and T_2 is short (figure 13.1A) the prebottleneck processing in task 2 can proceed without interference. This is illustrated by A_2 in figure 13.1A. When prebottleneck processing is completed, the processing of T_2 is ready to engage the mechanisms that constitute the processing bottleneck, but these mechanisms are busy with

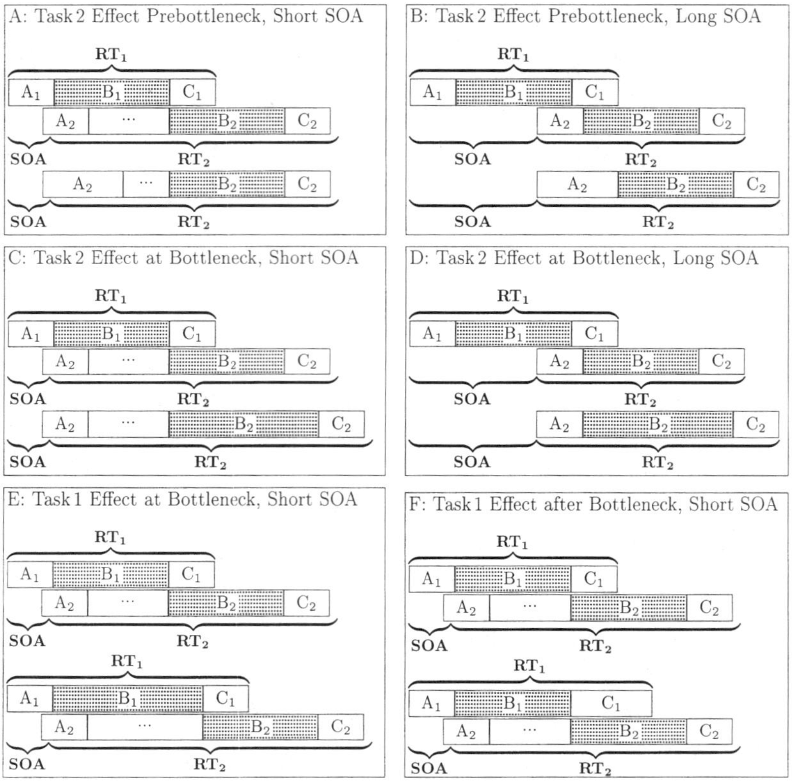

Figure 13.1 Stage diagrams showing the predicted task interactions in dual-task paradigms.

task 1. The result is a period of waiting, represented by three dots, during which no further processing of T_2 takes place. When task 1 no longer requires the bottleneck, processing of T_2 resumes. The initiation of bottleneck processing in task 2 (B_2) thus coincides with the termination of bottleneck processing in task 1 (B_1). RT_2 is the sum of stage durations plus the period of waiting (slack).

Now, suppose that a factor manipulated in task 2 increases the duration of a prebottleneck stage, as represented by an increase in the length of A_2. When the SOA is very short, as shown in figure 13.1A, this reduces the period of waiting before the initiation of bottleneck processing, but has no effect on RT_2. The effect of the factor is said to have been absorbed into the period of slack.

At longer SOAs (panel B), changing the duration of prebottleneck processing has the expected effect of increasing RT_2. Thus the effect of the factor should decrease as SOA is reduced. The resulting interaction is often described as underadditive with decreasing SOA (or with increasing task overlap).

If the factor manipulated in task 2 affects the duration of the bottleneck stage, additive effects of the factor and SOA are expected. As shown in figure 13.1C (short SOA) and 13.1D (long SOA). Although a period of waiting (slack) is created by the contention for the bottleneck at short SOA, the effects of the factor are not absorbed into slack because the factor affects a stage of processing that occurs after the period of waiting. The increased duration of processing through the bottleneck stage is fully and equally reflected in RT_2 at both short and long SOA. Additivity is also predicted if the factor affects the duration of processing after the bottleneck (not shown).

This analysis, developed by Pashler and Johnston (1989), can be used to interpret second-task factor effects on RT_2 in the PRP paradigm. If the factor effects are additive with SOA, the factor must be affecting a stage in or after the bottleneck. If the factor effects decrease in magnitude as SOA is reduced, then the factor must be affecting the duration of a stage before the bottleneck. This analysis is sometimes called "locus-of-slack logic," and strong support for the method has been provided in numerous studies (e.g., Pashler and Johnston 1989; McCann and Johnston 1992; see Pashler 1994a for a review).

Figure 13.1E–F illustrates another prediction of the postponement model of the PRP effect, concerning a factor manipulated in task 1. If the effect of the factor is to lengthen the duration of processing at the bottleneck (figure 13.1E) or before, then the effect should carry over to response times in task 2 as well. The longer bottleneck duration in the bottom pair postpones the onset of processing at the bottleneck stage in task 2, which results in a longer RT_2. At a very short SOA, as illustrated, the effect of the first-task factor should be the same on RT_1 and RT_2. Support for this prediction can be found in Smith 1967, Van Selst, Ruthruff, and Johnston 1999, Williams 1974, and Pashler, 1994b. In figure 13.1F, a first-task factor affecting processing after the bottleneck is assumed. While this factor would affect RT_1, no effect should be observed on RT_2 (see Pashler 1994b for some supporting evidence). The conclusion, therefore, is that a factor manipulated in task 1, whose effects carry over to RT_2, must affect a stage in or before—but not after—the bottleneck.

In general, effects of first-task variables on RT_2 such as the one in figure 13.1E are expected only at short SOAs. At longer SOAs, response times in task 2 are not predicted to be influenced by first-task variables because these effects are mediated by the competition for the bottleneck, and no such competition takes place if the SOA is long enough.

For the AB paradigm, we are concerned with the effects of factor manipulations in task 1 on performance in task 2. The analysis of factor effects shown in figure 13.1E–F also applies to the AB paradigm, although predictions are now made for accuracy in task 2, rather than for RT_2. In the AB paradigm, response times in task 2 are not measured; instead, the paradigm focuses on report accuracy to a masked target. For a wide range of possible models, including all extant models, a factor manipulated in task 1 of an AB experiment is not expected to affect accuracy in task 2 if the variable affects processing after the locus or loci of interference causing the AB effect. Therefore, if a first-task factor modulates the magnitude of the AB effect, which is measured as a change of accuracy in task 2, the factor must be affecting a stage of processing that is in or before the locus of AB interference (see Jolicœur 1998). Such effects are expected only at shorter SOAs; at longer SOAs, there is no competition for processing capacity, and thus no expected dual-task interactions. Clearly, some caution is required here because it is not difficult to think of events that could occur after the critical task interactions causing the AB effect that could cause a significant loss of information about T_2. The argument is sound, however, as long as the deficits in task 2 remain clearly time locked to the occurrence of T_1 at short SOAs, with a recovery to baseline conditions at long SOAs, and as long as we remain within the boundary conditions of the paradigms usually used to study the AB phenomenon.

To account for the AB effect using postponement models, we must also assume that there is a loss of information about T_2 during the period of waiting, with greater loss for longer waits (Jolicœur 1998). The results of Jolicœur (1999-b) and Giesbrecht and DiLollo (1998) suggest that such loss does not occur if T_2 is not masked, presumably because sensory persistence provides a form of storage of the information that can bridge the period of waiting.

Locus of Attentional Blink Interference Relative to the Psychological Refractory Period Bottleneck

Experiment 1 was a cross-modal AB experiment in which the first target was a pure tone and the second target a visually presented letter in an RSVP stream. The main factor manipulated in task 1 was the number of stimulus and response alternatives. This manipulation had a large effect on the magnitude of the AB effect. The conclusion is that this factor must have its effect in or before a locus of interference causing the AB phenomenon. In experiment 2, the same manipulation was performed in task 2 of a PRP experiment, and the effects were additive with SOA. The conclusion is that this factor must have its effects in or after the PRP bottleneck. Together, these results lead to the conclusion that at least one locus of AB interference must be in or after the PRP bottleneck.

13.2 EXPERIMENT 1

Experiment 1 used a cross-modal speeded AB paradigm. T_1 was a pure tone and task 1 was an immediate speeded-choice response based on pitch. T_2 was a visual stimulus, the letter *X* or *Y*, presented on every trial, embedded within an RSVP stream. Task 2 was a deferred and unspeeded discrimination between *X* and *Y* (see also Jolicœur 1998, 1999a, 1999c). The tone was presented concurrently with one of the letters in the RSVP stream.

There were two versions of task 1. One had two tone frequencies and two responses; the other, four frequencies and four responses. Based on earlier work, we hypothesized that the two-alternative task 1 would be associated with a shorter period of central processing than the four-alternative task 1 (Van Selst and Jolicœur 1997; Schubert 1999). According to Van Selst and Jolicœur (1997), this difference in processing should occur in or after the PRP bottleneck. This assumption is verified in experiment 2.

Strong claims have been made regarding the outcome of experiments like this. Pashler (1993) has influentially claimed that a speeded task will not produce an SOA-dependent deficit on a closely following unspeeded task. Duncan, Ward, and Shapiro (1994) have claimed that there is absolutely no cross-modal AB. Yet earlier work in our laboratory has shown that a speeded task 1 response to a tone can cause a significant AB effect in a concurrent visual encoding task 2 (Jolicœur 1999a; Jolicœur and Dell'Acqua 1999). Experiment 1 repeated Jolicœur's experiment (1999a) but also included two levels of first-task difficulty designed to influence the duration of central processing.

Subjects

Twenty-six undergraduates at the University of Waterloo participated for pay. All reported having normal or corrected-to-normal vision and normal hearing.

Stimuli

The auditory stimuli were pure tones presented well above threshold for 100 msec at a frequency of 200, 363, 660, or 1,200 Hz, using an internal computer speaker. The middle two frequencies (363 and 660 Hz) were used in the two-alternative condition.

The visual stimuli were white uppercase letters on a black background presented in RSVP at the center of a computer screen, at a rate of 10 letters/sec (100 msec each with no blank interstimulus interval).

Between 6 and 9 letters were presented prior to the letter concurrent with the tone, and 12 to 15 after the tone. The *X* or *Y* could occur with equal probability at positions 1, 3, 5, 7, 9, or 11, following the tone. Thus even the last target position had 1 to 4 letters following it, ensuring that T_2 was always effectively masked (Giesbrecht and Di Lollo 1998; Jolicœur 1999b).

On every trial, the background stream items were selected at random, without replacement, from the letters of the alphabet, excluding *H*, *S*, *X*, and *Y*. Each letter subtended about 1 degree of visual angle and had a luminance of about 25 cd/m^2 and CIE(x, y) coordinates of (0.278, 0.306). The background was black with a luminance of less than 1 cd/m^2.

Procedure

Each trial began with two symbols at the center of the screen, which provided both fixation markers and performance feedback for the previous trial. Pressing the space bar eliminated the fixation and feedback symbols and initiated the RSVP sequence. A tone (T_1) was presented on half of the trials. Trials with no tone served as control trials, in which preparation was equivalent to that in experimental trials. Tone-present and tone-absent trials were intermixed at random within each test session.

The experiment was divided into two sessions separated by a short break. In one session task 1 involved two choices (363 Hz → ">"; 660 Hz → "?"). In the other session, task 1 involved four choices (200 Hz → "M"; 363 Hz → "<"; 660 Hz → ">" 1,200 Hz → "?"). The response buttons were contiguous on the bottom right row of the keyboard, and responses were made with the index, middle, ring, and little fingers of the right hand. The index and middle fingers were used for the two-alternative task 1. The instructions were to press the correct response button as quickly as possible after hearing the tone, while keeping errors to a minimum. A message asking subjects to respond more quickly to the tone was presented if the RT to the tone was greater than 1,300 msec.

At the end of every trial, after the response to the tone, a prompt asked subjects to indicate which visual target had been shown (*X* or *Y*). The "X" key was used to respond "X," and the "C" key was used to respond "Y." This response was not speeded.

The two-alternative discrimination session consisted of one block of 48 practice trials, followed by 3 blocks of 96 trials. The four-alternative discrimination session consisted of two block of 48 practice trials, followed by 3 blocks of 96 trials. The order of sessions was counterbalanced across subjects.

Each block of experimental trials in the four-alternative task 1 contained a full crossing of T_1–T_2 SOA (100, 300, 500, 700, 900, or 1,100 msec), T_1 frequency, T_2 present versus absent, and T_2 identity (*X* versus *Y*). When T_1 was not presented, a corresponding position in the RSVP stream was selected nonetheless. This made it possible to create control trials in which the absolute position of T_2 in the RSVP stream was equated across T_1-present and T_1-absent trials. Each block of experimental trials in the two-alternative task 1 contained two full crossings of the experimental variables. Different random orders of the trials were used for each block and for each subject. Performance feedback was given in the form of a plus or minus sign for each response, at fixation, following each trial.

Results

The data from three subjects were eliminated. Two were less than 69% correct in the control condition for task 2 in one or both sessions, and one

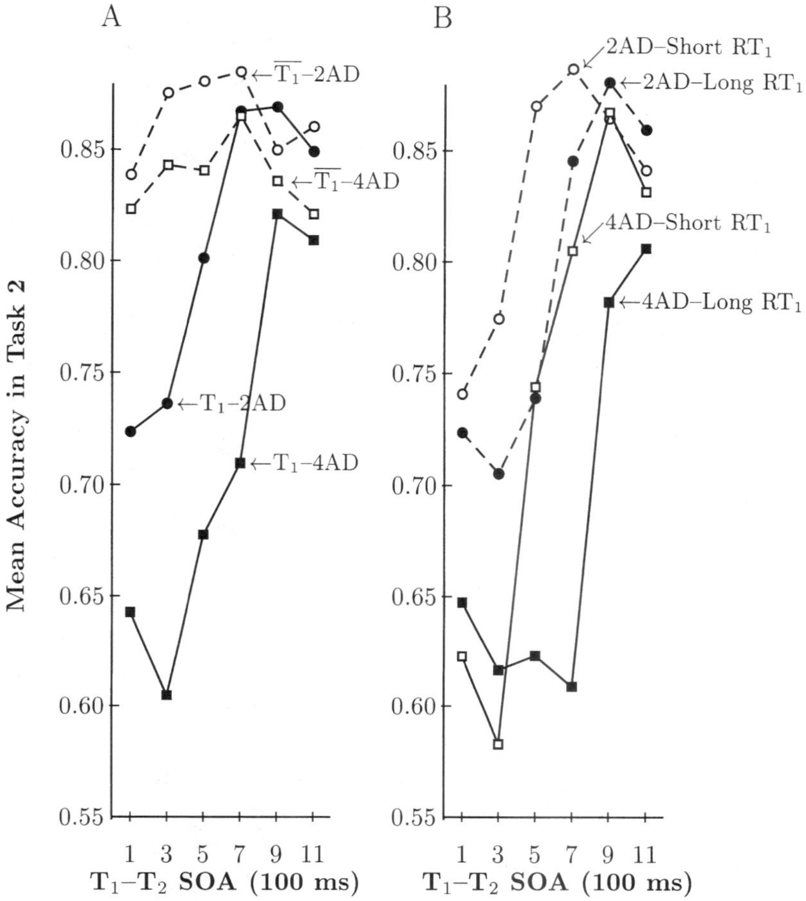

Figure 13.2 Mean proportion correct in task 2 of experiment 1. A. Results for each stimulus onset asynchrony (SOA), each level of first-task difficulty (two alternatives: circular symbols; four alternatives: square symbols), for trials on which T_1 was presented (filled symbols, solid lines), or omitted (unfilled symbols, dashed lines). B. Results for each SOA, each level of first-task difficulty (two alternatives: circular symbols, dashed lines; four alternatives: square symbols, solid lines), for trials in which T_1 was presented, depending on the duration of RT_1 (RT_1 below the median: open symbols; RT_1 above the median: filled symbols).

was correct on only 54% of four-alternative tone task trials. The analyses presented below are based on 13,248 pairs of responses generated by the remaining 23 subjects.

Correct trials in tone-present trials were screened for outliers using a close variant of the Van Selst and Jolicœur (1994) procedure (e.g., Jolicœur 1998, 1999a,b,c). Less than 1.9% of the trials were rejected. Analyses based on data that included outliers produced the same patterns of results.

Task 2 Figure 13.2A displays mean accuracy in task 2. Both versions of task 1 produced large AB effects. In an analysis of the two-alternative condition, the interaction between SOA and T_1 present or absent was highly significant: $F(5, 110) = 6.91$, $p < 0.0001$; as were the two main effects: $p < 0.0001$ in both cases. In a separate analysis of the four-alternative condition, the SOA by T_1 (present/absent) interaction was highly significant: $F(5, 110) = 10.94$, $p < 0.0001$; as were the two main effects; $p < 0.0001$ in both cases.

A separate analysis of the data from the T_1-present trials revealed a significant interaction between the number of first-task response alternatives and SOA: $F(5, 110) = 2.44$, $p < 0.04$. There was also a large main effect of number of Task 1 alternatives, $F(1, 22) = 43.53$, $p < 0.0001$, corroborating what can be seen in figure 13.2, namely, that accuracy in task 2 was lower when task 1 had four response alternatives than when it had only two. A companion analysis examining the control trials showed that the control conditions differed in overall levels of performance across the two- and four-alternative versions of task 1: $F(1, 22) = 9.47$, $p < 0.006$. However, there was no main effect of SOA and no interaction for the control conditions ($p > 0.05$ in both cases).

The three-way interaction between SOA, T_1 (present versus absent), and number of first-task responses was not significant in the omnibus analysis of variance (ANOVA), but the interaction between number of first-task responses and T_1 (present versus absent) was highly significant, reflecting the larger difference between the experimental and control conditions for the four-alternative than for the two-alternative first-task condition: $F(1, 22) = 24.42$, $p < 0.0001$. The AB effect was reliably larger in the four-alternative condition than in the two-alternative condition.

We also compared the difference between control and experimental performance during the blink versus after the blink, as a more direct test of the difference in AB effects across conditions (this more sensitive test is justified by a priori expectations; see Jolicœur 1998). The difference between average control performance and the average of the first four SOAs (during the blink) was contrasted with the difference between the control condition and the average of the last two SOAs (after the blink). This difference was significantly larger for the four-alternative condition than for the two-alternative condition: $F(1, 22) = 15.66$, $p < 0.0007$.

Task 1 As expected, mean RT_1 was longer for four alternatives (691 msec) than for two (530 msec); $F(1, 22) = 217.37$, $p < 0.0001$. Neither the main effect of SOA nor the interaction between SOA and number of first-task alternatives was significant: $F(5, 110) = 1.55$, $p > 0.18$; $F(5, 110) = 1.41$, $p > 0.22$, respectively.

First-task accuracy was higher for two alternatives (93.2%) than for four (77.4%); $F(1, 22) = 105.84$, $p < 0.0001$. There was no significant effect of SOA and no interaction between SOA and number of first-task alternatives ($p > 0.29$ in both cases).

Task 2 as a Function of RT_1 Accuracy in task 2 was also examined, as a function of the speed of processing in task 1. The trials on which T_1 was presented were divided into cells for each subject, each SOA, whether task 1 had two or four alternatives, and for each of the three blocks of trials within each session. For each of these cells, the trials were sorted further into two more cells depending on whether RT_1 was above or below the median RT_1 in that bin. (Mean RT_1 was 456 msec for faster two-alternative responses, 596 msec for slower two-alternative responses, 606 msec for faster four-alternative responses, and 792 msec for slower four-alternative responses.) For each resulting cell, the mean accuracy in task 2 was computed and submitted to an ANOVA with SOA, number of first-task alternatives, block, and short and long RT_1s as within-subject factors. The means are displayed in figure 13.2B.

Accuracy in task 2 was higher (0.786) when RT_1 was shorter than the median and lower (0.737) when RT_1 was longer than the median; $F(1, 22) = 23.09$, $p < 0.0001$. The interaction between short and long RT_1s and SOA was significant, $F(5, 110) = 5.03$, $p < 0.0003$. This effect was modulated by the number of response alternatives, as shown in figure 13.2B: $F(5, 110) = 2.78$, $p < 0.025$. The interaction between SOA and RT_1 appears to have the following form. Accuracy in task 2 is similar across short and long RT_1s at very short SOAs, accuracy for short and long RT_1s diverge for intermediate SOAs, with lower accuracy for long RT_1s, followed by a convergence of accuracy levels across short and long RT_1s at the longest SOAs.

The main effect of block was significant, with accuracy remaining about the same from block 1 (0.756) to block 2 (0.745), and then improving in block 3 (0.784): $F(2, 44) = 3.61$, $p < 0.0355$. The observed relationship between RT_1 and accuracy in task 2 was not an artifact of a general improvement in performance in both tasks, as subjects became more practiced, because lower accuracy in task 2 (for trials with a longer RT_1) was observed within each block or trials and because there was no overall increase in accuracy across blocks 1 and 2.

There was also little change in response times across blocks. Mean RT_1 was 619 msec in block 1, 613 msec in block 2, and 605 msec in block 3; and these means were not significantly different: $F(2, 44) = 1.56$, $p < 0.22$. Thus it is unlikely that the association between RT_1 and accuracy in task 2 could be due to correlated changes in overall performance levels with practice.

Discussion

The results were clear-cut. A larger AB effect was produced when task 1 involved four alternatives rather than two alternatives (figure 13.2). Furthermore, within each first-task condition, a larger and longer AB effect was found when processing of T_1 took longer. Both of these results support the hypothesis that a longer period of processing in task 1 in one or more stages of processing carried over into accuracy scores in task 2.

13.3 EXPERIMENT 2

The interpretation of the results of experiment 1 hinges critically on the locus of effect of the number of alternatives in task 1. On the one hand, Van Selst and Jolicœur (1997) and Schubert (1999) both provided evidence that the locus is in or after the PRP bottleneck, finding that number of alternatives (two or more), when manipulated in task 2 of a PRP experiment, produced additive effects with SOA. On the other hand, Schumacher et al. (1999) showed that underadditive interactions of number of alternatives and SOA can be found under certain conditions. The conditions used in task 1 of experiment 1 do not match exactly the conditions of any of these previous experiments, making it difficult to extrapolate from earlier work.

In experiment 2, the manipulation used in task 1 of experiment 1 was applied to task 2 of a PRP experiment. According to the locus-of-slack logic outlined earlier (figure 13.1), if the manipulation used in experiment 1 is at or after the stage of processing that causes PRP interference, then additive effects of this variable should be observed with SOA. If some or all of the effect is at an earlier stage, then an underadditive interaction with SOA would result, as SOA is reduced.

Subjects

Thirty-three undergraduates at the University of Waterloo participated for pay. All reported having normal or corrected-to-normal vision, and normal hearing.

Stimuli

The auditory stimuli were identical to those used in experiment 1. The visual stimuli were three letters, *H*, *O*, and *S*, presented at the same size and luminance as the stream items in experiment 1. The exposure duration of the letter was 100 msec. The letter was not masked.

Procedure

Pressing the space bar removed the fixation symbols and initiated the trial sequence. After a delay of 750 msec, a letter was presented, requiring a speeded, three-alternative choice response to be made with the left hand: *H* → ring finger ("Z" key); *O* → middle finger ("X" key); *S* → index finger ("C" key). The key mapping was described on a piece of paper that was at the top of the keyboard to facilitate learning during the practice trials.

After an SOA of 50, 200, 500, or 1,100 msec, chosen pseudorandomly at run time, the tone was presented. The frequencies and frequency response mappings (right hand) were identical to those in experiment 1.

Each subject was tested in two sessions separated by a short break, one for the two-alternative and one for the four-alternative version of task 2, in counterbalanced order. Each session began with 48 practice trials, followed by 6 blocks of 48 trials. Each block contained one or two full crossings of the independent variables. The order of trials was randomized anew for each block. Feedback was given in the form of a plus or minus sign at fixation, for each response, following each trial.

Results

The data from 6 subjects were excluded because of accuracy less than 50% in one or more cells in one or both tone tasks. The remaining 27 subjects produced 15,552 experimental response pairs. Prior to RT analyses, the correct trials were first screened for outliers using the same algorithm as in experiment 1. Screening based on RT_1 resulted in a loss of 2.8% of the trials. The surviving trials were then screened for outliers on RT_2, eliminating an additional 2.2%. Analyses performed on the unscreened data produced the same patterns of results as those reported below.

Task 2 The most important results concern mean RT_2, as a function of SOA and number of second-task alternatives, as shown in figure 13.3. The main effect of SOA was highly significant: $F(3, 78) = 248.96$, $p < 0.0001$; as was the main effect of number of second-task alternatives: $F(1, 26) = 157.99$, $p < 0.0001$. However, the interaction between these two factors was not significant: $F(3, 78) = 0.19$, $MS_E = 1698.24$, $p > 0.90$. The difference between the four-alternative condition and the two-alternative condition was 199 msec at 50 msec SOA, 210 msec at 200 msec SOA. 200 msec at 500 msec SOA, and 203 msec at 1,100 msec SOA.

Accuracy in task 2 varied slightly as SOA increased (0.845, 0.859, 0.874, and 0.866): $F(3, 78) = 5.22$, $p < 0.0024$. Accuracy was higher for two (0.933) than for four alternatives (0.789): $F(1, 26) = 160.65$, $p < 0.0001$. There was no interaction between these two variables, however: $F(3, 78) = 0.11$, MSE $= 0.001483$, $p > 0.95$.

Task 1 The mean RT_1 for each SOA and each number of alternatives in task 2 is also shown in figure 13.3. The main effect of SOA was significant: $F(3, 78) = 7.94$, $p < 0.0001$. Mean RT_1 was also longer when there were four second-task alternatives (590 msec) than when there were two (544 msec): $F(1, 26) = 27.05$, $p < 0.0001$. These two variables also interacted, as shown in the figure: $F(3, 78) = 13.66$, $p < 0.0001$. The mean RT_1 was constant

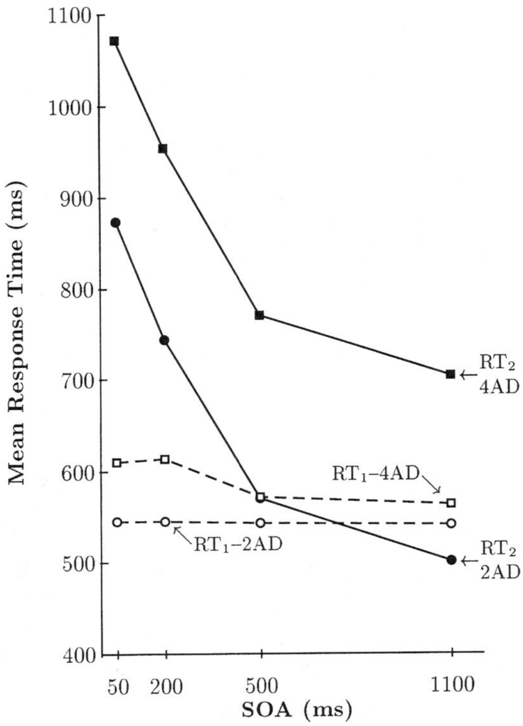

Figure 13.3 Results from experiment 2. Mean response time in task 2 (filled symbols, solid lines) and in task 1 (open symbols, dashed lines), for each stimulus onset asynchrony (SOA), and each level of second-task difficulty (two alternatives: circular symbols; four alternatives: square symbols).

across SOAs for the two-alternative second-task condition, but it declined with increasing SOA for the four-alternative second-task condition. It is not clear to what these effects on RT_1 should be attributed, but their small magnitudes suggest that, in the main, the assumptions of the postponement model of PRP were not badly violated.

Mean first-task accuracy was slightly higher when there were two alternatives in task 2 (0.962) than when there were four (0.950): $F(1, 26) = 11.43$, $p < 0.0025$. Neither the main effect of SOA nor the interaction between SOA and number of second-task alternatives was significant ($p < 1$ in both cases).

Discussion

The results were clear-cut. The effects of varying the number of alternatives in task 2 were additive with SOA. The implication is that this manipulation had an effect that was in or after the PRP bottleneck. Given that the manipulation in experiment 2 was identical to the one used in task 1 of experiment 1, it is reasonable to assume that effects of number of response alternatives in experiment 1 also took place at or after the stages of processing constituting the PRP bottleneck.

It was important to test directly whether the number of alternatives has its principal effect at or after the PRP bottleneck. A priori, one might have expected some of the effect to occur relatively early, for example, due to a greater difficulty of discrimination for four stimuli than for two stimuli. The frequencies were equally spaced on a log scale (200, 363, 660, and 1,200 Hz), in an attempt to produce approximately equal steps in perceived pitch; the two tones used in the two-alternative condition were adjacent (363 and 660 Hz) in the sequence, in an attempt to equate the degree of perceptual difficulty across the two conditions. It appears that, under these conditions, the degree of perceptual difficulty in the two conditions was very similar, such that the main difference between them was later in processing, perhaps at response selection.

13.4 GENERAL DISCUSSION

In experiment 1, varying the number of stimulus-response alternatives in task 1 of a speeded attentional blink paradigm produced a large modulation of the AB effect. Response times in task 1 were clearly longer when there were more response alternatives in task 1 than when there were fewer. Changes in the duration of the stages of processing affected by this manipulation carried over into accuracy in task 2. The manipulation therefore had its effects at or before the locus of dual-task interaction in the AB paradigm.

Experiment 2 showed that the manipulation of number of alternatives had its effect in or after the PRP bottleneck, given that number of alternatives in task 2 and SOA were additive in a PRP experiment (figure 13.1; Pashler and Johnston 1989; Pashler 1994a; McCann and Johnston 1992).

Together, these two experiments lead to the conclusion that at least one locus of interference contributing to the AB phenomenon is in or after the PRP bottleneck.

The results converge nicely with those of Jolicœur (1998, 1999a,b,c) in suggesting a close connection between the dual-task interference observed in the AB and PRP paradigms. In experiment 1, large AB effects were obtained using stimuli presented in different sensory modalities, replicating and extending those of Jolicœur (1999a).

Attentional Blink versus Task Switch Costs

Potter et al. (1998) argued that there are two distinct attentional deficits in serial target search tasks such as the one used in our experiment 1. One deficit, the attentional blink (AB) hypothesized by Raymond, Shapiro, and Arnell (1992), was claimed to occur only when both target stimuli are visual, and not when one or both are auditory. The other deficit, an amodal effect, was hypothesized to be caused by capacity demands of task switching (as discussed in several other chapters in this volume).

If Potter et al. (1998) are correct, then one could argue that the observed deficits in task 2 of experiment 1 were due to task switch costs, rather than to the within-modality AB effect studied by several researchers (e.g., Raymond, Shapiro, and Arnell 1992; Ward, Duncan, and Shapiro 1996). It could be that task switch costs take place later in the system than the distinct within-modality AB effect postulated by Potter et al. (1998). If so, the evidence provided in our experiments 1–2 may apply only to the amodal AB effect, and not to the within-modality AB effect.

The present results suggests that at least some component of the AB effect occurs relatively late in the information-processing stream (i.e., at or after the PRP bottleneck). Additional research will be required to determine whether our results apply to the within-modality AB effect, to the postulated amodal effect, or to both.

Attentional Blink and Short-Term Consolidation

Jolicœur and Dell'Acqua (1998) showed that encoding information into memory causes responses in a subsequent speeded task to be delayed. In their experiment 7, every trial began with the presentation of one or three letters exposed for 250 msec and followed by a pattern mask (100 msec). On "encode" trials, the letters had to be reported, without speed pressure, at the end of the trial. On "ignore" trials, the letters could be ignored. On every trial, the second stimulus was a tone to which the subjects made a speeded pitch discrimination response (two-alternatives). The SOA between the letter display and the tone was varied between 350 and 1,600 msec.

The response times to the tones are shown in figure 13.4 (solid lines, filled symbols). Responses to the tone were delayed as the SOA between the letters and the tone was reduced, but only when the information had to be encoded (top two functions). Minimal effects of SOA were found when the letters could be ignored (bottom function). A larger effect of SOA was found when more information had to be subjected to short-term consolidation (encode-3) than when less information had to be encoded (encode-1).

The results of Jolicœur and Dell'Acqua (1998) did not constrain the nature of the interference causing the delay in responses to the tone (i.e., postponement versus capacity sharing). Computer simulations, however, showed that the results could be approximated reasonably well by assuming that some stage of processing in the tone task (e.g., response selection) was postponed for some time while short-term consolidation of the letters was taking place (see figure 13.4; Jolicœur and Dell'Acqua 1998). These results support the view that the short-term consolidation of

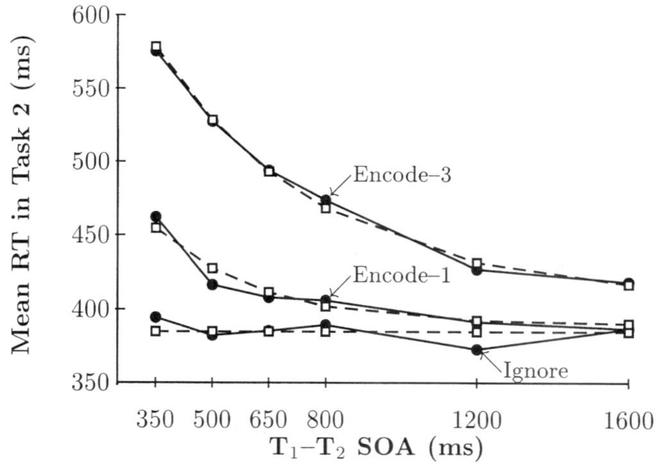

Figure 13.4 Cost of short-term consolidation. Results from experiment 7 of Jolicœur and Dell'Acqua 1998. Mean response time (RT_1) to the tone (in milliseconds) for each stimulus onset asynchrony (SOA), by whether the visual information was encoded (top two functions) or ignored (bottom function). The results from the encode condition are further split depending on the number of letters to be encoded (1, middle function; 3, top function). The unfilled symbols joined by dotted lines show the results of a simulation in which it was assumed that response selection in the tone task was postponed for some period of time while the short-term consolidation of the information to be remembered was taking place.

information into a durable form of memory is a capacity-demanding operation that can delay or slow down other cognitive processes. Chun and Potter (1995) and Jolicœur (1998; Jolicœur and Dell'Acqua 1998, 1999; Crebolder and Jolicœur forthcoming) argue that short-term consolidation is a likely locus of the dual-task interference causing the AB phenomenon.

Summary, Conclusions, and Implications for Control

We have presented two new experiments designed to provide constraints on possible loci of interference contributing to the AB phenomenon. Experiment 1 showed that a large AB effect in an RSVP scanning task with a deferred response can be caused by performing a speeded response to a pure tone (see also Crebolder and Jolicœur forthcoming; Jolicœur, 1998, 1999a,b; Jolicœur and Dell'Acqua forthcoming). Furthermore, larger effects resulted when the tone task had more stimulus-response alternatives. Also, within each version of task 1, a larger AB effect was found when RT_1 was longer. The manipulation of first-task difficulty carried over strongly onto task 2, as expected if the manipulation in task 1 had its effect in or before a locus of interference involved in the AB phenomenon.

In experiment 2, the number of stimulus-response alternatives used in experiment 1 (task 1) was now used in task 2 of a PRP experiment, and the effects were additive with SOA. Thus the manipulation had its effects in or after the PRP bottleneck. Consequently, at least one locus of interference causing the AB effect must be in or after the PRP bottleneck.

Crebolder and Jolicœur (forthcoming) performed a series of experiments that had the same logical structure as those in this chapter. Rather than manipulating number of alternatives in task 1, they varied the relative frequency of T_1 in AB experiments and of T_2 in PRP experiments. In the AB experiments, T_1 and T_2 were both letters, and less frequent T_1 signals caused larger AB effects. Hence the first-task manipulation carried over onto task 2. These effects were found both when task 1 was speeded and when task 1 was unspeeded. Furthermore, effects of the frequency of T_2 were additive with SOA in PRP experiments. These results show that the conclusions based on experiments 1–2 extend to within-modality AB paradigms, and to AB paradigms in which task 1 is deferred.

The results also suggest a closer connection between interference in the AB paradigm and in the PRP paradigm than has heretofore been supposed (e.g., Shapiro and Raymond 1994; Chun and Potter 1995; Ward, Duncan, and Shapiro 1996). At least one major source of AB interference appears to be at or after the same stage as the PRP bottleneck. Although a locus after the PRP bottleneck is logically possible, this alternative seems less likely than loci of interference that coincide in the two paradigms. While, this contention needs to be further tested, experiment 5 of Jolicœur 1999-b already provides evidence against a very late locus involving motor codes.

Additional evidence for a similarity between AB interference and PRP interference was also reviewed. The results of Jolicœur and Dell'Acqua (1998, forthcoming) suggest that the short-term consolidation of visual information into memory causes responses in a concurrent tone task to be delayed, suggesting that short-term consolidation requires central capacity-limited mechanisms.

We began this chapter by noting that the issue of control has figured prominently in some recent theoretical work on the PRP phenomenon (Meyer and Kieras 1997b), and of the AB phenomenon (Potter et al. 1998). Our results suggest that an effect at or after the PRP bottleneck also contributes substantially to the AB phenomenon. This effect, in the AB paradigm, could not be due to the need to control order of responses because the second response in that paradigm is not speeded. Given that there is good evidence ruling out late (motor-coding) accounts of the interference in such paradigms (e.g., Jolicœur 1999b), the most natural explanation of AB effects in the Meyer-Kieras framework would be to suppose that interference takes place early, in mechanisms that operate before central processing. Such an account, however, runs into difficulty given that a substantial component of the AB effect appears to be in or after the PRP bottleneck. A likely locus of effect for number of response alternatives (the main manipulation in our experiment 2) is response selection, and that locus is clearly beyond the early locus of interference that would be most easily incorporated into the Meyer-Kieras (1997) framework. This suggests to us that there may be more significant sources of structural central capacity limitations than are allowed for in that framework. Indeed, the results suggest to us that structural central capacity limitations, rather than the need to control response order, may be contributing causes of both AB and PRP dual-task interference.

The consolidation of information into memory is one important process required to perform task 2 in the AB paradigm (Chun and Potter 1995; Jolicœur 1998). The results of Jolicœur and Dell'Acqua (1998); (see figure 13.4) strongly suggest that short-term consolidation of letters causes dual-task slowing in a concurrent cross-modal speeded task. Jolicœur and Dell'Acqua (1997) showed that encoding a random polygon also causes dual-task slowing. This latter result is important because random polygons do not have names in long-term memory, thus ruling out explanations that hinge on the use of names to represent stimuli. The results of Jolicœur and Dell'Acqua (1997, 1998) show that dual-task slowing occurs even when only one response is speeded. This slowing cannot therefore be due to the need to control the order of output of two rapidly produced responses.

Although we argue that dual-task slowing in the Jolicœur and Dell'Acqua (1997, 1998) experiments was not caused by the consequences of strategic control, we want to highlight the important role of control for the results in that paradigm and in the AB paradigm. The key point is that short-term consolidation is not obligatory but under active control. Indeed, in many AB experiments, the control condition consisted of trials in which a salient target was shown, but could be ignored (e.g., Raymond, Shapiro, and Arnell 1992). These trials do not show the time-locked performance deficit that characterizes the AB effect. Similarly, "ignore" trials in the short-term consolidation experiments of Jolicœur and Dell'Acqua (1997, 1998; see figure 13.4) do not exhibit the dual-task slowing found when the information has to be consolidated.

The selection of information to be consolidated and the onset of the consolidation process itself are both controlled operations. Given that short-term consolidation appears to involve a significant cost in terms of the concomitant capacity demands, a key role for control processes is to minimize such costs by engaging capacity-demanding processes only when they are necessary.

Clearly, we are still quite far from having achieved a complete understanding of the AB and PRP phenomena. The present results and the evidence reviewed suggest, however, that a closer consideration of the similarities and differences between the patterns of interference in the AB paradigm and in the PRP paradigm is likely to provide useful constraints on theorizing in both domains.

NOTE

This work was supported by a research grant from the Natural Sciences and Engineering Research Council of Canada (NSERC). We thank Margaret Ingleton for technical assistance and Stephen Monsell, Molly Potter, and an anonymous reviewer for their very helpful comments and criticisms. Roberto Dell'Acqua received postdoctoral support from Fondation Fyssen, and Jacquelyn Crebolder was supported by an NSERC doctoral fellowship.

REFERENCES

Allport, D. A., Styles, E. A., and Hsieh, S. (1994). Shifting intentional set: Exploring the dynamic control of tasks. In C. Umiltà and M. Moscovitch (Eds), *Attention and Performance XV: Conscious and nonconscious information processing,* pp. 421–452. Cambridge, MA: MIT Press.

Broadbent, D. E., and Broadbent, M. H. P. (1987). From detection to identification: Response to multiple targets in rapid serial visual presentation. *Perception and Psychophysics, 42,* 105–113.

Chun, M. M., and Potter, M. C. (1995). A two-stage model for multiple target detection in rapid serial visual presentation. *Journal of Experimental Psychology: Human Perception and Performance, 21,* 109–127.

Crebolder, J., and Jolicœur, P. (Forthcoming). On the locus of the attentional blink bottleneck: Evidence from signal probability effects in the AB and PRP paradigms.

De Jong, R., and Sweet, J. B. (1994). Preparatory strategies in overlapping-task performance. *Perception and Psychophysics, 55,* 142–151.

Dell'Acqua, R., and Jolicœur, P. (1998). Evidence for dual-task interference on visual encoding. Poster presented at the Eighteenth International Symposium on Attention and Performance, Windsor Great Park, England.

Duncan, J., Ward, R., and Shapiro, K. L. (1994). Direct measurement of attentional dwell time in human vision. *Nature, 369,* 313–315.

Giesbrecht, B. L., and Di Lollo, V. (1998). Beyond the attentional blink: Visual masking by item substitution. *Journal of Experimental Psychology: Human Perception and Performance, 24,* 1454–1466.

Jolicœur, P. (1998). Modulation of the attentional blink by on-line response selection: Evidence from speeded and unspeeded task 1 decisions. *Memory and Cognition, 26,* 1014–1032.

Jolicœur, P. (1999a). Restricted attentional capacity between sensory modalities. *Psychonomic Bulletin and Review, 6,* 87–92.

Jolicœur, P. (1999b). Dual-task interference and visual encoding. *Journal of Experimental Psychology: Human Perception and Performance, 25,* 596–616.

Jolicœur, P. (1999c). Concurrent response selection demands modulate the attentional blink. *Journal of Experimental Psychology: Human Perception and Performance, 25,* 1097–1113.

Jolicœur, P., and Dell'Acqua (1997). Short-term consolidation of random polygons causes dual-task slowing. Paper presented at the annual meeting of the Psychonomic Society, Philadelphia.

Jolicœur, P., and Dell'Acqua, R. (1998). The demonstration of short-term consolidation. *Cognitive Psychology, 36,* 138–202.

Jolicœur, P., and Dell'Acqua, R. (1999). Attentional and structural constraints on visual encoding. *Psychological Research/Psychologische Forschung, 62,* 154–164.

Jolicœur, P., and Dell'Acqua, R. (Forthcoming). Selective influence of second target exposure duration and task 1 load effects in the attentional blink phenomenon. Psychonomic Bulletin and Review.

Luck, S. J., Vogel, E. K., and Shapiro, K. L. (1996). Word meaning can be accessed but not reported during the attentional blink. *Nature, 382,* 616–618.

McCann, R. S., and Johnston, J. C. (1992). Locus of the single-channel bottleneck in dual-task interference. *Journal of Experimental Psychology: Human Perception and Performance, 18,* 471–484.

Meyer, D. E., and Kieras, D. E. (1997a). A computational theory of executive cognitive processes and human multiple-task performance: 1. Basic mechanisms *Psychological Review, 104,* 3–65.

Meyer, D. E., and Kieras, D. E. (1997b). A computational theory of executive cognitive processes and human multiple-task performance: 2. Accounts of psychological refractory-period phenomena. *Psychological Review, 104,* 749–791.

Monsell, S. (1996). Control of mental processes. In V. Bruce (Ed.), *Unsolved mysteries of the mind: Tutorial essays in cognition,* pp. 93–148. Hove, U.K.: Erlbaum and Taylor and Francis.

Pashler, H. (1993). Dual-task interference and elementary mental mechanisms. In D. E. Meyer and S. Kornblum (Eds.), *Attention and Performance XIV: Synergies in experimental psychology, artificial intelligence, and cognitive neuroscience,* pp. 245–264. Cambridge, MA: MIT Press.

Pashler, H. (1994a). Dual-task interference in simple tasks: Data and theory. *Psychological Bulletin, 116,* 220–244.

Pashler, H. (1994b). Overlapping mental operations in serial performance with preview. *Quarterly Journal of Experimental Psychology, 47A,* 161–191.

Pashler, H., and Johnston, J. C. (1989). Chronometric evidence for central postponement in temporally overlapping tasks. *Quarterly Journal of Experimental Psychology, 41A,* 19–46.

Potter, M. C., Chun, M. M., Banks, B. S., and Muckenhoupt, M. (1998). Two attentional deficits in serial target search: The visual attentional blink and an amodal task-switch deficit. *Journal of Experimental Psychology: Learning, Memory, and Cognition, 24,* 979–992.

Raymond, J. E., Shapiro, and Arnell, K. M. (1992). Temporary suppression of visual processing in an RSVP task: An attentional blink? *Journal of Experimental Psychology: Human Perception and Performance, 18,* 849–860.

Raymond, J. E., Shapiro, K. L., and Arnell, K. M. (1995). Similarity determines the attentional blink. *Journal of Experimental Psychology: Human Perception and Performance, 21,* 653–662.

Rogers, R. D., and Monsell, S. (1995). Costs of a predictable switch between simple cognitive tasks. *Journal of Experimental Psychology: General, 124,* 207–231.

Schubert, T. (1999). Some more evidence for a central bottleneck in dual-task performance. *Journal of Experimental Psychology: Human Perception and Performance, 25,* 408–425.

Schumacher, E. H., Lauber, E. J., Glass, J. M., Zurbriggen, E. L., Gmeindl, L., Kieras, D. E., and Meyer, D. E. (1999). Concurrent response selection in dual-task performance: Evidence for adaptive executive control of task scheduling. *Journal of Experimental Psychology: Human Perception and Performance, 25,* 791–814.

Shapiro, K. L., Driver, J., Ward, R., and Sorensen, R. E. (1997). Priming from the attentional blink: A failure to extract visual tokens but not visual types. *Psychological Science, 8,* 95–100.

Shapiro, K. L., and Raymond, J. E. (1994). Temporal allocation of visual attention: Inhibition or interference? In D. Dagenbach and T. H. Carr (Eds.), *Inhibitory processes in attention, memory, and language,* pp. 151–188. San Diego, CA: Academic Press.

Shapiro, K. L., Raymond, J. E., and Arnell, K. M. (1994). Attention to visual pattern information produces the attentional blink in rapid serial visual presentation. *Journal of Experimental Psychology: Human Perception and Performance, 20,* 357–371.

Smith, M. C. (1967). The psychological refractory period as a function of performance of a first response. *Quarterly Journal of Experimental Psychology, 19,* 350–352.

Van Selst, M., and Jolicœur, P. (1994). A solution to the effect of sample size on outlier elimination. *Quarterly Journal of Experimental Psychology, 47A,* 631–650.

Van Selst, M., and Jolicœur, P. (1997). Decision and response in dual-task interference. *Cognitive Psychology, 33,* 266–307.

Van Selst, M., Ruthruff, E., and Johnston, J. C. (1999). Can practice eliminate the psychological refractory period effect? *Journal of Experimental Psychology: Human Perception and Performance, 25,* 1268–1283.

Ward, R., Duncan, J., and Shapiro, K. L. (1996). The slow time-course of visual attention. *Cognitive Psychology, 30,* 79–109.

Welford, A. T. (1952). The "psychological refractory period" and the timing of high-speed performance: A review and theory. *British Journal of Psychology, 43,* 2–19.

Williams, L. R. T. (1974). Effects of number of alternatives on the psychological refractoriness of an extended movement. *Journal of Motor Behavior, 6,* 227–234.

14 Intentional Reconfiguration and Involuntary Persistence in Task Set Switching

Thomas Goschke

ABSTRACT Switching between different tasks often increases response time compared to repeated performance of a task. This switch cost has been thought to reflect either an executive process of task set reconfiguration or proactive interference from competing task sets. This chapter tries to reconcile these views by showing that switch costs are influenced both by voluntary preparation and involuntary carry-over of inhibition and stimulus-response-bindings from the previous trial. Three experiments are reported in which participants switched between responding to the color and responding to the identity of letters. Switch costs were reduced when participants verbalized each task before the stimulus, compared to when they performed a verbal distractor task, suggesting that intention retrieval supported advance reconfiguration. Switch costs increased when the two stimulus dimensions activated incongruent responses and when task switches *followed* incongruent trials, indicating persisting activation of preceding task sets and persisting inhibition of irrelevant perceptual dimensions, S-R mappings, or both. Findings suggest that voluntary actions are not controlled by a unitary central executive, but emerge from the interaction of separable component processes, some maintaining intentions, others reconfiguring task sets. According to the proposed model, seemingly dysfunctional aspects of cognitive control are manifestations of adaptive mechanisms that have evolved to satisfy partially incompatible constraints on action control.

14.1 INTENTIONAL RECONFIGURATION AND COGNITIVE CONTROL

A remarkable property of willed action is its flexibility: by receiving an instruction or forming an intention, we can transiently couple almost any response to almost any stimulus or aspect of a stimulus, even when there are neither innate nor acquired connections between stimulus and response. For instance, if you, as a participant in a psychological experiment, are instructed to press a response key when the word "Green" is presented, or if you form the intention to lift your left index finger at the end of this sentence, your response dispositions are reconfigured such that your intended action is usually triggered by the stimulus condition specified in the instruction or intention. Although seemingly trivial, even such simple instances of voluntary action require that various processing systems be coordinated from moment to moment in novel ways, that new couplings between stimuli and action schemata be set into readiness, that skills be recombined into new behavioral sequences, and that a specific mode of interaction between various processing systems be established. Following similar proposals, I will use the term *task set* to denote such transient configurations of response dispositions and processing modes, and the term *intentional reconfiguration* to denote the processes underlying the formation and change of task sets (cf. Allport, Styles, and Hsieh 1994; Kuhl 1996; Meiran 1996; Monsell 1996; Rogers and Monsell 1995). Whereas in everyday life, "task sets" may involve long-term goals whose realization lies hours, days, or even years in the future, in this chapter, I restrict my analysis to much simpler task sets, ones that consist of transient couplings of elementary stimulus features and immediate behavioral responses.

Although the problem of cognitive control was acknowledged early in cognitive psychology (cf. Neisser 1967), as evidenced by the influential distinction between automatic and controlled (or control) processes (e.g., Atkinson and Shiffrin 1968; Posner and Snyder 1975; Schneider and Shiffrin 1977), until recently there has been little systematic research on the mechanisms underlying intentional reconfiguration. Whereas sophisticated models have been developed to account for performance in tasks such as naming, categorizing, and visual search, the question of how the cognitive system is configured for a given task in the first place still represents what Monsell (1996) has aptly called the "heart of darkness" of cognitive psychology.

This chapter reports three new experiments that use task switching to investigate processes underlying intentional reconfiguration. After a brief review of theoretical controversies in the task-switching literature (see also Allport and Wylie, chap. 2, De Jong, chap. 15, and Meiran, chap. 16, this volume; Monsell 1996; Pashler, chap. 12, this volume), I will try to show that the switch cost observed when individuals alternate between different tasks is influenced both by active preparatory processes (in particular, retrieval of verbal task representations) and by involuntary processes (in particular, *persisting activation* of the previously relevant task set and *persisting inhibition* of previously task-irrelevant perceptual dimensions, S-R mappings, or both). Finally, I will outline a theoretical framework, according to which voluntary control is a multiple constraint satisfaction problem, which affords a dynamic balance between maintaining and switching intentions, and between inhibition of distracting information and continuous background monitoring (Goschke 1996, 1997; cf. Allport 1989; Brandtstädter, Wentura, and Rothermund, forthcoming; Kuhl, 1985, 2000)

14.2 TASK SWITCHING AS A TOOL FOR STUDYING VOLUNTARY CONTROL

The experimental investigation of intentional reconfiguration has recently received renewed attention in the study of task switching, introduced as early as 1927 by Jersild, but seldom used (most notably by Spector and Biederman 1976) until recently (e.g., Allport, Styles, and Hsieh 1994; Allport and Wylie, chap. 2, De Jong, chap. 15, Keele and Rafal, chap. 28, this volume; Kluwe 1997; Mayr and Keele forthcoming; Meiran 1996, chap. 16, this volume; Rogers and Monsell 1995; Rubinstein, Meyer, and Evans forthcoming). The rationale of the method is to compare a condition or trials in which participants repeatedly perform the same task (for instance, subtracting 3 from successive digits in a list), with a condition or trials in which subjects have to alternate between different tasks (for instance, between subtracting 3 and adding 3 to successive digits in a list). Alternating between tasks usually results in a switch cost, that is, a prolonged response time compared to that for task repetition.

Task Switch Cost as a Manifestation of Proactive Interference

At first sight, the time cost incurred by a task switch may be taken to reflect the time required for executive control processes that configure the cognitive system for the new task. However, Allport, Styles, and Hsieh (1994, 436) have suggested that the switch cost does not directly reflect the duration of a stagelike executive process, but rather is due to proactive interference from previously executed task sets—"task set inertia" (TSI). In their experiment 4, participants first performed a block where they read color words printed in conflicting colors, named the digit in a stimulus such as "3333," or alternated between the two tasks. In a subsequent block, they had to perform different tasks with the same stimuli (naming the print color and counting the number of digits). Whereas in the first block, switch costs dissipated almost completely across 8 runs of trials, at the beginning of the second block they were significantly greater than in the first block and remained significant throughout the block. The authors interpreted this as evidence that the stimulus-response mappings from the first block persisted for at least some minutes and interfered with the tasks in the second block (p. 436). From this and other findings, they concluded that switch costs "cannot be understood as the reflection of a discrete processing stage that must be completed before execution of the next S-R- task can begin. Rather, ... they appear to represent the additional time needed for the system to settle to a unique response decision (or response retrieval) *after* the next imperative stimulus has arrived" (p. 436; see Allport and Wylie, chap. 2, this volume, for further elaboration of this view).

Task Switch Cost as a Manifestation of Advance Reconfiguration

That proactive interference influences switch costs does not exclude the possibility that there are endogenous executive processes as well, which may reconfigure processing systems before or after the stimulus.

Evidence for advance reconfiguration has been obtained by Rogers and Monsell (1995, exp. 3), who used an alternating-runs method in which two tasks were presented in a predictable sequence (AABB). The switch cost was reliably reduced when the response-stimulus interval (RSI) was increased from 150 to 1,200 msec, provided the RSI was constant throughout a block. Because the time for advance preparation was confounded with the temporal distance from the preceding response, one might suspect that fast decay of the previous task set was in part responsible for the switch cost reduction. This appears unlikely, however, because in their experiment 2, Rogers and Monsell found no reduction of the switch cost with a variable RSI. Nor can passive decay be easily reconciled with results reported by Meiran (1996), who presented subjects instructional cues before each stimulus and varied the response-cue interval and the cue-stimulus interval independently. The switch cost was reliably reduced when the cue-stimulus interval was increased from 216 to 1,716 msec, even when the RSI was held constant, which strongly suggests advance reconfiguration before the stimulus (see also De Jong, chap. 15, this volume; Rubinstein, Meyer, and Evans forthcoming).

14.3 OPEN QUESTIONS AND AIMS OF THE PRESENT STUDY

In the following sections, I defend the view that proactive interference and advance reconfiguration are not mutually incompatible explanations, but denote separable component processes influencing overall switch costs (see also Meiran 1996, chap. 16, this volume; Rogers and Monsell 1995; Rubinstein, Meyer, and Evans forthcoming). Three task-switching experiments were performed to elucidate processes underlying advance reconfiguration and to investigate the interaction of advance preparation and involuntary aftereffects of previous task sets. Three main issues were addressed and three corresponding hypotheses proposed.

Hypothesis 1: Advance Reconfiguration and Intention Retrieval

The first hypothesis states that an important component of advance reconfiguration is the retrieval of an abstract intention or task representation. It assumes that—at least in the case of nonautomatized actions—abstract intentions are preferentially represented in a verbal format, that is, in terms of self-instructions like "respond to the color" (cf. Goschke and Kuhl 1996; Kuhl and Kazén 1999). This assumption is consistent with the long-standing idea that the ability to represent intentions in a linguistic format and to generate self-instructions endogenously is an essential precondition for volitional self-control (Ach 1910; Luria 1961; Vygotski 1962). To test the task retrieval hypothesis in the following experiments, the length of the RSI was varied (14 versus 1,500 msec). Moreover, in conditions with the long RSI, participants either had to overtly verbalize the next task before each stimulus, or they had to say task-irrelevant words during the RSI in order to prevent them from retrieving the next task. The task retrieval hypothesis predicts a reduction of the switch cost when subjects retrieve the next task before the stimulus, compared to conditions in which task retrieval before the stimulus is prevented because the RSI is too short or a distractor task must be performed during the RSI.

Hypothesis 2: Persisting Activation of Task Set

The second hypothesis states that *persisting* activation of a previous task set can interfere with or facilitate a subsequent task switch, depending on whether it activates a response that is the same as or different from the response activated by the new task set. To test this hypothesis, task-relevant and -irrelevant stimulus dimensions were variously mapped to the same (congruent) or different (incongruent) responses. This manipulation allowed me to investigate possible interactions between advance reconfiguration and persisting task set activation, in particular, to determine whether proactive interference from a previous task set is suppressed when a new intention is retrieved.

Hypothesis 3: Persisting Inhibition of Task-Irrelevant Perceptual Dimensions or Stimulus-Response Mappings

The third hypothesis concerns the role of inhibitory processes in task switching. When a task requires responding to a particular stimulus dimension such as form, color, or location, one important function of task sets is presumably to enhance the sensitivity of task-relevant perceptual processing modules (cf. Hommel, chap. 11, this volume; Meiran, chap. 16, this volume). When, however, task-irrelevant stimulus features activate incompatible competing responses, it may also be necessary to inhibit or selectively decouple from action irrelevant perceptual information (Houghton and Tipper 1994; see also Mayr and Keele forthcoming). The third hypothesis states that the degree of inhibition is adjusted depending on the amount of response conflict evoked by a stimulus. More specifically, if one conceives of response selection in terms of a constraint satisfaction process, to settle into a maximally coherent state, the system will tend to suppress irrelevant information that imposes incompatible constraints on the activation of response codes. By contrast, no inhibition will be triggered when a stimulus imposes compatible constraints (cf. Houghton and Tipper 1994).

Two forms of inhibition will be distinguished. First, inhibition may affect stimulus *feature values* (for instance, when the task is to respond to the identity of the letter *A* printed in red, the color red may be inhibited). Inhibition of feature values should show up in increased switch costs when, on a subsequent trial, the stimulus feature to be responded to happens to have the same value as the task-irrelevant feature on the preceding trial, compared to switch trials on which a different feature value must be responded to (e.g., green). This form of inhibition is similar to the negative priming effect, that is, the increase in response time (RT) when one responds to a stimulus that was a distractor on the preceding trial (see Fox 1995; May, Kane, and Hasher 1995 for review). Second, inhibition may affect irrelevant stimulus *dimensions* (e.g., color) as a whole. According to hypothesis 3, dimensional inhibition should show up in longer RTs on task switch trials following incongruent than on those following congruent trials, whether or not specific feature values are repeated.

14.4 EXPERIMENT 1

Participants

Twelve undergraduates from the University of Osnabrück participated in the experiment.

Apparatus

Stimulus presentation and reaction time measurement were controlled by an IBM-compatible PC; presentation was synchronized with the vertical retrace signal of the monitor.

Procedure

Stimuli were the uppercase letters *A*, *B*, *C*, and *D*, which could appear in the colors red, green, blue, or yellow. Participants were instructed to respond to the color or to the identity of the letters as fast and accurately as possible by pressing one of two response keys with their left and right index fingers. For half of the participants, the letter *A* and the color red were mapped to the left key ("y"), and the letter *B* and the color green were mapped to the right key ("-"), whereas the other half received the reverse mapping. The remaining colors and letters were not mapped to any responses and occurred only as values of the irrelevant stimulus dimension.

Each trial started with a 200 Hz tone lasting 50 msec. After a delay of 500 msec, a letter was presented at the center of the screen and remained there until the participant pressed one of the response keys. After an RSI of either 14 or 1,500 msec the second letter was presented and remained on the screen until the second response was made. After a delay of 1,500 msec, the next trial started.

There were four different types of blocks, each consisting of 144 such trial pairs. Before each block, participants were informed about the task to be performed throughout the block. There were two task repeat blocks, in which participants either had to respond only to the color (task repeat "color") or to the identity of the letters (task repeat "letter") throughout the block. In task switch blocks, they either had to respond to the color of the first letter and the identity of the second letter in each trial pair (task switch "color-letter"), or to the identity of the first letter and the color of the second letter (task switch "letter-color"). Each participant performed each of the four blocks with both the long and the short RSI. Both the order of the RSI conditions and the order of the four types of blocks within each RSI condition were counterbalanced.

In one-third of the trials of each block, the task-relevant and task-irrelevant stimulus dimensions were mapped to the same response *(congruent trials)*; in one-third of the trials, the two stimulus dimensions required different responses *(incongruent trials)*; and in one-third of the trials, the value of the task-irrelevant dimension was not mapped to any response *(neutral trials)*. Within each experimental condition resulting from the orthogonal manipulation of task switch, RSI, and congruence, all possible combinations of colors and letters appeared equally often across subjects. For each combination of the experimental variables, half of the trial pairs required the same response to the two stimuli, and half involved a response switch.

Results

Reaction times (RTs) below 200 msec or more than 3 standard deviations above a participant's mean RT were discarded from the analyses (a stricter criterion for outliers did not substantively alter the results). Means of the remaining RTs for correct responses were computed for each participant and each experimental condition. Data from color-color and letter-letter trials were averaged to obtain mean RTs for task repeat trials, and trials from color-letter and letter-color trials were averaged to obtain mean RTs for switch trials.[1]

Effects of Task Switch, Response-Stimulus Interval, and Congruence Figure 14.1 (left panel) shows mean RT for correct responses (as well as error rates) on the second trial of each trial pair for the different experimental conditions. A $2 \times 2 \times 3$ repeated-measure analysis of variance (ANOVA) with the independent variables task switch, RSI, and congruence yielded a reliable effect of task switch, indicating that mean RT was longer on task switch than on task repeat trials: $F(1, 11) = 115.47$, $p < 0.001$. This main effect was qualified by a reliable interaction of task switch and RSI, indicating that the switch cost was markedly reduced

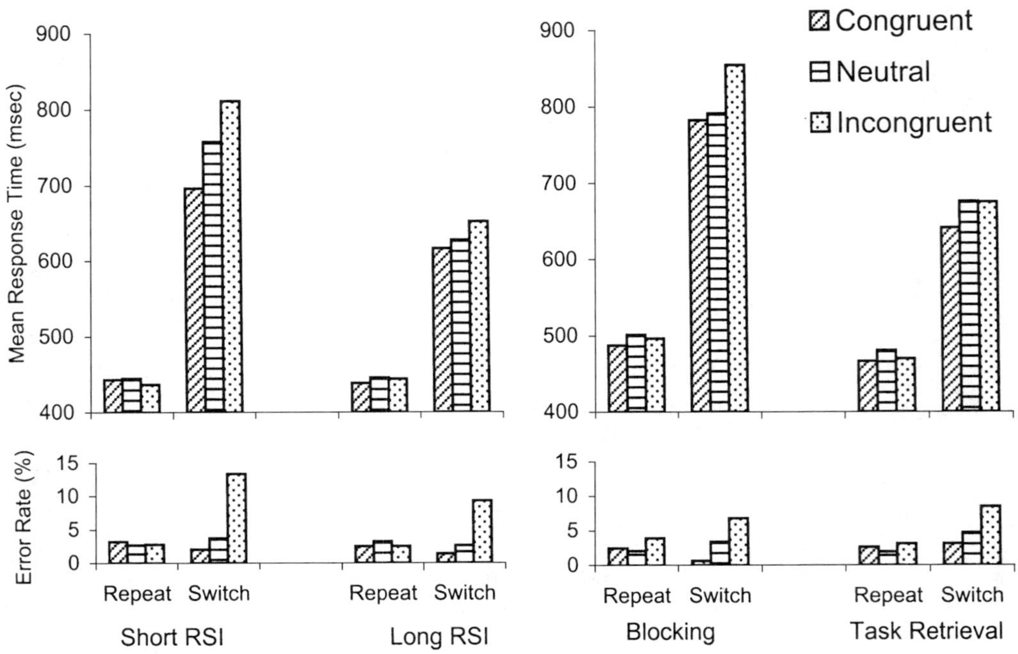

Figure 14.1 Mean response time for congruent, incongruent, and neutral task switch and task repeat trials in the short- and long-RSI conditions of experiment 1 (left panel), and in the blocking and task retrieval groups of experiment 2 (right panel).

after the long versus the short RSI: $F(1, 11) = 26.39$, $p < 0.001$. Even after the long RSI, however, there was still a reliable residual switch cost: $F(1, 11) = 199.68$, $p < 0.001$.

There was also a reliable effect of congruence: $F(2, 22) = 17.89$, $p < 0.001$, which was qualified by a reliable interaction with task switch: $F(2, 22) = 19.49$, $p < 0.001$. Congruence had a reliable effect on RT on task switch trials: $F(2, 22) = 20.25$, $p < 0.001$; but not on task repeat trials: $F < 1.1$, $p = 0.35$. Planned comparisons showed that RT on congruent, neutral, and incongruent nonswitch trials did not reliably differ from each other (all ps > 0.09), whereas congruent switch trials produced shorter RTs than neutral and incongruent switch trials (both ps < 0.001), and incongruent switch trials produced longer RTs than neutral trials ($p < 0.03$). The two-way interactions described thus far were further qualified by a reliable three-way interaction between task switch, RSI, and congruence: $F(2, 22) = 9.11$, $p < 0.001$. This reflects the fact that the interaction between task switch and congruence was reliable only for the short RSI: $F(2, 22) = 22.42$, $p < 0.001$; not for the long RSI: $F(2, 22) = 2.71$, $p = 0.09$. Thus the congruence effect on switch trials was strongly attenuated after the long RSI.

Error Rates Showing an analogous pattern, error rates increased on incongruent switch trials, especially after a short RSI. An ANOVA

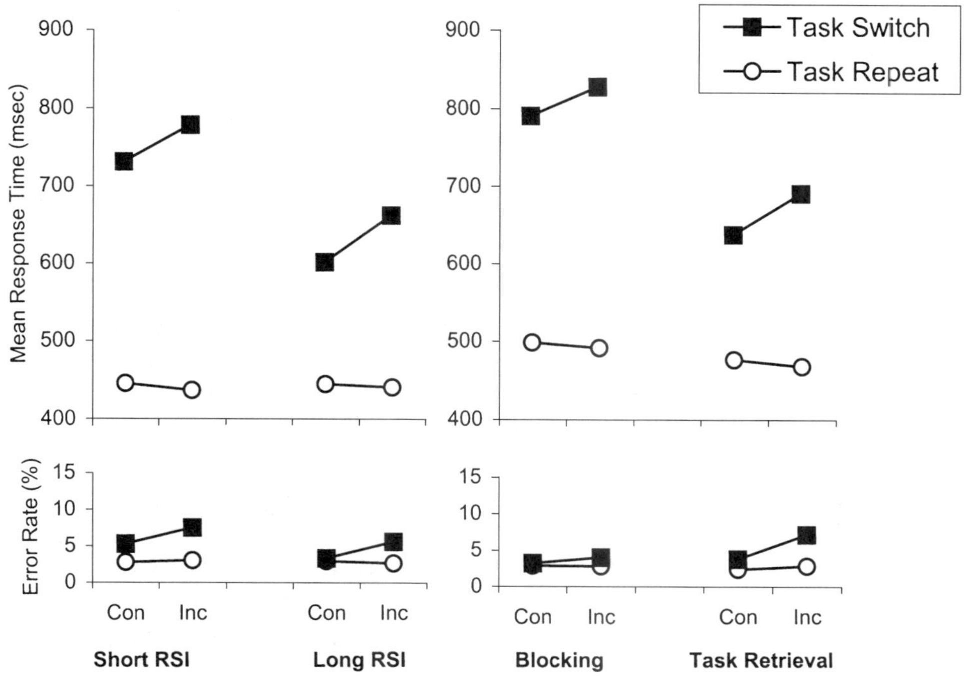

Figure 14.2 Mean response time for task switch (solid squares) and task repeat (circles) trials preceded by congruent (Con) and incongruent (Inc) trials in experiment 1 (left panel) and experiment 2 (right panel).

revealed reliable effects of task switch: $F(1, 11) = 20.04$, $p < 0.001$, congruence: $F(2, 22) = 11.90$, $p < 0.001$, and a reliable interaction of congruence and task switch: $F(2, 22) = 18.06$, $p < 0.001$.

Effects of Congruence on the Preceding Trial To investigate inhibition effects (hypothesis 3), all trial pairs were classified depending on whether the task-relevant dimension of the second stimulus had the same value as or a different value from that on the first trial (for instance, when color was task relevant on the second trial, trials were classified depending on whether the first and the second stimulus had the same or a different color). In addition, all trial pairs were classified depending on whether the first stimulus was congruent or incongruent. An ANOVA with the independent variables task switch, feature value repetition, previous congruence, and congruence on the second trial, and mean RT on the second trial as the dependent variable, yielded no evidence for feature-specific inhibition. The interaction of task switch and feature value repetition was not reliable: $F(1, 11) = 2.41$, $p = 0.15$. Mean RT for task switch trials on which participants responded to a feature value identical to the irrelevant feature value on the preceding trial was not longer than for switch trials in which the stimulus feature had changed (656 versus 668 msec).

There was, however, a highly reliable interaction of task switch and previous congruence: $F(1, 11) = 24.08$, $p < 0.001$. As can be seen in figure 14.2 (left panel), task switch trials following incongruent trials produced longer RTs (720 msec) than switch trials following congruent trials (666 msec): $F(1, 11) = 20.73$, $p < 0.001$. By contrast, mean RT on task repeat trials was slightly, though reliably *shorter* after incongruent than after congruent trials: $F(1, 11) = 5.98$, $p < 0.04$.

The analogous ANOVA for the error data yielded an almost reliable interaction of task switch and previous congruence: $F(1, 11) = 4.47$, $p = 0.058$, indicating that slightly more errors were made on task switch trials following incongruent trials than on those following congruent trials, whereas no such difference was present on task repeat trials.

Because incongruent first trials in task switch blocks produced longer RTs than did congruent trials (807 versus 721 msec), one might object that the effect merely reflects a tendency to produce slower responses following long RTs (RTs on first and second trials were indeed positively correlated: $r = 0.24$, $p < 0.001$). To address this objection, an analysis of covariance was performed at the level of individual trials with RT on switch trials as the dependent variable and with RT on the first trial as the covariate. Although this analysis yielded a reliable effect of the covariate, the effect of previous congruence also remained highly reliable: $F(1, 6{,}640) = 387.37$, $p < 0.001$; and $F(1, 6{,}640) = 15.72$, $p < 0.001$, respectively.

Discussion

Experiment 1 yielded three main findings. First, the task switch cost was reliably reduced, albeit not eliminated, after a long (1,500 msec) versus a short (14 msec) RSI. One possible explanation for this effect is that participants in the long-RSI condition had the opportunity to retrieve the next task prior to the stimulus. A majority of participants in fact reported that they had covertly said the words "color" or "letter" at least on a portion of trials with the long RSI. This interpretation was tested more directly in experiment 2.

Second, there was a reliable congruence effect. Switch costs were reliably greater on incongruent than on neutral trials, whereas they were smaller on congruent than on neutral trials, which indicates that the task set from the previous trial persisted in a state of residual activation (at least after a short RSI). It is noteworthy that Rogers and Monsell (1995, exps. 1 and 3) also obtained greater switch costs in mixed blocks, when congruent and incongruent stimuli were presented, than in pure blocks containing only neutral stimuli. Although incongruent stimuli produced longer RTs and higher error rates than congruent stimuli on switch trials, both congruent and incongruent trials produced longer RTs and greater switch costs than did neutral trials. The authors suggest that this may indicate that stimuli in mixed blocks not only activated an S-R-association defined by the recently performed task, but also evoked the complete competing task set, thus causing interference whether or not the irrelevant task set happened to trigger the same response as the relevant task set. Although not incompatible with this interpretation, the findings of my experiment 1 are evidence for more specific, trial-to-trial after-effects of recently activated task sets. Interestingly, in contrast to previous studies (e.g., Meiran 1996; Rogers and Monsell 1995), the congruence effect was almost completely attenuated after the long RSI, which may indicate that preparatory processes during the RSI helped to suppress the preceding task set. This possibility was further addressed in experiment 2.

Third, switch costs were reliably larger when task switches were preceded by incongruent versus congruent trials, whether or not task-relevant feature values were repeated. This effect did not reflect an unspecific slowing after long RTs, but was reliable even if response speed on the preceding trial was statistically controlled. Results are thus consistent with the interpretation that the task-irrelevant perceptual dimension was inhibited or selectively decoupled from the response system on incongruent trials. It is noteworthy that the persisting inhibition effect was not affected by the RSI. Inhibition of distracting perceptual information was obviously released only after the next imperative stimulus had been processed. Experiment 2 investigated whether inhibition persists until the next stimulus, even when task retrieval is explicitly induced.

14.5 EXPERIMENT 2

In addition to the questions noted above, experiment 2 addressed two obvious objections against the interpretation of the RSI effects in experiment 1. First, both the reduction of the switch cost and the attenuation of the congruence effect after the long RSI might have been due, not to active preparation, but merely to rapid dissipation of the previous task set. Second, although it may seem plausible that task retrieval is an important component of advance preparation, the results of experiment 1 provided no direct evidence for this. However, hypothesis 1 predicts that there should be no reduction of the switch cost even after a long RSI if task retrieval is prevented prior to the stimulus. To test this prediction in experiment 2, only a long (1,500 msec) RSI was used, and participants had either to verbalize the next task before the stimulus, or to perform a verbal distractor task during the RSI. According to hypothesis 1, verbalizing the task should produce the same reduced switch cost as observed with the long RSI in experiment 1, whereas a distractor task that prevents task retrieval should yield a switch cost of about the same magnitude as after the short RSI in experiment 1. If, on the other hand, the decrease in switch cost after a long RSI merely reflected passive decay of the previous task set, or if a previous task set is suppressed by *any* kind of intervening activity, there should be no differences between the task retrieval and blocking conditions.

Participants and Apparatus

Sixteen undergraduates from the University of Osnabrück participated in the experiment, which used the same equipment as in experiment 1.

Procedure

The procedure and response time analyses were the same as in experiment 1, with the following exceptions. Only the long RSI of 1,500 msec was used. Half of the participants were assigned at random to a *task retrieval* group; half were assigned to a *blocking* group. Participants in the task retrieval group were instructed to say either the word "color" or "letter" once during the interval between the warning signal and the first stimulus of each trial, and once during the RSI and prior to the second stimulus, depending on what the next task was. Participants in the blocking group were instructed to say one of two task-irrelevant words ("Monday" or "Tuesday") prior to each stimulus.

Results

Effects of Task Switch, Task Retrieval, and Congruence Means of the RTs for correct responses served as the dependent variable in a $2 \times 3 \times 2$ ANOVA with the independent variables: task switch, congruence, and group (task retrieval versus blocking). This analysis yielded a reliable effect of task switch: $F(1, 14) = 80.89$, $p < 0.001$; and a reliable interaction of task switch and group: $F(1, 14) = 4.73$, $p < 0.05$. Mean RT was markedly longer in task switch than in task repeat blocks (see figure 14.1, right panel). Most important, the switch cost was reliably smaller in the task retrieval than in the blocking group, although there was still a reliable residual switch cost in the task retrieval group: $F(1, 7) = 14.93$, $p < 0.01$. Planned comparisons showed that there was no reliable difference between the blocking and task retrieval groups for task repeat trials ($p > 0.40$), whereas RTs on task switch trials were reliably shorter in the task retrieval than in the blocking group: $t(14) = 1.94$, $p < 0.05$ (one-tailed test).

There was also a reliable main effect of congruence: $F(2, 28) = 15.48$, $p < 0.001$, as well as a reliable interaction between task switch and congruence: $F(2, 28) = 5.43$, $p < 0.01$. In task repeat blocks, incongruent and neutral trials differed only by a nonreliable -8 msec: $t(15) = -1.64$, $p = 0.12$; mean RT was 14 msec shorter on congruent than on neutral trials: $t(15) = -4.23$, $p < 0.01$. By contrast, in task switch blocks, mean RT was on average 31 msec longer on incongruent than on neutral trials: $t(15) = 2.54$, $p = 0.03$; and RT was 22 msec shorter on congruent than on neutral trials: $t(15) = -2.13$, $p = 0.05$.

As can be seen in figure 14.1 (right panel), the effect of congruence on the switch cost was greater in the blocking than in the task retrieval group. A 2×3 (group $\times$ congruence) ANOVA, with RT on switch trials as the dependent variable, yielded a reliable interaction of the two variables: $F(2, 28) = 4.51$, $p < 0.02$. Whereas the effect of congruence was highly reliable in the blocking group, it was at best marginally reliable in the task retrieval group: $F(2, 14) = 12.84$, $p < 0.001$ versus $F(2, 14) = 3.57$, $p = 0.06$. Analogous results were obtained when the switch cost served as the dependent variable: the effect of congruence was reliable in the blocking group, but not in the task retrieval group: $F(2, 14) = 8.82$, $p < 0.01$ versus $F(2, 14) = 1.48$, $p > 0.26$.

Error Rates Corresponding analyses of error rates yielded reliable effects of task switch: $F(1, 14) = 7.05$, $p < 0.02$; of congruence: $F(2, 28) = 12.43$, $p < 0.001$; and a reliable interaction of congruence and task switch: $F(2, 28) = 8.31$, $p < 0.001$. Error rates increased on task switch trials, and this increase was more pronounced on incongruent trials.

Effects of Congruence on the Preceding Trial The data were further analyzed depending on whether the relevant stimulus dimension on the second trial had the same value as on the first trial or a different value, and depending on whether the first trial was congruent or incongruent (see figure 14.2, right panel). As in experiment 1, there was no evidence for inhibition on the level of specific feature values. Mean RT for task switch trials on which participants responded to a feature value identical to the irrelevant feature value on the preceding trial was virtually identical to mean RT for task switch trials on which the stimulus feature value had changed (739 versus 735 msec). There was, however, a reliable interaction of task switch and previous congruence: $F(1, 14) = 15.72$, $p < 0.001$. Whereas RT on task switch trials was reliably longer after incongruent trials than after congruent trials (759 versus 714 msec), mean RT on task repeat trials was slightly, but reliably *shorter* after incongruent than after congruent trials: $F(1, 14) = 14.19$, $p < 0.002$, $F(1, 14) = 5.29$; $p < 0.04$.

Because RTs produced by the first and the second stimuli of the trial pairs were positively correlated in task switch blocks ($r = 0.33$; $p < 0.001$) the effect of previous congruence may again have been due merely to longer RTs on incongruent first trials. Although an analysis of covariance with RT on switch trials as the dependent variable and with RT on first trials as the covariate yielded a reliable effect of the covariate, the effect of previous congruence remained reliable: $F(1, 4{,}193) = 488.10$, $p < 0.001$; and $F(1, 4{,}193) = 4.19$, $p < 0.05$, respectively.

Error Rates Corresponding analyses of error rates yielded no reliable results.

Discussion

The results of experiment 2 replicate and extend the findings of experiment 1. When participants verbalized the next task before the stimulus, the switch cost was reliably smaller than in the blocking group, for whom task retrieval was interfered with by a verbal distractor task. In fact, the magnitude of the switch cost in the task retrieval group (192 msec) was almost identical to that in the long-RSI condition of experiment 1 (189 msec), whereas the switch cost in the blocking group (315 msec) was practically identical to that in the short (14 msec)-RSI condition of experiment 1 (313 msec), despite the long (1,500 msec) RSI.

There was again a reliable congruence effect, as indicated by greater switch costs on incongruent than on neutral or congruent trials. This effect was reliable only in the blocking group, but not in the task retrieval group, which speaks against an interpretation in terms of passive decay of the previous task set. The preceding task set neither decayed in a passive manner as a function of the length of the RSI, nor was it deactivated by an unrelated intervening activity; it was suppressed only by retrieval of a new intention.

Finally, switch costs were again reliably greater after incongruent than after congruent trials, whereas previous congruence had a small reverse effect on task repeat trials. This further supports the assumption that the task-irrelevant perceptual dimension was inhibited when it activated an incompatible response. It is noteworthy, that—in contrast to the congruence effect—the dimensional inhibition effect persisted even after the new task was retrieved.

Dimensional Inhibition or Episodic Stimulus-response Binding? Up to this point, I have interpreted the effect of previous congruence as evidence for inhibition of task-irrelevant percepual dimensions (or the decoupling of perceptual dimensions from the response system). There is, however, an alternative interpretation that deserves consideration. With the two-choice reaction tasks used, it was inevitable that previous congruence was confounded with particular combinations of switches and repetitions of the response and the task-relevant stimulus feature. Consider the case in which the previous trial $n-1$ is congruent and both stimulus dimensions are mapped to the same response. On a following task switch trial n, either the task-relevant stimulus feature will have the same value as on trial $n-1$ and the response must be repeated, or both the stimulus feature and the response will switch. Consider now an incongruent trial $n-1$, in which the two stimulus dimensions are mapped to different responses. When on a following task-switch trial n the rele-

Table 14.1 Example of Different Stimulus Combinations on Two Successive Trials

			Trial $n-1$: Task = COLOR	
			Congruent $Red_L A_L$	Incongruent $Red_L B_R$
Trial n: Task = LETTER	Con	$Red_L A_L$	S=R=	S≠R=
	Inc	$Green_R A_L$	S=R=	S≠R=
	Con	$Green_R B_R$	S≠R≠	S=R≠
	Inc	$Red_L B_R$	S≠R≠	S=R≠

Note: Stimuli are letters (*A*, *B*) with different colors (green, red). The task on trial $n-1$ is to respond to the color, the task on trial n is to respond to the letter. Subscripts (L, R) attached to stimulus values denote the response (left, right) associated with a given stimulus value. Symbols S = and S ≠ denote whether the task-relevant stimulus value on trial n is or is not repeated from trial $n-1$; symbols R = and R ≠ denote whether the response on trial n is or is not repeated from trial $n-1$.

vant stimulus feature has the same value as on trial $n-1$, it will require a response switch, whereas a switch of the stimulus feature will be accompanied by a response repetition (see table 14.1 for an illustration).

When RT on switch trials was analyzed, not in terms of previous congruence, but in terms of the orthogonal combination of stimulus feature switch and response switch, this yielded in both experiments a highly reliable interaction of the two variables: $F(1, 11) = 20.73$; $p < 0.001$, for experiment 1; $F(1, 14) = 14.19$; $p < 0.002$, for experiment 2. The effect of previous congruence may thus alternatively be explained in terms of *episodic bindings* of stimulus and response codes (cf. Hommel 1998, chap. 11, this volume). According to this explanation, task-relevant and -irrelevant stimulus features together with the current response will be encoded as an integrated episode. If the task-relevant feature on the following switch trial is repeated, the previous S-R configuration will be reevoked. This will facilitate the task switch when the same response is produced as on the preceding trial, whereas it will interfere with the production of a different response, which requires an unbinding of the previously established S-R configuration. If, on the other hand, the task-relevant stimulus feature is different from that on the preceding trial, this should facilitate a switch to a different response, one not previously bound to a different stimulus feature, whereas it should interfere with a repetition of the response, which again requires an unbinding of the previously established S-R episode (see Hommel 1998, chap. 11, this volume, for empirical evidence for automatic stimulus-response bindings).

14.6 EXPERIMENT 3

Experiment 3 was performed to unconfound previous congruence from the effect of particular stimulus-response bindings. This was achieved by using four-choice instead of two-choice reaction tasks, so that there could be task switch trials preceded by congruent and incongruent trials, in both cases accompanied by a switch of the relevant stimulus feature and a switch of the response. If the effect of previous congruence is due to the confounding described above, it should disappear under these conditions.

Participants and Apparatus

Sixteen undergraduates from the University of Osnabrück participated in the experiment, which used the same equipment as in experiment 1.

Procedure

Participants had to respond to the color or identity of four uppercase letters (*A, B, C, D*), which could appear in four colors (red, green, blue, yellow), by pressing one out of four response keys on the computer keyboard ("y", "x", ":", "-"). In contrast to the experiments 1 and 2, the two tasks appeared in a computer-generated pseudorandom sequence of 500 trials. Each trial started with a blank screen for 250 msec, followed by an instructional cue (the word "color" or "letter") at the center of the screen. After a cue-stimulus interval of 1,500 msec, the imperative stimulus appeared and remained on the screen until a response was made. Half of the trials were task repeat trials; half required a task switch. After 250 trials, participants were given a brief rest. The first three trials after the break were not included in the data analyses. Prior to the main block, participants performed 40 practice trials to become familiar with the task and the S-R mapping.

Results and Discussion

Trimmed mean RTs for correct responses were computed as in the previous experiments. The analyses included only those trials on which both the response and the value of the task-relevant stimulus dimension differed from the immediately preceding trial (there were too few data points to analyze other possible combinations). This selection did not result in any confoundings of previous congruence with some other variable. In particular, previous congruence and congruence on the current trial were orthogonal. A $2 \times 2 \times 2$ ANOVA with the independent variables task switch, congruence on the current trial, and congruence on the preceding trial yielded a reliable effect of task switch, indicating that RT was longer on task switch than on task repeat trials (844 versus 815 msec): $F(1, 15) = 4.63$; $p < 0.05$. The switch cost was smaller than in experiments 1 and 2, which presumably reflects beneficial effects of the instructional cues and the fact that the randomized presentation of tasks unconfounded task switch from intention memory load, which may increase switch costs in a blocked design (cf. Rogers and Monsell 1995). There was also a reliable effect of current congruence, indicating that RT was longer on incongruent trials than on congruent trials (870 versus 788 msec): $F(1, 15) = 30.71, p < 0.001$. Most important, there was a reliable interaction of task switch and previous congruence (no other main effects or interactions were reliable): $F(1, 15) = 5.35$, $p < 0.04$. RT was longer on task switch trials preceded by incongruent trials than on task switch trials preceded by congruent trials (866 versus 822 msec): $F(1, 15) = 7.12$, $p < 0.02$, By contrast, no such difference was present on task repeat trials (810 versus 819 msec): $F < 1$. Given that the two categories of trials were both accompanied by a response and a stimulus feature switch, this shows that the dimensional inhibition effect cannot be accounted for in terms of episodic S-R binding.

It should be noted, however, that there was also evidence suggesting an effect of episodic S-R binding. Task switch trials that required a response switch produced longer RT when accompanied by a stimulus feature repetition (889 msec) than when accompanied by a stimulus feature switch (866 msec), whereas task switches accompanied by a response repetition produced longer RT when the stimulus feature was switched (883 msec) than when it was repeated (843 msec). Although the interaction of response switch and stimulus feature switch was only marginally reliable: $F(1, 15) = 3.63$, $p < 0.08$, the present results suggest that dimensional inhibition and episodic S-R binding constitute separate influences on task switching.

Error Rates Corresponding analyses of error rates yielded a reliable interaction of current congruence and task switch: $F(1, 15) = 16.28$, $p < 0.001$. Error rates for congruent and incongruent trials were 1.5% versus 5.3% for task repeat trials and 5.3% versus 8.8% for task switch trials.

14.7 GENERAL DISCUSSION: ACTION CONTROL AS A MULTIPLE CONSTRAINT SATISFACTION PROBLEM

The present results have shown that task switch costs are influenced by various separable processes, including advance preparation in the form of task retrieval, proactive interference from recently activated task sets, persisting inhibition of distracting perceptual dimensions, and episodic stimulus-response bindings. In discussing implications of these findings for the interaction of intentional and involuntary processes, this final section outlines a tentative theoretical framework according to which seemingly dysfunctional aspects of cognitive control, such as proactive interference, can be seen as manifestations of an adaptive design, evolved to cope with partially incompatible constraints in the control of action.

On the Nature of Advance Reconfiguration

One aim of the present study was to provide evidence for advance reconfiguration in terms of retrieval of verbal task representations. The most serious objection against the present interpretation is that the reduction of the switch cost in the task retrieval group could merely have reflected fast dissipation of task set inertia. This objection deserves serious consideration, given that the time for advance preparation was confounded with the temporal distance from the previous response. But it is not easily reconciled with the complete absence of a switch cost reduction in the blocking group. Obviously, neither the length of the RSI nor the presence of an intervening task as such was responsible for the switch cost reduction, but rather the *content* of what was verbalized. This conclusion fits with other evidence against a passive decay account of the reduction of switch costs with a long RSI (Meiran 1996; Rogers and Monsell 1995). It is also consistent with the suggestion that the endogenous aspect of task switching consists in the deletion of old and insertion of new goals in a declarative working memory before activation of specific condition action rules (Rubinstein, Meyer, and Evans forthcoming; see also Kieras et al., chap. 30).

Given that the results demonstrate active preparation, one may further ask whether preparation actually consisted in the retrieval of a *verbal* task representation. One might argue that the distractor task in the blocking group need not have specifically interfered with retrieval of a verbal task representation, but may rather have impaired other, yet-to-be-specified nonverbal executive processes. This interpretation, however, raises the question of why such nonverbal executive processes were completely blocked by saying the words "Monday" and "Tuesday," while they were not at all impaired by saying the words "color" and "letter." Again, it was not that participants said *something* during the RSI, but whether they verbalized the next task, that accounts for the results.[2] Converging evidence for the role of verbal processes in task switching has recently been reported in a neuropsychological study (Mecklinger et al. 1999). Although patients with left-brain damage showed greater switch costs than patients with right-brain damage, this difference was exclusively due to a subgroup of left-brain-damaged patients suffering from central speech disorders, who showed disproportionately great switch costs. The authors suggest that articulatory processes may be important for suppressing interference from previously activated task sets, which fits nicely with the present finding that task retrieval attenuated the congruence effect.[3]

This brings us to why and how task retrieval facilitated task switching. At first sight, one might interpret the difference in the RT cost of a switch between the task retrieval group and the blocking group as a measure of the time it takes to retrieve a task representation, time that augments the

RT if task retrieval can only be initiated after the stimulus. On the other hand, as has been noted by Allport and Wylie (1999), the switch cost reduction caused by a process performed before the stimulus need not be a direct measure of the duration of that process, but may reflect additional *effects* of this process on subsequent response selection. Loading an intention into working memory presumably has a number of such effects, in particular (1) it may increase in an anticipatory way the sensitivity of task-relevant perceptual processing modules (cf. Corbetta et al. 1990; Cohen, Dunbar, and McClelland 1990; Posner and Peterson 1990; see also Meiran, chap. 16, this volume); (2) it may set specific stimulus-response connections into readiness; and (3) it may suppress representations of competing intentions.

Varieties of Involuntary Priming in Task Set Reconfiguration

While the present results suggest that retrieval of an intention into working memory constitutes a strong top-down constraint for subsequent processing and response selection, they also demonstrated involuntary aftereffects of preceding task sets. These findings are consistent with other reports of involuntary priming in task switching (e.g., Allport and Wylie, chap. 2, this volume). Taken together, these findings suggest that competing task sets may influence response selection for a number of reasons, in particular, because they were recently activated, because they must be maintained in a state of readiness, or because they were consistently associated with the same stimuli in the past. It will be important in future research to investigate differences and commonalities between different sources of interference and cross talk. For instance, while slowly dissipating task set inertia effects may result from competing stimulus-task associations (Allport, Styles, and Hsieh 1994), much shorter-lived aftereffects of recently executed tasks as observed in the present experiments may reflect more transient changes in the activation level of task sets. In this respect it is also noteworthy that the present results showed that persisting task set activation was attenuated by task retrieval, whereas the inhibition (or decoupling from the response system) of task-irrelevant perceptual dimensions persisted even after task retrieval. Although this dissociation clearly needs to be replicated, it suggests that different kinds of facilitatory and inhibitory aftereffects of task sets differ in their resistance to top-down control (cf. Mayr and Keele forthcoming). It should be noted, however, that other studies have reported no reduction of congruence effects with an increasing opportunity for preparation (e.g., Meiran 1996; Rogers and Monsell 1995). At least with respect to experiment 2, this discrepancy may reflect the fact that in the present experiment subjects were forced to retrieve the next task before the stimulus on each trial; in other experiments, merely providing the opportunity to prepare may not have been sufficient to motivate subjects to prepare on each trial (see De Jong, chap. 15, this volume). On the other hand, such an account cannot explain the reduction of the congruence effect in experiment 1, and further research is needed to clarify the discrepancy.

From a more general perspective, the foregoing conclusions are consistent with the view that automatic and controlled (intentional) processes do not constitute an either-or distinction. Rather than conceiving of automatic processes as necessarily triggered by a stimulus, and of controlled processes as directly initiated by conscious intentions, we should see intentions rather as constituting constraints that set the stage for later processing and that modulate the readiness of responses to be activated more or less "automatically" by subsequent stimuli (Cohen, Dunbar, and McClelland 1990; Exner 1873; Gollwitzer 1996; Goschke 1996, 1997; Hommel, chap. 11, this volume; Neumann 1984, 1987; Neumann and Prinz 1987). Thus intentions modulate or "configure" automatic processes for voluntary action, whereas the selection of responses, though dependent on prior intentions, is influenced by various forms of involuntary priming.

Control Dilemmas and Adaptive Constraints: Toward a Functional Analysis of Action Control

In a sense, the present results may appear to reveal the suboptimal design of the cognitive system. Humans neither switch between tasks without a cost nor inhibit competing intentions efficiently, but are obviously prone to various kinds of interference from irrelevant information or competing task sets. I propose, however, that these seemingly dysfunctional features are manifestations of an adaptive design and reflect competing mechanisms, which have evolved to satisfy partially incompatible constraints on intelligent action. I have described these constraints as "control dilemmas" (Goschke 1996, 1997, 1998; see also Kuhl 2000 for a related view) and will briefly relate some of them to the problem of task switching.

The Selection-Orienting Dilemma On the one hand, an acting organism should select intention-relevant information to specify parameters of immediate action and should inhibit irrelevant information to avoid cross talk (Allport 1989). On the other, the organism should continuously monitor the environment for potentially significant information, even if this information is not directly relevant for the ongoing action. For this reason, it would not be adaptive if attentional selection operated so efficiently as to suppress irrelevant information completely (cf. Allport 1989; Houghton and Tipper 1994). Ignored information should be processed to a level at which threats or affordances relevant for higher-level goals or vital needs can be recognized (e.g., the smell of fire while working on an important paper). Thus what is considered interference and cross talk in the light of the current intention is a necessary by-product of continuous *background monitoring* and thus a precondition for flexible reorientation (Goschke 1996, 1997; cf. Allport 1989; Dibbelt 1996; Brandtstädter, Wentura, and Rothermund forthcoming; Kuhl and Goschke 1994).

The Persistence-Interruption Dilemma On the one hand, the system should shield a current intention against competing intentions and motivational tendencies in order to persist in pursuing long-term goals (Kuhl 1985). On the other, an organism must be able to interrupt an ongoing action and to switch to a different action if necessary. Indeed, animals incapable of responding to the sudden appearance of, say, a predator with a fast switch from the ongoing activity (e.g., eating) to a very different behavior (e.g., flight) are most probably not numbered among our evolutionary ancestors. From this perspective, task set inertia and the related finding that uncompleted intentions persist automatically in a state of high activation in long-term memory (Goschke and Kuhl 1993, 1996) may be manifestations of an inherent tendency of intentions to persist in the face of distractions. Although this persistence promotes the realization of a selected intention, it incurs a cost when fast and flexible switching is required.

The Stability-Flexibility Dilemma On the one hand, the system should incrementally strengthen fixed stimulus-response and stimulus-task associations in order to respond to invariant or recurrent situations with well-established habits (Goschke 1998). On the other, the system should be able to flexibly reconfigure response dispositions from moment to moment. From this perspective, long-lasting task set inertia effects after prolonged performance of competing tasks, as observed by Allport, Styles, and Hsieh (1994), may reflect the formation of relatively stable stimulus-task associations. Although such associations will allow for efficient responding under invariant conditions, they will interfere when reconfiguration of response dispositions is required.

It is beyond the scope of this chapter to describe these dilemmas in greater detail. Suffice it to say that the foregoing analysis supports a view of action control as an optimization problem, which requires a dynamic, context-sensitive balance between competing constraints (Goschke 1996, 1997; cf. Allport 1989; Brandtstädter, Wentura, and Rothermund forthcoming; Kuhl 2000; Kuhl and Goschke 1994). Insofar as these constraints pose functionally incompatible demands, they presumably promoted the evolution of a functional architecture in which different control operations are subserved by separable competing and cooperating subsystems (Baars 1988; Goschke 1996; Hayes-Roth 1985; Kieras et al., chap. 30, this volume), as opposed to being controlled top-down by a unitary central executive (central processor, intention system, or operating system). Although admittedly speculative, this account receives support from recent neuropsychological and brain-imaging studies suggesting that the prefrontal cortex, long considered to be the anatomical locus of executive control, appears to exhibit an unanticipated degree of functional specialization (for reviews, see Della Sala and Logie 1993; Fuster 1989; Goldman-Rakic 1995; McCarthy and Warrington 1990; Robbins 1998; Roland 1984; Shallice and Burgess 1998). Although our knowledge about the neurocognitive systems underlying cognitive control is still very restricted, a functional analysis of adaptive constraints on action control may serve as a fruitful framework for further experimental explorations.

NOTES

This research was supported by German Science Foundation grant Go-720/2-1. I thank Frauke Bastians and Kristina Gräper for their assistance in running the experiments. I thank Jon Driver, Bernhard Hommel, Steve Keele, Uli Mayr, Nachshon Meiran, Stephen Monsell, Dirk Vorberg, and an anonymous reviewer for their valuable comments on a previous version of this chapter.

1. Responses to the first stimulus of each trial pair showed a pattern of results similar to responses to the second stimulus. Because, however, reaction times to the first stimuli are uninformative concerning the effect of the response-stimulus interval, I will report results only for second responses.

2. This is not to say that *overt* verbalization is crucial; covert task retrieval should produce similar effects. In addition, it should be noted that, while the present results show that retrieval of a verbal task representation is sufficient to facilitate preparation for the next task, it is an open question whether verbal task retrieval is also *necessary* for intentional reconfiguration.

3. It may be asked why a reliable residual switch cost was observed in the task retrieval group. One possibility is that the blocked presentation of task switch and repeat trials confounded task switching with the number of stimulus-response mappings in working memory (Rogers and Monsell 1995). Consistent with this possibility, in experiment 3, where switch and repeat trials were intermixed, the residual switch cost was much smaller. A second possibility is that further reconfiguration processes (for instance, the activation of specific condition action rules) can only be completed after the imperative stimulus (cf. Rogers and Monsell 1995), or that participants do not initiate these processes before the stimulus due to a lack of motivation (De Jong, chap. 15, this volume).

REFERENCES

Ach, N. (1910). *Über den Willensakt und das Temperament*. Leipzig: Quelle and Meyer.

Allport, D. A. (1989). Visual attention. In M. I. Posner (Ed.), *Foundations of cognitive science*, pp. 631–681. Cambridge, MA: MIT Press.

Allport, D. A., Styles, E., and Hsieh, S. (1994). Switching intentional set: Exploring the dynamic control of tasks. In C. Umiltà and M. Moscovitch (Eds.), *Attention and Performance XV: Conscious and nonconscious information processing*, pp. 421–452. Cambridge, MA: MIT Press.

Allport, D. A., and Wylie, G. (1999). Task-switching: Positive and negative priming of task-set. In G. W. Humphreys, J. Duncan, and A. M. Treisman (Eds.), *Attention, space, and action: Studies in cognitive neuroscience*. Oxford: Oxford University Press.

Atkinson, R. C., and Shiffrin, R. M. (1968). Human memory: A proposed system and its control processes. In K. W. Spence (Ed.), *The psychology of learning and motivation*, vol. 2, pp. 89–195. New York: Academic Press.

Baars, B. J. (1988). *A cognitive theory of consciousness*. Cambridge: Cambridge University Press.

Brandtstädter, J., Wentura, D., and Rothermund, K. (Forthcoming). Intentional self-development through adulthood and later life: Tenacious pursuit and flexible adjustment of goals. In J. Brandstädter and R. M. Lerner (Eds.), *Action and self-development: Theory and research through the life-span*. Thousand Oaks, CA: Sage.

Cohen, J. D., Dunbar, K., and McClelland, J. L. (1990). On the control of automatic processes: A parallel distributed processing account of the Stroop effect. *Psychological Review, 97,* 332–361.

Corbetta, M, Miezin, F. M., Dobmeyer, S., Shulman, G. L., and Petersen, S. E. (1990). Attentional modulation of neural processing of shape, color, and velocity in humans. *Science, 248,* 1556–1559.

Della Sala, S., and Logie, R. H. (1993). When working memory does not work: the role of working memory in neuropsychology. In F. Boller and J. Grafman (Eds.), *Handbook of neuropsychology*, vol. 8, pp. 1–61. Amsterdam: Elsevier.

Dibbelt, S. (1996). Wechseln und Beibehalten als Grundfunktionen der Handlungskontrolle. Ph.D. University of Osnabrück.

Exner, S. (1873). Experimentelle Untersuchungen der einfachsten psychischen Processe: 1. Die persönliche Gleichung. *Archiv für die gesammte Physiologie des Menschen und der Thiere, 7,* 601–660.

Fox, E. (1995). Negative priming from ignored distractors in visual selection: A review. *Psychonomic Bulletin and Review, 2,* 145–173.

Fuster, J. M. (1989). *The prefrontal cortex*. 2d ed. New York: Raven Press.

Goldman-Rakic, P. S. (1995). Architecture of the prefrontal cortex and the central executive. In J. Grafman, K. J. Holyoak, and F. Boller, F. (Eds.), *Structure and functions of the human prefrontal cortex* Reprinted in *Annals of the New York Academy of Sciences, 769,* 71–83.

Gollwitzer, P. (1996). The volitional benefits of planning. In P. M. Gollwitzer and J. A. Bargh (Eds.), *The psychology of action. Linking cognition and motivation to behavior,* pp. 287–312. New York: Guilford Press.

Goschke, T. (1996). Wille und Kognition: Zur funktionalen Architektur der intentionalen Handlungssteuerung. In J. Kuhl and H. Heckhausen (Eds.), *Enzyklopädie der Psychologie.* Series 4, vol. 4, *Motivation, Volition und Handeln,* pp. 583–663. Göttingen: Hogrefe.

Goschke, T. (1997). Zur Funktionsanalyse des Willens: Integration kognitions-, motivations- und neuropsychologischer Perspektiven. *Psychologische Beiträge, 39,* 375–412.

Goschke, T. (1998). Implicit learning of perceptual and motor sequences: Evidence for independent learning systems. In M. Stadler and P. Frensch (Eds.), *Handbook of implicit learning,* pp. 401–444. Thousand Oaks, CA: Sage.

Goschke, T., and Kuhl, J. (1993). The representation of intentions: Persisting activation in memory. *Journal of Experimental Psychology: Learning, Memory, and Cognition, 19,* 1211–1226.

Goschke, T., and Kuhl, J. (1996). Explicit and implicit memory for intentions: Empirical findings and a theoretical framework. In M. Brandimonte, G. Einstein, and M. McDaniel (Eds.), *Prospective memory: Theory and applications,* pp. 53–91. Hillsdale, NJ: Erlbaum.

Hayes-Roth, B. (1985). A blackboard architecture for control. *Artificial Intelligence, 26,* 251–321.

Hommel, B. (1998). Event files: Evidence for automatic integration of stimulus-response episodes. *Visual Cognition, 5,* 183–216.

Houghton, G., and Tipper, S. P. (1994). A model of inhibitory mechanisms in selective attention. In D. Dagenbach and T. Carr (Eds.), *Inhibitory processes in attention, memory, and language,* pp. 53–112. San Diego, CA: Academic Press.

Jersild, A. T. (1927). Mental set and switch. *Archives of Psychology,* no. 89.

Kluwe, R. (1997). Intentionale Steuerung kognitiver Prozesse. *Kognitionswissenschaft, 6,* 1–17.

Kuhl, J. (1985). Volitional mediators of cognitive-behavior consistency: Self-regulatory processes and actions versus state orientation. In J. Kuhl and J. Beckmann (Eds.), *Action control: From cognition to behavior,* pp. 101–128. Berlin: Springer.

Kuhl, J. (1996). Wille und Freiheitserleben: Formen der Selbststeuerung. In J. Kuhl and H. Heckhausen (Eds.), *Enzyklopädie der Psychologie.* Series 4, vol. 4, *Motivation, Volition und Handlung,* pp. 665–765. Göttingen: Hogrefe.

Kuhl, J. (2000). A functional-design approach to motivation and self-regulation: The dynamics of personality systems interactions. In M. Boekaerts, P. R. Pintrich, and M. Zeidner (Eds.), *Self-regulation: Directions and challenges for future research,* pp. 111–169. Academic Press.

Kuhl, J., and Goschke, T. (1994). A theory of action control: Mental subsystems, modes of control, and volitional conflict-resolution strategies. In J. Kuhl and J. Beckmann (Eds.), *Volition and personality: Action versus state orientation,* pp. 93–124. Göttingen and Toronto: Hogrefe.

Kuhl, J., and Kazén, M. (1999). Volitional facilitation of difficult intentions: Joint activation of intention memory and positive affect removes Stroop interference. *Journal of Experimental Psychology: General, 128,* 382–399.

Luria, A. R. (1961). *The role of speech in regulation of normal and abnormal behavior.* London: Pergamon.

May, C. P., Kane, M. J., and Hasher, L. (1995). Determinants of negative priming. *Psychological Bulletin, 118,* 35–54.

Mayr, U., and Keele, S. (Forthcoming). Changing internal constraints on action: The role of backward inhibition. *Journal of Experimental Psychology: General.*

McCarthy, R. A., and Warrington, E. K. (1990). *Cognitive neuropsychology. A clinical introduction.* San Diego, CA: Academic Press.

Mecklinger, A., von Cramon, D. Y., Springer, A., and Matthes-von Cramon, G. (1999). Executive control functions in task switching: Evidence from brain injured patients. *Journal of Clinical and Experimental Neuropsychology, 21,* 606–619.

Meiran, N. (1996). Reconfiguration of processing mode prior to task performance. *Journal of Experimental Psychology: Learning, Memory, and Cognition, 22,* 1423–1442.

Monsell, S. (1996). Control of mental processes. In V. Bruce (Ed.), *Unsolved mysteries of the mind,* pp. 93–148. Hove, U.K.: Erlbaum.

Neisser, U. (1967). *Cognitive psychology.* New York: Appleton Century Crofts.

Neumann, O. (1984). Automatic processing: A review of recent findings and a plea for an old theory. In W. Prinz and A. F. Sanders (Eds.), *Cognition and motor processes,* pp. 255–294. Berlin: Springer.

Neumann, O. (1987). Beyond capacity: A functional view of attention. In H. Heuer and A. F. Sanders (Eds.), *Perspectives on perception and action*, pp. 361–394. Hillsdale, NJ: Erlbaum.

Neumann, O., and Prinz, W. (1987). Kognitive Antezedenzien von Willkürhandlungen. In H. Heckhausen, P. M. Gollwitzer, and F. E. Weinert (Eds.), *Jenseits des Rubikon: Der Wille in den Humanwissenschaften*, pp. 195–215. Berlin: Springer.

Posner, M. I., and Peterson, S. E. (1990). The attention system of the human brain. *Annual Review of Neuroscience, 13*, 25–42.

Posner, M. I., and Snyder, C. R. (1975). Attention and cognitive control. In R. L. Solso (Ed.), *Information processing and cognition*, pp. 55–85. Hillsdale, NJ: Erlbaum.

Robbins, T. W. (1998). Dissociating functions of the prefrontal cortex. In A. C. Roberts, T. W. Robbins, and L. Weiskrantz (Eds.), *The prefrontal cortex: Executive and cognitive functions*, pp. 117–130. Oxford: Oxford University Press.

Rogers, R. D., and Monsell, S. (1995). Costs of a predictable switch between simple cognitive tasks. *Journal of Experimental Psychology: General, 124*, 207–221.

Roland, P. E. (1984). Metabolic measurement of the working frontal cortex in man. *Trends in Neuroscience, 7*, 430–435.

Rubinstein, J., Meyer, D. E., and Evans, J. E. (Forthcoming). Executive control of cognitive processes in task switching.

Schneider, W., and Shiffrin, R. W. (1977). Controlled and automatic human information processing: 1. Decision, search, and attention. *Psychological Review, 84*, 1–66.

Shallice, T., and Burgess, P. (1998). The domain of supervisory processes and the temporal organization of behavior. In A. C. Roberts, T. W. Robbins, and L. Weiskrantz (Eds.), *The prefrontal cortex: Executive and cognitive functions*, pp. 22–35. Oxford: Oxford University Press.

Spector, A., and Biederman, I. (1976). Mental set and mental switch revisited. *American Journal of Psychology, 89*, 669–679.

Vygotski, L. S. (1962). *Thought and language*. Cambridge, MA: MIT Press.

15 An Intention-Activation Account of Residual Switch Costs

Ritske De Jong

ABSTRACT *Residual switch costs* are performance costs associated with a shift of task that persist even when there is ample time to prepare in advance for the new task. I present a mixture-model approach for evaluating the contributions of two possible causes of residual switch costs: (1) failures to take advantage of opportunities for advance preparation, and (2) limitations to the completeness of task-set reconfiguration attainable by fully endogenous means. The proposed intention-activation hypothesis of failures to engage in advance preparation is shown to provide a coherent account of the influences of a variety of factors on residual switch costs. Two new experiments tested predictions of the hypothesis regarding the effects of task duration and of time on task on the incidence of preparatory failures.

Although people can perform an almost endless variety of tasks, they are limited in the number of tasks they can perform concurrently, and they generally devote themselves to just one task at any moment. As pointed out by Simon (1994), serial organization of activities should perhaps be viewed not as the result of resource scarcity prohibiting a presumably more efficient parallel organisation, but as an efficient solution to the problem of getting a powerful parallel processing device, the human brain, to support coherent behavior in environments that provide multiple affordances for action. The division of labor by time segments, with processing resources devoted, in turn, to satisfying successive goals, requires signaling and attention control mechanisms to establish priorities, to protect task performance in progress from interference, to update priorities, and to switch from one task to another.

The *task-switching* paradigm provides a simple experimental framework for systematic study of the control processes underlying our ability to switch from one task goal to another and to reconfigure the processing system for engaging in another task. This chapter presents the approach we have developed for detailed analysis and modeling of task-switching performance (De Jong et al. forthcoming) and outlines the novel perspective our approach provides on the causes of performance limitations in task switching. It reports the findings of two new experiments investigating the effects of time on task and of expected task duration on task-switching performance.

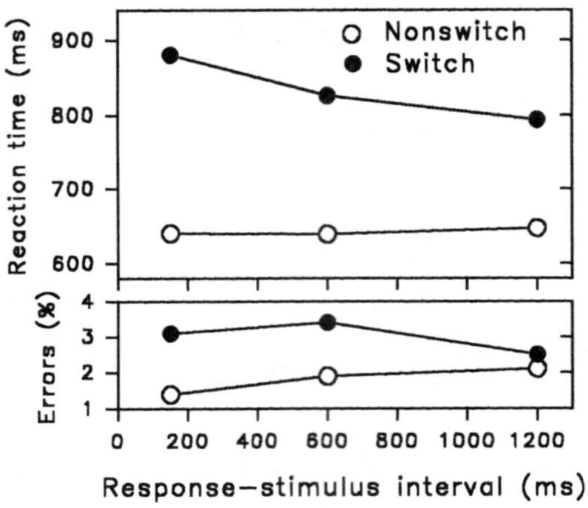

Figure 15.1 Mean correct reaction time and error rate as a function of trial type and response-stimulus interval.

15.1 THE TASK-SWITCHING PARADIGM

In the task-switching paradigm, the task to be performed on each trial is selected from a set of alternative tasks, usually choice reaction time (RT) tasks. In the "explicit cue" version, tasks are presented in an unpredictable order. At the start of each trial a cue or instruction signal signals the task to be performed (e.g., Meiran 1996), followed by the presentation of the imperative stimulus after a fixed or random delay, called the "preparation interval." In the "implicit cue" version, the tasks are presented in a predictable order, either in a simple alternating order (e.g., Allport, Styles, and Hsieh 1994) or in a more complex pattern (e.g., Rogers and Monsell 1995), with the response-stimulus interval (RSI) serving as the preparation interval.

There are two basic types of trials. On *nonswitch trials*, the task to be performed is the same as that on the previous trial, and the task set remains in place. On *switch trials*, the task changes, and the task set must be reconfigured. Longer preparation intervals provide more time for advance preparation, that is, for the selection and configuration of the relevant task set before the imperative stimulus is presented. Thus *switch costs*, defined as the difference in performance between switch and nonswitch trials, are expected to diminish gradually as the preparation interval is prolonged.

Residual Switch Costs

In this chapter, I will focus on *residual switch costs*, defined as switch costs at long preparation intervals, that should provide ample time for advance preparation to be completed. Figure 15.1 presents a representative example. The stimuli in the experiment (De Jong et al. forthcoming, exp. 1) were red or blue letters. Both tasks involved the same two keypress response alternatives but in one task, the response was to the color of the letter, and in the other, to its category (consonant or vowel). The experiment used the implicit cue version of the task-switching paradigm, with tasks alternating across trials according to a fixed AABB scheme. Following Rogers and Monsell (1995), clockwise cycling of the position of successive stimuli in a 2×2 grid was used to help subjects keep track of the tasks. Subjects were required to perform one task when the stimulus was displayed in one of the two top positions, and the other, when it was displayed in one of the two bottom positions. Mean initial switch costs at the shortest preparation interval (150 msec) were 240 msec; switch costs declined to 143 msec at the longest interval (1,200 msec). Virtually all task-switching studies have yielded residual switch costs, although the magnitude of such costs relative to initial switch costs varies widely across studies, ranging from very large (e.g., Allport, Styles, and Hsieh 1994; Rogers and Monsell 1995, exp. 2) to very small (e.g., De Jong et al., forthcoming, exp. 3; Meiran 1996, chap. 16, this volume).

Two basic accounts of residual switch costs have been proposed (a third account, by Allport and Wylie, chap. 2, this volume, will be considered later). The first one, which I refer to as the "additional process" (AP) hypothesis, is best exemplified by Rogers and Monsell (1995), who argued that, while the endogenous component of task set reconfiguration could be carried out during the preparation interval, completion of the reconfiguration process had to await triggering by a task-relevant stimulus. The duration of this exogenous component results in residual switch costs. The second account, which I refer to as the "failure-to-engage" (FTE) hypothesis, starts from the notion that advance preparation is optional. Advance preparation is useful because it promotes fast responding to the imperative stimulus, but postponing task set reconfiguration until the arrival of the imperative stimulus still suffices to ensure an accurate, albeit slow, response. According to this perspective, residual switch costs are due to intermittent failures to engage in advance preparation, rather than to a fundamental inability to attain a complete reconfiguration of task set during the preparation interval (i.e., by fully endogenous means). Some broader implications of the FTE hypothesis will be discussed later (see also De Jong, Berendsen, and Cools 1999). First, I will present the modeling approach that we have developed to evaluate the relative merits of these alternative hypotheses (De Jong et al., submitted).

15.2 A MIXTURE MODEL OF RESIDUAL SWITCH COSTS

According to the AP hypothesis, residual switch costs should be manifest on all switch trials. In contrast, the FTE hypothesis holds that such costs should be concentrated within that subset of switch trials on which, for

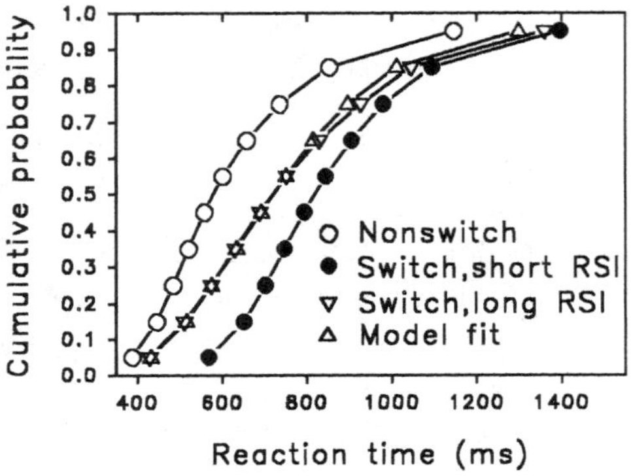

Figure 15.2 Cumulative distribution functions as a function of trial type and response-stimulus interval (RSI). The fit was produced by the restricted mixture model with $\alpha = 0.51$.

reasons to be discussed later, subjects failed to prepare in advance for the change of task. This suggests that, to distinguish between these hypotheses, entire RT distributions, rather than only their means, should be considered. We therefore computed cumulative distribution functions (CDFs) by dividing the rank-ordered RTs for each subject, for each condition into deciles (10% bins) and then computing the mean RT for each decile. Figure 15.2 shows these functions averaged across subjects, collapsed across the different preparation intervals for nonswitch trials, and at the shortest and longest intervals for switch trials.

The most striking feature of the figure concerns the shape of the CDF for switch trials at the longest preparation interval. In the fast-response range this function approaches that for nonswitch trials, whereas at the slow-response range it approaches the function for switch trials at the shortest preparation interval. This feature is consistent with the FTE hypothesis that responses on switch trials at the longest preparation interval consist of a mixture of two basic types. When advance preparation is carried out, the long preparation interval should provide ample time to attain a suitably reconfigured task set. Responses in this *prepared* state should be relatively fast and have the same RT distribution as those on nonswitch trials, where, by definition, a properly configured set is assumed to be in place. When advance preparation fails to be triggered, the system will remain unprepared throughout the preparation interval. Responses in this *unprepared* state should be relatively slow and have about the same distribution as responses on switch trials at the shortest preparation interval, where preparation may be assumed to have hardly gotten under way when the imperative stimulus arrives.

The FTE hypothesis holds that switch trials at a long preparation interval should yield a mixture of outcomes, with task set reconfiguration either being completed or not having been attempted by the end of the interval. This hypothesis can be formalized in terms of CDFs by the following equation:

$$F_{\text{switch, long PI}}(t) = \alpha\, F_{\text{prepared}}(t) + (1 - \alpha)\, F_{\text{unprepared}}(t), \tag{15.1}$$

where $_{\text{switch, long PI}}$ is the CDF for switch trials and a long preparation interval, F_{prepared} and $F_{\text{unprepared}}$ the theoretical CDFs for the prepared and unprepared state, and α the probability that preparation is carried out and completed during the long preparation interval, which I refer to as the "mixing probability." From the definition of residual switch costs (in RT) as the RT difference between switch and nonswitch trials at the longest preparation interval, it follows that $F_{\text{nonswitch, long PI}}$ provides the proper empirical estimate of F_{prepared}. The best available estimate of $F_{\text{unprepared}}$ is provided by $F_{\text{switch, short PI}}$. However, this latter estimate may well be somewhat biased. For one thing, we cannot exclude the possibility that a significant amount of preparation might be carried out within the shortest preparation interval (see De Jong et al. forthcoming for a discussion of other potential problems regarding this estimate). As shown in the chapter appendix, this difference in a priori appropriateness of the two estimates can be effectively dealt with in the model-testing procedure. Substitution of these estimates into equation 15.1 gives a testable version of the mixture model (i.e., the FTE hypothesis) in terms of a relation between the CDFs for three experimental conditions:

$$F_{\text{switch, long PI}}(t) = \alpha\, F_{\text{nonswitch, long PI}}(t) + (1 - \alpha)\, F_{\text{switch, short PI}}(t). \tag{15.2}$$

This mixture model can be generalized to allow also for a possible contribution of any additional exogenous component of task set reconfiguration to residual switch costs. Let δ represent the average duration of this hypothetical exogenous component. Even when advance preparation is carried out during the long preparation interval, with probability α, a response on switch trials should then yet incur an average time cost of δ, as compared to a response on nonswitch trials. Incorporating this hypothetical time cost in equation 15.2, we arrive at the following expression for the generalized mixture model for residual switch costs:

$$F_{\text{switch, long PI}}(t) = \alpha\, F_{\text{nonswitch, long PI}}(t - \delta) + (1 - \alpha)\, F_{\text{switch, short PI}}(t). \tag{15.3}$$

which assumes that the duration of the exogenous component is invariant. Although this simplifying assumption imposes some restrictions on the generality of our approach, I suggest that it is unlikely to compromise our main objective, for two reasons. First, the assumption yields a first-order approximation that should give the generalized mixture model a substantial advantage over the restricted model ($\delta = 0$) whenever an exogenous component of appreciable mean duration actually contributes to residual switch costs. Second, the approximation may in fact be quite close. A consistent finding in our experiments has been that the relation between the two basis distributions of the mixture model, $F_{\text{nonswitch, long PI}}$ and $F_{\text{switch, short PI}}$, is captured quite accurately by a simple shift on the time axis (e.g., figures 15.2, 15.4, and 15.6). If the RT distributions for the two most extreme preparatory states are related through such a shape-conserving shift, it would seem reasonable to assume that the distributions for completely and partially prepared states are similarly related.

With $\delta = 0$, the generalized mixture model reduces to the pure FTE hypothesis; with $\alpha = 1$, it reduces to the pure AP hypothesis. The intermediate cases, $0 < \alpha < 1$ and $\delta > 0$, comprise a range of models in which various proportions of residual switch costs are attributed to failures to engage in advance preparation and to an exogenous component of task set reconfiguration. We used the "multinomial maximum likelihood method" (MMLM), developed by Yantis, Meyer, and Smith (1991), to determine maximum likelihood estimates of α for the restricted mixture model ($\delta = 0$), and of α and δ for the generalized or full model, and to compute goodness-of-fit statistics for the two models. (Details of the model-testing procedure are given in the chapter appendix.)

Application of this procedure to the present data set (De Jong et al. forthcoming, exp. 1) gave the following results. The average estimate of δ was a nonsignificant $12(\pm 8)$ msec: $t(15) = 1.48$, $p > 0.15$. The fit of the restricted mixture model was fairly good: $G^2(48) = 64.8$, $p > 0.05$, $\alpha = 0.51$ (average estimate). Thus the residual switch costs can be adequately explained by the hypothesis that subjects engaged in advance preparation on only about 50% of the switch trials. Figure 15.2 depicts the corresponding fit of $F_{\text{switch,long PI}}$, which can be seen to be close in the fast-response range, but less so in the slow-response range. This progressive worsening of the fit over the slower RT range can plausibly be attributed to the likelihood, discussed above, that $F_{\text{switch,short PI}}$ provides a biased estimate of $F_{\text{unprepared}}$ (De Jong et al. forthcoming).

These results lend credence to the hypothesis that failures to engage in advance preparation were the predominant cause of residual switch costs in our first experiment. Although they do not completely rule out a possible contribution of an exogenous component of task set reconfiguration, they indicate that this contribution must have been at best a minor one (see De Jong et al. forthcoming for discussion of the power of the MMLM analyses).

Corroborating evidence for this conclusion was obtained in two other experiments. In our second experiment (De Jong et al. forthcoming), we contrasted the implicit cue version of the task-switching paradigm with the explicit cue version. In both versions, the vertical position of the imperative stimulus determined the relevant task. In the implicit cue version, the position of the next stimulus could be easily predicted from the clockwise cycling of stimulus position in a 2×2 grid. In the explicit cue version, the cue, consisting of a square above or below a horizontal midline, was presented, followed by the display of the stimulus within that square after the preparation interval had elapsed. Although initial switch costs were very similar for the two versions, residual switch costs were about twice as large for implicit as for explicit cues. Mixture model analyses attributed this latter difference to the finding that failures to engage in advance preparation were twice as likely with implicit as with explicit cues. The difference can be easily understood in terms of the prompting effect of the explicit cue on triggering advance preparation.

In our third experiment (De Jong et al. forthcoming), we succeeded in eliminating residual switch costs altogether by using a combination of explicit cues and short trial blocks. That residual switch costs can be eliminated under suitable conditions is consistent with the notion that such costs have a strategic origin and do not arise from the fundamentally limited effectiveness of endogenously initiated preparation.

To summarize, mixture model analyses of residual switch costs have provided consistent support for the hypothesis that such costs are primarily, if not exclusively, due to failures to engage in advance preparation. I would like to stress, however, that these results were obtained for experiments that used young college students as subjects and pairs of relatively simple, speeded tasks. Thus preparatory limitations of the sort assumed by the AP hypothesis may yet prove to make a substantial contribution to residual switch costs in different populations or with pairs of more complex tasks associated with more intricately structured task sets. Indeed, elderly subjects have already been found to exhibit marked preparatory limitations, at least in initial stages of practice (De Jong et al. forthcoming).

15.3 THE ORIGIN OF TRIGGER FAILURES IN TASK SWITCHING

What might cause intermittent failures to engage in anticipatory preparation in the task-switching paradigm? As pointed out, advance preparation is optional, serving primarily to optimize performance on switch trials. Effective use of the option requires (1) that an explicit goal or intention to engage in advance preparation be added to the basic goal structure that governs performance in the task-switching paradigm; and (2) that this intention be retrieved and carried out at the proper time, namely, at the start of the preparation interval. We can thus see a marked correspondence in this aspect of performance between the task-switching paradigm and prospective memory tasks requiring subjects to carry out their intentions at a future time. This suggests that an answer to the question above may be informed by current ideas regarding the factors and mechanisms that determine success and failure in prospective memory tasks.

In prospective memory tasks, people may be assumed to form an associative encoding of a target cue-action pairing and to hold this representation in a state of extra activation (Goschke and Kuhl 1993; Yaniv and Meyer 1987). Success in subsequently retrieving the intention or action depends on the joint influence of two factors: (1) the activation level of the associative encoding; and (2) the characteristics of the target cue (Mantyla 1996). The application of these ideas to the case of advance preparation in task switching is straightforward.

One possible reason for failures to trigger advance preparation might be that the associative encoding of a cue-action pair, with advance preparation as the action, was never formed in the first place. This might happen if subjects failed to understand or appreciate the benefits to be gained by advance preparation, and it should be associated with complete and consistent failures (i.e., $\alpha = 0$). In our experiments, we have encountered such cases only rarely, presumably because our instructions explicitly pointed out such benefits and generally emphasized speed of responding. On the other hand, this factor may have played a role in studies that have found little or no reduction of switch costs with preparation interval (Allport, Styles, and Hsieh 1994; Rogers and Monsell 1995, exp. 2).

Another reason for trigger failures might be that the activation level or strength of the cue-action representation was too low for the cue to reliably trigger its associated action. Several factors may influence the activation level of the cue-action representation. A prominent factor would be the subjective utility of the expected benefits of the action, a low utility being associated with reduced activation of the representation. Because enhanced response speed is the primary benefit to be gained by advance preparation, we can predict that trigger failures should be especially prevalent when response speed is assigned low priority. Two pieces of evidence bear out this prediction. First, manipulation of speed-accuracy instructions in the task-switching paradigm strongly affects both the magnitude of residual switch costs and the estimated mixing probability α, which is much smaller when instructions emphasize accuracy over speed than when speed and accuracy were equally emphasized (De Jong, Schellekens, and Meyman in preparation). Second, correlational analysis of individual differences in task switching within a group of college students has yielded a strong negative correlation (-0.72) between estimated α and mean RT on nonswitch trials (De Jong et al. forthcoming). On the assumption that differences in mean RT reflect, at least in part, differences in priority assigned to response speed, this result nicely corroborates the evidence from explicit manipulation of this priority.

An important factor that may influence the activation level of the cue-action representation is the ability or capacity to generate and maintain goals or intentions in working memory. This ability has been held by some to be a primary determinant of success and failure in prospective memory tasks (Duncan et al. 1996) and in other tasks requiring organization and management of a hierarchy of goals (Anderson, Reder, and Lebiere 1996; Carpenter, Just, and Shell 1990). Following reports of a relation between this ability and "general intelligence" or Spearman's *g*

(Carpenter, Just, and Shell 1990; Duncan et al. 1996), we can predict a similar relation between general intelligence and α. The results of a recently completed study are generally consistent with this prediction (Cools 1998). Like high-g normals, low-g normals performed with high accuracy in the task-switching paradigm, indicating they were able to switch between tasks according to instructions. Moreover, estimates of δ did not significantly differ from zero for either group, although estimated α was much lower for low-g than for high-g normals, especially in the implicit cue version of the paradigm.

The intention-activation account also provides a ready explanation for the higher incidence of trigger failures in the implicit cue than in the explicit cue version of the paradigm. The implicit cue version requires subjects to anticipate or predict a change of task on the basis of the regular ordering of tasks; failure to do so would obviously prevent the triggering of advance preparation. This potential cause of trigger failures does not apply to the explicit cue version. Moreover, the commandlike nature of an explicit cue may be assumed to make it a particularly powerful trigger of preparatory activities.

Finally, holding the cue-action representation at a high level of activation may require substantial effort—effort that can be maintained for only brief periods of time. This suggests that failures to engage in advance preparation may become more prevalent as a function of task duration or time on task, a possibility investigated in the following two experiments.

15.4 EXPERIMENT 1: EFFECTS OF TASK DURATION AND TIME ON TASK ON TASK SWITCHING

It may take considerable effort to hold the intention to engage in advance preparation at a sufficiently high level of activation to ensure that advance preparation will be successfully triggered. If people are able to sustain this effort for only brief periods of time, trigger failures should be expected to be more prevalent during long than during short blocks of trials. There is some evidence to support this conjecture. The only experiment finding residual switch costs to be virtually eliminated among individual college students used short blocks of trials (De Jong et al. forthcoming, exp. 3), although procedural details other than block length may have been responsible for this exceptional finding. The two experiments reported here were designed to provide more definitive evidence on this issue.

The experiments addressed two related questions. First, does block length exert reliable effects on the incidence of trigger failures in the task-switching paradigm? Second, if it does, are such effects present right from the start of the block or do they gradually emerge during the course of long blocks? The former possibility would suggest that people pace themselves, setting intention-activation at a level that they expect to be able to sustain for the duration of the block. The latter possibility would suggest that, irrespective of known block length, people initially set the activation at a high level that they cannot sustain for prolonged periods of time. The first experiment required subjects to alternate between blocks of 12 or 48 trials, with subjects being informed about the block length at the beginning of a block. The second experiment used a between-subjects design, with blocks of 12 trials being used for one group of subjects and blocks of 96 trials for the other.

Subjects

Eight students from the University of Groningen, 3 women and 5 men between 19 and 24 years of age, were paid to participate in the experiment.

Apparatus and Procedure

Subjects sat approximately 70 cm in front of a VGA color monitor of an IBM-compatible PC. A white 2×2 grid, consisting of a 6 cm square subdivided into four 3 cm squares, was displayed continuously at the center of the display against a black background. On each trial, the stimulus was a red- or blue-colored letter displayed at the center of one of the small squares; on the next trial, the stimulus was presented in the next square clockwise. Half of the subjects were instructed to perform the letter-classification task when the stimulus appeared in either of the two top squares and the color-classification task when the stimulus appeared in either of the two bottom squares; for the other half, the assignment of tasks to positions was reversed. Because stimulus position cycled in a clockwise fashion, the task changed predictably on every second trial, according to an AABB scheme.

Letters were displayed in an uppercase sansserif font, 1.0 cm wide and 1.4 cm tall. On each trial, the letter was sampled randomly from the set *A*, *E*, *Y*, *U*, *G*, *K*, *M*, and *R*, and its color was sampled randomly from the set red and blue. The stimulus remained on the screen until a response was registered or until 5,000 msec had elapsed. After a response was registered and the stimulus extinguished, the next stimulus appeared after a response-stimulus interval (RSI) with a randomly determined duration of 150, 600, or 1,500 msec.

Subjects received written instructions, which also told them to minimize RT while avoiding errors, and that, to do so, they should make effective use of the RSI to prepare for the upcoming task. An abbreviated version of the instructions appeared for 5,000 msec on the screen at the beginning of a new trial block, after which the first stimulus appeared in the top left square.

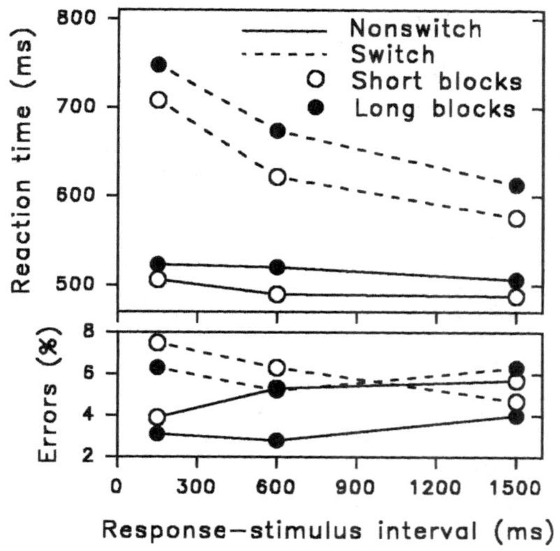

Figure 15.3 Experiment 1: Mean correct reaction time and error rate as a function of trial type, block length, and response-stimulus interval.

Design

The experiment consisted of a single session lasting about two hours. The first three trial blocks consisted of 60 trials each and were used for training. Subjects practiced the individual letter and color tasks in the first two blocks and then practiced the task switch condition in the third block. They subsequently completed 124 experimental blocks, with blocks of 12 trials and blocks of 48 trials randomly intermixed in a 4:1 ratio. At the start of a new block, a message on the screen informed them about the length of the block. Because subjects had to start a new block by pushing the space bar, they had ample opportunity to take short breaks in between blocks and were encouraged to do so.

There were two responses: a *left* response, made by pressing the "v" key of the computer keyboard with the left index finger; and a *right* response, made by pushing the "n" key with the right index finger. For the letter task, vowels required one response and consonants the other. For the color task, red letters required one response and blue letters the other. The four possible stimulus-response mapping combinations (two possible mappings for each task) were counterbalanced across subjects.

Results

Reaction Time and Errors Figure 15.3 shows mean correct RT and error rate for switch and nonswitch trials as a function of RSI and block length. Although switch costs decreased with RSI, sizable residual switch costs were obtained for both short and long trial blocks. Responses were somewhat faster in short than in long blocks, especially on switch trials.

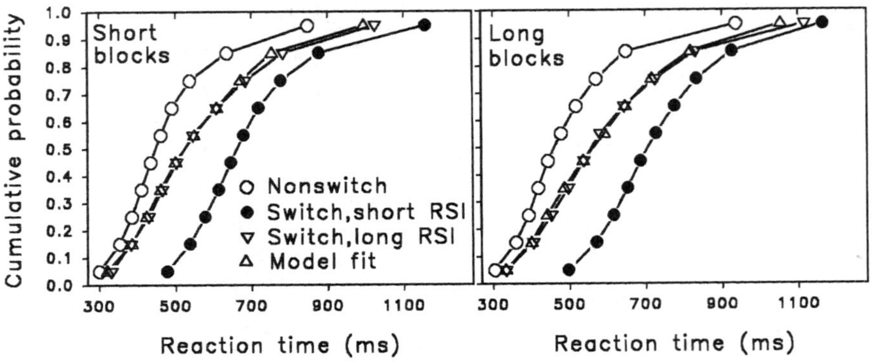

Figure 15.4 Experiment 1: Cumulative distribution functions for short and long trial blocks, as a function of trial type and response-stimulus interval (RSI). The fits were produced by the restricted mixture model with $\alpha = 0.64$ (short blocks) and $\alpha = 0.58$ (long blocks).

An ANOVA with block length, RSI, and trial type (switch/nonswitch) as within-subjects factors yielded, for RT, main effects of trial type: $F(1,7) = 28.0$, $p < 0.001$; of RSI: $F(2,14) = 14.0$, $p < 0.001$; and of block length: $F(1,7) = 20.9$, $p < 0.003$. These effects were qualified by interactions of trial type and RSI: $F(2,14) = 120.9$, $p < 0.001$; and of trial type and block length: $F(1,7) = 12.5$, $p < 0.01$. No other effects on RT approached significance. Analysis of error rate yielded a significant effect only of trial type: $F(1,7) = 6.7$, $p < 0.05$. Mean error rate was 4.1% for nonswitch trials and 6.1% for switch trials.

Reaction Time Distributions and Modeling Results Figure 15.4 shows averaged CDFs of RT for the relevant conditions (nonswitch trials at the longest RSI and switch trials at the shortest and longest RSIs) separately for short and long trial blocks. Average estimates of δ were 10 (± 11) msec and 13 (± 9) msec for short and long blocks, respectively; neither value differed significantly from zero ($p\text{s} > 0.15$). The restricted mixture model ($\delta = 0$) gave very good fits for short blocks, $G^2(24) = 25.0$, $p > 0.40$, as well as for long blocks, $G^2(24) = 26.1$, $p > 0.35$. The average estimate of α was 0.64 for short and 0.58 for long blocks; although this difference was in the predicted direction, it was not significant: $F(1,7) = 2.6$, $p > 0.20$. The corresponding fits of $F_{\text{switch,long RSI}}$ produced by the restricted model are depicted in figure 15.4.

Time on Task To assess possible time-on-task effects, the data for the long trial blocks were reanalyzed with the factor block half (the first versus the last 24 trials) included. Though this test is admittedly crude, the limited number of trials did not permit a more precise decomposition. For RT, this analysis yielded as new results a main effect of block half: $F(1,7) = 14.0$, $p < 0.01$; and an interaction of block half and trial type:

$F(1, 7) = 9.1$, $p < 0.02$. Mean RT in the second half of the block was longer than that in the first half by 25 msec on nonswitch trials and by 51 msec on switch trials. Error rates did not differ between the first and second half. The average estimates of α were 0.61 and 0.57 for the first and second half, respectively; this difference did not approach significance.

Discussion

Substantial residual switch costs were obtained for both short and long trial blocks. Replicating previous findings (De Jong et al. forthcoming), the modeling results indicated that these residual costs could be attributed almost exclusively to failures to engage in advance preparation, rather than to an additional poststimulus component of task set reconfiguration. Contrary to predictions, however, there was only a nonsignificant tendency for such failures to be more prevalent in long trial blocks. Instead, responses tended to be somewhat slower in long blocks, especially on switch trials, and this effect appears to have been largely due to a decline in response speed from the first to the second half of long blocks. Because these effects of block length did not interact with RSI, we suggest that they may reflect a gradual slowing of both preparation and poststimulus task execution during the course of a long block.

Although these results seem to refute our assumption that people may have trouble sustaining for prolonged periods the effort needed to keep the intention to prepare in advance highly activated, an alternative interpretation is possible. Faced with a mixture of short and long blocks, subjects may have adopted a conservative, worst-case strategy, and have set intention-activation at a level that they could sustain for the duration of the long blocks. Though admittedly ad hoc, this interpretation receives some support from several recent studies that suggest a marked lack of flexibility in adjusting control settings in response to different instructions or task requirements (Los 1996; Strayer and Kramer 1994). Experiment 2 was designed to address these remaining uncertainties.

15.5 EXPERIMENT 2

Using a between-subjects design, one group of subjects was exposed only to short trial blocks and the other group only to long blocks. If the suggested interpretation of the absence of clear effects of block length on α in experiment 1 is correct, then two predictions follow for experiment 2. First, a clear effect of block length on α should now be present. Second, if the intermixing of short and long blocks indeed caused subjects to adopt an overly conservative level of intention-activation for the short blocks of experiment 1, then α for the short blocks should be larger than that in experiment 1.

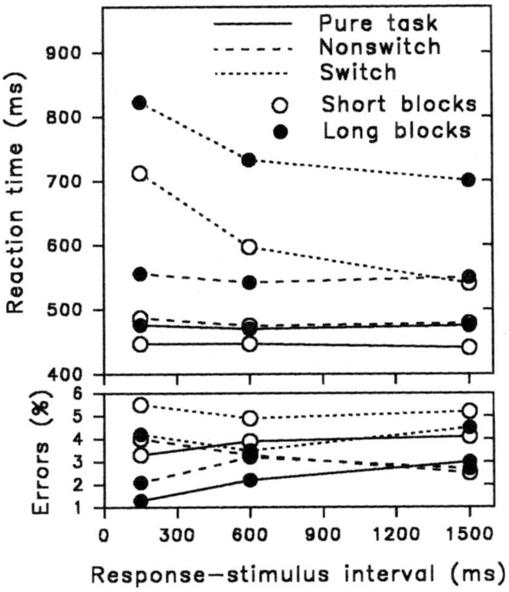

Figure 15.5 Experiment 2: Mean correct reaction time and error rate as a function of trial type, block length, and response-stimulus interval.

Method

There were 20 new paid participants, 10 male and 10 female, all students at the University of Groningen between 19 and 26 years of age. The apparatus and procedure were the same as in experiment 1, as was the design, with two important exceptions. After three practice blocks, half the subjects completed 100 blocks of 12 trials whereas the other half completed 12 blocks of 96 trials. Every fourth block was a pure-task block exactly similar to the mixed-task blocks except that only one of the two alternative tasks was relevant throughout the block.

Results

RT and Errors Figure 15.5 shows mean correct RT and error rate for pure task, nonswitch, and switch trials as a function of RSI and block length. As in experiment 1, switch costs declined with RSI but sizable residual switch costs were obtained, especially in long blocks. Responses were substantially faster in short than in long blocks, especially on switch trials.

An ANOVA with block length as a between-subjects factor and with RSI and trial type (switch/nonswitch) as within-subject factors yielded, for RT, main effects of trial type: $F(1, 18) = 93.6$, $p < 0.001$; of RSI: $F(2, 36) = 59.7$, $p < 0.001$; and of block length: $F(1, 18) = 9.8$, $p < 0.01$. These effects were qualified by interactions of trial type and RSI: $F(2, 36) = 109.4$, $p < 0.001$; and of trial type and block length: $F(1, 18) = 6.8$,

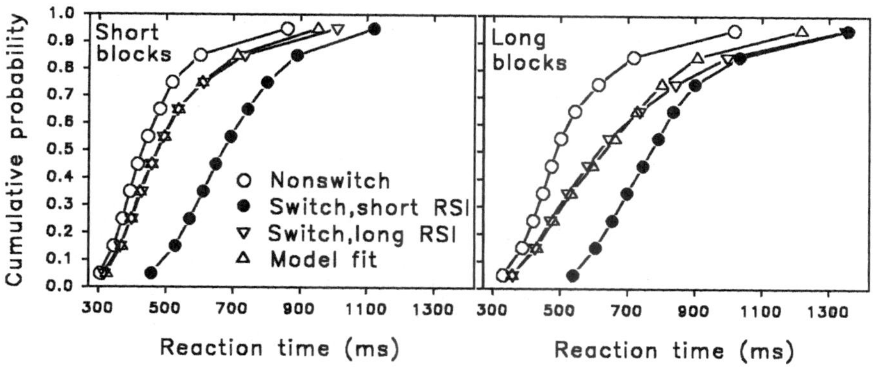

Figure 15.6 Experiment 2: Cumulative distribution functions for short and long trial blocks, as a function of trial type and response-stimulus interval (RSI). The fits were produced by the restricted mixture model with $\alpha = 0.80$ (short blocks) and $\alpha = 0.57$ (long blocks).

$p < 0.02$. No other effects on RT approached significance. Analysis of error rates yielded a significant effect only of trial type: $F(1, 18) = 12.1$, $p < 0.01$. Mean error rate was 3.0% for nonswitch trials and 4.6% for switch trials.

We conducted a separate analysis of the difference between pure task and nonswitch trials. For RT, this analysis yielded main effects of trial type: $F(1, 18) = 58.2$, $p < 0.001$; and of block length: $F(1, 18) = 4.9$, $p < 0.05$; and a significant interaction of trial type and block length, reflecting a larger pure task/nonswitch RT difference in long than in short blocks: $F(1, 18) = 8.9$, $p < 0.01$. No other effects on RT approached significance. Mean error rate in pure task blocks was 2.6%.

Reaction Time Distributions and Modeling Results Figure 15.6 shows averaged CDFs of RT for the relevant conditions separately for short and long trial blocks. Average estimates of δ were -1 (± 14) msec and 4 (± 11) msec for short and long blocks, respectively; neither value differed significantly from zero (ps > 0.25). The restricted mixture model gave excellent fits for both short and long blocks: $G^2(30) = 24.6$, $p > 0.70$ and $G^2(30) = 29.2$, $p > 0.50$, respectively. The average estimate of α was 0.80 for short blocks and 0.57 for long blocks; this difference was highly significant: $F(1, 18) = 10.9$, $p < 0.01$. The corresponding fits of $F_{\text{switch, long RSI}}$ are depicted in figure 15.6.

Time on Task In order to assess time-on-task effects, the data for the long blocks were reanalyzed with the factor block half included. For RT, this analysis yielded as the only new result a significant main effect of block half, reflecting an increase in RT of 26 msec from the first to the second half: $F(1, 9) = 7.1$, $p < 0.05$, Error rates did not differ between the two halves. The average estimate of α was 0.55 and 0.61 for the first and second half, respectively; this difference did not approach significance.

Comparison with Experiment 1 The average estimate of α for short blocks in experiment 1 was 0.64, as compared to 0.80 in experiment 2; this difference was significant: $F(1, 16) = 5.7$, $p < 0.05$.

Discussion

The two key predictions for this experiment were confirmed. First, differences in estimated α indicate that failures to engage in advance preparation were about twice as prevalent in long as in short trial blocks. Moreover, the absence of negative time-on-task effects on α in long blocks suggests that subjects paced themselves, setting intention-activation at an initial level they could sustain for the duration of the block. Second, trigger failures were more prevalent in short blocks when such blocks were intermixed with long blocks (experiment 1) than when only short blocks were administered (experiment 2). This provides further evidence for pacing because such a difference could have occurred only if subjects did take block length into account in setting the initial level of intention-activation. It also lends credence to the idea that the intermixing of short and long trial blocks in experiment 1 led subjects to adopt a compromise setting, rather than to adjust the setting for each of the two block lengths.

Consistent with the intention-activation account of residual switch costs, the combined results from the two experiments indicate that holding the intention to engage in advance preparation at a high level of activation requires considerable effort, and that, as in distance running, people can adaptively manage these requirements to maintain a steady level of performance in a prospective, rather than only a reactive manner. On the other hand, the results also suggest clear limits to the flexibility with which people adjust the level of intention-activation on the basis of expected task duration in the task-switching paradigm.

Finally, responses in pure task blocks were found to be considerably faster than those on nonswitch trials, especially in long trial blocks. As has been emphasized by Allport and Wylie (chap. 2, this volume), this finding suggests that task set reconfiguration could not have been optimal or complete even on nonswitch trials. The implications of this for the present theoretical approach will be discussed in section 15.6.

15.6 CONCLUSIONS

The mixture model approach has yielded consistent support for the FTE hypothesis that residual switch costs stem from intermittent failures to take advantage of opportunities for advance preparation. The all-or-none conception of advance preparation implicit in this hypothesis should be taken quite literally. For instance, consider the alternative hypothesis that the degree of advance preparation has, on a trial-to-trial basis, a continuous and smooth distribution with 0% and 100% as extremes and with α

representing its central tendency. While such a continuous conception of advance preparation would seem perfectly plausible, it can be shown to be incompatible with the small and nonsignificant estimates of δ that have consistently been obtained (De Jong et al. forthcoming). Clearly, although we cannot exclude the possibility that some other, yet-unspecified, model may offer an equally precise account of residual switch costs, at this point, the FTE hypothesis seems to come close to identifying the actual primary cause of this intriguing empirical phenomenon.

This conclusion, it must be stressed, is based exclusively on evidence from experiments that used young college students as subjects and pairs of simple tasks, and should therefore not be generalized beyond such cases at this point. Rather, it should provide a clear point of reference for the evaluation and interpretation of residual switch costs in other populations or for pairs of more complex tasks, where limitations to the completeness of task set reconfiguration attainable by fully endogenous means might well be present. Indeed, the mixture model approach should be most useful, perhaps even indispensable, when residual switch costs are jointly due to such preparatory limitations and to intermittent failures to engage in advance preparation, and it becomes important to assess the relative contributions of these different causes (De Jong et al. forthcoming).

An intention-activation account of intermittent failures to engage in advance preparation was proposed, based on a marked correspondence between this aspect of task-switching performance and prospective memory performance. The account was argued to provide a coherent explanation of the influence of a variety of factors on the incidence of such failures, including the effects of task duration and time on task in the two experiments reported here. Admittedly, pertinent empirical evidence is still scant and potentially important factors, such as the predictive validity of the task cue, task complexity, and training, remain to be explored. Nevertheless, these initial results are encouraging and suggest that the intention-activation account may provide a versatile theoretical framework for future studies of strategic control in the task-switching paradigm.

Residual Switch Costs and Nonswitch/Pure Task Differences

The intention-activation account may also shed light on another intriguing finding in the recent task-switching literature. Responses on nonswitch trials are usually considerably slower than those in pure task blocks, and Stroop-like interference by the competitor task is usually observed on nonswitch trials (see Allport and Wylie, chap. 2, this volume). These findings indicate that previously relevant task sets are generally not fully disengaged on nonswitch trials. It is important to note that this does not undermine the all-or-none conception of advance preparation embodied by the FTE hypothesis. The FTE hypothesis holds that the task set in place on nonswitch trials can also be attained on switch trials by fully endogenous means, and will be if advance preparation is carried out and completed. It does not assume or require this set to be fully reconfigured, with competing task sets fully disengaged, or to be the same as the task set in pure task blocks. Thus there is no logically necessary relation between residual switch costs and performance differences between pure task and nonswitch trials. Yet the question of whether and how these two phenomena might be related is an important one.

Allport and Wylie (chap. 2, this volume) outlined an interesting theoretical perspective on this issue. They suggest that incomplete disengagement of prior task sets is caused by involuntary residual priming. Residual priming of prior task sets can retard the system's settling to a unique response. Such proactive interference can be long-lasting, which explains differences between pure task and nonswitch trials, but is typically largest for the first trial of a run (the "restart" effect), which explains residual switch costs. On the other hand, the notion that residual switch costs are due largely to involuntary residual priming of prior task sets is clearly incompatible with the present evidence that such costs reflect, not fundamental preparatory limitations, but inconsistent use of preparatory capabilities.

I would like to argue that an integrative account should probably be based on the notion that, like residual switch costs, pure task/nonswitch differences depend on the control strategies that subjects adopt. Even though capable of completely disengaging prior task sets, subjects might opt not to fully exercise this capability when, for instance, these sets may need to be reinstalled shortly. The effort requirements for executive control might be significantly reduced by such a conservative control strategy, but at the expense of suboptimal task performance and potential interference effects.

The hypothesis that pure task/nonswitch performance differences may reflect a strategic compromise between minimizing control effort and maximizing task performance closely resembles the intention-activation account of residual switch costs. Combining the two accounts, we can predict that greater effort invested in executive control should have the dual effect of enhancing α and reducing pure task/nonswitch performance differences. The previously mentioned strong negative correlation between α and nonswitch RT is consistent with this prediction. Two more specific predictions can be made. First, experimental factors that affect α by influencing the level of intention-activation should also affect pure task/nonswitch differences. This prediction is borne out by the finding in experiment 2 that α was substantially larger and the pure task/nonswitch RT difference substantially smaller in short blocks than in long trial blocks. Note also that the very presence of an effect of block length on the pure task/nonswitch RT difference would seem to argue against the notion that this difference is due to involuntary persistence of the competing task set. Second, factors that affect α by influencing trigger strength should leave the pure task/nonswitch difference unaffected. This is the pattern obtained when implicit and explicit cues were contrasted: estimated α was substantially larger for explicit cues whereas the pure task/nonswitch RT difference was the same for the two types of cue (De Jong et al. forthcoming, exp. 2). These considerations would seem to provide reason to take seriously the possibility that pure task/nonswitch performance differences are not an inevitable result of involuntary persistence of competing task sets but, like residual switch costs, have a largely strategic origin.

APPENDIX

The multinomial maximum likelihood method (MMLM) for testing mixture models requires grouping of rank-ordered RTs into a finite number of bins (Yantis, Meyer, and Smith 1991). In our analyses, we used five bins, with bins 1 to 4 comprising the consecutive first four 8% portions of RTs in the mixture condition and bin 5 comprising the remaining 68% slowest RTs. This choice of bins served to reduce unwanted effects of a possibly biased estimate of $F_{\text{unprepared}}$ on goodness-of-fit statistics (see De Jong et al. forthcoming for details).

The log likelihood ratio statistic G^2 served as the goodness-of-fit statistic. For a valid restricted mixture hypothesis ($\delta = 0$) and with five bins, this statistic has an asymptotic X^2 distribution with 3 degrees of freedom (Yantis, Meyer, and Smith 1991). Because the generalized (full) mixture model is not linear in its parameters α and δ, the asymptotic distribution of the G^2 statistic for a valid model is not known a priori and, from Monte Carlo simulations, depends on such factors as the true value of α and the degree of overlap between the two basis distributions. This complicates the application of the common likelihood ratio procedure to test for a significant improvement of fit by the generalized model. Though this technical problem is not insurmountable, we (De Jong et al. forthcoming) used an alternative and less complex procedure that sufficed for the analysis of the experimental data presented in this chapter. In the first step of the procedure, maximum likelihood estimates of the models' parameters were computed for each subject. In the second step, we tested whether the average estimate of δ differed significantly from zero across subjects. This test is based on the notion that a significantly improved fit by the generalized model would be meaningful only if accompanied by consistently positively valued estimates of δ. Because the null hypothesis of $\delta = 0$ could not be rejected for any of the experimental conditions presented in this chapter, precise assessment of the relative adequacy of the generalized model was unnecessary. In the third and final step, the overall adequacy of the restricted model was therefore assessed by summing the individual G^2 values and their associated degrees of freedom for subjects as a group.

REFERENCES

Allport, D. A., Styles, E. A., and Hsieh, S. (1994). Shifting attentional set: Exploring the dynamic control of tasks. In C. Umiltà and M. Moscovitch (Eds.), *Attention and Performance XV*, pp. 421–452. Cambridge, MA: MIT Press.

Anderson, J. R., Reder, L. M., and Lebiere, C. (1996). Working memory: Activation limitations on retrieval. *Cognitive Psychology, 30*, 221–256.

Carpenter, P. A., Just, M. A., and Shell, P. (1990). What one intelligence test measures: A theoretical account of the processing in the Raven Progressive Matrices Test. *Psychological Review, 97*, 404–431.

Cools, R. (1998). Frontal deficits and symptoms in schizophrenia. Master's thesis, University of Groningen.

De Jong, R., Berendsen, E., and Cools, R. (1999). Goal neglect and inhibitory limitations: Dissociable causes of interference effects in conflict situations. *Acta Psychologica, 101*, 379–394.

De Jong, R., Emans, B., Eenshuistra, R., and Wagenmakers, E-J. (Forthcoming). Structural and strategical determinants of intentional task control.

De Jong, R., Schellekens, J., and Meyman, T. F. (In preparation). Ansatz, Aufgabe, and Einstellung in task switching.

Duncan, J., Emslie, H., Williams, P., Johnson, R., and Freer, C. (1996). Intelligence and the frontal lobe: The organization of goal-directed behavior. *Cognitive Psychology, 30*, 257–303.

Goschke, T., and Kuhl, J. (1993). Representation of intentions: Persisting activation in memory. *Journal of Experimental Psychology: Learning, Memory, and Cognition, 19*, 1211–1226.

Los, S. A. (1996). On the origin of mixing costs: Exploring information processing in pure and mixed blocks of trials. *Acta Psychologica, 94*, 145–188.

Mantyla, T. (1996). Activating actions and interrupting intentions: Mechanisms of retrieval sensitization in prospective memory. In M. A. Brandimonte, G. O. Einstein, and M. A. McDaniel (Eds.), *Prospective memory: Theory and applications*, pp. 241–275. Hillsdale, NJ: Erlbaum.

Meiran, N. (1996). Reconfiguration of processing mode prior to task performance. *Journal of Experimental Psychology: Learning, Memory, and Cognition, 22*, 1423–1442.

Rogers, R. D., and Monsell, S. (1995). Costs of a predictable switch between simple cognitive tasks. *Journal of Experimental Psychology: General, 124*, 207–231.

Simon, H. A. (1994). The bottleneck of attention: Connecting thought with motivation. In D. Spaulding (Ed.), *Integrative views on motivation, cognition, and emotion: The Nebraska Symposium on Motivation*, vol. 41, pp. 1–21. Lincoln, NE: University of Nebraska Press.

Strayer, D. L., and Kramer, A. F. (1994). Strategies and automaticity: 2. Dynamic aspects of strategy adjustment. *Journal of Experimental Psychology: Learning, Memory, and Cognition, 20*, 342–365.

Yaniv, I., and Meyer, D. E. (1987). Activation and metacognition of inaccessible stored information: Potential bases for incubation effects in problem solving. *Journal of Experimental Psychology: Learning, Memory, and Cognition, 13*, 187–205.

Yantis, S., Meyer, D. E., and Smith, J. E. K. (1991). Analyses of multinomial mixture distributions: New tests for stochastic models of cognition and action. *Psychological Bulletin, 110*, 350–374.

16 Reconfiguration of Stimulus Task Sets and Response Task Sets during Task Switching

Nachshon Meiran

ABSTRACT A tentative model of task switching was tested in two experiments. The model accounts for the switching costs observed in previous experiments by attributing them to multivalent task elements, in the present paradigm bivalent stimuli (relevant for both tasks) and bivalent responses (used in both tasks). It assumes that stimulus task sets enable nearly univalent mental representations of bivalent stimuli, and that response task sets enable nearly univalent mental representations of bivalent responses. Results support two novel predictions of the model: (1) the residual switching cost is substantial with bivalent responses, but negligible with univalent responses; and (2) the preparatory cost is substantial when bivalent target stimuli follow bivalent stimuli, but negligible when either the current target stimulus or the previous one is univalent. Hence there is an approximate one-to-one mapping between preparatory cost and reconfiguration of stimulus task set, on the one hand, and between residual switching cost and reconfiguration of response task set, on the other.

Despite its obvious importance to the study of cognitive control, task switching was barely studied until recently. Furthermore, what used to be the dominant experimental paradigm (i.e., Jersild 1927) suffers from serious shortcomings (see Pashler, chap. 12, this volume), limiting the usefulness of most previous results. Although two better-controlled paradigms were developed, the alternating-runs paradigm (Fagot 1994; Rogers 1993; Rogers and Monsell 1995; Stablum et al. 1994) and the cuing paradigm (e.g., De Jong 1995; Meiran 1996; Shaffer 1965; see also Sudevan and Taylor 1987), extensive work with these paradigms is so recent that our understanding of the phenomena remains rudimentary, and models based on them should be regarded as first approximations. The present chapter introduces such a model, which accounts successfully for previous results and two of whose novel predictions were tested in two experiments.

16.1 THE EXPERIMENTAL PARADIGM

Two and sometimes more different tasks were performed over a long series of trials; in most of the experiments, the tasks required locating a target stimulus within a 2×2 grid (figure 16.1). Subjects were instructed to indicate either the vertical position (the *up-down* task) or the horizon-

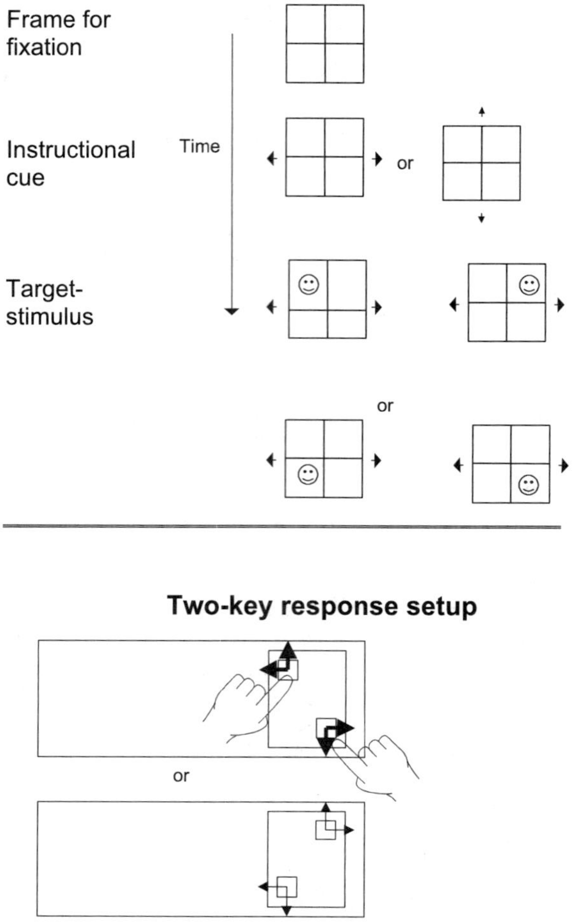

Figure 16.1 Experimental paradigm.

tal position of the target stimulus (the *right-left* task). Two keys were used to indicate the four possible nominal responses. For example, the upper left key indicated either *up* or *left*, depending on the task, while the lower right key indicated *down* or *right*.

This paradigm had several critical features:

1. The tasks were of similar difficulty level. This creates a relatively simple experimental situation by avoiding strategies such as being preferentially prepared for more difficult tasks (e.g., De Jong 1995).

2. The tasks varied randomly from trial to trial. Hence the subjects needed to be instructed on each trial which task to perform, and the effect of switching tasks was estimated by comparing performance on *switch trials*, where the task was different from that on the previous trial, to performance on *nonswitch* trials, where the task was the same.

3. In most instances, the instructional cues were uninformative with respect to which of the two responses would be required on the upcom-

ing trial, which target stimulus would be presented, or when exactly the target onset would occur.

4. With the two-key response setup (figure 16.1), some trials were *congruent*, where the same keypress was appropriate whichever task was being performed (e.g., the correct response to the upper left target stimulus was indicated by pressing the upper left key for both the up-down and the right-left tasks). Other trials were *incongruent*, where different keypresses were appropriate for different tasks (e.g., the correct response to the upper right target stimulus was indicated by pressing the upper left key in the up-down task, where it indicated *up*, and the lower right key in the *right-left* task, where it indicated *right*).

5. The use of instructional cues allowed control over two intervals: the *cue-target interval* (CTI), the time allowed for any preparation for the task; and the *response-cue interval* (RCI), the time during which the subject waited for the instructional cue for the next trial.

Because the trials were ordered randomly, subjects were unlikely to prepare for a switch during the RCI. In fact, the results for switching costs were virtually unaffected by a manipulation in which task repetitions exceeded task switches by a ratio of 2 : 1. The manipulation presumably discouraged attempts to prepare for a task switch during the RCI (Meiran, Chorev, and Sapir forthcoming). A third interval, the *response-target interval* (RTI), is simply the sum of RCI and CTI.

Because of its ability to manipulate CTI and RCI, the cuing paradigm offers an advantage over the alternating-runs paradigm (Rogers and Monsell 1995), where the point in time when task preparation begins is not as tightly controlled.

Previous Results

Components of Task-Switching Cost Probably the most prominent finding in previous studies is that task switching is associated with a reaction time (RT) cost (switch RT > nonswitch RT). The present chapter concerns the trial-by-trial switching costs revealed in the alternating-runs and the cuing paradigms. (For a comparison between nonswitch trials from a task alternation block and pure task blocks, see, for example, Fagot 1994; Kray and Lindenberger forthcoming; Mayr and Liebscher forthcoming.)

Manipulating the CTI and RCI reveals three components of the trial-by-trial task-switching cost. Relevant results from two illustrative experiments (Meiran, Chorev, and Sapir forthcoming) are presented in figure 16.2.

The abscissa in figure 16.2 is the response-target interval, allowing the presentation of the two experiments on the same graph. In our first experiment, the RCI was manipulated, and the CTI was fixed at 117 msec, a

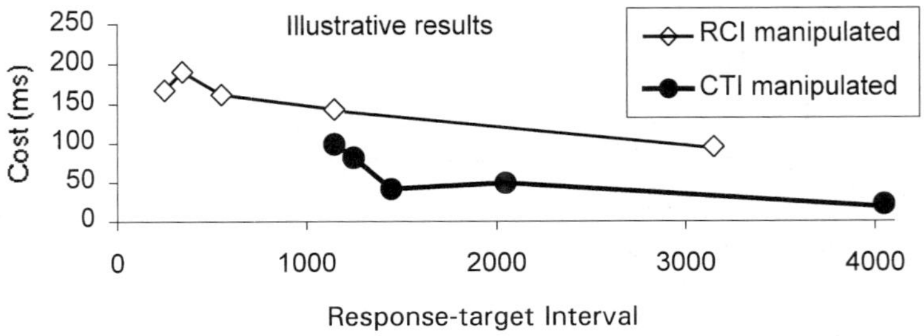

Figure 16.2 Illustrative results from Meiran et al. forthcoming. CTI = cue-target interval; RCI = response-cue interval.

period presumably sufficient for cue encoding but not for task preparation.[1] We found that the task-switching cost first increased and then declined as the RCI increased. The rate of decline was initially fast, but slowed when the RCI exceeded 0.5–1 sec. In our second experiment, the RCI was fixed at 1,016 msec (the time at which the decline in switching cost associated with an increase in the RCI becomes slow), and the CTI was manipulated. The results indicate a sharp decline in the task-switching cost following the presentation of the instructional cue, as the CTI increased. Based on the results of our first experiment, we know that the decline in the cost in our second experiment could not be attributed to the increased remoteness from the previous response, hence must be attributed to processes evoked by the instructional cue. As can be seen in figure 16.2, even when the CTI was relatively long, switching tasks was still associated with a small cost. On the basis of these results and suggestions by Fagot (1994) and Rogers and Monsell (1995), we argued that the task-switching cost has components, of which we identify three: (1) a *waiting component*, related to the effects of the RCI on the cost; (2) a *preparatory component*, related to the effects of CTI on the cost; and (3) a *residual component*, reflecting a portion of the task-switching cost that seems relatively resistant to increases of either interval.

Residual Costs De Jong (chap. 15, this volume) argues that the residual cost reflects a failure to take advantage of the advance information provided in the cue, possibly because of lack of motivation. He proposes that the residual cost results from a mixture of two types of trials: some associated with complete preparation, and others where no preparation took place. Although I believe that motivation may influence the size of the residual cost, it seems that under specific circumstances and without extensive practice, subjects are faced with a genuine limitation in their ability to be fully prepared for task switching. Furthermore, this limitation does not necessarily reflect a lack of motivation to prepare. For example, in previous work (Meiran 1996, exp. 3) two groups of subjects were compared. In the first group, for whom 80% of the trials were incongruent, subjects must have processed the instructional cues to have reached a reasonable error rate. In the second group, for whom 80% of the trials were congruent, subjects could have ignored the instructional cues and still have made only 10% errors. Presumably, the subjects in the first group were more strongly motivated to pay attention to the instructional cues than the subjects in the second group. Nonetheless, the findings indicated a significantly *larger* residual cost in the first ("motivated") condition than in the second ("less motivated") condition—just the opposite to what De Jong's model would have predicted. Furthermore, as explained in "General Discussion" (section 16.3), De Jong's model, at least in its purest form, cannot explain the present results concerning residual costs.

Empirical Dissociations The argument that the trial-by-trial switching costs comprise three components is not merely a summary of the results. It is based on empirical dissociations, suggesting that the components reflect different underlying processes.

Empirical dissociations are indicated when variables selectively affect one component but not another. We found, for example, that the time spent on task reduced the size of the preparatory component of the task-switching cost but affected neither the residual component (Meiran 1996; Meiran, Chorev, and Sapir forthcoming) nor the waiting component of switching cost (Meiran, Chorev, and Sapir forthcoming). Old age (Meiran, Gotler, and Perlman, forthcoming) did not affect the preparatory component of the cost (see also Hartley, Kieley, and Slabach 1990; Kray and Lindenberger, forthcoming; Mayr and Liebscher, forthcoming) but did affect the waiting component. With young and elderly subjects alike, an increase in the RCI led to an initial rise in the switching cost, followed by a gradual decline. On the other hand, the initial rise in the cost among the elderly subjects came later and the rate of the subsequent decline in the cost was slower than among the young. We (Chorev and Meiran 1998) also manipulated phasic alertness by presenting an uninformative highlighted grid before presenting the instructional cue or the target stimulus. In both instances, this alerting manipulation led to faster and more accurate responses, as would be expected from the literature (e.g., Posner and Boies 1971). Interestingly, alertness did not modulate the effect of CTI on the switching cost, although it reduced the residual cost.[2] Finally, in most of the experiments in our lab, congruency affected the residual component of the cost (larger when incongruent), but did not affect the preparatory component of the cost (e.g., Meiran 1996; see also Rogers and Monsell 1995 for a similar effect). The results to be presented in the present chapter constitute additional empirical dissociations.

A Processing Model

Although empirical dissociations strongly suggest that different underlying processes are responsible for the three components of task-switching costs, they do not indicate what these processes might be. The present model describes the underlying processes. I shall outline the model informally (for a formal mathematical description, see Meiran forthcoming). The model has five free parameters, and was fit to explain results from an experiment including 24 conditions, yielding $R = 0.994$ between the predicted mean RT for a given condition and the observed mean RT for that condition. The 24 conditions resulted from orthogonal manipulation of congruency (2), task switch (2), response repetition (2), and CTI (3).

In line with Allport, Styles, and Hsieh 1994 and Rogers and Monsell 1995, our proposed model assumes that task sets have several facets. What is novel about the model, however, is the explicit claim that the various facets of a task set are reconfigured independently of one another, and, under specific constraints, are adopted at specified (and different) points in time. In other words, the model holds that task set reconfiguration cannot be identified with the activation of a unitary algorithm (Dixon 1981) or schema (Norman and Shallice 1986). Moreover, it makes three other critical assumptions.

First, it assumes that task-switching costs arise because the target stimuli, the responses, and possibly other task facets are multivalent with respect to the tasks at hand. In the two experiments to be presented, the target stimuli were bivalent because they had values associated with responses in both tasks. Similarly, the responses were bivalent because they signaled two different properties of the stimulus.

Thus, to execute the correct task, subjects need to recruit task sets, which enable a nearly univalent mental representation of the target stimuli, the responses, or both. *Stimulus task sets* control the representation of the target stimuli, so that the relevant stimulus dimension is emphasized relative to the irrelevant dimension. Similarly, *response task sets* control the representation of the available responses. The suppression of irrelevant information, the activation of relevant information, or both may achieve selective representation.

Second, our model assumes that task-switching costs arise because task sets maintain their configuration until the next trial. This causes interference if the next trial involves a task switch, and hence requires a different configuration of these sets (cf. Allport, Styles, and Hsieh 1994; Allport and Wylie, chap. 2, this volume). Furthermore, if subjects are prewarned of a task switch, some reconfiguration can take place before task execution proper, which results in less interference and smaller task-switching cost.

And third, our model assumes that the stimulus task set can be adopted relatively quickly and efficiently, and hence is usually the one to be reconfigured before task execution proper, that is, during the CTI. In contrast, the response task set is adopted relatively slowly and inflexibly, and hence its reconfiguration is usually completed only *after* responding.

The assumptions listed above lead to an approximate one-to-one mapping between cognitive processes and two of the three components of the task-switching cost. This mapping is the heart of the model. Specifically, it is suggested that the preparatory component of the task-switching cost reflects the reconfiguration of the stimulus task set before task execution proper. In contrast, the residual task-switching cost component is (mainly) attributed to the delayed reconfiguration of the response task set.

Details and Rationale An important characteristic of the model is that response selection is achieved via the interaction of stimulus and response codes.[3] Specifically, response activation is a function of the similarity between the stimulus code and the response code, weighted according to the current status of the stimulus task set and the response task set. To give an example, in the context of the up-down task, an almost fully reconfigured stimulus task set might imply that the vertical dimension is assigned a weight of, say, 0.8, while the horizontal dimension is assigned a weight of 0.2. Consequently, upper right is coded so that the weights for *up* and *right* are 0.8 and 0.2, respectively. The (weighted) stimulus code then interacts with the two response codes, *up-left* and *down-right*. Let us assume, for simplicity, that the response task set is not reconfigured, meaning that neither the vertical dimension nor the horizontal dimension is emphasized in the response task set. This is represented by equal weights (0.5) for the two features in the response code. As a result of the interaction, the stimulus attribute *up* activates the upper left keypress, while the stimulus attribute *right* activates the lower right keypress. Nonetheless, the upper left keypress is more strongly activated (and is thus selected) because *up* is more heavily weighted than *right*.[4]

Congruency effects arise because the irrelevant dimension is represented in the response codes, and because the stimulus task set, although strongly biased, also includes the irrelevant features. This results in the wrong response (e.g., the lower right keypress) being activated, although not selected. The example above also demonstrates why correct responding can be entirely based on the reconfiguration of the stimulus task set.

Another critical assumption is that the response task set is (usually) adjusted after responding. This assumption is based on Hommel's "action-coding theory" (1997), according to which responses are coded (also) in terms of their outcomes. We assume that subjects are more inclined to code their responses (adjust the response task set) when response outcomes are available, that is, after responding.

In the present paradigm, a given response is associated with at least two outcomes. In the first, a key is pressed at a particular position; when this happens, either the vertical dimension or the horizontal dimension is attended, depending on whether the up-down task or the right-left task was executed.

In the second outcome, a key is pressed to express a nominal response. In our experiments, the instructions describe the keypresses as means to express nominal responses. Pressing the key presumably links the motor response to the respective nominal response. Regardless of which outcome is more important, pressing the key results in emphasizing one of its interpretations (e.g., *up*) over the other (e.g., *left*).

In task switching, however, coding responses in terms of their outcomes is counterproductive, and subjects do better if they do not reconfigure the response task set at all. The reason is that the postresponse reconfiguration of the set results in suboptimal response codes in the case of a task switch and, consequently, in a switching cost. One piece of evidence that task set reconfiguration is usually completed after responding is the initial rise in the task-switching cost as a result of increasing the RCI (figure 16.2). The reasoning goes as follows. With sufficiently long RCIs, response codes are determined by the preswitch trial. Hence the response task set is appropriately reconfigured for a task repetition and inappropriately reconfigured for a task switch. If the RCI is extremely short, there is insufficient time to permit response recoding. Consequently, response codes are determined in the trial preceding the preswitch trial. Given the random ordering of tasks, the codes are predicted to be appropriate in 50% of the trials, irrespective of task switching. Hence, with very short RCIs, response recoding does not contribute to the task-switching costs. When the RCI slightly increases, this permits response recoding and increases the overall switching costs.

Accounting for Previous Results

Congruency-Related Effects Switching costs were larger in the incongruent condition than in the congruent condition, indicating that the irrelevant task rule was not completely suppressed, although congruency effects on switching costs did not decrease systematically as preparation time increased (e.g., Fagot 1994; Goschke, chap. 14, this volume;[5] Meiran 1996; Rogers and Monsell 1995). In one exception to this rule, the fifth experiment of Allport, Styles, and Hsieh 1994, preparation did not significantly affect the switching costs.

The aforementioned pattern of results indicates that the reduction in switching costs by task preparation is not usually due to the selection or bias of stimulus-response (S-R) rules, as many researchers seem to believe. If this were the case, task preparation would be accompanied by a reduction in congruency effects in the switch condition. Because this is not usually found, it is suggested that in many circumstances subjects keep all S-R rules active, which is represented by nearly equal weights given to the two attributes (e.g., *up, left*) of each response. Controlling responses is achieved by selectively attending to the relevant stimulus dimension (stimulus set reconfiguration), that is, by controlling the input into S-R rules.[6] Accordingly, the preparatory component of the switching costs reflects the process of selecting the relevant stimulus dimension. Because this precedes response selection, preparation is not reflected in a reduction in congruency effects.

The residual component reflects the delayed, hence counterproductive, incremental change in the response task set and response codes (analogous to reweighting S-R rules). Consequently, in the nonswitch condition, the relevant response codes are primed, whereas in the switch condition, the irrelevant response attributes are primed. Priming the irrelevant response features after a task switch results in an increased congruency effect in that condition.

Interference Due to Response Repetition A surprising finding is that, in the switch condition, response repetition results, not in facilitation, but in interference, slower responses, or a higher error rate (Fagot 1994; Meiran 1996; Rogers and Monsell 1995). This is easily explained if we assume that responses are coded after responding. Consider the following example, where the task is up-down and subjects press the upper left key. As a result of the keypress, the code for that response is adjusted, giving more emphasis to task-relevant features (e.g., assigning the weights 0.6 and 0.4 to the features *up* and *left*, respectively). However, because the lower right key was not pressed, its code is either adjusted more moderately (e.g., 0.55 and 0.45) or not adjusted at all. After switching to the right-left task, pressing the upper left key again would be more difficult than pressing the lower right key. This is because *left* is more strongly deemphasized in the response code (0.4 in the example) than *right* (0.45 or 0.5). Rogers and Monsell (1995, 226) offered several explanations for the effect, one of which is quite similar to the present suggestions.

In summary, the model suggests that, in the present paradigm at least, there is an approximate one-to-one mapping between the task set facet (stimulus or response) and the two components of the task-switching cost. Although the model accounts successfully for basic findings, as shown in the several examples given above,[7] like other models, it should be judged mainly by its ability to generate novel and nontrivial predictions.

Novel Predictions

The assumptions regarding approximate one-to-one mapping between switching cost components and the facets of the task set lead to three straightforward predictions:

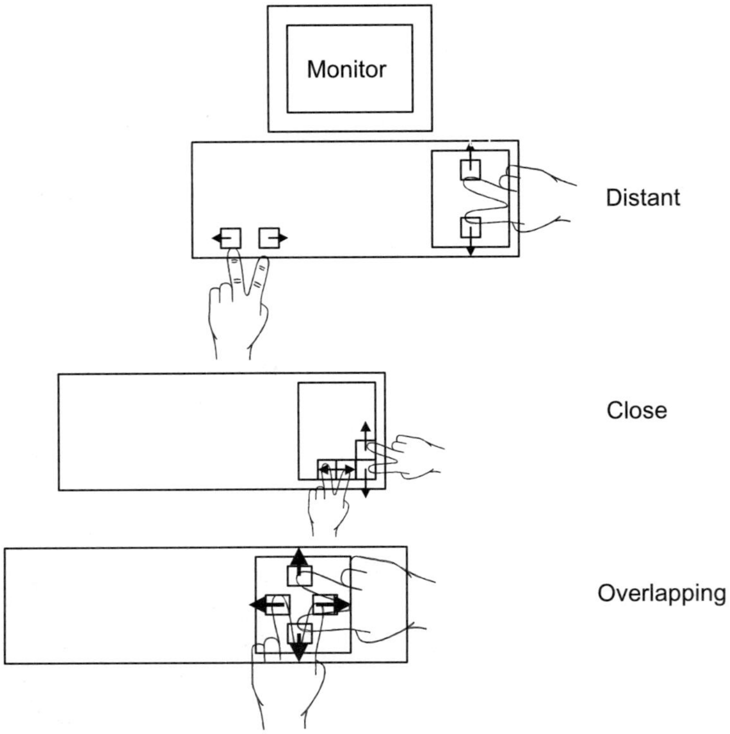

Figure 16.3 Response setups for experiment 1.

1. When the target stimuli are bivalent, but the responses are univalent, the preparatory component of the trial-by-trial cost will be present, whereas the residual task-switching cost will be absent or nearly absent.

2. When the responses are bivalent but the target stimuli are univalent, the residual cost will be present, whereas the preparatory cost will be absent or nearly absent.

3. When both the target stimuli and the responses are univalent there will be no trial-by-trial task-switching cost at all.

Prediction 3 was not tested because it is not unique to the present model.

16.2 EXPERIMENT 1: BIVALENT TARGET STIMULI AND UNIVALENT RESPONSES

The target stimuli were bivalent (figure 16.1), and several response setups were compared. In the standard two-key setup (figure 16.1), the responses were bivalent, as explained above, and both a preparatory switch component and a residual component were predicted for this condition. The two-key setup was compared to three different orthogonal four-key setups: distant, close, and overlapping (figure 16.3), in which the responses were univalent. The prediction was that the task-switching cost in these setups would be eliminated or nearly eliminated by preparation (long CTI), in other words, only the preparatory component would be found, but the residual component would be negligible.

On the basis of previous experiments (e.g., Moulden et al. 1998), it was already known that the residual task-switching cost is abolished in the four-key setup, but there were several problems associated with the interpretation of the results. First, only the distant four-key setup was used, and RT was much faster than in the standard two-key setup. This leaves open the possibility that general speeding led to the reduction of all experimental effects, including the task-switching cost. Second, the two-key setup and the four-key setup were compared across experiments.

The three orthogonal four-key setups differed from one another with respect to perceptual factors. Three different setups were tried because, based on previous literature (e.g., Reeve et al. 1992) it was predicted that proximity and overlap would slow responses and produce average RTs comparable to those in the two-key setup. This, of course, is not the only difference between these setups, which differ in motor aspects as well. The crucial prediction was that, despite all these differences, the three four-key setups would yield similar patterns of switching costs.

Subjects

Twenty-four undergraduate subjects from Ben-Gurion University and the affiliated Achva College participated in this experiment as part of a course requirement. Six subjects were assigned to each group according to order of entry into the experiment.

Apparatus and Stimuli

All testing was performed in front of an IBM PC clone with a 14-inch monitor. The stimuli were drawn in white on black and included a 2×2 grid that subtended approximately 3.4 degrees (width) $\times$ 2.9 degrees (height). The target stimulus subtended approximately 0.3 degree (width) $\times$ 0.5 degree (height). The arrowheads subtended approximately 0.3×0.3 degree, and were positioned 0.7 degree from the end of the grid.

Procedure

After the instructions, there was a short warm up block (20 trials) followed by five identical blocks of 96 trials, all in a 1-hour session. The subjects were encouraged to stretch a little between blocks. The keyboard, used to collect responses, was positioned so that its center (distant four-key setup group) or its keypad (for the remaining groups) was aligned with the center of the computer monitor. Each trial consisted of (1) the

Table 16.1 Mean Reaction Times and Error Rates for Experiment 1

				Four-keys (univalent)								
	Two-keys (bivalent)			Distant			Close			Overlapping		
CTI (msec)	S	NS	Cost	S	NS	Cost	S	NS	Cost	S	NS	Cost
166 Reaction time (msec)	1,028	750	278	735	622	113	999	861	138	808	719	89
Error rate	0.081	0.035	0.046	0.067	0.012	0.055	0.040	0.003	0.037	0.085	0.018	0.067
366 Reaction time (msec)	879	748	131	565	538	27	822	747	75	688	656	32
Error rate	0.060	0.036	0.024	0.009	0.006	0.003	0.018	0.000	0.018	0.035	0.009	0.026
716 Reaction time (msec)	846	776	70	532	522	10	759	756	110	646	608	38
Error rate	0.052	0.039	0.013	0.009	0.000	0.009	0.011	0.003	0.008	0.012	0.003	0.009
1616 Reaction time (msec)	827	716	111	520	532	–12	730	727	3	604	628	–24
Error rate	0.054	0.031	0.023	0.006	0.009	–0.003	0.014	0.008	0.006	0.011	0.003	–0.008

Note: CTI = cue-target interval; S = switch; NS = nonswitch.

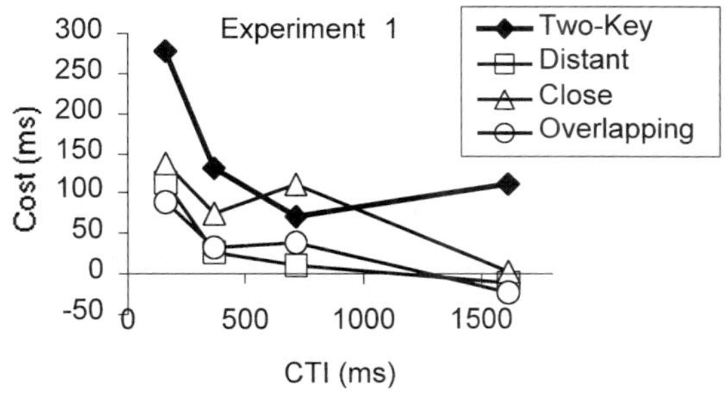

Figure 16.4 Task-switching costs: Experiment 1. CTI = cue-target interval.

presentation of an empty grid for a constant RCI of 1,532 msec; (2) the presentation of an instructional cue for a variable CTI (166, 366, 716, 1,616 msec); and (3) the presentation of the target stimulus along with the instructional cue until the response. A 50 msec 400 Hz beep signaled an error.

Results and Discussion

In the two-key setup, the mean RT was 744 msec, which compares to 555, 763, and 642 msec in the distant, close, and overlapping four-key setups, respectively (see table 16.1 and figure 16.4). The fact that mean RT was similar in the two-key setup and in one of the four-key setups permits a safer interpretation of the results concerning switching costs.

Responses preceded by errors or by RTs longer than 3 sec were discarded. Responses that were either inaccurate or associated with an excessively long RT (3 sec) were included in the error score, but not in the estimate of mean RT. Each cell was represented by the mean, after trimming values exceeding 2 standard deviations (SDs) from the untrimmed mean. Because space is limited and errors were relatively rare, formal statistical analyses of errors are not reported. However, as can be seen in the tables, the critical RT effects do not reflect a speed-accuracy trade-off. The alpha level was 0.05.

Because the assignment of trials to conditions was partly random, the number of analyzable responses per condition was not identical and ranged from 47 to 59. Two focused comparisons were conducted; mean square errors were taken from an analysis of variance, with CTI, task switch, and group as the independent variables. In one analysis, the two-key setup was compared to the three groups with the orthogonal four-key setup. The group main effect was insignificant, while the interaction of CTI and Group just missed significance: $F(3, 60) = 2.74$, $p = 0.051$; and the triple interaction was significant: $F(3, 60) = 2.85$. On the other hand, there was a significant main effect of task switch: $F(1, 20) = 24.40$; a significant interaction between CTI and task switch: $F(3, 60) = 24.56$; and most important, a significant interaction between group and task switch:

$F(1, 20) = 11.20$. The simple interaction of group and task switch at the longest cue-target interval was also significant: $F(1, 20) = 21.46$, reflecting a significant residual cost in the two-key setup: $F(1, 20) = 9.50$, compared to a residual cost that was negative in two of the four-key setups, and was 3 msec in the third group. The significant triple interaction indicates that the group differences in the task-switching cost were somewhat larger in the short CTI compared with the long CTI. In the second analysis, where the three four-key setups were compared to one another, the main effect of group was significant: $F(1, 20) = 5.43$; but none of the interactions involving group approached significance, $F < 1$.

One could argue that the two-key setup yielded larger costs only because it involved an incongruent condition and task-switching costs are known to be larger in that condition. This was not the case, however, because the residual costs (at the longest CTI) were 143 and 93 msec for the incongruent and congruent conditions, respectively.[8] Namely, the costs in the congruent condition were considerably larger than the costs in any of the four-key setups. An alternative explanation is based on Monsell et al. 1998, which showed that switching costs were larger when the responses were incompatible with the stimuli (e.g., pressing a key in response to the words "left" and "right") as compared to a compatible setting (reading the words). One might argue that this is the reason why residual costs were larger in the two-key setup, where the incongruent condition was also incompatible in that the relative position of the target stimulus (e.g., upper *right*) was opposite to the relative position of the response along one dimension (e.g., upper *left*). However, the congruent condition in the two-key setup was highly compatible because the response key occupied the same relative position as the target stimulus. The four-key setups were associated with an intermediate level of S-R compatibility because the response key never occupied the same relative position as the target stimulus, although it was never opposite to it. Nonetheless, the residual cost in the congruent condition (two-key setup) was much larger than in the less compatible four-key setups. Hence compatibility cannot explain the differences in the residual costs in the present case.

The results of experiment 1 generally support the predictions by showing that when the responses were univalent, the residual task-switching cost was eliminated. The small triple interaction may indicate that while most of the preparation applied to the stimulus task set (common to all four response setups), a little preparation also applied to the response task set. The findings therefore indicate an empirical dissociation, namely, response valence affects residual cost, although its effect on the preparatory cost was much smaller. The findings also support the predicted (approximate) one-to-one mapping between response task set reconfiguration and the residual component of the task-switching cost.

Figure 16.5 Univalent target stimuli.

16.3 EXPERIMENT 2: UNIVALENT TARGET STIMULI AND BIVALENT RESPONSES

The responses were bivalent (the two-key setup was used), but half of the target stimuli were univalent and could be classified only in one manner (figure 16.5). There were two reasons for this manipulation. First, this condition constitutes a replication of the standard conditions using the two-key setup of experiment 1 (figure 16.1). Second, it was hoped that intermixing bivalent and univalent target stimuli in an unpredictable order would encourage subjects to maintain the same strategy they used when both the stimuli and the responses were bivalent. Including only, or too many, univalent target stimuli could potentially lower subjects' motivation to reconfigure the stimulus task set during the CTI because that set would often not be needed. Furthermore, under these conditions, it would make more sense to change strategy and prepare for a task by reconfiguring the response task set during the CTI. This was probably the case in De Jong 1995 and in Rogers and Monsell 1995, exp. 4.

Rogers and Monsell (1995, exp. 3) mixed univalent and bivalent target stimuli. Nonetheless, they did not include the status of the target (univalent, bivalent) in the previous trial as a variable in their analyses. Including that variable allows one to distinguish between two scenarios, as elaborated below. The subjects were assumed to reconfigure the stimulus task set on every trial because, when the instructional cue was presented, they were unable to predict whether the upcoming target stimulus would be univalent or bivalent. On the other hand, using the stimulus set for responding depended on the nature of the target stimulus as univalent or bivalent. The reason is that correct responding depended on the stimulus task set only when the target stimulus was bivalent, where the set enabled univalent representation.

One possible scenario is that the stimulus task set remains roughly unchanged after being reconfigured. In that case, it would not matter if the previous trial involved a bivalent or a univalent target stimulus because in both cases the stimulus task set was reconfigured. This scenario predicts that the presence of a preparatory cost component depends only on the status of the current target stimulus, present when bivalent and absent when univalent. The reason is that the reconfiguration of the stimulus task set may be skipped once the subject realizes that the target stimulus is univalent.

A second possible scenario is that although the stimulus task set is reconfigured during the CTI, if not used (that is, with univalent target stimuli), it returns quickly to its previous or to a neutral state. In either case, this would result in zero preparatory cost on the following trial. Hence this scenario predicts that the preparatory cost would be missing if the previous target stimulus, the current target stimulus, or both were univalent. The preparatory cost would be present only when both trials involved bivalent target stimuli.

Subjects

Twenty students from the Negev College, affiliated with Ben-Gurion University, served as subjects in this experiment. Half were assigned to each of the two possible two-key combinations.

Stimuli

The stimuli were the same as in experiment 1, except for the inclusion of the 4 univalent target stimuli that were identical in size to the target stimuli used in experiment 1.

Procedure

The only changes from experiment 1 were that all the subjects used the two-key setup (figure 16.1) for responses. The CTIs were 166, 516, and 2,516 msec. When the target stimulus was univalent, it was always one that matched the task. For example, when the task was up-down, the target stimuli were either *up* or *down*, but neither *right* nor *left*. The task switch condition, target, target type (bivalent, univalent), and CTI were randomly selected with equal probabilities in each trial. The warm-up block included 25 trials, and each of the 5 experimental blocks included 96 trials.

Results and Discussion

There were between 18 and 20 observations per condition (see table 16.2 and figure 16.6). The triple interaction between target type combination

Table 16.2 Mean Reaction Times and Error Rates for Experiment 2

	Previous target bivalent						Previous target univalent					
	Bivalent target			Univalent target			Bivalent target			Univalent target		
CTI (msec)	S	NS	Cost	S	NS	Cost	S	NS	Cost	S	NS	Cost
166 Reaction time (msec)	971	756	215	744	660	84	906	844	62	720	642	78
Error rate	0.053	0.008	0.045	0	0	0	0.024	0.014	0.010	0.002	0.005	–0.003
516 Reaction time (msec)	827	688	139	673	624	49	812	774	38	660	604	56
Error rate	0.033	0.015	0.018	0.007	0	0.007	0.013	0.010	0.003	0.003	0	0.003
2516 Reaction time (msec)	749	686	63	667	626	41	742	717	25	626	609	17
Error rate	0.012	0.006	0.006	0.005	0	0.005	0.017	0.010	0.007	0	0	0

Note: CTI = cue-target interval; S = switch; NS = nonswitch.

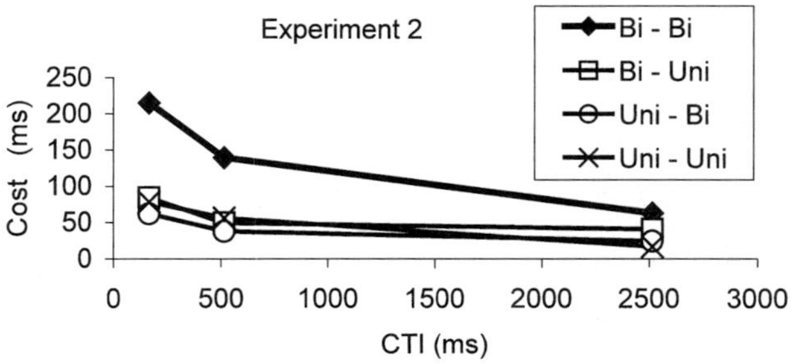

Figure 16.6 Task-switching costs in experiment 2. Bi = bivalent; Uni = univalent; CTI = cue-target interval.

(bivalent-bivalent, bivalent-univalent, univalent-univalent, univalent-bivalent), CTI, and task switch was significant: $F(6, 116) = 2.25$. It resulted mainly from the difference between the bivalent-bivalent combination and the remaining three conditions, $F(2, 38) = 4.44$; and not from the differences among the remaining three conditions: $F < 1$. An increase in CTI was associated with a significant reduction in the task-switching cost in the bivalent-bivalent condition: $F(2, 38) = 12.67$. Nonetheless, there was a small preparatory component even in the remaining conditions, seen in the fact that an increase in the CTI led to a reduction in the task-switching cost even when one or both of the targets were univalent: $F(2, 38) = 4.79$. It was much smaller, however, than that obtained in the bivalent-bivalent condition because task preparation reduced the cost by only 27–61 msec, as compared to 152 msec.[9] It is important to note that there was a significant residual cost even when either the previous or the current trial involved a univalent target stimulus, as seen in the effects of task switch in the longest CTI: $F(1, 19) = 5.90$. Thus including any univalent task element is insufficient to eliminate the residual costs in the present paradigm. The univalent task element must be the responses.

The results may be summarized as follows. When either the current or the previous target stimulus, or both, were univalent, the task-switching cost was relatively small, and barely influenced by the CTI. In other words, the cost comprised mainly the residual component. In contrast, when both the current target stimulus and the preceding target stimulus were ambivalent, the task-switching cost was larger, mainly in the short CTIs. In other words, both the residual component and the preparatory component were present in that condition. In terms of the model, if a stimulus task set was used in the preceding trial, and not merely reconfigured, this made it difficult to adopt a new stimulus task set. In that respect, the current findings support the suggestion of Allport, Styles, and Hsieh (1994) that the task-switching cost results from interference from the task set in the previous trial.

The results of experiment 2 also indicate an empirical dissociation. Namely, the combination of current and previous target valence affected the preparatory component more strongly than they affected the residual component. As in experiment 1, there was an indication that the response task set is slightly prepared during the CTI. The reasoning is that reconfiguring the stimulus task set was unlikely to help when the target was univalent. Finally, the results may also explain why Rogers and Monsell (1995, exp. 3) did not find that stimulus valence affected the preparatory cost: the valence of the previous target stimulus was not included in the analyses. A relevant comparison is between their experiments 3 and 4. In experiment 3, univalent and bivalent stimuli were mixed, and the results indicated that preparation reduced the cost from 207 to 115 msec (a preparatory component of 92 msec). This is probably an underestimation because the experiment included trials in which either the current or previous target stimulus was univalent. In comparison, when there were only univalent target stimuli (experiment 4), the reduction was from 67 to 42 msec (25 msec difference), which is probably an overestimate because having nothing else to prepare, the subjects probably reconfigured the response set, which explains the modest decline in the switching costs. In other words, Rogers and Monsell's results also indicate that target stimulus valence affects the preparatory component of the switching costs more strongly than it affects the residual component.

An unexpected finding was that responses in the nonswitch condition were slower when the current target was bivalent, especially when the previous target was also bivalent (table 16.2). This may have reflected the fact that the bivalent condition included incongruent trials. Although one could argue that this slowing of responses in the bivalent-bivalent condition caused an increase in switching costs, even if switching costs are represented as proportional increases in RT relative to the nonswitch condition, the picture remains essentially unchanged. In the bivalent-bivalent condition, preparation reduced the proportional switching cost by 19.2% (from 28.4% to 9.2%). This value compares to a reduction of 6.2% (12.7% to 6.5%) in the bivalent-univalent condition, 3.8% (7.3% to 3.5%) in the univalent-bivalent condition, and 9.3% (12.1% to 2.8%) in the univalent-univalent condition.

General Discussion

Our proposed model serves as a reasonable first approximation in describing subjects' performance in a particular task-switching paradigm. Like other models, the present model should be judged, not only by its ability to account for previous findings, but more important, by its ability to generate new, nontrivial, and testable predictions. Although alternative explanations may apply to the present results, to the best of my knowledge, none of the existing models could predict these results. Several relevant issues are discussed below.

De Jong's Model According to De Jong's model (chap. 15, this volume), residual costs represent lack of motivation to prepare. When preparation time is short (short CTI), there should thus be no difference between "motivated" and "unmotivated" trials. The difference between the two types of trials should be evident given sufficient preparation time. One may argue that the near-zero residual costs in experiment 1 were due to a higher motivation to prepare with four-key setups. This explanation, besides being ad hoc, leads to the prediction that the switching costs in the two-key and the four-key setups would be similar when CTI was very short, so that the motivation to prepare did not yet affect the switching costs. The results are clearly inconsistent with that prediction, showing a *larger* difference between the setups in the shortest CTI compared to the longest CTI. (A similar argument applies to the results of experiment 2.) In summary, lack of motivation to prepare is not the only reason why residual costs exist.

Applicability to Other Switching Paradigms At the heart of the model is the assumption that task sets are adopted, and hence cause interference, because several facets of the task are multivalent with respect to the tasks at hand. In the present paradigm, both the target stimuli and the responses were bivalent. Certainly, additional task facets may be multivalent and contribute to the task-switching costs in other paradigms. Furthermore, the nature of the approximate one-to-one mapping between task set facets and the two components of the task-switching cost may be specific to the present tasks and the very explicit instructional cues that were used. This may have made it easier to reconfigure the stimulus task set than the response task set. Consequently, the subjects adopted a strategy of preparing by reconfiguring the stimulus task set.

Despite the peculiar aspects, two general principles emerge. First, the task-switching cost should not be treated as a single phenomenon. Within a given paradigm, the components of the switching cost reflect different underlying processes. This general principle allows for some variability. For example, in one paradigm, subjects might prepare by reconfiguring the stimulus task set, whereas, in another paradigm, they might prepare by reconfiguring the response task set, or a rule task set. Thus the processes underlying the preparatory component would not be the same across the two paradigms.

Following the models of other researchers, our model holds that the trial-by-trial switching costs resulted from the multivalence of task elements. The second general principle to emerge is that separate task sets are required to deal with each multivalent task element, and that these task sets need not be adopted at the same time. Using valence-related manipulations, one can determine that task set facet is reconfigured and when. A valence-related manipulation that affects the preparatory switching cost component indicates that the related task set is reconfigured during the CTI. For example, in experiment 2, stimulus valence affected the preparatory cost, indicating that the stimulus task set was reconfigured during the CTI. In contrast, valence-related manipulations that affect the residual cost indicate that the respective task set is reconfigured sometime after target stimulus presentation. For example, in experiment 1, response valence affected the residual cost, which supported the present claim regarding the relatively delayed reconfiguration of the response task set.

NOTES

This research was supported by a grant from the Israel Science Foundation. I wish to thank Meirav Levi and Eldad Weisbach for running the experiments.

1. This presumption can be defended on the basis of a study which employed high-density event-related potential (ERP) recording (Moulden et al. 1998). In that study, the first (cue-locked) switch related component was revealed 200 msec after cue presentation, and the locus of its generator was bioccipital. Based on the commonly accepted assumption that the occipital lobes are involved in encoding visual information, this result suggests that about 200 msec are required to encode the present type of instructional cues.

2. This finding may be specific to the present paradigm. Using a different method to alert their subjects, Rogers and Monsell (1995, exp. 5) did not find that alertness reduced the cost, although the effect of the alerting stimulus on RT was very weak in that study (10–21 msec).

3. My choice of the term *response codes* instead of *S-R rules* allows a natural link to selective attention theories and theories of response coding (Hommel 1997); moreover, it fits well into current cognitive theorizing. Specifically, most cognitive psychologists would agree that S-R rules do not relate physical stimuli to physical responses, instead, they relate stimulus representations to response representations. They would also agree that mental representations are influenced by selective attention.

4. The present formulation may be extended to situations in which a translation must apply to the stimulus code. For example, if subjects switch between odd versus even judgments and larger versus smaller than 5 judgments, the code of a given target digit (e.g., 7) needs to be first translated to either "high" or "odd". This requires a translation phase between stimulus encoding and response activation. If we assume only two responses (e.g., Sudevan and Taylor 1987), the responses may be coded as *high-odd*, and *low-even*, with one set of attributes (e.g., *high*, *low*) being emphasized relative to the other set of attributes (e.g., *odd*, *even*). Once the digit "7" is coded as *high*, this would result in the activation of the response that contains *high* in its code.

5. I am referring here to Goschke's comparison of two conditions. In the first condition, RCI was short and CTI long (short-long); in the second, RCI was long and CTI short (long-short). These conditions are equal with respect to the time allowed for the dissipation of the previous task set, and differ with respect to task preparation only (Meiran 1996). In Goschke's experiment, congruency effects declined with task preparation (from short-long to long-short), but more or less equally in switch trials and nonswitch trials.

6. This partly explains the advantage of pure task blocks (where only one S-R rule is active) over task repetitions within a task alternation block.

7. In the model, it is possible to eliminate residual costs by adopting specific strategies, although subjects rarely employ these strategies. One such strategy is total biasing of the stimulus task set (assigning a weight of 1 to the relevant dimension, and a weight of 0 to the

irrelevant dimension). Another strategy is learning not to reconfigure the response set after responding. In neither case would the irrelevant stimulus dimension activate the wrong response. The model also predicts for these strategies that the two-way interaction between congruency and task switch would be eliminated. The most common strategy, and the one on which the predictions were based, is to sufficiently bias the stimulus task set before selecting the response. A fuller description of the strategy may be found in Meiran forthcoming.

8. As one may notice, the average, 118 msec, is not identical to the residual cost reported in table 16.1, 111 msec. This is because values exceeding 2 SDs were trimmed, and including congruency as a variable changed cell means and SDs. When untrimmed arithmetic means were used, the pooled residual cost in the two-key setup was 105 msec, which reflected a cost of 113 msec in the incongruent condition and 97 msec in the congruent condition. These values were compared with −10 , 8, and −23 msec (based on arithmetic means) in the distant, close, and overlapping setups, respectively.

9. There is no agreed-upon method to compute the reduction in the costs by preparation. I tried two methods: the first based on raw costs (figure 16.6); the second based on the proportional reduction in raw cost, that is, switch RT minus nonswitch RT in milliseconds. The reduction in the bivalent-bivalent condition was 71% (raw cost was reduced from 215 to 63 msec). This value is compared to a reduction of 78% in the univalent-univalent condition, 51% in the bivalent-univalent condition, and 60% in the univalent-bivalent condition. Although the last analysis may suggest that the efficiency of preparation does not depend on target stimulus valence, if the same logic were applied to the results of experiment 1, the conclusion would be that using univalent responses resulted in complete or close to complete reduction in switching cost (~ 100 %). Thus the present results indicate a dissociation of response valence and stimulus valence, regardless of the computational method. Specifically, univalent responses resulted in improving the proportional reduction in switching costs (experiment 1). On the other hand, univalent responses did not result in such improvement (experiment 2). The reasons to prefer the computational method used is that it is the one most commonly used. Moreover, the emergent picture fits the predictions of a model successfully fit to RT results (Meiran forthcoming). The last statement holds, of course, as long as there is no alternative model that can account for the results concerning proportional effects on switching costs.

REFERENCES

Allport, D. A., Styles, E. A., and Hsieh, S. (1994). Shifting intentional set: Exploring the dynamic control of tasks. In C. Umiltà and M. Moscovitch (Eds.), *Attention and Performance XV*, pp. 421–452. Hillsdale, NJ: Erlbaum.

Chorev, Z., and Meiran, N. (1998). Phasic arousal affects the residual task switching cost. Poster presented at the Tenth Congress of the European Society for Cognitive Psychology, Jerusalem, September.

De Jong, R. (1995). Strategical determinants of compatibility effects with task uncertainty. *Acta Psychologica, 88,* 187–207.

Dixon, P. (1981). Algorithms and selective attention. *Memory and Cognition, 9,* 177–184.

Fagot, C. (1994). Chronometric investigations of task switching. Ph.D. diss., University of California, San Diego.

Hartley, A. A., Kieley, J. M., and Slabach, E. H. (1990). Age differences and similarities in the effects of cues and prompts. *Journal of Experimental Psychology: Human Perception and Performance, 16,* 523–537.

Hommel, B. (1997). Toward an action-concept model of stimulus-response compatibility. In Hommel and W. Prinz (Eds.), *Theoretical issues in stimulus-response compatibility,* pp. 281–320. Amsterdam: Elsevier.

Jersild, A. T. (1927). Mental set and shift. *Archives of Psychology,* no. 89.

Kray, J., and Lindenberger, U. (Forthcoming). Adult age differences in task-switching. *Psychology and Aging.*

Mayr, U., and Liebscher, T. (Forthcoming). Is task-set disengagement an active process? Evidence from cognitive aging.

Meiran, N. (1996). Reconfiguration of processing mode prior to task performance. *Journal of Experimental Psychology: Learning, Memory, and Cognition, 22,* 1423–1442.

Meiran, N. (Forthcoming). Modeling cognitive control in task switching. *Psychological Research.*

Meiran, N., Chorev, Z., and Sapir, A. (Forthcoming). Component processes in task switching. *Cognitive Psychology.*

Meiran, N., Gotler, A., and Perlman, A. (Forthcoming). Old age is associated with a pattern of relatively intact and relatively impaired task set switching abilities.

Monsell, S., Azuma, R., Eimer, M., Le Pelley, M., and Strafford, S. (1998). Does a prepared task switch require an extra (control) process between stimulus onset and response selection? Poster presented at the Eighteenth International Symposium on Attention and Performance, The Great Park, Windsor, England, July.

Moulden, D. J. A., Picton, T. W., Meiran, N., Stuss, D. T., Riera, J. J., and Valdes-Sosa, P. (1998) Event-related potentials when switching attention between task sets. *Brain and Cognition, 37,* 186–190.

Norman, D. A., and Shallice, T. (1986). Attention to action: Willed and automatic control of behavior. In R. J. Davidson, G. E. Schwartz, and D. Shapiro (Eds.), *Consciousness and self-regulation, Vol 4,* pp. 1–18. New York: Pellenum.

Posner, M. I., and Boies, S. J. (1971). Components of attention. *Psychological Review, 78,* 391–408.

Reeve, T. G., Proctor, R. W., Weeks, D. J., and Dornier, L. (1992). Salience of stimulus and responses features in choice reaction tasks. *Perception and Psychophysics, 52,* 453–460.

Rogers, R. D. (1993). The costs of switching between cognitive tasks: A performance analysis. Ph.D. diss., University of Cambridge.

Rogers, R. D., and Monsell, S. (1995). The cost of a predictable switch between simple cognitive tasks. *Journal of Experimental Psychology: General, 124,* 207–231.

Shaffer, L. H. (1965). Choice reaction with variable S-R mapping. *Journal of Experimental Psychology, 70,* 284–288.

Stablum, F., Leonardi, G., Mazzoldi, M., Umiltà, C., and Morra, S. (1994). Attention and control deficits following closed head injury. *Cortex, 30,* 603–618.

Sudevan, P., and Taylor, D. A. (1987). The cueing and priming of cognitive operations. *Journal of Experimental Psychology: Human Perception and Performance, 13,* 89–103.

17 Task Switching in a Callosotomy Patient and in Normal Participants: Evidence for Response-Related Sources of Interference

Richard B. Ivry and Eliot Hazeltine

ABSTRACT We examined multitask coordination in neurologically healthy subjects and in a callosotomy patient. Subjects in two new experiments responded to two successive stimuli separated by a variable stimulus onset asynchrony (SOA), with the left hand to the first stimulus and with the right hand to the second. The task-relevant dimension for the two stimuli was the same for both hands or required a change in task set. For all subjects, reaction time to the second stimulus was inversely related to SOA, an effect referred to as the "psychological refractory period" (PRP). For control subjects, the effect of switching set was either additive or overadditive with SOA, whereas, for the callosotomy patient, no difference was observed between the same- and different-set conditions, even when the stimuli were presented along the vertical meridian and presumably available to both hemispheres. These results indicate that the primary locus of interference associated with task switching arises at processing stages associated with response preparation, selection, or initiation. Unlike the control subjects, the split-brain patient was able to maintain separate stimulus-response mappings in the two hemispheres.

17.1 COORDINATION OF BEHAVIOR AFTER CALLOSOTOMY

Callosotomy (split-brain) patients provide a unique opportunity for exploring the organization of our cognitive architecture (Sperry 1982; Gazzaniga 1995). Most of the split-brain work has addressed issues related to hemispheric specialization, focusing on identifying the basic capabilities of the two cerebral hemispheres in the areas of perception, memory, and language, although researchers have also studied these patients to learn how information processing is integrated and coordinated between the hemispheres. In general, perceptual studies have demonstrated that, while each hemisphere in isolation is capable of deriving perceptually and semantically rich representations, the integration of this information is dependent on callosal fibers (see Corballis 1995). Attention studies have provided even more impressive evidence of the general competence of each hemisphere (Luck et al. 1994). For example, Holtzman and Gazzaniga (1985) found that split-brain patients were able to monitor two lateralized stimulus sequences without interference, whereas control subjects showed extensive cross talk between the two sequences, suggesting an inability to segregate the two sources of information.

Results such as these might suggest that the callosotomy operation functionally splits an individual into two separate halves. And yet the very success of this operation argues against such an extreme conclusion. Indeed, from simple observation, it is impossible to discern any indication that the hemispheres are operating in isolation of one another (Bogen 1993). The actions of split-brain patients are quite coherent: they move about, talk, and use tools like neurologically intact individuals.

While the actions of split-brain patients continue to manifest at least some integration (Sergent 1987), this does not mean that the selected actions following callosotomy are the result of integrated processing between the two hemispheres (see Kingstone and Gazzaniga 1995). Rather, each hemisphere may independently control the actions of the contralateral limbs. For example, in a bimanual drawing task, split-brain patients performed the same when the component movements entailed orthogonal spatial trajectories as when the movements entailed parallel trajectories (Franz et al. 1996), whereas normal subjects, showed severe interference (see also Franz et al. 1991). It appears that for tasks such as these, the subjects must generate two spatial plans, one associated with movement of the right hand and a second with movement of the left. In normal subjects, cross talk between these two representations produces interference when the representations entail conflicting spatial trajectories and goals. Rather than reflecting the operation of an integrated control operation, this interference presumably involves communication across the corpus callosum.

These results suggest that separate response plans can be generated and selected in the isolated cerebral hemispheres, although further examination of the patients' performance on these drawing tasks indicates that the initiation of these responses continues to be severely constrained (Franz et al. 1996). In contrast to their spatial uncoupling, the movements of the right and left hands remain tightly coupled in the temporal domain (see also Tuller and Kelso 1989). Thus there is a striking dissociation between two well-documented constraints on bimanual movements. The callosotomy operation leads to spatial uncoupling, yet has minimal effect on temporal coupling.

17.2 DUAL-TASK PERFORMANCE AFTER CALLOSOTOMY

We have recently examined the dissociation of spatial and temporal constraints in a very different context (Ivry et al. 1998). In our dual-task study, subjects made two successive speeded responses to two different stimuli, the psychological refractory period (PRP) paradigm, in which one stimulus always appears first and subjects are instructed to respond as quickly as possible to this event (RT1). The second stimulus appears after a variable stimulus onset asynchrony (SOA), and also requires a speeded response (RT2). Across a wide range of studies, the time required to respond to the second stimulus is longer when the SOA between S1 and S2 is short than when it is long (see reviews by Pashler 1994; chap.

12, this volume; Meyer and Kieras 1997). The inverse relationship between RT2 and SOA has been dubbed the "psychological refractory period" or "PRP effect."

The analysis of the PRP effect has been useful for examining the architecture of human cognition, seeking to determine the limitations in multitask coordination (see Pashler, chap. 12, Jolicoeur, Dell'Acqua, and Crebolder, chap. 13, and Kieras et al., chap. 30, this volume). In Pashler's influential model, perceptual analysis and response execution are assumed to be independent processing stages for the two tasks. The critical limitation in dual-task performance, according to Pashler, is associated with response selection, which cannot occur in parallel for the two tasks. Rather, it is assumed that there is a unitary response selection process that must be accessed successively, first for task 1 and then for task 2. With short SOAs, response selection for task 2 must be delayed until this process is completed for task 1.

Pashler et al. (1994) tested three split-brain patients on a PRP task. The design involved the lateralized presentation of two up-down spatial discrimination tasks, with the onset of the tasks separated by a variable SOA. The first stimulus was presented to the left visual field (right hemisphere), and the subjects indicated the position of this stimulus by pressing one of two keys with the left hand. The second stimulus was presented to the right visual field (left hemisphere) and, correspondingly, was responded to with the right hand. The results convincingly demonstrated a robust PRP effect for all of the callosotomy patients.

Given our evidence that split-brain patients could maintain separate spatial plans in the two hemispheres (Franz et al. 1996), we sought to examine the persistent PRP effect in greater detail (Ivry et al. 1998, exps. 2 and 3). We used the same spatial discrimination tasks as Pashler et al. However, in separate blocks, the consistency between the two S-R mappings was manipulated (e.g., the spatial S-R mapping for the two hands was either symmetric or reversed). Because similar manipulations have been shown to affect response selection processes (McCann and Johnston 1992), we expected that the consistency manipulation would produce additive effects with SOA for RT2 (see also Duncan 1979).

As predicted, the consistency manipulation had a substantial effect on the performance of the control subjects. A PRP effect, additive or overadditive with SOA, was found for both consistent and inconsistent S-R pairings. There was substantial slowing of RT1 in the inconsistent condition, even though subjects were instructed to give priority to this task. The results were strikingly different for patient J.W. (Ivry et al. 1998). While the PRP effect was again present, the consistency manipulation was underadditive with SOA and there was no cost on RT1. That is, the split-brain patient responded as fast to stimulus 1 when the two S-R mappings were inconsistent with one another as when they were consistent.

These results provide further confirmation of spatial uncoupling after callosotomy. The split-brain patient showed no cost attributable to the maintenance of inconsistent spatial S-R mappings in the two hemispheres. Moreover, the patient showed underadditivity between the effects of the S-R mappings and SOA, suggesting that the effect of the consistency manipulation influenced processing in or before the bottleneck. Thus whatever processing limits may persist following callosotomy, they do not appear to be associated with the same limitation on response selection identified in PRP studies with control subjects.

On the other hand, the split-brain patient did show a persistent delay in RT2 at short versus long SOAs, indicating that the two hemispheres were not completely independent. The source of this interference remains unclear, although, given the pattern of underadditivity, it arises at a relatively late stage of processing. One possibility is that the bottleneck for the split-brain patient is associated with a subcortical process associated with response implementation, a process accessed by action commands from the two hemispheres. There is evidence that such a limitation in response implementation also exists for normal participants, but is not typically evident because they bottleneck at an earlier stage of processing (De Jong 1993; Ruthruff, Johnston, and Van Selst forthcoming). Another hypothesis is that the persistent PRP effects reflect a strategy adopted by the split-brain patients to comply with the task instructions to make two successive responses (Meyer and Kieras 1997).

17.3 TASK SWITCHING AFTER CALLOSOTOMY

Many real-world situations require highly flexible behavior. For example, when approaching an exit on the highway, you may note the fuel gauge on your car and start looking for a gasoline station. As you turn off, however, the clamoring of the children, as well as the growls of your stomach may redirect your action toward the fast-food restaurant for a quick lunch. This fluctuation as to the goals of behavior is termed *task switching* (Allport, Styles, and Hsieh 1994; Rogers and Monsell 1995; Spector and Biederman 1976), reflecting the change in the salience of different stimulus properties as well as the viable responses.

In the typical task-switching experiment, subjects are required to switch between two tasks, each involving distinct S-R mappings. For example, for one task, subjects judge whether a digit is odd or even; for a second, they judge whether the digit is greater or less than 5 (Allport, Styles, and Hsieh 1994). Or they may be presented with bivalent stimuli and have to alternate between responding on the basis of the shape or color (Hayes et al. 1998). Switching costs are evident from the fact that reaction times are longer when the task set changes (e.g., color to shape) than when the task set remains constant (e.g., shape to shape). These costs are assumed to reflect the time required to retrieve and instantiate a new task set (see Pashler, chap. 12, Goschke, chap. 14, Meiran, chap. 16, and Kieras et al., chap. 30, this volume). Moreover, competition is also likely at the time of the switch between the old and new sets (Allport and Wylie, chap. 2, this volume; Mayr and Keele forthcoming).

We were interested in whether task-switching costs would be evident after callosotomy when the two tasks were associated with different hemispheres. Task-switching costs reflect limitations in our ability to maintain multiple goals and coordinate the processes required to achieve these goals. In Ivry et al. 1998, we had observed that, unlike control subjects, the split-brain patient J.W. could maintain separate, and even conflicting, stimulus-response mappings for his two hands. Thus, when two tasks were assigned to separate hemispheres, he did not show a limitation evident in normal individuals.

The generality of this claim may be limited, however. First, tasks that have shown independence between hemispheres after callosotomy have generally been spatial in nature (e.g., Franz et al. 1996; Luck et al. 1989; Holtzman and Gazzaniga 1985). It is unknown whether such independence would be observed with nonspatial tasks. Second, in Ivry et al. 1998, the task sets remained constant for each block of trials. Thus neither hemisphere was ever required to switch set. In the following studies, we examine what happens when the mappings need to be continually modified from trial to trial. Specifically, will a split-brain patient show signs of interference between the two hemispheres when the task requires continuous task switching?

Experiment 1

We employed a hybrid task that combined features of task switching and PRP experiments. On each trial, the subjects were required to make two successive responses. For the first task, a colored shape, a blue or green square or triangle, was presented in the left visual field; subjects made a speeded response with the left hand, identifying in separate blocks, either its shape or color. For the second task, one of four univalent stimuli was presented in the right visual field requiring a second speeded response with the right hand (see figure 17.1). This stimulus could be defined either by its color (a blue or green circle) or by its shape (a white square or triangle). A 4:2 mapping was used for the right hand, with one color and one shape assigned to each of two response keys. Stimulus onset asynchronies of 50, 150, 400, and 1,000 msec separated the presentation of the two stimuli. Task-switching costs were expected on trials in which the second stimulus was defined on a dimension different from that used to define the first stimulus. For example, if task 1 required the identification of shape, then task-switching costs would be evident by comparing response latencies when task 2 also required a shape judgment to those when task 2 required a color judgment.

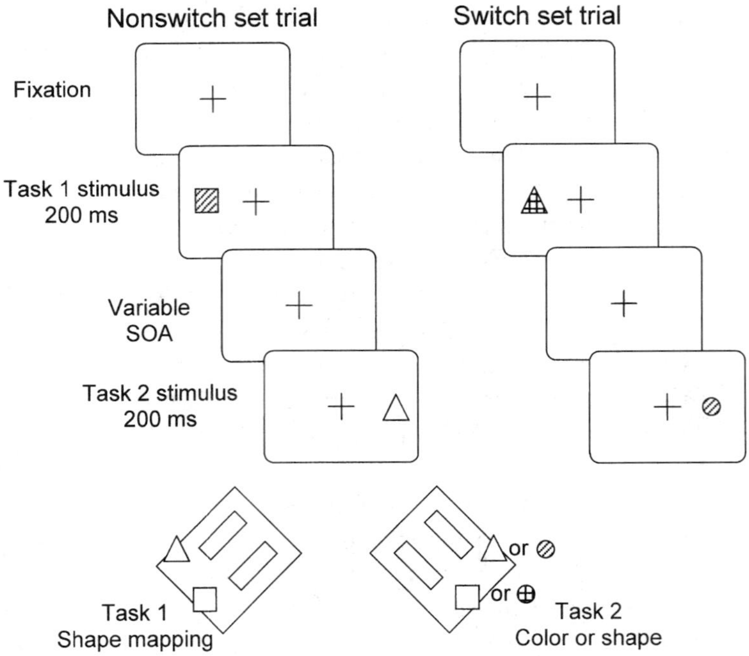

Figure 17.1 Sequence of events in each trial for block in which the relevant dimension for task 1 was shape. A same-set trial (nonswitch) is shown on the left and a different-set trial (switch) is shown on the right. Two keyboards were position below the monitor with the left keyboard used to make responses for task 1 and the right used to make responses for task 2. The S-R mappings for each keyboard are shown below the keys. The diagonal texture indicates the color green and the grid texture indicates the color blue.

This design entailed two significant differences from typical task-switching experiments. first, the stimuli were lateralized and the two responses made with different effectors, to assess whether task-switching costs would persist after callosotomy. Second, the trials were always presented as pairs of events with varying SOAs between stimulus 1 and stimulus 2, to evaluate switching costs in terms of the process models that have been developed for analyzing PRP data (Pashler 1994, chap. 12, this volume), a manipulation that, to our knowledge, has not been applied in previous task-switching studies. If the effect of task switching reflects a bias in perceptual set, then we would expect the cost to be underadditive with SOA because the change in perceptual set could be achieved during the refractory period (Pashler and Johnston 1989). On the other hand, if the effect of task switching is due to the establishment of a different stimulus-response mapping, then the cost would be additive or overadditive with SOA (McCann and Johnston 1992; Ivry et al. 1998).

Note that making the stimuli for task 1 bivalent, even though only one dimension was relevant for each block of trials, ensured that the stimuli for tasks 1 and 2 were perceptually different on all trials. It also provided another means for assessing interactions between the sets adopted by the two hands. For example, we could look at filtering costs by comparing response latencies for task 1 on trials where the irrelevant dimension was congruent with the relevant dimension, to those where the irrelevant dimension was incongruent with the relevant dimension.

Subjects Our split-brain patient was again J.W., who had participated in Ivry et al. 1998, a right-handed 44-year-old male who underwent a two-stage callosotomy operation in 1979 for the treatment of intractable epilepsy. MRI scans reveal that all of the fibers of the corpus callosum and posterior commissure were sectioned and that the anterior commissure is intact. J.W. continues to take antiseizure medication, and seizure activity has been minimal postoperatively. His recovery has been excellent, and he has no difficulty in everyday activities (Gazzaniga 1998; Gazzaniga et al. 1984).

J.W. was administered the Wechsler Adult Intelligence Scale (WAIS-R) postoperatively, scoring 97 and 95 on the verbal and performance subtests, respectively. He has participated in neuropsychological studies for almost twenty years now (Gazzaniga 1995). He is able to comprehend language in both hemispheres; indeed, even when performing tasks where the input is restricted to the right hemisphere and responses are made with the fingers on the left hand, verbal instructions are sufficient. J.W. is adept in using his hands, as evidenced by his two favorite hobbies, drawing and building model cars, but does show mild clumsiness in finger movements with the left hand.

Three control subjects were tested; one male, aged 40, and two females, aged 48 and 42. All were right-handed and, based on self-report, had no known neurological disturbance. One of the authors (R.I.) served as a control and was aware of the hypotheses under study. The other two control subjects were naive as to the purposes of the experiment.[1]

Procedure The experiment was conducted with a PC-based computer system. Two customized response boards were used, one for the left hand and one for the right hand, with participants using the index and middle finger of each hand to press low-resistance response keys measuring 10×1.9 cm. To maximize the participants' comfort, the keyboards were oriented at 45 degrees with respect to the frontal plane.

A cross, 2 degrees on a side, was present at the center of the monitor at all times, and subjects were instructed to focus their eyes on this fixation marker. Each trial began with the 200 msec presentation of a bivalent stimulus in the left visual field. This stimulus was either a triangle or square, colored green or blue. The side of either object subtended a visual angle of 2.2 degrees and the center-to-center distance from the object to the fixation marker was approximately 9 degrees. On shape blocks, subjects were to respond with the left hand, pressing with the middle finger if the stimulus was a triangle, and with the index finger if the stimulus was a square. On color blocks, the middle finger was to be used if the stimulus was green, and the index finger if the stimulus was blue. Subjects were told to ignore the value on the irrelevant dimension.

After an SOA of 50, 150, 400, or 1,000 msec, the stimulus for task 2 appeared for 200 msec, 9 degrees to the right of fixation. Unlike first-task stimuli, second-task stimuli were univalent, either a green or blue circle or a white square or triangle. Subjects were taught a 4:2 mapping and responded on one of two response keys with the right hand. To maintain a consistent mapping with task 1, the middle finger for the right hand was used to respond to the triangle and green circle, and the index finger was used to respond to the square and blue circle. A 3 sec window was provided during which the participant could complete the two responses before the next trial began. A 2 sec intertrial interval separated the response to the second-task stimulus on trial n from the onset of the first-task stimulus on trial $n + 1$, with the fixation marker present during this intertrial interval.

The subjects were repeatedly instructed to maintain fixation at the center of the screen. Although we did not monitor eye movements, J.W. has participated in many similar experiments and is quite good at maintaining fixation. While the subjects were encouraged to make fast and accurate responses for both tasks, they were explicitly told that their primary responsibility was to respond as quickly as possible to the first stimulus. They were informed that variable delays would occur between the two stimuli and that they should not wait for the second stimulus before making their first response. These instructions were repeated before each block of trials.

Subjects were tested in test blocks of 64 trials formed by the factorial combination of four color-shape combinations for stimulus 1, four univalent values for stimulus 2, and four SOAs. Four test blocks were completed in which the relevant dimension for task 1 was shape, and another four in which the relevant dimension for task 1 was color. Within each block, there were 32 trials on which the relevant dimension for task 2 was the same as for task 1 (nonswitch) and 32 trials on which the relevant dimension for task 2 was different (switch). Practice blocks, consisting of 16 trials were performed before the first test block for each of the first-task color and shape conditions. Visual feedback, presented at the center of the display, was provided after errors on either task during the practice blocks; this was repeated until the experimenter judged subjects to have learned the stimulus-response mappings and generally involved 2–3 repetitions. No on-line feedback was given during the test blocks, although the percent correct and mean reaction times for tasks 1 and 2 were displayed at the end of each block.

J.W. was tested on first-task color and shape conditions on separate days, with a four-day break between the sessions. Due to technical problems, the data were lost for the final two test blocks in the color condition. Thus, J.W.'s results are based on six blocks, two color and four shape. The control subjects each completed four test blocks of the color and shape conditions in a single session with a 15 min break separating the two conditions. Testing began with the first-task color condition for half of the controls and with the first-task shape condition for the other half.

Results and Discussion Our primary goal in this experiment was to determine whether the split-brain patient would show evidence of cross talk between the two tasks. In particular, would a task-switching cost be observed when the relevant dimension for task 2 differed from the relevant dimension for task 1. This question could be addressed in within-subject analyses because the design entailed both nonswitch and switch trials. Thus we used within-subject repeated-measure analyses of variance (ANOVA), with test block treated as the repeating measure.

A three-way ANOVA was used to measure task switching, with the variables task 1 (color, shape), set (nonswitch, switch), and SOA (50, 150, 400, 1,000 msec). Only trials on which both response 1 and response 2 were correct were included. Figure 17.2 shows the mean response latencies for tasks 1 and 2 on these trials. Because there were no systematic differences in performance whether the first task required responses based on color or shape, the data in the figure are combined over the color and shape blocks. While there were some main effects and interactions involving the task variable for the control participants, these effects were inconsistent. Furthermore, given that task and order were confounded for each individual, it is unclear whether these effects are related to idiosyncratic differences in discriminability between the color and shape stimuli or to practice effects. Given this, we collapsed data over this variable to examine the other variables, verifying that all of the basic conclusions were essentially the same regardless of the relevant dimension for task 1.

A PRP effect is seen for all of the participants, with the main effect of SOA reliable at the 0.001 level in all of the analyses. The current findings of a robust PRP effect in a split-brain patient are in accord with previous studies (Pashler et al. 1994; Ivry et al. 1998) and indicate a persistent source of constraint in the timing or scheduling of the two responses. Most relevant for the present study, second-task response latencies are similar in the nonswitch and switch conditions for J.W. Although his latencies were slower on both tasks for the shortest SOA, neither the effect of set, nor the interaction of set and SOA were significant: $F(1, 4) = 1.0$ and $F(3, 12) = 1.1$, respectively. Thus J.W. does not appear to show a cost in changing response set when the two tasks are associated with different hemispheres.

A very different picture is evident for the control subjects. The effect of set was significant for all of the controls—R.I.: $F(1, 6) = 51.0$, $p < 0.001$; M.S.: $F(1, 6) = 27.8$, $p < 0.005$; A.L.: $F(1, 6) = 187.6$, $p < 0.001$. The interac-

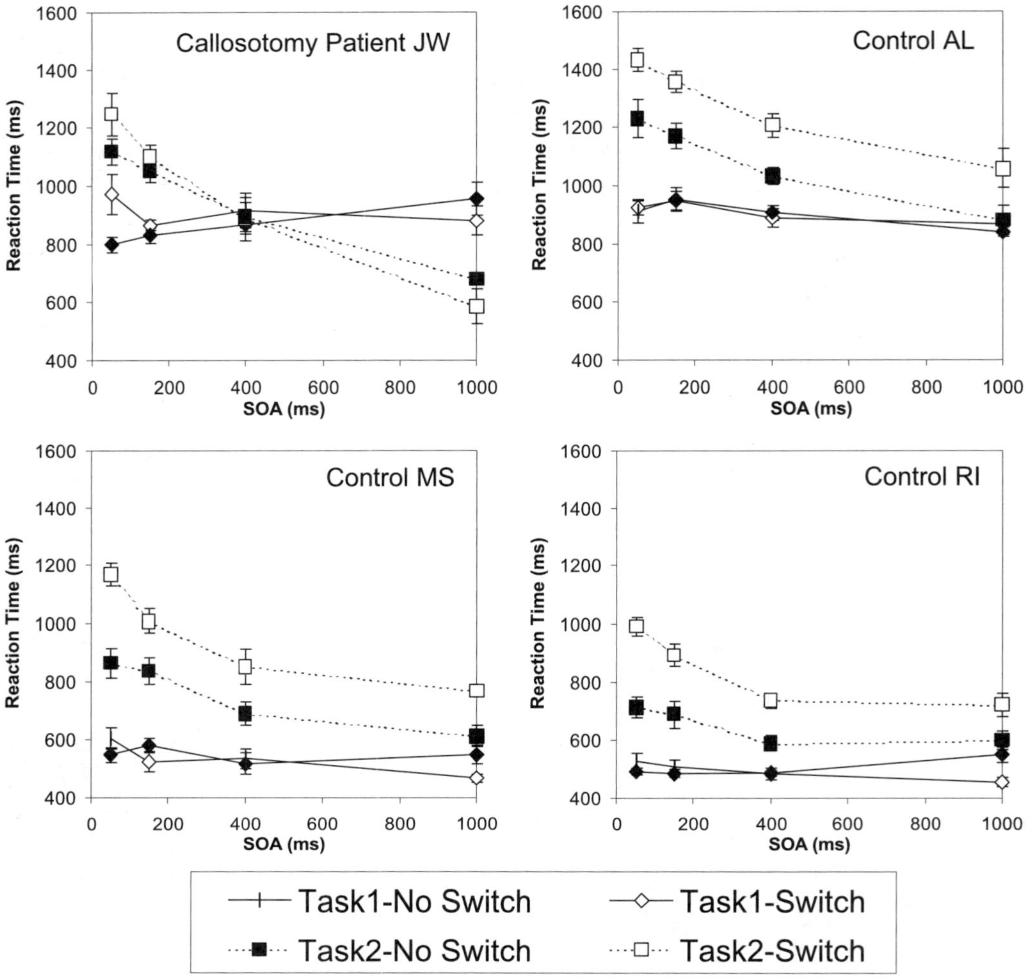

Figure 17.2 Mean response latencies for the two tasks as a function of stimulus onset asynchrony (SOA) in experiment 1.

tion of set and SOA was significant for two of the controls—R.I.: $F(3, 18) = 3.9$, $p < 0.05$; M.S.: $F(3, 18) = 3.4$, $p < 0.05$. For the other control, interaction of the set and SOA was not reliable—A.L.: $F(3, 18) < 1$. The results for the control subjects are consistent with the hypothesis that the task-switching cost is associated with a relatively late stage of processing, such as response selection (Pashler, chap. 12, this volume). There is no indication of underadditivity, for A.L., M.S. or R.I., the hallmark of a manipulation that influences processing prior to the bottleneck stage. Rather, the cost of switching set was either additive or overadditive.[2]

Given our instructions emphasizing that priority should be given to task 1, we expected response latencies on task 1 to be relatively invariant across SOA. J.W.'s performance matched this expectation: he did not show any reliable differences on task 1 as a function of task, SOA, or set, nor any interactions between these variables. On the other hand, the control participants were less successful in keeping the two tasks segregated. RT1 tended to decrease with SOA, with a main effect for SOA observed for M.S. and A.L.: $F(3, 18) = 3.4$, $p < 0.05$ and $F(3, 18) = 4.2$, $p < 0.05$, respectively. For R.I., this factor interacted with set, due to slower responses to the first stimulus when the second stimulus required a shift in set at the short SOAs: $F(3, 18) = 4.7$, $p < 0.05$.

In the preceding analyses, the same set trials are composed of two very different types of conditions. In one condition, the task-relevant stimulus value is identical for tasks 1 and 2, and correspondingly, the two responses entail homologous effectors. For example, in the color condition, the task-relevant stimulus might be green for both tasks, requiring successive keypresses with the middle finger of the left and right hands. In the other condition, even though the set remains the same, the task-relevant stimulus values differ and the responses are made with nonhomologous effectors. For example, the task-relevant stimuli for tasks 1 and 2 might be blue and green, respectively, requiring successive responses with the index and middle fingers of the left and right hands. It is important to assess whether the results observed for the control subjects might reflect repetition benefits, either on the stimulus or response end, rather than task-switching costs. This is especially relevant given that the stimuli for task 2 were univalent.

We evaluated the effects of repetition benefits by performing additional analyses in which we focused on whether the two responses were made with homologous or nonhomologous fingers. For these analyses, we defined a new variable, "response relationship" (homologous, nonhomologous), and examined the effects of this variable by itself and as a function of whether the set remained the same or switched. Because the number of observations per condition is relatively small, we combined the data over SOA. Note that when the set remains the same, responses with homologous effectors also entail a repetition of the task-relevant stimulus value, allowing some insight into the contributions of both stimulus and response repetition. In addition, a measure of switching cost relatively uncontaminated by repetition benefits can be gleaned from these analyses. By focusing solely on those trials where the two responses are made with nonhomologous effectors (and thus involve different stimuli), one can compare latencies on task 2 conditions where the set remains the same to those where the set changes.

Figure 17.3 summarizes the key findings of these analyses. Consider first the results for the split-brain patient. None of the main effects or interactions was significant. While there is a trend for J.W.'s responses to be faster when the two tasks require successive responses with homologous fingers, this difference was not reliable: $F(1, 4) = 3.8$, $p = 0.12$. Focusing on the uncontaminated measure of switching cost (the right side of each panel in figure 17.3), his mean latencies for same and differ-

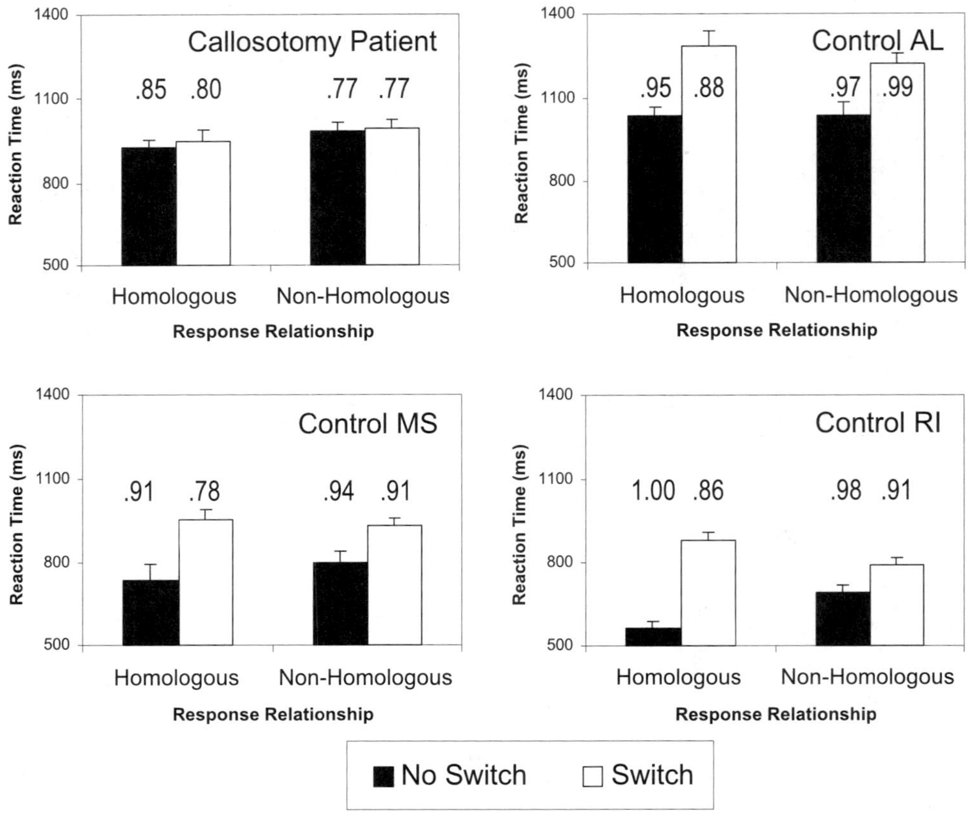

Figure 17.3 Mean response latencies for task 2 in experiment 1 as a function of whether the fingers used for task 1 and task 2 were homologous (repeated) or nonhomologous (different). Note that nonswitch task data involve the repetition of the task relevant stimulus. The switch response trials provide an uncontaminated measure of task-switching effects.

ent set trials were 984 and 993 msec, respectively. Moreover, unlike the control participants, J.W. did not show an advantage on trials in which the set and response remained the same, conditions in which benefits from stimulus repetition would be observed. Thus, as in the primary analysis, there is no evidence of cross talk between the two tasks in the split-brain patient.

The picture is more complex for the control subjects. The interaction of set and response relationship was significant for two of the controls—R.I.: $F(1, 6) = 82.8$, $p < 0.001$; M.S.: $F(1, 6) = 6.8$, $p < 0.05$. First, consider the uncontaminated measure of switching cost, the comparison restricted to nonhomologous responses in same- and different-set conditions. Post hoc analyses confirmed a significant switching cost for all of the controls, with a mean increase on different set trials of 150 msec across the three controls. These findings provide the most compelling evidence that this task entailed a task-switching cost. Second, the control subjects are substantially faster on trials in which the task-relevant stimulus value is the same in both visual fields. We suspect that this effect is due to an interfield stimulus repetition benefit, although these trials also entail successive responses with homologous fingers.

One final assay of cross talk between the two tasks centers on the value of the irrelevant dimension for task 1. Although the subjects were aware of the relevant dimension for task 1, the value on the irrelevant dimension was one of the possible targets for task 2. Thus a different form of repetition effect is possible on the switch trials. For example, when task 1 is color, blue square followed by square would involve a repetition on the response and value of the shape dimension. Blue triangle followed by triangle would involve a repetition on the value of the shape dimension, but here the two successive responses would involve nonhomologous responses. In the former case, color and shape are consistent in terms of their S-R mapping (i.e., both blue and square are mapped to the index finger), and in the latter, the color and shape are inconsistent. To determine whether there was an effect of filtering the irrelevant dimension for task 1 (see also, Goschke, chap. 14, this volume), we compared latencies on trials where the value on the irrelevant dimension was consistent with the target value for task 1 to those on trials where the value on the irrelevant dimension was inconsistent with the target value. The variable "filter" (consistent, inconsistent) was added to the ANOVAs reported for the repetition effects, analyzed here in terms of its effect on both the first and second responses.

Effects of filtering on RT1 were minimal and nonsignificant for all of the participants, including J.W. On average, consistent trials were responded to 9 msec faster than inconsistent for the controls and 7 msec slower for J.W. However, the consistency of the values of stimulus 1 influenced the latencies to stimulus 2 for R.I., who was faster on RT2 when the two values of stimulus 1 were consistent: $F(1, 6) = 10.7$, $p < 0.02$. The means for the other two controls were in the same direction. Thus the controls appear to show another source of interference from task 1 to task 2. When values for stimulus 1 are inconsistent (in terms of their S-R mapping for task 2), slower responses are observed to stimulus 2. Importantly, the value on the irrelevant dimension for task 1 did not influence the magnitude of the switching cost.[3]

The accuracy data were, in general, in accord with the latency results (table 17.1, left half). The tasks were challenging for J.W.: both responses were correct on only 73% of the trials. On task 1, J.W. responded correctly on 88% of the trials. For task 2, his performance dropped to 80%. Although the mean error rate appears higher for J.W. on switch trials, neither the main effect nor the interaction approached significance (both $Fs < 1$). He did make more errors as SOA increased on task 2, with 90%, 83%, 95%, and 64% correct across the four SOAs. It is not clear why his performance was so poor at the 1,000 msec SOA. The short reaction times in this condition suggest a speed-accuracy trade-off, perhaps reflecting a

Table 17.1 Accuracy across the Four Stimulus Onset Asynchronies as a Function of Task Switching for Experiments 1 and 2

	Experiment 1				Experiment 2			
SOA	50	150	400	1000	50	150	400	1000
Callosotomy patient J.W.								
Same RT1	0.90	0.90	0.88	0.83	0.89	0.83	0.81	0.86
Switch RT1	0.88	0.83	0.85	0.96	0.81	0.81	0.83	0.86
Same RT2	0.96	0.85	0.90	0.58	0.69	0.80	0.81	0.69
Switch RT2	0.83	0.81	0.81	0.69	0.72	0.88	0.66	0.73
Control A.L.								
Same RT1	0.97	0.97	0.98	0.95	0.97	0.98	1.00	1.00
Switch RT1	0.89	0.95	0.98	1.00	0.94	0.94	0.98	1.00
Same RT2	0.97	0.95	1.00	0.94	0.98	1.00	1.00	1.00
Switch RT2	0.97	0.91	0.97	0.91	0.89	0.88	0.92	0.94
Control M.S.								
Same RT1	0.98	0.97	0.98	0.92	0.98	1.00	0.95	0.95
Switch RT1	0.97	0.95	0.94	0.98	0.94	0.98	0.91	0.98
Same RT2	0.92	0.95	0.92	0.80	0.94	0.98	0.91	0.95
Switch RT2	0.84	0.77	0.86	0.91	0.64	0.81	0.83	0.91
Control R.I.								
Same RT1	0.98	0.97	0.97	0.97	0.97	0.92	1.00	0.92
Switch RT1	0.98	0.95	0.92	0.98	0.94	0.97	0.98	0.98
Same RT2	1.00`	0.97	1.00	0.97	0.95	0.98	1.00	0.98
Switch RT2	0.84	0.91	0.91	0.89	0.81	0.91	0.92	0.95

difficulty in withholding the second response for a long interval after the first response.

Overall, the control subjects responded correctly to both stimuli on 89% of the trials. No significant effects were observed in the accuracy data for task 1. On task 2, a main effect of set was found for R.I.: $F(1, 6) = 60.5$, $p < 0.001$; and M.S. an interaction of showed set and SOA: $F(3, 18) = 5.8$, $p < 0.01$. In both cases, accuracy declined when the set changed, in correspondence with the latency data. In all cases, there is no indication of a speed-accuracy trade-off.

In summary, experiment 1 provides further insights into changes in multitask performance that occur after callosotomy. Across a variety of measures, the split-brain patient J.W. failed to show any sign of cross talk between two tasks, one lateralized to the right hemisphere and the other to the left. These findings extend those reported in Ivry et al. 1998, where J.W. was found able to maintain inconsistent S-R mappings within each hemisphere. In the current study, J.W. exhibited neither a task-switching cost on task 2, nor repetition effects across the hemispheres, nor any costs associated with processing the irrelevant dimension of stimulus 1. Thus his ability to maintain separable S-R mappings is not limited to the spatial domain, and holds even when the task-relevant S-R mapping for the second response fluctuates from trial to trial.

As in the other PRP studies with split-brain patients, J.W. continued to exhibit a pronounced PRP effect. Consistent with the findings of Ivry et al. 1998, the PRP effect after callosotomy appears to be quite different from that observed in healthy control participants. We expect that the PRP effect for J.W. results from his compliance with the task instructions, reflects the operation of a late bottleneck associated with response execution, or both.

While the control subjects also follow these generic instructions, their performance suffers from task-specific sources of interference: they exhibit task-switching costs as well as other signs of cross talk between the two tasks. By applying the PRP logic to a task-switching experiment, we were able to examine the locus of interference. The patterns of additivity and overadditivity indicate that the costs associated with changing set are not related to processes involved in perceptual identification, but rather arise at a later stage of processing, one likely involved in the retrieval of the task-relevant S-R mappings or the selection of the appropriate response codes. Previous task-switching studies have typically entailed a single response system for both tasks (e.g., the right hand with a 4:2 mapping). The current study demonstrates similar costs when the two tasks are associated with different hands (see also Rogers and Monsell forthcoming). Presumably, this reflects the unity of these response processes, at least when the callosal fibers are intact.

The hybrid task used in experiment 1 combined elements of task switching with the PRP paradigm. Although our focus was on the performance of the split-brain patient, this manipulation also proved insightful in terms of the performance of the control participants, specifying the locus of interference associated with task switching. The results of experiment 1 suggest a late stage of interference: for all of the control subjects, the switching effect was either additive or overadditive with SOA. This points to a locus of interference associated with response preparation, selection, or initiation (McCann and Johnston 1992; Ivry et al. 1998). Such a hypothesis is consistent with the notion that task-switching costs are associated with the retrieval and instantiation of new S-R mappings.

Experiment 2

To test this hypothesis, we conducted a second experiment with the same pair of tasks as in experiment 1, except that two stimuli were now presented along the vertical meridian. The first-task stimulus was presented above fixation and, after a variable SOA, the second-task stimulus was presented below fixation. We expected that each stimulus would be available to both hemispheres. If some or all of the various costs observed for the controls in experiment 1 were due to intermingling of stimulus information, then we should observe these in J.W. On the other hand, because the results of experiment 1 suggested that task switching produced interference at a later stage of processing, we expected again to find no switching cost for J.W.: the successive responses were to be made with different hands, and we assumed the S-R mappings for each hand would still be restricted to the contralateral hemisphere.

Subjects J.W. and the same three age-matched control subjects as in experiment 1 were tested.

Procedure The only modification to the procedure was in the placement of the stimuli. On each trial, two stimuli were presented, a colored triangle or square above fixation followed by one of four univalent stimuli below fixation. The center-to-center distance between the fixation cross and the stimuli was 4.2 degrees of visual angle. The participants were instructed to respond to the upper stimulus with the left hand, and to the lower stimulus with the right. The stimuli were presented for only 200 msec to discourage eye movements, and the cross was always present to provide a fixation marker.

Each subject completed eight test blocks, four in which the relevant dimension for task 1 was color and four in which the relevant dimension was shape. For all blocks, the second-task stimulus could be either a target shape or color, with a neutral value (circle or white) used for the irrelevant dimension. The instructions were as in experiment 1 with special emphasis now given to respond first to the upper stimulus. J.W. was tested six months after completing experiment 1, whereas the control participants were tested only a day or two after completing experiment 1.

Results and Discussion Overall, the pattern of results was similar to that observed in experiment 1. Presenting the stimuli at the vertical meridian, and thus making them accessible to both hemispheres, was not sufficient to induce interference between the two tasks for the split-brain patient. Figure 17.4 presents the latency data for RT1 and RT2, collapsing over the color and shape conditions. For all of the participants, RT2 decreases with SOA. This is the only variable that is significant for J.W. In contrast, all of the controls show an effect of set—R.I.: $F(1, 6) = 127.3$, $p < 0.001$; M.S.: $F(1, 6) = 11.5$, $p < 0.05$; A.L.: $F(1, 6) = 376.9$, $p < 0.001$. The interaction of set and SOA is significant for two of the control participants—R.I.: $F(3, 18) = 6.6$, $p < 0.01$; M.S.: $F(3, 18) = 7.8$, $p < 0.01$. As in experiment 1, the interaction is one of overadditivity, with the switching cost most evident at the shortest SOA.

Whereas J.W.'s latencies on task 1 did not differ across conditions, all of the controls showed an effect of SOA—R.I.: $F(3, 18) = 5.0$, $p < 0.05$; A.L.: $F(3, 18) = 8.8$, $p < 0.001$; or an of set and SOA interaction—M.S.:

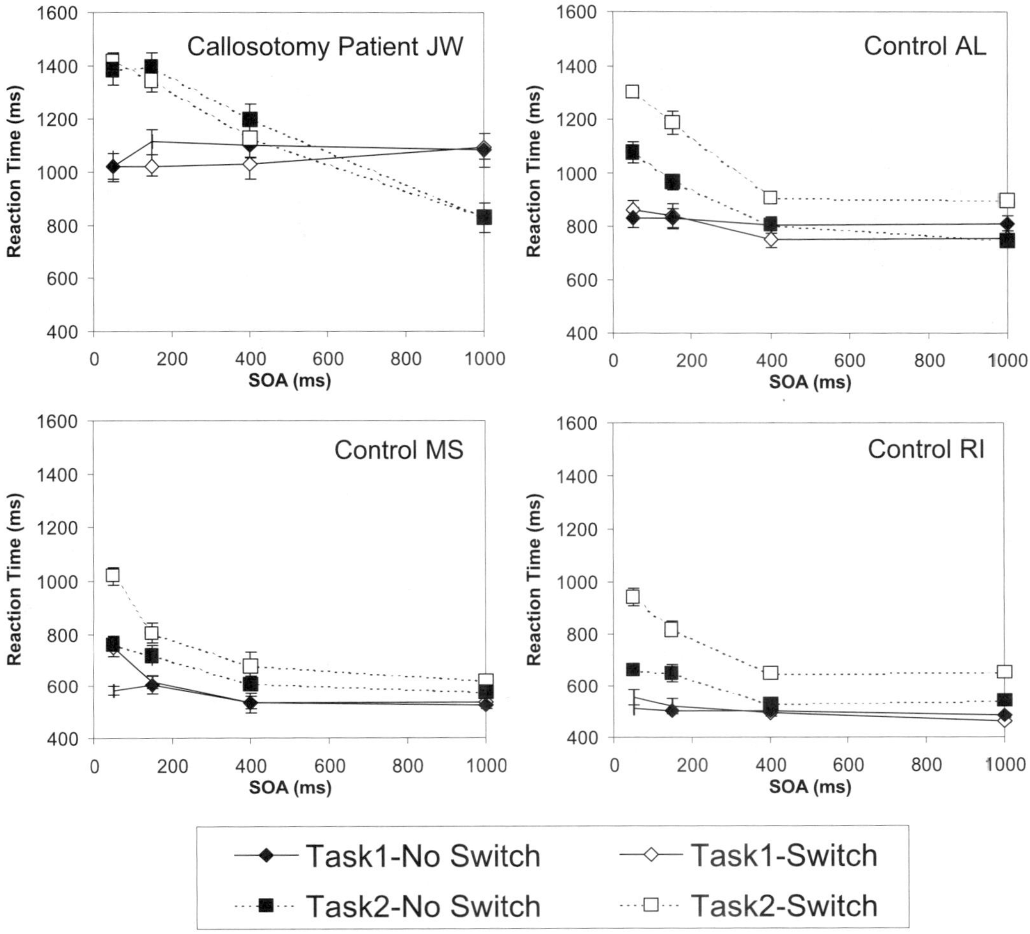

Figure 17.4 Mean response latencies for the two tasks as a function of stimulus onset asynchrony (SOA) in experiment 2.

$F(3, 18) = 7.8$, $p < 0.01$. The controls responded more quickly to the first stimulus at the long SOAs.

J.W. did not show evidence of cross talk between the two tasks on the additional measures of multitask interference (figure 17.5). There was no evidence of either response or set repetition benefits for RT2: $F(1, 7) < 1$. Nor did J.W. show any filtering effects related to the value of the irrelevant dimension for stimulus 1 on either RT1 or RT2: $F(1, 7) = 1.9$, $p = 0.21$; $F(1, 7) < 1$, respectively. The interaction of response relationship and set was reliable for two of the controls—R.I.: $F(1, 6) = 12.3$, $p < 0.05$; A.L.: $F(1, 6) = 18.9$, $p < 0.01$. For R.I. and A.L., the interaction indicates that the switching cost was greatest when the two responses involved homologous effectors. Nonswitch trials here correspond to trials in which the task-relevant stimulus value remained identical across the two hemifields. For M.S., only the main effect of response relationship approached significance: $F(1, 6) = 5.3$, $p = 0.06$.

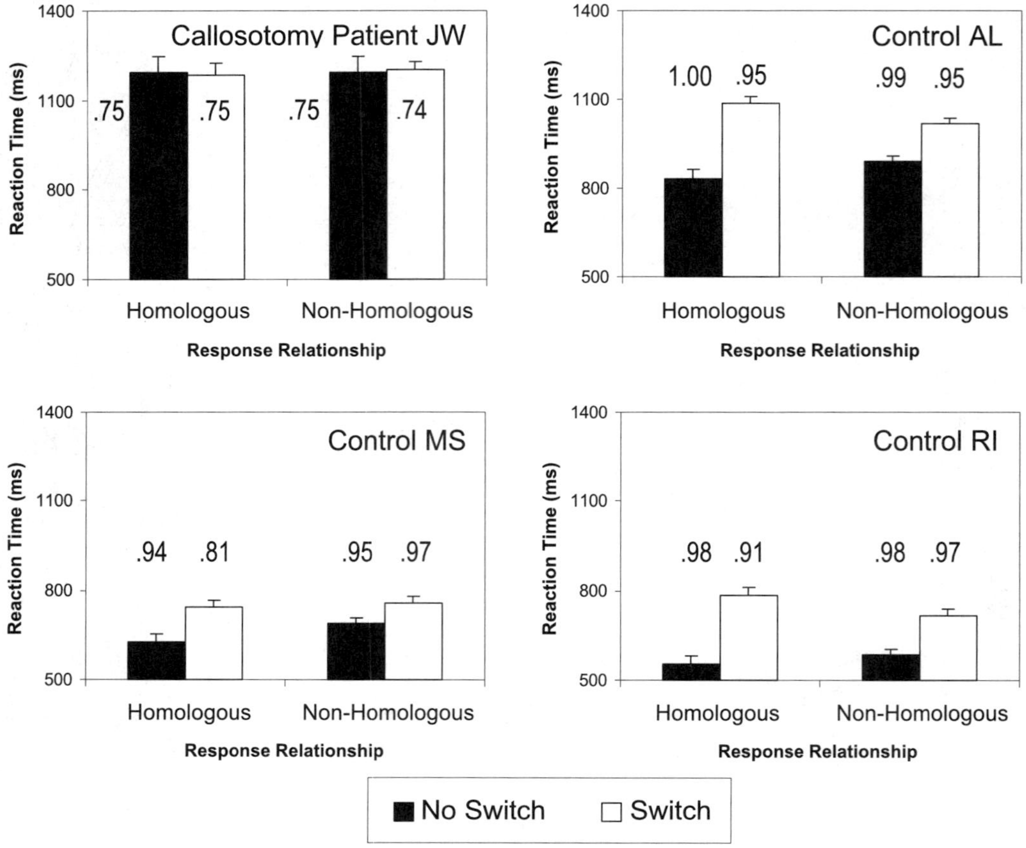

Figure 17.5 Mean response latencies for task 2 in experiment 2, presented as in figure 17.4.

In terms of filtering costs, R.I. was slower on RT1 when the value on the irrelevant dimension for stimulus 1 was inconsistent than when it was consistent: $F(1, 6) = 10.7$, $p < 0.05$. R.I. and A.L. also showed a carryover filtering effect on RT2—R.I.: $F(1, 6) = 35.2$, $p < 0.01$; A.L.: $F(1, 6) = 17.7$, $p < 0.01$—but this did not interact with set or response relationship, indicating that the switching costs were similar regardless of the relationship between the values on the irrelevant dimension for task 1 and relevant dimension for task 2.

J.W. was correct on 85% of his responses to stimulus 1 and 75% of his responses to stimulus 2 (table 17.1, right half); the control participants were generally more accurate, with mean values of 97% and 92% for the two tasks. Nonetheless, the accuracy data are in accord with the latency data in terms of interference between the two tasks. J.W.'s accuracy scores were similar for the same and switch trials. The controls consistently exhibited higher error rates on switch trials, for RT1—A.L.: $F(1, 6) = 8.0$, $p < 0.05$; M.S.: $F(1, 6) = 8.0$, $p < 0.05$; and for RT2—R.I.: $F(1, 6) = 40.3$, $p < 0.001$; A.L.: $F(1, 6) = 42.8$, $p < 0.001$; M.S.: $F(1, 6) = 69.6$, $p < 0.001$. The accuracy data in experiment 2 also revealed another difference between the performance of the split-brain patient and that of the controls. Whereas J.W.'s accuracy was relatively constant across SOA, the control subjects became more accurate as SOA increased, especially on RT2—R.I.: $F(3, 18) = 5.9, p < 0.01$; M.S.: $F(3, 18) = 12.9, p < 0.001$.

The results of experiment 2 provide new evidence that the costs observed in task-switching experiments are associated with stages of processing closely linked to response processes rather than to perceptual analysis. Even though the stimuli were presented along the vertical meridian, the performance of the split-brain patient again indicated that the two tasks were effectively segregated. While we assume that each hemisphere had access to information related to both the upper and lower stimuli, it nonetheless appears that the processing of each stimulus is essentially restricted to the hemisphere required for generating the responses: there was no evidence that the stimulus or response codes for the two tasks interacted.

17.4 IMPLICATIONS FOR MODELS OF EXECUTIVE CONTROL

Task-switching experiments have been used to study control processes associated with the coordination of performance in multitask situations. The concept of task switching has been used to capture the idea that our behavior is not simply exogenously guided, but also reflects the interaction of the stimulus information with our internal goals. Indeed, it is this interaction that allows human behavior to be so flexible and adaptive (see Gotschke, chap. 14, this volume). Although we can exert some control over which information to attend to, and respond in a way that will help achieve our current goals, this control comes at a cost. Adopting a particular task set limits the speed with which we can alter our behavior should the environmental conditions suddenly change, or should the task requirements mandate a new set of candidate actions. This cost has been interpreted as reflecting limitations in our ability to integrate perceptual, cognitive, and response processes to meet the behavioral requirements of the moment (Allport, Styles, and Hsieh 1994; Rogers and Monsell 1995; Rogers et al. 1998).

An important component operation of task switching involves the establishment and maintenance of S-R mappings. In our previous studies (Franz et al. 1996; Ivry et al. 1998), we observed that callosotomy patients fail to exhibit interactions between spatial codes represented in each hemisphere. The current study was designed to examine whether the lack of interaction would also be evident in a nonspatial task as well as under conditions in which the S-R mappings, at least in one hemisphere, had to be dynamically reorganized from trial to trial.

As expected, the neurologically healthy control participants exhibited numerous manifestations of interference between the two tasks: intermanual task-switching costs, repetition effects, and filtering costs associated with the value of the task-irrelevant dimension of the first stimulus.

While stimulus repetition benefits were found in a few situations, the results suggest that the prominent source of interference was associated with processes involved in response preparation and selection. In particular, the PRP analysis indicated that the effect of task switching was additive or overadditive with the interval between the two stimuli, a pattern indicative of a source of interference downstream from processes associated with perceptual identification. We have argued that the task-switching interference arises from the operation of processes involved in the establishment of task-relevant S-R mappings, a hypothesis similar to the response selection bottleneck hypothesis promoted by Pashler (1994; chap. 12, this volume).

The fact that the task-switching cost is found even when the successive responses are performed with different hands indicates that this operation occurs at a relatively abstract level (see also Rogers and Monsell forthcoming). Although consistent with previous findings in the motor literature that, at higher levels, S-R codes are not linked to particular effectors, this finding is better conceptualized in terms of a goal-based representation (e.g., Hommel 1993; MacKay 1982). The unity of goal-oriented representations would provide a locus for the interference between the two tasks. Nonetheless, it seems likely that under certain conditions, different sets could be associated with distinct effectors. For example, when driving, we do not find ourselves pushing on the steering wheel when we go to engage the clutch. In this condition, there does not appear to be any cross talk between the actions produced by the hands and feet. On the other hand, the sets associated with effector system in such situations are well learned and relatively invariant. In task-switching experiments, the context and thus mapping are in constant flux, placing high demands on control processes (Norman and Shallice 1986). We expect that the cost of switching set would be as great within an effector as between effectors in such conditions.

In sharp contrast to the control subjects, the split-brain patient did not exhibit any evidence of task-specific interference in the two experiments. He was just as fast to respond to the second stimulus when the task-relevant dimension changed as when the task-relevant dimension remained the same. Moreover, he did not exhibit repetition effects between the two hemispheres, nor did he show any costs associated with the value of the task-irrelevant dimension for the first stimulus. These results suggest that processes involved in the establishment and maintenance of S-R codes can be independently supported in the two cerebral hemispheres. The interhemispheric task-switching costs found in the normal participants are likely to involve interactions across the corpus callosum rather than to arise from a single control process localized to one hemisphere. Interestingly, the lack of cross talk was also evident even when the stimuli were projected bilaterally in experiment 2. This finding provides additional evidence that the source of interference from changing set is linked to response processes rather than to perceptual processes.

The present experiments provide specificity to the putative operations underlying executive function. An important component of flexible behavior is the ability to create transient representations of S-R codes to achieve task-relevant goals. The costs associated with task switching reflect the fact that, when the task changes, new goals must be instantiated, leading to the activation of new S-R codes. These codes, at least in normal individuals, are generically available to all response systems. Although this may create interference when the task requirements change, the activation of abstract response codes should be adaptive in promoting goal-oriented behavior.

NOTES

This work was supported by National Institutes of Health grants NS30256 and NS17778. We are grateful to Kristi Hiatt, Emily Levins, Fredrik Nilsen, and Michael Miller for their assistance in collecting data and preparing this chapter, and to Michael Gazzaniga for his many insightful comments.

1. We also tested 12 college-age controls for experiment 1 and 19 controls for experiment 2. The pattern of results for these groups was quite consistent with that of the age-matched controls with the exception that the switching costs were smaller for the college students in experiment 2.

2. We also tested a fourth age-matched control subject, H.A. Like the split-brain patient, J.W., this person did not exhibit a task-switching cost, although his null result is likely due to very different reasons. H.A. was extremely slow in performing task 2, with mean latencies collapsed over SOA of 1,765 msec for the same-set trials and of 1,708 msec for the different-set trials. Indeed, even at the longest SOA of 1,000 msec, the mean latencies for H.A. were slower than for any of the other participants at all SOAs. Given that H.A. did not appear to heed the instructions to respond rapidly, we do not report his data in the main text. Such data suggest, however, that control subjects can avoid a task-switching cost in this task only by making the second response very slowly.

3. There are other ways in which repetition effects can be analyzed with these data. For example, on nonswitch trials, homologous responses entailed a repetition of the relevant feature value (e.g., with shape relevant for both tasks, a blue square would be followed by a square), and nonhomologous responses precluded the repetition of the relevant feature (e.g., blue square followed by triangle). That is, the relevant feature for task 2 was always present in the stimulus for task 1 when the two tasks required homologous responses and never present when the two tasks required nonhomologous responses. In contrast, for the switch trials, homologous responses included repetitions and nonrepetitions of second-task-relevant features with equal frequency. To assess the importance of different repetition factors (stimulus, response, set), however, would require more than two values for each dimension.

REFERENCES

Allport, D. A., Styles, E. A., and Hsieh, S. (1994). Shifting intentional set: Exploring the dynamic control of tasks. In C. Umiltà and M. Moscovitch (Eds.), *Attention and Performance XV*, pp. 421–452. Hillsdale, NJ: Erlbaum.

Bogen, J. E. (1993). The callosal syndromes. In K. M. Heilman and E. Valenstein (Eds.), *Clinical neuropsychology*, 3d ed., pp. 337–407. New York: Oxford University Press.

Corballis, M. C. (1995). Line bisection in a subject with complete forebrain commissurotomy. *Neuropsychology, 9,* 147–156.

De Jong, R. (1993). Multiple bottlenecks in overlapping task performance. *Journal of Experimental Psychology: Human Perception and Performance, 19,* 965–980.

Duncan, J. (1979). Divided attention: the whole is more than the sum of its parts. *Journal of Experimental Psychology: Human Perception and Performance, 5,* 216–228.

Eliassen, J. C., Baynes, K., and Gazzaniga, M. S. (1996). Spatial independence of bilateral upper limb movement is mediated by the posterior callosum. Paper presented at the Third Annual Meeting of the Cognitive Neuroscience Society, San Francisco, April.

Franz, E. A., Eliassen, J. C., Ivry, R. B., and Gazzaniga, M. S. (1996). Dissociation of spatial and temporal coupling in bimanual movements of callosotomy patients. *Psychological Science, 7,* 306–310.

Franz, E. A., Zelaznik, H. N., and McCabe, G. (1991). Spatial topological constraints in a bimanual task. *Acta Psychologia, 77,* 137–151.

Gazzaniga, M. S. (1995). Principles of human brain organization derived from split-brain studies. *Neuron, 14,* 217–228.

Gazzaniga, M. S. (1998). The split brain revisited. *Scientific American, 279,* 50–55.

Gazzaniga, M. S., Nass, R., Reeves, A., and Roberts, D. (1984). Neurologic perspectives on right hemisphere language following surgical section of the corpus callosum. *Seminars in Neurology, 4,* 126–135.

Hayes, A. E., Davidson, M. C., Keele, S. W., and Rafal, R. D. (1998). Toward a functional analysis of the basal ganglia. *Journal of Cognitive Neuroscience, 10,* 178–198.

Holtzman, J. D., and Gazzaniga, M. S. (1985). Enhanced dual task performance following corpus commissurotomy in humans. *Neuropsychologia, 23,* 315–321.

Hommel, B. (1993). Inverting the Simon effect intention: Determinants of direction and extent of effects of irrelevant spatial information. *Psychological Research/Psychologische Forschung, 55,* 270–279.

Hsieh, S., Lee, C.-Y., and Tai, C.-T. (1995). Set-shifting aptitude in Parkinson's disease: External vs. internal cues. *Psychological Reports, 77,* 339–349.

Ivry, R. B., Franz, E. A., Kingstone, A., and Johnston, J. C. (1998). The psychological refractory period effect following callosotomy: Uncoupling of lateralized response codes. *Journal of Experimental Psychology: Human Perception and Performance, 24,* 463–480.

Kingstone, A., and Gazzaniga, M. S. (1995). Subcortical transfer of higher order information: More illusory than real? *Neuropsychology, 9,* 321–328.

Luck, S., Hillyard, S. A., Mangun, G. R., and Gazzaniga, M. S. (1989). Independent attentional scanning in the separated hemispheres of split-brain patients. *Nature, 342,* 543–545.

Luck, S. J., Hillyard, S. A., Mangun, G. R., and Gazzaniga, M. S. (1994). Independent hemispheric attentional systems mediate visual search in split-brain patients. *Journal of Cognitive Neuroscience, 6,* 84–91.

MacKay, D. M. (1982). Ourselves and our brains: Duality without dualism. *Psychoneuroendocrinology, 7,* 285–294.

Mayr, U., and Keele, S. W. (Forthcoming). Changing internal constraints on action: The role of backward inhibition. *Journal of Experimental Psychology: General.*

McCann, R. S., and Johnston, J. C. (1992). Locus of the single-channel bottleneck in dual-task interference. *Journal of Experimental Psychology: Human Perception and Performance, 2,* 471–484.

Meyer, D. E., and Kieras, D. E. (1997). A computational theory of executive cognitive processes and multiple-task performance: 1. Basic mechanisms. *Psychological Review, 104,* 3–65.

Norman, D. A., and Shallice, T. (1986). Attention to action: Willed and automatic control of behavior. In R. J. Davidson, G. E. Schwartz, and D. Shapiro (Eds.), *Consciousness and self-regulation,* vol. 4, pp. 1–18. New York: Plenum Press.

Owen, A. M., Roberts, A. C., Polkey, C. E., Sahakian, B. J., and Robbins, T. W. (1991). Extradimensional versus intradimensional set-shifting performance following frontal lobe excisions or amygdala-hippocampectomy in man. *Neuropsychologia, 10,* 993–1006.

Owen, A. M., Roberts, A. C., Hodges, J. R., Summers, B. A., Polkey, C. E., and Robbins, T. W. (1993). Contrasting mechanisms of impaired attentional shifting in patients with frontal lobe damage or Parkinson's disease. *Brain, 116,* 1159–1175.

Meyer, D. E., and Kieras, D. E. (1997). A computational theory of executive cognitive processes and multiple-task performance: 2. Accounts of psychological refractory-period phenomena. *Psychological Review, 104,* 749–791.

Pashler, H. (1994). Dual-task interference in simple tasks: Data and theory. *Psychological Bulletin, 116,* 220–244.

Pashler, H., and Johnston, J. C. (1989). Chronometric evidence for central postponement in temporally overlapping tasks. *Quarterly Journal of Experimental Psychology, 1,* 19–45.

Pashler, H., Luck, S. J., Hillyard, S. A., Mangun, G. R., O'Brien, S., and Gazzaniga, M. S. (1994). Sequential operation of the disconnected cerebral hemispheres in "split-brain" patients. *NeuroReport, 5,* 2381–2384.

Preilowski, B. (1990). Intermanual transfer, interhemispheric interaction, and handedness in man and monkeys. In C. B. Trevarthen (Ed.), *Brain circuits and functions of the mind: Essays in honor of Roger W. Sperry,* pp. 168–180. New York: Cambridge University Press.

Rogers, R. D., and Monsell, S. (1995). The cost of a predictable switch between simple cognitive tasks. *Journal of Experimental Psychology: General, 124,* 207–231.

Rogers, R. D., and Monsell, S. (Forthcoming). Task-switching within and between response modalities.

Rogers, R. D., Sahakian, B. J., Hodges, J. R., Polkey, C. E., Kennard, C., and Robbins, T. W. (1998). Dissociating executive mechanisms of task control following frontal lobe damage and Parkinson's disease. *Brain, 121,* 815–842.

Ruthruff, E., Johnston, J. C., and Van Selst, M. (Forthcoming). Why practice reduces dual-task interference. *Journal of Experimental Psychology: Human Perception and Performance.*

Sergent, J. (1987). A new look at the human split brain. *Brain, 110,* 1375–1392.

Spector, A., and Biederman, I. (1976). Mental set and mental shift revisited. *American Journal of Psychology, 89,* 669–679.

Sperry, R. W. (1982). Some effects of disconnecting the cerebral hemispheres. *Science, 217,* 1223–1226.

Tuller, B., and Kelso, J. A. S. (1989). Environmentally specified patterns of movement coordination in normal and split-brain patients. *Experimental Brain Research, 75,* 306–316.

IV Control of Multistep Tasks

18 The Organization of Sequential Actions

Glyn W. Humphreys, Emer M. E. Forde, and Dawn Francis

ABSTRACT Successful performance on many everyday tasks depends on the ability to recognize the objects involved, stored knowledge about the steps in the action, the ability to organize the steps in the correct temporal order, and maintenance of information about the steps completed. Insights into these abilities and their interrelations can be gained by studying the breakdown of performance, either after brain damage or when people perform tasks with forms of processing load imposed. We report data from patients with the neuropsychological deficit "action disorganization syndrome" (ADS) and from normal subjects under dual-task conditions, demonstrating the fractionation of some of these abilities. ADS can entail deficits in long-term knowledge for the component steps and their orders, in actions, and in inhibiting component actions already completed. Moreover, the problems ADS patients have in maintaining the steps within a complex action sequence interact with learned knowledge about familiar object usage, which can be invoked directly by objects. Qualitatively similar effects can be observed in control subjects under dual-task conditions. We discuss the implications of the results for understanding how sequential behavior is organized in complex tasks.

Many everyday tasks, such as making a cup of tea or writing a letter, consist of several component actions. To accomplish the tasks successfully, we must recognize the objects involved, recall the component actions and their sequence, and, as the component actions are being carried out, we must maintain a record of our current position and not repeat steps already completed. Perhaps because of the complexity of the processes involved, there are few detailed accounts of how such sequential behaviors are conducted. In this chapter, we explore some of these processes by studying performance breakdown in neuropsychological patients and in control subjects under dual task conditions.

18.1 PREVIOUS MODELS

General models of behavior in complex tasks have been outlined by Norman and Shallice (1986; see also Cooper and Shallice forthcoming) and Grafman (1995). Norman and Shallice distinguish between a "contention-scheduling system," concerned with routine complex behaviors, and a "supervisory attentional system," used in the control of novel actions. Associations between individual objects and actions lead to the activation of "action units," which must be output in a certain order for a complex behavior to succeed. For familiar tasks, this is accomplished by means of the contention-scheduling system, which regulates activation so that the correct actions are made in the correct order. The supervisory attentional system is required to modulate the contention-scheduling system when a less familiar variation of a task must be conducted (e.g., make lemon tea) and a more familiar action overruled (e.g., do not use the milk jug in the task). Shallice (1988) suggests that the supervisory attentional system is associated with frontal lobe structures in the brain.

Grafman (1995) does not distinguish between different systems for routine and for novel actions, proposing instead that complex sequential actions depend on the activation of "structured event complexes," represented hierarchically at different levels of abstraction. At the lowest level, structured event complexes represent information about particular motor skills (e.g., how to use chopsticks). A number of structured event complexes can then become associated to form higher-level "managerial knowledge units," which can be linked either to particular contexts (e.g., how to eat in a Chinese restaurant) or, at yet higher levels, to more general contexts (e.g., how to behave in a restaurant). According to this account, the "supervisory attentional system" (Norman and Shallice 1986) is a set of abstracted managerial knowledge units that guide behavior in underspecified (perhaps unfamiliar) conditions.

For both models, selective damage to either the supervisory attentional system or to high-level managerial knowledge units should disrupt performance on novel but not on routine tasks. Neither model, however, provides a detailed specification of the nature of the memory representations that mediate routine tasks, or how such memory representations are accessed by stimuli. For example, are memories for the component actions integrated with memories for the temporal order of action? Are the memories activated to the same degree irrespective of how stimuli are presented? Furthermore, exactly how do the structures that maintain the goal and instructions for a given task interact with the procedures that determine the production of familiar action?

18.2 NEUROPSYCHOLOGICAL STUDIES

It has long been known that lesions to the frontal lobes can produce severe disturbances in a large variety of tasks, particularly those involved in reasoning in new situations or in applying novel task instructions (see Duncan 1986; Luria 1973; Shallice 1988). On the other hand, disorders can also be found in complex everyday tasks when detailed measures are taken. Schwartz and colleagues (Schwartz 1995; Schwartz et al. 1991, 1993, 1995; Schwartz and Buxbaum 1997) introduced the term *action disorganization syndrome* (ADS) to describe patients who make many errors even on familiar multistep tasks. For instance, using standardized measures of performance on everyday tasks, they have reported that such patients make abnormal numbers of sequence and omission errors (performing component actions in the wrong sequence or failing to perform them at all), as well as addition and semantic errors (inserting an extra component action incorrectly or using an object as another semantically related one). These impairments can arise even though patients can show good recognition and appropriate use of individual objects, thus not showing signs of either ideomotor or ideational apraxia as conventionally defined (Buxbaum, Schwartz, and Carew 1997).

Schwartz and Buxbaum (1997) propose that such problems on everyday tasks reflect a joint impairment, not only to a high-level system controlling behavior in unfamiliar circumstances (the supervisory attentional system or managerial knowledge units) but also to more basic knowledge representations for familiar actions (the contention-scheduling system or structured event complexes). They suggest that an impairment to a high-level system alone would not disrupt everyday tasks, as intact high-level processes should allow problem-solving strategies to be constructed "on the fly" to accomplish tasks even without supportive lower-level knowledge (as when the task is unfamiliar). Thus, by demonstrating deficits on familiar tasks, patients with ADS reveal impairments in both lower- and higher-order procedures.

Although the above studies indicate how performance on everyday tasks can be analyzed, they do not directly test whether any disorders reflect impaired stored representations for familiar actions, as opposed to, say, problems in reviewing performance as it proceeds. Furthermore, if there are impaired stored representations, does this reflect poor knowledge of the actions or of their temporal sequencing? And what is the form of interaction between higher- and lower-level procedures in this? Does the disruption of higher-level procedures, required for novel behavior, lead to poor activation of the lower-level (routine) procedures, to poor control of this activation (e.g., failure to inhibit inappropriate habitual procedures), or indeed to both?

18.3 A PRELIMINARY STUDY

Our own first investigation of these issues involved a study of four "patients," two of whom were diagnosed as having ADS due to their severe problems with everyday tasks, and two "control" patients (Humphreys and Forde 1998). One ADS patient, H.G. (aged 78), had damage to the right frontal and parietal lobes following a stroke. The other, F.K. (aged 29), had bilateral medial frontal and temporal lobe damage due to carbon monoxide poisoning. H.G. and F.K. had a variety of neuropsychological problems in addition to their deficits with routine multistep actions (see Humphreys and Forde 1998 for details). For example, they had deficits in a variety of "executive" behaviors (e.g., in suppressing incongruent overlearned responses in Stroop color naming), and they had poor episodic memory (e.g., on the Wechsler Memory Scale). The control patients, F.L. and D.S., were matched to F.K. and H.G. on these ancillary deficits (Humphreys and Forde 1998). F.L. (aged 61) was severely amnesiac as a result of carbon monoxide poisoning and showed poor performance on episodic memory tasks. D.S. (aged 64) had sustained a large left frontal infarction and had problems at least as severe as those found with H.G. and F.K. on the tests of excutive behaviors. If problems in episodic memory or executive behaviors alone produce ADS, then problems in everyday tasks should occur in the control patients as well.

The patients carried out 7 everyday tasks, and performance was assessed relative to "norms" from these tasks collected from 45 non-brain-damaged control subjects. We took as the correct "basic" actions, and the correct sequence, the action and sequence lists generated by 80% or more of the controls (and any action reported by less than 80% of the subjects was not counted as "basic"). For example, the basic actions listed by subjects for the task "write and post a letter" were as follows: (1) write the letter; (2) sign the letter; (3) fold the letter, (4) put the letter in the envelope; (5) seal the envelope; (6) write the address on the envelope; (7) lick the stamp; (8) stick the stamp on the envelope (here a few participants listed actions such as "pick up a pen," but these were not included in our list of basic actions because they were not reported frequently enough). It is interesting to note that there was strikingly good agreement across controls as to the basic actions involved for each task and the sequence in which the actions should be carried out.[1] This high level of agreement suggests that basic actions and their sequence of production are stored in our long-term knowledge of familiar tasks.[2] The behavior of the patients, videotaped as they performed the tasks, was scored according to whether they produced the basic actions in the standard order. The tasks were carried out twice, once with only the objects for the tasks placed in front of the patients, and once with three additional distractors present (distractors were semantically related to objects in the tasks). The distractors did not affect performance greatly, and the data presented here are summed over the two test occasions.

The number of errors made by each patient, for each task, are shown in table 18.1. The data were analyzed by treating each task as a subject and each patient as a level, in a repeated-measure analysis of variance (ANOVA). There was a significant effect of patient: $F(3, 18) = 8.13$, $p < 0.001$. H.G. and F.K. performed worse than the two control patients ($p < 0.01$ for all comparisons; Newman-Keuls test), consistent with their diagnosis as having ADS. Stepwise regressions showed significant effects of the number of steps in each action, but no effects of the number of target objects present, for both H.G. and F.K.: $F(1, 6) = 35$ and 17, respec-

Table 18.1 Number of Errors Made on Each Task, along with Number of Basic Steps and Target Objects

Task	Number of steps	Number of objects	Patient H.G.	F.K.	D.S.	F.L.
Write letter	8	4	20	18	2	4
Wrap gift	8	5	29	16	7	7
Make sandwich	7	7	15	21	6	3
Make tea	6	7	11	13	3	4
Make toast	6	6	14	7	8	4
Paint wood	5	4	7	2	1	4
Eat cereal	3	4	0	0	0	0
Total			96*	77*	27	26

* If repeated perseverations are discounted, then H.G. made a total of 78 errors and F.K. a total of 71 errors.

tively, both $p<0.01$. The patients made more errors in the tasks where a larger number of steps were required.[3]

Mistakes were predominantly sequence errors, omissions, and perseverations, though some were semantic (e.g., drinking from a teapot), addition (adding a new step into a task) or "quality/spatial" errors (e.g., filling the cup with more milk than tea).[4] Interestingly, differences did emerge in the kinds of perseverative errors produced by H.G. and F.K. F.K. tended to repeat earlier actions later in the sequence (e.g., in making a cup of tea, F.K. placed teabag in teapot, poured water from a kettle into teapot, poured from teapot into cup, poured milk into cup, and placed teabag in teapot again); in contrast, H.G. made many perseverative errors in which he repeatedly performed an action (e.g., in wrapping a present, H.G. repeatedly cut the wrapping paper until it was far too small for the present—despite remarking that the paper was now too small!). For H.G., proportionately more of his perseverative errors were immediate repeats (18/25) relative to F.K. (6/17): chi-square (1) $=4.17$, $p<0.05$. This difference between the patients suggests that perseveration errors can reflect different factors—a problem in inhibiting the action last produced (particularly for H.G.) and a problem in preventing a more distantly completed action from recurring (in F.K.). We return to this point in section 18.7.

We also assessed performance as a function of the order of the steps in the tasks (using the 4 tasks with either 6 or 8 steps; see table 18.1). F.K. successfully completed more steps in the first than the second half of the sequence: 23/28 versus 14/28; chi-square (1) $=6.5$, $p<0.01$. This did not hold for H.G. (14/28 versus 19/28; control patients were at ceiling). On the other hand, H.G. alone produced more "overt" errors in the final steps of the tasks (by "overt" we refer to perseverations, sequence or quality/spatial errors): 22/28 versus 8/28; chi-square (1) $=12.13$,

$p < 0.01$.[5] For F.K., any effect may have been obscured by omissions in the second half of the task. Thus there was some evidence that these two ADS patients failed to retrieve or maintain actions and their sequence, or both, as they proceeded through the more complex tasks, so that they performed worse on the second halves.

Subsequently, we tested for long-term knowledge of the actions and the action sequences comprising the everyday tasks. In one case, patients were asked to give verbal descriptions of how each task should be completed. In another, the basic actions were written on cards, and patients had to order the cards in the correct sequence. The description task required knowledge of both the component actions and their sequence. The ordering task required only sequence knowledge. H.G. and F.K. were impaired in both tasks. In the description task, H.G. produced only 33% and F.K. only 28% of the basic actions generated by the controls, even when responses were scored only according to whether the basic actions were produced (irrespective of their order). In the ordering task, both H.G. and F.K. placed only 26% (12/46) of the component actions in the correct consecutive order. The chance level of ordering component actions in correct pairwise relations was 20% (9/46). Neither patient was better than chance and in no instance were the two patients able to order correctly all the actions in any single task. H.G. and F.K. were also tested in control-sequencing tasks, which required that they sequence items based either on stored knowledge (e.g., ordering sets of letters and numbers, tested with both patients) or on perceptual information (e.g., the sizes of circles; tested with F.K. only). Both patients performed better when sequencing other stereotyped orders than when sequencing actions (see Humphreys and Forde 1998), and F.K. performed at ceiling when sequencing with perceptual information. These data indicate that H.G. and F.K. had some problems in accessing their long-term knowledge for the basic actions in everyday tasks, including their knowledge for the sequential order of the actions. The poor retrieval of information about action sequences was over and above any general deficit in sequencing information. Although other investigators have argued for a separate loss of sequence information, with the knowledge of component actions being preserved (Sirigu et al. 1996), we found no such dissociation: H.G. and F.K. were impaired with both action and sequence knowledge.

18.4 CONTROL PERFORMANCE UNDER DUAL-TASK CONDITIONS

The patients with ADS studied by Humphreys and Forde (and also by Schwartz and colleagues) had problems on general measures of "executive functions" as well as on everyday tasks.[6] From such cases alone, it is difficult to judge whether a deficit in executive functions is sufficient to generate the observed problems with familiar sequential actions. Note, however, that one of the control patients, D.S., performed as poorly as the patients with ADS on tests of executive function. This suggests that a deficit in executive functions is not sufficient to cause problems in everyday actions. This issue can also be addressed by assessing performance on the everyday tasks by control subjects under dual-task conditions. Do problems in performance arise when dual tasks load supposed "executive structures" within working memory (cf. Baddeley 1986)? Diary studies suggest that "action errors" occur in everyday life under conditions in which people are distracted or "thinking of something else" (see Reason 1990), that is, perhaps when working memory is otherwise occupied. This was tested experimentally here.

To load working memory, 10 young (aged 18–24) control subjects carried out the Trails Test (Heaton, Grant, and Mathews 1991) while simultaneously performing the everyday tasks; 10 others were given a simple verbal rehearsal task (repeating the word "the" aloud as quickly as possible). Our version of the Trails Test involved the experimenter naming an arbitrary letter and number pair (e.g., "D7") and asking subjects to continuously shift both the letter and the number in sequence ("E8," "F9," etc.) while concurrently carrying out the everyday tasks. Subjects were required to say the numbers and letters aloud when shifting each sequence, and to do this fluently, without pausing. When performed in this way, the Trails Test can be considered to demand both verbal and "central executive" components of working memory (e.g., keeping track of the last letter and number produced; cf. Baddeley 1986). Subjects in the articulatory suppression condition should only use the verbal component of working memory. The contrast between the two conditions should inform us of the contribution of central executive processes to performance on familiar multistep tasks. The behavior of each subject was videotaped and both primary and secondary task behavior scored.

Several interesting results arose. One is that, despite having to perform a secondary task, the controls made far fewer errors than the patients. Using the same scoring procedure as applied to the patients, there were 35 errors in the Trails Test condition (summed over subjects), and 13 in the condition with articulatory suppression. Summing across the two dual-task conditions, there were step omissions (38) but few additions (4), perseverations (5) or quality/spatial errors (1), and no semantic errors. In addition, the controls made a form of error we had not observed in the ADS patients; namely, they sometimes inappropriately reached for an object, but then discontinued the action (26 in the Trails Test condition and 8 in the articulatory suppression condition). Thus controls appeared to suppress activated actions prior to their completion. Errors also tended to be linked to the secondary task. In particular, subjects made a total of 48 errors in the Trails Test, which were in all cases immediately self-corrected (e.g., "F9," "G10," "E11," ... "H11," "I12," etc.). Thirty-seven of the errors on the everyday tasks (typically omissions or discontinued action errors) occurred on the next step after the one where the mistake arose in the Trails Test. Considering the probabilities of errors per step in the action tasks and Trails Tests by themselves, 6.8 coincident errors could be expected by chance. The proportional number of coincident to non-coincident errors in the everyday actions tasks is considerably higher than would be expected by chance: chi-square (1) = 35.8, $p < 0.001$. Very few errors were detectable in the articulatory suppression task, making it difficult to judge the relations between performance on this task and on the primary everyday task.

The coincident errors in the everyday tasks and the Trails Test suggest an association between the executive component of working memory, which is challenged by the Trails Test, and everyday task performance. Several accounts of this association are possible. One is that the goal state for the task must be maintained in working memory to ensure both that the correct components actions are produced, and the correct action sequence. An explicit account along these lines, based on a competitive queuing network, is outlined in section 18.7. Temporary loss of this goal state, due to working memory being used in self-corrections on the Trails Test, may lead to (1) the loss of activation for component actions and (2) competition from other objects in the scene for actions from other parts of the sequence. Omission errors and reaching for incorrect objects result. In normal subjects, recovery of the goal state in working memory may nevertheless be sufficiently rapid to self-correct reaching for incorrect objects. This account differs from theories that assign familiar task performance to a system operating independently of working memory (e.g., the contention-scheduling system; Shallice 1988). An alternative account is that executive processes are involved solely in error monitoring, whereas the multistep actions are generated by another, autonomous system. However, when executive processes are occupied, errors go unnoticed (or are only noticed after an incorrect action has been initiated, in the case of discontinued errors). This account presupposes there is some nonnegligible probability that errors arise within the routine procedures involved in generating familiar action sequences, but are normally prevented by an active error-monitoring process. This leaves unexplained how the "error monitor" knows that a misreach is being made, or a step omitted, unless it has its own model of how the task should be performed.

18.5 NOVEL TASK PERFORMANCE

In this section, we assess the relations between working memory and long-term knowledge of actions when novel tasks are conducted. With novel tasks, we might again expect that task goals and instructions, held in working memory, would modulate activation in systems carrying out learned actions with objects (see section 18.7). In this case, however, inhibition from working memory may be required to prevent overlearned actions from being generated in place of the novel ones (see Kimberg and Farah 1993). It follows that, as the load on working memory increases, there may be difficulty in inhibiting overlearned actions.

We tested these ideas with patient F.K., who was presented with 6 objects on a desk, 3 sets of 2 related objects from the tests of everyday action, all of which he could identify. On each trial he had to carry out a novel action involving 2 unrelated objects from the set (e.g., objects: teapot, teabag, cheese, plate, cellophane, scissors; task: "Put the teabag on the scissors"). There were either 1, 2, or 4 instructions, presented auditorily, which F.K. was asked to repeat back immediately after performing the task (in the two- and four-instruction condition, instructions were given as a list before F.K. initiated any action). There were 3 different arrays of objects and up to 4 different instructions per array. In one session the four-instruction condition was carried out once with each array, the two-instruction condition twice, and the one-instruction condition 4 times. There were two sessions. The actions for the one- and two-instruction conditions were the same as those for the four-instruction condition (see chapter appendix). After each trial (with 1, 2, or 4 instructions), the table was cleared and the next trial chosen at random. Performance was not time limited and F.K.'s behavior was videotaped. Performance in the two- and four-instruction conditions was scored according to whether each individual action was performed correctly.

F.K. completed 14/24 one-instruction trials correctly, 6/24 two-instruction trials and 0/24 four-instruction trials. There was a clear effect of the number of instructions: chi-square (2) = 34, $p < 0.001$. Many of the errors were "standard actions," where F.K. used the two related objects together (e.g., task: "Put the teabag on the scissors," given the array of objects listed above; F.K. put the teabag in the teapot). We compared the likelihood of a "standard action" error relative to all errors. Standard actions increased as a proportion of the errors as the instructions increased. A standard action occurred on 4/10 error trials in the one-instruction condition, on 14/18 error trials in the two-instruction condition, and on 18/24 error trials in the four-instruction condition: chi-square (2) = 27, $p < 0.001$.

The above effect did not occur, simply because on trials where F.K. failed to remember the instructions, he carried out a learned action with objects. F.K. recalled 23/24 of the instructions correctly on one-instruction trials; 12/24 on two-instruction trials; 0/24 on four-instruction trials. When he recalled the instructions, he completed the actions correctly 19 times, but on a further 16 trials he repeated back the instruction correctly *after having first made an action error*. Twelve of these trials involved a standard action error. Thus, on these trials at least, verbal working memory dissociated from the system that maintained or applied task instructions to action. Because there was just one occasion when F.K. performed the action correctly but then failed to remember the instruction, it is difficult to assess whether the opposite dissociation might also occur (good action along with poor verbal memory). We obtained essentially similar results when F.K. was given written instructions available throughout the trial, to which he could refer back if he had forgotten the task.

These results indicate that (1) as the working-memory load (the number of instructions) increased, so the likelihood increased that F.K.'s performance was determined by a learned rather than an instructed relationship between objects; and (2) this was not always due to poor maintenance of task instructions in a verbal component of working memory. The application of novel goal states and instructions for action can dissociate from the ability to maintain the instructions verbally. Again, different accounts of these results are possible. One account is that verbal working memory is disconnected from the memory representations responsible for familiar actions. F.K. fails to *apply* the instructions. Alternatively, F.K. lacks a nonverbal working-memory state that maintains novel goals and instructions and that modulates the activation of stored memories. F.K. fails to *maintain* the instructions in the critical manner. This nonverbal system is distinct from verbal working memory. Whichever the case, weakening the goal instruction state, by giving more instructions, increases the propensity for errors to be based on stored action routines.

18.6 MODALITY EFFECTS: DIRECT EFFECTS FROM VISION

In a final study, we further evaluated the procedures involved when standard action errors are made, asking whether actions generated in the same way when stimuli are presented in different modalities. Although many theories hold that actions are driven from conceptual knowledge about stimuli, abstracted from the modality of stimulus presentation (e.g., Roy and Square 1985), evidence also suggests that actions can be evoked directly from seen objects, without conceptual mediation (see Riddoch, Humphreys, and Edwards, chap. 27, this volume). F.K.'s tendency to make "standard" rather than instructed, novel actions might be most pronounced when seen objects activate a familiar action directly. We tested this by repeating the novel instruction task but using cards with the names of the objects written on them. We assumed that actions to names are conceptually mediated.

We examined one- and two-instruction trials with a subset of the original actions.[7] Maintaining the arrays and the combinations of instructions from the object study, there were 3 two-instruction trials and 6 one-instruction trials. The procedure was otherwise the same as that used for objects with aural instructions. F.K. was tested on two occasions. Subsequently, we returned to test F.K.'s performance with objects, and verified that this remained at the same level as when first tested.

Summing across the one- and two-instruction trials, F.K. scored 18/24; on the equivalent trials with objects, F.K. made 7 correct responses. Performance was better with words than with objects: chi-square (1) = 8.35, $p < 0.01$. Only 1/6 errors with words involved standard actions, whereas 11/17 of the errors with objects took this form.

F.K. performed better with words than with objects, and he was better able to refrain from making standard actions with words than with objects. F.K. was able to name both the words and the objects, making it unlikely that there were differences in accessing concepts for the stimuli (especially because object naming is usually thought to operate via semantics). Rather, the results suggest that there is stronger activation of learned actions from objects than from words, which exacerbates F.K.'s difficulty in overruling learned actions in favor of novel instructed actions. This is consistent with a direct route from objects to actions (Riddoch, Humphreys, and Edwards, chap. 27, this volume).

18.7 GENERAL DISCUSSION

We have shown that performance on multistep tasks can break down in various ways. Patients with ADS can fail to retrieve familiar component actions and their temporal sequence in familiar tasks. There can also be contrasting deficits in preventing both immediate and earlier actions from recurring. Patients with a common impairment in activating stored knowledge can have different deficits in modulating behavior over time, to prevent different types of perseverative response.

Errors also occur when control subjects perform familiar multistep tasks, particularly when a dual task is imposed that challenges executive processes in working memory. Mistakes on the load task are associated with transitory errors in action, suggesting some link between working memory and task performance.

Finally, the working-memory load of the instructions affects the ability of patients with ADS to perform novel tasks. In particular, behavior becomes increasingly driven by learned rather than instructed actions as the instruction load increases. Nevertheless, standard action errors arise even when verbal memory representations are maintained. Thus either (1) verbal working memory can be disconnected from the procedures that modulate the activation of familiar actions; or (2) there is a deficit in a nonverbal component of working memory that maintains goal and instruction sets. Problems in novel tasks are also exacerbated when standard actions are strongly activated in a bottom-up fashion, from visually presented objects rather than from words. This is consistent with a direct route to action from seen objects (see Riddoch, Humphreys, and Edwards, chap. 27, this volume).

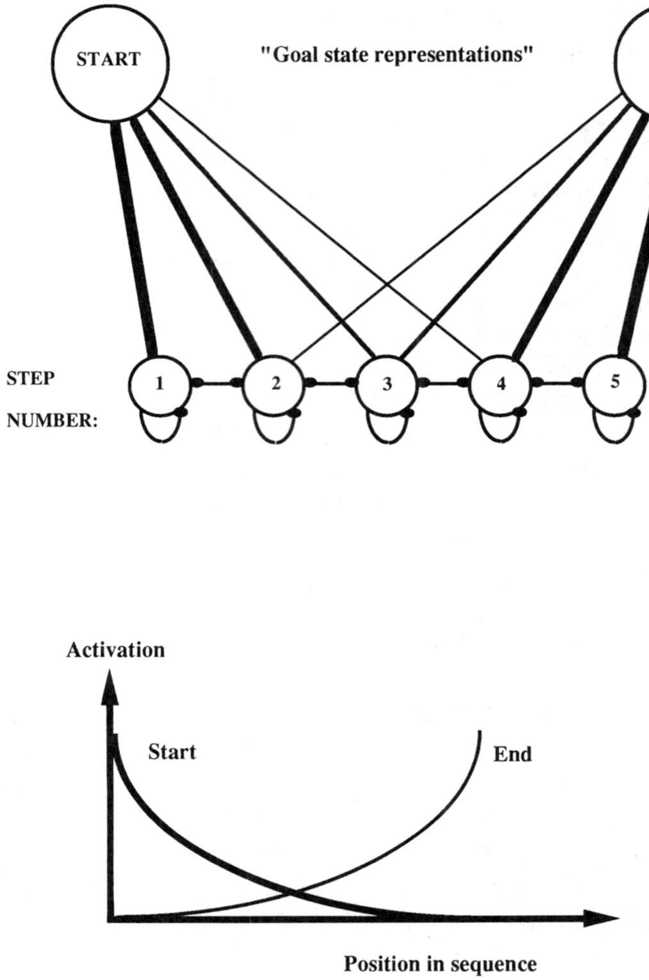

Figure 18.1 Competitive queuing framework for the production of familiar, sequential tasks. Goal state representations generate gradients of activation for the start and end of each sequence, and units corresponding to steps in the sequence compete for output in a winner-take-all fashion. Following output, the unit for a given step is immediately inhibited to allow other steps to take place.

A Suggested Framework

One way of conceptualising the impairments we have reported is in terms of theories of serial behavior that use "competitive queuing" mechanisms (e.g., Houghton 1990). In such theories, a temporal gradient of activation (from high to low) is applied to a set of processing nodes from an initiation or "goal state" unit. Nodes compete for output, with the most strongly activated node being output at any given time. Activation of a node will be affected by the temporal gradient of goal-related activation and also by bottom-up cues from objects. After output, the most active node is immediately inhibited, allowing the next-most-activated node to win the competition, and the steps in behavior to emerge. The nodes represent the component actions in a task. The weights from the goal state unit to the nodes that determine the temporal gradient represent stored knowledge of the sequence. This is illustrated in figure 18.1. Moreover, when actions are performed with real objects, there will also be bottom-up activation of component actions from the objects present, which can compete with activation patterns imposed top-down, by stored knowledge of action sequences.

Damage to the goal state, to the weights representing temporal sequence information, or to the component action nodes would lead to problems in everyday tasks. Poor top-down activation, in particular, should also lead to behavior in which component actions are overdetermined by bottom-up object-action associations, as we have observed in our studies on the reproduction of novel tasks. Decreased top-down activation of component actions will also produce particular problems later in a sequence because the gradient of activation typically decreases across later steps (Houghton 1990), and because there is competition from earlier actions following their initial inhibition. "Distant" perseverations (from actions completed some steps back) may be expected under these circumstances. A separate problem, in the immediate application of inhibition, could cause the types of repeated perseverations we observed with patient H.G. In normal subjects, temporary loss of the goal state (under secondary task conditions) could also lead to transitory decreases in the activation of particular component actions, so that errors then occur. With novel actions, separate goal states may need to inhibit those for familiar actions. Poor maintenance of these novel states, especially when coupled with strong bottom-up activation of action nodes, leads to standard action error. A model of this form may provide an articulated framework for accounting for disorders in both familiar and novel multistep tasks. Within the model, retrieval of the component actions within an action sequence is intimately bound to retrieval of the temporal order of the actions: the ADS patients we observed had problems in retrieving both forms of information. Whether knowledge of component actions can be dissociated from knowledge of their order is a question for future research.

APPENDIX

The stimuli for the test of novel action were derived from three everyday tasks and used the following objects.

Make a cup of tea: teapot, spoon, teabag, saucer, sugar, cup

Make a sandwich: plate, bread, knife, sandwich bag, cheese, butter

Wrap a gift: bow, wrapping paper, cellophane, scissors, gift, label

The novel arrays were

Array 1: cup, saucer, sandwich bag, bread, bow, wrapping paper

Array 2: teapot, teabag, cheese, plate, cellophane, scissors

Array 3: spoon, sugar, knife, butter, gift, label

The novel instructions were

Array 1:

a. Pour the cup on the sandwich bag

b. Wrap the bread with the wrapping paper

c. Put the bow on the bread*

d. Put the saucer on the wrapping paper*

Array 2:

a. Wrap the cellophane around the cheese

b. Cut the cheese with the scissors

c. Put the teabag on the scissors*

d. Put the teapot on the plate*

Array 3:

a. Put the gift on the spoon*

b. Put the label on the butter*

c. Cut the sugar with the knife

d. Put the sugar on the label

Note: In the two-instruction condition, actions *a* and *b* and *c* and *d* were paired together. Asterisked actions were used in the study with words.

NOTES

This work was supported by a grant from the Medical Research Council (U.K.) to Glyn W. Humphreys. We thank Hayley Watson and Amber Wilcox for their assistance with data collection and Jane Riddoch for her helpful comments.

1. Basic actions were placed in the same sequence by a minimum of 80% of our population for each task. The only significant disagreement was found for the task of making a cup of tea, and this concerned the order in which milk should be put in the tea. For this reason, responses by the patients were scored as correct when they put the milk in either before or after the tea.

2. Note that there is not necessarily a physical constraint on the order with which component actions need to be conducted. In our example of writing a letter, the stamp could be placed on the envelope at the beginning of the action. However, the vast majority of normal subjects list this as the last, not the first, action performed when writing a letter. This suggests that temporal sequence information is stored and not simply computed "on the fly," constrained by the physical situation.

3. It might be argued that the effects of the number of steps arose because individual actions were harder to accomplish in the tasks with more component actions. However the difficulty of individual actions is unlikely to be a major factor. In control studies, the patients were able to conduct all individual actions in response to an immediate instruction and had

no particular difficulty with any one action in a task. Also, our normal subjects listed the basic actions and their orders as consistently for the longer as for the shorter tasks. Hence the shorter steps do not appear any more stereotypic on this measure.

4. These errors are not necessarily independent of one another. For instance, an omission error would preclude any other type of error occurring on the step omitted. For this reason, the numbers of errors made in a task were simply summed together for analysis.

5. For this analysis, omission errors were not included; the assessment of steps successfully completed showed that H.G. made no more omissions on the second than on the first half of the steps in the tasks.

6. We use "executive functions" here to describe a clinical pattern across a set of tests designed to tap novel problem solving. We do not take a view on whether these functions are served by a single, central processing component or by a set of dissociable processes.

7. Only a subset of actions could be assessed because "wrapping," "pouring," "sticking," and "cutting" actions could only be performed with the real objects: they could not be conducted using the cards.

REFERENCES

Baddeley, A. D. (1986). *Working memory*. Oxford: Oxford University Press.

Buxbaum, L. J., Schwartz, M. F., and Carew, T. G. (1997). The role of semantic memory in object use. *Cognitive Neuropsychology, 14,* 219–254.

Cooper, R., and Shallice, T. (Forthcoming). Modelling the selection of routine actions: Exploring the criticality of parameter values.

Duncan, J. (1986). Disorganisation of behaviour after frontal lobe damage. *Cognitive Neuropsychology, 3,* 271–290.

Grafman, J. (1995). Similarities and distinctions among current models of prefrontal functions. *Annals of the New York Academy of Sciences, 769,* 337–368.

Heaton, R. K., Grant, I., and Mathews, C. G. (1991). *Comprehensive norms for expanded Halstread-Reitan battery*. Odessa, FL: Psychological Assessment Resources.

Houghton, G. (1990). The problem of serial order: Opponent mechanisms in sequencing and selective attention. In R. Dale, C. Mellish, and M. Zock (Eds.), *Current research in natural language generation*, pp. 287–319. London: Academic Press.

Humphreys, G. W., and Forde, E. (1998). Disordered action schema and action disorganization syndrome. *Cognitive Neuropsychology, 15,* 771–811.

Kimberg, D. Y., and Farah, M. J. (1993). A unified account of cognitive impairments following frontal lobe damage: The role of working memory in complex, organized behaviour. *Journal of Experimental Psychology: General, 122,* 411–428.

Luria, A. R. (1973). *The working brain*. New York: Basic Books.

Norman, D. A., and Shallice, T. (1986). Attention to action: Willed and automatic control of behavior. In R. J. Davison, G. E. Schwartz, and D. Shapiro (Eds.), *Consciousness and self-regulation,* vol. 4, pp. 1–8. New York: Plenum Press.

Reason, J. (1990). *Human error*. London: Cambridge University Press.

Roy, E. A., and Square, P. A. (1985). Common considerations in the study of limb, verbal and oral apraxia. In E. A. Roy (Ed.), *Neuropsychological studies of apraxia and related disorders,* pp. 111–162. Amsterdam: Elsevier.

Schwartz, M. F. (1995). Re-examining the role of executive functions in routine action production. *Annals of the New York Academy of Sciences, 769,* 321–335.

Schwartz, M. F., and Buxbaum, L. J. (1997). Naturalistic action. In L. Rothi and K. Heilman (Eds.), *Apraxia: The neuropsychology of action,* pp. 269–290. London: Psychology Press.

Schwartz, M. F., Mayer, N. H., Fitzpatrick De Salme, E. J., and Montgomery, M. W. (1993). Cognitive theory and the study of everyday action disorders after brain damage. *Journal of Head Trauma Rehabilitation, 8,* 59–72.

Schwartz, M. F., Montgomery, M., Fitzpatrick De Salme, E. J., Ochipa, C., Coslett, H. B., and Mayer, N. H. (1995). Analysis of a disorder of everyday action. *Cognitive Neuropsychology, 12,* 863–892.

Schwartz, M. F., Reed, E. S., Montgomery, M., Palmer, C. and Mayer, N. H. (1991). The quantitative description of action disorganisation after brain damage: A case study. *Cognitive Neuropsychology, 8,* 381–414.

Shallice, T. (1988). *From neuropsychology to mental structure.* Cambridge: Cambridge University Press.

Sirigu, A., Zalla, T., Pillon, B., Grafman, J., Agid, Y., and Dubois, B. (1996). Encoding of sequence and boundaries of scripts following prefrontal lesions. *Cortex, 32,* 297–310.

19 Cognitive Control of Multistep Routines: Information Processing and Conscious Intentions

Richard A. Carlson and Myeong-Ho Sohn

ABSTRACT The procedural frame hypothesis, derived from a theoretical analysis of cognitive control (Carlson 1997), suggests that instantiated goals correspond to intentions to apply operators and provide procedural frames to which operands are assimilated. It predicts that participants will perform multistep mental routines more quickly when operators can be processed before than after operands. Participants in four experiments solved running arithmetic or spatial path construction problems. Performance in both arithmetic and spatial tasks was more fluent when operator information preceded operand information, regardless of whether the information was displayed or held in working memory, supporting the procedural frame hypothesis. We consider several alternative accounts, and discuss the possibility that operator-operand structure is a general feature of cognitive control.

This chapter focuses on understanding the control of multistep mental routines by examining the information-processing dynamics of component skills embedded in cascaded sequences. We consider a hypothesis about the role of instantiated goals in such processes, for both symbolic and spatial tasks and when goals are specified by information in the environment or in working memory. Evidence that the dynamic structure of intentions is similar across these contexts contributes to a general account of the processing functions of goals, and provides a basis for linking information-processing descriptions with hypotheses about the informational structure of conscious agency.

Four new experiments tested the prediction that participants will perform mental routines more quickly when operators can be processed before than after operands. Participants in the first two experiments performed mental arithmetic routines when both operators and operands were displayed step by step (experiment 1) and when either operators or operands were held in working memory (experiment 2). Experiments 3 and 4 repeated the designs of the first two experiments, extending their logic from the symbolic arithmetic task to a spatial task with a large perceptual component and addressing some alternative explanations of the results of the first two experiments. We found consistent support for the major hypothesis, as opposed to several alternative accounts.

The present experiments and theoretical discussion are motivated by Carlson's theoretical analysis (1997) of control by conscious intentions, based in part on the idea, also expressed in some production system theories (e.g., Anderson and Lebiere 1998), that fine-grained explicit goals are central to the moment-by-moment control of cognition. The analysis also suggests that the dynamic structure of intentions will be the same across symbolic and perceptual-motor domains, whether the information specifying immediate goals is represented externally or internally. The central idea examined here is that forming an intention—instantiating a goal specified by information about an operator, available to perception or in working memory—provides a frame to which operands are assimilated to perform mental activities. This "procedural frame hypothesis" (Sohn and Carlson 1998) contrasts with views that do not attribute internal structure to intentions but suggest a common role for operator and operand symbols as retrieval cues. We therefore predict that performance of multistep mental routines will be most fluent when information that specifies operators is available before information specifying operands. We also predict that this operator-operand structure can be mapped to spatial analogues of the arithmetic tasks. The distinctive roles hypothesized for operators and operands can be mapped to an account of how the information specifying actions and their objects contributes to the conscious experience of agency (Carlson 1997). Exploring this mapping in detail is beyond the scope of the present chapter, but we return to it briefly in section 19.7.

19.1 COMPONENT SKILLS IN MULTISTEP MENTAL ROUTINES

Multistep mental activities are characterized by hierarchical goal structures in which some steps may be cascaded. By "hierarchical goal structure," we mean that some steps are performed to accomplish higher-level goals. This is, of course, a familiar idea in the problem-solving literature (e.g., Anderson 1983). By "cascaded," we mean both that (1) the result of one step may serve as data (as operand or premise) for a subsequent step, as in multistep inference (e.g., Schum 1977); and (2) a step may begin—for example, a goal may be activated—before the prior step is completed, corresponding to the sense of "cascaded" used by McClelland (1979).

Individuals performing multistep routines must assemble component skills, weaving together the cascaded steps specified by a goal structure. To fluently perform the series of steps that constitute a multistep routine, individuals must coordinate the instantiation of goals with the availability of results from previous steps and of new operands picked up by perception (Carlson 1997). According to the procedural frame hypothesis, at each step the instantiation of a goal precedes processing of new operands. In Sohn and Carlson 1998, we provided evidence for this hypothesis in the context of single-step tasks, using both conventional arithmetic and newly acquired symbolic skills. To test this prediction in the context of multistep tasks, we examined performance of multistep mental tasks in a paradigm where information is displayed briefly, but pacing of the task is under the participant's control.

Operator-Operand Structure

Conventional notation for arithmetic explicitly distinguishes symbols for operators (e.g., the plus sign "+") and for operands (numerals). This distinction seems to correspond to the cognitive structure of arithmetic computation, and some authors (e.g., Crosby 1997) have described such notation as a breakthrough in the development of calculation skills. At least for the schooled cognitive skill of mental arithmetic, it seems likely that the operator-operand distinction also characterizes the structure of mental processes and their informational support.

In cognitive theory, the term *operator* refers to basic actions that accomplish single steps of mental activity (Newell and Simon 1972; Bovair and Kieras 1991). We hypothesize that these basic actions may be represented at a level of abstraction that distinguishes operator and operand—for example, that the appropriate representation of an intention to add 3 to a current result treats "add" as the action to be performed and "3" as a parameter of that action, rather than treating "add 3" as a unitary, basic action. Although the latter representation is possible (cf. Anderson and Lebiere 1998), it would not naturally account for phenomena such as the benefit of a consistent sequence of operators in learning multistep arithmetic routines when operands vary from trial to trial (e.g., Carlson and Lundy 1992). This distinction is also implicit in common conceptions of task switching (see Allport and Wylie, chap. 2, Goschke, chap. 14, De Jong, chap. 15, and Meiran, chap. 16, this volume; Sohn and Carlson forthcoming). What is of interest is the ability to switch judgments or operations, not changes in data or operands judged.

Goals and Operators

While researchers generally assume that experimental participants adopt goals that orient them to tasks and provide a context for the aspects of control to be investigated, the relation between these goals and the moment-to-moment control of mental activity is seldom described or investigated in detail. A significant exception is the role of goals in Anderson's ACT-R theory (1993; Anderson and Lebiere 1998), in which cognition is serial at the level of individual production rules, each including a goal clause and requiring up to several hundred milliseconds for execution. If we assume that these goals reflect the finest grain of deliberate cognition (also see Newell 1990), their role provides a basis for linking information-processing description with theoretical descriptions of conscious control (Carlson 1997).

Establishing this link requires briefly considering the concept of *goal,* which has been used in two quite different ways. In one sense, a goal refers to desired outcomes or final problem states (Austin and Vancouver 1996; Newell and Simon 1972); in a second sense, which better captures the role of goals in controlling activity, a goal corresponds to an intention to achieve an outcome by taking a particular action. As Mandler (1984, 82) wrote, there are "no goals without means." This sense of goal may also capture its role in ACT-R—Anderson and Lebiere (1998) describe goal clauses using action verbs (e.g., "to add"). In the context of tasks like mental arithmetic, actions are applications of operators. Of course, goals may be represented declaratively and considered as objects of thought even when they are not currently controlling activity. We therefore use "instantiated goals" to make clear that we are referring to goals currently active as intentions controlling cognition.

Procedural Frame Hypothesis

This brief consideration of the structure of component skills suggests that symbols specifying operators and those specifying operands play distinctive roles in the control of mental activities such as arithmetic. The procedural frame hypothesis is that an operator symbol supports the instantiation of a goal, which provides a procedural frame to which operands are assimilated. This hypothesis contrasts with those (e.g., Siegler 1988) in which operator and operand symbols play a uniform role as retrieval cues. The procedural frame hypothesis predicts that performance will be faster when operators appear before operands because the instantiated goal provides a basis for interpreting the operand. In contrast, the uniform role hypothesis predicts that any benefit of advance information will be equivalent for the two types of information, or will depend on factors other than the hypothesized processing roles. (Other implications of these hypotheses are considered in greater detail in Carlson 1997; Sohn and Carlson 1998.)

19.2 EXPERIMENT 1: MENTAL ARITHMETIC WITH ON-LINE ACQUISITION OF OPERATORS AND OPERANDS

The first empirical question addressed here is whether the solution time advantage of operator-first displays observed with single-step arithmetic tasks (Sohn and Carlson 1998) will also be present in a multistep, cascaded tasks. A major difference between single- and cascaded multistep tasks is that intermediate results must be carried forward from step to step to serve as operands.

We therefore began by asking subjects to solve multistep arithmetic problems in which operator and operand information were available on-line, displayed at each problem step. This task, which minimizes the

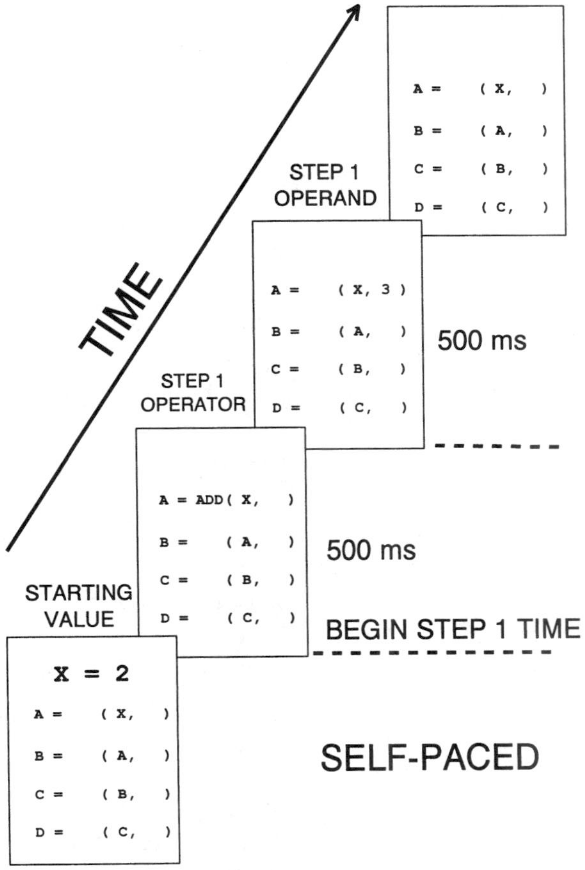

Figure 19.1 Starting display and time course of events on each step of the arithmetic task used in experiment 1. Operand-first displays were identical, except that the order of operator and operand displays on each step was reversed.

demands on working memory, might be seen as involving primarily exogenous control because information specifying the action to be performed at each step is perceptually available.

Experimental Task

Subjects in experiment 1 solved running arithmetic problems, in which a value was updated at each step. Each problem comprised four steps, and at each step a new operator and operand appeared in a computer display. Four operations were possible, indicated by the three-letter abbreviations "ADD," "DIF," "MIN," and "MAX." "ADD" represented the familiar operation of addition, "DIF" represented obtaining the absolute difference between the current value and the displayed operand, and "MIN" and "MAX" represented choosing the smaller or larger of the current value or displayed operand as the new value. Each problem included one step with each operation, appearing in a new random order on each trial.

The problem began with a starting value and display of the problem frame (figure 19.1). Subjects updated results at each step on the basis of an operation and a new single-digit operand (2, 3, 4, or 5). Subjects controlled the presentation by pressing the space bar to request the display of each step. Each operator or operand was visible for only 500 msec. The time course of events on each step is illustrated in figure 19.1. Because subjects controlled the onsets of successive steps, we could infer something about the pacing of mental processes. At the end of four steps, subjects entered answers using the computer keyboard.

For some subjects, the order in which operator and operand appeared at each step was constant over steps within problems, and varied from problem to problem. For others, the order varied from step to step within problems. The purpose of this manipulation was to rule out alternative explanations in terms of optional strategies. If the solution time benefit of operator-first presentation is due to a strategy specialized for a particular display order, this benefit might be reduced or eliminated in the constant-order condition because subjects could adopt the appropriate strategy for each problem. If the benefit is instead due to the structure of component skills, as suggested by the procedural frame hypothesis, it should be apparent in both conditions.

Subjects

Twenty-seven students from introductory psychology classes at Pennsylvania State University participated in exchange for course credit. All reported normal or corrected-to-normal vision.

Design and Procedure

We manipulated two factors, the order in which operator and operand appeared at each step, and whether this order was constant throughout a problem or varied from step to step. Fourteen subjects were randomly assigned to the constant-order display condition, and thirteen to the varied-order display condition. In the constant-order display condition, operators preceded operands, or vice versa, on every step of a problem, and this display order varied randomly from problem to problem. In the varied-display condition, operators preceded operands on two of the four steps of each problem, with the order reversed on the other two steps. The sequence of these display orders was random within problems.

Each problem began with a starting value chosen randomly from the range 1 to 6. The four operands 2–5 were assigned randomly to the steps in each problem. The final solution to each problem was always a single digit. The uncertainty concerning operators and operands was thus equivalent. We instructed subjects to solve problems as quickly as possible while maintaining a high level of accuracy.

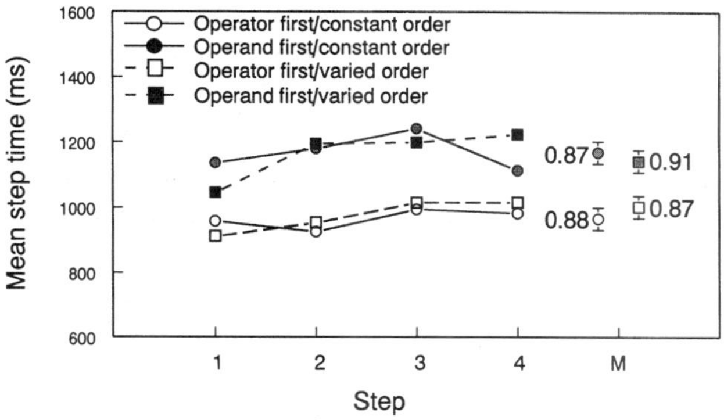

Figure 19.2 Mean step-by-step latency in experiment 1 as a function of experimental conditions. Markers with error bars at the right of the figure show means over steps together with standard errors; mean accuracy for each condition is displayed next to the corresponding marker.

Results and Discussion

Subjects gave correct answers to approximately 89% of problems, and this proportion did not vary as a function of whether display order was constant (87%) or varied (90%) within problems: $t(25) = 0.98$. The remaining analyses focus on latency data for correct trials. In all of the experiments reported here, latency for each step is measured from the point at which all information needed for that step is available, the onset of operand information for operator-first cases, and of operator information for operand-first cases.

The average time per step for correct responses was 1,069 msec. The effects of display order and step are depicted in figure 19.2. As expected, subjects were faster when operators appeared first than when operands appeared first: $F(1, 25) = 53.4$, $p < 0.001$. No other effects were significant in this analysis, all $p > 0.19$. In particular, it made no difference whether display order was constant (M = 1,067 msec, SE = 94 msec) or varied (M = 1,072 msec, SE = 110 msec) from step to step within problems.

Times to initiate problems (M = 1,930 msec) and to enter answers (M = 740 msec) did not differ between groups, providing a check on random assignment: $t(25) = 1.34$ and 0.54. We also examined the effects of operator and of operand value (because these effects, depicted in figure 19.2, were present in all cases, they are not reported here).

As predicted by the procedural frame hypothesis, participants performed steps more fluently when operator information appeared before operand information, consistent with the results of single-step arithmetic studies and in contrast to the uniform role hypothesis (Sohn and Carlson 1998). This result held for both constant- and varied-display orders, consistent with the claim that it should be attributed to the structure of component skills, rather than to problem-specific strategies.

19.3 EXPERIMENT 2: MENTAL ARITHMETIC WITH WORKING-MEMORY PRELOAD OF OPERATORS OR OPERANDS

Although on-line pickup of information specifying both operators and operands provides a useful paradigm for studying the information-processing dynamics of control, one might wonder whether conclusions from an on-line paradigm also apply to the common circumstance under which the information supporting intentions is available in memory rather than in the environment. That is, we commonly form intentions without immediate prompting from the environment. This question is also relevant to recent discussions about the role of endogenous and exogenous sources of information in control (e.g., Rogers and Monsell 1995; Meiran, chap. 16, this volume). The procedural frame hypothesis suggests that the information-processing structure of component skills should not depend on whether the information supporting goal instantiation is held in working memory or available to perception, consistent with other evidence that control structures are the same whether information to be processed is selected from working memory or perception (Carlson, Wenger, and Sullivan 1993). We therefore again expected to observe the operator-first advantage in solution time seen in experiment 1.

Experiment 2 used the same running arithmetic task as experiment 1, with one important difference. On each trial, either the series of operator symbols or the series of operand symbols was presented in advance, requiring subjects to hold this information in working memory while stepping through the problem. At each step, the remaining piece of information—either the operator or the operand—appeared briefly, as in experiment 1. As in some previous work (e.g., Carlson, Sullivan, and Schneider 1989; Carlson et al. 1990), subjects had to hold and manipulate a substantial task-relevant working-memory load in order to solve the problems. This experiment thus extends the paradigm in experiment 1 in two ways: the information supporting goal instantiation was sometimes held in working memory, and component skills were performed in a relatively demanding context.

Subjects

Sixteen students from introductory psychology classes at Pennsylvania State University, who had not participated in experiment 1, participated in exchange for course credit. All reported normal or corrected-to-normal vision. Two subjects failed to reach a criterion of 60% correct, and their data were excluded from analysis.

Design and Procedure

Each trial began with a ready signal. When subjects pressed the space bar to initiate the problem, the series of four operator labels or four operands

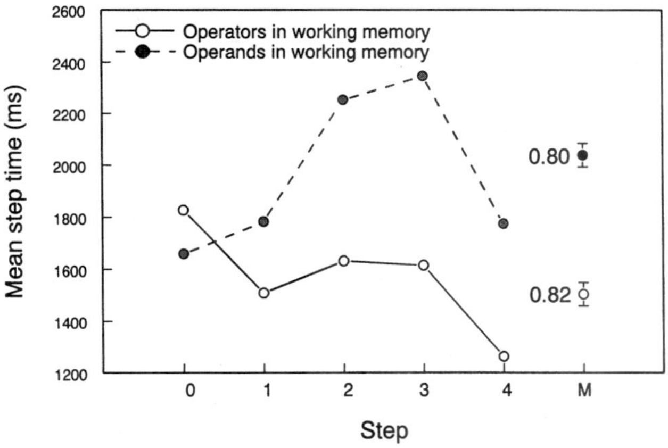

Figure 19.3 Mean step-by-step latency in experiment 2 as a function of experimental conditions. Markers with error bars at the right of the figure show means for steps 1–4 together with standard errors.

appeared at a rate of two per second above the problem frame (in the same location as the starting value; figure 19.1), constituting the memory preload set for that problem. Following the final item of the memory set, the starting value appeared, and subjects initiated the problem by pressing the space bar to request the remaining piece of information for the first step. The procedure thus allowed self-paced preparation of information in working memory (for example, by establishing a verbal rehearsal loop).

Type of working-memory preload—operators or operands—varied randomly problem by problem. The experimental design was thus entirely within subjects, and display order was not manipulated within problems. In all other respects, the design and procedure was identical to that of experiment 1.

Results and Discussion

Subjects answered an average of 81% of problems correctly, and this average did not vary as a function of display order: $t(13) = 0.62$. Step-by-step latencies and initiation times are displayed in figure 19.3. As with on-line acquisition of information, subjects were faster when operator rather than operand information was available in advance: $F(1, 13) = 98.8$, $p < 0.001$. This analysis included only steps 1–4 because initiation time includes time to prepare the memory load, but not calculation time. As the figure shows, the effect of type of advance information was greater for later steps within problems. Both the interaction of step and type of advance information and the main effect of step were significant: $F(3, 39) = 10.2$, $p < 0.001$ and $F(3, 39) = 12.8$, $p < 0.001$, respectively.

The time to initiate problems (measured from the display of the starting value) averaged 1,829 msec on operator-first trials and 1,660 msec on operand-first trials, a marginally significant effect: $t(13) = 2.0$, $p = 0.07$. Assuming a rehearsal rate of 3 to 4 syllables per second, this would allow one or two complete rehearsals of the memory set. The additional time used on operator-first trials might reflect time to encode the operators procedurally, but we have no independent evidence of this. Times to enter answers averaged 738 msec and did not vary as a function of type of advance information: $t(13) = 0.56$.

The central result in experiment 2 is that preparation time was marginally longer, but subsequent steps substantially faster, when the sequence of operators rather than the sequence of operands is held in working memory. Overall, solution time was substantially longer than in experiment 1, presumably reflecting the additional activity needed to maintain and select from information in working memory. This difference was especially pronounced when operands were held in working memory, reflecting the difficulty of maintaining numbers representing both operands and intermediate results, presumably stored in the same format. The additional time required by operand-first presentation increases substantially after step one, when participants must first coordinate storage of an intermediate result with retention of the remaining operands.

19.4 EXPERIMENT 3: SPATIAL PATH CONSTRUCTION WITH ON-LINE ACQUISITION OF DIRECTIONS AND DISTANCES

Experiment 3 extends the logic of experiment 1 to a spatial task, for two principal reasons. First, although experiments 1 and 2 support the predictions of the procedural frame hypothesis in multistep arithmetic tasks, it might be argued, that arithmetic is a special case. The operator-operand distinction is embodied in a conventional notation system, the results of manipulating operands are symbolic values that do not necessarily refer to anything, and individuals probably perform the task using serial, verbal coding to remember both operators and operands. Examining our hypotheses in the spatial domain thus provides some useful generality.

Second, theories of working memory (e.g., Baddeley 1986) suggest that people's distributed capacities for working memory are divided along verbal or spatial lines. The comparison of similar tasks in verbal-symbolic and spatial domains may therefore help us understand how control interacts with working-memory strategies. The spatial task used operands corresponding to those used in the arithmetic task; the operations, starting values (locations), and intermediate results were spatial in nature. This extension addresses another possible criticism of experiments 1 and 2: that interference due to the similarity between new operands and intermediate results made it more difficult for participants to begin each step when operands appeared first. Note, that, according to the procedural frame hypothesis, such interference in the on-line case (experiment 1) would result primarily because in the operand-first case participants must form a working-memory representation of the operand rather than immediately assimilate it to a procedural framework. In Experiments 3 and 4, however, it is unlikely that participants retained intermediate results (locations) and new operands (distances) in the same format.

A critical point is the mapping of operator and operand to the spatial domain. In the mapping we chose, we identified operator with direction and operand with distance, was based in part on the formal structure of arithmetic and spatial domains. As Piaget (e.g., 1954), noted, integer arithmetic and spatial displacements share a set of structural characteristics known as "group structure." For example, both are characterized by *composition under closure*—any sequence of operations results in a position (location or number) also belonging to the system. Directed moves (or vectors of unspecified length) may therefore correspond to operators (and indeed addition and subtraction can be seen as directed moves in the unidimensional space of integer arithmetic). Given this mapping, numbers serve analogous roles as operands in both domains. We have, of course, placed some restrictions on the representation of each domain; for example, limiting intermediate results to single-digit numbers or locations within the displayed grid, and using only four directions. We also attributed particular orientations to subjects; for example, we assumed that subjects considered single digits as representing numbers rather than categorical labels, and we used a constant viewer-centered frame of reference for specifying direction.

Experimental Task

We therefore designed an experimental task that provided a close analogue to the arithmetic task in terms of organization and control demands, but required construction of a spatial path rather than arithmetic calculation. Each problem required subjects to mentally move around a checkerboard grid, using the four directions "UP," "DOWN," "LEFT," and "RIGHT" (defined with relation to the subjects' point of view). Each problem began with the display of the grid and a starting location (figure 19.4). Each step required an imagined movement in one of the four directions, for a distance of 2–5 squares. The intermediate locations were not marked on the screen, and thus had to be maintained mentally. Subjects paced the presentation of steps by clicking a mouse key. After four steps, they used the mouse to move a check mark to the ending location of the path they had mentally constructed, clicking to indicate their answers.

The structure of this task closely parallels that of the arithmetic task. At each step, one of four possible operations (the four directions of movement) and one of four possible operands (the numbers 2–5) appears. Note, however, that the starting values and intermediate results to be maintained in working-memory—locations on the grid—are of a com-

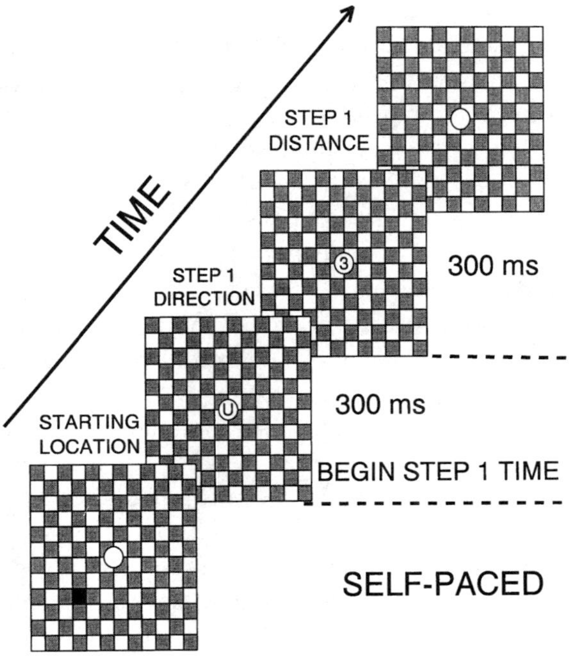

Figure 19.4 Starting display and time course of events on each step in experiment 3. Distance-first displays were identical, except that the order of direction and distance displays on each step was reversed.

pletely different type than the numerical values in the arithmetic task. For each step, then, one operand is a location and one is a number representing distance.

Subjects

Thirty-four students from introductory psychology classes at Pennsylvania State University, none of whom had participated in experiments 1 or 2, participated in exchange for course credit. All reported normal or corrected-to-normal vision.

Design and Procedure

The design was the same as that for experiment 1. For a randomly selected half of subjects, the display order was constant from step to step within problems, and for the other half, it varied randomly from step to step.

Each problem included the four directions and four distances, randomly sampled without replacement for assignment to steps within each problem. Based on pilot studies, the display times were slightly shorter than in experiment 1, with direction and distance cues each appearing for

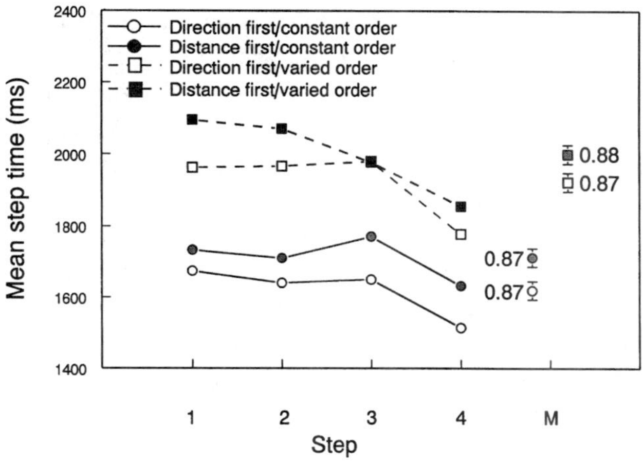

Figure 19.5 Mean step-by-step latency in experiment 3 as a function of experimental conditions. Markers with error bars at the right of the figure show means over steps together with standard errors; mean accuracy for each condition is displayed next to the corresponding marker.

300 msec. Starting locations were chosen randomly, with the constraint that the starting location never appeared in the innermost or outermost cells of the grid (although the imagined path could pass through these cells). As shown in figure 19.4, the grid was a 12 × 12 square of alternating light and dark cells. Each cell was approximately 1.1 cm square, which was also the diameter of the center circle. Subjects were instructed not to use their fingers to touch or point at the screen. In all other respects, the procedure was the same as in experiment 1.

Results and Discussion

Subjects answered approximately 87% of problems correctly, and this proportion did not vary as a function of whether display order was constant or varied within problems: $t(32) = 0.05$. The remaining analyses focus on latencies for correct trials.

As with the arithmetic task used in experiments 1 and 2, steps were performed more quickly when directions (indicating operators) appeared before distances (operands): $F(1, 32) = 17.5$, $p < 0.001$. This effect is depicted in figure 19.5, together with the effect of step within problem. In contrast to experiment 1, the effect of step was significant: $F(3, 96) = 10.7$, $p < 0.001$. Also in contrast to experiment 1, participants were marginally faster with constant display orders (M = 1,666 msec, SE = 98) than with display orders that varied from step to step (M = 1,964 msec; SE = 141): $F(1, 32) = 3.5$, $p = 0.07$. No interactions approached significance, all $p > 0.15$.

Average times to initiate problems from the "ready" display (838 msec) and to enter answers (798 msec) provided a check on random assignment. Subjects who saw constant-display orders within problems were slightly faster to initiate problems and slightly slower to enter answers, compared to those who saw varied-display orders, but neither difference was significant: $t(32) = 0.70$ and 1.6. Mean step times varied with direction and distance, but because the effects of display order were apparent in all cases, these results are not reported here.

Experiment 3 replicated the major results of experiment 1. Subjects completed steps more quickly when operators (directions) were available before operands (distances), regardless of whether display order was constant or varied within problems. The substantial, though marginally significant, effect of whether display order was constant or varied was likely due to the need to use different strategies for coordinating spatial operations and for maintaining intermediate locations as a function of display order.

19.5 EXPERIMENT 4: SPATIAL PATH CONSTRUCTION WITH WORKING-MEMORY PRELOAD OF DIRECTIONS OR DISTANCES

Experiment 4 investigated whether the parallel between arithmetic and spatial tasks established by experiments 1 and 3 would extend to the case in which subjects held the series of directions or of distances in working memory. If the parallel held, performance should be more fluent when directions are held in working memory than when distances are held. We expected that subjects would hold intermediate results not by verbal rehearsal but by spatial strategies, such as fixing their gaze or visual attention on the appropriate cell of the grid, or by coding spatial relations, such as a vector relating the location to the central circle in the display. Pilot research demonstrated that subjects could pick up operator and operand information from the central circle while maintaining intermediate locations, and no pilot subject reported a verbal strategy for maintaining intermediate locations. If this was the case in experiment 4, and the interaction of step and type of advance information observed in experiment 2 (figure 19.3) was due to the need to hold intermediate results and the memory preload in the same format, that interaction should not be observed here.

The experimental task was like that in experiment 3, except that subjects saw the series of directions or distances prior to the start of each problem, and the remaining piece of information appeared at each self-paced step.

Subjects

Nineteen students from introductory psychology classes at Pennsylvania State University, who had not participated in experiments 1–3, participated in exchange for course credit. All reported normal or corrected-to-

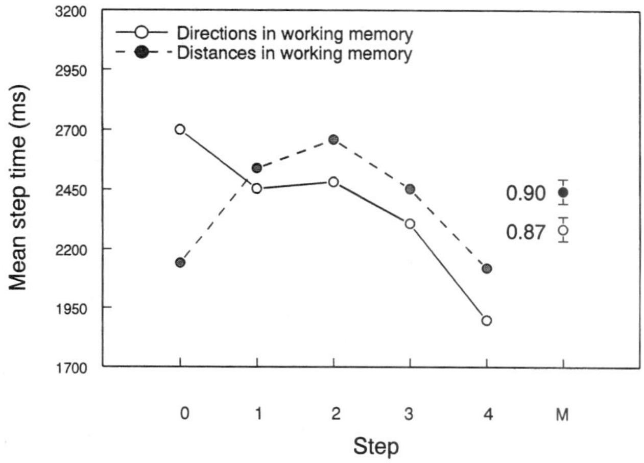

Figure 19.6 Mean step-by-step latency in experiment 4 as a function of experimental conditions. Markers with error bars at the right of the figure show means for steps 1–4 together with standard errors.

normal vision. Three subjects failed to reach a criterion of 70% correct, and their data were excluded from analysis.

Design and Procedure

The experimental design and procedure were the same as those used in experiment 2, except for the modifications based on using the spatial task described earlier. Here the elements to be held in working memory appeared for 500 msec each in the center of the grid, separated by 100 msec, when the center of the grid was blank. After the last memory item appeared, the starting location was displayed until subjects initiated the problem by pressing the space bar to request the remaining information for the first step.

Results and Discussion

Subjects answered 87% of the distance-first problems and 90% of the direction-first problems correctly, a nonsignificant difference: $t(15) = 1.15$, $p > 0.2$. The remaining analyses focus on latency for correct problems.

The latency data supported our predictions. Figure 19.6 displays initiation times and mean latencies as a function of step within problem. Subjects took longer to begin the problem when holding the sequence of directions in working memory than when holding the sequence of distances: $t(15) = 3.62$, $p < 0.005$. On subsequent steps, they were faster when directions rather than distances were in working memory: $F(1, 15) = 20.4$, $p < 0.001$. As in experiment 2, this analysis included only steps $1-4$, because initiation time includes time to prepare the memory load, but not calculation time. As suggested by figure 19.6, latency varied across steps:

$F(3, 45) = 22.3$, $p < 0.001$; but the effects of type of advance information and step did not interact: $F(3, 45) = 0.95$. Participants entered their answers an average of 805 msec after the information for the final step appeared, and this did not vary with type of information in working memory: $t(15) = 1.21$, $p > 0.2$. Step times also varied as a function of direction and distance, but because the effect of type of advance information was apparent in all cases, these results are not reported here.

Experiment 4 replicated the major result observed in experiments 1 – 3. Subjects performed steps more quickly when operator rather than operand information was available in advance. The longer preparation time between receiving the memory load and initiating the problem was more prominent than in experiment 2, possibly reflecting procedural encoding of the directions. Consistent with this speculation, some subjects reported imagining a path (presumably using procedures similar to those involved in constructing a path step by step) in order to remember the sequence of directions. In contrast to experiment 2, there was no interaction between type of advance information and step, reflecting the availability of different working-memory strategies for maintaining intermediate results and the memory load of directions or distances.

19.6 A BIGGER PICTURE: LEARNING TO CONTROL MENTAL ROUTINES

The present experiments provide evidence for the operator-operand processing sequence suggested by the procedural frame hypothesis, and for parallel control structures for symbolic and spatial problem solving. They also demonstrate the interaction of control and storage requirements, when storage must be updated dynamically to manage both intermediate results and operators or operands held in working memory. Subjects took, on average, 600–700 msec longer per step when using information from working memory than when information was available on-line, demonstrating the value of control by "just-in-time" pickup of information (Ballard, Hayhoe, and Pelz 1995; Carlson et al. 1990).

These results were obtained with tasks that explicitly distinguished operators and operands in order to realize our experimental manipulations. On the other hand, operator-operand structure may be a general characteristic of the control of mental activity. In particular, this structure may make possible the fluent performance achieved by overlapping sequential steps, for example, picking up or retrieving information that specifies an operator for the next step while calculating the result of the current step.

Several lines of evidence support this conjecture. First, learners speed up more with practice when sequences of operators are consistent rather than varied, even if operands vary from problem to problem (Carlson and Lundy 1992). Second, the opportunity to preview upcoming operators results in faster performance than a no-preview condition (Carlson and Shin 1996). Both results suggest that fluency depends in part on early instantiation of goals for problem-solving steps. Third, subjects in a task-switching experiment may prepare for an upcoming step during performance of the previous step (Sohn and Carlson forthcoming). When foreknowledge of task switches is available on a global timescale (i.e., a block of trials), responses on preswitch steps are slowed, while responses on switch steps are speeded, relative to cases in which no foreknowledge is available.

Instantiating a goal in advance may require individuals to anticipate the time course of their own mental processes. For example, subjects in Carlson, Shin, and Wenger 1994 performed a running arithmetic task, pressing the space bar to request a display for each step. The time between this keypress and the display was either 200 or 1,000 msec, manipulated between subjects. Early in practice, there was no difference between groups in stepwise latency, suggesting that subjects simply pressed the key after completing each step. With practice, however, latencies became shorter for subjects with the longer delays, suggesting that they anticipated when they would be ready for the next step. Subjects performing a more complex arithmetic task adjusted the rate of performance to the time required for individual steps (Sohn and Carlson in preparation). Step time was manipulated by varying the values of operands (based on the problem size effect for simple arithmetic; e.g., LeFevre, Sadesky, and Bisanz 1996). In a large-digit version of the task, subjects who learned the routine with small digits performed more quickly than those who had practiced with large digits all along, demonstrating that performance speed was a learned parameter rather than simply a consequence of other factors.

These findings speak to how learners weave multiple steps together into fluent sequences in later stages of skill acquisition, when performance is being adjusted at the fine-grained level of individual steps or transitions between steps. During earlier stages, learners find ways to organize component skills to accomplish tasks, ways that satisfy both cognitive and situational constraints. These earlier stages of skill acquisition are beyond the scope of this chapter.

19.7 DISCUSSION

The major results of these experiments supported the prediction of the procedural frame hypothesis: that subjects would complete multistep routines more quickly when information specifying operators was available before information specifying operands. This was true for both arithmetic and spatial tasks, and for information both acquired on-line and held in working memory. The consistent pattern of results provides support for the procedural frame hypothesis and, regarding cognitive control, for the suggested parallels between the points of view of information-processing dynamics and of conscious intentions. Let us examine these results and their implications from the perspective of alternative accounts.

Linguistic Habits

The processing sequence we hypothesize corresponds to the word order of English-language imperatives (e.g., "Add three."), or more generally to the standard verb-object syntax of English. One possible alternative account for the outcome we observed is that individuals more fluently process information in an order that corresponds to the syntax of their native language. If our results reflect a linguistic phenomenon rather than a more fundamental property of cognitive control, they might be reversed in native speakers of non-Indo-European languages (e.g., Korean) where the syntactic order is the reverse of that for English.

There are at least four reasons to doubt this alternative account. First, in the arithmetic task, participants were faster on operator-first trials both for tasks with lexicalized operators (e.g., "Add") and for those not usually lexicalized as single words (e.g., "Take the MINimum of"). Second, the similar results for spatial and arithmetic tasks cast doubt on the linguistic account because operator information in the spatial task leaves the verb ("move") implicit. Third, with single-step arithmetic problems, performance was faster when the operator symbol came before both operands, for example, +(2,3), rather than appearing in its conventional middle position, for example, 2 + 3 (Sohn and Carlson 1998), suggesting that conventional reading order is not responsible for these results. And fourth, the finding that performance on such a cognitive task is strongly determined by a language-specific grammatical feature would be dramatic evidence for a Whorfian hypothesis, which is at odds with most of the literature on linguistic influences on cognition (e.g., Hunt and Agnoli 1991). Nevertheless, it would be useful to repeat these experiments with native speakers of a language that uses verb-final structures for imperative sentences.

Strategy and Memory Effects

Several alternative accounts are based on assumptions about the strategies participants might apply to hold and use operator and operand information in working memory. For example, a participant might use knowledge of the possible operands (the numbers 2–5 in these experiments) together with a just-displayed operator to generate the four possible answers, then select among those answers when operand information becomes available. Again, there are several reasons to doubt this alternative. First, in all of these experiments the number of possible operators (four) was the same as the number of possible operands, so that such a strategy would be possible with either kind of advance information. Second, subjects in a single-step arithmetic study (Biederman 1973) did not use this strategy when more than two operators could appear. Third, given the relatively short step times observed, it is unlikely that our subjects applied such a strategy.

Another alternative account depends on presumed interference in working memory. For example, one reason arithmetic performance was slower when operands appeared first may be that operands are more likely to interfere with retention or retrieval of intermediate results held in working memory. Although the results of experiment 2 demonstrate that such interference is possible, as already argued, the parallel results for the arithmetic and spatial tasks weigh against this possibility as a general account.

Evidence from studies of task switching (e.g., Allport and Wylie, chap. 2, this volume) suggests that there are long-term proactive interference effects from prior tasks. Given that different operators were applied at each step, such effects may be present in our tasks. For example, if an intention from a prior step is still active—as it might be in the operand-first case—one might expect interference based on an aborted application of that now-inappropriate operator to the new operand. Given that humans are almost always doing *something*, and that moment-to-moment intentions adopted by experimental participants are embedded in a hierarchical goal structure, such effects may be practically irreducible, and truly neutral baselines difficult or impossible to find. Understanding such effects is, however, relevant to understanding the effects of context on goal instantiation.

Structure of Tasks

One way of viewing the present studies is as a fine-grained examination of how most effectively to communicate to experimental subjects the information needed to construct and complete mental procedures. Research on procedural instructions (e.g., Bovair and Kieras 1991) has addressed similar questions, typically at a somewhat larger grain size and with more complex instructions. One possibility, however, is that the *use order principle*—which states that the best order for presenting procedural information is the order in which it will be used—could account for our data (Bovair and Kieras 1991). For example, in our spatial tasks, subjects might first move in a direction (specified by an operator), then decide where to stop (specified by an operand). As Bovair and Kieras note, at the micro level of individual steps, applying a use order analysis depends critically on appropriately identifying those steps. The operator-operand analysis—based on a theoretical link between intentions and basic actions—provides one basis for such identification, and the use order principle at the micro level of analysis might correspond to the structure of conscious intentions.

Procedural Frames and Agency

Research on cognitive control is informed by viewing ourselves and our research subjects as agents pursuing deliberate goals. This can be seen in choices of experimental tasks and paradigms for studying control, the selection of patients thought to show disorders of control, and the commonsense way in which theoretical constructs and empirical observations are described. In particular, scientists studying cognitive control must, to interpret their data, assume (and occasionally verify) that subjects have understood instructions and adopted specific intentions. It is therefore important to bring our understanding of conscious intentions into contact with theoretical proposals about cognitive control.

Some of the earliest presentations of the computational approach to cognition addressed control issues (e.g., Miller, Galanter, and Pribram 1960; Newell, Shaw, and Simon 1958), and substantial computational resources are currently available for considering control issues (e.g., Anderson and Lebiere 1998; Meyer and Kieras 1997; Kieras et al., chap. 30, this volume; Newell 1990). On the other hand, these approaches generally regard conscious experience of agency—insofar as they regard it all—as an "extra" problem, to be addressed after computational theorizing is done (e.g., Jackendoff 1987). Thus computational theorizing about cognitive control has not been adequately linked with research either on how children and adults think about agency (e.g., Hauser and Carey 1998; Vallacher and Wegner 1987) or on consciousness (e.g., Carlson 1997). One of our goals in the research program from which the current experiments are drawn is to establish such links.

Three specific links are supported by the present results. First, just as experimental instructions are effective when they are adopted as intentions to perform tasks, so goals control cognitive activity when they are instantiated as intentions to apply operators. Second, just as intentions or other psychological modes (e.g., believe, see) that participate in the structure of conscious states specify perspectives from which objects are considered or viewed (Carlson 1997), so operators and the intentions they support specify frames to which operands are assimilated. And third, just as similar formal structures relating self, action, and object can be identified in both symbolic and perceptual awareness (Carlson 1997) that are supported by information from memory or from the environment, so similar information-processing dynamics can be observed in both symbolic and spatial tasks, based on information in memory or currently displayed. Agency in multistep mental activities, then, may be described in information-processing terms as series of instantiated goals that both control activity and constitute points of view on one's own actions and the objects of those actions. Establishing links such as these will, in our view, lead to a convincing solution to the homunculus problem.

NOTE

This work was supported in part by National Science Foundation grant NSF RED-9554504. We thank Lori Forlizzi, Lynn Liben, David Rosenbaum, and two anonymous reviewers for their helpful comments on earlier drafts.

REFERENCES

Anderson, J. R. (1983). *The architecture of cognition*. Cambridge, MA: Harvard University Press.

Anderson, J. R. (1993). *Rules of the mind*. Hillsdale, NJ: Erlbaum.

Anderson, J. R., and Lebiere, C. (1998). *The atomic components of thought*. Mahwah, NJ: Erlbaum.

Austin, J. T., and Vancouver, J. B. (1996). Goal constructs in psychology: Structure, process, and content. *Psychological Bulletin, 120*, 338–375.

Baddeley, A. D. (1986). *Working memory*. Oxford: Oxford University Press.

Ballard, D. H., Hayhoe, M. M., and Pelz, J. B. (1995). Memory representations in natural tasks. *Journal of Cognitive Neuroscience, 7*, 66–80.

Biederman, I. (1973). Mental set and mental arithmetic. *Memory and Cognition, 1*, 383–386.

Bovair, S., and Kieras, D. E. (1991). Toward a model of acquiring procedures from text. In R. Barr, M. L. Kamil, P. Mosenthal, and P. D. Pearson (Eds.), *Handbook of reading research*, vol. 2, pp. 206–229. White Plains, NY: Longman.

Carlson, R. A. (1997). *Experienced cognition*. Mahwah, NJ: Erlbaum.

Carlson, R. A., Khoo, B. H., Yaure, R. G., and Schneider, W. (1990). Acquisition of a problem solving skill: Levels of organization and use of working memory. *Journal of Experimental Psychology: General, 119*, 193–214.

Carlson, R. A., and Lundy, D. H. (1992). Consistency and restructuring in learning cognitive procedural sequences. *Journal of Experimental Psychology: Learning, Memory, and Cognition, 18*, 127–141.

Carlson, R. A., and Shin, J. C. (1996). Practice schedules and subgoal instantiation in cascaded problem solving. *Journal of Experimental Psychology: Learning, Memory, and Cognition, 22*, 157–168.

Carlson, R. A., Shin, J. C., and Wenger, J. L. (1994). Timing and the control of fluent cognitive sequences. Paper presented at the Thirty-fifth Annual Meeting of the Psychonomic Society, St. Louis, November 1994.

Carlson, R. A., Sullivan, M. A., and Schneider, W. (1989). Practice and working memory effects in building procedural skill. *Journal of Experimental Psychology: Learning, Memory, and Cognition, 15*, 517–526.

Carlson, R. A., Wenger, J. L., and Sullivan, M. A. (1993). Coordinating information from perception and working memory. *Journal of Experimental Psychology: Human Perception and Performance, 19*, 531–548.

Crosby, A. W. (1997). *The measure of reality: Quantification and Western society, 1250–1600.* Cambridge: Cambridge University Press.

Hauser, M., and Carey, S. (1998). Building a cognitive creature from a set of primitives: Evolutionary and developmental insights. In D. D. Cummins and C. Allen (Eds.), *The evolution of mind,* pp. 51–106. New York: Oxford University Press.

Hunt, E., and Agnoli, F. (1991). The Whorfian hypothesis: A cognitive psychology perspective. *Psychological Review, 98,* 377–389.

Jackendoff, R. (1987). *Consciousness and the computational mind.* Cambridge, MA: MIT Press.

LeFevre, J., Sadesky, G. S., and Bisanz, J. (1996). Selection of procedures in mental addition: Reassessing the problem size effect in adults. *Journal of Experimental Psychology: Learning, Memory, and Cognition, 22,* 216–230.

Mandler, G. (1984). *Mind and body.* New York: Norton.

McClelland, J. L. (1979). On the time relations of mental processes: An examination of systems of processes in cascade. *Psychological Review, 86,* 287–330.

Meyer, D. E., and Kieras, D. E. (1997). A computational theory of executive cognitive processes and multiple-task performance: 1. Basic mechanisms. *Psychological Review, 104,* 3–65.

Miller, G. A., Galanter, E., and Pribram, K. H. (1960). *Plans and the structure of behavior.* New York: Holt, Rinehart, and Winston.

Newell, A. (1990). *Unified theories of cognition.* Cambridge, MA: Harvard University Press.

Newell, A., Shaw, J. C., and Simon, H. A. (1958). Elements of a theory of human problem solving. *Psychological Review, 65,* 151–166.

Newell, A., and Simon, H. A. (1972). *Human problem solving.* Englewood Cliffs, NJ: Prentice-Hall.

Piaget, J. (1954). *The construction of reality in the child.* New York: Basic Books.

Rogers, R. D., and Monsell, S. (1995). Costs of a predictable switch between simple cognitive tasks. *Journal of Experimental Psychology: General, 124,* 207–231.

Schum, D. A. (1977). The behavioral richness of cascaded inference models: Examples in jurisprudence. In N. J. Castellan, D. B. Pisoni, and G. R. Potts (Eds.), *Cognitive theory,* vol. 2, pp. 149–173. Hillsdale, NJ: Erlbaum.

Siegler, R. S. (1988). Strategy choice procedures and the development of multiplication skill. *Journal of Experimental Psychology: General, 117,* 258–275.

Sohn, M.-H., and Carlson, R. A. (1998). Procedural frameworks for simple arithmetic skills. *Journal of Experimental Psychology: Learning, Memory, and Cognition, 24,* 1052–1067.

Sohn, M.-H., and Carlson, R. A. (Forthcoming). Effects of foreknowledge on disengagement and engagement in task set reconfiguration.

Sohn, M.-H., and Carlson, R. A. (In preparation). Coordinative modes and the transfer of cognitive skill.

Vallacher, R. R., and Wegner, D. M. (1987). What do people think they're doing: Action identification and human behavior. *Psychological Review, 94,* 3–15.

20 Real-World Multitasking from a Cognitive Neuroscience Perspective

Paul W. Burgess

ABSTRACT This chapter examines the demands made by multitasking situations in the real world, and argues that the human brain systems critical in dealing with them may be surprisingly circumscribed. Four kinds of evidence are considered: single-case studies of patients with selective multitasking problems; group studies of the relationship between multitasking failures and other cognitive control problems; the neuroanatomical locus of multitasking deficits according to group lesion studies, and evidence from functional imaging. These studies suggest three distinct brain systems are involved in supporting the retrospective memory, prospective memory and planning demands of multitasking, and tentative suggestions for the neuroanatomical correlates of these systems are proposed.

In a recent television program, the U.S. astronaut Jerry Linenger described his experiences aboard the Mir space station: "We had many system failures and they were in need of your constant attention. Many days I'd start an experiment in the morning and then I'd run over and help hacksaw through a pipe and plug the ends and then run back to my experiments. I'd have three or four watches on with alarms set to different things that I'd have to run back to. So I was multitasking in order to try to get everything accomplished."

Although, at first sight, Jerry Linenger's use of the term *multitasking* accords well with the *Compact Oxford English Dictionary* definition: the "ability to perform concurrent tasks or jobs by interleaving," his account suggests something more complex than interleaving tasks in a multiple-task sense. The situation he faced also required further mental activities, such as prioritization, planning, and prospective memory (i.e., the realization of a delayed intention; Ellis 1996).

The ability to deal with such complex situations is clearly important to effectiveness in everyday life. Neurological patients who have lost this ability are severely handicapped, especially in work situations. However, although the present volume is testament to recent advances in understanding many situations which have some relevance to aspects of multitasking (e.g. dual- or multiple-task paradigms, task switching etc), more complex situations akin to those faced by Jerry Linenger have been rarely studied within an experimental psychology or cognitive neuroscience framework. Indeed, the complexity of such situations would seem to make them poor candidates for scientific investigation. However, recent findings, principally from human neuropsychology, suggest that, to the contrary, such multitasking makes demands on a relatively discrete set of resources, and thus may be experimentally tractable. Before examining these findings, let us briefly review the characteristics of these situations.

20.1 THE DEMANDS OF REAL-WORLD MULTITASKING

Although the multitasking situation that faced Jerry Linenger was highly atypical in its setting and its seriousness, its actual characteristics were not unlike those of situations commonly faced in everyday life:

1. *Numerous tasks:* A number of discrete and different tasks have to be completed.

2. *One task at a time:* Due to physical or cognitive constraints, it is not possible to perform more than one task at a time.

3. *Interleaving required:* Performance on these tasks must be dovetailed; the most time-effective course of action is not to completely finish one task before moving to another, but to switch between them as appropriate.

4. *Delayed intentions:* The time for a switch or return to a task is not signaled directly by the situation. Jerry Linenger adopts the use of watch alarms in order to reduce this problem.

In addition, most busy everyday multitasking situations will share three further characteristics:

5. *Interruptions:* Occasionally, interruptions and unforeseen circumstances will occur.

6. *Differing task characteristics:* Tasks usually differ in terms of priority, difficulty, and the length of time they will take.

7. *No feedback:* People decide for themselves what constitutes adequate performance, and there is no minute-by-minute performance feedback of the sort that participants receive in, for instance, a typical "psychological refractory period" (PRP) dual-task experiment, where errors are apparent.

Although not every multitasking situation will have all these characteristics, it is arguably easier to think of generic everyday activities lasting several minutes or more (e.g., cooking, shopping) that have these characteristics than it is to think of ones that do not.

20.2 SINGLE-CASE STUDIES: PATIENTS WITH SELECTIVE MULTITASKING IMPAIRMENTS

The assertion that there may be discrete brain systems supporting performance in these situations is initially based on neurological patients with "strategy application disorder" (Shallice and Burgess 1991;

Goldstein et al. 1993; Levine et al. 1998), a cluster of symptoms whose cardinal feature is an impairment that manifests itself particularly in multitasking situations of the kind just outlined. Shallice and Burgess (1991) described three patients, all of whom had suffered frontal lobe damage, but who had superior IQs and no significant deficits in language, memory, or visual-perceptual functions, and at least one of whom was unimpaired on a wide range of cognitive tests traditionally considered sensitive to frontal lobe lesions (e.g., Wisconsin Card-Sorting Test, Tower of London, Cognitive Estimates, Verbal Fluency). Despite their lack of apparent disability on traditional psychometric examination, all three had made unsuccessful attempts to return to work, with employers complaining of tardiness, disorganization, and inability to meet deadlines or to finish lengthy projects.

Shallice and Burgess demonstrated these patients' problems by constructing two multitasking tests. The first, called the "Multiple Errands Test" (MET) was a real-world shopping task, where the subjects also had to follow a series of rules such as "No shop should be entered other than to buy something" or "On leaving a shop you must always inform an experimenter what you have bought there" while purchasing a series of items, finding out some information (e.g., Where was the coldest place in Britain yesterday?), and meeting the experimenters at a certain place at a prespecified time.

In the second multitasking test, designed for use in the laboratory and called the "Six Element Test" (SET), subjects were faced with three different tasks, (describing memorable events; writing the answers to simple arithmetic sums; and writing the names of items shown in simple line drawings), each of which is split into two sections, A and B. Subjects were told that they had 15 minutes to score as many points as they could, given that (1) within each section, earlier items scored more points than later ones and (2) they were not permitted to perform section A of a particular task directly followed by section B of that same task.[1] The subjects were told that otherwise they were free to organize their performance in any way they liked, and they were not given any other information (e.g., about the exact "point value" of items). In this way, their tasks met all the characteristics of everyday multitasking situations outlined above except characteristic 5 (unforeseen interruptions).

Shallice and Burgess's frontal lobe patients (1991) all showed impairments on both these multitasking tests, compared with age- and IQ-matched controls. Of especial interest was the finding that their work rates on the SET were normal: their difficulties consisted of failures to switch tasks and to follow the simple task rules. Similar cases have been reported by Penfield and Evans (1935); Eslinger and Damasio (1985); Goldstein et al. (1993) and Duncan, Burgess, and Emslie (1995; see also Levine et al. 1998).

20.3 GROUP STUDY: DYSEXECUTIVE PATIENTS

If tests like the Six Element Test measure processes specific to multitasking, one should be able to demonstrate their discriminative validity by finding stronger relationships between performances on these tests and everyday multitasking problems than occurs with other measures, such as memory or IQ tests or even other executive tests (e.g., Wisconsin Card-Sorting Test, Verbal Fluency) traditionally associated with frontal lobe damage.[2] In a study of this kind (Burgess et al. 1998), the caregivers or close relatives of 92 neurological patients of mixed etiology were asked to rate the frequency of occurrence of twenty of the most common dysexecutive symptoms in the patients they knew well. When the results were subjected to factor analysis (orthogonal rotation), five factors appeared: inhibition (deficits in response suppression and disinhibition); intentionality (deficits in planning, plus distractibility and poor decision making that could be expected to interfere particularly with real-world multitasking); executive memory (e.g., confabulation, perseveration); positive affective changes; and negative affective changes. Of all the tests given, which included measures of intelligence, memory, language, and visual perception, as well as ten measures of executive function, only one—the Six Element Test—correlated significantly with the factor scores for intentionality: $r = 0.46$, $p < 0.001$ criterion. This occurred despite many significant relationships between the other neuropsychological tests and the inhibition and executive memory factors. Thus it would seem that the Six Element Test measures something not measured by other neuropsychological tests and that this function is relevant to intentionality in everyday life. A related finding is that multitasking deficits are not necessarily accompanied by other symptoms of the dysexecutive syndrome (e.g. confabulation, perseveration).

20.4 GROUP STUDIES OF PATIENTS WITH CIRCUMSCRIBED CEREBRAL LESIONS

Together, the results of these single-case and group studies provide strong evidence that multitasking impairments can be seen independently of other neuropsychological impairments and of other problems in everyday life. They do not explain, however, why the multitasking impairments are occurring or indicate the lesion locations causing them.

Burgess et al. (2000) have examined these issues directly by administering a multitasking test (closely resembling the Six Element Test) to 60 patients with circumscribed cerebral lesions to isolate the particular stage or stages of failure in the patients, and to see whether different lesion locations were associated with decrement at different stages.

First, before the task was attempted, we measured the speed and accuracy with which the subjects learned the task rules. Subjects were then asked how they intended to perform the task, and the appropriateness and complexity of the plan they produced was scored. Next, they performed the test itself, and this was scored as the number of task switches minus the number of rule breaks. A measure of "plan following" was derived by comparing actual test performance with the reported plan. Finally, after the task was completed, subjects were asked to recall (1) what they had done (a measure of autobiographical recollection) and (2) what the task rules were (delayed recall). In this way, it was possible to examine the relative contributions to multitasking performance of task learning and remembering, planning, plan following, and remembering one's actions.

Lesions to the left posterior cingulate and regions in the vicinity of the forceps major gave deficits on all measures except planning. Remembering task contingencies after a delay was also affected by lesions to the left anterior cingulate, and rule breaking and failures of task switching were additionally found in patients with lesions affecting the medial aspects of Brodmann's areas 8, 9, and 10 in the left frontal lobe. Planning deficits were associated with lesions to right dorsolateral prefrontal cortex. Examination of the relationship between the individual task components by structural equation modeling of the data from the patients and 60 age- and IQ-matched healthy controls suggested that there are three primary constructs that underpin multitasking: retrospective memory, prospective memory, and planning.

The data further suggested that the second and third draw on the products of the first. The left anterior and posterior cingulates (plus regions surrounding and the forceps major) appear to play some part in the retrospective memory demands of multitasking (e.g., learning and remembering task rules), whereas prospective memory (e.g., rule following and task switching) makes demands on the processes supported by left frontal areas 8, 9, and 10, with the right dorsolateral prefrontal cortex playing a critical part in planning.

20.5 FUNCTIONAL IMAGING STUDIES

Although current functional imaging technology cannot examine entire multitasking performance on tests with the complexity and duration of the Six Element Test, it can examine specific contributory components in isolation, and a recent study of this kind in our laboratory shows promising concordance with the lesion studies already outlined.

We (Burgess, Quayle, and Frith forthcoming) used positron-emission tomography (PET) to examine the brain regions involved in maintaining and realizing a delayed intention (known as "prospective memory"). The behavioral analogues in the Six Element Test would be plan following, rule following, and task switching. In this study, eight healthy subjects were given four different prospective memory tasks under two randomized conditions. In the "expectation condition," subjects were expecting to see a prospective memory (PM) stimulus, but during the PET scanning period one never occurred. In the "realization condition," subjects were expecting a PM stimulus, and it did occur. In both conditions, subjects were engaged in a foreground task of sufficient difficulty to prevent conscious intention rehearsal; a baseline condition involving only the foreground task was also given.

For the expectation condition, relative to the baseline, regional blood flow (rCBF) increased in Brodman's area 10 of the frontal lobes bilaterally, right dorsolateral prefrontal cortex (RDLPFC), precuneus, and inferior regions of the right parietal lobe. In the realization condition, relative to the expectation condition, rCBF increased in the thalamus and decreased in RDLPFC. The findings for area 10 and RDLPFC are concordant with data from our group lesion study described in the previous section. We concluded that these regions are involved in the creation and maintenance of intentions, with other regions, such as thalamus, anterior and posterior cingulates, and forceps major, supporting retrospective and prospective memory (see Burgess and Shallice 1996 for discussion of the relationship between prospective and retrospective memory).

20.6 CONCLUSIONS

Although the apparent complexity of multitasking would seem to make scientific investigation of this human activity problematic, recent results from cognitive neuroscience suggest that this may not be the case. This chapter has reviewed a series of investigations observations of behavior in real-world situations, covering the development and validation of experimental tasks designed to make similar demands, examination of the brain regions that, when damaged, lead to poor multitasking performance and their relative roles in performance, and (functional imaging) results that show promising cross-method concordance. The two principal conclusions to emerge from all of this are (1) the control processes involved in multitasking may be usefully seen as distinct from many other control and general cognitive functions; and (2) there may be a more straightforward mapping between these processes and the activity of specific brain regions than might initially be supposed.

There are, however, many aspects of multitasking in ill-structured situations which would be most appropriately investigated by the methods of cognitive and experimental psychology. The present chapter is intended as an appeal to my colleagues in this field to consider them scientifically tractable.

NOTES

This work was supported by Wellcome Trust grant 049241/Z/96/Z/MRE/HA/JAT.

1. In the version of SET now in common clinical use (Burgess et al. 1996), the test period is 10 minutes, and the first rule is simplified: "You must attempt at least some of all the six subtasks."

2. The terms *executive tests* or *tests of executive function* are used in the neuropsychological literature to designate tests that have a strong "cognitive control" component (e.g., response suppression, planning tests). Although such tests were often referred to as "frontal lobe tasks" because deficits on them were most often seen in patients with frontal lobe damage, Baddeley and Wilson (1986) pointed out that doing so confused anatomical and psychological descriptions. They proposed the alternative, now more common "executive tasks,"; patients (usually with frontal lobe damage) who show a range of executive control deficits are referred to as "dysexecutive."

REFERENCES

Baddeley, A. D., and Wilson, B. A. (1986). Amnesia, autobiographical memory, and confabulation. In D. C. Rubin (Ed.), *Autobiographical memory*, pp. 225–252. Cambridge: Cambridge University Press.

Burgess, P. W. (1997). Theory and methodology in executive function research. In P. Rabbit (Ed.), *Methodology of frontal and executive functions*, pp. 81–116. Hove, U.K.: Psychology Press.

Burgess, P. W., Alderman, N., Evans, J., Emslie, H., and Wilson, B. A. (1998). The ecological validity of tests of executive function. *Journal of the International Neuropsychological Society, 4*, 547–558.

Burgess, P. W., Alderman, N., Evans, J. J., Wilson, B. A., Emslie, H., and Shallice, T. (1996) The simplified six element test. In B. A. Wilson, N. Alderman, P. W. Burgess, H. Emslie, and J. J. Evans (Eds.), *Behavioural assessment of the dysexecutive syndrome*, Bury St. Edmunds, U.K.: Thames Valley Test Company.

Burgess, P. W., Frith, C. D. and Quayle, A. (Forthcoming). Brain regions involved in prospective memory according to positron-emission tomography.

Burgess, P. W., and Shallice, T. (1997). The relationship between prospective and retrospective memory: Neuropsychological evidence. In M. A. Conway (Ed.), *Cognitive models of memory*, pp. 247–272. Hove, U.K.: Psychology Press.

Burgess, P. W., Veitch, E., de Lacy Costello, A. and Shallice, T. (2000) The cognitive and neuroanatomical correlates of multitasking. *Neuropsychologia, 38*, 848–863.

Duncan, J., Burgess, P. W., and Emslie, H. (1995). Fluid intelligence after frontal lobe lesions. *Neuropsychologia, 33*, 261–268.

Goldstein, L. H., Bernard, S., Fenwick, O., Burgess, P. W., and McNeil, J. E. (1993). Unilateral frontal lobectomy can produce strategy application disorder. *Journal of Neurology, Neurosurgery and Psychiatry, 56*, 271–276.

Ellis, J. (1996). Prospective memory or the realization of delayed intentions: A conceptual framework for research. In M. Brandimonte, G. O. Einstein, and M. A. McDaniel (Eds.), *Prospective memory: Theory and applications*, pp. 1–22. Mahwah, NJ: Erlbaum.

Eslinger, P. J., and Damasio, A. R. (1985) Severe disturbance of higher cognition after bilateral frontal lobe ablation: Patient E.V.R. *Neurology, 35*, 1731–1741.

Kimberg, D. Y., and Farah, M. J. (1993). A unified account of cognitive impairments following frontal lobe damage: The role of working memory in complex, organized behavior. *Journal of Experimental Psychology: General, 122,* 411–428.

Levine, B., Stuss, D. T., Milberg, W. P., Alexander, M., Schwartz, M., and Macdonald, R. (1998). The effects of focal and diffuse brain damage on strategy application: Evidence from focal lesions, traumatic brain injury and normal ageing. *Journal of the International Neuropsychological Society, 4,* 247–264.

Penfield, W., and Evans, J. (1935) The frontal lobe in man: A clinical study of maximum removals. *Brain, 58,* 115–133.

Shallice, T., and Burgess, P. W. (1991). Deficits in strategy application following frontal lobe damage in man. *Brain, 114,* 727–741.

V The Neural Substrate of Control

21 Functioning of Frontostriatal Anatomical "Loops" in Mechanisms of Cognitive Control

Trevor W. Robbins and Robert D. Rogers

ABSTRACT The neurobiological and functional organization of the prefrontal cortex and the striatum is reviewed in the context of parallel, functionally segregated anatomical 'loops'. Although cortical input converges to some extent within the striatum, particular striatal sectors project back to a subset of their cortical inputs via relays in the globus pallidus and thalamus. The control of striatal outflow by direct and indirect pathways and their modulation by striatal dopamine are described, and recent attempts to provide neurocomputational models of the striatum briefly reviewed. The possible functions of cortico-striatal loops in the formation, maintenance, and shifting of cognitive set, as well as in reversal learning, are investigated using a paradigm related to the Wisconsin Card-Sorting Test (WCST) in variants for patients with frontal lobe lesions or basal ganglia disorders, for monkeys with selective and excitotoxic lesions, and for normal humans in functional imaging studies.

Making inferences about function from structure that can be applied, with all due caution, to the study of brain and behavior interrelationships can be seen as a heuristic device for constraining theory, and even for defining mechanism. This device may help resolve the conundrum of the executive functions of the prefrontal cortex and its associated structures, such as the basal ganglia. To use it, we must of course embrace an evolutionary perspective: much of what we know about the anatomical connectivity of the frontal lobes is derived from information obtained in other species. We must also be mindful, however, of the dangers of such theorizing from structural evidence: while it may stimulate research, it can never replace conclusions arrived at from well-designed behavioral experiments in humans and other species.

The general notion that the prefrontal cortex plays an important role in higher cognitive functions, including the ill-defined category of "executive functions" that serve to optimize performance, is hardly controversial. It is, for example, consistent with the steady increase in size of this region within the primate order from 8.5% of the total cortex in lemurs, to 11.5% in macaques, to 17% in chimpanzees, and to 29% in humans (Brodmann 1912). More problematic, however, is the nature and organization of executive functions supported by the prefrontal cortex (Norman and Shallice 1980; Baddeley 1986: Duncan 1986; Passingham 1993; Burgess 1997; Damasio 1998; Goldman-Rakic 1998; Petrides 1998). These variously include the scheduling of multitask performance, working-memory functions involving "on-line processing" (i.e., maintaining stimulus representations for further processing after the eliciting stimulus is no longer present), behavioral inhibition, attentional control, and the application of task or somatic markers (i.e., bodily feedback derived from previous experiences that evoke previous outcomes—in common parlance, "gut feelings"). Some theoretical positions (e.g., Goldman-Rakic 1998) have argued for unitary processes of working memory that subordinate mechanisms of inhibition and response selection to that of maintaining stimulus representations across time within anatomically discrete prefrontal cortical "modules" (see Kimberg and Farrah, chap. 32, this volume). By contrast, other positions have stressed the serial (Petrides 1998; Rushworth and Owen 1998) or the parallel or hierarchical (Dias, Robbins, and Roberts 1996; Wise, Murray, and Gerfen 1996) organization of processing routes within the prefrontal cortex. It is hardly surprising that this evident heterogeneity of function is matched by the anatomical complexity of the frontal cortex itself. These processes, which can be collectively referred to as "executive mechanisms of attentional control," including the coordinated control over both input and output mechanisms, comprise several distinct operations with probably distinct anatomical substrates.

21.1 PREFRONTAL CORTEX EXECUTIVE FUNCTION: CLUES FROM ANATOMY

Anatomical Subdivisions of the Prefrontal Cortex

Although detailed analysis of the complex anatomy of the prefrontal cortex is well beyond the scope of this chapter (see Petrides and Pandya 1994; Barbas and Pandya 1989), one pragmatic anatomical nomenclature (Wise, Murray, and Gerfen 1996) divides it into six main regions: the orbitofrontal cortex (Pfo), the ventrolateral prefrontal cortex (Pfvl), the dorsolateral prefrontal cortex (Pfdl; mainly defined as the banks of the sulcus principalis); the dorsal prefrontal cortex (Pfd); the medial prefrontal cortex (Pfm); and the frontal pole (Pfp). These six regions lie anterior to the other main components of the frontal lobes, which can be labeled as the "motor" and "premotor" (including supplementary motor area) cortex (Brodmann's areas 4 and 6). A rough mapping of some cytoarchitectonic regions in the primate prefrontal cortex is shown in figure 21.1.

Studies of neocortical development have shown that the prefrontal cortex arises from at least two moieties or "trends," the archicortical (dorsal) and paleocortical (ventral), which derive from the cingular or parahippocampal and from the parapiriform cortical regions, respectively, to meet in the anterior cortex on its dorsolateral aspect (Barbas and Pandya

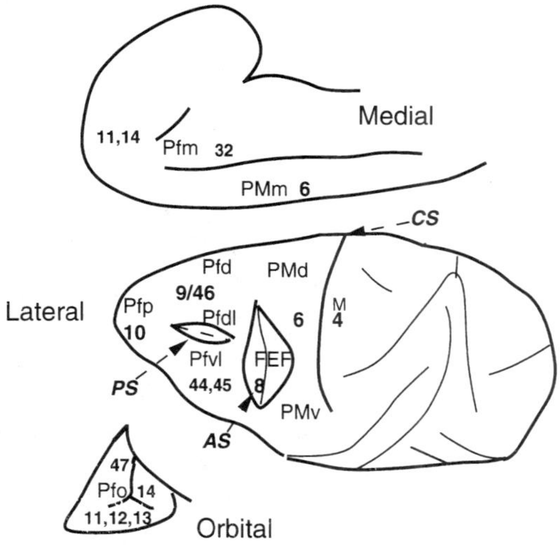

Figure 21.1 Regions and cytoarchitectonic areas (numbered Brodmann's areas) of the prefrontal cortex in the nonhuman primate. Pfm = medial prefrontal cortex; Pfd = middorsal prefrontal cortex; Pfdl = dorsolateral prefrontal cortex; Pfp = frontal pole; Pfvl = ventrolateral prefrontal cortex; Pfo = orbital prefrontal cortex; PMm = medial premotor region; PMd = dorsal premotor region; PMv = ventral premotor region; M = motor cortex, FEF = frontal eye fields; *CS* = central sulcus; *AS* = arcuate sulcus; *PS* = principal sulcus. Modified version of figure from Wise et al. 1996.

1989).[1] The main cytoarchitectonic regions contributing to the archicortical and paleocortical trends are shown in figure 21.2. What is striking is the relatively specific nature of the interconnectivity existing between the different cytoarchitectonic regions and the two trends themselves. Direct communication between these trends through interconnected neurons is not highly evident; most of the interactions appear to occur within different regions of area 8, for example, or between areas 9 and 12. This may explain why it has proven relatively easy to show double dissociations of function between Pfo and Pfd or Pfdl in human and nonhuman primates (Fuster 1989): between, say, different aspects of shifting of responding (Dias et al. 1996) or between dorsolateral working-memory functions and orbitofrontal decisional processes (Bechara et al. 1998).

Anatomical Connectivity of the Prefrontal Cortex

Some of the main interconnections of the prefrontal cortex to other brain systems are indicated in figure 21.3 (see Goldman-Rakic 1987 for an exhaustive review; see also Pandya and Yeterian 1998). First, there are the reciprocal projections to posterior cortical structures, such as the temporal and parietal cortex, as well as the parahippocampal gyrus, that can be assumed to play modulatory roles in the processing of information in these posterior cortical processing modules (see Desimone et al. 1995). Second, there are projections to the brain stem and hypothalamus that

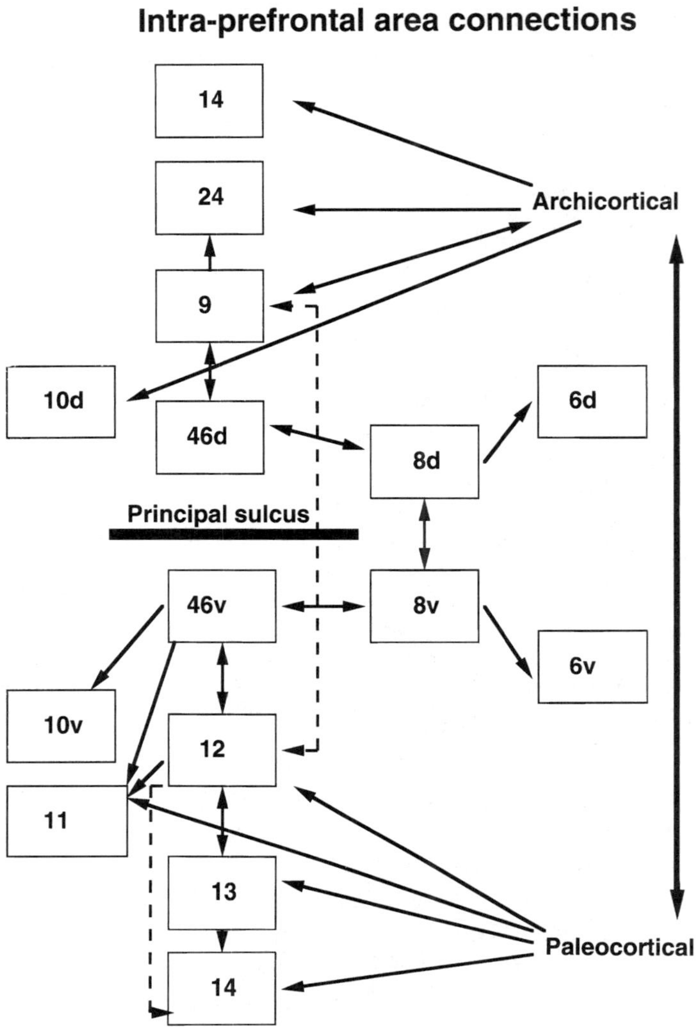

Figure 21.2 Known interconnections between cytoarchitectonic regions (Brodmann's areas) in the prefrontal cortex, showing the archicortical and paleocortical trends in development (d = dorsal; v = ventral). Data derived from Barbas and Pandya 1989; Pandya and Yeterian 1998.

indicate important functions of the prefrontal cortex in control over even basic vegetative and reflexive mechanisms (Goldman-Rakic 1987).[2] Finally, there are important connections between different regions of the prefrontal cortex and the striatum, which also include relays in the globus pallidus and thalamus. Usually discussed under the rubric of "cortico-striatal loops," these interconnections are presumed to play important roles in the executive control of output because of the well-known role of the basal ganglia in the control of action. From the bald neuroanatomical facts, two rather obvious conclusions follow: (1) the extensive neural connectivity of the prefrontal cortex with other brain regions potentially

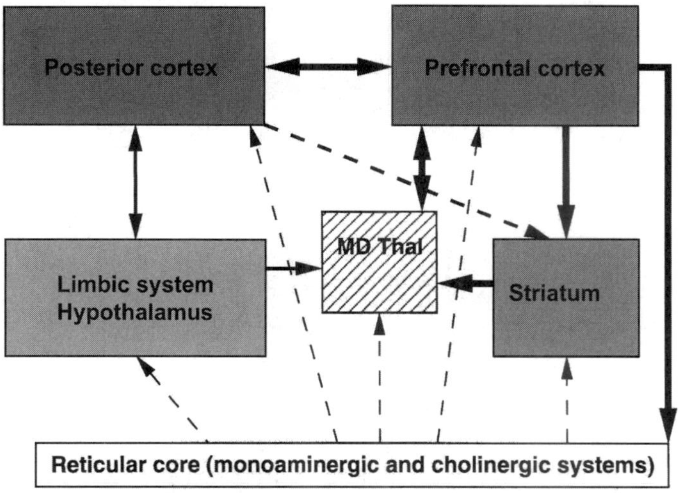

Figure 21.3 Schematic diagram of the main anatomical connections of the prefrontal cortex. MD Thal = mediodorsal nucleus of the thalamus.

enables it to exert executive control over many types of function; and (2) the often quoted paradox of patients with a "dysexecutive syndrome" and no apparent dysfunction of the prefrontal cortex may simply result from alterations in function in brain structures distal from, and yet intimately connected to, the prefrontal cortex. This appears especially to be the case for the basal ganglia, as it is apparent that many of the effects of basal ganglia damage in humans or monkeys reflect the types of executive or cognitive impairments seen following damage to the frontal cortex itself (e.g., see Divac et al. 1967; Owen et al. 1992).

Links with the Basal Ganglia

The old assumption that the basal ganglia, incorporating the neostriatum (itself consisting of the putamen and caudate nucleus) as well as the nucleus accumbens, the globus pallidus, and the substantia nigra of the midbrain, exclusively control motor function has long been superseded by evidence from a variety of sources. This includes findings from single-cell recording in nonhuman primates that even movement-related firing generally depends on environmental context, for example, on which instructional cues are present to elicit responding (Mink 1996; see also "Corticostriatal Loops Targeting the Prefrontal Cortex" in section 21.2). Moreover, classical neurodegenerative diseases of the basal ganglia, such as Parkinson's and Huntington's diseases, as well as progressive supranuclear palsy, are associated not only with motor dysfunction but with a wider range of cognitive deficits, at least some of which are likely to involve damage to particular portions of the striatum or, in the case of Parkinson's disease, their dopaminergic innervation. A pressing question

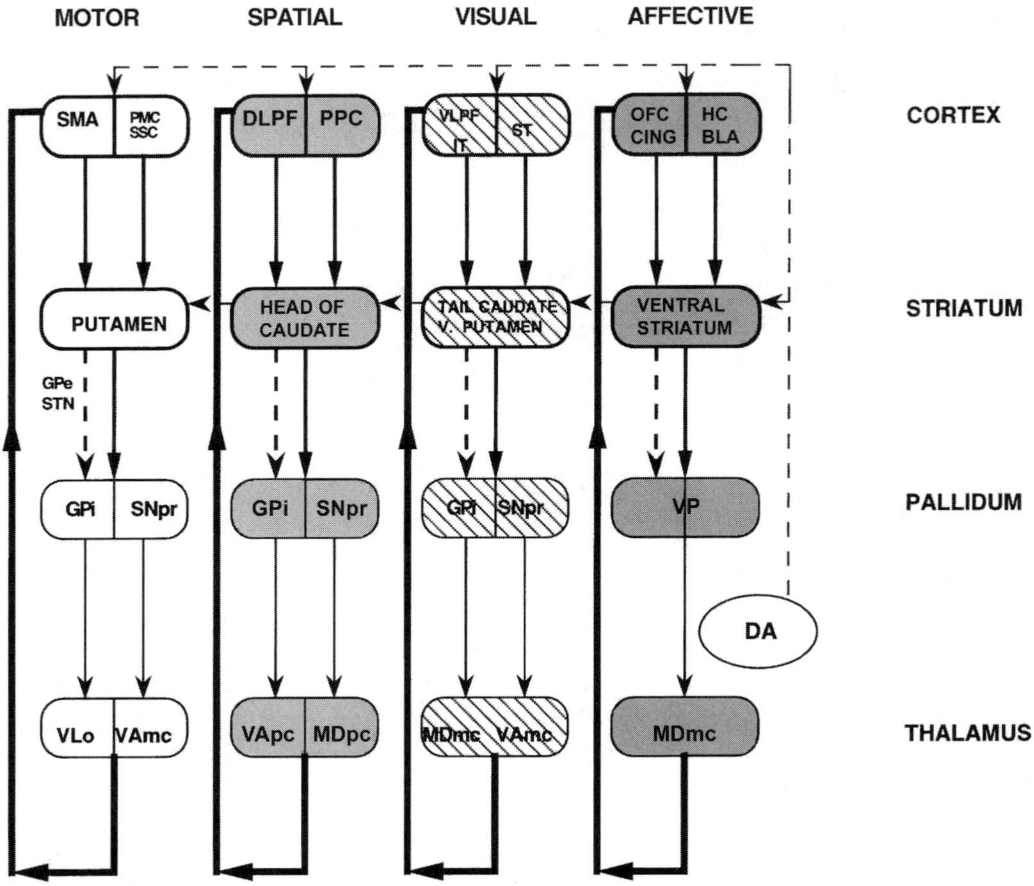

Figure 21.4 Corticostriatal loops, modified from the original scheme of Alexander, DeLong, and Strick (1986). Four of the putative segregated, parallel loops are shown with possible functions. SMA = supplementary motor area; PMC = premotor cortex; SSC = somatosensory cortex; DL-PF = dorsolateral prefrontal cortex; PPC = posterior parietal cortex; VL-PF = ventrolateral prefrontal cortex; ST = superior temporal gyrus; IT = inferotemporal cortex; OFC = orbitofrontal cortex; CING = anterior cingulate; HC = hippocampus; BLA = basolateral amygdala; V PUTAMEN = ventral putamen; Gpi = internal segment of the globus pallidus; SNpr = substantia nigra pars reticulata; VP = ventral pallidum; VLo = ventrolateral thalamus; VA = ventral anterior thalamus; MD = mediodorsal thalamus; STN = subthalamic nucleus; GPe = external segment of globus pallidus; o = pars oralis; pc = parvocellularis; mc = magnocellularis. Thick dashed lines indicate net opposed influences of the indirect over the direct striatal output pathway. Thin dashed lines reflect the modulatory influences of the mesocortical and mesostriatal dopamine (DA) systems originating in the midbrain (from, e.g., the substantia nigra pars compacta).

is whether these diverse motor and cognitive symptoms reflect a unitary processing operation of the striatum.

Anatomical advances over the past fifteen years or so have given rise to some remarkable generalizations about the relationship between the frontal lobes and the basal ganglia. The striatum via the globus pallidus (GP) and substantia nigra pars reticulata (SNr; see figure 21.4) can clearly gain access to brain stem structures such as the pedunculopontine nucleus and the tectum, which are components of the supraspinal motor system, and which have been implicated, from an early phylogenetic stage, in orientational responses of the eyes and head, as well as in forward locomotion. On the other hand, at some evolutionary stage, presumably coincidental with neocortical development, other output pathways became available to the striatal system that mainly, though not exclusively, target the frontal lobe. Evidence from neuroanatomy, electrophysiology, and some functional studies indicates that there is a systematic set of anatomical and presumed functional relationships between certain cortical and striatal regions such as the putamen and caudate, as first noted in the landmark article Alexander, DeLong, and Strick 1986, which consolidated the "parallel, segregated corticostriatal loop" hypothesis.

21.2 CORTICOSTRIATAL LOOP HYPOTHESIS

Convergence of the Corticostriatal Inputs

Most of the cerebral cortex projects to the striatum where excitatory synaptic contacts are made with medium spiny neurons, which constitute by far the greatest proportion of striatal cells. The spiny cells, so called because of the huge number of synaptic spines on their long dendrites, send inhibitory striatal outflow exclusively to the globus pallidus (GP) and the substantia nigra (SN). Anatomical studies have revealed striking patterns of projections by cortical regions onto the striatum. Kemp and Powell (1970) suggested a corticostriatal topography by which the more posterior cortical areas project to the tail of the caudate nucleus and the caudal putamen, whereas frontal regions project to the head of the caudate and the rostral putamen. By contrast, Yeterian and van Hoesen (1978) suggested a convergence of separate inputs from those cortical areas with shared functions, such as spatial processing or the control of eye movements or other aspects of motor function. That inputs from the representations of homologous body parts in the primary somatosensory and primary motor cortex have been shown to overlap in small zones in the putamen (Flaherty and Graybiel 1993) lends support to the notion of convergence.

A possible reconciliation of the Kemp-Powell and the Yeterian–van Hoesen modes of organization is provided by recent detailed anatomical studies. Corticostriatal axons make few synaptic contacts with any particular medium spiny cell but do synapse on many different striatal neurons (Wilson 1995), which means that a given cortical region such as the prefrontal cortex may project, not only densely to a specific region of the striatum, but also more diffusely to other regions that are main targets of other cortical inputs. It also means that, in order to be fired, a striatal spiny cell has to receive input simultaneously from many different cortical inputs.

Corticostriatal "Loops"

A further important feature is that regions of the striatum project back to a limited subset of the cortical regions that initially projected to them (figure 21.4). This mode of organization represents the origin and termination of seemingly partially closed corticostriatal loops in which information is fed from a number of regions to a common striatal sector, to return via the GP or SN and thalamus to a particular region of the frontal cortex. The original view was that the putamen was mainly concerned with motor functions and formed a circuit or loop with the motor cortex via the supplementary motor area, thalamus, and GP. By contrast, the caudate nucleus was hypothesized to have cognitive functions reflected in its independent complex loop completed by reentrant circuitry to the prefrontal cortex (DeLong and Georgopoulos 1981). The classic synthesis by Alexander, DeLong, and Strick (1986) extended this view by defining five such corticostriatal loops (four of which are shown in figure 21.4) for the primate brain. Although the most detailed evidence is available for a "motor loop" comprising inputs from sensorimotor regions of the cortex to the putamen and thus back to the supplementary motor area and premotor cortex, four other possible loops are identified, including one (not shown) apparently specialized for the control of eye movements and projecting to frontal eye field regions (see Rafal et al., chap. 6, this volume) one involving the parietal, Pfd or Pfdl, and other systems feeding output to the Pfo and the anterior cingulate cortex. A similar analysis has been made for the rat (Pennartz, Groenewegen, and Lopes da Silva 1994) focusing on the loops that involve the nucleus accumbens or ventral striatum. While these loops in the primate brain are probably critical for governing motivational influences over action, this review will focus on dorsal striatal loops in primates.

Although several generalizations can be made about corticostriatal organization, the picture may be radically altered by new findings and subsequent reinterpretations of the previous organizational principles. For example, it is no longer possible to make the generalization that the frontal lobe is the only cortical target of striatal outflow: a recent study has identified a projection back to the temporal cortex (Middleton and Strick 1995). Nevertheless, the close relationships between striatal outflow and the prefrontal cortex are potentially of considerable functional significance for the control of action, and perhaps even cognitive output (Wise, Murray, and Gerfen 1996).

Parallel and Segregated Nature of the Corticostriatal Loops: The Motor Loop

A yet stronger claim is that the corticostriatal loops are segregated throughout the course of their trajectory from the cortical regions to the striatum, globus pallidus, and thalamus before reentry to the cortex. Evidence used to establish such specific relationships with respect to the motor loop includes electrophysiological recording data and anatomical findings derived from the imaginative use of anatomical tracers. Important principles established by single-unit recording studies in behaving primates are (1) the specificity of the neuronal responses to active movements versus passive manipulation of individual body parts; (2) somatotopic organization of such movement neurons throughout the circuit (Alexander, DeLong, and Strick 1986); and (3) the lack of precise relationship of single-cell firing to most parameters of movement, except for its direction (Mink 1996).

A combination of techniques has been used to examine multiple stages in the organization of striatal outflow via the GP with the motor areas of the cerebral cortex (Strick, Dum, and Picard 1995). After injection of a strain of the herpes virus that is transported transneuronally in an anterograde direction into the "arm" area of the primary motor cortex, virus was transported first to neurons of the putamen and then to neurons in the external (GPe) and internal (GPi) globus pallidus. The labeled neurons in the GPe were restricted mainly to its caudal third, where one can record from neurons with activity changes related specifically to arm movements. These results suggest that a pathway from the "arm" area of the primary motor cortex targets the GPe through a specific region of the putamen. When a second herpes virus strain, one transported transneuronally in a *retrograde* manner, was injected into the "arm" area of the primary motor cortex, densely labeled neurons were initially found in subdivisions of the ventrolateral thalamus known to innervate the primary motor cortex. A few days later, virus was found to have been transported to neurons in GPi, again to specific regions with neurons exhibiting activity specifically related to arm movements. Parallel experiments in which the virus is targeted to other motor regions of the frontal cortex, such as defined "arm" regions of the premotor and supplementary motor areas, show that there are topographical projections from the GPi to these cortical regions spatially related to, yet distinct from, the projection to the motor cortex. Such findings support the proposal that there are specific input "channels" in the motor portion of the striatum, also referred to by Houk (1995) as "striatal modules."

The striatal input channels or modules may correspond to histologically defined regions referred to as "matrisomes" (matrix compartment) or possibly "striosomes" (patch compartment). For all of the corticostriatal loops, including the "motor" loop, these striatal compartments reflect largely independent channels of information processing with relatively little apparent intercommunication. They are distinguished by differences in neurochemical composition, input-output relationships, and topographic distribution within the striatum (Graybiel 1990). This anatomical complexity, superimposed on the fundamental corticostriatal loop organization, is reminiscent of a similar heterogeneity representing independent channels of information processing within the extrastriate visual cortex, where it is often suggested that visual input is analyzed in parallel streams before being subjected to "binding" operations. While there may be a similar principle at work on the very different information handled by the frontal cortex–basal ganglia circuitry, we are not yet in a position to understand the full functional significance of the patch and matrix mode of organization.

There is an apparent segregation of movement channels within the GPi, as shown by single-unit recording studies in awake trained primates. Cells preferentially active during the performance of remembered sequences of movement were located dorsally in the GPi, in the "channel" normally innervating the supplementary motor area (SMA). By contrast, those pallidal neurons that were preferentially active during the performance of sequences of movement guided by external cues were located in the ventral GPi, which gains access to the premotor areas (see Strick et al. 1995). These observations further support the notion that there is a further "nesting" of channels within each loop, in this case specialized for different types of motor processing.

Corticostriatal Loops Targeting the Prefrontal Cortex

A similar type of analysis can be made of the segregated and nested functions of the other postulated corticostriatal loops as initiated by Alexander, DeLong, and Strick (1986), although the evidence is still far less well developed than for the "motor" loop. In brief, the frontal eye fields (Brodmann's area 8) project to the central (body) portion of the caudate nucleus that also receives projections from the Pfdl and posterior parietal cortex, regions also implicated in the control of eye movements. The central body of the caudate (as distinct from its head or tail regions) projects back to the frontal eye fields via specific regions of the GPi and SNr. The precise nature of the organization within the "oculomotor" loop is not known, although it is plausible that it might contain separate channels arranged according to an appropriate coordinate system for controlling different eye movements.

From available anatomical evidence for the prefrontal loops incorporating the dorsolateral and orbitofrontal sectors in primates, the electrophysiological properties of different sectors of the striatum appear to reflect, at a functional level, their distinct profile of inputs from the cortex. An important generalization for the findings from single-unit studies is that the activity of striatal neurons is not invariably associated with specific stimuli or motor responses, but is often context dependent, in the sense of being tied to particular configurations of environmental factors, such as particular task contingencies and potentially including internal states (see Wise, Murray, and Gerfen 1996 for a review). While this electrophysiological evidence has been informative for modeling striatal functions, evidence of neural connectivity has largely depended on some exceptionally innovative, but now quite old, behavioral studies in monkeys that compared the effects of lesions to different parts of the neocortex, with the effects of electrolytic bilateral lesions to those parts of the striatum to which the cortical regions project (Divac, Rosvold, and Szarcbart 1967). Thus visual discrimination learning was impaired by lesions to the tail of the caudate nucleus, presumably reflecting its input from the inferotemporal (visual association) cortex (see figure 21.4); the spatial working-memory task of delayed alternation was impaired by lesions of the anterodorsal head of the caudate nucleus, which receives input from Pfdl; and a test of object reversal was impaired by damage to the ventrolateral head of the caudate, whose major input comes from the orbitofrontal cortex, which is also implicated in reversal learning (Jones and Mishkin 1972; Rolls 1998).

These commonalities of effects of cortical and striatal lesions, do not, by themselves, establish the operation of a serial circuitry in control of specific forms of behavior. To do this, crossed asymmetric lesion procedures are necessary, in which damage is inflicted on different nodes of a putative common system on opposite sides of the brain; a deficit obtained with asymmetric lesions compared to the effects of lesions confined to one side only of the brain provides some evidence of a common neural system. This type of logic has been used in the rat to establish some commonality in the effects of lesions of the amygdala and nucleus accumbens on reward-related preferences within the "ventral striatal loop circuitry" (Everitt et al. 1991). Although, in theory, it might be possible to make inferences about the separate contributions of the different nodes of the cortical and striatal circuitry by comparing the effects of lesions in these different nodes on performance of the same task, this route has provided surprisingly little definitive information thus far, whether in experimental animals or in humans with cortical or striatal lesions, a point to which we will return in section 21.3.

Evidence for Limited Convergence and Interloop Interactions

Although there is a strong view that information processing remains segregated in the pallidal circuitry (Alexander et al. 1986), other anatomical data indicate that this can only be partially true in view of the large reduction in numbers of neurons between the striatum and pallidum, suggesting at least some convergence of information processing there (Joel and Weiner 1994). This is supported by the existence of large, disk-shaped dendritic fields of pallidal neurons that might play such a role by receiving input from many striatal sources (Percheron, Yelnik, and Francois 1984). The other main challenge to the parallel, segregated corticostriatal loop hypothesis comes from observations of additional anatomical features that suggest that the partially closed loops, are in fact more open than is generally considered (Joel and Weiner 1994). The Alexander et al. (1986) position is that each striatal region projects to both the GPi and the SNr, which in turn project to different regions of the thalamus, before reconvergence within the same frontocortical region. An alternative position is that each striatal region innervates either the SNr or the GPi and then continues via the thalamus to different frontocortical regions (Parent and Hazrati 1993; Joel and Weiner 1994). This led Joel and Weiner (1994) to postulate the existence of "split" or partially open loops by which a striatal region might input to a cortical area that is *not* the source of innervation to this striatal zone. The functional significance of such split circuits is that they would allow some degree of interaction between the parallel segregated loops defined by Alexander, DeLong, and Strick (1986), including putative feedback and feedforward functions.

The Role of the Direct and Indirect Striatal Output Pathways and Striatal Dopamine in the Modelling of Corticostriatal Functions

How exactly is striatal outflow to the globus pallidus and substantia nigra pars reticulata controlled? The striatum has two distinct routes to the palladium, the "direct" and "indirect" pathways (see figure 21.4), which appear to arise from different pools of neurons within the matrix compartment of the striatum. Although they both use the inhibitory neurotransmitter gamma-aminobutryric acid (GABA), the two pathways are associated with different neuropeptides as products of gene expression, modulated by different types of dopamine receptor, and are opposed in their influence on the inhibitory functions of the globus pallidus. Activation of the loop therefore depends on activation of the inhibitory direct pathway, which effectively disinhibits an excitatory drive to the thalamocortical circuitry (Chevalier and Deniau 1990; see figure 21.4). The indirect pathway is routed via the GPe and the subthalamic nucleus (STN), which has an excitatory link to the GP and SN. A consecutive, double inhibitory relay to the STN delivers an excitatory influence to the GPi and SN (figure 21.4). The balance of activity in the direct and indirect pathways determines the degree of disinhibition of the pallido-thalamocortical path, and hence the level of thalamofrontocortical activity (DeLong 1990). It should be noted that the STN, which probably has a pivotal role in regulating basal ganglia output (see Berns and Sejnowski 1996) also receives important projections directly from the frontal lobe.

The two output pathways are regulated by the release of dopamine (DA) within the striatum from neurons originating in the substantia nigra pars compacta (SNc; not indicated explictly in figure 21.4). The DA system itself is regulated not only by pre-synaptic inputs from corticostriatal neurons, but also by feedback from the striosomal or patch compartment to the SNc. The DA system exerts its effects via two different types of receptor, called "D1" and "D2." These receptors may preferentially control responses of the striatal output cells, through the direct pathway (via the D1 receptor) and indirect pathway (via the D2 receptor) (see Gerfen, 1992), although this sharp dichotomy is somewhat controversial. Because the dopamine receptor predominantly associated with the indirect pathway (D2) has an *inhibitory* action, enhanced dopaminergic activity not only enhances the inhibitory influence of the direct pathway on the GP and SN via the D1 receptor, but also *reduces* the excitatory influence of the indirect pathway on pallidal output via the D2 receptor (Gerfen 1992). Both of these actions promote behavioral disinhibition (e.g., dyskinesias of the limbs). Therefore cortical and dopaminergic inputs to the striatum modulate the balance between the direct and indirect pathways in different ways. Dopamine facilitates behavior, whereas the effect of corticostriatal activity will depend on its precise profile of activity and the balance between the direct and indirect pathways. This is consistent with a role for dopamine in the motivational modulation of action (particularly in the ventral striatum; see Robbins and Everitt 1992); in "rule potentiation"—an exaggeration of a behavioral tendency or disposition established by prior training (Wise, Murray, and Gerfen 1996); and also in neural plasticity in the striatum conferred, for example, by reinforcement learning (Schultz et al. 1995). Houk (1995) considers that striatal dopamine plays an essential part in the reinforcement or synaptic strengthening that accompanies the "recognition" of various contexts, as defined by a profile of corticostriatal activity within a particular striatal channel or module (see figure 21.5). Because dopamine innervates both the patch and matrix compartments, it is possible that striatal activity operates in a successive or cascadelike manner, of significance for the transfer of representations across different sectors of the striatum, as may be important, for example, in sequence learning.

One way of conceptualizing the outcome of computations realized by striatal outflow is that it acts as a "winner loses all" mechanism for response selection (Berns and Sejnowski 1996). Because the inhibitory GPi and SNr neurons are tonically active, a GPi-SNr unit has to be "turned off" to allow disinhibition of the corresponding thalamocortical circuitry. A "winning" excitatory signal emanating from the corticostriatal circuitry achieves this by delivering inhibition to a restricted portion of the GPi and SNr via the direct pathway and by focusing that inhibition still further via the indirect pathway (Mink 1996), possibly by cancelling or inhibiting the effect of a given constellation of stimuli or context ("context negation"). This may explain why STN lesions in humans produce

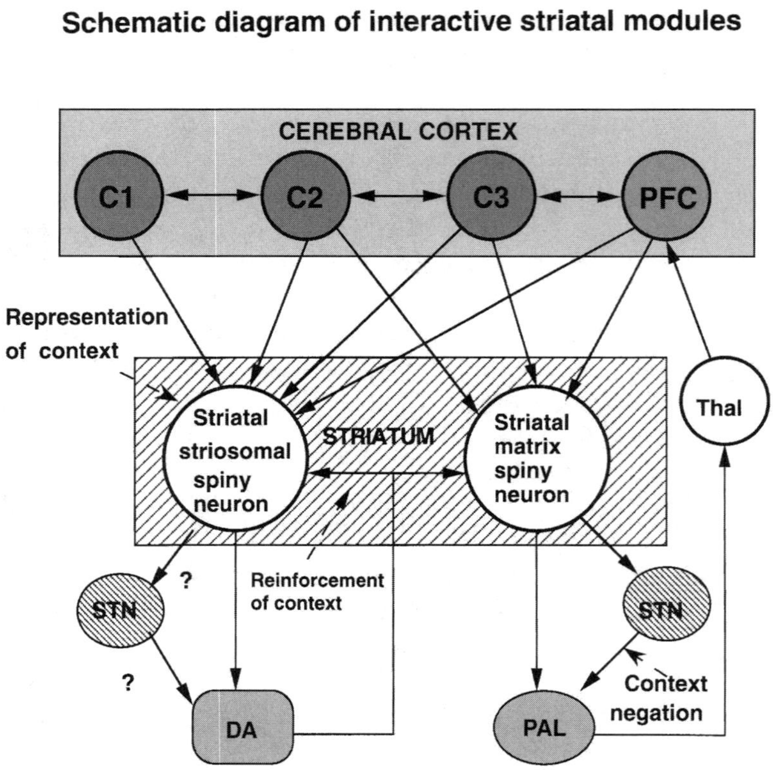

Figure 21.5 Convergence of cortical (C) inputs from different regions of neocortex to the medium spiny neurons of the striatum, striosome, and matrix compartments provides a hypothetical context for striatal output to influence mechanisms of response selection. Note how the matrix compartment targets the prefrontal cortex (PFC) via the striatopallidal-thalamic (PAL-Thal) loop. Hypothetical roles for the midbrain dopamine systems and the indirect pathway via the subthalamic nucleus (STN) are shown in reinforcement and context negation, respectively. The question marks indicate doubt that the indirect pathway from the striosomal compartment has been conclusively shown to exist. Modified version of figure from Houk, Adams, and Barto 1995.

an excess of movements, as in hemiballismus, although, again, one must beware of making too many assumptions about anatomical connectivity at this level of analysis: the very existence of the indirect pathway to the STN from the striosomal compartment (figure 21.5) remains in some doubt. For the time being, it may be sensible to relate the behavioral evidence to more macromolar aspects of the anatomy, such as the corticostriatal loops themselves.

The arrangement described above is generally consistent with behavioral and electrophysiological evidence that the striatum may play an important role in mediating behavioral set, namely, the predisposition to respond in a particular way sustained over a delay. The activity of some dorsal striatal neurons may be related to a particular set: they fire, for example, in the delay period after an instructional cue has cued the required direction of a particular motor response (see Mink 1996). Midbrain dopamine cells, projecting to both the ventral and the dorsal striatum, fire in response to conditioned stimuli that predict food (Schultz et al. 1995). These set-related functions contribute to response selection at some level, for example, in terms of response preparation or reward expectancy. The electrophysiological evidence of set-related activity is consistent with the effects of striatal lesions or dopaminergic manipulations of the striatum in simple and choice reaction times in the rat (Robbins and Brown 1990; Brown and Robbins 1991). Thus DA depletion from the dorsal striatum blocked the normal progressive speeding of reaction time that occurred as the imperative cue to respond became more likely (equally so for both choice and simple reaction time). On the other hand, because similar set-related activity appears to occur in the cortex before the striatum (Mink 1996), we cannot assume that the formation of set is a specifically striatal function.

We have laid out in some detail the anatomical and electrophysiological nature of the corticostriatal loops, focusing especially on the functioning of motor portions of this circuitry and relatively simple forms of behavior. Three main patterns have emerged: the convergence of cortical information to relatively segregated functional circuitries that appear to operate according to broadly similar principles; the impact of activity from midbrain dopamine neurons, which evidently helps to sharpen and reinforce certain patterns of neural activity within the striatum, presumably supporting some form of response selection function; and the apparent importance of the striatum in preparatory processes such as set that optimize responding in particular rule-governed contexts. In section 21.3, we consider whether the parallel nature of the circuitry between motor and cognitive regions of the corticostriatal pathways means that set-related and other processes mediated by the striatum are relevant to the control of cognitive as well as motor output.

21.3 FUNCTIONS OF THE CORTICOSTRIATAL LOOPS

Discrimination Learning within the Corticostriatal Loops?

The putative context recognition function of the striatum shown in figure 21.5, where a context is defined by the unique convergence of inputs from different cortical processing domains, is ideally suited to certain forms of learning and performance where combinations of specific stimuli in a certain context evoke the performance of specific responses. Such stimulus-response mappings can be essentially arbitrary and hence of the "conditional discrimination learning" type in which, for example, one cue "instructs" one response, whereas a different cue "instructs" an alternative response. Conditional or rule learning is well known to involve the frontal lobes (Petrides 1985; Passingham 1993) and is an essential component of tasks requiring the ordering of sequences of responses, as may occur in tests of planning ability such as the Tower of London (Owen et al. 1990). In the form we use, subjects consider two arrangements of colored balls hanging in stockings on a touch-screen computer monitor. Subjects have to rearrange the balls in the bottom arrangement to match the top, "goal" arrangement, in a defined, minimum number of moves. The balls are moved by touching them and their desired destination according to a few simple rules. Each move in the sequence, including appropriate "subgoals," entails a separate conditional choice evoked by the relative positions of the test stimuli in relation to the required goal configuration. These conditional choices have to be visualized as part of a precise solution involving candidate move sequences held in working memory.

Conditional learning itself is complex, involving distinct forms of associative learning between discriminative stimuli, different responses, and their associations with certain outcomes, which may be mediated by distinct corticostriatal loops. For example, conditional discrimination learning may involve action-outcome associations, as well as contemporaneous stimulus-response ("habit") learning, which is ultimately independent of the goal (Dickinson and Balleine 1994). Action-outcome and stimulus-response learning have been associated with different sectors of the striatum in the rat (e.g., White 1989; Robbins and Everitt 1992), and thus presumably, at different stages of learning, recruit either distinct corticostriatal loops or the striosomal and matrisomal compartments of the striatum as depicted in figure 21.5. An obvious correlate of stimulus-response learning in experimental animals is skill or "procedural" learning in humans. Its possible striatal basis has been investigated both in patients (Butters et al. 1985) and (using PET) in normal volunteers (e.g., Jenkins et al. 1994) but remains hard to specify. Wise, Murray, and Gerfen (1996) postulate that the rule learning occurs at the level of the frontal cortex (e.g., premotor areas) rather than the striatum itself, which acts mainly to modulate or "potentiate" this rule learning. This may be consistent with the set-related activity of the striatum that Wise, Murray, and Gerfen (1996) and Robbins and Brown (1990) regard as contributing to procedural or rule learning at the cortical level. Unlike patients with prefrontal or medial temporal lesions, however, patients with Parkinson's or Huntington's disease do have impairments in certain forms of probabilistic discrimination learning (Knowlton, Mangels, and Squire 1996; Knowlton et al. 1996).

The relationship of elementary forms of response selection, such as set, to more complex cognitive functions is important for defining the unique contributions of striatal and frontal cortical nodes within the corticostriatal loop circuitry. Cognitive theories of prefrontal cortical functioning like that of Shallice (1982) emphasize the importance of a "supervisory attentional system" that facilitates selection among "schemata" that often consist of no more than innate or learned stimulus-response mappings. The circumstances under which this system is required have included novel situations and the presence of stressors, as well as conflicts between alternative response options and changes in the contingencies that link particular responses to particular outcomes (Shallice 1982). These situations may engage several mechanisms of executive control including the capacity to override particular associations or, at a more abstract level, attentional biases to specific classes of information. The Wisconsin Card-Sorting Test has long been a popular clinical test of such functions because it stresses the capacity to establish, maintain, and, most crucially, alter cognitive set. The remainder of the review focuses on the suitability of this class of tasks for functional analysis at both the cognitive and neural levels, within the anatomical and theoretical framework introduced above.

Attentional Set Shifting, Reversal Learning, and Corticostriatal Loops

The Wisconsin Card-Sorting Test (WCST) requires subjects to sort cards that vary in three perceptual dimensions (color, form, and number of stimuli) according to rules that have to be learned through trial-and-error feedback. The cardinal sign of frontal lobe damage is not the inability to learn the original rule, but to shift from a learned rule to an alternative one (Milner 1964). This task has also been used in various forms to demonstrate deficits in patients with neurodegenerative diseases of the basal ganglia, including Parkinson's disease (Bowen et al. 1975) and Huntington's disease (Josiassen, Curry, and Mancall 1983). While the frontal deficit is thought mainly to consist of perseverative responding to the formerly reinforced category, in the early stages of unmedicated Parkinson's disease, there have been some indications of impairments even in learning the first rule, that is, in compound discrimination learning itself, where subjects have to distinguish between stimuli that comprise more than one perceptual dimension (Cooper et al. 1991).

The psychological and neural substrates of the frontal deficit on the WCST have attracted considerable controversy. Inflexible or perseverative responding could in theory arise from an executive failure at any of a number of different levels of processing (see Sandson and Albert 1984), for example, in attentional, decisional or response-related aspects of processing, and might additionally implicate working-memory failure (cf. Goldman-Rakic 1987). However, what is often forgotten about the WCST (though see Dehaene and Changeux 1991) is that it engages associative processes, as well as working memory, in a series of visual discriminations over multidimensional compound stimuli in which several different stimulus features are reinforced at different stages of the test.

This associative analysis has driven a functional decomposition of the WCST via a series of different types of visual discrimination effected on a

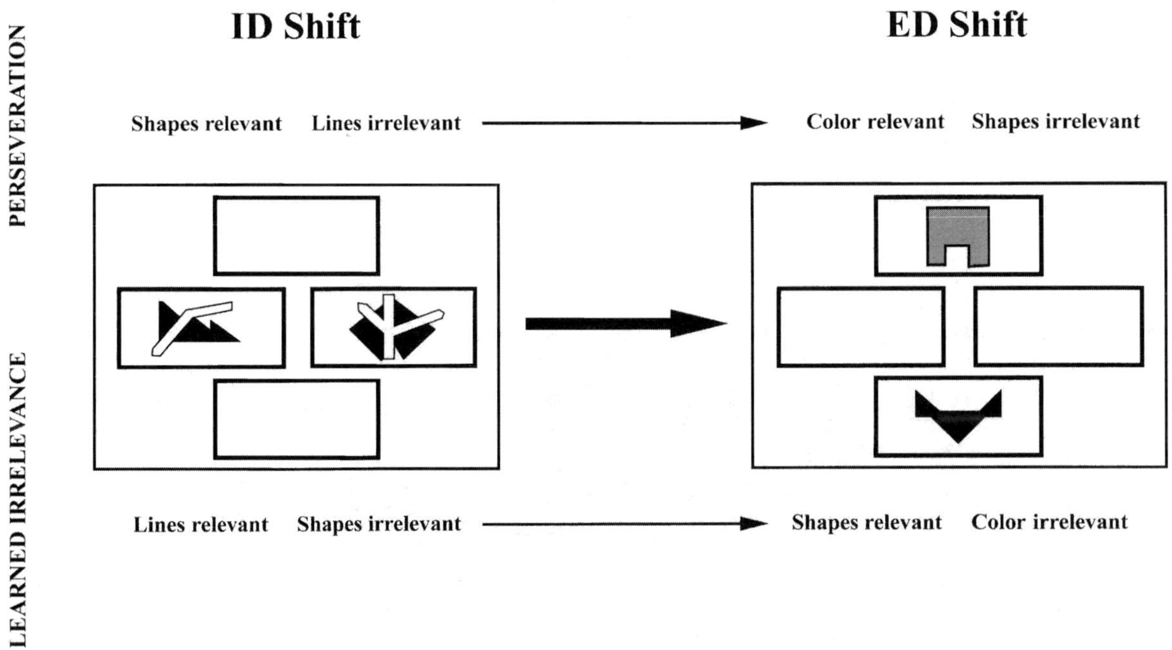

Figure 21.6 *Left*. Basic arrangement of the intra-/extradimensional (id/ed) shift task, as displayed on the video monitor. Different elements of perceptual dimensions ('Shapes' or 'Lines') can occur as compound stimuli, in any combination (i.e. as two pairs) on different trials, and at any two of the four boxes shown on the screen. In the standard task (e.g., Downes et al. 1989), novel exemplars from the same two perceptual dimensions are used in the transition from the previous compound discrimination phase to the intradimensional shift phase, and thence again in the transition to the extradimensional shift. In the intradimensional shift phase, subjects are confronted with novel exemplars of both stimulus dimensions, to which they have been responding in the preliminary training stages. They have to maintain discriminated

responding to the previously reinforced dimension. The subject may have previously had to respond to one Shape and not to another, with both Lines being irrelevant. In the intra-dimensional shift stage, Shape continues to be the relevant dimension and the subject has to learn which is the correct shape and which is the incorrect one, by trial and error, on the basis of the feedback received. When the standard extradimensional shift occurs, again, novel exemplars of Shapes and Lines appear (not shown in this figure), but this time the subject has to shift responding to the previously irrelevant dimension, 'Lines'. Thus one of the lines is correct, the other incorrect, and Shapes are now irrelevant.

Right. Transitions between intra- and extradimensional shifts, as used in Owen et al. 1993. In the perseveration extradimensional shift condition (top right), the previously irrelevant dimension, 'Lines' is discarded for a novel dimension, 'Color,' which becomes relevant to reinforcement in the perseveration condition, thus testing whether subjects can disengage from responding to the previously reinforced dimension ('Shapes'). In the learned irrelevance extradimensional shift condition (bottom right), the previously relevant dimension, 'Lines' is discarded and replaced by the novel dimension, 'Color,' but 'Shapes' now becomes relevant, thus testing whether subjects can overcome inhibition caused by learned irrelevance and learn to transfer responding to the previously irrelevant dimension, 'Shapes.' See Owen et al (1993) for further details.

touch-sensitive screen in a format suitable for testing experimental animals as well as human subjects (Roberts, Robbins, and Everitt 1988). To analyze the neural as well as the cognitive basis of WCST performance, the test, whose selectivity for prefrontal dysfunction has come increasingly into question (Anderson et al. 1991) must be refined. From the perspective of Houk, Adams, and Barto (1995), the different visual dimensions of the WCST can be seen as engaging groups of different striatal modules (see "Parallel and Segregated Nature of Corticostriatal Loops: The Motor Loop" in section 21.2) within a particular corticostriatal loop responsible for visual discrimination learning, by which specific exemplars are linked to specific responses. Given the apparent specificity of many cells within the striatum to particular stimuli (Rolls 1994), this seems entirely plausible. New learning would involve a degree of conflict among the different striatal modules for the control of response output, which may be partly resolved by frontal input or through the direct and indirect pathways (see figure 21.4), thus leading to an alteration of behavioral set.

One form of the visual discrimination learning test requires subjects to discriminate between two exemplars from either a "shape" or "lines" dimension (see figure 21.6). Following attainment of criterion on this simple discrimination, subjects are exposed to a reversal of these contingencies, in which the previously reinforced exemplar is not reinforced and vice versa. This form of learning requires the inhibition of previous associations during new stimulus-reinforcement learning, hence executive control. Subsequently, an irrelevant stimulus dimension (either lines or shapes) is added, initially as a distractor (see figure 21.6). Subjects are then exposed to two kinds of shifts. In the *intradimensional shift*, which corresponds to the *learning set* of comparative psychology, novel exemplars of shapes and lines are used, subjects merely have to keep responding to the previously reinforced dimension. In the *extradimensional shift*, a core component of the WCST, novel stimuli are again introduced, but subjects have now to respond to the alternative dimension. A final stage examines the reversal of contingencies between the two exemplars within the shifted dimension. The decomposition of the standard WCST into discrimination learning and reversal and into intra- and extradimensional shifts is consistent with theoretical analyses of discrimination learning based on animal learning theory (Sutherland and Mackintosh 1971) as well as with studies of human discrimination learning and its computational modeling (Kruske 1996), which show that it is difficult, if not impossible, to reduce reversal learning and extradimensional shifting to a single associative mechanism. It is not immediately clear whether these two forms of shifting behavior would be controlled by hierarchically nested circuits or modules within the same corticostriatal loop, on the one hand, or by different loops engaged to varying degrees in visual discrimination learning, on the other.

This suite of visual discrimination tests, widely used in clinical neuropsychology, has confirmed that patients with neurosurgical excisions of the frontal (but not the temporal) cortex, and patients with basal ganglia disorders are impaired, especially at the extradimensional stage (Downes et al. 1989; Owen et al. 1991; Lawrence et al. 1996; Lawrence et al. 1998). There is a tendency for some Parkinson's disease patients to fail the test even at the earlier visual discrimination learning stage (Owen et al. 1992). Patients in the advanced stages of Huntington's disease fail the test at the early stage of reversal learning, blatantly continuing to respond in a perseverative manner to the stimulus from the simple discrimination stage previously associated with reinforcement (Lange et al. 1995). These results suggest, not only that perseverative responding is a product of several different forms of processing (Sandison and Albert 1984), but also that there may be a neural substrate for the extradimensional and reversal learning impairments, which coincide with the progession of the disease from dorsal to more ventral portions of the caudate nucleus (Hedreen and Folstein 1995). On the other hand, it is entirely possible that the progression in cognitive deficit arises from progressively greater impairment of one sector of the striatum. This calls into question the relative psychometric sensitivity of the reversal task, which may simply be less difficult than the extradimensional shift; thus the apparent progression in deficit may reflect increasing intellectual deficit rather than a relationship to neuropathology. The psychometric sensitivity argument is somewhat blunted, however, by recent findings (Rahman et al. 1999) of selective deficits in the reversal learning, rather than the extradimensional shifting, components of the task in patients with lobar atrophy or dementia of the frontal type. Here the pathology is known to begin in orbitofrontal portions of the prefrontal cortex, which, as noted above, projects to the more ventral regions of the striatum (see figure 21.4). Notwithstanding these results, the neural basis of any such deficit is particularly difficult to assess in patients, particularly given the known cortical pathology with increasing progression of Huntington's disease. We will therefore further address this issue (1) in experiments with monkeys bearing specific lesions of frontal cortical and striatal circuitry; and (2) in studies with human volunteers with functional neuroimaging. A brief account will be given of these projects, both in their early stages.

Neural Substrates of Extradimensional Shifting and Reversal Learning

Discrete lesions using excitotoxic methods of lesioning that target cell bodies and not fibers of passage have been made in different regions of the marmoset prefrontal cortex, to the Pfdl or Pfvl and Pfo areas (Dias, Robbins, and Roberts 1996, 1997) after pretraining on compound discrimination learning with lines and shapes, similar to the stimuli used in human subjects. The lesioned and sham-operated animals were subsequently exposed to intra- and extradimensional shifts, and to a reversal after the extradimensional shift, using the same stimuli. The results clearly dissociated deficits on the reversal and extradimensional shifting task to different regions of the prefrontal cortex; reversal learning was impaired by lesions to Pfo, but not to Pfdl or Pfvl, whereas extradimensional shifting was selectively impaired by lesions of Pfdl or Pfvl. The anatomical locus of the reversal learning deficit is consistent with previous evidence in monkeys (see Rolls 1998 for a review). The extradimensional shift deficit may be consistent with previous suggestions regarding the anatomical basis of the WCST deficit in human patients (Milner 1964), although this is controversial.

These results pose interesting problems for understanding the organization of prefrontal cortical function. Wise, Murray, and Gerfen (1996), for example, believe they reflect the control of lower-order (reversal learning) and higher-order (abstract rule–shifting) processes, rather than contrasting shifting processes per se. The psychological basis of the deficits has been investigated in a follow-up study, where monkeys with similar lesions were found not to be impaired on compound discrimination learning itself, which probably entails a working-memory load similar to that of the extradimensional shift in the number of previous episodes of informative reinforcing feedback that must be processed to solve the discrimination (Dias, Robbins, and Roberts 1997). The deficits on extradimensional shifting may thus have more to do with inhibitory control than with holding stimuli "on-line." The reversal learning deficit following Pfo lesions was also obtained for both simple discriminations (i.e., where exemplars vary in a single stimulus dimension) and compound discriminations (where exemplars vary in at least two perceptual dimensions), thus confirming its generality. The most surprising result, however, was the lack of any deficit whatsoever in reversal learning or extradimensional shifting, provided the animals had been previously exposed, after surgery, to the same types of discrimination, reversal, and shifting with similar stimulus dimensions and exemplars (Dias, Robbins, and Roberts 1997). The finding suggests that performance of such shifts depends less on the prefrontal cortex when the shifts are no longer novel. It raises the possibility that other circuitry, possibly including the striatum, mediates the performance of the discriminations and their associated shifts when these have become more routine. The reported effects of selective prefrontal lesions on performance of the novel tasks are consistent with the proposed neural course of the deficit within the striatum in patients with Huntington's disease, although we have not yet completed studies with selective striatal lesions in monkeys to test the hypothesis further.

The monkey studies allow us to predict a priori the likely substrates for the various forms of shifting and reversal learning in the normal human subjects of functional imaging studies. The use of multistage visual discriminations lends itself well to the subtractive paradigm because several contrasts are possible, for example, between intra- and extradimensional shifts. We have recently completed such a study using positron-emission tomography with labeled oxygen to measure regional cerebral blood flow (Rogers et al. 2000). To date, the results indicate that extradimensional shift learning, relative to intradimensional shift learning, produced significant changes in exclusively prefrontal regions, including the left anterior prefrontal cortex and right Pfdl (Brodmann's areas 9, 10, and 46). By contrast, reversal learning, also relative to intradimensional shifting, produced activations not in the prefrontal cortex but in the ventral part of the left caudate nucleus. The contrast between cortical and subcortical effects for these different forms of shifting clearly bears on general attempts to distinguish cortical and striatal contributions to different forms of cognitive flexibility (Eslinger and Grattan 1993). In this case, the relatively complex shifting requirement of the extradimensional shift is associated with cortical function, whereas the lower-level process (cf. Wise, Murray, and Gerfen 1996) of reversal learning is linked with striatal activation.

The ventral portion of the caudate nucleus is anatomically connected to the orbitofrontal cortex, a structure we have seen associated with reversal impairments in monkeys (Jones and Mishkin 1972; Dias, Robbins, and Roberts 1996) and in humans (Rolls 1998). Taking into account these data, together with the evidence from Huntington's disease patients reviewed above, it may therefore be most parsimonious to conclude that the functional neuroimaging results are consistent with the engagement of different corticostriatal loops. On the other hand, we have yet to test this directly by examining the effects of relevant excitotoxic lesions of the striatum in monkeys. The comparison of striatal and frontal lesions is crucial for two reasons: (1) to determine the involvement of a given corticostriatal loop in a particular form of discrimination learning, reversal, or shifting, thus following the lead of Divac, Rosvold, and Szarcbart 1967, which demonstrated a congruence of lesion effects in different sectors of the prefrontal cortex and in the targets of their striatal projections; and (2) to infer the relative contributions of the cortical and striatal region in different aspects of processing by comparing the precise types of deficit obtained after each lesion. Some preliminary data bear on this issue. Dopamine depletion from the caudate nucleus and lateral prefrontal cortical lesions in marmosets have different effects on familiar versus novel extradimensional shifts. Striatal dopamine depletion impairs only the familiar shifts, and Pfdl or Pfvl lesions, only the novel shifts (Dias, Robbins, and Roberts 1997). One interpretation of this is that the prefrontal cortex is especially engaged during novel shifts, whereas the striatum is especially engaged during familiar shifts, between competing, "active" sets.

The Psychological Nature of the Shifting and Reversal Deficits

The mapping of processes of attentional control at the neural level will have to proceed in parallel and in conjunction with a more intensive psychological analysis. For example, a notable feature of the contrast between extra- and intradimensional shifts in the imaging study mentioned above (Rogers et al. 2000) was that the regions of activation did not include regions associated with verbal aspects of working memory, apparently unlike the WCST (Berman, Zee, and Weinberger 1995; Konishi et al. 1998). This suggests that the visual discrimination paradigm is less susceptible to verbal rehearsal strategies than the WCST (Dunbar and Sussman 1995). Moreover, the working-memory load, as inferred from the number of errors made (deriving presumably in part from episodes of particular associations of different stimuli with reinforcing feedback over previous trials) is not monotonically related to the degree of activation of the right Pfdl, further strengthening the argument made above that the extradimensional shift requires processes over and above working memory in the sense of 'holding stimuli on-line' (Rogers et al. 2000). While these additional "executive" processes are assumed to include inhibitory control over responding, the psychological locus of inhibition is unclear; it could include inhibition of perceptual processing, inhibition of response outflow, or both.

Manipulating the perceptual dimensions of the exemplars used at the extradimensional shift stage has made it possible to distinguish between impairments produced by shifting *from* a previously reinforced dimension, and those produced by shifting *to* a previously reinforced dimension (Owen et al. 1993, see figure 21.6). Neurosurgical excisions of the prefrontal cortex affect the former much more than the latter, whereas patients with Parkinson's disease have deficits in both forms of shifting.[3] Because the subjects have to overcome an inhibitory bias to respond to a previously unreinforced dimension, the data remain in line with the "rule potentiation" function posited by Wise, Murray, and Gerfen (1996). In terms of the analysis offered earlier, the lack of reinforcement of the previously irrelevant dimension, in the relative absence of striatal DA, may promote the enhanced "learned irrelevance" observed in the Parkinson's patients.

Alternatively, the "disengagement" of responding from one dimension to allow responding to the other could be mediated by inhibitory control operating within the indirect pathway of the striatal outflow systems or by top-down control from the prefrontal cortex. Failures at either of these levels would lead to perseverative responding, although we have to date seen no significant evidence for changes in regional cerebral blood flow in the caudate itself during an extradimensional shift versus reversal learning or an intradimensional shift (Rogers et al. 2000). In fact, more striking deactivations were found in the left visual cortex and right inferotemporal cortex, suggesting that the process of overriding an acquired attentional set or responding to a particular stimulus dimension depends in part on a nonstriatal system, specifically on transcortical pathways through prefrontal, temporal, and occipital cortex. On the other hand, these data must eventually be reconciled with evidence that patients with basal ganglia lesions are very susceptible to failing on extradimensional shifts and on related forms of set-shifting tasks (Hayes et al. 1998).

Performance on WCST can also benefit from a problem-solving or hypothesis-testing approach, which depends on experience. Once a subject has "cracked" the WCST, it provides little further challenge to mechanisms that promote cognitive flexibility, becoming, in effect, a well-learned routine. This is consistent with the short-lived nature of the shifting deficits seen in monkeys (Dias, Robbins, and Roberts 1997). To analyze the cognitive mechanisms underlining set shifting further, they must be disconfounded from those allied to new learning. As we saw above (Dias, Robbins, and Roberts 1997), the problem of learning is also present in the intra- versus extradimensional shift paradigm. If shifting or switching deficits could be identified in well-trained subjects, this would serve to isolate such impairments from those of learning per se. The question can be addressed using task set switching paradigms derived from human experimental psychology (Jersild 1927; Allport, Styles, and Hsieh 1994; Rogers and Monsell 1995), where subjects are required to switch responses between consistent stimulus-response mappings that have been previously well learned.

Task Set Switching

Rogers and Monsell (1995) developed a task-switching procedure that continuously compares switching and nonswitching between task sets. The basic design involves the subject switching between two stimulus-response mappings (or tasks) such as naming printed letters or digits, when letters and digits are always presented in pairs together, as compound stimuli, so that either task could, in principle, be performed on any trial. The sequence is arranged so that two trials with the letter mapping (task A) are followed immediately by two trials with the digit mapping (task B), with this basic AABB design being repeated throughout a number of trial blocks. The key measure is the cost of switching, measured in terms of reaction time or errors, calculated as differences between AB or BA transitions relative to AA and BB transitions. This paradigm was adapted for use in patients with frontal lobe excisions or medicated patients with mild Parkinson's disease (Rogers et al. 1998). Subjects saw a pair of characters on each trial with a 1 sec interval between the response and the onset of the next stimulus. For the first task, one character was a letter and subjects had to name it. For the second task, one character was a digit and subjects had to name it. The irrelevant character might, or might not, be a member from the other category. The experimental design provided for several other manipulations to test different aspects of the control process required for task set reconfiguration.

As has been suggested, the level of control over the stimulus-response mapping may be important (e.g., the internal-external dichotomy; Brown and Marsden 1988; see also Robbins and Brown 1990). Hence we employed two ways of specifying the relevant category: (1) direct word cuing of the required category on the screen (i.e., "letters" or "digits"); and (2) indirect color cuing, where the arbitrary learned cue for the relevant category was provided by the color of the display.

Because, in patients with frontal damage, the ability to switch may be influenced by interference from other cues or tasks that "capture" particular responses (cf. Shallice 1982), the design manipulated the possibility of cross talk between tasks (i.e., whether the irrelevant character was a member of the other category, thus also afforded a potentially competing naming response). Rogers and Monsell (1995) found that, in normal subjects, the time cost incurred by a switch is greater in the presence of a stimulus that activates the currently inappropriate task. In our study, this was achieved by including blocks of trials where the currently relevant stimulus was combined with and without the currently irrelevant stimulus. These two conditions were crossed to form blocks of trials constituting four conditions (color cue/no cross talk; color cue/cross talk; word cue/no cross talk; word cue/cross talk). The principal finding was that the time costs of a task switch were significantly increased in those patients with left-sided damage frontal lobe damage (LF) versus patients with right-sided frontal lobe damage (RF) and control subjects in the color cue/cross talk condition only (i.e., when there was cross talk between the tasks and only arbitrary task cues were used). By contrast, RF patients and mild, medicated Parkinson's disease patients were unimpaired in this condition (see table 21.1). Moreover, no subject group showed any task-switching deficits in the absence of cross talk, or when the stronger task cues (i.e., the name of the relevant category) were printed inside the display. Although these results are consistent with the changes in regional cerebral blood flow following task set switching reported by Meyer et al. (1998), they appear to contrast with the findings of Keele and Rafal (chap. 28, this volume), who found no differences for LF patients in a task set switching paradigm where sequences of trials before a switch were longer than in the present study. There are several such procedural differences between the two studies that might explain the discrepant findings. In reaching their essentially negative conclusions, however, Keele and Rafal also ignored higher error rates on switch versus nonswitch trials in LF patients, rates that might confound any assessments of switch RT costs in their study.

Table 21.1 Task-Switching Performance in Groups of Patients

Subject groups	Early performance of task-switching procedure	Effects of task-specific cross talk on task switching	Effects of stronger task cues on task switching	RT effects of prior processing on switch trials
LF patients	RTs: ↑ Errors: ↑	Time cost: ↑ Error cost: —	Time cost: ↓ Error cost: —	Enhanced
RF patients	RTs: ↑ Errors: —	Time cost: — Error cost: —	Time cost: ↓ Error cost: —	Unchanged
PD patients	RTs: — Errors: —	Time cost: — Error cost: —	Time cost: ↓ Error cost: —	Unchanged

Note: LF patients had focal damage to the left frontal cortex; RF patients, to the right frontal cortex. PD patients had mild medicated Parkinson's disease. ↓ = increased relative to controls; ↓ = decreased relative to controls.

Our results suggest that LF damage is associated with deficits in dynamic switches between tasks. The impairment may well be independent of other deficits in the *acquisition* of conditional aspects of simple stimulus-response tasks. Although both LF and RF groups exhibited slow and disorganized performance in the initial training blocks of the color cue/cross talk condition, only the LF group manifested a persistent slowing of reaction time on trials requiring a task switch. Moreover, patients in the LF group were also more sensitive to both the inhibition and the facilitation of task switching that resulted from processing on the trial immediately before, which suggests that the left prefrontal cortex may modulate the priming effects of task set reconfiguration. We were able to find only very minor changes in the switch costs shown by patients with mild, medicated Parkinson's disease, and then only under specific circumstances (i.e. toward the end of blocks of trials with cross talk; see Rogers et al. 1998). Studies with patients in later stages of the disease, or no longer medicated, might show stronger effects.

In keeping with roles for the striatum and frontal cortex in different aspects of attentional control, the results demonstrate that task set shifting is a function of the prefrontal cortex, and one relatively independent of learning; they are quite consistent with a role for the left prefrontal cortex in extradimensional shifting, given the specific increase in regional cerebral blood flow shown in the left Pfp, although, activations were also seen in right prefrontal cortex. We must emphasize, however, that the results are preliminary and certainly do not, as yet, establish differential roles for the prefrontal cortex and the striatum in different aspects of cognitive control. The Parkinson's disease patients used in our study were quite stable on dopaminergic medication, thus any deficits may have been essentially remediated. Downes et al. (1989) observed just such an effect when comparing unmedicated and medicated Parkinson's disease patients on the extradimensional shifting task. Moreover, there are some recent data relevant to task set control for Parkinson's disease (Hayes et al. 1998) that actually show evidence of impairments in shifting task set among Parkinson's disease patients, partly influenced by the efficacy of medication. The effects of striatal damage per se will also be important to assess, allowing a direct comparison between the effects of striatal damage and DA depletion. One study with Huntington's disease patients (Sprengelmeyer, Lange, and Homberg 1995) showed enhanced costs of task set shifting under conditions that cannot be directly compared with our own. Thus the hypothesis that the striatum is involved in task set shifting, independently of learning, is still viable.

21.4 SUMMARY AND CONCLUSIONS

We have described what is known about the anatomical and physiological organization of a major interaction of the prefrontal cortex with the striatum and its associated structures. There is currently much interest in relating the neurobiological organization of corticostriatal loops to possible motor and cognitive functions, whether based on neuropsychological studies of patients and experimental animals, on functional neuroimaging investigations in normal human subjects, or on neurocomputational modeling (see Braver and Cohen, chap. 31, this volume). Because the prefrontal cortex is assumed to have a role in aspects of executive function, including processes of response selection and attentional control, it seems likely that the neurobiological approach could interface quite well with experimental psychology approaches that seek to define the nature of the underlying cognitive processes. Although we have cautioned about the dangers of making too much of information from a neuronal, as distinct from a neural systems, analysis at this stage, there seems little doubt that detailed psychological analyses will constrain neurobiological models to such a degree that their precise applicability to molar cognitive functions will become more apparent. Some signs of this are already to be seen in the growing interest of cognitive scientists in psychopharmacological investigations (e.g., D'Esposito, and Postle, chap. 26, this volume), which promise to interface with the cognitive neuropsychological approach at several levels, including that of treatment.

The main results we have reviewed in the second half of this chapter concern a class of executive operations that we have designated as the establishing, maintaining, and shifting of cognitive set. The corticostriatal systems seem to play a major role in such functions, as shown especially by the performance of patients with basal ganglia lesions, and the parallels between cognitive set and the motor set–related activity of single neurons in these systems. We have shown, for example, from effects of lesions in nonhuman primates, from functional neuroimaging studies of normal subjects, and more indirectly, from reports of various neurological patient groups, that different aspects of shifting, for example, between stimulus-response mappings associated with different perceptual dimensions that make up a single compound stimulus (i.e., extradimensional shifting) or between previously reinforced and unreinforced preferences (i.e., reversal learning), are governed by different corticostriatal loops. Although we may have been less successful in defining precisely what the striatum does that is different from frontal cortical mechanisms in these operations, and how exactly they interact with basic associative processes, we have provided evidence that the role of prefrontal cortical mechanisms in certain task-switching operations is not confined to novel situations, where learning is required. The effects of more localized lesions of prefrontal cortex and of the striatum itself and results from well-designed studies using functional neuroimaging in normal human subjects should help us to further elucidate the role of specific elements of the corticostriatal loops. While the successful application of these methodologies will undoubtedly depend on theoretical advances in understanding the cognitive architecture of executive processes, it may also help to shape some of those advances.

NOTES

The work reported in this chapter was supported by a program grant from the Wellcome Trust to Trevor W. Robbins, Barry J. Everitt, Angela C. Roberts, and Barbara J. Sahakian. We thank our colleagues for their collaboration, T. Roehling for providing figure 21.2, and M. Easter for manuscript preparation.

1. A third trend has been suggested by Sanides (see Fuster 1989) to arise in the premotor regions of the prefrontal cortex and to move anteriorly toward the others.

2. The prefrontal cortex is one of the few neocortical regions to project back to the chemically identified neurotransmitter systems of the reticular core of the brain. There is burgeoning evidence that the prefrontal cortex can exert regulatory effects on noradrenergic, serotoninergic, dopaminergic, and cholinergic systems, each of which gives rise to diffuse and general projections to different regions of the forebrain, where they exert important neuromodulatory influences such as arousal, alertness, and (most probably) mood (Robbins and Everitt 1995).

3. This result may be more pertinent to the functions of the nigrostriatal DA system than to any difference between the striatum and the prefrontal cortex: the performance of patients having Huntington's disease (hence damage to the striatum itself) resembles that of patients having prefrontal cortex lesions (Lawrence 1997).

REFERENCES

Alexander, G. E., DeLong, M. R., and Strick, P. L. (1986). Parallel organization of functionally segregated circuits linking basal ganglia and cortex. *Annual Review of Neuroscience, 9,* 357–381.

Allport, A., Styles, E. A., and Hsieh, S. (1994). Shifting intentional set: Exploring the dynamic control of tasks. In C. Umiltà and M. Moscovitch (Eds.), *Attention and Performance XV,* pp. 421–452. Cambridge, MA: MIT Press.

Anderson, S. W., Damasio, H., Dallas Jones, R., and Tranel, D. (1991). Wisconsin Card-Sorting Test performance as a measure of frontal damage. *Journal of Clinical and Experimental Neuropsychology, 13,* 909–922.

Baddeley, A. D. (1986). *Working memory.* Oxford: Clarendon Press.

Barbas, H., and Pandya, D. N. (1989). Architecture and intrinsic connections of the prefrontal cortex in the rhesus monkey. *Journal of Comparative Neurology, 286,* 353–375.

Bechara, A., Damasio, H., Tranel, D., and Anderson, S. W. (1998). Dissociation of working memory from decision making within the human prefrontal cortex. *Journal of Neuroscience, 18,* 428–437.

Berman, K. F., Zee, R. F., and Weinberger, D. R. (1995). Physiologic dysfunction of dorsolateral prefrontal cortex in schizophrenia: 2. Role of neuroleptic treatment, attention and mental effort. *Archives of General Psychiatry, 43,* 126–135.

Berns, G. S., and Sejnowski, T. J. (1996). How the basal ganglia make decisions. In A. R. Damasio and V. Christen (Eds.), *Neurobiology of decision-making,* pp. 101–113. Berlin: Springer.

Bowen, F. P., Kamienny, R. S., Burns, M. M. and Yahr, M. D. (1975). Parkinsonism: Effects of levodopa on concept formation. *Neurology, 25,* 701–704.

Brodmann, K. (1912). Neue Ergebneisse über die vergleichende histologische Lokalisation der Gross-hirnrinde mit besonderer Berücksichtigung des Stirnhirns. *Anatomischer Anzeiger* (suppl.) *41,* 157–216.

Brown, R. G., and Marsden, C. D. (1988). Internal versus external cues and the control of attention in Parkinson's disease. *Brain, 111,* 323–345.

Brown, V. J., and Robbins, T. W. (1991). Simple and choice reaction time performance following unilateral striatal dopamine depletion in the rat: Impaired motor readiness but preserved response preparation. *Brain, 114,* 513–525.

Burgess, P. (1997). Theory and methodology in executive function research. In P. Rabbitt (Ed.), *Methodology of frontal and executive functions,* pp. 81–111. Hove, U.K.: Erlbaum.

Butters, N., Wolfe, J., Martone, M., Granholm, E., and Cermak, L. S. (1985). Memory disorders associated with Huntington's disease: Verbal recall, verbal recognition, and procedural memory. *Neuropsychology, 23,* 729–743.

Chevalier, G., and Deniau, J. M. (1990). Disinhibition as a basic process in the expression of striatal functions. *Trends in Neuroscience, 13,* 277–280.

Cooper, J. A., Sagar, H., Jordan, N., Harvey, N. S., and Sullivan, E. V. (1991). Cognitive impairment in early, untreated Parkinson's disease and its relationship to motor disability. *Brain, 114,* 2095–2122.

Damasio, A. R. (1998). The somatic marker hypothesis and the possible functions of the prefrontal cortex. In A. C. Roberts, T. W. Robbins, and L. Weiskrantz (Eds.), *The prefrontal cortex: Executive and cognitive functions,* pp. 36–50. Oxford: Oxford University Press.

Dehaene, S., and Changeux, J. P. (1991). The Wisconsin Card-Sorting Test: Theoretical analysis and simulation of a reasoning task in a model neuronal network. *Cerebral Cortex, 1,* 62–79.

DeLong, M. R. (1990). Primate models of movement disorders of basal ganglia origin. *Trends in Neuroscience, 13,* 281–285.

DeLong, M. R., and Georgopoulos, A. P. (1981). Motor functions of the basal ganglia. In J. M. Brookhart, V. B. Mountcaste, and V. B. Brooks (Eds.), *Handbook of physiology,* pp. 1017–1062. Bethesda, MD, American Physiological Society.

Desimone, R., Miller, E. K., Chelazzi, L., and Lueschow A. (1995). Multiple memory systems in the visual cortex. In M. S. Gazzaniga (Ed.), *The cognitive neurosciences*, pp. 475–486. Cambridge MA: MIT Press.

Dias, R., Robbins, T. W., and Roberts, A. C. (1996). Dissociation in prefrontal cortex of affective and attentional shifts. *Nature, 380*, 69–72.

Dias, R., Robbins, T. W., and Roberts, A. C. (1997). Dissociable forms of inhibitory control within prefrontal cortex with an analogue of the Wisconsin card sort test: Restriction to novel situations and independence from "on-line" processing. *Journal of Neuroscience, 17*, 9285–9297.

Dickinson, A., and Balleine, B. (1994). Motivational control of goal-directed action. *Animal Learning and Behavior, 22*, 1–18.

Divac, I., Rosvold, H. E., and Szarcbart, M. K. (1967). Behavioral effects of selective ablation of the caudate nucleus. *Journal of Comparative and Physiological Psychology, 63*, 184–190.

Downes, J. J., Roberts, A. C., Sahakian, B. J., Evenden, J. L., and Robbins, T. W. (1989). Impaired extra-dimensional shift performance in medicated and unmedicated Parkinson's disease: Evidence for a specific attentional dysfunction. *Neuropsychologia, 27*, 1329–1344.

Dunbar, K., and Sussman, D. (1995). Toward a cognitive account of frontal lobe function: Simulating frontal lobe deficits in frontal normal subjects. In J. Grafman, K. J. Holyoke, F. Boller (Eds.), *Structure and functions of the human prefrontal Cortex, Annals of the New York Academy of Sciences, 769*, 289–304.

Duncan, J. (1986). Disorganisation of behaviour after frontal lobe damage. *Cognitive Neuropsychology, 3*, 271–290.

Eslinger, P. J., and Gratton, L. M. (1993). Frontal lobe and fronto-striatal substrates for different forms of human cognitive flexibility. *Neuropsychologia, 31*, 17–28.

Everitt, B. J., Morris, K., O'Brien, A., and Robbins, T. W. (1991). The amygdala-ventral striatal system: Role in reward-related processes. *Neuroscience, 42*, 1–18.

Flaherty, A. W., and Graybiel, A. M. (1993). Two input systems for body representations in the primate striatal matrix: Experimental evidence in the squirrel monkey. *Journal of Neuroscience, 13*, 1120–1137.

Fuster, J. M. (1989). *The Prefrontal Cortex: Anatomy, physiology, and neuropsychology of the frontal lobe*. 2d ed. New York: Raven Press.

Gerfen, C. R. (1992). The neostriatal mosaic: Multiple levels of compartment organization in the basal ganglia. *Annual Review of Neuroscience, 15*, 285–320.

Goldman-Rakic, P. S. (1987). Circuitry of primate prefrontal cortex and regulation of behavior by representational memory. In F. Plum (Ed.), *Handbook of physiology*, pp. 373–417. Bethesda, MD: American Physiological Society.

Goldman-Rakic, P. S. (1998). The prefrontal landscape: Implications of functional architecture for understanding human mentation and the central executive. In A. C. Roberts, T. W. Robbins, and L. Weiskrantz (Eds.), *The prefrontal cortex: Executive and cognitive functions*, pp. 87–102. Oxford: Oxford University Press.

Graybiel, A. M. (1990). Neurotransmitters and neuromodulators in the basal ganglia. *Trends in Neurosciences, 13*, 244–254.

Hayes, A. E., Davidson, M. C., Keele, S. W., and Rafal, R. D. (1998). Toward a functional analysis of the basal ganglia. *Journal of Cognitive Neuroscience, 10*, 178–198.

Hedreen, J. C., and Folstein, S. E. (1995). Early loss of neostriatal striosome neurons in Huntington's disease. *Journal of Neuropathology and Experimental Neurology, 54*, 105–120.

Houk, J. C. (1995). Information processing in modular circuits linking basal ganglia and cerebral cortex. In J. C. Houk, J. L. Davis, and D. G. Beiser (Eds.), *Models of information processing in the basal ganglia,* pp. 3–10. Cambridge, MA: MIT Press.

Houk, J. C., Adams, J. L., and Barto, A. G. (1995). A model of how the basal ganglia generate and use neural signals that predict reinforcement. In J. C. Houk, J. L. Davis, and D. G. Beiser (Eds.), *Models of information processing in the basal ganglia,* pp. 249–270. Cambridge, MA: MIT Press.

Jenkins, I. H., Brooks, D. J., Nixon, P. D. Frackowiak, R. S., Passingham, R. E. (1994). Motor sequence learning: A study with positron emission tomography. *Journal of Neuroscience, 14,* 3775–3790.

Jersild, A. T. (1927). Mental set and shift. *Archives of Psychology,* no. 89.

Joel, D., and Weiner, I. (1994). The organization of the basal ganglia–thalamocortical circuits: Open interconnected rather than close segregated. *Neuroscience, 63,* 363–379.

Jones, B., and Mishkin, M. (1972). Limbic lesions and the problem of stimulus-reinforcement associations. *Experimental Neurology, 36,* 362–377.

Josiassen, R. C., Curry, L. M., and Mancall, E. L. (1983). Development of neuropsychological deficits in Huntington's disease. *Archives of Neurology, 40,* 791–796.

Kemp, J. M., and Powell, T. P. S. (1970). The cortico-striate projection in the monkey. *Brain, 93,* 525–546.

Kish, S. J., Shannak, K., and Hornykiewicz, O. (1988). Uneven patterns of dopamine loss in the striatum of patients with idiopathic Parkinson's disease: Pathophysiologic and clinical implications. *New England Journal of Medicine, 318,* 876–880.

Knowlton, B. J., Mangels, J. A., and Squire, L. R. (1996). A neostriatal habit learning-system in humans. *Science, 273,* 1399–1402.

Knowlton, B. J., Squire, L. R., Paulsen, J. S., Swedlow, N. S., Swenson, M., and Butters, N. (1996). Dissociations within non-declarative memory in Huntington's disease. *Neuropsychology, 10,* 538–548.

Konishi, S., Nakajima, K., Uchida, I., Kameyama, M., Nakahara, K., Sekihara, K., and Miyashita, Y. (1998). Transient activation of inferior prefrontal cortex during cognitive set shifting. *Nature Neuroscience, 1,* 80–84.

Kruske, J. K. (1996). Dimensional relevance shifts in category learning. *Connection Science, 8,* 225–247.

Lange, K. W., Sahakian, B. J., Quinn, N. P., Marsden, C. D., and Robbins, T. W. (1995). Comparison of executive and visuospatial function in Huntington's disease and dementia of the Alzheimer-type matched for degree of dementia. *Journal of Neurology, Neurosurgery, and Psychiatry, 58,* 598–606.

Lawrence, A. D. (1997). Executive and memory functions in Huntington's disease. Ph.D. diss., University of Cambridge.

Lawrence, A. D., Hodges, J. R., Rosser, A. E., Kershaw, A., fFrench-Constant, C., Rubinsztein, D. C., Robbins, T. W., and Sahakian, B. J. (1998). Evidence for specific cognitive deficits in preclinical Huntington's disease. *Brain, 121,* 1329–1341.

Lawrence, A. D., Sahakian, B. J., Hodges, J. R., Rosser, A. E., Lange, K. W., and Robbins, T. W. (1996). Executive and mnemonic functions in early Huntington's disease. *Brain, 119,* 1633–1645.

Meyer, D. E., Evans, J. E., Lauber, E. J., Gmeindl, L., Rubinstein, J., Junck, L. and Koeppe, R. A. (1998). The role of dorsolateral prefrontal cortex for executive cognitive processes in task switching. Poster presented to the Cognitive Neuroscience Society, San Francisco, April.

Middleton, F. A., and Strick, P. L. (1995). The temporal lobe is a target of output from the basal ganglia. *Proceedings of the National Academy of Sciences, U.S.A., 93,* 8383–8387.

Milner, B. (1964). Some effects of frontal lobectomy in man. In J. M. Warren and K. Akert (Eds.), *The frontal granular cortex and behavior,* pp. 410–444. New York: McGraw-Hill.

Mink, J. W., (1996). The basal ganglia: focused selection and inhibition of competing motor programs. *Progress in Neurobiology, 50,* 381–425.

Norman, D. A., and Shallice, T. (1980). *Attention to action: Willed and automatic control of behavior.* Technical Report No. 99. San Diego: University of California, Center for Information Processing.

Owen, A., Downes, J. J., Sahakian, B. J., Polkey, C. E., and Robbins, T. W. (1990). Planning and spatial working memory following frontal lobe lesions in man. *Neuropsychologia, 28,* 1021–1034.

Owen, A. M., Roberts, A. C., Polkey, C. E., Sahakian, B. J., and Robbins, T. W. (1991). Extra-dimensional versus intradimensional set shifting performance following frontal lobe excision, temporal lobe excision or amygdalo-hippocampectomy in man. *Neuropsychologia, 29,* 993–1006.

Owen, A. M., James, M., Leigh, P. H., Summers, B. A., Marsden, C. D., Quinn, N. P., Lange, K. W., and Robbins, T. W. (1992). Fronto-striatal cognitive deficits at different stages of Parkinson's disease. *Brain, 115,* 1727–1751.

Owen, A. M., Roberts, A. C., Hodges, J. R., Summers, B. A., Polkey, C. E., and Robbins, T. W. (1993). Contrasting mechanisms of impaired attentional set-shifting in patients with frontal lobe damage or Parkinson's disease. *Brain, 116,* 1159–1179.

Pandya, D. N., and Yeterian, E. H. (1998). Comparison of prefrontal architecture and connections. In A. C. Roberts, T. W. Robbins, and L. Weiskrantz (Eds.), *The prefrontal cortex: Executive and cognitive functions,* pp. 51–66. Oxford: Oxford University Press.

Parent, A., and Hazrati, L. N. (1993). Anatomical aspects of information processing in primate basal ganglia. *Trends in Neuroscience, 16,* 111–116.

Passingham, R. E. (1993). *The frontal lobes and voluntary action.* Oxford: Oxford University Press.

Passingham, R. E. (1998). Attention to action. In A. C. Roberts, T. W. Robbins, and L. Weiskrantz (Eds.), *The prefrontal cortex: Executive and cognitive functions,* pp. 131–143. Oxford: Oxford University Press.

Pennartz, C. M. A., Groenwegen, H. J., and Lopes de Silva, F. (1994). The nucleus accumbens as a complex of functionally distinct neuronal ensembles: An integration of behavioural, electrophysiological and anatomical data. *Progress in Neurobiology, 42,* 719–761.

Percheron, G., Yelnik, J., and Francois, C. (1984). The primate striato-pallido-nigral system: An integrative system for cortical information. In J. S. McKenzie, R. E. Kemm, and L. N. Wilcock (Eds.), *The basal ganglia: Structure and function,* pp. 87–105. New York: Plenum Press.

Petrides, M. (1985). Deficits on conditional associative tasks after frontal and temporal-lobe lesions in man. *Neuropsychologia, 23,* 601–614.

Petrides, M. (1998). Specialized systems for the processing of mnemonic information within the primate frontal cortex. In A. C. Roberts, T. W. Robbins, and L. Weiskrantz (Eds.), *The prefrontal cortex: Executive and cognitive functions,* pp. 103–116. Oxford: Oxford University Press.

Petrides, M., and Pandya, D. N. (1994). Comparative architectonic analysis of the human and macaque frontal cortex. In F. Boller and J. Grafman (Eds.), *Handbook of neuropsychology,* vol. 9, pp. 17–58. Amsterdam: Elsevier.

Rahman, S., Sahakian, B. J., Hodges, J. R., Rogers, R. D., and Robbins, T. W. (1999). Specific cognitive deficits in mild frontal variant frontotemporal dementia. *Brain, 122,* 1469–1493.

Robbins, T. W., and Brown, V. J. (1990). The role of the striatum in the mental chronometry of action: A theoretical review. *Reviews in the Neurosciences, 2,* 181–213.

Robbins, T. W., and Everitt, B. J. (1992). Functions of dopamine in the dorsal and ventral striatum. In T. W. Robbins (Ed.), *Seminars in the neurosciences,* vol. 4, pp. 119–127. London: Saunders.

Robbins, T. W., and Everitt, B. J. (1995). Arousal systems and attention. In M. Gazzaniga (Ed.), *The cognitive neurosciences,* pp. 703–720. Cambridge, MA: MIT Press.

Roberts, A., Robbins, T. W., and Everitt, B. J. (1988). Extra- and intra-dimensional shifts in man and marmoset. *Quarterly Journal of Experimental Psychology, 40B,* 321–342.

Rogers, R. D., Andrews, T. C., Grasby, P. M., Brooks, D. J., and Robbins, T. W. (2000). Contrasting cortical and subcortical activations produced by reversal learning and attentional-set shifting in humans. *Journal of Cognitive Neuroscience, 12,* 142–162.

Rogers, R. D., and Monsell, S. (1995). Costs of a predictable switch between simple cognitive tasks. *Journal of Experimental Psychology: General, 124,* 207–231.

Rogers, R. D., Sahakian, B. J., Hodges, J. R., Polkey, C. E., Kennard, C. and Robbins, T. W. (1998). Dissociating executive mechanisms of task control following frontal lobe damage and Parkinson's disease. *Brain, 121,* 815–842.

Rolls, E. T. (1994). Neurophysiology and cognitive functions of the striatum. *Revue Neurologique* (Paris), *150,* 648–660.

Rolls, E. T. (1998). The orbitofrontal cortex. In A. C. Roberts, T. W. Robbins, and L. Weiskrantz (Eds.), *The prefrontal cortex: Executive and cognitive functions,* pp. 67–86. Oxford: Oxford University Press.

Rushworth, M. F. S., and Owen, A. M. (1998). The functional organisation of the lateral frontal cortex: Conjuncture or conjecture in the electrophysiology literature? *Trends in Cognitive Sciences, 2,* 46–53.

Sandison, J., and Albert, M. L. (1984). Varieties of perseveration. *Neurology, 22,* 715–732.

Schultz, W., Romo, R., Ljundberg, T., Mirenowicz, J., Hollerman, J. R., and Dickinson, A. (1995). Reward-related signals carried by dopamine neurons. In J. C. Houk, J. L. Davis, and D. G. Beiser (Eds.), *Models of information processing in the basal ganglia,* pp. 223–248. Cambridge, MA: MIT Press.

Selemon, L. D., and Goldman-Rakic, P. S. (1985). Longitudinal topography and interdigitation of corticostriatal projections in the rhesus monkey. *Journal of Neuroscience, 5,* 776–6–794.

Shallice, T. (1982). Specific impairments of planning. *Philosophical Transactions the Royal Society of London B298,* 199–209.

Sprengelmeyer, R., Lange, H., and Homberg, V. (1995). The pattern of attentional deficits in Huntington's disease. *Brain, 118,* 145–152.

Strick, P. L., Dum, R. P., and Picard, N. (1995). Macro-organization of the circuits connecting the basal ganglia with the cortical motor areas. In J. C. Houk, J. L. Davis, and D. G. Beiser

(Eds.), *Models of information processing in the basal ganglia,* pp. 117–130. Cambridge, MA: MIT Press.

Sutherland, N. S., and Mackintosh, N. J. (1971). *Mechanisms of animal discrimination learning.* New York: Academic Press.

White, N. M. (1989). A functional hypothesis concerning the striatal matrix and patches: Mediation of S-R memory and reward. *Life Science, 45,* 1943–1957.

Wilson, C. J. (1995). The contribution of cortical neurons to firing pattern of striatal spiny neurons. In J. C. Houk, J. L. Davis, and D. G. Beiser (Eds.), *Models of information processing in the basal ganglia,* pp. 29–50. Cambridge, MA: MIT Press.

Wise, S. P., Murray, E. A., and Gerfen, G. R. (1996). The frontal cortex–basal ganglia system in primates. *Critical Reviews in Neurobiology, 10,* 317–356.

Yeterian, E. J., and Van Hoesen, G. W. (1978). Cortico-striate projections in the rhesus monkey: The organization of certain cortico-caudate connections. *Brain Research, 139,* 43–63.

22 The Neural Basis of Top-Down Control of Visual Attention in the Prefrontal Cortex

Earl K. Miller

ABSTRACT The prefrontal (PF) cortex is thought to be a major source of top-down signals that modulate processing in other brain regions. Recent neurophysiological studies in monkeys illustrate that PF neurons have properties ideal for providing the top-down signals that control the allocation of attention. By selecting and integrating information from diverse sources, these neurons can convey information about items to be attended: they can maintain activity about the items in the face of distractions; they seem to play a role in acquiring and representing behavior-guiding rules that dictate what is relevant and needs attending.

Many views of cognition (e.g., Baddeley 1986; Norman and Shallice 1986; Johnson and Hirst 1991) posit the existence of top-down signals that select and coordinate information. These signals are thought to enhance the representations that underlie our conscious perceptions, thoughts, and plans of actions, while inhibiting irrelevant or inappropriate information. Many brain processes can work without top-down control: well-learned, habitual behaviors can be executed automatically, and unexpected events can automatically grab our attention and enter our awareness. Top-down control is necessary, however, when we need to ignore distractions or to inhibit reflexive, prepotent responses and when habitual behaviors cannot be used, as in novel or difficult situations.

Perhaps the best understood example of top-down control is selective visual attention, that is, the ability to voluntarily focus our awareness on certain portions of a visual scene. This ability is critical because our capacity for visual processing is severely limited; at any given moment, we can only fully process a small portion of a scene. For example, when briefly presented with a scene containing multiple objects, we can typically process and remember only 4–5 items (Luck and Vogel 1997). Intelligent behavior thus depends on suppressing reflexive orientation to physically salient inputs and on selectively gathering inputs that are behaviorally relevant.

Various accounts of selective attention have held that it can be focused on relevant visual field locations and objects, and that processing of relevant visual attributes is enhanced, whereas processing of irrelevant attributes is suppressed. Desimone and Duncan (1995) have recently proposed a biologically plausible model to explain these phenomena. According to biased competition, neurons in the extrastriate visual cortex that represent different visual field locations and objects are mutually inhibitory. Top-down signals are excitatory and represent the item to be attended. These bias signals increase activity of neurons that process the relevant information and, by virtue of the mutual inhibition, suppress activity of neurons processing irrelevant information. Top-down signals are thought to derive from maintained activity of the task-relevant information, activity that conveys information about the sought-after item. In this chapter, I will discuss neurophysiological studies relevant to top-down attentional selection by biased competition, focusing on the properties of neurons in the prefrontal cortex, a brain region thought to be involved in top-down control and to provide the bias signals that mediate attentional selection.

22.1 PREFRONTAL CORTEX AND TOP-DOWN SIGNALS

The prefrontal (PF) cortex is associated with a wide range of "executive" functions critical for complex behavior, such as problem solving, planning, selecting action, and working memory (Milner and Petrides 1984; Petrides 1990; Duncan et al. 1996; Burgess and Shallice 1996a,b; Humphreys, Forde, and Francis, chap. 18, Petrides, chap. 23, Frith, chap. 24, D'Esposito and Postle, chap. 26, Riddoch, Humphreys, and Edwards, chap. 27, Kimberg and Farah, chap. 32, this volume). Consistent with an "executive" role in brain function are the extensive interconnections between the PF cortex and many other brain regions (Pandya and Yeterian 1990; Pandya and Barnes 1987; Cavada and Goldman-Rakic 1989; Preuss and Goldman-Rakic 1989; Webster, Bachevalier and Ungerleider 1994). It should be noted, however, that the prefrontal cortex is unlikely either to be the only region involved in top-down control nor to act alone. For example, some studies implicate frontostriatal loops in top-down control of attention (Robbins and Rogers, chap. 21, this volume).

Selective attention has long been thought to be an important prefrontal function. Damage to the PF cortex in humans can cause deficits in sustained attention and detection of novel events (Knight 1984, 1991; Stuss and Benson 1986). Further, deficits on complex tasks after PF damage have been thought to reflect a dysfunction in switching attention between different visual features of a task, between different sets of abstract behavior-guiding rules, or both (Owen et al. 1991). Similarly, PF lesions in monkeys can result in deficits in shifting attention between different stimulus dimensions (Dias, Robbins, and Roberts 1997).

According to the biased competition model, the role of the prefrontal cortex in visual attention is to provide activity that biases competition in the visual cortex in favor of neurons representing that information. The PF cortex is thought to provide "attentional templates" by maintaining activity that conveys information about the sought-after item. This ability is typically studied in delay tasks in which a single stimulus is presented as a cue and then, after a delay, monkeys make a response based on that cue. During the delay, many PF neurons show high levels of often cue-specific activity (Fuster and Alexander 1971; Kubota and Niki 1971; Funahashi, Bruce, and Goldman-Rakic 1989; Quintana and Fuster 1992; Wilson, O'Scalaidhe, and Goldman-Rakic 1993; Miller, Erickson, and Desimone 1996; Rainer, Asaad, and Miller 1998a; Rainer, Asaad, and Miller 1998b). Human imaging studies also indicate high levels of sustained PF activity during such tasks (Cohen et al. 1997; Courtney et al. 1997).

This "delay activity" can convey information about stimulus identity and location, and thus might play a role in directing attention to relevant form or color attributes, and or to particular locations. Other properties of prefrontal neurons also seem ideal for a role in voluntarily directing attention. They can select and integrate information from diverse sources and can maintain activity about this information in the face of distractions. Further, the PF cortex seems to play a central role in the "executive" brain functions that determine what is relevant and needs attending. This may be mediated by PF mechanisms that acquire and represent behavior-guiding rules and behavioral context. Physiological evidence for these claims is reviewed below.

Selection of Information for Bias Signals: Sensory Information

In monkeys, many tests of attention are identical to delay tasks used to test active short-term, or working, memory. For example, in visual search the animal is briefly shown a cue object. Then, after a delay (during which the cue must be held in memory), two or more test objects are presented; the animal must find the test object that matches the cue (the target) and ignore the test objects that do not match (the distractors). Prefrontal delay activity that maintains information about the cue is thought to provide the attentional template that guides selection when the test objects appear. To do so, PF activity must be more than a simple visual buffer, recording any incoming visual input. It must have the ability to selectively represent only the information needed to guide selection (e.g., the cue) and not maintain other, irrelevant, information that also happens to fall on the retina. In other words, delay activity must be "pure"; it must reflect only the item to be attended, for only the visual representations of the target to be enhanced. There has been little or no testing of this ability. In almost all studies of PF delay activity in monkeys, the cue to be remembered has been presented in isolation from other stimuli. Thus it was not known whether irrelevant information can be filtered from PF delay activity.

We addressed this issue by training monkeys on a visual search task that required them to view a cluttered scene and remember only one object from it (Rainer, Asaad, and Miller 1998a). Monkeys were first cued

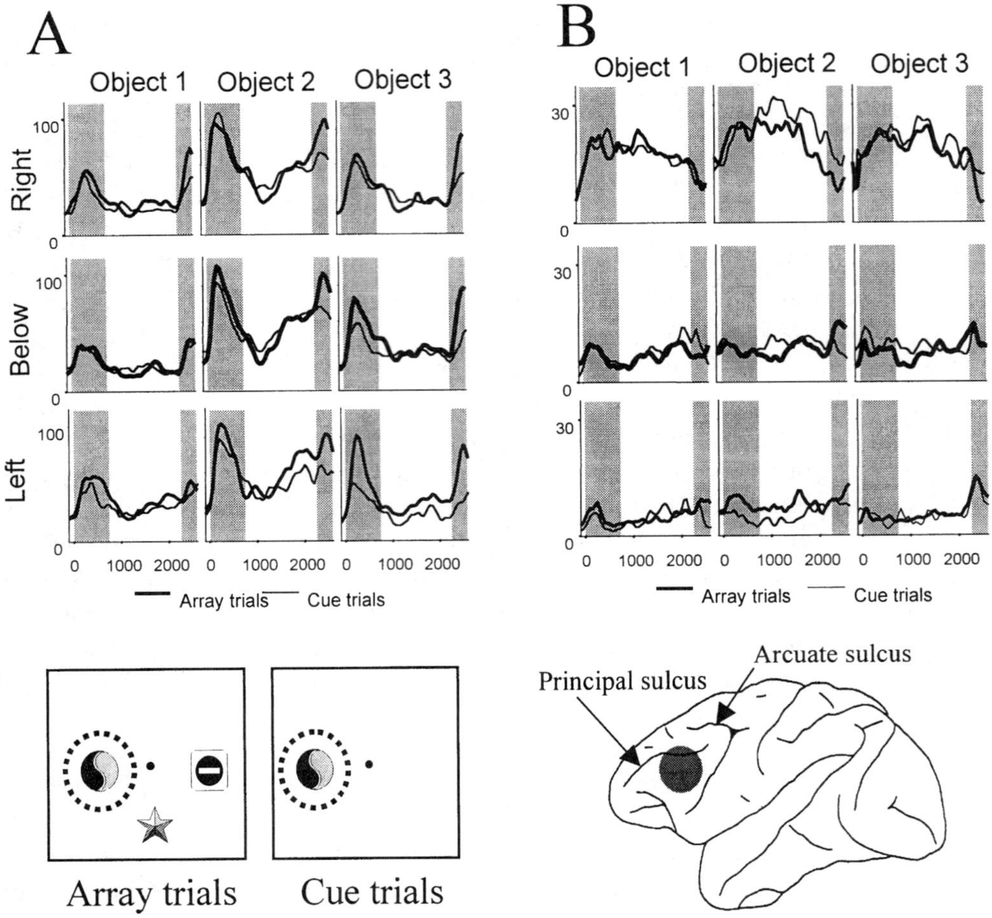

Figure 22.1 Activity from single prefrontal neurons that varied with target object (A) and target location (B). Gray bar on left of each histogram indicates time of sample presentation; gray bar on right indicates presentation of the test array. The x-axis represents time and the y-axis firing rate in spikes per second. Column labels refer to target object in array trials or to sample object on cue trials. Row labels refer to target location on array and cue trials. Across a given row, the arrays were physically identical but differed in relevant target object. Bin width, 40 msec. The line drawing shows a lateral view of the left side of a macaque brain. The recording sites for this experiment (shaded) were around the principal sulcus region (Brodmann's area 46) and on the inferior convexity below it (Brodmann's area 12). Most of the neurons (85% of 97 tested) showed attentional effects in one or more task epoch. Adapted from Rainer et al. 1998b.

to a relevant (target) object by cue trials in which the target appeared alone. On "array trials," the monkeys had to find the target in a sample array of three objects: they needed to remember its position in the sample array over a brief delay. Then, an array of three test objects appeared. Monkeys released a lever if the target appeared in the same location as it had in the sample array. Three objects were used and each was used as a cue and target on a different block of trials.

On array trials, PF delay activity reflected only information about the target object, its location, or both; activity related to the nontargets was almost completely filtered out. For example, figure 22.1A shows examples of two PF neurons, each of which selectively represents behaviorally relevant information from the sample array. The neuron in figure 22.1A showed a high level of activity on array trials (thick lines) when object "B" was the target (middle column), but lower activity when object "A" or "C" was the target. The rows show the position of the target. Note that this neuron showed similar activity to the target in each location (i.e., it was object, but not location, selective). We also found neurons selective for the target's location (figure 22.1B) and for both the target object and its location. For each neuron, irrelevant information had little or no effect on neural activity. The thin lines show each neuron's activity on corresponding "cue trials" in which the target appeared alone in the same position. Note that activity on array and cue trials was strikingly similar. Thus on array trials the neurons responded as if the target were presented alone (as it was on cue trials). These results show that PF delay activity does not merely reflect what the animal had just seen (the whole sample array); it conveys only the information needed to make the decision about test array (the target).

One striking aspect of this study was that most of the neurons (85%) showed attention effects. In studies from our laboratory we do not prescreen neurons for task-related responses. Instead, we advance each electrode until the activity of one or more neurons is well isolated, and then begin to collect data. This procedure is used to ensure an unbiased estimate of prefrontal activity. That many neurons in this study (and in others discussed below) show task-related properties suggests that the PF cortex becomes "tuned" to the task through the months of training needed to teach it to the monkeys. Bichot, Schall, and Thompson (1996), for example, showed that neurons in prefrontal area 8 acquire color selectivity through training. Rather than create altogether new mechanisms, training may enhance preexisting mechanisms, much as training monkeys to perform sensory discriminations results in expanded representations in sensory cortex (Recanzone et al. 1992; Recanzone, Schreiner, and Merzenich 1993). This ability to adapt to current task demands may be critical to the role of the PF cortex in guiding complex behavior, a point we will revisit.

Selective encoding by prefrontal neurons has been observed in other studies as well. Sakagami and Niki (1994a) showed that many PF neurons selectively encoded the dimension of a stimulus (e.g., color versus shape) that was currently relevant for behavior. We have observed similar effects during a task that required monkeys to remember first the identity and then the location of an object (Rao et al. 1997). The activity of many PF neurons mirrored these task demands. The monkeys appeared to "switch modes," from being highly object selective in the first half of the trial to being purely location selective in the second half. In other words, PF neurons only reflected information about the object features that were currently relevant. This ability to selectively convey task-relevant information suggests that PF delay activity reflects an active encoding and maintenance process rather than a passive buffering of sensory information. Studies of patients with PF dysfunction (e.g., D'Esposito and Postle, chap. 26, this volume) also suggest a disruption of an active rehearsal process.

Thus prefrontal neurons can selectively represent behaviorally relevant information, a prerequisite for providing bias signals. The PF cortex is not the only region where selection is evident. Indeed, selection is a hallmark of attention and it is evident throughout the visual cortex (Bushnell, Goldberg, and Robinson 1981; Moran and Desimone 1985; Chelazzi et al. 1993; Motter 1993, 1994; Gottlieb, Kusunoki, and Goldberg 1998). We found, however, that selection in the PF cortex during visual search occurred very rapidly; information about the location of a target object appeared in PF activity 135–140 msec after onset of the sample array (Rainer, Asaad, and Miller 1998a). In another object-based attention task, Chelazzi et al. (1993) found that information about a target was not reflected in the inferior temporal cortex (an extrastriate visual area) until 175–200 msec after stimulus onset. Although caution must be taken in interpreting this difference (the experiments used different monkeys with different training histories), the early selection of the target in the PF cortex is consistent with its role as the source of bias signals that mediate selection.

Selection of Information for Bias Signals: Recall of Stored Information

In most neurophysiological studies of attention, attentional templates can be derived from sensory information. For example, in a typical visual search task, a monkey is shown a cue object shortly before it must find the object in a cluttered display. Outside the laboratory, however, bias signals often cannot be derived from available sensory information; they must be derived from information stored in long-term memory. For example, our missing keys are not available to us as a cue shortly before we begin looking for them. Instead, we must recall what they look like. A region that provides bias signals thus needs access to stored knowledge.

In studies of animal cognition, the process of bringing to mind the information from long-term memory is referred to as "prospective memory" (Honig and Thompson 1982); in human studies, "prospective memory" refers to the slightly different process of remembering to execute an action in the future. Memory is "prospective" when recall occurs in anticipation of an upcoming event or action, such as when we bring an image of our keys to mind shortly before we begin to search for them. By contrast, the mechanisms that preserve recent sensory inputs are called "retrospective memory." Working memory contains both retrospective and prospective mechanisms (Honig and Thompson 1982). In many delay tasks, though, prospective mechanisms seem to dominate, particularly when the information needed after a delay is different from that seen before a delay (e.g., Gaffan 1977; Roitblat 1993; Colombo and Graziano 1994).

We (Rainer, Rao, and Miller 1999) demonstrated this, training monkeys on a go/no-go symbolic delayed match-to-sample (SDMS) task. A sample object was briefly presented at the center of gaze. This was followed, after a brief delay, by a single test object. The monkeys had to release a lever if the test object was "correct." SDMS differs from identity (standard) delayed match-to-sample (IDMS) in that the correct test object is different from the sample object. The monkeys learned through months of training that when, for example, object S1 was the sample, they had to select choice object C1. In this situation, monkeys can use either a retrospective or a prospective strategy. A retrospective strategy would involve holding the sample (e.g., S1) in memory over the delay and, when the test object appeared, querying long-term memory to determine whether the current test object was the correct one. By contrast, a prospective strategy would involve immediately recalling the correct choice (e.g., C1) shortly after the sample was presented, holding that stimulus, rather than the sample, in memory over the delay, and, when the test object appeared, querying the representation currently in active memory.

To determine whether monkeys were using a prospective strategy, we used the three pairs of sample-correct choice objects to form a "confusion matrix" (figure 22.2A). Of the three sample objects, two were similar and one was dissimilar from the other two. The three choice objects also included two that were similar and one that was dissimilar. The two similar sample stimuli were associated with two dissimilar choice stimuli, and the two dissimilar sample objects, with similar choice objects (figure 22.2A). The monkeys' choice errors were prospective in nature. That is, the errors reflected the similarity of the choice objects, not the samples (figure 22.2B). This suggests that upon seeing the sample object, the monkeys used a prospective strategy of "thinking ahead" to the choice object during the delay. Other studies in monkeys (Gaffan 1977; Erickson and Desimone 1996; Colombo and Graziano 1994) and in a variety of other species (Honig and Thompson 1982; Roitblat 1993) have also found evidence for prospective coding. Thus working memory not only maintains sensory inputs; it can also be used to maintain information prospectively recalled from long-term storage.

In principle, delay activity observed in this task could reflect either the object the animal just saw or the object the animal anticipated choosing at the end of the delay. To distinguish between these possibilities, we also trained the monkeys on a standard IDMS task, where they chose the test object that matched the sample (e.g., if object C1 was the sample, it was

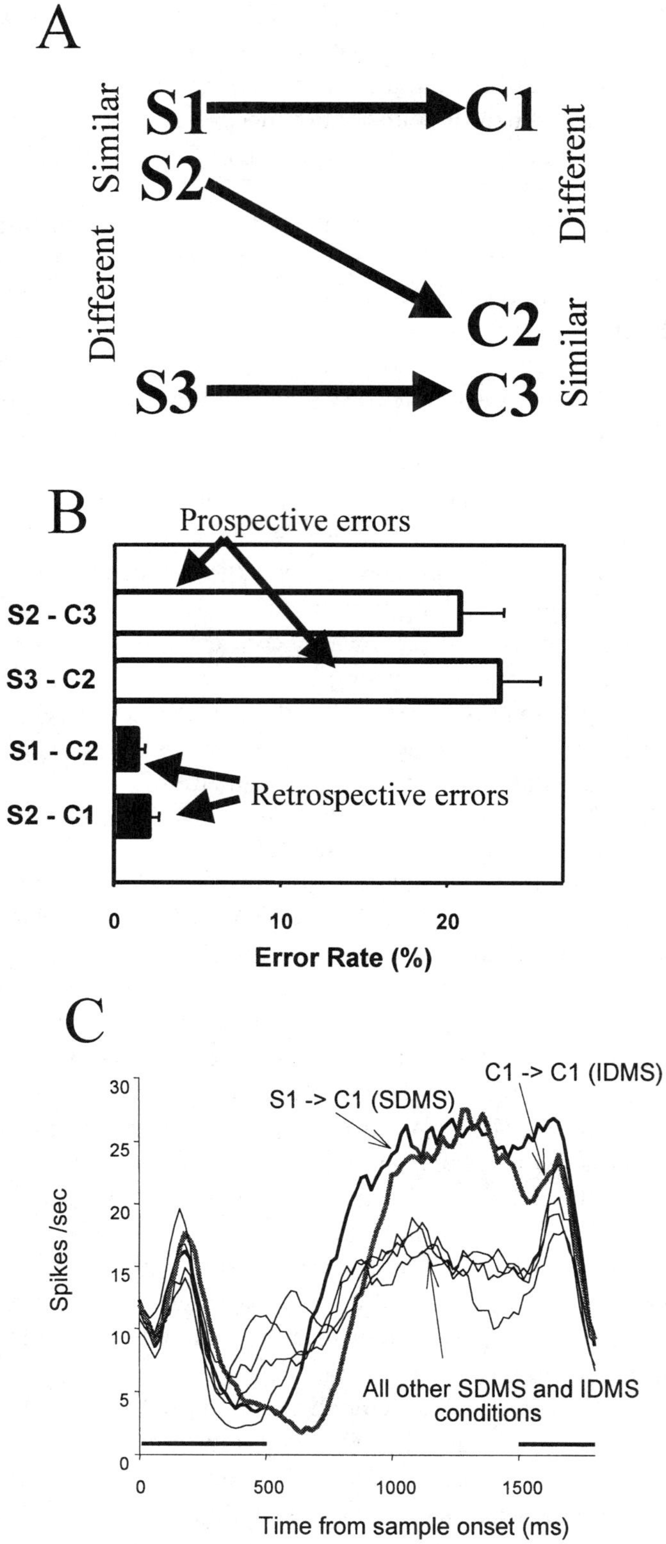
A
Similar
S1
S2
Different
S3
C1
C2
C3
Different
Similar
B
Prospective errors
S2 - C3
S3 - C2
S1 - C2
S2 - C1
Retrospective errors
0
10
20
Error Rate (%)
C
C1 -> C1 (IDMS)
S1 -> C1 (SDMS)
All other SDMS and IDMS conditions
Spikes /sec
0
5
10
15
20
25
30
0
500
1000
1500
Time from sample onset (ms)

also the correct choice). By comparing delay activity during the SDMS task to that during the IDMS task, we could determine whether delay activity during the SDMS task conveyed the sample or its anticipated associate. This revealed that many prefrontal neurons showed properties consistent with prospective coding. That is, their delay activity reflected the anticipated choice object regardless of whether the monkey had been cued with its paired associate (SDMS task) or with the object itself (IDMS task; figure 22.2C). PF neurons seem to generate prospective codes for other types of information as well. Watanabe (1996) demonstrated that they can convey information about expected rewards. Quintana and Fuster (1992) found that PF activity reflects the probability that a given response will be required at the end of a delay, which suggests prospective coding for actions. Thus PF neurons have another property critical for providing attentional templates: they can maintain representations of information recalled from long-term memory.

Integration of Diverse Information: Attention to Conjunctions of Features

We have seen that prefrontal neurons can maintain information about object identity and location and have access to diverse sensory inputs and stored memories. Complex behavior typically requires coordinating and integrating diverse information to serve common behavioral goals. Visual attention, for example, is rarely directed only to objects or only to certain locations. Take searching for a coffee cup. We have in mind not only what the cup looks like but also where it is likely to be.

It was unclear whether PF neurons can form attentional templates that combine object and spatial information. The visual cortex has been proposed to contain two "streams" or pathways that separate processing of object and spatial information (Ungerleider and Mishkin 1982; Maunsell and Newsome 1987).

Figure 22.2 Prospective memory effects. A. Schematic representation of the stimulus relationships used in this experiment. Distance on vertical axis represents the relative degree of physical similarity between stimuli in the same column. B. Prospective and retrospective error rates for one monkey. The y-axis lists the type of error. Note that monkeys made more errors confusing similar test objects (C2 and C3) than similar samples (S1 and S2). C. Activity of a single prefrontal neuron involved in recall of a long-term memory. The figure shows activity during performance of a symbolic delayed match-to-sample (SDMS) task and an identity delayed match-to-sample (IDMS) task. The small horizontal line on the left of the graph shows the time of sample presentation and the small horizontal line on the right shows when the choice objects were presented. Note that this neuron showed a high level of delay activity on IDMS trials when the monkey remembered "C1" over the delay. It showed a similar level of delay activity when, on SDMS trials, the sample was S1's paired associate, C1. Thus on SDMS trials delay activity seemed to reflect the object anticipated at the end of the delay (C1), which needed to be recalled from long-term memory. Recording sites were in the inferior convexity (Brodmann's area 12) and Brodmann's area 46.

This raises the question of how and where object and spatial information come together. Because the separation between the visual system pathways is relative, not absolute, the two kinds of information are likely to be integrated to some extent within the visual system. There are interconnections within the visual cortex that can bring together object and spatial information (Maunsell and Van Essen 1983; Boussaoud, Ungerleider, and Desimone 1990; Van Essen, Anderson, and Fellman 1992); moreover, visual cortical areas thought to be relatively specialized for processing either object or spatial information also have neurons that select for, or are modulated by, the other kind of information (Moran and Desimone 1985; Ferrera, Rudolph, and Maunsell 1994; McAdams and Maunsell 1997; Sereno and Maunsell 1998). Other studies (e.g., Goodale and Haffenden 1998) indicate that the two cortical visual streams may separate processing of perceptual information from that of information needed for action. In this model, the object and spatial information used for perception are not separate but instead are integrated within the ventral visual pathway. Indeed, object-selective ventral pathway neurons do carry spatial information (Gross, Rocha-Miranda, and Bender 1972; Desimone et al. 1984; Schein and Desimone 1990). Integration of disparate information is likely to occur within the prefrontal cortex as well. While inputs to the PF cortex from different sensory systems only partly overlap, there are extensive interconnections between different PF regions that could integrate information from these inputs (Barbas and Pandya 1989, 1991). Few neurophysiological studies have addressed this issue, however. Most have studied how PF neurons convey object or spatial information alone.

To explore whether single prefrontal neurons have access to both object and spatial information, we (Rao, Rainer, and Miller 1997) trained monkeys on a task that required them to remember first an object and then its location. They were shown a sample object they needed to remember. After a delay, two objects were simultaneously and briefly presented. One of the objects matched the sample; the other did not. The monkeys needed to remember the location of the match because, after another delay, they directed a saccadic eye movement to its remembered location. Thus they needed to find a specific object and then, ultimately, to direct action to its location. Although some of the PF neurons were specialized for object or spatial working memory, about half were able to link objects with their locations, conveying information first about the identity of the sample and then about the location of the match.

This suggests that many prefrontal neurons have access to both object and spatial information. Of course, a top-down bias signal would often need to simultaneously convey both kinds of information. In the Rainer, Asaad, and Miller 1998a study described above, many neurons did just that. In Rainer, Asaad, and Miller 1998b, we explored the receptive fields

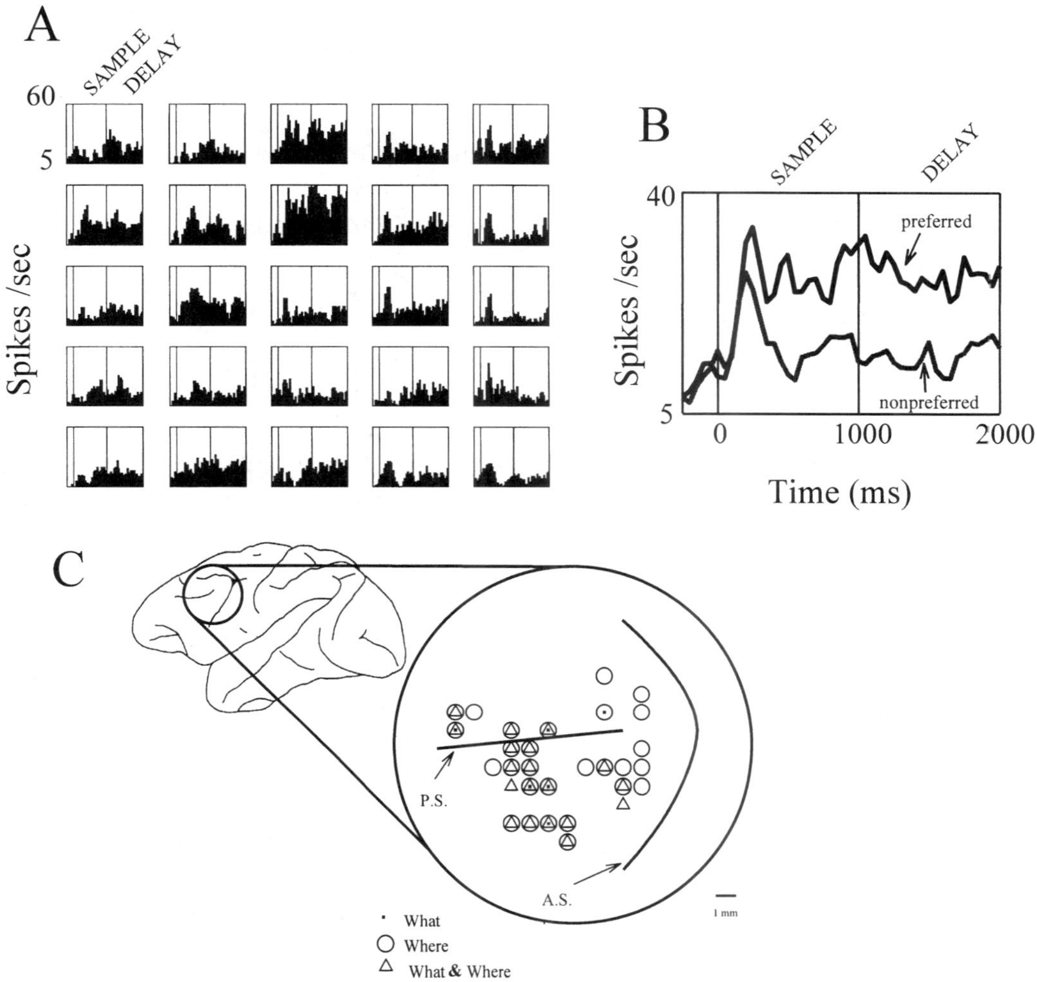

Figure 22.3 A. Histograms of a single prefrontal neuron's activity to an object appearing at each of the 25 tested locations. The vertical line to the left of each histogram shows time of sample onset and the vertical line in the middle denotes sample offset. Bin width, 40 msec. The timescale for each histogram is identical to that shown in figure 22.2C. Note that this neuron is highly spatial selective. It only shows sustained activation when the object appears at two extrafoveal locations. The remaining locations may elicit brief bursts of activity at sample onset, but they do not elicit robust sustained activity. B. Average activity of the same neuron to a preferred and nonpreferred object appearing at the two locations that elicited delay activity. Note that this neuron is also highly object selective. C. Recording sites. Each symbol represents a recording site where neurons with object-selective delay activity ("What"), location-selective delay activity ("Where"), or both object- and location-selective delay activity ("What" and "Where") were found. Typically, several neurons were found at the same site. About half of the 149 neurons with task-related properties showed activity selective for both "What" and "Where." Adapted from Rainer, Asaad, and Miller 1998.

of PF neurons. Monkeys were trained on a go/no-go delayed match-to-object-place task that required them to remember, over a brief delay, which of 2–5 sample objects had appeared in which of 25 visual field locations. They released a lever when a test object matched a sample in both identity and location. During the delay, about half of the neurons simultaneously conveyed information about the identity of the sample object and its precise location (figure 22.3). In fact, the average diameter of the receptive field derived from delay activity (i.e., "memory fields," or MFs) of these neurons was only about 9 degrees. Further, unlike inferior temporal neurons, object-selective PF neurons did not emphasize central vision. Rather, they seemed well suited to the task demand to remember an object throughout a wide portion of the visual field. Many object- and location-selective neurons had MFs that were entirely extrafoveal and many were maximally activated by peripheral locations. Thus, across the population, these neurons could simultaneously identify and localize objects throughout a wide area of the visual field, both near the fovea and in the periphery.

These results are consistent with other neurophysiological studies that have found an intermixing of prefrontal neurons that process object and spatial information within the same PF regions (Watanabe 1981; Fuster, Bauer, and Jervey 1982). Similarly, functional imaging studies in humans have found that similar, often identical, regions of the PF cortex are activated during object memory tasks and spatial memory tasks (Owen et al. 1996, 1998; Oster et al. 1997; Courtney et al. 1998; Cullen et al. 1998; Postle and D'Esposito 1998). Even studies that find some separation of PF regions activated by object and spatial processing also find regions of overlap (Courtney et al. 1998). In fact, some functional imaging and behavioral studies (e.g., Duncan and Owen, chap. 25, Petrides, chap. 23, this volume) suggest that the PF cortex is organized by the type of processing required rather than by the nature of the information processed (e.g., object or location). There may, however, be some regional biases in the representations of object and spatial information (Wilson, O'Scaladaihe, and Goldman-Rakic 1993; O'Scalaidhe, Wilson, and Goldman-Rakic 1997; Courtney et al. 1998). Cells specialized for processing facial information appear to be highly localized within the ventral PF cortex, much as they are highly localized within the temporal cortex (O'Scalaidhe, Wilson, and Goldman-Rakic 1997).

Thus prefrontal neurons can provide bias signals that convey both object and spatial information, a characteristic useful for guiding attention based on conjunctions of attributes. They may play a role in integrating more diverse information. The lateral PF cortex receives converging visual, auditory, and somatosensory information; some of its neurons have multimodal responses. Watanabe (1992), for example, has found that many PF neurons will respond to both visual and auditory stimulation when they have similar behavioral significance.

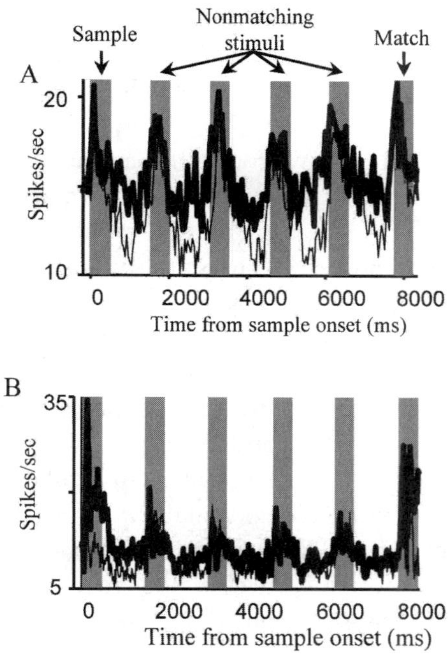

Figure 22.4 Average histograms of a population of prefrontal neurons (A) and inferotemporal neurons (B) following preferred and nonpreferred sample objects. Responses are shown separately for trials in which a "preferred" or "nonpreferred" stimulus was used as a sample. The gray bars show the time of stimulus presentation. Bin width, 40 msec. Prefrontal recordings were from the inferior convexity (Brodmann's area 12) and Brodmann's area 46. Inferotemporal recordings were primarily from the perirhinal cortex. Adapted from Miller, Erickson, and Desimone 1996.

Maintenance of Signals in the Face of Distractions

Once an attentional template is formed, it needs to be maintained until attention is successfully directed to the visual field item of choice. Most studies of prefrontal delay activity have not addressed this issue; they have used tasks that employed a "blank" delay interval, in which no stimuli intervene between the sample and the choice phases of the task. In the real world, however, bias signals need to be maintained across intervening sensory inputs; our retention intervals are often filled with new stimuli entering the visual system. In visual search, for example, we need to hold an attentional template in mind while we inspect the visual environment. If delay activity were disrupted each time we inspected a new portion of a scene, it would be useless as an attentional template.

We (Miller, Erickson, and Desimone 1996) tested the ability of prefrontal neurons in monkeys to convey information about a given stimulus across intervening inputs, using a delayed match-to-sample task with intervening stimuli. After they were presented a sample object, the monkeys viewed a sequence of one to five test objects; they were rewarded for releasing a lever when one of the test objects matched the sample. There was a short (1 sec) delay between each stimulus presentation, and the monkeys could not predict when the match would appear in the sequence.

Consistent with other studies, we found that in the delay immediately following the sample, many prefrontal neurons maintained sample-specific delay activity. The intervening stimuli in the delay revealed that this activity was robust. Figure 22.4A shows the average activity of a population of PF neurons when a preferred or nonpreferred object was the remembered sample. While sample-specific activity is temporarily disrupted during stimulus presentation (gray bars), there is more activity following a preferred object in each delay. Thus the neural representation of the sample was maintained throughout the trial across intervening objects. diPelligrino and Wise (1993) also found a similar maintenance of PF delay activity across intervening visual inputs. This ability is not unique to the PF cortex. Suzuki, Miller, and Desimone (1997) found that some neurons in the entorhinal cortex, another region critical for visual memory, also maintain sample-specific delay activity across intervening stimuli.

By contrast, at least some extrastriate visual areas responsible for analysis of sensory information do not appear to have this property. Object-specific delay activity has been reported in the inferior temporal (IT) cortex and neurons in the posterior parietal (PP) cortex have delay activity selective for spatial locations (Miyashita and Chang 1988; Fuster and Jervey 1981; Gnadt, Bracewell, and Andersen 1991; Miller, Li, and Desimone 1993; Constantinidis and Steinmetz 1996), although delay activity in these areas is labile and easily disrupted by intervening inputs (Miller, Li, and Desimone 1993; Miller, Erickson, and Desimone 1996; Constantinidis and Steinmetz 1996). This can be seen for IT neurons in figure 22.4B, where sample-specific activity in the delay immediately following the sample is attenuated by the first intervening stimulus and abolished after the second intervening stimulus.

To summarize, prefrontal neurons appear to have properties ideal for attentional templates that bias competition in extrastriate visual cortex in favor of behaviorally relevant visual field items. PF neurons can form attentional templates by selecting relevant sensory inputs and stored knowledge and by integrating diverse information to meet current attentional demands. They can maintain the templates across distracting inputs so that they are available until attention is successfully focused.

But how do we determine what is relevant? This is perhaps the central question in top-down control and the most difficult to study. The prefrontal cortex has long been thought to be important for such "executive decisions." In the next subsection, we will examine some of the neural mechanisms that may mediate them.

Determining Relevance: Prefrontal Cortex and Rule Representation

Complex behavior is typically rule based. Our previous experiences arm us with sets of behavior-guiding scripts, or rules, that relate events to possible outcomes and consequences. They specify the conditions and behaviors needed for achieving a goal (Abbott, Black, and Smith 1985; Barsalou and Sewell 1985; Norman and Shallice 1986). Behavior-guiding rules not only dictate what behaviors are likely to be rewarding or appropriate, but also which visual features are likely to be important and worth attending.

Rules are also important for monkeys. Indeed, to perform any of the tasks described here, monkeys must have some internal representation of the task rules. Models of prefrontal function by Wise, Murray, and Gerfen (1996) and Passingham (1993) based on animal studies argue that rule learning and representation are cardinal PF functions, and that the pattern of deficits seen after PF damage reflects a loss in these functions. PF mechanisms for acquiring, representing, and selecting among behavior-guiding rules may correspond to Norman and Shallice's "supervisory attention system" (1986; thought to be located in the PF cortex) that switches attention to important sensory information and actions.

Understanding how rules are engendered by prefrontal neural activity is central to understanding directed attention in particular and cognition in general. Cohen and colleagues (Cohen and Servan-Schreiber 1992; Braver and Cohen, chap. 31, this volume) have suggested a biologically plausible model. They posit that cognitive control emanates from a PF representation of context, the constellation of information needed to mediate an appropriate behavior. One prediction of this model is that many PF neurons should have complex, multimodal responses that represent, not simply single stimuli, but also conjunctions of behaviorally related information. In other words, their response to a stimulus should also reflect the behavioral context in which the stimulus appears.

To explore the effects of behavioral context on prefrontal activity and the neural mechanisms involved in rule-learning, we (Asaad, Rainer, and Miller 1998) used a conditional visuomotor task. Studies in humans and monkeys (Petrides 1982, 1986, 1990; Passingham 1993; Gaffan and Harrison 1988; Eacott and Gaffan 1992; Parker and Gaffan 1998) suggest that the PF cortex is involved in a wide variety of conditional learning tasks, including conditional visuomotor learning. In conditional tasks, a set of rules must be learned. In our task, monkeys learned to associate each of two initially novel cue objects with either a saccade to the left or a saccade to the right (e.g., A → go right; B → go left). While the monkeys maintained fixation of a fixation target, one of the objects was presented at the center of gaze. Then, after a 1 sec delay, the fixation point was extinguished and two choice dots were presented to the left and right of

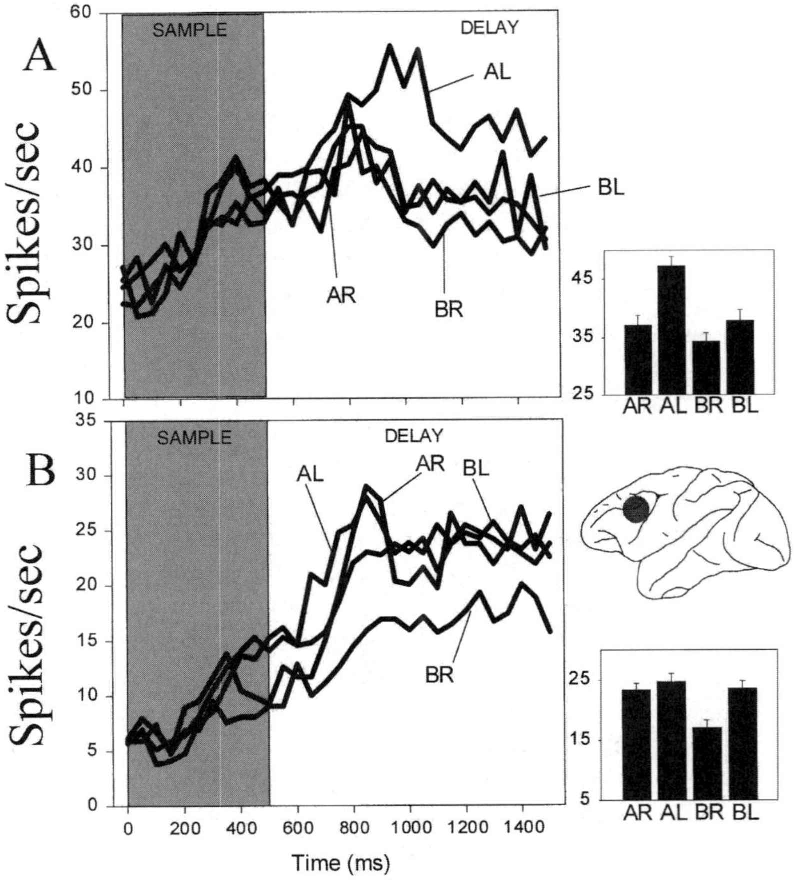

Figure 22.5 Histograms of two single prefrontal neurons tuned for object-spatial associations. AR = object A associated with "go right"; AL = object A associated with "go left"; and so on. Gray bars indicate the times of sample presentation. Small bar graphs show the average activity in the delay for each of the conditions; error bars show the standard errors of the mean. Note that neuron in panel A shows a high level of delay activity when sample object A is associated with a saccade to the left. By contrast, its delay activity is lower when the same object is associated with a saccade to the right or when sample object B is associated with a saccade to the left or to the right. Note also that neuron in panel B shows lower activity when object B is associated with a saccade to the right. The line drawing shows the brain recording sites (see figure 22.1 for conventions). About half (47%) of the 254 cells studied showed activity tuned for both objects and saccade direction. Adapted from Asaad, Rainer, and Miller 1998.

fixation. Monkeys made a saccade to one of the dots depending on which object had been the cue. After the monkeys learned the initial object-direction pairings, the associations were reversed (now A → go left; B → go right). Once the reversals were learned, the associations were reversed again, and again, for six or more reversals. This allowed us to explore how the conditional rules were represented by PF activity. The reversals allowed us to avoid confounding object and spatial information, that is, by not exclusively associating a specific object with a specific saccade direction, we could determine the relative effects of object and spatial information on neural activity.

After the object-saccade pairings were learned, many PF neurons seemed to explicitly represent them, showing activity that depended on both the sample object and the direction of the forthcoming saccade. For most of these cells, however, object and spatial information combined in a nonlinear fashion. For example, the neuron depicted in figure 22.5A showed the highest level of activity in the second half of the delay whenever sample object A instructed a saccade to the leftward location. By contrast, lower activity was apparent for the other associations. This neuron was not merely tuned to object A because "A → go right" did not elicit the same level of activity as "A → go left." Nor was it merely tuned to "go left" because "B → go left" also did not produce the same activity as "A → go left." This neuron thus seemed to be tuned to the combination of "A" and "go left." Another type of "nonlinear" neuron (figure 22.5B) showed weaker activity to the combination of "B → go right" than for all other combinations.

Activity reflecting these stimulus-response pairings was not as evident before learning. When, at first, the monkeys were "guessing" which response was correct for each cue, spatial activity related to the impending response only appeared just before the saccade was made. During learning, however, location-selective activity appeared progressively earlier within each successive trial. By the time the pairings were well learned, many neurons showed object and spatial activity that overlapped throughout most of the trial. These results suggest that many prefrontal neurons play a role in acquiring and representing the stimulus-response associations the animals used to guide their behavior. This ability to represent conjunctions of disparate behaviorally related information has been observed in other studies. Sakagami and Niki (1994b) found that many PF neurons responded differently to a visual stimulus depending on whether that stimulus currently required an immediate or delayed release of a response lever. Watanabe (1990, 1992) found that many neurons responded differentially to a sensory stimulus depending on whether it signalled that a reward would be delivered on that trial; indeed, many single PF neurons were tuned to the associative significance of both visual and auditory cues (Watanabe 1992). In a particularly relevant example, White and Wise (1997) trained monkeys to attend to a particular location by teaching them a spatial rule (attend to the location where a cue had appeared) or a conditional rule (attend to the location associated with the cue, e.g., "object A → attend right"). They found that the activity of many PF neurons reflected not only the relevant location but also which rule the animal had followed.

Thus prefrontal activity does not merely reflect a stimulus or a response but also conveys information about behavioral context. It can convey information about conjunctions of related sensory events, actions, and their expected consequences, such as reward. These properties are what we would expect from a region involved in acquiring and representing rules. Indeed, a wealth of behavioral evidence indicates that the PF cortex is central to these processes (Shallice 1982; Burgess and Shallice 1996a; Duncan et al. 1996). The ability to acquire and choose among rules is important for flexible, intelligent behavior, particularly in novel situations when we must apply generalizations from our previous experiences to solve a new problem. More to the point, behavior-guiding rules can convey information about which visual features are, or are likely to be, important and need attending.

22.2 CONCLUSIONS

Competition plays an important role in visual processing. Inhibitory interactions between neurons are thought to play a central role in sensory processing by, for example, enhancing contrast representation and by segmenting figure from ground. The neurons that "win" the competition and remain active incur a higher level of activity than those with which they share inhibitory interactions. The model of biased competition posits that visual attention exploits these mechanisms (Desimone and Duncan 1995). Competitive advantage can result from physical properties of the stimulus; a stimulus that is different from its surroundings seems to automatically "pop out" and grab our attention. In voluntary shifts of attention, however, a competitive advantage must often be incurred, not from the stimulus, but from top-down signals related to its behavioral relevance. These bias signals must originate from brain regions that are not exclusively visual; information about what is relevant and needs attending requires multimodal, abstract sources of information.

The prefrontal cortex seems ideally suited for this role. It is interconnected with virtually all of the brain's sensory systems, with neural structures critical for storing knowledge and with cortical and subcortical structures critical for voluntary behavior (Pandya and Barnes 1987; Barbas and Pandya 1991). Its interconnections with virtually all of extrastriate visual cortex place PF cortex in an ideal position for modulating visual processing (Barbas 1988; Ungerleider, Gaffan, and Pelak 1989; Pandya and Yeterian 1990; Webster, Bachevalier, and Ungergleider 1994). Evidence for such interactions comes from observations that cooling the PF cortex modulates activity in IT cortex, causing cells to be less selective (Fuster, Bauer, and Jervey 1985).

Consistent with their multivariate connections, the activity of prefrontal neurons reflects behavioral context, the constellation of behaviorally relevant information associated with stimuli such as associated behavioral responses, reward value, and expected events. These associations may develop from past experience at achieving a particular goal or similar goals. As a result, sensory inputs to the PF cortex may evoke a neural representation of the behavioral context associated with those inputs, including the conjunction of relevant visual features that need attending to achieve the current goal. This attentional template may then feed back to the visual cortex, enhancing the activity of neurons sensitive to features that match the template and thus biasing competition in their favor. Knight (1997) found evidence for this process, observing that patients with PF damage do not show attention-related enhancement of extrastriate scalp potentials during attention tasks. Of course, the PF cortex is unlikely to be the sole source of feedback signals pertaining to behavioral relevance. Other regions share at least some properties with PF cortex (Suzuki, Miller, and Desimone 1997), and structures interconnected with the PF cortex, such as the striatum, are likely to be important (Robbins and Rogers, chap. 21, this volume). Given its central role in organizing complex behavior, however, the PF cortex is likely to be a major source of top-down bias signals.

Finally, it is worth noting that the principles of biased competition are unlikely to be limited to attention. Indeed, the neural architecture on which biased competition rests (local inhibitory interactions, long-range excitatory influences) is common in the brain (White 1989), and a wide variety of functions may exploit them. Indeed, mechanisms similar to biased competition have been proposed to play a role in the highest levels of cognition. For example, in Norman and Shallice's model (1986), conflicting thoughts and actions are mutually inhibitory and compete for control of behavior. Excitatory influences from a supervisory attention system (thought to be located in the PF cortex) enhance appropriate representations, which then inhibit their competitors (Shallice 1982; Norman and Shallice 1986). Studies of visual attention may offer a tractable means for understanding the principles that determine which perceptions, thoughts, and actions reach our awareness and guide our behaviors.

NOTE

I thank my collaborators on these experiments: Wael Asaad, Robert Desimone, John Duncan, Cynthia Erickson, S. Chenchal Rao, and Gregor Rainer. I also thank Wael Asaad, Richard Wehby, and Marlene Wicherski for their valuable comments on the manuscript.

REFERENCES

Abbott, V., Black, J. B., and Smith, E. E. (1985). The representation of scripts in memory. *Journal of Memory and Language, 24,* 179–199.

Asaad, W. T., Rainer, G., and Miller E. K. (1998). Neural activity in the primate prefrontal cortex during associative learning. *Neuron, 21,* 1399–1407.

Baddeley, A. (1986). *Working memory.* Oxford: Clarendon Press.

Barbas, H. (1988). Anatomic organization of basoventral and mediodorsal visual recipient prefrontal regions in the rhesus monkey. *Journal of Comparative Neurology, 276,* 313–342.

Barbas, H., and Pandya, D. N. (1989). Architecture and intrinsic connections of the prefrontal cortex in the rhesus monkey. *Journal of Comparative Neurology, 286,* 353–375.

Barbas, H., and Pandya, D. N. (1991). Patterns of connections of the prefrontal cortex in the rhesus monkey associated with cortical architecture. In H. S. Levin, H. M. Eisenberg, and A. L. Benton (Eds.), *Frontal lobe function and dysfunction,* pp. 35–58. New York: Oxford University Press.

Barsalou, L. W., and Sewell, D. R. (1985). Contrasting the representation of scripts and categories. *Journal of Memory and Language, 24,* 646–665.

Bichot, N. P., Schall, J. D., and Thompson, K. G. (1996). Visual feature selectivity in frontal eye fields induced by experience in mature macaques. *Nature, 381,* 697–699.

Boussaoud, D., Ungerleider, L. G., and Desimone, R. (1990). Pathways for motion analysis: Cortical connections of the medial superior temporal and fundus of the superior temporal visual areas in the macaque. *Journal of Comparative Neurology, 296,* 462–495.

Burgess, P. W., and Shallice, T. (1996a). Bizarre responses, rule detection and frontal lobe lesions. *Cortex, 32,* 241–259.

Burgess, P. W., and Shallice, T. (1996b). Response suppression, initiation and strategy use following frontal lobe lesions. *Neuropsychologia, 34,* 263–272.

Bushnell, M. C., Goldberg, M. E., and Robinson, D. L. (1981). Behavioral enhancement of visual responses in monkey cerebral cortex: 1. Modulation in posterior parietal cortex related to selective vision. *Journal of Neurophysiology, 46,* 755–771.

Cavada, C., and Goldman-Rakic, P. S. (1989). Posterior parietal cortex in rhesus monkey: 2. Evidence for segregated corticocortical networks linking sensory and limbic areas with the frontal lobe. *Journal of Comparative Neurology, 287,* 422–445.

Chelazzi, L., Miller, E. K., Duncan, J., and Desimone, R. (1993). A neural basic for visual search in inferior temporal cortex. *Nature, 363,* 345–347.

Cohen, J. D., Perlstein, W. M., Braver, T. S., Nystrom, L. F., Noll, D. C., Jonides, J., and Smith, E. E. (1997). Temporal dynamics of brain activation during a working memory task. *Nature, 386,* 604–608.

Cohen, J. D., and Servan-Schreiber, D. (1992). Context, cortex, and dopamine: A connectionalist approach to behavior and biology in schizophrenia. *Psychological Review, 99,* 45–77.

Colombo, M., and Graziano, M. S. (1994). Effects of auditory and visual interference on auditory-visual delayed matching to sample in monkeys (*Macaca fascicularis*). *Behavioral Neuroscience, 108,* 636–639.

Constantinidis, C., and Steinmetz, M. A. (1996). Neuronal activity in posterior parietal area 7a during the delay periods of a spatial memory task. *Journal of Neurophysiology, 76,* 1352–1355.

Courtney, S. M., Petit, L., Maisog, J. M., Ungerleider, L. G., and Haxby, J. V. (1998). An area specialized for spatial working memory in human frontal cortex. *Science, 279,* 1347–1351.

Courtney, S. M., Ungerleider, L. G., Keil, K., and Haxby, J. V. (1997). Transient and sustained activity in a distributed neural system for human working memory. *Nature, 386,* 608–611.

Cullen, C. M., Bucci, J., Snow, M., Miller, E. K., and Corkin, S. (1998). FMRI activation during performance on object and spatial components of a working memory task. *NeuroImage, 8,* S867.

Desimone, R., Albright, T. D., Gross, C. G., and Bruce, C. (1984). Stimulus-selective properties of inferior temporal neurons in the macaque. *Journal of Neuroscience, 4,* 2051–2062.

Desimone, R., and Duncan, J. (1995). Neural mechanisms of selective visual attention. *Annual Review of Neuroscience, 18,* 193–222.

DiPellegrino, G., and Wise, S. P. (1993). Effects of attention on visuomotor activity in the premotor and prefrontal cortex of a primate. *Somatosensory Motor Research, 10,* 245–262.

Dias, R., Robbins, T. W., and Roberts, A. C. (1997). Dissociable forms of inhibitory control within prefrontal cortex with an analogue of the Wisconsin Card Sorting Test: Restriction to novel situations and independence from "on-line" processing. *Journal of Neuroscience, 4,* 2051–2062.

Duncan, J., Emslie, H., Williams, P., Johnson, R., and Freer, C. (1996). Intelligence and the frontal lobe: The organization of goal-directed behavior. *Cognitive Psychology, 30,* 257–303.

Eacott, M. J., and Gaffan, D. (1992). Inferotemporal-frontal disconnection: The uncinate fascicle and visual associative learning in monkeys. *European Journal of Neuroscience, 4,* 1320–1332.

Erickson, C. A., and Desimone, R. (1996). Neural correlates of visual associative learning in inferior temporal cortex of rhesus macaque. *Society for Neuroscience Abstracts, 21,* 1616.

Ferrera, V. P., Rudolph, K. K., and Maunsell, J. H. (1994). Responses of neurons in the parietal and temporal visual pathways during a motion task. *Journal of Neuroscience, 14,* 6171–6186.

Funahashi, S., Bruce, C. J., and Goldman-Rakic, P. S. (1989). Mnemonic coding in visual space in the monkey's dorsolateral prefrontal cortex. *Journal of Neurophysiology, 61,* 331–349.

Fuster, J. M., and Alexander, G. E. (1971). Neuron activity related to short-term memory. *Science, 173,* 652–654.

Fuster, J. M., Bauer, R. H., and Jervey, J. P. (1982). Cellular discharge in the dorsolateral prefrontal cortex of the monkey in cognitive tasks. *Experimental Neurology, 77,* 679–694.

Fuster, J. M., Bauer, R. H., and Jervey, J. P. (1985). Functional interactions between inferotemporal and prefrontal cortex in a cognitive task. *Brain Research, 330,* 299–307.

Fuster, J. M., and Jervey, J. P. (1981). Inferotemporal neurons distinguish and retain behaviorally relevant features of visual stimuli. *Science, 212,* 952–955.

Gaffan, D. (1977). Response coding in recall of colours by monkeys. *Quarterly Journal of Experimental Psychology, 29,* 597–605.

Gaffan, D., and Harrison, S. (1988). Inferotemporal-frontal disconnection and fornix transection in visuomotor conditional learning by monkeys. *Behavioural Brain Research, 31,* 149–163.

Gnadt, J. W., Bracewell, R. M., and Andersen, R. A. (1991). Sensorimotor transformation during eye movements to remembered visual targets. *Vision Research, 31,* 693–715.

Goodale, M. A., and Haffenden, A. (1998). Frames of reference for perception and action in the human visual system. *Neuroscience and Biobehavioral Reviews, 22,* 161–172.

Gottlieb, J. P., Kusunoki, M., and Goldberg, M. E. (1998). The representation of visual salience in monkey parietal cortex. *Nature, 391,* 481–484.

Gross, C. G., Rocha-Miranda, C. E., and Bender, D. B. (1972). Visual properties of neurons in inferotemporal cortex of the macaque. *Journal of Neurophysiology, 35,* 96–111.

Honig, W., and Thompson, R. K. R. (1982). Retrospective and prospective learning processes in animal working memory.In G. H. Bower (Ed.), *The psychology of learning and motivation: Advances in research and theory,* pp. 239–283. New York: Academic Press.

Johnson, M. K., and Hirst, W. (1991). Processing subsystems of memory. In R. G. Lister and H. J. Weingartner (Eds.), *Perspectives in cognitive neuroscience,* pp. 3–16. New York: Oxford University Press.

Knight, R. T. (1984). Decreased response to novel stimuli after prefrontal lesions in man. *Clinical Neurophysiology, 59,* 9–20.

Knight, R. T. (1991). Evoked potential studies of attention capacity in human frontal lesions. In H. S. Levin, H. M. Eisenberg, and A. L. Benton (Eds.), *Frontal lobe function and dysfunction,* pp. 139–153. New York: Oxford University Press.

Knight, R. T. (1997). Distributed cortical network for visual attention. *Journal of Cognitive Neuroscience, 9,* 75–91.

Kubota, K., and Niki, H. (1971). Prefrontal cortical unit activity and delayed alternation performance in monkeys. *Journal of Neurophysiology, 34,* 337–347.

Luck, S. J., and Vogel, E. K. (1997). The capacity of visual working memory for features and conjunctions. *Nature, 390,* 279–281.

Maunsell, J. H., and Newsome, W. T. (1987). Visual processing in monkey extrastriate cortex. *Annual Review of Neuroscience, 10,* 363–401.

Maunsell, J. H., and Van Essen, D. C. (1983). The connections of the middle temporal visual area (MT) and their relationship to a cortical hierarchy in the macaque monkey. *Journal of Neuroscience, 3,* 2563–2586.

McAdams, C. J., and Maunsell, J. H. (1997). Spatial attention and feature-directed attention can both modulate responses in macaque area V4. *Society for Neuroscience Abstracts, 23,* 2062.

Miller, E. K., Erickson, C. A., and Desimone, R. (1996). Neural mechanisms of visual working memory in prefrontal cortex of the macaque. *Journal of Neuroscience, 16,* 5154–5167.

Miller, E. K., Li, L., and Desimone, R. (1993). Activity of neurons in anterior inferior temporal cortex during a short-term memory task. *Journal of Neuroscience, 13,* 1460–1478.

Milner, B., and Petrides, M. (1984). Behavioural effects of frontal-lobe lesions in man. *Trends in Neuroscience, 7,* 403–407.

Miyashita, Y., and Chang, H. S. (1988). Neuronal correlate of pictorial short-term memory in the primate temporal cortex. *Nature, 331,* 68–70.

Moran, J., and Desimone, R. (1985). Selective attention gates visual processing in the extrastriate cortex. *Science, 229,* 782–784.

Motter, B. C. (1993). Focal attention produces specially selective processing in visual cortical areas V1, V2, and V4 in the presence of competing stimuli. *Journal of Neuroscience, 70,* 909–919.

Motter, B. C. (1994). Neural correlates of feature selective memory and pop-out in extrastriate area V4. *Journal of Neuroscience, 14,* 2190–2199.

Norman, D. A., and Shallice, T. (1986). Attention to action: Willed and automatic control of behavior. In R. J. Davidson, G. E. Schwartz, and D. Shapiro (Eds.), *Consciousness and self-regulation: Advances in research and theory.* New York: Plenum Press.

O'Scalaidhe, S. P., Wilson, F. A., and Goldman-Rakic, P. S. (1997). Areal segregation of face-processing neurons in prefrontal cortex. *Science, 278,* 1135–1138.

Oster, M. N., Snow, M., Miller, E. K., and Corkin, S. (1997). FMRI activation during performance on object and spatial components of a working memory task. *Society for Neuroscience Abstracts, 23,* 2110.

Owen, A. M., Milner, B., Petrides, M., and Evans, A. C. (1996). Memory for object features versus memory for object location: A positron-emission tomography study of the encoding and retrieval processes. *Proceedings of the National Academy of Sciences, U.S.A., 93,* 9212–9217.

Owen, A. M., Roberts, A. C., Polkey, C. E., Sahakian, B. J., and Robbins, T. W. (1991). Extradimensional versus intra-dimensional set shifting performance following frontal lobe excisions, temporal lobe excisions, or amygdalo-hippocampectomy in man. *Neuropsychologia, 29,* 993.

Owen, A. M., Stern, C. E., Look, R. B., Tracey, I., Rosen, B. R., and Petrides, M. (1998). Functional organization of spatial and nonspatial working memory processing within the human lateral frontal cortex. *Proceedings of the National Academy of Sciences, U.S.A., 95,* 7721–7726.

Pandya, D. N., and Barnes, C. L. (1987). Architecture and connections of the frontal lobe. In E. Perecman (Ed.), *The frontal lobes revisited,* pp. 41–72. New York: IRBN Press

Pandya, D. N., and Yeterian, E. H. (1990). Prefrontal cortex in relation to other cortical areas in rhesus monkey: Architecture and connections. *Progress in Brain Research, 85,* 63–94.

Parker, A., and Gaffan, D. (1998). Memory after frontal/temporal disconnection in monkeys: Conditional and non-conditional tasks, unilateral and bilateral frontal lesions. *Neuropsychologia, 36,* 259–271.

Passingham, R. (1993). *The frontal lobes and voluntary action.* Oxford: Oxford University Press.

Petrides, M. (1982). Motor conditional associative learning after selective prefrontal lesions in the monkey. *Behavioural Brain Research, 5,* 407–413.

Petrides, M. (1986). The effect of periarcuate lesions in the monkey on the performance of symmetrically and asymmetrically reinforced visual and auditory, go, no-go tasks. *Journal of Neuroscience, 6,* 2054–2063.

Petrides, M. (1990). Nonspatial conditional learning impaired in patients with unilateral frontal but not unilateral temporal lobe excisions. *Neuropsychologia, 28,* 137–149.

Postle, B. R., and D'Esposito, M. (1998). Analysis of spatial and object-delayed response with event-related fMRI. *Proceedings of the Cognitive Neurology Society, 5,* 85.

Preuss, T. M., and Goldman-Rakic, P. S. (1989). Connections of the ventral granular frontal cortex of macaques with perisylvian premotor and somatosensory areas: Anatomical evidence for somatic representation in primate association cortex. *Journal of Comparative Neurology, 282,* 293–316.

Quintana, J., and Fuster, J. M. (1992). Mnemonic and predictive functions of cortical neurons in a memory task. *NeuroReport, 3,* 721–724.

Rainer, G., Asaad, W. F., and Miller, E. K. (1998a). Selective representation of relevant information by neurons in the primate prefrontal cortex. *Nature, 393,* 577–579.

Rainer, G., Asaad, W. F., and Miller, E. K. (1998b). Memory fields of neurons in the primate prefrontal cortex. *Proceedings of the National Academy of Sciences, U.S.A. 95,* 15008–15013.

Rainer, G., Rao, S. C., and Miller, E. K. (1999). Prospective coding for objects in the primate prefrontal cortex. *Journal of Neuroscience, 19,* 5493–5505.

Rao, S. C., Rainer, G., and Miller, E. K. (1997). The integration of what and where in the primate prefrontal cortex. *Science, 276,* 821–824.

Recanzone, G. H., Merzenich, M. M., Jenkins, W. M., Grajski, K. A., and Dinse, H. R. (1992). Topographic reorganization of the hand representation in cortical area 3b owl monkeys trained in a frequency-discrimination task. *Journal of Neurophysiology, 67,* 1031–1056.

Recanzone, G. H., Schreiner, C. E., and Merzenich, M. M. (1993). Plasticity in the frequency representation of primary auditory cortex following discrimination training in adult owl monkeys. *Journal of Neuroscience, 13,* 87–103.

Roitblat, H. L. (1993). Representations and processes in working memory. In T. R. Zentall (Ed.), *Animal cognition*, pp. 175–192. Hillsdale, NJ; Erlbaum.

Sakagami, M., and Niki, H. (1994a). Encoding of behavioral significance of visual stimuli by primate prefrontal neurons: Relation to relevant task conditions. *Experimental Brain Research, 97,* 423–436.

Sakagami, M., and Niki, H. (1994b). Spatial selectivity of go/no-go neurons in monkey prefrontal cortex. *Experimental Brain Research, 100,* 165–169.

Schein, S. J., and Desimone, R. (1990). Special properties of V4 neurons in the macaque. *Journal of Neuroscience, 10,* 3369–3389.

Sereno, A. B., and Maunsell, J. H. (1998). Shape selectivity in primate lateral intraparietal cortex. *Nature, 395,* 500–503.

Shallice, T. (1982). Specific impairments of planning. *Philosophical Transactions of the Royal Society of London, B298,* 199–209.

Stuss, D. T., and Benson, D. F. (1986). *The frontal lobes.* New York: Raven Press.

Suzuki, W. A., Miller, E. K., and Desimone, R. (1997). Object and place memory in the macaque entorhinal cortex. *Journal of Neurophysiology, 78,* 1062–1081.

Ungerleider, L. G., Gaffan, D., and Pelak, V. S. (1989). Projections from inferior temporal cortex to prefrontal cortex via the uncinate fascicle in rhesus monkeys. *Experimental Brain Research, 76,* 473–484.

Ungerleider, L. G., and Mishkin, M. (1982). Two cortical visual systems. In J. Ingle, M. A. Goodale, and R. J. W. Mansfield (Eds.), *Analysis of visual behavior,* pp. 549–586. Cambridge, MA: MIT Press.

Van Essen, D. C., Anderson, C. H., and Felleman, D. J. (1992). Information processing in the primate visual system: An integrated systems perspective. *Science, 255,* 419–423.

Watanabe, M. (1981). Prefrontal unit activity during delayed conditional discriminations in the monkey. *Brain Research, 225,* 51–65.

Watanabe, M. (1990). Prefrontal unit activity during associative learning in the monkey. *Experimental Brain Research, 80,* 296–309.

Watanabe, M. (1992). Frontal units of the monkey coding the associative significance of visual and auditory stimuli. *Experimental Brain Research, 89,* 233–247.

Watanabe, M. (1996). Reward expectancy in primate prefrontal neurons. *Nature, 382,* 629–632.

Webster, M. J., Bachevalier, J., and Ungerleider, L. G. (1994). Connections of inferior temporal areas TEO and TE with parietal and frontal cortex in macaque monkeys. *Cerebral Cortex, 4,* 470–483.

White, E. L. (1989). *Cortical circuits: Synaptic organization of the cerebral cortex: Structure, function, and theory.* Boston: Birkhauser.

White, I. M., and Wise, S. P. (1997). Rule-dependent neuronal activity in the prefrontal cortex. *Society for Neuroscience Abstracts, 23,* 1303.

Wilson, F. A. W., O'Scalaidhe, S. P., and Goldman-Rakic, P. S. (1993). Dissociation of object and spatial processing domains in primate prefrontal cortex. *Science, 260,* 1955–1958.

Wise, S. P., Murray, E. A., and Gerfen, C. R. (1996). The frontal-based ganglia system in primates. *Critical Reviews in Neurobiology, 10,* 317–356.

23 Middorsolateral and Midventrolateral Prefrontal Cortex: Two Levels of Executive Control for the Processing of Mnemonic Information

Michael Petrides

ABSTRACT According to the proposed hypothesis, the middorsolateral prefrontal cortex (areas 46, 9/46, and 9) is a specialized system for the monitoring and manipulation of information within working memory, whereas the midventrolateral prefrontal cortex (areas 47/12 and 45) is involved in the active retrieval of information from posterior cortical association areas. Data are presented that support this two-level hypothesis. In the monkey, lesions restricted to the middorsolateral region of the prefrontal cortex yield a severe impairment in the performance of tasks that require monitoring within working memory, this impairment appearing against a background of normal performance on several basic mnemonic tasks. In functional activation studies with normal human subjects, specific changes in activity within the middorsolateral region of the prefrontal cortex are observed with respect to monitoring of information within working memory. In the midventrolateral prefrontal region, changes in activity are observed with respect to the active retrieval of information from memory.

Although it is generally agreed that the prefrontal cortex plays an important role in memory, precise characterization of this role has proved elusive. Patients with damage to the lateral prefrontal cortical region perform well on several tests that are sensitive measures of the well-established memory disorder that follows damage to the medial temporal region of the brain (for review, see Petrides 1989). For instance, performance can be normal on standard tests of basic recognition memory and story recall. When a severe memory disorder is reported after frontal lesions, there is often involvement of the caudal orbito-medial limbic region of the frontal lobe and the immediately adjacent basal forebrain region (e.g. septal area, nucleus basalis of Meynert, etc.) or there is additional damage outside the frontal cortex (see Petrides 1989).

Nevertheless, damage to the lateral prefrontal cortex in both the human and the monkey brain can severely impair mnemonic performance under certain testing conditions. To characterize the essential nature of the specific contribution of the human lateral prefrontal cortex to mnemonic processing, I developed a working-memory task that required monitoring of earlier selections from a set of stimuli for successful performance (Petrides and Milner 1982). On this self-ordered working-memory task, we observed severe impairments after lateral prefrontal lesions, although these same patients could perform well on

Figure 23.1 Experimental arrangement in the self-ordered task. Subjects face a stack of cards on which the same designs are presented in different arrangements. Subjects have to select one of the designs and touch it, then turn to the next card and touch another design, until all designs have been touched once.

several other memory tests, such as those for recognition memory, digit span, and story recall.

In the self-ordered working-memory task, the subjects are presented with different arrangements of the same set of stimuli and, on each trial, they have to select a different stimulus until all have been selected once. For instance, they may be presented with a stack of cards bearing the same stimuli (e.g., a set of abstract designs) but in a different arrangement on each card (figure 23.1). The subjects are told to touch one stimulus per card until all stimuli have been touched without repeating any stimuli. Successful performance therefore encourages the subjects to compare carefully stimuli they have already selected with those they have yet to select. In other words, events in working memory must be closely monitored.

When normal human subjects perform the self-ordered task, they tend to sort the stimuli into subjective categories and therefore reduce their memory load. Patients with frontal lesions are less likely to adopt such strategies and this is clearly one source of their problem. However, this poorer organization does not account for the entire deficit exhibited by patients with lateral frontal lesions. In analyses in which I obtained and covaried the effect of an organization score, I found that the patients with prefrontal lesions were still severely impaired (Petrides, unpublished work). Similarly, to obtain a purer measure of monitoring, subjects in a recent study (Petrides, unpublished work) were specifically instructed not to adopt any organization strategies and to make random choices in performing the self-ordered task. Patients with lateral prefrontal lesions still exhibited a very severe impairment in comparison with normal controls.

Thus, work with patients and monkeys has shown that monitoring the information, in the sense that each selection must be marked in the subjects' minds and simultaneously considered in relation to the others that still remain to be selected, is an important source of impairment on the self-ordered working-memory task. Monitoring within working memory must not be confused with simple attention to a stimulus held in memory. For instance, there are many situations (e.g., recognition memory, meaningful story recall) in which attention is directed to a particular stimulus in memory, but the other stimuli are not in the center of current awareness. These situations do not challenge monitoring within working memory in the sense used here, although they demand attention to the stimulus being remembered.

23.1 ROLE OF MIDDORSOLATERAL PREFRONTAL CORTEX IN THE MONITORING OF EVENTS HELD IN WORKING MEMORY

My work in the monkey (Petrides 1991, 1995) has demonstrated that the impairment on working-memory tasks after lesions restricted to the middorsolateral prefrontal cortex (i.e., areas 46, 9 and 9/46; see figure 23.2) can be attributed to these tasks' monitoring requirements rather than to the maintenance of the information per se. This work led to the identification of the middorsolateral prefrontal cortex as the critical region for monitoring information held within working memory. The evidence is based on the following facts. In the monkey, lesions restricted to the middorsolateral prefrontal cortex result in a severe impairment on tasks having requirements comparable to those of the self-ordered working-memory tasks used with patients and on the related, externally ordered working-memory tasks. The externally ordered tasks provided a particularly striking illustration of the role of the middorsolateral prefrontal cortex in monitoring information held within working memory. Here subjects were trained to expect a certain set of stimuli to occur. During testing, a subset of these stimuli was presented, and subjects had to monitor carefully their occurrence to detect stimuli that had not been presented. For instance, in the case illustrated in figure 23.3, subjects knew, on the basis of previous training, that the particular three objects constituted the set to be monitored. On a given test session, two of these three

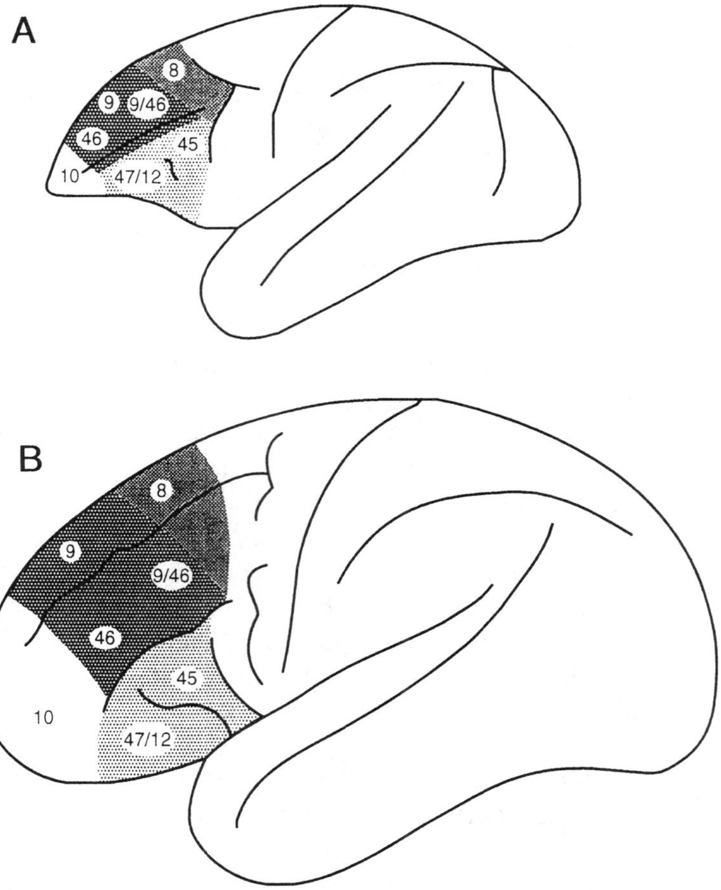

Figure 23.2 Lateral surface of the macaque monkey (panel A) and the human (panel B) cerebral hemisphere illustrating the middorsolateral prefrontal region (areas 46, 9/46, and 9) and the midventrolateral prefrontal region (areas 45 and 47/12). The term *middorsolateral prefrontal cortex* is used to distinguish this region from the frontopolar cortex (i.e., area 10) and the posterior dorsolateral frontal cortex (i.e., area 8 and rostral area 6).

expected objects were randomly selected and presented alone (A and B) and then, on the critical test trial, all three objects were presented together, and subjects had to select the object not previously presented. On such trials, monkeys with middorsolateral prefrontal lesions were severely impaired (Petrides 1995). On the other hand, these monkeys performed as well as normal control animals if the same sequence of testing events proceeded with stimuli, whether novel or familiar, that did not belong to an expected set. For instance, performance was normal if the animals were shown objects A and B, which were then presented together with object C, and the animals had to select object C (Petrides 1995, exp. 6).

My explanation of these results is as follows. The middorsolateral prefrontal cortex is a specialized region where stimuli or events, first interpreted and maintained in posterior association cortical areas, can be recoded to monitor expected acts or events (Petrides 1991, 1994). This

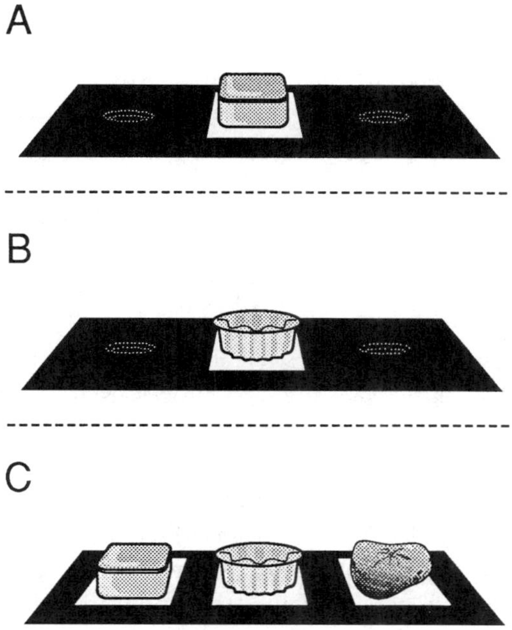

Figure 23.3 Schematic illustration of the experimental arrangement in the externally ordered tests with monkeys. Two objects, randomly selected from a set of three expected objects, are presented on the central food well on trials A and B. On the critical test trial C, all three objects are presented and the animal has to displace the object that had not been previously presented. If the animal responds correctly, it will find a reward in the food well.

region of the prefrontal cortex evolved, not to maintain information for short periods of time, but rather to hold coded representations of events expected to occur, so as to mark their occurrence or nonoccurrence (i.e., monitor their relative status in relation to each other and the intended or expected set of events). If the monkeys expect a set of stimuli to occur, as in the externally ordered tasks, coded representations of the stimuli will be activated in the middorsolateral prefrontal cortex, and when some of these stimuli are presented, those pools of neurons coding them will be marked (i.e., their neural response will be modulated). Thus monitoring within working memory is carried out by marking the occurrence or nonoccurrence of an expected set of stimuli or planned acts. On the other hand, as pointed out above, if the testing does not require this careful monitoring of events within working memory, the monkeys can demonstrate considerable short-term memory capacity even after a middorsolateral prefrontal lesion. Indeed, performance can be normal if the animals, presented with a number of stimuli, can base their choice on simply remembering the stimuli that they have recently seen (Petrides 1991, 1995). Similarly, monkeys with such lesions perform well on delayed match-to-sample tasks in which they have to recognize which one of two constantly recurring stimuli was most recently presented (Passingham 1975), and on delayed object alternation tasks in which they have to alternate their responses between two stimuli following the imposed delay period (Mishkin et al. 1969; Petrides 1995). On these tasks, normal performance requires the capacity to discriminate which one of two frequently occurring stimuli was more recently presented (e.g., on the delayed match-to-sample task) or selected (e.g., on the delayed object alternation task).

In summary, the monkeys with middorsolateral prefrontal lesions can perform normally when required to make a choice based on memory of which stimuli were previously seen and which were not (i.e., recognition memory) or based on the relative recency of frequently recurring stimuli. By contrast, these animals are severely impaired when performance cannot be based on these basic memory processes alone (e.g., familiarity, primacy, or recency judgments) but requires that the status of multiple events in working memory be monitored, as in the self-ordered and externally ordered tasks described above.

The above findings led me to propose that the middorsolateral prefrontal cortex serves as a specialized region where representations of stimuli or events can be maintained on-line and their relative status marked with regard to various requirements set by the task at hand (Petrides 1991, 1994). The essential characteristic of the specialized contribution of this region is the coded representation in memory of an expected set of acts or events (stimuli) and the accompanying marking signals that define the status of these events vis-à-vis each other.

Manipulation of information in working memory requires precise coding of the current status of a given event in memory vis-à-vis the other events so that a transformation in that relative status (i.e., a manipulation) can be effected. I have argued that the capacity to manipulate information in working memory, and therefore to carry out complex plans of action so characteristic of primate behavior, emerged largely because of the specialized computational capacity of the middorsolateral prefrontal region, which permits marking and monitoring within memory of the relative status of multiple intended acts or expected occurrences (Petrides 1991, 1994).

Recent functional neuroimaging studies have extended the animal findings discussed above to the human brain. In the first study to address the role of middorsolateral prefrontal cortex in human memory (Petrides et al. 1993a), the distribution of regional cerebral blood flow (rCBF; a marker of local neuronal activity) was measured by positron-emission tomography in normal human subjects as they performed a nonspatial visual self-ordered task, a visual matching control task, and a visual conditional task. The same eight visual stimuli (abstract designs) were used in all three tasks, and these eight stimuli were presented in a different random arrangement on each trial. Subjects were required to indicate their response by pointing to particular stimuli; the only difference between the three tasks lay in their cognitive requirements. In the self-ordered

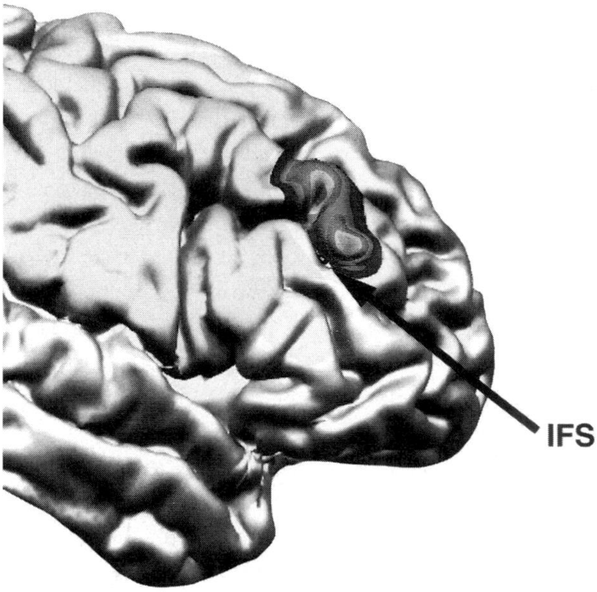

Figure 23.4 Increased activity within the middorsolateral prefrontal cortex during the performance of a self-ordered task. Note that the activity is located on the middle frontal gyrus above the inferior frontal sulcus (IFS), that is, in areas 46 and 9/46.

task, subjects were required to select a different stimulus on each trial until all had been selected, thus to consider actively (i.e., to monitor) their earlier selections as they were preparing their next response. The matching control task, in which subjects had to search and find the same stimulus on each trial, involved the same visual stimuli and searching behavior as the self-ordered task, but did not require that subjects consider their earlier responses in relation to the current one. In the conditional task, before being scanned, subjects had learned associations between the stimuli and particular color cues. During scanning, they were required to select the stimulus appropriate for the color cue presented. Thus, although the searching among the stimuli was the same as in the self-ordered task, because the stimulus to be selected was completely determined by the color cue presented on each trial, no monitoring within working memory of prior selections was required.

Performance of the self-ordered task, in comparison with either the matching control or the conditional task, resulted in significantly greater activity within the middorsolateral prefrontal cortex (areas 46 and 9/46), particularly within the right hemisphere (figure 23.4). There was no greater activity in this region when rCBF in the conditional task was compared with that in the control task, although there was now significant activity within area 8 of the posterior dorsolateral prefrontal cortex, an area known to be critical for visual conditional learning (see Petrides 1987). The contrast in the activation patterns between the self-ordered and the conditional tasks emphasizes the specificity of activation within the middorsolateral frontal cortex in relation to the monitoring requirements of the self-ordered task.

A related study (Petrides et al. 1993b) demonstrated bilateral increase in activity in the middorsolateral prefrontal cortex in relation to the performance of a verbal self-ordered task and a verbal externally ordered working-memory task. With regard to spatial working memory, activation of either the ventrolateral prefrontal cortex (areas 47/12; Jonides et al. 1993) or the middorsolateral prefrontal cortex (area 46; McCarthy et al. 1994) has been reported. In Owen, Evans, and Petrides, 1996, we showed that the occurrence of activity in the middorsolateral prefrontal cortex depends on whether monitoring of the spatial information within working memory is taxed. Thus increased activity within the middorsolateral prefrontal cortex occurs whenever the monitoring requirements are greater than those of the control task, regardless of the nature of the stimulus material (e.g., visual spatial, visual nonspatial, auditory).

The demonstration that the middorsolateral prefrontal cortex shows increased activity whenever monitoring of information within working memory is required (Petrides et al. 1993a,b) has now been repeatedly confirmed (for reviews, see Owen 1997 and D'Esposito et al. 1998). For instance, in one variation of the externally ordered monitoring tasks, the subjects were required to monitor, not the whole set of stimuli, but only a subset of them. In these *n*-back working-memory tasks, subjects are typically presented with a series of stimuli and must respond upon reappearance of a stimulus presented a specified number of steps earlier (e.g., two steps back). Thus, instead of monitoring all the items in short-term memory (as in the original externally ordered tasks), subjects need monitor only the last few items presented. As would be predicted from the lesion studies with monkeys (Petrides 1991, 1995) and the first functional neuroimaging studies with tasks requiring monitoring of information within working memory (Petrides et al. 1993a,b), all studies that have used *n*-back monitoring tasks (e.g., Cohen et al. 1994; Braver et al. 1997; Owen et al. 1998) have observed increases in the middorsolateral prefrontal cortex. D'Esposito et al. (1995) observed increased activity in the middorsolateral prefrontal cortex when subjects were performing two concurrent tasks, even though neither task resulted in increased activity in this region when performed alone. Successful dual-task performance requires that multiple items of information be simultaneously attended to (e.g., recent information in tasks 1 and 2) and thus challenges monitoring of information within working memory in the sense defined above.

23.2 ROLE OF MIDVENTROLATERAL PREFRONTAL CORTEX IN ACTIVE JUDGMENTS ON MNEMONIC INFORMATION HELD IN POSTERIOR ASSOCIATION CORTICAL REGIONS

There is a fundamental difference between the middorsolateral and the midventrolateral prefrontal cortex in terms of their involvement in mem-

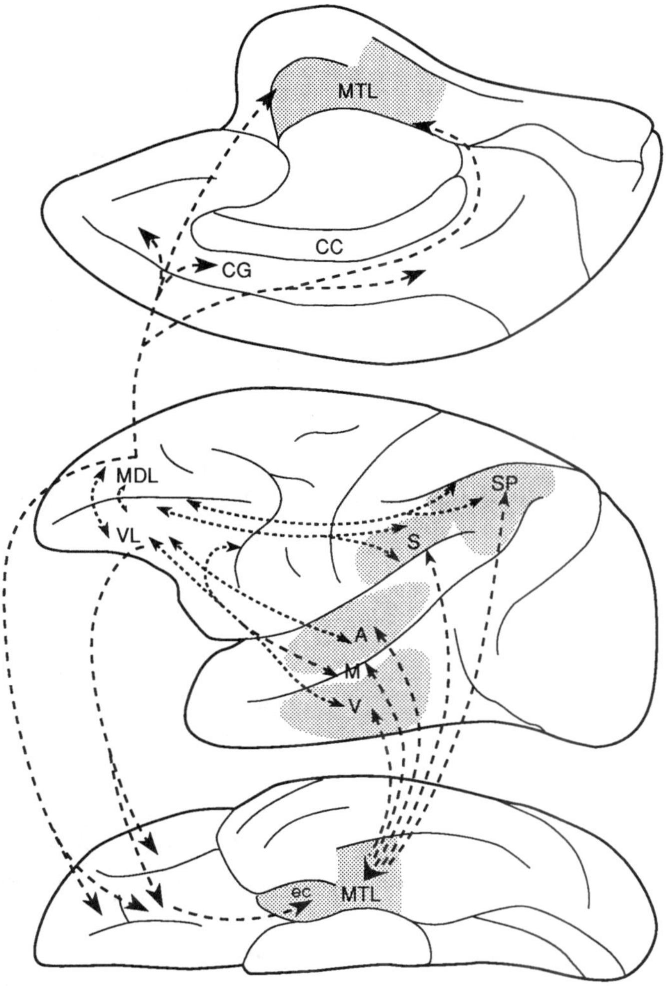

Figure 23.5 Schematic diagram of the brain of the macaque monkey to illustrate some of the functional interactions postulated by the two-level hypothesis of the role of the lateral prefrontal cortex to mnemonic processing. Somatosensory (S), spatial (SP), auditory (A), visual (V), and some aspects of multimodal (M) information are processed in posterior association cortex. CC = corpus callosum; CG = cingulate gyrus; ec = entorhinal cortex; MDL = middorsolateral frontal cortex; MTL = medial temporal lobe; VL = ventrolateral frontal cortex.

ory (Petrides 1994). According to the two-level hypothesis proposed, the midventrolateral prefrontal cortex, in interaction with posterior cortical association areas, subserves the expression within memory of various first-order executive processes, such as active selection, comparison, and judgment of stimuli held in short-term and long-term memory (figure 23.5; see Petrides 1994 for details). This type of interaction is necessary for active (explicit) encoding and retrieval of information, processes initiated under effort by subjects and guided by their plans and intentions. By contrast, as stated above, the middorsolateral prefrontal cortex (areas 46,

9/46, and 9) constitutes another level of interaction with mnemonic information and is involved when several pieces of information in working memory need to be monitored and manipulated on the basis of the task's requirements or the subjects' current plans. It must be emphasized that the two levels of mnemonic executive processing posited above are likely to be involved in several tasks, often at the same time. The successful demonstration of the specific contribution of different regions will therefore depend on selective lesion studies in nonhuman primates, where impaired performance on certain mnemonic tasks is contrasted with normal performance on other similar tasks, as well as on neuroimaging studies with normal human subjects in which experimental tasks are differentially loaded with requirements thought to involve one or the other area.

A distinction must be made between active (strategic) retrieval, which requires the ventrolateral prefrontal cortex, and automatic retrieval, which does not (Petrides 1994). Automatic retrieval is the by-product of the triggering of stored representations in the posterior cortical association regions either by incoming sensory input that matches preexisting representations or by recalled events that trigger stored representations of related information on the basis of strong preexisting associations or other relations, such as thematic context. This kind of *automatic* retrieval is mediated by connections between posterior temporal and parietal association areas and subcortical structures. When, however, *active* retrieval of specific information held in posterior association areas is required, the midventrolateral prefrontal cortex interacts with these posterior association areas via strong bidirectional connections. By "active retrieval," I mean effortful retrieval of specific items of information that is guided by the subjects' intentions and plans. This attempt at retrieval may be self-generated or initiated by the instructions given to the subjects in an experiment.

The above hypothesis of the role of the prefrontal cortex (middorsolateral and midventrolateral) explains why performance on several standard memory tests can be normal after lateral prefrontal lesions. For instance, in memory tasks where recognition of previously presented information is required, performance can be adequate when the reexposure to the stimuli triggers existing representations in posterior association cortex, and these reactivated representations are the basis of the knowledge that the stimulus has been experienced before. Thus performance on several basic recognition tasks that simply require awareness of familiarity of the stimuli can be normal after lateral frontal lesions. Similarly, in recalling a narrative story previously read or heard, the thematic relations between the various components of the story automatically trigger related information in posterior association cortical areas and, to a large extent, can guide recall of the story; the prefrontal cortex becomes critical to the extent that strong thematic relations are not sufficient for adequate recall. Thus free-recall tasks on which subjects are asked to recall specific pieces of information not automatically triggered either by current sensory input or by thematic or other strong relations demonstrate the clearest impairments in patients with prefrontal lesions (see Petrides 1989). Under these circumstances, an active planned search must be initiated to retrieve the particular pieces of information. According to the two-level hypothesis presented above, this type of search depends on interactions between the ventrolateral prefrontal cortex and the posterior temporal and parietal association cortex: the ventrolateral prefrontal cortex can exert top-down control on posterior cortical association circuits and thus enable the retrieval of specific pieces of information in posterior cortical areas that cannot be automatically triggered either by strong preexisting associations or by thematic context.

In Petrides, Alivisatos, and Evans 1995, we tested the prediction from the above hypothesis that the midventrolateral prefrontal cortex, in the left hemisphere, is involved in the active, strategic retrieval of verbal information from long-term memory. The main experimental condition during scanning involved the free recall of a list of arbitrary words that had been studied before scanning. Performance on such a free-recall task cannot be simply the result of recognizing familiar words that are presented again, nor can it be the result of retrieving information by thematic relatedness, as in a logical story. Free recall under these conditions is the result of active strategic retrieval processes because subjects are now asked to recall from their lexicons a specific set of arbitrary words that were presented on a particular recent occasion under particular conditions, namely, the words studied just before scanning.

Because any recall task will require some degree of monitoring within working memory of the output from long-term memory, during the performance of the above free-recall task, there should be significant activity in the middorsolateral region of the frontal cortex, in addition to any ventrolateral activity that might be observed. Note that in our earlier work with positron-emission tomography (Petrides et al. 1993a, b), the middorsolateral prefrontal cortex, but not the midventrolateral, was shown to be specifically activated in relation to monitoring information within working memory. Two control scanning conditions were therefore employed to reveal any specific contribution of the left midventrolateral prefrontal cortex to the active retrieval of verbal information. One of these control conditions required the simple repetition of auditorily presented words and was designed to control for processes involved in listening to, understanding, and producing words. The other involved verbal retrieval significantly easier than retrieval on the free-recall task, but required that, the retrieved verbal output be monitored within working memory at about the same level as on the free-recall task. For this purpose, a verbal paired-associate task was used on which the pairs were well learned before scanning and therefore easy to retrieve in comparison with the free-recall task.

In relation to the repetition control task, the free-recall task resulted in greater activation within both the midventrolateral and middorsolateral prefrontal cortex because both active retrieval and monitoring of the retrieved output within working memory were greater in the free-recall task. Comparison of the free-recall (difficult retrieval) and the highly learned paired-associate (easy retrieval) tasks revealed significantly greater activity in the left midventrolateral prefrontal cortex in the free-recall task, but no difference between the two tasks in the middorsolateral prefrontal cortex (Petrides, Alivisatos, and Evans 1995).

In agreement with the above results, Fletcher et al. (1996) reported increased activity in left prefrontal cortex in cued recall of nonimageable versus imageable pairs. Although the authors described this focus as being in dorsolateral prefrontal cortex, the coordinates provided clearly indicate the activity to be in ventrolateral prefrontal area 45, the same area that showed increased activity in Petrides, Alivisatos, and Evans 1995. Buckner et al. (1996) also observed increased activity in left, as well as right, ventrolateral prefrontal cortex in their studies of verbal episodic retrieval when comparing paired-associate word recall with word repetition or with rest.

Fletcher et al. (1998) have provided results consistent with our proposal (Petrides, Alivisatos, and Evans 1995) that the activity observed in the middorsolateral prefrontal cortex in episodic retrieval tasks reflects, not retrieval per se, but rather monitoring of information within memory. In Fletcher et al. 1998, subjects retrieved verbal material under two conditions: one that required monitored memory search and one that did not require monitored search, as retrieval was externally driven. The middorsolateral prefrontal cortex showed greater activity when monitoring demands were emphasized, whereas the midventrolateral region showed greater activity in the externally driven condition. Buckner et al. (1998) and MacLeod et al. (1998) have also concluded that the right anterior prefrontal activity observed in episodic retrieval may reflect monitoring processes.

In conclusion, the data reviewed above show that within the midlateral part of the prefrontal cortex, two systems can be distinguished: one centered on the middorsolateral prefrontal cortex and the other on the midventrolateral prefrontal cortex. The fundamental distinction between these two regions of the frontal lobe is shown to involve the nature of the executive processing carried out, rather than the modality (e.g., spatial versus nonspatial) of the information processed. While this does not exclude the possibility that, *within* the dorsolateral and the ventrolateral prefrontal regions, there may be some specialization according to the sensory modality of the information being processed, the fundamental principle of organization between the dorsolateral and ventrolateral prefrontal regions cannot be reduced to one of modality specificity.

NOTE

This work was supported by grants from the National Sciences and Engineering Research Council of Canada and from the Medical Research Council of Canada.

REFERENCES

Braver, T. S., Cohen, J. D., Nystrom, L. E., Jonides, J., Smith, E. E., and Noll, D. C. (1997). A parametric study of the prefrontal cortex involvement in human working memory. *NeuroImage, 5,* 49–62.

Buckner, R. L., Koutstaal, W., Schacter, D. L., Dale, A. M., Rotte, M., and Rosen, B. R. (1998). Functional-anatomic study of episodic retrieval: II. Selective averaging of event-related fMRI trials to test the retrieval success hypothesis. *NeuroImage, 7,* 163–175.

Buckner, R. L., Raichle, M. E., Miezin, F. M., and Petersen, S. E. (1996). Functional anatomic studies of memory retrieval for auditory words and visual pictures. *Journal of Neuroscience, 16,* 6219–6235.

Cohen, J. D., Forman, S. D., Braver, T. S., Casey, B. J., Servan-Schreiber, D., and Noll, D. C. (1994). Activation of the prefrontal cortex in a nonspatial working memory task with functional MRI. *Human Brain Mapping, 1,* 293–304.

D'Esposito, M., Aguirre, G. K., Zarahn, E., Ballard, D., Shin, R. K., and Lease, J. (1998). Functional MRI studies of spatial and nonspatial working memory. *Cognitive Brain Research, 7,* 1–13.

D'Esposito, M., Detre, J. A., Alsop, C. D., Shin, R. K., Atlas, S., and Grossman, M. (1995). The neural basis of the central executive of working memory. *Nature, 378,* 279–281.

Fletcher. P. C., Shallice, T., Frith, C. D., Frackowiak, R. S. J., and Dolan, R. J. (1996). Brain activity during memory retrieval. *Brain, 119,* 1587–1596.

Fletcher, P. C., Shallice, T., Frith, C. D., Frackowiak, R. S. J., and Dolan, R. J. (1998). The functional roles of prefrontal cortex in episodic memory. *Brain, 121,* 1249–1256.

Jonides, J., Smith, E. E., Koeppe, R. A., Awh, E., Minoshima, S., and Mintun, M. A. (1993). Spatial working memory in humans as revealed by PET. *Nature, 363,* 623–625.

Macleod, A. K., Buchner, R. L., Miezin, F. M., Petersen, S. E., and Raichle, M. E. (1998). Right anterior prefrontal cortex activation during semantic monitoring and working memory. *NeuroImage, 7,* 41–48.

McCarthy, G., Blamire, A. M., Puce, A., Nobre, A. C., Bloch, G., Hyder, F., Goldmann-Rakic, P., and Shulman, R. G. (1994). Functional magnetic resonance imaging of human prefrontal cortex activation during a spatial working memory task. *Proceedings of the National Academy of Sciences, U.S.A., 91,* 8690–8694.

Mishkin, M., Vest, B., Waxler, M., and Rosvold, H. E. (1969). A re-examination of the effects of frontal lesions on object alternation. *Neuropsychologia, 7,* 357–363.

Owen, A. M. (1997). The functional organization of working memory processes within the human lateral frontal cortex: The contribution of functional neuroimaging. *European Journal of Neuroscience, 9,* 1329–1339.

Owen, A. M., Evans, A. C., and Petrides, M. (1996). Evidence for a two-stage model of spatial working memory processing within the lateral frontal cortex: A positron-emission tomography study. *Cerebral Cortex, 6,* 31–38.

Owen, A. M., Stern, C. E., Look, R. B., Tracey, I., Rosen, B. R., and Petrides, M. (1998). Functional organization of spatial and nonspatial working memory processing within the human lateral frontal cortex. *Proceedings of the National Academy of Sciences, U.S.A., 95,* 7721–7726.

Passingham, R. E. (1975). Delayed matching after selective prefrontal lesions in monkeys (*Macaca mulatta*). *Brain Research, 92,* 89–102.

Petrides, M. (1987). Conditional learning and the primate frontal cortex. In E. Perecman (Ed.), *The frontal lobes revisited,* pp. 91–108. New York: IRBN Press.

Petrides, M. (1989). Frontal lobes and memory. In F. Boller and J. Grafman (Eds.), *Handbook of neuropsychology,* vol. 3, pp. 75–90. Amsterdam: Elsevier.

Petrides, M. (1991). Monitoring of selections of visual stimuli and the primate frontal cortex. *Proceedings of the Royal Society of London B246,* 293–298.

Petrides, M. (1994). Frontal lobes and working memory: Evidence from investigations of the effects of cortical excisions in nonhuman primates. In F. Boller and J. Grafman (Eds.), *Handbook of neuropsychology,* vol. 9, pp. 59–82. Amsterdam: Elsevier.

Petrides, M. (1995). Impairments on nonspatial self-ordered and externally ordered working memory tasks after lesions of the middorsal part of the lateral frontal cortex in the monkey. *Journal of Neuroscience, 15,* 359–375.

Petrides, M., Alivisatos, B., and Evans, A. C. (1995). Functional activation of the human ventrolateral frontal cortex during mnemonic retrieval of verbal information. *Proceedings National Academy of Sciences U.S.A., 92,* 5803–5807.

Petrides, M., Alivisatos, B., Evans, A. C., and Meyer, E. (1993a). Dissociation of human middorsolateral frontal cortex in memory processing. *Proceedings of the National Academy of Sciences, U.S.A., 90,* 873–877.

Petrides, M., Alivisatos, B., Meyer, E., and Evans, A. C. (1993b). Functional activation of the human frontal cortex during the performance of verbal working memory tasks. *Proceedings of the National Academy of Sciences U.S.A., 90,* 878–882.

Petrides, M., and Milner, B. (1982). Deficits on subject-ordered tasks after frontal- and temporal-lobe lesions in man. *Neuropsychologia, 20,* 249–262.

24 The Role of Dorsolateral Prefrontal Cortex in the Selection of Action as Revealed by Functional Imaging

Chris Frith

ABSTRACT Functional imaging studies reveal that the dorsolateral prefrontal cortex (DLPFC) is more active when we select one from a number of possible responses. The same region is activated whether the choice is between limb movements or words. The magnitude of the activity does not increase with increasing rate of response selection, although the activity decreases when performance starts to break down at high rates. In a sentence completion task, the more constrained the response is by the sentence, the less activity is seen in DLPFC. These observations suggest that DLPFC biases possible responses top-down, thereby creating an arbitrary and temporary category of responses appropriate to the task in hand. This biasing depends on interactions between DLPFC and more posterior brain regions where responses are represented; the location of these regions depends on response modality, and their activity varies with response rate.

The development of functional imaging techniques seemed to place within our grasp the possibility of fractionating the prefrontal cortex and identifying specific roles for separate components of this large region of the brain. In practice, progress in identifying such roles has been remarkably slow. There was a time when every task seemed to activate dorsolateral prefrontal cortex (DLPFC), and every experimenter was happy to define a different role for this region. For example, it was proposed that DLPFC was critical for willed action (Frith et al. 1991), for working memory (Petrides et al. 1993), or for semantics (Petersen et al. 1988). The tasks used in these studies were complex and involved many processes. Inevitably, the selection of one of these processes to be associated with DLPFC was somewhat arbitrary. If we are to specify a precise role for DLPFC and other frontal regions, we need evidence from a whole range of tasks and from studies where the parameters of one task have been systematically varied. In this chapter, I will present data from a series of studies that provide convergent evidence about the role of DLPFC in the control of action selection. My working assumption is that it will be possible to characterize a single function associated with activity in DLPFC.

24.1 WORD GENERATION STUDIES

The task of word generation has been more widely used than any other in functional imaging. Some experiments involve the traditional verbal

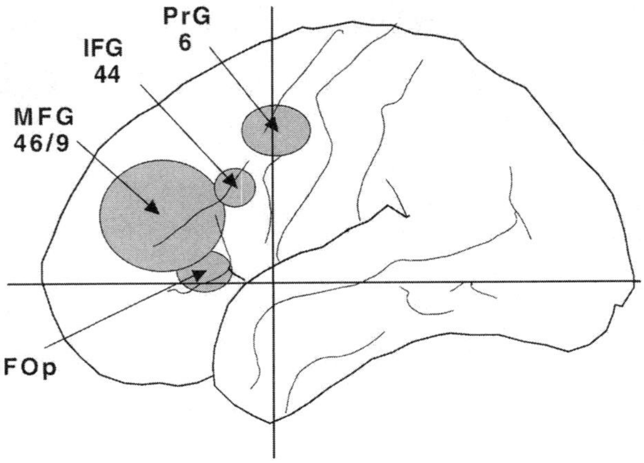

Figure 24.1 Frontal areas activated in common in seven studies of word generation. The center of each ellipse represents the mean Talairach coordinates across the studies (see table 24.1B). The periphery of each ellipse is two standard deviations from the centre. Brain region and likely Brodmann's areas are indicated. PrG-precentral gyrus; MFG-middle frontal gyrus; IFG-inferior frontal gyrus; FOp-frontal operculum. Data from Friston et al. 1993; Frith et al. 1991; Spence (personal communication); Warburton et al. 1996. Although, in most studies, activity was seen only on the left as shown, activity from left and right frontal regions is combined in this and the following figures. These same activations are shown again in subsequent figures to allow comparison with maximal activations in other tasks.

fluency tasks used by neuropsychologists (e.g., "Produce as many words beginning with *S* or as many animals as possible"; Benton 1968) while others use versions of the "verb for noun" task introduced by Petersen et al. (1988) where subjects must generate a verb that goes with a noun (e.g., cake—eat, knife—cut). The pattern of activation produced by these tasks, when compared to baselines such as word repetition, is relatively robust. Increased activity is typically seen in left DLPFC, Broca's area, and anterior cingulate cortex. The precise pattern of activity will, of course, depend on the control task used for comparison. For example, when compared to rest, word generation is associated with an increase in temporal lobe areas, whereas, when compared to word repetition, there is a relative decrease in these areas (e.g., Warburton et al. 1996, exp. 4). On the other hand, the pattern of activity in more anterior regions of the brain seems to be less affected by the nature of the control task.

In figure 24.1 and table 24.1 I have summarized data from seven studies of word generation using the same PET camera and the same method of analysis, "statistical parametric mapping" (SPM; Friston et al. 1996). In the four experiments described by Warburton et al. (1996) subjects silently generated words. In experiments 1–3 (the verb generation task), subjects heard six concrete nouns per minute and generated as many verbs as possible for each noun (e.g., apple—eat, pick, slice, peel). In experiment 3 (the noun generation task), they generated basic level

Table 24.1A Five Distinct Brain Regions Identified on the Basis of the Coordinates of the Peak Activations in Frontal Cortex Listed in Seven Independent Experiments on Word Generation

Region	Findings
Anterior cingulate cortex–supplementary motor area (Brodmann's areas 32/6)	All peaks were within 12 mm from the midline ($\|x\| < 12$). All other activations were at least 26 mm from the midline.
Frontal operculum	All peaks were inferior to 10 mm above the line joining the anterior and posterior commissures ($z < 10$). All other activations were superior to this level.
Precentral gyrus (Brodmann's area 6)	All peaks were less than 5 mm in front of the origin defined by the anterior commissure ($y < 5$). All other activations were more anterior.
Inferior frontal gyrus (Brodmann's area 44)	All peaks lay between 5 and 18 mm in front of the origin ($18 > y > 5$).
Middle frontal gyrus (Brodmann's areas 46/9)	All peaks lay more than 20 mm in front of the origin ($y > 18$).

Sources: Warburton et al. 1996; Frith et al. 1991; Friston et al. 1993; and Spence (personal communication).

Table 24.1B Five Distinct Brain Regions: Mean Talairach Coordinates for Locations of Peak Activations and Number of Studies Where Activations Were Observed

	Mean Talairach coordinates*			
Region	x	y	z	Number of studies
Anterior cingulate–supplementary motor area	–3 (6)	15 (8)	46 (7)	6
Frontal operculum	–39 (4)	21 (4)	4 (3)	6
Precentral gyrus	–39 (4)	1 (5)	44 (4)	3
Inferior frontal gyrus	–38 (4)	12 (3)	28 (3)	5
Middle frontal gyrus	–38 (7)	32 (7)	21 (8)	6

* Standard deviations in parentheses.

Sources: Warburton et al. 1996; Frith et al. 1991; Friston et al. 1993; and Spence (personal communication).

nouns appropriate to superordinate nouns (e.g., furniture—table, chair, stool, cabinet). In experiment 4, German subjects carried out the verb generation task in German. In all these experiments, control data were available for rest. Additional comparison tasks included detecting verb-noun matches (experiment 1), listening to nouns (experiment 2), and subvocal repetition of heard pseudo-words (experiment 4). In the studies by Frith et al. (1991), Friston et al. (1993), and Spence (personal communication), subjects generated words out loud, beginning with specified letters cued at a fixed rate (one word for each cue heard). In the baseline task, subjects repeated the letter cues rather than generating new words. In the Friston et al. and Spence studies, there were data from six generation scans and six repetition scans for each volunteer. This summary is restricted to activity observed in the frontal lobes.

The size of the regions shown in figure 24.1 is determined by the standard deviations from the mean peak of activity across the studies. Each axis of the ellipse is 4 SDs. The large region centered on Brodmann's areas 46/9 probably includes distinct subregions, but these could not be resolved on the basis of the studies discussed here. On the basis of lesion studies, we would expect the frontal operculum to have a specific role in the production of speech (Dronkers 1996). The regions listed as being in Brodmann's areas 44 and 6 are part of premotor cortex, therefore likely to be involved in high-level aspects of movement production (Passingham 1997). The large area listed as being in Brodmann's areas 46/9 is the region of dorsolateral prefrontal cortex widely believed to have a key role in planning and executive control (Luria 1966; Goldman-Rakic 1987; Fuster 1989). Anterior cingulate cortex (ACC) has also been assigned a high-level role, although more specifically related to the control of action than to planning (Posner and Dehaene 1994) While this chapter will concentrate on DLPFC, I will indicate under which circumstances the pattern of activity in ACC diverges from that seen in DLPFC. If DLPFC has a role in high-level executive function, we would expect activations of this region for response generation tasks, whatever the modality of the response.

24.2 RESPONSE MODALITY

Frith et al. 1991 included a separate experiment in which subjects generated a sequence of random finger movements by lifting either the first or second finger of the right hand in response to a tactile pacing signal. This task was characterized as involving "willed action": subjects decided for themselves which finger to lift on each trial. This task was contrasted with one on which the choice of response was determined by an external signal: on each trial subjects simply lifted the finger that was touched. The "willed action" task produced activations in DLPFC (Talairach coordinates: −35, 39, 21) close to the area activated during word generation (see figure 24.2). Several other studies have also shown that DLPFC is activated when volunteers have to select for themselves among different hand and arm movements. Deiber et al. 1991 compared selecting between four different movements of a joystick to repeating the same movement on every trial and observed activation in DLPFC. Jueptner et al. 1997 compared selecting four different button presses to a well-learned sequence of presses and also observed activity in DLPFC. Jahanshahi et al. 1995 showed that DLPFC was also active when subjects had to select when to make a movement rather than which movement to make.

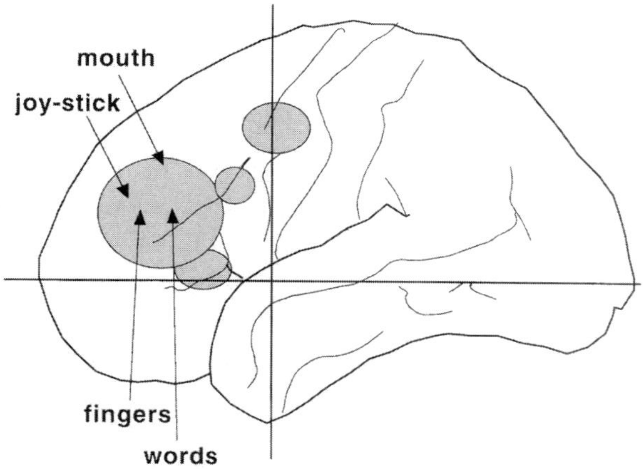

Figure 24.2 Comparison of responses generated in different modalities. Mouth and joystick locations from Spence et al. 1998; finger and word locations from Frith et al. 1991.

Frith et al. 1991 thus implies that left DLPFC activation during self-generated response selection may arise regardless of response modality, although the two tasks used did not differ only in response modality. In the finger-lifting task, only two basic responses are possible, whereas, in the word generation task, a different response must be produced on every trial. Spence et al. (1998) looked for any effect of response modality in two much more comparable tasks. The first was a standard joystick task in which subjects had to produce a series of movements using the right hand in four different directions in response to a pacing tone. In the second, subjects had to produce a series of mouth movements by saying the two syllables "lah" and "bah" in random order, again in response to a pacing tone. In both paradigms, the control tasks were to produce a prespecified stereotyped sequence of joystick or mouth movements. For both response modalities, an area of activity was seen in left DLPFC (hand coordinates: −38, 32, 36; mouth coordinates: −30, 42, 24; see figure 24.2). When both tasks were entered into the same analysis, there was a main effect of condition (self-generated versus stereotyped sequences) in DLPFC (coordinates: −40, 30, 32; $Z=3.8$; $p<0.001$), but no interaction with response modality. Activity was seen in anterior cingulate cortex (ACC) for self-generated sequences in all four response modalities.

The pattern of brain activity in more posterior regions differed markedly between the tasks. For example, the joystick task generated activity in the parietal cortex, whereas the mouth task did not. The only difference in the frontal regions, however, was that both joystick and finger movements were associated with bilateral activations of DLPFC and premotor cortex, whereas mouth movements and word generation were associated with activity solely in the left DLPFC and the left frontal operculum.

These results suggest that DLPFC and ACC have a general role in tasks involving the generation of response sequences, one independent of response modality. Although, at first sight, it might also seem that activity in this region is also independent of the number of responses available for selection, this is probably a false impression. When instructed to produce a long, random sequence of two finger movements, we probably would choose, not just one response at a time, but rather a short subsequence of movements that passes some criterion for randomness. The number of possible such subsequences could be quite large. For example, if we are choosing from two finger movements, there are sixteen different sequences of four movements. We also need to keep track of where we are in the current subsequence and which subsequences have been produced already, just as we need to keep track of the words produced thus far in a word generation task. These considerations indicate at least four possible roles for DLPFC: (1) generating candidate responses or response sequences; (2) checking suitability of responses; (3) keeping track of what has happened thus far; and (4) coordinating all these different task components. Role 3 (keeping track of what has happened thus far) is one of the important roles of working memory and one many believe is instantiated in DLPFC (see Petrides, chap. 23, this volume).

24.3 RESPONSE RATE

If the role of DLPFC is to generate or check responses, we would expect there to be a transient increase in neural activity associated with each response and less activity in the gaps between responses. If, on the other hand, the region is more concerned with keeping track of what has happened, we would expect to see the activity sustained across the gaps between responses. (For a similar argument, in relation to studies of working memory, Cohen et al. 1997.) We can infer that transient increases of activity are occurring in conjunction with stimuli or responses by examining the effect of changing stimulus or response rate on regional cerebral blood flow (rCBF). Because rCBF is integrated across the scanning window, the more bursts of transient activity that have occurred during the window, the greater the total rCBF will be. This effect is manifest as a linear relationship between rCBF and stimulus or response rate (Price et al. 1992; Rees and Frith 1998). We examined the effects of rate in a word generation task (Frith and Friston, in preparation). Responses were cued by spoken letter names that the subjects either repeated or used to generate a word beginning with the same letter. Responses could not be prepared in advance because the subjects could not predict which letter would be spoken next. In both "repeat" and "generate" conditions, there were very marked linear effects of rate in auditory cortex bilaterally and in the cerebellum. These effects presumably reflect transient responses to the auditory stimuli and the motor

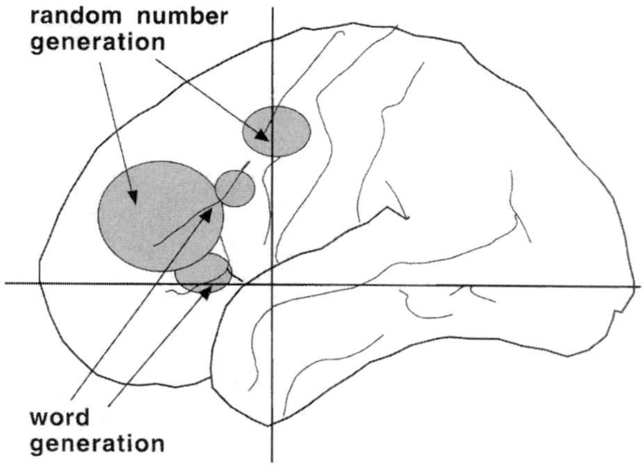

Figure 24.3 Regions where there were increases in activity associated with response generation independent of rate of responding (main effect of task). Random number generation from Jahanshahi et al. forthcoming; word generation from Frith and Friston in preparation.

movements. Activity in DLPFC, frontal operculum, and ACC was significantly greater when subjects generated than when they repeated words (see figure 24.3), although there was no detectable effect of rate in any of these areas. These results suggest that activity was sustained across trials rather than occurring transiently in association with each response.

Jahanshahi et al. (forthcoming) also looked at the effects of rate in a response generation task. Subjects were required to count aloud or to generate random sequences of numbers at six different rates. Here again marked linear effects of rate were seen in auditory cortex and in motor cortex and cerebellum for both conditions. During random number generation, there was greater activity in DLPFC and premotor cortex (bilaterally) (see figure 24.3), as well as in anterior cingulate cortex. On the other hand, activity in these areas did not increase with increasing rate. Indeed, at the two highest rates (1 per second and 2 per second), there was a significant *decrease* in activity in DLPFC, although not in ACC (see figure 24.4A).

These results suggest to me that there is no transient increase in activity in DLPFC associated with the generation and checking of each response. Rather activity is sustained across trials, but cannot be maintained at the highest rate of responding in the random number generation task. Sustained activity in DLPFC could be related to keeping in mind what has happened across the sequence of responses or to some form of high-level task set concerned with the overall goals and rules of the task. I do not believe, however, that a high-level executive role is compatible with the reduction of activity in this region seen at the highest rates of responding. This reduction in activity is associated with a decrease in the randomness of the response sequence (see figure 24.4B). Jahanshahi et al. (forthcoming) interpret this as reflecting interference between the task of

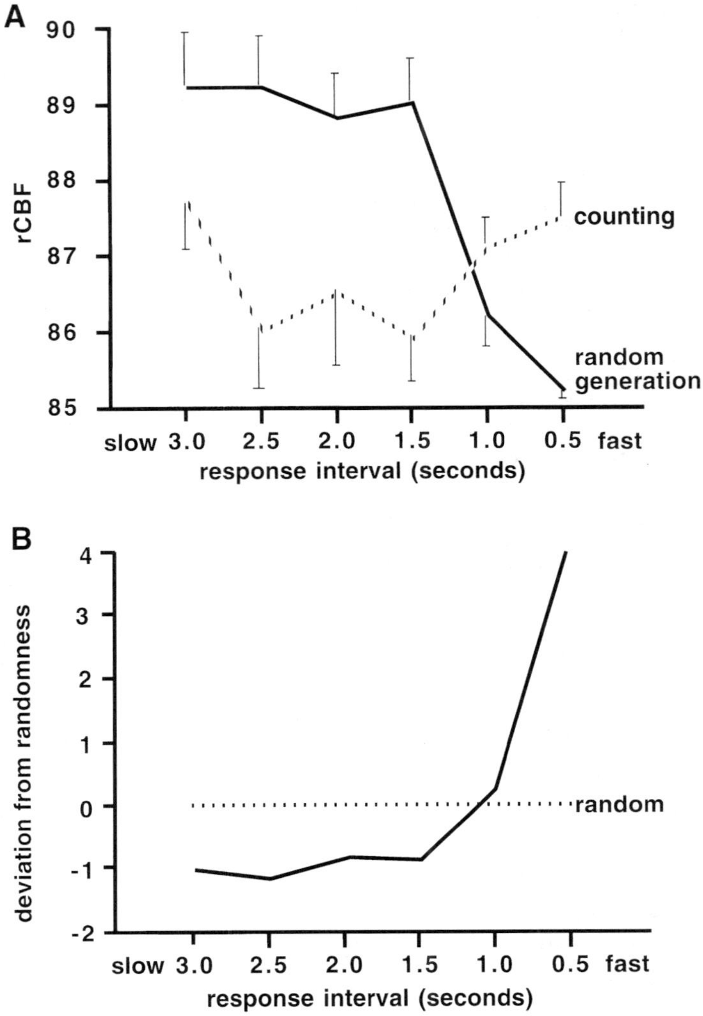

Figure 24.4 A. Regional cerebral blood flow (rCBF) in left DLPFC is higher during random number generation than during counting except at the highest rates of production. At high rates, rCBF decreases during random generation but not during counting. B. A similar pattern is seen in a performance measure of randomness. Performance is less random at the two highest rates of production. From Jahanshani et al. forthcoming.

random generation and the need to produce responses rapidly. Similar reduction of activity in DLPFC has been observed in some explicit studies of dual task interference, as described later.

24.4 DUAL-TASK INTERFERENCE

On the "random" number generation task (Jahanshahi et al. forthcoming), the reduction in the randomness of the response sequence at the highest rates of performance (see figure 24.4B) took the form of an

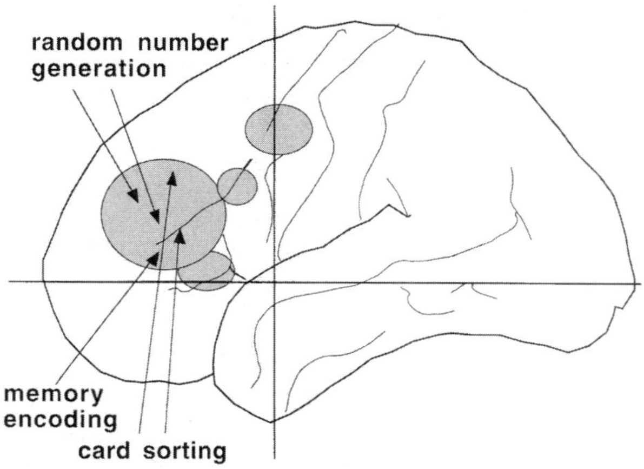

Figure 24.5 Regions where there was a decrease in activation associated with dual-task interference. Random number generation from Jahanshahi et al. forthcoming. There was a significant interaction between task and rate, with less activity at high rates for the random number generation task only in left and right DLPFC (see figure 24.4). Card sorting from Goldberg et al. 1998. Memory encoding from Fletcher et al. 1995.

increase in the number of response pairs (consecutive numbers: 1-2, 5-6, etc.) and a reduction in the number of response pairs (numbers 2 apart: 1-3, 5-7, etc.). This effect was manifest in brain activity as an interaction between task and rate (i.e., a decrease in activity at high rates for the random number generation task versus no change in the counting task). This interaction effect was seen in DLPFC (see figure 24.4A; coordinates: −34, 40, 24; $Z = 3.2$, $p < 0.001$), but not in any of the other frontal areas associated with random number generation. There was also a negative correlation between "randomness" (as defined above) and activity in left DLPFC (coordinates: −52, 34, 18; $Z = 3.8$; $p < 0.001$).

Goldberg et al. (1998) have shown a similar effect with dual-task interference on the Wisconsin Card-Sorting Task. Performance of a shadowing task while sorting produced an impairment of performance and a reduction of the activity in DLPFC associated with sorting (see figure 24.5). This effect was revealed as a significant interaction in DLPFC (card sorting − control) > ((card sorting + shadowing) − (control + shadowing)); coordinates: −52, 28, 16; $Z = 3.6$; $p < 0.001$. The same effect was also observed by Fletcher et al. (1995) in a study of memory acquisition. A secondary choice reaction time task impaired memory performance and reduced the activity in left DLPFC associated with memory acquisition. Here again there was a significant interaction in left DLPFC (memory − control) > ((memory + RT task) − RT task)); coordinates: −48, 34, 8; $Z = 2.7$; $p < 0.01$. Different results were obtained by D'Esposito et al. (1995), who observed increases in activity in DLPFC when two tasks had to be performed at once, although the decrements in performance were not large and may not have been significant. There was no detectable activity in DLPFC in this study when the two tasks were performed separately. In Jahanshahi et al. forthcoming and in Goldberg et al. 1998, activity in ACC did not decrease during the interfering conditions, whereas in Fletcher et al. 1995, ACC was the only area to show an increase in activity during the dual-task condition, evidence that it has a function distinct from that of DLPFC.

When several processes are competing for limited resources, there may be a need for some higher-level executive system to make appropriate allocations. The greater the competition, the harder this executive will have to work. The observation of reduced activity in DLPFC coupled with impaired performance may suggest that this region is concerned with a lower-level process that receives insufficient resources at high levels of competition. Only in ACC can we observe a pattern of activity that would be consistent with a higher level of executive function. What sort of low-level executive process might be subserved by DLPFC? My experience of trying to generate random numbers at too high a rate is as follows. At the moment that I have to make the next response, I have not had time to think of an item that I consider sufficiently random. I am forced, therefore, to produce one of the unsuitable items that happens to be available. This is likely to be a recently emitted item or one that has been primed by the last response (i.e., the next number in a counting sequence). I do not totally give up, however. I continue to try and find a "random" response for each trial and the sequence I produce does not become completely stereotyped. Perhaps activity in DLPFC is necessary to prevent the production of inappropriate responses and has to be reduced when an inappropriate response has to be emitted because there is not time to complete the selection process. Another low-level process that might be instantiated in DLPFC could be keeping track of the responses selected thus far. In the random number generation task, a reduction in randomness would occur if subjects could not keep track of recent selections. In this case, they would not be aware that their responses were not random. As the last set of studies I will discuss shows, DLPFC is active even when there is no requirement to keep track of responses.

24.5 CONTEXTUAL CONSTRAINT

Nathaniel-James, Fletcher, and Frith (1997) studied word generation using the sentence completion task developed by Burgess and Shallice (1996). On this task, subjects are shown a sentence with the last word missing. In one version of the task they must generate the word that best fits the sentence; in the other, they must generate a word that does *not* fit the sentence. Both versions of this task, especially the latter, are performed badly by patients with frontal lobe lesions (Burgess and Shallice 1996). When normal subjects perform this task, much activity is observed

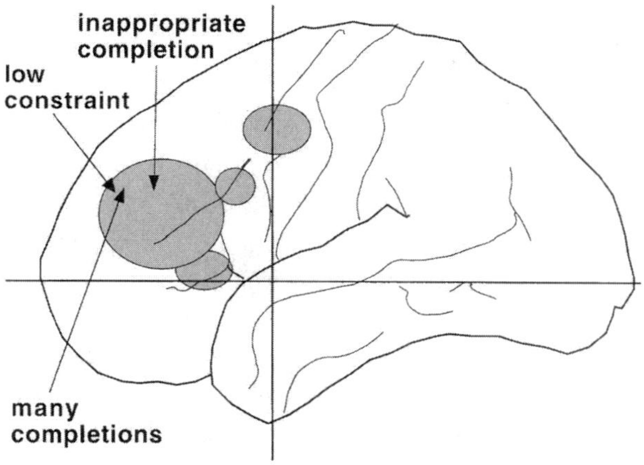

Figure 24.6 Regions where activity was associated with less constraint on response selection. Inappropriate completion and low constraint from Nathaniel-James and Frith in preparation. Many completions from Desmond, Gabrieli, and Glover 1998.

in left DLPFC for both versions compared to rest or to reading sentences in which the last word is supplied. Nathaniel-James and Frith (in preparation) have examined the effect of the constraint supplied by the sentence on the pattern of activity. Six levels of constraint were derived from the Bloom and Fischler 1980 sentence completion norms. An example of high constraint would be "He posted the letter without a ———" (99% of subjects said "stamp") and an example of low constraint would be "The police had never seen a man so ———" (14 different responses were given; the most frequent, "drunk," was given by 9% of subjects). Subjects were asked to give an appropriate or an inappropriate completion, making a total of twelve different conditions. When the six inappropriate completions were compared with the six appropriate completions, activity was observed in left DLPFC (see figure 24.6). There was no effect of constraint in the inappropriate condition, but when subjects had to give an appropriate completion there was more activity in DLPFC under conditions of low constraint, which appeared as a significant interaction between task and the linear component of constraint in left DLPFC (coordinates: −32, 58, 26; $Z = 4.2$; $p < 0.001$).

For both low-constraint appropriate and any inappropriate completions, it was necessary to select between a number of possible responses. However, because these three conditions were also more difficult than the high-constraint appropriate condition, subjects took longer to produce their responses. Is it the lack of constraint that leads to the activity in DLPFC or is it simply the difficulty of the task? In Desmond, Gabrieli, and Glover 1998, subjects had to generate words on the basis of word stems, with fMRI used to contrast activity elicited by stems with many or few possible completions. In contrast to the sentence completion task, this word stem completion task is more difficult when the stem has few versus many possible completions. Nevertheless, the many completions condition (i.e., the less constrained condition) produced greater activity in the left DLPFC (see figure 24.6). Clearly it is the lack of constraint and not the difficulty of the task that leads to activity in DLPFC.

In these completion tasks, because there is no requirement to keep track of the sequence of responses, this component of working memory is not involved. In addition, subjects cannot prepare and hold their response in advance. Clearly, working memory is required to keep in mind the current instructions (whether the response should be appropriate or inappropriate, for example), although this is required for both conditions. In combination with the various studies previously considered, the results from these studies of the effects of response constraint strongly suggest that DLPFC activation is greater in situations where subjects must select one from many rather than few alternatives. One possible formulation of the common feature of all the tasks reviewed here would be the need to create and sustain an arbitrary category of responses appropriate for the task in hand. This process may include the requirement to suppress responses outside the arbitrary category. For example, when generating words that start with a certain letter, it may be necessary to suppress semantic associations, and when generating random sequences of finger movements, it may be necessary to suppress sequences such as LLLL or LRLR. In these tasks, it is the "sculpting" of the response space normally achieved by external context that has to be self-generated. In this regard, Braver and Cohen (chap. 31, this volume) also assign DLPFC a role in sustaining contextual information.

24.6 OTHER CHARACTERIZATIONS OF THE FUNCTION OF DORSOLATERAL PREFRONTAL CORTEX

One important component of the executive system likely to be involved in response generation tasks is monitoring. Although I have suggested that DLPFC is not involved in monitoring in the sense of keeping track of the responses produced thus far, monitoring might be required before each individual response to ensure that the appropriate response is going to be selected. Such a process might have to work harder when many rather than few responses are available for selection. I suspect that monitoring in this sense is closely related to "response sculpting" because both are mechanisms for ensuring that the correct response is selected, although it might be possible to choose between these formulations by studying what happens when response selection breaks down at high rates of performance. If this was due to the failure of a monitor system, then presumably subjects would not be able to indicate which responses were inappropriate. On the other hand, if the "response sculpting" mechanism failed, then subjects would know that incorrect responses were being made (see section 24.4).

Although this review has concentrated largely on tasks involving response selection, it is well established that activation in DLPFC is also elicited by working-memory tasks. Such tasks involve many processes, leaving open the question of which particular process is relevant to the activity in DLPFC. Petrides (chap. 23, this volume) argues persuasively that DLPFC is not required simply to maintain items in working memory, but is involved when items have to be manipulated in working memory. A working-memory task popular with brain imagers that requires such manipulation is the *n*-back task, in which subjects see a sequence of letters and have to detect whether the currently presented item is the same as the item presented *n* trials previously. To do this, subjects must keep the last *n* items in memory and continuously update which is the target item. It is only the target items to which the subjects must respond. Thus the continuous updating involves creating new and arbitrary stimulus-response relationships. Clearly, the manipulation of items in working memory involved in the *n*-back task can be seen in terms very similar to the "response sculpting" process that I suggest is required for the performance of response generation tasks. A more detailed analysis of the various working-memory tasks that activate DLPFC, supplemented by the use of new tasks concerned with particular components of working memory, will be needed to determine whether my formulation of the function of DLPFC in response generation can also be applied to other working-memory tasks.

24.7 ACHIEVEMENT OF RESPONSE SELECTION

How does DLPFC influence response selection across different modalities? We have previously suggested (Friston et al. 1991) that DLPFC modulates activity in those posterior brain regions where responses relevant to the task are represented. In word generation tasks, activity is seen in the temporal lobe (Warburton et al. 1996), whereas in tasks requiring the movement of a joystick (Spence et al. 1998) or the fingers (Frith et al. 1991), activity is seen in parietal cortex. Whether this activity represents an increase or a decrease seems to depend critically on the control task. With word generation, there is an increase of activity in temporal lobe relative to rest, but a decrease relative to word repetition. This is the case even when the generation and repetition is "silent" (see Warburton et al. 1996, exp. 4), and thus the activity does not reflect a response to external inputs from the subjects' own voice. The situation is much less clear in relation to the parietal activity seen in the limb movement tasks. Decreases were seen when self-generated finger movements were compared to repetition (Frith et al. 1991), whereas increases were seen for the equivalent comparison on the joystick task (Spence et al. 1998).

We (Friston et al. 1991) have suggested that the relative decreases seen in temporal cortex during word generation reflect a modulation of the region by signals from DLPFC that permit self-generated response selection to occur. The appropriate responses emerge through the suppression of the very much greater number of inappropriate responses, leading to an overall reduction of activity. In terms of my formulation in section 24.5, the decrease could represent the self-generated "sculpting of the response space" imposed by DLPFC. The manipulation of response rate during word generation sheds some light on the precise nature of the modulation. We observed a highly significant effect of rate on activity in temporal cortex, an effect observed in a number of studies and one, we have suggested, reflects the transient increase in activity associated with each trial. This activity is presumably associated with stimulus analysis, response production, or both (Rees and Frith 1998). Because the activity is located in Wernicke's area, I suggest that the rate effect observed in temporal cortex in the word generation task reflects transient activity associated with the selection of each word. Moreover, there was an effect of task on the activity in this area (see figure 24.7). Although the effect did not alter the slope of the line relating activity and task rate (i.e., there was no significant difference in the slopes of rCBF against rate between the two conditions), it shifted the intercept down so that there was a general reduction of activity in the word generation task compared to the word repetition task.

Because there is no significant change in slope, the transient activity associated with each response was not detectably affected by the task. The change in intercept suggests that there was a tonic change in activity, which implies activity was reduced in this area during the word generation task even when no responses were being generated. This is consistent with a mechanism whereby a form of bias is imposed on the relevant population of cells by the task set, analogous to the bias proposed to underlie stimulus selection in the attentional model of Desimone and Duncan (1995; see also Miller, chap. 22, this volume). Similar ideas are also put forward by Braver and Cohen (chap. 31, this volume).

24.8 CONCLUSIONS

By identifying a series of different circumstances under which DLPFC is activated in association with response selection, I have tried to derive a single cognitive function for this region. The evidence suggests, first, that DLPFC is not at the apex of an action control system because the process instantiated there competes for resources with other processes. Second, it appears that DLPFC is not solely involved in keeping track of response sequences because it is activated in tasks where keeping track of responses is not required (e.g., the stem completion task of Desmond, Gabrieli, and Glover 1998). I conclude that DLPFC is most likely involved in defining a set of responses suitable for the task and biasing these for selection when external inputs achieve such selection to only a very limited degree. This function resembles that component of Shallice's

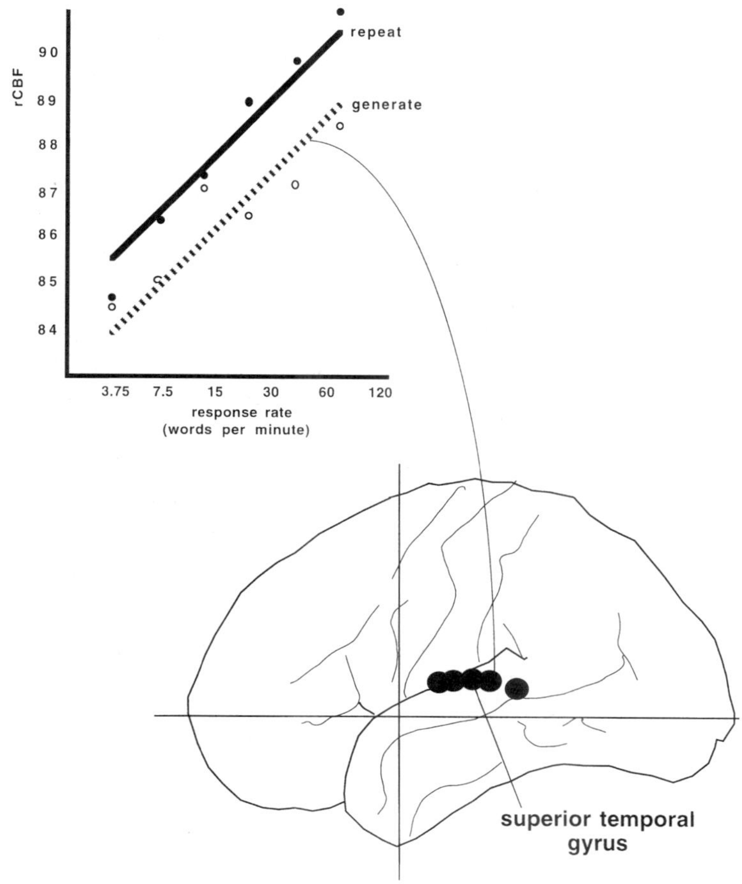

Figure 24.7 The effects of task (repeat versus generate) and response rate on activity in the temporal lobe. Data from Frith and Friston in preparation. The lines shown in the inset are the best fit straight lines given that there was no significant difference between the conditions in the slope of the lines relating rate and regional cerebral blood flow (rCBF).

"supervisory attentional system" (1988, chap. 14) which modulates his proposed lower-level contention-scheduling system. By breaking this executive system into such components, it may eventually lose its mystical and homuncular nature. Although over the next few years it should be possible to associate these various components of the executive system to particular frontal regions, the success of this program will depend on converging evidence from many imaging and lesion studies employing a variety of tasks.

NOTE

My thanks to Sean Spence, Marjan Jahanshahi, and David Nathaniel-James for allowing me to present some of their data prior to full publication; my apologies to Karl Friston for not having presented his data earlier. I am grateful to Tim Shallice for our many discussions about the contribution of imaging to our understanding of executive processes.

REFERENCES

Benton, A. L. (1968). Differential behavioural effects in frontal lobe disease. *Neuropsychologia, 6,* 53–60.

Bloom, P. A., and Fischler, I. (1980). Completion norms for 329 sentence contexts. *Memory and Cognition, 8,* 631–642.

Burgess, P. W., and Shallice, T. (1996). Response suppression, initiation and strategy use following frontal lobe lesions. *Neuropsychologia, 34,* 263–273.

Cohen, J. D., Braver, T. S., and O'Reilly, R. C. (1996). A computational approach to the prefrontal cortex. *Philosophical Transactions of the Royal Society of London B351,* 1515–1527.

Cohen, J. D., Perlstein, W. M., Braver, T. S., Nystrom, L. E., Noll, D. C., Jonides, J., and Smith, E. E. (1997). Temporal dynamics of brain activation during a working memory task. *Nature, 386,* 604–608.

Deiber, M.-P., Passingham, R. E., Colebatch, J. G., Friston, K. J., Nixon, P. D., and Frackowiak, R. S. J. (1991). Cortical areas and the selection of movement. *Experimental Brain Research, 84,* 393–402.

Desimone, R., and Duncan, J. (1995). Neural mechanisms of selective visual attention. *Annual Review of Neuroscience, 18,* 193–222.

Desmond, J. E., Gabrieli, J. D. E., and Glover, G. H. (1998). Dissociation of frontal and cerebellar activity in a cognitive task: Evidence for a distinction between selection and search. *NeuroImage, 7,* 368–376.

D'Esposito, M., Detre, J. A., Alsop, D. C., Shin, R. K., Atlas, S., and Grossman, M. (1995). The neural basis of the central executive system of working memory. *Nature, 378,* 279–281.

Dronkers, N. F. (1996). A new brain region for coordinating speech articulation. *Nature, 384,* 159–161.

Fletcher, P., Frith, C. D., Grasby, P., Shallice, T., Frackowiak, R. S. J., and Dolan, R. J. (1995). Brain systems for encoding and retrieval of auditory-verbal memory. *Brain, 118,* 401–416.

Friston, K. J., Frith, C. D., Liddle, P. F., and Frackowiak, R. S. J. (1991). Investigating a network model of word generation with positron emission tomography. *Proceedings of the Royal Society of London, B244,* 101–106.

Friston, K. J., Frith, C. D., Liddle, P. F., and Frackowiak, R. S. J. (1993). Functional connectivity: the principal-component analysis of large (PET) data sets. *Journal of Cerebral Blood Flow and Metabolism, 13,* 5–14.

Friston, K. J., Holmes, A., Poline, J.-B., Price, C. J., and Frith, C. D. (1996). Detecting activations in PET and fMRI: levels of inference and power. *NeuroImage, 4,* 223–235.

Frith, C. D., Friston, K. J., Liddle, P. F., and Frackowiak, R. S. J. (1991). Willed action and the prefrontal cortex in man: A study with PET. *Proceedings of the Royal Society of London, B244,* 241–246.

Fuster, J. M. (1989). *The prefrontal cortex.* Raven Press: New York.

Goldberg, T. E., Berman, K. F., Fleming, K., Ostrem, J., Van Horn, J. D., Esposito, G., Mattay, V. S., Gold, J. M., and Weinberger, D. R. (1998). Uncoupling cognitive workload and prefrontal cortical physiology *Neuroimage, 7,* 296–303.

Goldman-Rakic, P. S. (1987). Circuitry of primate prefrontal cortex and regulation of behaviour by representational knowledge. In V. B. Mountcastle (Ed.), *Handbook of physiology: The nervous system,* vol. 5, pp. 373–417. Baltimore: Williams and Wilkins.

Jahanshahi, M., Jenkins, I. H., Brown, R. G., Marsden, C. D., Passingham, R. E., and Brooks, D. J. (1995). Self-initiated vs internally triggered movements. *Brain, 118,* 913–933.

Jahanshahi, M., Dirnberger, G., Fuller, R., and Frith, C. D. (Forthcoming). The role of dorsolateral prefrontal cortex in random number generation.

Jueptner, M., Stephan, K. M., Frith, C. D., Brooks, D. J., Frackowiak, R. S. J., and Passingham, R. E. (1997). Anatomy of motor learning: 1. Frontal cortex and attention to action. *Journal of Neurophysiology, 77,* 1313–1324.

Luria, A. R. (1966). *Higher cortical functions in man.* London: Tavistock.

Nathaniel-James, D. A., Fletcher, P., and Frith, C. D. (1997). The functional anatomy of verbal initiation and suppression using the Hayling test. *Neuropsychologia, 35,* 559–566.

Passingham, R. E. (1997). Functional organization of the motor system. In R. S. J., Frackowiak, K. J., Friston, C. D., Frith, R. J., Dolan, and J. C., Mazziotta, (Eds.), *Human brain function,* pp. 243–274. San Diego, CA: Academic Press.

Petersen, S. E., Fox, P. T., Posner, M. I., Mintun, M., and Raichle, M. E. (1988). Positron-emission studies of the cortical anatomy of single-word processing. *Nature, 331,* 585–589.

Petrides, M., Alivisatos, B., Meyer, R., and Evans, A. C. (1993). Functional activation of the human frontal cortex during the performance of verbal working memory tasks. *Proceedings of the National Academy of Science, U.S.A., 90,* 878–882.

Posner, M. I., and Dahaene, S. (1994). Attentional networks. *Trends in the Neurosciences, 17,* 75–79.

Price, C. J., Wise, R. J. S., Ramsay, S., Friston, K. J., Howard, D., Patterson, K., and Frackowiak, R. S. J. (1992). Regional response differences within the human auditory cortex when listening to words. *Neuroscience Letters, 146,* 179–182.

Rees, G., and Frith, C. D. (1998). How do we select perceptions and actions? Human brain imaging studies. *Philosophical Transactions of the Royal Society of London, B353,* 1283–1293.

Shallice, T. (1988). *From neuropsychology to mental structure.* Cambridge: Cambridge University Press.

Spence, S. A., Hirsch, S. R., Brooks, D. J., and Grasby, P. M. (1998). Prefrontal cortex activity in people with schizophrenia and control subjects. *British Journal of Psychiatry, 172,* 316–323.

Warburton, E., Wise, R. J. S., Price, C. J., Weiller, C., Hadar, U., Ramsay, S., and Frackowiak, R. S. J. (1996). Noun and verb retrieval by normal subjects. *Brain, 119,* 159–179.

25 Dissociative Methods in the Study of Frontal Lobe Function

John Duncan and Adrian M. Owen

ABSTRACT In principle, the specialization of function within prefrontal cortex can be shown by double dissociation using any of a variety of neuroscientific methods, including functional imaging, comparison of lesion groups, and single-cell electrophysiology. In practice, full dissociation designs are rarely used, and when they are, clear dissociations are hard to obtain. Taken together, neuroimaging, lesion, and electrophysiological results suggest that well-defined regions of frontal cortex—middorsolateral, midventrolateral, and dorsal anterior cingulate—have somewhat dynamic functions, adapting themselves to solution of a broad range of cognitive problems. In neuroimaging, for example, these regions are activated by many different increases in cognitive demand, including response conflict, task novelty, working-memory load, and even perceptual difficulty. At the same time, these regions can be distinguished from much of medial and orbital frontal cortex, perhaps more concerned with affective and motivational processes. We suggest that refinement of this rather coarse subdivision of frontal functions will require a substantial strengthening of commitment to full-scale double-dissociation methodology.

As Frith (chap. 24, this volume) observes, a long-standing goal of research into frontal lobe functions has been "fractionating the prefrontal cortex and identifying specific roles for separate components of this large region of the brain." As Frith goes on to say, "In practice progress in identifying such roles has been remarkably slow." In this commentary, we shall make some remarks about the methodology of fractionation experiments, as illustrated by the contributions to this volume, and what conclusions are indicated by current results.

What is the explanation for the "remarkably slow" progress that Frith describes? In principle, the basic methodology of fractionation has been well understood at least since Teuber's introduction (1955) of the "double dissociation." As conceived by Teuber, the double-dissociation experiment requires investigation of (at least) two tasks or cognitive operations in the context of (at least) two brain areas. By some means, it is shown that brain area X is more involved in operation A than in operation B, whereas area Y is more involved in B than in A. Although Teuber developed this principle for lesion studies—operation A is more affected by lesions to area X than to area Y, whereas operation B is more affected by lesions to Y than to X—the basic idea of double dissociation applies equally to other methods. Here we shall discuss its application in functional imaging, in animal and human lesion studies, and in single-unit electrophysiology.

25.1 FUNCTIONAL IMAGING

The recent functional imaging literature contains many proposals concerning specialized function within prefrontal cortex. At first sight, the typical basis for such proposals is strong double dissociation. In one experiment or comparison, task A is compared with a control. Significant activation in frontal area X suggests that X is important in task A. In a second experiment or comparison, task B is compared with (possibly the same) control. Significant activation in frontal area Y suggests that Y is important in task B. Because X and Y are different, the apparent conclusion is that the operations of A and B are dissociated within frontal cortex.

The problem with this inference is statistical noise. Suppose that both A and B in fact activate very much the same, broad region of frontal cortex, including both the regions X and Y. In any given experiment, some particular part of this broad region will by chance be measured as *most* significantly active. Indeed, limited statistical power makes it likely that only a subset of this region will be measured as "significantly" active at all. Under these conditions, it is more or less inevitable—simply through statistical noise—that any two comparisons (one for task A minus control, the other for task B minus control) will indicate somewhat different "most active" prefrontal areas.

Some results from the literature may help to make this point more concrete. As Frith (chap. 24, this volume) points out, proposals for specialization of function within prefrontal cortex provide only one theme in current neuroimaging work; a contrasting theme is a strong impression of rather substantial prefrontal regions activated over and over again, in studies using widely different tasks designed to investigate quite separate cognitive domains. To examine this impression more systematically, we (Duncan and Owen forthcoming) have recently carried out a comparison of studies manipulating different forms of cognitive "demand" or task difficulty, asking whether these different demands are associated with the same or different regions of frontal activation. For this purpose, we defined "activation" as increased activity with increased cognitive demand, decreases being inconsistently analyzed and reported in the studies reviewed. Of the various demand factors covered in our analysis, five may be considered here. First, to capture the process of inhibiting prepotent response tendencies, we combined the results from seven studies of response confiict, including four studies of the Stroop effect (Bench et al. 1993; Carter, Mintun, and Cohen 1995; George et al. 1994; Pardo et al. 1990), two of incompatible stimulus-response mappings (Sweeney et al. 1996; Taylor et al. 1994), and one of reversing previously learned stimulus-response associations (Paus et al. 1993). Second, to capture the response to task novelty, we combined results from four studies comparing initial and practiced performance in an assortment of cognitive learning contexts (Jenkins et al. 1994; Jueptner et al. 1997; Klingberg and Roland 1998; Raichle et al. 1994). Third, turning to the role of frontal cortex in working memory, we combined results from three studies varying the number of elements to be tracked in *n*-back comparison tasks (Braver et al. 1997; Carlson et al. 1998; Cohen et al. 1997). In these tasks, a sequence of stimuli is presented one after the other. Subjects must respond when the current stimulus matches the item preceding it by *n* steps, requiring constant updating and reorganization of working memory as the sequence progresses (cf. the "monitoring" concept of Petrides, chap. 23, this volume). Fourth, to examine more passive aspects of short-term memory, we combined results from three studies with simple manipulations of delay between stimulus presentation and test (Barch et al. 1997; Goldberg et al. 1996; Smith et al. 1995). Finally, as a demand factor normally thought to be somewhat unrelated to executive functions, we combined results from three studies of perceptual difficulty, including two studies of stimulus degradation (Barch et al. 1997; Grady et al. 1996) and one of usual versus unusual views (Kosslyn et al. 1994). To obtain as clean a set of contrasts as possible, we included only studies that had manipulated the specific demand variable (e.g., presence of response conflict, length of delay) in an otherwise identical task. From each study, we listed all reported activations within the frontal lobe (coordinates of peak increase in activity with increased demand), excluding only those judged to lie in primary motor (Brodmann's area 4) or premotor (Brodmann's area 6) cortex.

The results are summarized in figure 25.1. In this figure, peak activations from all selected studies have been plotted together onto the standard brain of the SPM96 software (Wellcome Department of Cognitive Neurology, London). Each reported peak is plotted as a letter, different letters distinguishing the five different demand manipulations. Six different views are shown, including lateral and medial views of each hemisphere, and at the bottom of the figure, views of the whole brain from above and below.

The first and perhaps most noteworthy result is the remarkable specificity of the activated region in this diverse group of studies. Though studies were carried out in many different laboratories using many different tasks and methods of analysis, activated points are seen only within a compact region of frontal cortex. On the medial surface, this region is entirely restricted to the dorsal part of the anterior cingulate, excluding the surrounding cortex and the whole orbitomedial region of each hemisphere. On the lateral surface, points cluster within the middorsolateral and midventrolateral regions discussed by Petrides (chap. 23, this volume; for closely similar activations see also Frith, chap. 24, this volume),

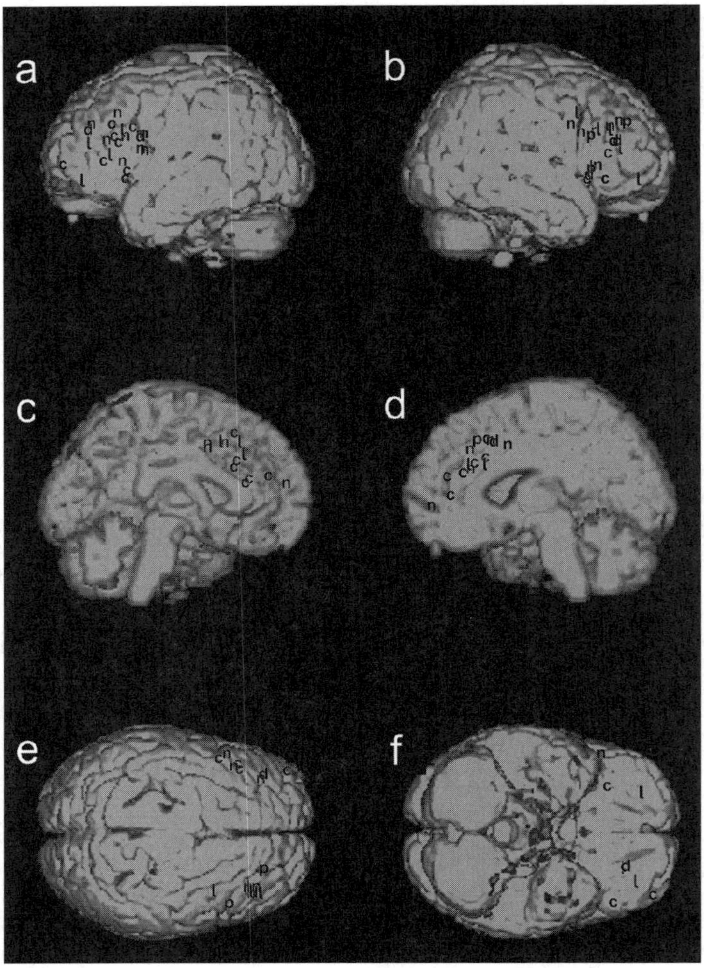

Figure 25.1 Activations in prefrontal cortex associated with increased cognitive demand. Foci of peak activation from studies of five different demand factors are plotted on six surface views of a standard brain: lateral views, left and right hemisphere (*a* and *b*); medial views, left and right hemisphere (*c* and *d*); and dorsal and ventral views (*e* and *f*). Activation peaks are plotted as letters: c = response conflict; *l* = learning; n = *n*-back; d = delay; p= perceptual difficulty.

with occasional further scattered points toward the frontal pole. The dorsal view of the brain shows with particular clarity how much of the lateral surface is excluded, including the whole strip of cortex running down the midline to the frontal pole.

The second important result is the lack of differentiation between the five aspects of cognitive demand contributing to the analysis. For each demand, activations are distributed broadly throughout the middorsolateral, midventrolateral, and dorsal anterior cingulate regions. The only real suggestion of differentiation is a preponderance of right-hemisphere activations associated with perceptual demand. Though the active frontal region is compact and specific anatomically, these results give little clue of specificity in cognitive function. Instead they reveal a region that is activated rather generally by any increase in task "demand" or difficulty.

Of course, there are reasons for caution with respect to this kind of exercise. One risk is that apparently large regions of activation may be produced by inappropriately combining data from contrasts that, though superficially similar (e.g., multiple studies of working-memory delay), in fact have rather different cognitive components. Against this, as we have said, we were extremely strict in including only studies with the purest possible manipulations of our chosen demand factors. A second possibility is that fine-grained specializations within the active region of figure 25.1 are concealed by the restricted spatial precision of current imaging methods. Even as they stand, however, the data warn that it would be all too easy to overinterpret any single study as showing a specific relation between some manipulated cognitive demand and some activated frontal region. When information is combined across studies, many different cognitive manipulations may produce rather similar results.

How can such overinterpretation be avoided? To show a strong double dissociation, it is not enough just to show that area X is significantly active in a comparison of task A with control, while area Y is significantly active in a comparison of task B with control. For neither area do such tests actually show a difference between A and B themselves: as any student of statistics is taught, a demonstration that A differs significantly from control, whereas B does not, is not at all a demonstration that A differs from B. A supplementary analysis is needed to show that, in area X, task A gives significantly more activation than task B, whereas in area Y, task B gives significantly more activation than task A (see, for example, Courtney et al. 1998; Fletcher et al. 1998). It is entirely possible that many of the apparent "dissociations" reported within the active region shown in figure 25.1 would fail this more appropriate test. If the truth is that both of two regions are somewhat activated in both tasks A and B—but that, by chance, one region is most significant in the task A versus control comparison; the other in the task B versus control comparison—then opposite, significant differences in the two regions should not appear in a direct A versus B comparison.

Of course, it would be unjustified to conclude that all proposed dissociations are indeed the spurious consequence of noise in individual sets of results. As Petrides (chap. 23, this volume) discusses, for example, a number of converging experiments have suggested a separation between directed information retrieval operations in midventrolateral cortex, and more complex information manipulation in middorsolateral cortex (see also Owen 1997; D'Esposito et al. 1998). This would be one promising candidate for direct statistical test in the way that we have suggested.

Before leaving the neuroimaging literature, it is worth noting one broad dissociation already indicated by current results. As figure 25.1

shows, much of frontal cortex, including most of the medial surface and the whole orbitomedial region, does not increase its activity with increased cognitive demands of the sort we have been considering. Indeed, a recent comparison of assorted active tasks with passive, no-task controls suggests that increased task demand may often *decrease* activity in much of this region (Shulman et al. 1997). At the same time, activations within this general region have been associated with emotional (Lane et al. 1997), social (Fletcher et al. 1995), and motivational (Elliott, Frith, and Dolan 1997) manipulations. As often suggested (e.g., Walsh 1978), there may indeed be a general division between more cognitive and more affective aspects of frontal function.

25.2 LESION STUDIES

At least as remarkable as the paucity of strong double dissociations within the functional imaging literature is their paucity in lesion studies. With only occasional exceptions, double dissociation in the sense defined by Teuber (1955) has not been the basis for proposing specialization of function in different frontal regions.

In the human literature, for example, it is widely accepted that very different consequences follow dorsolateral and orbitomedial lesions. Plausible though this is in light of both animal work (see below) and the imaging results we have reviewed, the human lesion evidence comes largely from striking single cases (e.g., Eslinger and Damasio 1985), rather than dissociative group studies (for a partial exception, see Bechara et al. 1998). Beyond this, double dissociations in human lesion studies are all but restricted to a few suggestions of hemispheric specialization (e.g., Milner 1971).

Of course, technical difficulties, in particular the difficulty of sorting naturally occurring lesions into anatomical groups, may make dissociations hard to demonstrate in the human case. Complete double dissociations, however, are also a rarity in the monkey literature. Consider the influential proposal that spatial and object tasks are respectively associated with more dorsal and more ventral divisions of the lateral frontal surface (Goldman-Rakic 1988). Certainly, important experiments have shown spatial deficits after dorsal lesions (e.g., Funahashi, Bruce, and Goldman-Rakic 1993), and object deficits after ventral lesions (e.g., Mishkin and Manning 1978). The full-scale double dissociation is more elusive, however. Although suggestive results were obtained by Passingham (1975), most studies have not used the full dissociation design, and indeed, spatial tasks can be impaired by ventral lesions (e.g., Passingham 1975), and object tasks by dorsal lesions (Petrides 1995).

As we have said, dissociations between lateral and orbital frontal functions are among the most robust in the monkey literature (Butter 1969; Robbins and Rogers, chap. 21, this volume). In a study by Dias, Robbins, and Roberts 1996, monkeys with lateral frontal lesions were impaired on an extradimensional shift task, an impairment attributed to disordered "attentional selection." Monkeys with orbital lesions, contrastingly, were impaired in reversal learning, an impairment attributed to "the ability to alter behaviour in response to fluctuations in the emotional significance of stimuli" (p. 69). The full-scale double dissociation was shown by significant, opposite differences between the two frontal groups on the two measures. Again, such results may suggest a rather general distinction between more cognitive and more affective frontal functions.

25.3 SINGLE-CELL RECORDING

Strong specialization of function within frontal cortex has also been inferred from single-cell recording studies. Again elegant studies of spatial function in dorsolateral neurons provide an outstanding example. Funahashi, Bruce, and Goldman-Rakic (1989), for example, recorded cells in the region of the principal sulcus during a task designed to tap spatial short-term memory. In this task, monkeys were shown a brief target stimulus, positioned at one of several locations around the fixation point. After a delay period, they were required to make an eye movement to the location where the target had been seen. During this delay period, individual cells showed activity tuned to the remembered target location, suggesting a specific role in spatial working memory.

From the perspective of double dissociation, such results—obtained in a single task at a single recording site—raise two questions. What would these same neurons be doing in other cognitive contexts? And what would neurons elsewhere in prefrontal cortex be doing in the same delayed-saccade task? As amply documented by Miller (chap. 22, this volume), both questions receive surprising answers. First, the exact same neuron can carry very different information in different contexts, even different phases of the same trial (Rao, Rainer, and Miller 1997). For example, when the monkey remembers the identity of a target object, a neuron may be selective for what that object was, but when the monkey switches to remembering where the target occurred, selectivity for identity is replaced by selectivity for location. Second, in any particular task, neurons with very similar properties are found throughout a large region of both dorsolateral and ventrolateral frontal cortex. Again, the conclusion must be that much of frontal cortex is not dedicated to extremely specific cognitive functions; instead, neurons throughout a large area have the plasticity to acquire response properties dependent on current behavioral significance. Indeed, substantial tuning by distinctions relevant to current context is implied by the simple observation that, at least after training, a high proportion of recorded frontal units show selectivity for some aspect of events in whatever task it is that has been trained (e.g., Rao, Rainer, and Miller 1997).

25.4 CONCLUSIONS

Scant evidence for functional dissociations within prefrontal cortex could be explained in two ways. First, there could be far less regional specialization than is commonly presumed. Second, current methods could be inadequate to demonstrate such specialization. Our suspicion is that both of these factors contribute. Taken together, neuroimaging and single-unit results indicate regions of prefrontal cortex with substantial ability to adapt themselves to the solution of widely different cognitive problems. At the same time, some functional boundaries are already apparent, and in all probability, more are waiting to be discovered. Establishing such functional specializations, however, requires a far stronger commitment to double dissociation methodology—whether in neuroimaging, in lesion studies, or in single-unit electrophysiology—than is currently the rule.

NOTE

We are endebted to Matthew Brett for use of his image generation software.

REFERENCES

Barch, D. M., Braver, T. S., Nystrom, L. E., Forman, S. D., Noll, D. C., and Cohen, J. D. (1997). Dissociating working memory from task difficulty in human prefrontal cortex. *Neuropsychologia, 35,* 1373–1380.

Bechara, A., Damasio, H., Tranel, D., and Anderson, S. W. (1998). Dissociation of working memory from decision making within the human prefrontal cortex. *Journal of Neuroscience, 18,* 428–437.

Bench, C. J., Frith, C. D., Grasby, P. M., Friston, K. J., Paulesu, E., Frackowiak, R. S. J., and Dolan, R. J. (1993). Investigations of the functional anatomy of attention using the Stroop test. *Neuropsychologia, 31,* 907–922.

Braver, T. S., Cohen, J. D., Nystrom, L. E., Jonides, J., Smith, E. E., and Noll, D. C. (1997). A parametric study of prefrontal cortex involvement in human working memory. *NeuroImage, 5,* 49–62.

Butter, C. M. (1969). Perseveration in extinction and in discrimination reversal tasks following selective frontal ablations in *Macaca mulatta. Physiology and Behavior, 4,* 163–171.

Carlson, S., Martinkauppi, S., Rämä, P., Salli, E., Korvenoja, A., and Aronen, H. J. (1998). Distribution of cortical activation during visuospatial *n*-back tasks as revealed by functional magnetic resonance imaging. *Cerebral Cortex, 8,* 743–752.

Carter, C. S., Mintun, M., and Cohen, J. D. (1995). Inteference and facilitation effects during selective attention: An $H_2{}^{15}O$ PET study of Stroop task performance. *NeuroImage, 2,* 264–272.

Cohen, J. D., Peristein, W. M., Braver, T. S., Nystrom, L. E., Noll, D. C., Jonides, J., and Smith, E. E. (1997). Temporal dynamics of brain activation during a working memory task. *Nature, 386,* 604–608.

Courtney, S. M., Petit, L., Ma. Maisog, J., Ungerleider, L. G., and Haxby, J. V. (1998). An area specialized for spatial working memory in human frontal cortex. *Science, 279,* 1347–1351.

D'Esposito, M., Aguirre, G. K., Zarahn, E., and Ballard, D. (1998). Functional MRI studies of spatial and non-spatial working memory. *Cognitive Brain Research, 7,* 1–13.

Dias, R., Robbins, T. W., and Roberts, A. C. (1996). Dissociation in prefrontal cortex of affective and attentional shifts. *Nature, 380,* 69–72.

Duncan, J., and Owen, A. M. (Forthcoming). Common regions of the human frontal lobe recruited by diverse cognitive demands.

Elliott, R., Frith, C. D., and Dolan, R. J. (1997). Differential neural response to positive and negative feedback in planning and guessing tasks. *Neuropsychologia, 35,* 1395–1404.

Eslinger, P. J., and Damasio, A. R. (1985). Severe disturbance of higher cognition after bilateral frontal lobe ablation: Patient E.V.R. *Neurology, 35,* 1731–1741.

Fletcher, P. C., Happe, F., Frith, U., Baker, S. C., Dolan, R. J., Frackowiak, R. S. J., and Frith, C. D. (1995). Other minds in the brain: A functional imaging study of "theory of mind" in story comprehension. *Cognition, 57,* 109–128.

Fletcher, P. C., Shallice, T., Frith, C. D., Frackowiak, R. S. J., and Dolan, R. J. (1998). The functional roles of prefrontal cortex in episodic memory: 2. Retrieval. *Brain, 121,* 1249–1256.

Funahashi, S., Bruce, C. J., and Goldman-Rakic, P. S. (1989). Mnemonic coding of visual space in the monkey's dorsolateral prefrontal cortex. *Journal of Neurophysiology, 61,* 331–349.

Funahashi, S., Bruce, C. J., and Goldman-Rakic, P. S. (1993). Dorsolateral prefrontal lesions and oculomotor delayed-response performance: Evidence for mnemonic "scotomas." *Journal of Neuroscience, 13,* 1479–1497.

George, M. S., Ketter, T. A., Parekh, P. I., Rosinsky, N., Ring, H., Casey, B. J., Trimble, M. R., Horwitz, B., Herscovitch, P., and Post, R. M. (1994). Regional brain activity when selecting a response despite interference: An $H_2{}^{15}O$ PET study of the Stroop and an emotional Stroop. *Human Brain Mapping, 1,* 194–209.

Goldberg, T. E., Berman, K. F., Randolph, C., Gold, J. M., and Weinberger, D. R. (1996). Isolating the mnemonic component in spatial delayed response: A controlled PET ^{15}O-labeled water regional cerebral blood flow study in normal humans. *NeuroImage, 3,* 69–78.

Goldman-Rakic, P. (1988). Topography of cognition: Parallel distributed networks in primate association cortex. *Annual Review of Neuroscience, 11,* 137–156.

Grady, C. L., Horwitz, B., Pietrini, P., Mentis, M. J., Ungerleider, L. G., Rapoport, S. I., and Haxby, J. V. (1996). Effect of task difficulty on cerebral blood flow during perceptual matching of faces. *Human Brain Mapping, 4,* 227–239.

Jenkins, I. H., Brooks, D. J., Nixon, P. D., Frackowiak, R. S. J., and Passingham, R. E. (1994). Motor sequence learning: A study with positron emission tomography. *Journal of Neuroscience, 14,* 3775–3790.

Jueptner, M., Stephan, K. M., Frith, C. D., Brooks, D. J., Frackowiak, R. S. J., and Passingham, R. E. (1997). Anatomy of motor learning: 1. Frontal cortex and attention to action. *Journal of Neurophysiology, 77,* 1313–1324.

Klingberg, T., and Roland, P. E. (1998). Right prefrontal activation during encoding, but not during retrieval, in a non-verbal paired-associates task. *Cerebral Cortex, 8,* 73–79.

Kosslyn, S. M., Alpert, N. M., Thompson, W. L., Chabris, C. F., Rauch, S. L., and Anderson, A. K. (1994). Identifying objects seen from different viewpoints: A PET investigation. *Brain, 117,* 1055–1071.

Lane, R. D., Reiman, E. M., Bradley, M. M., Lang, P. J., Ahern, G. L., Davidson, R. J., and Schwartz, G. E. (1997). Neuroanatomical correlates of pleasant and unpleasant emotion. *Neuropsychologia, 11,* 1437–1444.

Milner, B. (1971). Interhemispheric differences in the localization of psychological processes in man. *British Medical Bulletin, 27,* 272–277.

Mishkin, M., and Manning, F. J. (1978). Nonspatial memory after selective prefrontal lesions in monkeys. *Brain Research, 143,* 313–323.

Owen, A. M. (1997). The functional organization of working memory processes within human lateral frontal cortex: The contribution of functional neuroimaging. *European Journal of Neuroscience, 9,* 1329–1339.

Pardo, J. V., Pardo, P. J., Janer, K. W., and Raichle, M. E. (1990). The anterior cingulate cortex mediates processing selection in the Stroop attentional conflict paradigm. *Proceedings of the National Academy of Sciences, U.S.A., 87,* 256–259.

Passingham, R. (1975). Delayed matching after selective prefrontal lesions in monkeys (*Macaca mulatta*). *Brain Research, 92,* 89–102.

Paus, T., Petrides, M., Evans, A. C., and Meyer, E. (1993). Role of the human anterior cingulate cortex in the control of oculomotor, manual, and speech responses: A positron emission tomography study. *Journal of Neurophysiology, 70,* 453–469.

Petrides, M. (1995). Impairments on nonspatial self-ordered and externally ordered working memory tasks after lesions of the mid-dorsal part of the lateral frontal cortex of the monkey. *Journal of Neuroscience, 15,* 359–375.

Raichle, M. E., Fiez, J. A., Videen, T. O., MacLeod, A. K., Pardo, J. V., Fox, P. T., and Petersen, S. E. (1994). Practice-related changes in human brain functional anatomy during non-motor learning. *Cerebral Cortex, 4,* 8–26.

Rao, S. R., Rainer, G., and Miller, E. K. (1997). Integration of what and where in the primate prefrontal cortex. *Science, 276,* 821–824.

Shulman, G. L., Fiez, J. A., Corbetta, M., Buckner, R. L., Miezin, F. M., Raichle, M. E., and Petersen, S. E. (1997). Common blood flow changes across visual tasks: 2. Decreases in cerebral cortex. *Journal of Cognitive Neuroscience, 9,* 648–663.

Smith, E. E., Jonides, J., Koeppe, R. A., Awh, E., Schumacher, E., and Minoshima, S. (1995). Spatial vs. object working memory: PET investigations. *Journal of Cognitive Neuroscience, 7,* 337–358.

Sweeney, J. A., Mintun, M. A., Kwee, S., Wiseman, M. B., Brown, D. L., Rosenberg, D. R., and Carl, J. R. (1996). Positron emission tomography study of voluntary saccadic eye movements and spatial working memory. *Journal of Neurophysiology, 75,* 454–468.

Taylor, S. F., Kornblum, S., Minoshima, S., Oliver, L. M., and Koeppe, R. A. (1994). Changes in medial cortical blood flow with a stimulus-response compatibility task. *Neuropsychologia, 32,* 249–255.

Teuber, H.-L. (1955). Physiological psychology. *Annual Review of Psychology, 6,* 267–296.

Walsh, K. W. (1978). *Neuropsychology: A clinical approach.* New York: Churchill Livingstone.

VI Disorders of Control

26 Neural Correlates of Processes Contributing to Working-Memory Function: Evidence from Neuropsychological and Pharmacological Studies

Mark D'Esposito and Bradley R. Postle

ABSTRACT Theoretical and empirical investigations of the prefrontal cortex (PFC) have provided evidence that this region mediates both mnemonic (e.g., storage and rehearsal) and non-mnemonic (e.g., shifting attention, inhibition, updating) processes, each of which may be implicated in control of behavior. To understand the contributions of PFC to these components of working memory, we performed (1) a meta-analytic review of behavioral studies of patients with focal PFC lesions performing working-memory tasks; (2) behavioral studies of patients with PFC and dopaminergic dysfunction—traumatic brain injury and Parkinson's disease, respectively; and (3) pharmacological studies of traumatic brain injury patients. The results of these studies reveal anatomical, pharmacological, and functional dissociations of processes that contribute to the short-term retention and on-line manipulation of information, and that may underlie control processes. We propose a model in which posterior retention and storage processes contribute to working-memory capacity, whereas prefrontostriatal regions, interacting with dopamine, contribute to rehearsal and control processes.

The concept of working memory in mammalian cognition was first introduced by Pribram (Miller, Galanter, and Pribram 1960; Pribram et al. 1964), who appropriated the term from the artificial intelligence work of Newell, Simon, and Feigenbaum (1958, 1961). The current understanding of working memory among students of human and nonhuman primate memory and cognition is strikingly reminiscent of Pribram's description (Pribram et al. 1964, p. 48) of a system that can accomplish "temporary storage [of] a flexible set of temporary instructions" in the service of problem solving. Pribram noted that, for the monkey as well as for the computer, "[t]his temporary storage must take place not in the computer's permanent memory where it would do little good, but in the instruction program itself."

For our part, we view working memory, not as a dedicated "system" composed of neurally and computationally interrelated modules, like the visual system, but rather as a capacity of the nervous system, conceptually analogous to motor control. To maintain and manipulate information when that information is not accessible in the environment, the brain needs (1) a *storage process*; (2) *rehearsal processes*, to prevent the contents of the storage system from decaying; and (3) *control processes*, to perform manipulations on the mnemonic representations of the information being stored and rehearsed (for a similar view see Smith and Jonides 1998). We consider the first two of these to be mnemonic processes, and the third to be non-mnemonic in nature. The interaction of these three systems gives rise to the behavioral phenomenon of working memory.

In this chapter, we will articulate a functional neuroanatomical model of the mnemonic processes that contribute to working memory and will shed some light, at the psychological and at the neural level, on the functional organization of the non-mnemonic processes that contribute to performance on complex working-memory tasks. These nonmnemonic processes, we believe, are central to the issues of control that are the focus of this symposium. Our efforts to determine the dependence of purely mnemonic contributors to working-memory function (i.e., storage and rehearsal) on the prefrontal cortex (PFC) are detailed in section 26.1, a meta-analysis of the effects of PFC lesions on working-memory storage and rehearsal. We believe that the results of our review impose important constraints on the interpretation of neuroimaging studies of working memory. Our investigations of the non-mnemonic processes that contribute to working-memory function are presented in section 26.2, a series of behavioral studies of patient groups with PFC or dopaminergic dysfunction, and in section 26.3, a behavioral study of the effects of dopamine manipulation on patients with PFC dysfunction resulting from traumatic brain injury.

The mnemonic components of working memory can be organized into two classes of processes: storage and rehearsal (Awh et al. 1996; Baddeley 1990; Longoni, Richardson and Aiello 1993; Schweickert, Guentert, and Hersberger 1990). Storage is measured in terms of capacity, and can be indexed by span tasks (Baddeley 1990): digit span for verbal working memory (Wechsler 1945) and block span for visuospatial working memory (Milner 1971). It is important to note, however, that because these span tests also recruit rehearsal processes, they are not "pure" tests of storage. This is manifest in the "articulatory suppression effect" (Levy 1971; Murray 1968) and the "word length effect" (Baddeley, Thomson, and Buchanan 1975), experimental manipulations believed to tie up articulatory rehearsal resources, and whose effect is to decrease memory span. Such results are reasonable evidence that rehearsal processes contribute to performance on a span test. Nevertheless, patients with intact articulatory abilities, and thus intact rehearsal, can have severely circumscribed spans (Vallar and Baddeley 1984), suggesting that storage processes make a critical contribution to span performance. Many researchers have used immediate serial recall as an index of working-memory capacity. Among widely used clinical and experimental measures of working memory, digit and block span tests are the most likely to minimize rehearsal processes because subjects repeat the remembered information immediately following presentation.

"Rehearsal" refers to the processes necessary to refresh and maintain information held in working memory. Tests of delayed response are often used to measure rehearsal processes (Awh et al. 1996; Paulesu, Frith, and Frackowiak 1993) because such tests tax subjects' ability to maintain information over a period of time. The typical delayed-response task presents one or a few stimulus items to be remembered at the beginning of a trial, conceals them during a delay period, and then probes memory for them at the end of the trial. In contrast to span tasks, delayed-response tasks rarely require memory of a large number of items, and thus do not provide a measure of working-memory storage capacity. Conversely, because delayed-response tasks *always* require that subjects maintain information across intervals exceeding the passive decay threshold of working-memory storage, and because such tasks often challenge subjects with distraction, they necessarily index rehearsal. Thus, throughout this chapter, memory span measures serve as acceptable approximate indices of working-memory *storage* processes, whereas delayed-response measures serve as acceptable approximate indices of working-memory *rehearsal* processes.

Although the starting point for many cognitive neuroscientific investigations of mental phenomena in humans is research in nonhuman primates, there are few empirical assessments of working-memory capacity in monkeys, and therefore scant data addressing the neural substrates of working-memory storage. Electrophysiological (e.g., Funahashi, Bruce, and Goldman-Rakic 1989; Fuster and Alexander 1971; see Miller, chap. 22, this volume) and lesion (e.g., Funahashi et al. 1993) studies of monkeys performing delayed-response tasks, on the other hand, have established lateral PFC as an important neural substrate of information maintenance during the delay portion of delayed-response tasks. Moreover, the performance of monkeys with PFC lesions on tasks such as conditional associative learning (Petrides 1982, 1985) and self-ordered choosing (Collins et al. 1998; Petrides 1991, 1994) suggests that the non-mnemonic contributors to working memory in the monkey are also dependent on PFC. The non-mnemonic processes measured by these tasks may include shifting attention, monitoring responses, inhibiting behaviorally salient responses, and formulating strategies. In the empirical studies presented in sections 26.2 and 26.3, we will use a dual-task paradigm as an index of analogous control processes that can contribute to working-memory performance in humans.

The advent of neuroimaging technologies in human research has given rise to mounting empirical evidence of the contribution of many cortical regions, including PFC, to working-memory performance (for review, see D'Esposito et al. 1998). Two features of such studies, however, impose constraints on their inferential power with respect to the mnemonic and non-mnemonic processes that contribute to working-memory function. First, many of these studies employ complex working-memory tasks that render them unsuitable for a detailed examination of isolated cognitive processes. Second, neuroimaging studies, by their very nature, support inferences about the *engagement* of a particular brain system by a cognitive process, but not about its *necessity* to that process (Sarter, Bernston, and Cacioppo 1996). Which is to say, neuroimaging studies cannot, alone, tell us whether the activation of a neural system represents a neural substrate of a specific function or a nonessential process associated with that function.[1] Examples of such nonessential processes might include monitoring and detecting errors, regulating attention or vigilance, inhibiting other processes that could potentially compete for the same resources as the process in question, or even affective responses to a particular stimulus or behavior. The inference of necessity cannot be made without demonstrating that the inactivation of a brain system disrupts the function in question. This chapter will therefore emphasize studies in patients with brain lesions to provide the evidence needed to test our hypotheses.

26.1 ANALYSIS OF STUDIES OF PATIENTS WITH FOCAL FRONTAL LESIONS

To determine the contribution of prefrontal cortex to the mnemonic components of working memory, we (D'Esposito and Postle 1999) analyzed the performance of patients with lesions of the lateral PFC on tests of working memory, focusing on published reports of group studies with simple span and delayed-response tasks. We selected these measures because, as stated above, we believe that they offer reasonably direct measures of working-memory storage and rehearsal: they are unconfounded by non-mnemonic cognitive processes that fall under the rubric of "executive control processes."[2] We therefore considered span measures to be indices of the ability to *store* information temporarily, and thus of the capacity or load of working memory; and delayed-response measures, to be indices of the ability to *rehearse* information in an active state over a short period of time. Because non-mnemonic cognitive processes are more likely to be engaged by more complex working-memory tasks such as *n*-back, self-ordered pointing, and sentence and computation span, these tasks are not useful behavioral measures for isolating the mnemonic role of the PFC.

Our review, of the literature from 1960 to 1997 uncovered eight group studies reporting digit span results that used the standardized procedures of the Wechsler Adult Intelligence Scale-Revised (WAIS-R; Wechsler 1981). None of the reports of digit span reported a statistically significant deficit in patients with frontal lobe lesions (total number of patients from the eight studies = 115), as compared to normal control subjects. We also found four studies reporting results on the block span task that was developed by Corsi (Milner 1971) as a spatial analogue of the digit span test. None of these reports of block span reported a statistically significant deficit in patients with frontal lobe lesions (total number of patients from the four studies = 61). It can be seen from figure 26.1 that

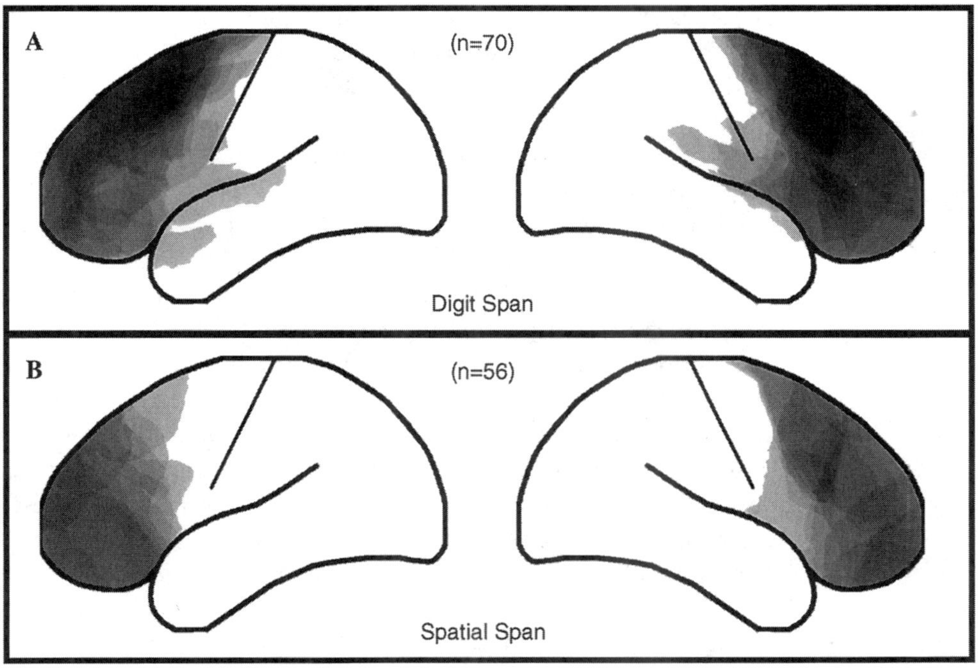

Figure 26.1 Composite diagrams illustrating extent of prefrontal cortex lesions of patients showing no deficit in span performance from: A. four studies of digit span (Canavan et al. 1989; Mangels et al. 1996; Pigott and Milner 1994; and Stuss et al. 1994); and B. three studies of spatial span (Canavan et al. 1989; Miotto et al. 1996; and Owen et al. 1990). To generate these diagrams, we digitized each published individual lesion diagram and superimposed it onto a brain hemisphere template with the other diagrams from the same study, creating two composite diagrams for each study (one for each hemisphere). Each lesion was drawn in a low saturation shade of gray, and thus regions representing overlapping lesions appeared darker than those representing a lesion in one subject. Each composite diagram was then transformed to a two-dimensional brain template in standard stereotaxic space (Talairach and Tournoux, 1988) using a linear scaling procedure (Morph 2.0, Gryphon Software Corporation).

the locations of the lesions in these studies do not appear to spare any portion of the PFC. Thus the consistently spared performance on span tasks cannot be linked reliably to any one spared region of PFC. Importantly, one of the eight studies reporting the span performance of PFC patients also reported data from patients with posterior cortical lesions that spared PFC (Ghent, Mishkin, and Teuber 1962). The patients with posterior lesions, in contrast to patients with prefrontal lesions, were impaired on the test of digit span, with the left-hemisphere group demonstrating the largest impairment. This result is consistent with reports linking impaired digit span performance with lesions of left inferior parietal lobe (Vallar and Papagno 1995).

Our review of studies of performance on delayed-response tasks in patients with PFC lesions encompassed six reports, many featuring multiple experiments that varied stimulus materials, and some that filled the delay interval with distracting stimuli. Patients were significantly impaired relative to normal control subjects in only 3 of the 9 experiments that employed undistracted delay periods, versus 4 of the 6 experiments that featured distraction during the delay period. Thus, in contrast with the span results, our review of the delayed-response literature indicated that there are conditions under which PFC lesions disrupt delayed-response performance.

Our findings also suggest that the dependence of delayed-response performance on prefrontal cortex may increase with distraction during delay periods, perhaps reflecting an increase in information-processing demands. That is, the rehearsal processes that suffice to support undistracted delayed-response performance may require the mediation of PFC-supported processes when distraction during the delay interval presents a source of interference or attentional salience. These PFC-supported processes may include inhibition of prepotent responses (Diamond 1988; Roberts, Hager, and Heron 1994); gating behaviorally irrelevant stimuli (Chao and Knight 1995); shifting attention among stimuli, among different components of a task, or among both (Postle and D'Esposito 1998; Rogers and Monsell 1995); maintaining or refreshing information in a noisy environment (Johnson 1992); and selection among competing responses (Thompson-Schill et al. 1997).

We interpret the findings we reviewed to indicate that working-memory storage processes are independent of PFC integrity, whereas working-memory rehearsal processes can demonstrate greater dependence on PFC integrity. The data reviewed thus far, however, represent only a single dissociation. It could be that delayed-response tasks are simply more difficult than span tests, and therefore more sensitive to PFC damage. On the other hand, the three studies of delayed response that included patients with posterior lesions found no evidence of delayed-response impairment in these patients (Chao and Knight 1995; Ghent, Mishkin, and Teuber 1962; Verin et al. 1993). These results, paired with the report of impaired digit span performance in posterior-lesioned patients (Ghent, Mishkin, and Teuber 1962), form a functional and neuroanatomical double dissociation of storage and rehearsal processes, strengthening our claim that span performance exhibits greater dependence on posterior cortex, whereas delayed-response performance exhibits a greater dependence on PFC.

The finding that humans with PFC lesions can be impaired on delayed-response tasks is consistent with the monkey literature (for review, see Fuster 1997), although among the studies we reviewed, there were several in which humans with PFC lesions were *not* impaired on certain delayed-response tasks (Baldo 1997; Ghent, Mishkin, and Teuber 1962; Prisko 1963; Ptito et al. 1995). There are several possible explanations for this observation. First, this disparity may reflect important differences in the role of PFC in working-memory function across species. Second, it may be that the non-mnemonic demands of the delayed-response task, such as attentional and strategic demands, rely to a greater extent on PFC mediation in the monkey than they do in the human. Third, methodological differences across tests may have contributed to the variability across studies. Finally, differences between studies in the site of the PFC lesions in patients may explain a great deal of the variance in the delayed-response results reviewed here.

Our review of the literature leads us to conclude that working-memory function is not a unitary process, but is comprised of dissociable processes subserved by distinct neural circuitry. We established that working-memory storage is not dependent on PFC function, whereas rehearsal and executive control processes can depend on PFC. Converging evidence from neuropsychological and neuroimaging research is consistent with the model of a functional neuroanatomical dissociation of storage and rehearsal processes that has emerged from our meta-analysis. For example, patients with focal parietal lesions demonstrate markedly reduced performance on digit span tests (Vallar and Papagno 1995; Shallice and Vallar 1990), indicating that short-term storage of verbal material is likely mediated by left inferior parietal cortex. This view is consistent with the results of neuroimaging studies indicating that the storage components of verbal working memory are associated with activation in inferior parietal cortex, whereas the rehearsal components are associated with activation in ventral PFC (Awh et al. 1996; Jonides et al. 1998; Paulesu, Frith, and Frackowiak 1993).

Other studies have observed that working-memory tasks that place demands on the processing or manipulation of information (i.e., control processes) often elicit greater activation in dorsolateral prefrontal cortex (Brodmann's areas 9 and 46), than those that do not place demands on such processes (D'Esposito et al. 1999; Postle, Berger, and D'Esposito 1999; Owen, Evans, and Petrides 1996). These empirical data are consistent with a model of PFC organization as originally proposed by Petrides (1994; see Petrides, chap. 23, this volume).

26.2 NEUROPSYCHOLOGICAL STUDIES OF WORKING MEMORY

We assembled this set of empirical studies to test some of the hypotheses about working-memory storage and rehearsal articulated in section 26.1, as well as to begin probing the non-mnemonic, executive control processes that can contribute to working-memory performance. The measures of storage and rehearsal employed in these studies differ little from those discussed in the previous section. To investigate control processes that can contribute to working memory, we employed a dual-task paradigm. Previous neuroimaging research showed that simultaneous performance of two non-mnemonic tasks (a mental rotation task, and a semantic judgment task), but not performance of either task alone, elicited activation of dorsolateral prefrontal cortex (D'Esposito et al. 1995). We hypothesized that this PFC activation reflected a neural correlate of the operation of the control processes needed to coordinate the successful performance of two tasks simultaneously. Moreover, we assumed that such control processes may also contribute to performance of working-memory tasks that require shifts of attention and coordination among competing behavioral demands (e.g., delayed response with a secondary distracting task, or *n*-back task). The experiments presented in this and the following section permitted us to assess the dependence of these non-mnemonic processes on PFC, and on the neurotransmitter dopamine, and to compare their neural substrates with those of the mnemonic processes of storage and rehearsal.

We studied three groups of subjects: patients with frontal lesions, patients with Parkinson's disease (PD), and normal healthy elderly. PD can affect dopamine in two ways: by disrupting the nigrostriatal system (thereby reducing dopamine delivery to the neostriatum) and by disrupting the mesocortical dopamine systems (thereby reducing delivery of dopamine directly to prefrontal cortex). Nonhuman primate studies have demonstrated that there is a high concentration of dopamine, dopamine receptors, and dopamine-containing terminals in lateral PFC (Brown, Crane, and Goldman 1979), and converging evidence suggests that neurochemical alterations in the dopaminergic neurotransmitter system can cause frontal lobe dysfunction. For example, dopamine depletion (Sawaguchi and Goldman-Rakic 1991) and pharmacological dopamine blockade (Brozoski et al. 1979) cause difficulty with spatial working-memory tasks. Performance on PFC-mediated tasks can also be disrupted by lesions in the caudate nucleus (Battig, Roswold, and Mishkin 1960; Dean and Davis 1959; Divac, Roswold, and Szwarcbart 1967), the neostriatal structure anatomically linked with PFC. Interestingly, several neuropsychological studies have demonstrated cognitive impairments in PD patients that are similar to those found in patients with PFC dysfunction (Owen et al. 1992; Taylor, Saint-Cyr, and Lang 1986), although the cognitive impairments of these groups clearly differ (Owen et al. 1993; see Robbins and Rogers, chap. 21, this volume). Recent event-related fMRI studies in our laboratory (Postle and D'Esposito 1999) have also implicated a role for the caudate nucleus in spatial working-memory function. Because normal aging also decreases dopamine receptor levels in the PFC (de Keyser, Ebinger, and Vauquelin 1990; Rinne, Lonnberg, and Marjamaki 1990; Wong et al. 1984) and has been reported to impair spatial working memory in monkeys (Arnsten et al. 1995), the age of healthy control subjects participating in this study was also treated as an independent variable of interest.

Subjects

We studied two groups of patients and two groups of normal control subjects (NCS) on a range of behavioral tasks. Patients with Parkinson's disease (PD) were recruited from the University of Pennsylvania Medical Center and were all in the earliest stage of their disease (Hoehn and Yahr stage I or II; Hoehn and Yahr 1967). PD patients in the earliest stages of their disease were chosen to avoid patients with dementia, namely, patients having mini–mental state scores of less than 26 (Folstein, Folstein, and McHugh 1975), significant motor disturbance, or both. PD patients (mean age: 66.1 years; mean education: 14.6 years; $n = 25$) and NCS_{PD} (mean age: 67.6 years; mean education: 14.2 years; $n = 25$) were matched for age and education. Patients with traumatic brain injury (TBI) were recruited from Moss Rehabilitation Hospital and were studied at least four weeks after their injury (range: 1 month to 10 years; mean: 34.1 months). TBI was defined as a brain injury causing concussion with loss of consciousness (Glasgow Coma Scale < 8). All TBI patients had evidence of frontal confusions, based on clinical brain scans. (Data from a subset of these patients have been published in McDowell, Whyte, and D'Esposito 1997.) TBI patients (mean age: 34.0 years; mean education: 13.7 years; $n = 30$) and NCS_{TBI} (mean age: 35.4 years; mean education: 15.0 years; $n = 30$) were matched for age and education. NCS participating in this study were recruited from spouses and friends of the patients, as well as from the Philadelphia community at large.

To examine the effects of normal aging, we selected young and old subgroups from the normal controls, by dividing the controls into two equal groups of 30 by age (greater or less than 60 years), and then by dropping young subjects with high span scores and elderly subjects with low span scores from each group until we had young and elderly subgroups, each of 22 subjects, matched in mean span (young group—mean span: 6.4 ± 0.7; mean age: 37.9 years; mean education: 16.4 years; elderly group—mean span: 6.3 ± 0.6; mean age: 72.3 years; mean education: 16.2 years). This was done to ensure that any differences between the two groups on the dependent measures of principal interest could not be ascribed to differences in working-memory capacity.

Behavioral Tasks

Three behavioral paradigms were studied: digit span (Wechsler 1981), spatial delayed-response, and dual-task.

Delayed-Response Task The delayed-response spatial location task was designed to be similar to that developed by Funahashi, Bruce, and Goldman-Rakic (1989) in monkeys. Subjects were required to recall the location of a black dot on a computer monitor after a brief delay. While the subjects were observing a central fixation point, a visual stimulus appeared for 0.2 sec at a peripheral location on an imaginary circle on the screen. This stimulus was presented within 10 degrees of the fixation point (to avoid the subjects' blind spot), excluding locations of 0, 90, 180, and 270 degrees (to avoid referencing to the exact vertical or horizontal). Following the presentation of the stimulus, the screen was blank for 8 sec. An auditory tone signaled end of the delay and prompted subjects to identify the location occupied previously by the stimulus by moving a cursor to that location. Error was assessed as the distance in pixels between the stimulus and the response. Testing was performed in a single block of 40 trials.

Dual-Task Paradigm Subjects first performed a simple visual reaction time task (the primary task), then performed it concurrently with each of two other tasks. In the primary task, subjects responded with a keypress to a target on the computer screen. The target appeared following one of four possible interval delays after the previous response, each used randomly 25% of the time: 0.5 sec., 1.0 sec., 1.5 sec., and 2.0 sec. The target was a sharply demarcated black dot that appeared in one of sixteen dot positions, evenly spaced on the perimeter of two concentric squares, in pseudorandom order, with location counterbalanced. The dot remained on the screen until the subject responded. Performance was measured as the mean reaction time across 64 trials. One secondary task required subjects to count aloud from 1 to 10 repeatedly, at a self-selected rate. This task was selected to make minimal demands on control processes. The other secondary task was an oral digit span task, which was expected to make greater demands on control processes. The number of items used was varied across subjects to match each subject's own "100% correct" digit span, determined as the largest digit span that the subject was able to perform correctly three times consecutively. In this way difficulty of the span task was calibrated across subjects.

Results

Span Performance Compared to their own set of NCS, neither PD nor TBI patients were impaired on digit span: $t(\mathrm{df}, 48) = 1.08$ and $t(\mathrm{df}, 58) = 1.34$, respectively. This is illustrated in the top row of figure 26.2.

Delayed-Response Performance PD patients were not impaired on the delayed-response task compared to NCS_{PD} whereas TBI patients were significantly impaired compared to NCS_{TBI}: $t(\mathrm{df}, 48) = 1.46$ and $t(\mathrm{df}, 58) = -3.54$, $p = 0.001$, respectively. The elderly NCS did not differ in spatial delayed-response performance from the young NCS with whom they were matched for span performance: $t(\mathrm{df}, 40) = -0.46$. These results are illustrated in the middle row of figure 26.2.

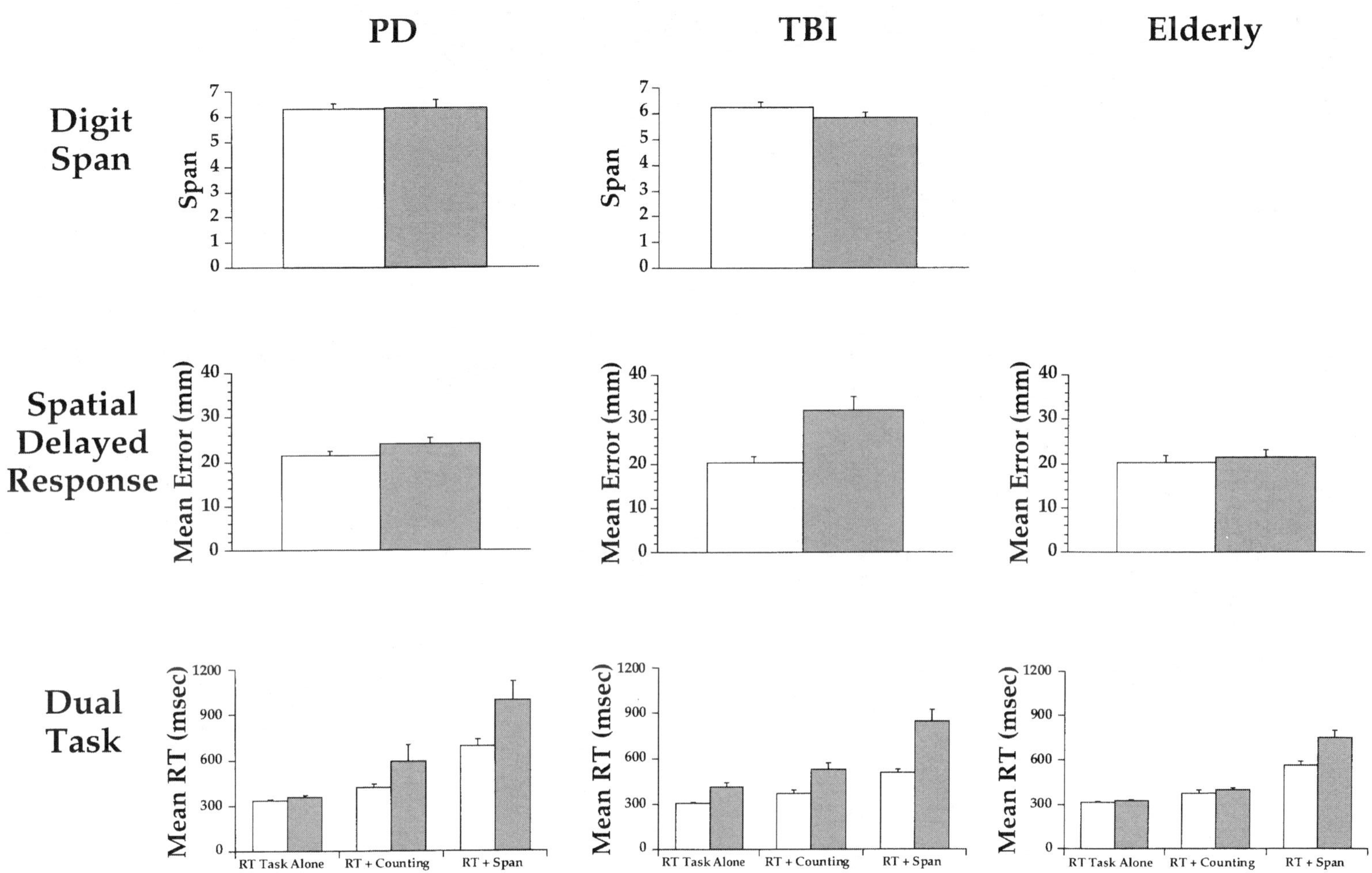

Figure 26.2 Results from behavioral studies of working memory of three experimental groups: Parkinson's disease, traumatic brain injury, and healthy elderly subjects. White bars represent matched control subjects for each group; dark bars represent patient and elderly data. The Elderly and Young control group were matched for digit span (Elderly: 6.3, Young: 6.4), not shown.

Dual-Task Performance The PD, TBI, and elderly groups were compared to the appropriate control groups on the *primary* task performed alone to determine if there was discrepancy in the level of difficulty of this task between groups. There was no significant difference in baseline performance between PD and NCS_{PD} groups or between the elderly and young NCS groups: $t(\text{df}, 48) = -0.99$ and $t(\text{df}, 42) = -1.18$, respectively. There were, however, differences in baseline performance between the TBI and NCS_{TBI} groups: $t(\text{df}, 58) = -3.17$; $p < 0.002$. Thus a repeated-measure analysis of variance (ANOVA) was performed to test for differences in the proportional decrement in primary task performance at the three levels of the dual-task paradigm (alone, concurrent counting, concurrent digit span).

Group $\times$ condition (single-task, dual-task counting, dual-task span) ANOVAs performed for each comparison (i.e., PD versus NCS_{PD}; TBI versus NCS_{TBI}; elderly versus young NCS), using performance on primary task as the dependent measure, revealed significant main effects of group—PD: $F(1, 48) = 5.86$, $p = 0.02$; TBI: $F(1, 58) = 15.96$, $p = 0.0002$; elderly: $F(1, 40) = 7.15$, $p = 0.01$); and of condition—PD: $F(1, 48) = 33.9$, $p = 0.01$; TBI: $F(1, 58) = 78.39$, $p = 0.01$; elderly: $F(1, 48) = 122.84$, $p < 0.0001$); and a significant interaction effect—PD: $F(1, 48) = 8.37$, $p = 0.01$; TBI: $F(1, 58) = 11.55$, $p < 0.0001$; elderly: $F(1, 40) = 8.72$, $p = 0.0004$. Planned *t*-tests revealed that PD patients, TBI patients, and elderly subjects had a significantly greater decrement in performance during concurrent performance of span—PD: $t(48) = -2.51$, $p = 0.02$; TBI: $t(58) = -2.22$, $p = 0.03$; elderly: $t(40) = -2.99$, $p = 0.005$; but not during concurrent articulation—PD: $t(48) = -1.78$; TBI: $t(58) = -0.93$; elderly: $t(40) = -1.08$. These results are illustrated in the bottom row of figure 26.2.

Because TBI patients performed worse than NCS_{TBI} on the primary task, subgroups of TBI patients and NCS_{TBI} ($n = 20$ in each subgroup) were matched for performance on this task. A significant interaction of group and condition was still observed: $F(1, 38) = 5.71$; $p = 0.005$. Planned *t*-tests revealed that TBI patients had a significantly greater decrement in performance during concurrent articulation and concurrent performance of span: $t(38) = 2.42$, $p = 0.02$; $t(38) = 2.54$, $p = 0.01$, respectively.

Performance of digit span concurrent with the primary task (as assessed by number of errors) was worse for PD, TBI, and elderly groups, as compared to their respective control groups, but only one of these comparisons reached statistical significance—PD: $t(48) = 1.60$; TBI: $t(58) = 2.69$, $p = 0.01$; elderly: $t(40) = -1.75$.

Conclusions

The results of performance for TBI patients (i.e., patients with frontal lesions) on the digit span task are consistent with the conclusion of our meta-analysis (section 26.1) that working-memory storage may be independent of prefrontal cortex integrity. Moreover, the performance of Parkinson's disease patients suggests that working-memory storage may not rely on the dopaminergic system (at least at the level of depletion in the early stages of PD). Whether span performance may be independent of the dopaminergic system was tested further in the pharmacological studies described in section 26.3.

Our investigation of spatial delayed-response performance revealed a single dissociation between Parkinson's disease patients (spared) and traumatic brain injury patients (impaired). The disruption of spatial delayed-response performance in the TBI group, contrasted with its intact digit span performance, is consistent with our proposal that working-memory storage and the working memory rehearsal processes required for delayed-response performance are supported by distinct, neuroanatomically dissociable processes. Because however, these two tasks tested different types of information (spatial and verbal, respectively), the performance differences we found may also reflect the difference in stimulus material. We believe that this alternative possibility is unlikely because our meta-analysis (section 26.1) revealed a process-specific pattern of results, with storage of both spatial and verbal information (indexed by span tests) spared in PFC patients, and rehearsal of spatial and nonspatial material (indexed by delayed-response tests) impaired by PFC damage.

The sparing of spatial delayed-response performance in PD patients is at odds with several previous reports of impaired spatial working memory in PD patients (Bradley, Welch, and Dick 1989; Morris et al. 1988; Taylor, Saint-Cyr, and Lang 1986), including three reports of a selective impairment in spatial working memory, as contrasted with preserved nonspatial working memory (Owen et al. 1997; Postle, Jonides, et al. 1997; Postle, Locascio, et al. 1997). One possible source of this discrepancy is methodology. Our task may have been considerably easier than more traditional delayed-response designs (e.g., Postle, Jonides, et al. 1997; Taylor, Saint-Cyr, and Lang 1986) because it did not incorporate a forced-choice decision, but merely required pointing to a location. Delayed-response tasks featuring a forced-choice component may introduce additional discrimination and decision requirements not present in a simple pointing task, such as used in the present study. Another possibility is that the PD groups selected in the different studies are at different stages of the disease. Many of the earlier studies reporting impaired spatial working memory included patients who were at a more advanced stage of the disease than were those in our sample. Regardless of the discrepancy between the present results and those of previous studies, the implication of the present findings for the present line of inquiry is that the maintenance and rehearsal processes engaged by our spatial delayed-response task may be more dependent on PFC integrity than on the integrity of the dopaminergic system.

Viewed in isolation, the dual-task results from this section tell us little about the control processes we assume are engaged by this task, other than to demonstrate that performance on such a task can be impaired even when maintenance processes and storage are normal. For example, PD patients and their corresponding subset of elderly NCS (who were matched on span performance with young NCS) were not impaired on tests of delayed response but were impaired on the dual-task experiment. Thus the control processes engaged by the dual-task paradigm are, themselves, dissociable from working-memory storage and working-memory maintenance processes. Section 26.3 presents experiments intended to help us refine our model of the functional and neural bases of these control processes.

26.3 PHARMACOLOGICAL STUDIES OF WORKING MEMORY

Administration of dopamine receptor agonists, which stimulate dopamine receptors in the same way that dopamine does, also provides a method for examining the role of dopaminergic systems in higher cognitive functions in humans. Most dopamine receptor agonists are relatively selective for a particular receptor subtype, the two most common being the D1 and the D2, although the selectivity of these drugs for cognitive functions is poorly understood. The studies described below employed bromocriptine, a drug relatively selective for the D2 receptor subtype and approved for human use. (Pergolide, another drug used to study human cognition, affects both D1 and D2 subtypes.)

As mentioned earlier, dopamine receptors are found in high densities in the prefrontal cortex. D2 dopamine receptors are present in much lower concentrations in the cortex than D1 receptors, and are localized primarily within the striatum (Camps et al. 1989), whereas D2 receptors are at their highest concentrations in PFC in layer V, which makes them especially well placed to influence PFC function (Goldman-Rakic, Lidow, and Gallager 1990). D1 receptors have also been implicated in mnemonic functions in monkeys (Arnsten et al. 1994), and evidence from animal studies (Arnsten, et al. 1995) suggests that that some PFC functions may depend on a synergistic interaction between these two dopamine receptor subtypes.

We have studied the effects of bromocriptine on the performance of TBI patients on the measures of working-memory storage, rehearsal, and executive control processes presented in section 26.2. We (McDowell, Whyte, and D'Esposito 1998) administered these tests to twenty-four TBI patients two times, on and off bromocriptine, in a double-blind procedure. Because this was a repeated-measure design, we did not test a group of NCS in this experiment. In addition to digit span, spatial delayed-response, and dual-task tests, we administered several traditional clinical measures of executive function including the Stroop test, in

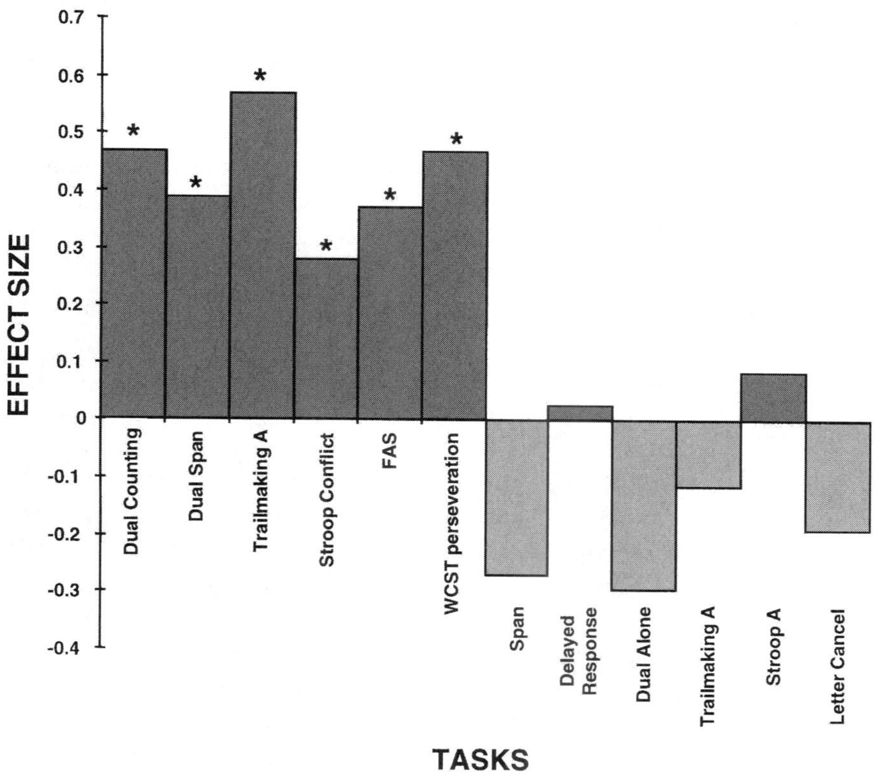

Figure 26.3 For each task, this chart shows the effect size of the performance change with bromocriptine of traumatic brain injury patients. A positive bar (value > 0) indicates improved performance on the drug. Because most dependent measures are time related, where greater values indicate worse performance, the difference for these was calculated as the placebo measurement minus the drug measurement; this difference was reversed (drug measurement minus placebo measurement) for the FAS test and the spatial delayed-response task for which greater values indicate better performance, to maintain consistency of direction with the other measures on this chart. For this graph, upward bars indicate drug benefit. An asterisk represents a significant effect of the administration of bromocriptine. Note that "Dual Alone" (i.e., single task), "Trails A" (nonswitching task), and "Stroop A" (nonconflict task) refer to the control conditions of these tasks.

which subjects are presented with an array of color names printed in different colored inks and are asked to name the ink colors or read the words (Stroop 1935); the Wisconsin Card-Sorting Test (WCST), in which subjects are given a series of cards and asked to sort them according to three different attributes (Nelson 1976; Grant 1948); the Trailmaking Test (Lezak 1995), which requires subjects to alternate between connecting letters and numbers in sequential order; and a verbal fluency test that requires subjects to produce words beginning with the letters *F*, *A*, and *S* (the "FAS" test; Lezak 1995).

We found that bromocriptine, as compared to placebo, improved performance on all measures requiring executive control processes, including the dual-task paradigm, the conflict condition of the Stroop task, the Wisconsin card-sorting task, the switching condition of the Trailmaking

Test, and the FAS test (see figure 26.3). In contrast, performance on the spatial delayed-response and digit span tasks did not improve with bromocriptine (see figure 26.3). Similarly, performance on the biletter cancellation control task and in the baseline conditions of the clinical measures of executive function (i.e. the nonconflict condition of the Stroop task, the nonswitching condition of the Trailmaking Test, the single-task condition of the dual-task paradigm) that assess basic attentional and sensorimotor processes was not improved with bromocriptine (McDowell, Whyte, and D'Esposito 1998). Performance on some tasks (the biletter cancellation test, the Stroop test, and the dual-task paradigm with concurrent digit span) can be measured in terms of time and accuracy, and thus either could be affected by the medication. To make certain that the beneficial effect of bromocriptine on speed for these tasks was not due to a speed-accuracy trade-off, the effect of medication on task accuracy was also assessed. Accuracy was not significantly affected by bromocriptine for any of these tasks, and the nonsignificant changes in function that did occur with bromocriptine were also in the direction of improvement.

These findings demonstrate a selective effect of bromocriptine on tasks that seem to engage executive control processes, as opposed to tasks that do not. The insensitivity of spatial delayed-response performance to bromocriptine is consistent with the result reported in section 26.2, that PD patients did not differ from NCS on this task. These two findings provide converging evidence that rehearsal processes engaged by this particular task may be relatively insensitive to dopaminergic neurotransmission. In a previous study (Kimberg, D'Esposito, and Farah 1997) with young normal subjects, we also found that bromocriptine did not improve performance on the same delayed-response and span tasks. In contrast, the sensitivity of dual-task performance and of the other clinical executive measures to dopamine manipulation indicates that executive control processes recruited by these tasks are sensitive to both PFC integrity and to dopamine levels. Again, our finding that patients with PFC lesions and PD patients are impaired on the same dual-task paradigm provides converging evidence to support this claim.

Another possible explanation for the results of McDowell and colleagues is that the tasks that did not show improvement with bromocriptine (i.e., delayed-response and span) were less sensitive in detecting differences. This is unlikely, however, because the range of performance on these tasks by patients was quite broad. For example, on the spatial delayed-response task, difference scores between sessions ranged from an improvement in spatial error of 22.7 pixels to a decrement of 12.8 pixels (raw data ranged from 8.8 to 68.4 pixels). Likewise, the span task difference scores ranged from an improvement of 6 correct words to a decrement of 10 words (raw data ranged from 16 to 52 words recalled). According to another alternative interpretation, our patients may have been more impaired on tasks that responded to bromocriptine (because they were more difficult), as compared to tasks that did not. Such an explanation seems unlikely because subjects were impaired on *all* of the tasks we administered relative to NCS, except for the Wisconsin card-sorting task, and performance even on this task showed improvement with bromocriptine (McDowell, Whyte, and D'Esposito 1997).

26.4 GENERAL DISCUSSION

The empirical data presented in sections 26.2 and 26.3, which are broadly consistent with the data from patients with prefrontal cortex lesions performing the working-memory tasks we reviewed in section 26.1, encourage us to draw six additional conclusions about the neural bases of processes underlying the functional components of working memory:

1. *Working memory storage, as assessed by simple span performance, is not dependent on PFC integrity nor on the neurotransmitter dopamine.* This conclusion is supported by the observation that neither the traumatic brain injury patients (representative of frontal injury) nor the Parkinson's disease patients (representative of dopamine depletion) were significantly impaired on span tasks, and because administration of a dopamine agonist to TBI patients did not improve span performance. Thus working-memory storage seems likely to be supported by neural networks located in posterior cortex, independent of PFC integrity, and relatively insensitive to manipulations of the neurotransmitter dopamine.

2. *The rehearsal processes engaged by delayed-response tasks, but not by span tasks, are dependent on PFC integrity, but not on the neurotransmitter dopamine.* Traumatic brain injury patients were impaired on a delayed-response task, whereas Parkinson's disease patients were not. Furthermore, administration of a dopaminergic agonist to TBI patients, as compared to a placebo, did not lead to improved delayed-response performance.

3. *The non-mnemonic control processes engaged by dual-task performance are dependent on PFC integrity and on the neurotransmitter dopamine.* This conclusion is supported by the observation that dual-task performance was impaired in both traumatic brain injury and Parkinson's disease patients. Moreover, administration of a dopaminergic agonist to TBI patients, as compared to a placebo, improved dual-task performance. Importantly, bromocriptine did not affect performance on either of the two component tasks of the dual-task paradigm when these tasks were performed individually. It therefore follows that the dopamine dependence can be ascribed to the non-mnemonic control functions critical to dual-task performance. The discrepancy in dopamine dependence between delayed-response performance and dual-task performance indicates a

dissociation between working-memory rehearsal and these nonmnemonic control processes.

Converging evidence for conclusion 3 can also be found from studies of PD patients "off" and "on" their dopaminergic medications, which have also revealed dopamine dependency in measures sensitive to PFC function. For example, PD patients "on" their medication (as compared to "off" it) have been shown to perform better on a wide range of executive function tasks such as the Wisconsin Card-Sorting Test, verbal fluency, and the Tower of London task (Bowen et al. 1975; Cooper et al. 1992; Lange et al. 1992, 1995). One study in particular revealed a behavioral dissociation reminiscent of some of the results presented earlier in this chapter: PD patients "on" their medication displayed improvement on executive measures, but not on tests of working-memory span for verbal and spatial information (Lange et al. 1995).

Thus far, we have interpreted the insensitivity of delayed-response performance to dopamine manipulations as evidence of the independence of working-memory rehearsal processes from dopaminergic neurotransmission. An alternative view of the dissociation of TBI and PD delayed-response performance might be that it merely reflects the graded effects of increasingly severe lesions. That is, a direct lesion to prefrontal cortex (TBI) might be expected to have a greater impact on *any* PFC-dependent process than a depletion of dopamine in the frontostriatal system (PD). The insensitivity of delayed-response performance in TBI patients administered a dopaminergic agonist, however, paired with the sensitivity of TBI dual-task performance to this same manipulation, is difficult to reconcile with an explanation based on graded difficulty. Nevertheless, the relationship between dopamine and PFC function is clearly complex: other investigators have found a relationship between dopamine administration and delayed-response performance in normal human subjects (Luciana and Collins 1997; Luciana et al. 1992; Müller, Pollman, and van Cramon 1998) as well as impaired performance on spatial delayed-response tasks in PD patients (Freedman and Oscar-Berman 1986; Postle, Jonides, et al. 1997).

Related to conclusions 1–3 are three others central to our proposed model:

4. *Storage processes are dissociable from rehearsal processes in working memory.*

5. *Storage processes are dissociable from executive control processes in working memory.*

6. *Rehearsal processes are dissociable from executive control processes in working memory.*

The data presented in this chapter are consistent with a model of working memory that emphasizes the distributed nature of the cognitive and neural architecture of the processes underlying working memory. Complex interactions between these anatomically, pharmacologically, and functionally dissociable processes enable the short-term retention and manipulation of information, functions that contribute importantly to control.

NOTES

This work was supported by National Institutes of Health grants NS01762 and AG13483 and by the American Federation for Aging Research.

1. This observation applies equally to all methods of physiological measurement, including single- and multiunit electrophysiology, EEG, MEG, hemodynamic measures, and measures of glucose metabolism.

2. We distinguish *non-mnemonic* control processes, as discussed in sections 26.2 and 26.3, from the control processes that are likely involved directly in the support of span and delayed-response performance (Kieras et al. 1999).

REFERENCES

Arnsten, K. T., Cai, J. X., Murphy, B. L., and Goldman-Rakic, P. S. (1994). Dopamine D1 receptor mechanisms in the cognitive performance of young adult and aged monkeys. *Psychopharmacology, 116,* 143–151.

Arnsten, A. F. T., Cai, J. X., Steere, J. C., and Goldman-Rakic, P. S. (1995). Dopamine D2 receptor mechanisms contribute to age-related cognitive decline: The effects of quinpirole on memory and motor performance in monkeys. *Journal of Neuroscience, 15,* 3429–3439.

Awh, E., Jonides, J., Smith, E. E., Schumacher, E. H., Koeppe, R. A., and Katz, S. (1996). Dissociation of storage and rehearsal in verbal working memory: evidence from PET. *Psychological Science, 7,* 25–31.

Baddeley, A. D. (1990). *Human Memory: Theory and Practice.* London: Lawrence Erlbaum.

Baddeley, A. D., Thompson, N., and Buchanan, M. (1975). Word length and the structure of short-term memory. *Journal of Verbal Learning and Verbal Behavior, 14,* 575–589.

Baldo, J. V. (1997). Spatial and object working memory in patients with lateral prefontal cortex lesions. Ph.D. diss., University of California, Berkeley.

Battig, K., Rosvold, H. E., and Mishkin, M. (1960). Comparison of the effects of frontal and caudate lesions on delayed response and alternation in monkeys. *Journal of Comparative and Physiological Psychology, 53,* 400–404.

Bowen, F. P., Kamienny, R. S., Burns, M. M., and Yahr M. D. (1975). Parkinsonism: Effects of levodopa treatment on concept formation. *Neurology, 25,* 701–704.

Bradley, V. A., Welch, J. L., and Dick, D. J. (1989). Visuospatial working memory in Parkinson's disease. *Journal of Neurology, Neurosurgery, and Psychiatry, 52,* 1228–1235.

Brown, R. M., Crane, A. M., and Goldman, P. S. (1979). Regional distribution of monoamines in the cerebral cortex and subcortical structures of the rhesus monkey: Concentrations and in vitro synthesis rates. *Brain Research, 168,* 133–150.

Brozoski, T. J., Brown, R. M., Rosvold, H. E., and Goldman, P. S. (1979). Cognitive deficit caused by regional depletion of dopamine in prefrontal cortex of rhesus monkey. *Science, 205,* 929–932.

Camps, M., Cortés, R., Gueye, B., Probst, A., and Palacios, J. M. (1989). Dopamine receptors in human brain: Autoradiographic distribution of D1 sites. *Neuroscience, 28,* 275–290.

Canavan, A. G. M., Passingham, R. E., Marsden, C. D., Quinn, N., Wyke, M., and Polkey, C. E. (1989). Sequencing ability in Parkinsonians, patients with frontal lobe lesions and patients who have undergone unilateral temporal lobectomies. *Neuropsychologia, 27,* 787–798.

Chao, L. L., and Knight, R. T. (1995). Human prefrontal lesions increase distractibility to irrelevant sensory inputs. *NeuroReport, 6,* 1605–1610.

Collins, P., Roberts, A. C., Dias, R., Everitt, B. J., and Robbins, T. W. (1998). Perseveration and strategy in a novel spatial self-ordered sequencing task for nonhuman primates: Effects of excitotoxic lesions and dopamine depletions of the prefrontal cortex. *Journal of Cognitive Neuroscience, 10,* 332–354.

Cooper, J. A., Sagar, H. J., Doherty, M., Jordan, N., Tidswell, P., and Sullivan, E. V. (1992). Different effects of dopaminergic and anticholingergic therapies on cognitive and motor function in Parkinson's disease. *Brain, 115,* 1701–1725.

Dean, W. H., and Davis, G. D. (1959). Behavior changes following caudate lesions in rhesus monkey. *Journal of Neurophysiology, 22,* 525–537.

de Keyser, J., Ebinger, G., and Vauquelin, G. (1990). Age-related changes in the human nigrostriatal dopaminergic system. *Annals of Neurology, 27,* 157–161.

D'Esposito, Aguirre, G. K., Zarahn, E., and Ballard, D. (1998). Functional MRI studies of spatial and non-spatial working memory. *Cognitive Brain Research, 7,* 1–13.

D'Esposito, M., Detre, J. A., Alsop, D. C., Shin, R. K., Atlas, S., and Grossman, M. (1995). The neural basis of the central executive system of working memory. *Nature, 378,* 279–281.

D'Esposito, M., and Postle, B. R. (1999). The dependence of span and delayed-response performance on the prefrontal cortex. *Neuropsychologia, 37,* 1303–1315.

D'Esposito, M., Postle, B. R., Ballard, D., and Lease, J. (1999). Maintenance versus manipulation of information held in working memory: An event-related fMRI study. *Brain and Cognition, 41,* 66–86.

Diamond, A. (1988). Differences betweeen adult and infant cognition: Is the crucial variable presence or absence of language? In L. Weiskrantz (Ed.), *Thought without language,* pp. 337–370. Oxford: Oxford University Press.

Divac, L., Rosvold, L. E., and Szwarcbart, M. K. (1967). Behavioral effects of selective ablation of the caudate nucleus. *Journal of Comparative and Physiological Psychology, 63,* 184–90.

Duncan, J., Burgess, P., and Emslie, H. (1995). Fluid intelligence after frontal lobe lesions. *Neuropsychologia, 33,* 261–268.

Feigenbaum, E. A., and Simon, H. A. (1961). Performance of a reading task by an elementary perceiving and memorizing program. *The RAND Corporation Paper, P-2358.*

Funahashi, S., Bruce, C., and Goldman-Rakic, P. (1993). Dorsolateral prefrontal lesions and oculomotor delayed-response performance: Evidence for mnemonic "scotomas." *Journal of Neuroscience, 13,* 1479–1497.

Folstein, M., Folstein, S., and McHugh, P. (1975). "Mini mental state": A practical method for grading the cognitive state of patients for the clinician. *Journal of Psychiatric Research, 12,* 189–198.

Freedman, M., and Oscar-Berman, M. (1986). Selective delayed response deficits in Parkinson's and Alzheimer's disease. *Archives of Neurology, 43,* 886–890.

Funahashi, S., Bruce, C. J., and Goldman-Rakic, P. S. (1989). Mnemonic coding of visual space in the monkey's dorsolateral prefrontal cortex. *Journal of Neurophysiology, 61,* 331–349.

Fuster, J. (1997). *The prefrontal cortex: Anatomy, physiology, and neuropsychology of the frontal lobes.* 3d ed. Raven Press: New York.

Fuster, J. M., and Alexander, G. E. (1971). Neuron activity related to short-term memory. *Science, 173,* 652–654.

Ghent, L., Mishkin, M., and Teuber, H.-L. (1962). Short-term memory after frontal lobe injury in man. *Journal of Comparative and Physiological Psychology, 5,* 705–709.

Godfrey, H., and Knight, R. (1985). Cognitive rehabilitation of memory functioning in amnesic alcoholics. *Journal of Clinical and Experimental Neuropsychology, 8,* 292–312.

Goldman-Rakic, P. S., Lidow, M. S., and Gallager, D. W. (1990). Overlap of dopaminergic, adrenergic, and serotoninergic receptors and complementarity of their subtypes in primate prefrontal cortex. *Journal of Neuroscience, 10,* 2125–2138.

Grant, D. A., and Berg, E. A. (1948). A behavioral analysis of degree of reinforcement and ease of shifting to new responses in a Weigl-type card-sorting problem. *Journal of Experimental Psychology, 38,* 404–411.

Hoehn, M. D., and Yahr, M. M. (1967). Parkinsonism: Onset, progression, and mortality. *Neurology, 17,* 427–442.

Johnson, M. K. (1992). MEM: Mechanisms of recollection. *Journal of Cognitive Neuroscience, 4,* 268–280.

Jonides, J., Schumacher, E. H., Smith, E. E., Koeppe, R. A., Awh, E., Reuter-Lorenz, P. A., Marshuetz, C., and Willis, C. R. (1998). The role of parietal cortex in verbal working memory. *Journal of Neuroscience, 18,* 5026–5034.

Kieras, D. E., Meyer, D. E., Mueller, S., and Seymour, T. (1999). Insights into working memory from the perspective of the EPIC architecture for modeling skilled perceptual-motor and cognitive human performance. In A. Miyake and P. Shah (Eds.), *Models of Working Memory: Mechanisms of Active Maintenance and Executive Control* (pp. 183–223). Cambridge, UK: Cambridge University Press.

Kimberg, D., D'Esposito, M., and Farah, M. (1997). Effects of bromocriptine on human subjects depend on working memory capacity. *NeuroReport, 8,* 3581–3585.

Lange, K. W., Paul, G. M., Naumann, M., and Gsell, W. (1995). Dopaminergic effects of cognitive performance in patients with Parkinson's disease. *Journal of Neural Transmission Supplement, 46,* 423–432.

Lange, K. W., Robbins, T. W., Marsden, C. D., James, M., Owen, A. M., and Paul, G. M. (1992). L-dopa withdrawal in Parkinson's disease selectively impairs cognitive performance in tests sensitive to frontal lobe dysfunction. *Psychopharmacology, 107,* 394–404.

Levy, B. A. (1971). The role of articulation in articulatory and visual short-term memory. *Journal of Verbal Learning and Verbal Behavior, 10,* 123–132.

Lezak, M. (1995). *Neuropsychological assessment.* 3d ed. New York: Oxford University Press.

Longoni, A. M., Richardson, J. T. E., and Aiello, A. (1993). Articulatory rehearsal and phonological storage in working memory. *Memory and Cognition, 21,* 11–22.

Luciana, M., and Collins, P. F. (1997). Dopaminergic modulation of working memory for spatial but not object cues in normal humans. *Journal of Cognitive Neuroscience, 9,* 330–347.

Luciana, M., Depue, R. A., Arbisi, P., and Leon, A. (1992). Facilitation of working memory in humans by a D2 dopamine receptor agonist. *Journal of Cognitive Neuroscience, 4,* 58–68.

Mangels, J. A., Gershberg, F. B., Shimamura, A. P., and Knight, R. T. (1996). Impaired retrieval from remote memory in patients with frontal lobe damage. *Neuropsychology, 10,* 32–41.

McDowell, S., Whyte, J., and D'Esposito, M. (1997). Working memory impairments in traumatic brain injury: Evidence from a dual-task paradigm. *Neuropsychologia, 35,* 1341–1353.

McDowell, S., Whyte, J., and D'Esposito, M. (1998). Differential effect of a dopaminergic agonist on prefrontal function in traumatic brain injury patients. *Brain, 121,* 1155–1164.

Miller, G. A., Galanter, E., and Pribram, K. H. (1960). *Plans and the structure of behavior.* New York: Henry Holt.

Milner, B. (1971). Interhemispheric differences in the localization of pyschological processess in man. *British Medical Bulletin, 27,* 272–277.

Miotto, E. C., Bullock, P., Polkey, C. E., and Morris, R. G. (1996). Spatial working memory and strategy formation in patients with frontal lobe excisions. *Cortex, 32,* 613–630.

Morris, R. G., Downes, J. J., Sahakian, B. J., Evenden, J. L., Heald, A., and Robbins, T. W. (1988). Planning and spatial working memory in Parkinson's disease. *Journal of Neurology, Neurosurgery, and Psychiatry, 51,* 757–766.

Müller, U., Pollmann, S., and von Cramon, D. Y. (1998). D1 versus D2-receptor stimulation of visuo-spatial short-term memory. *Journal of Neuroscience, 18,* 2720–2728.

Murray, D. J. (1968). Articulation and acoustic confusability in short-term memory. *Journal of Experimental Psychology, 78,* 679–684.

Nelson, H. E. (1976). A modified card sorting test sensitive to frontal lobe defects. *Cortex, 12,* 313–324.

Newell, A., Shaw, J. C., and Simon, H. A. (1958). Elements of a theory of human problem solving. *Psychological Review, 65,* 151–166.

Owen, A. M., Downes, J. J., Sahakian, B. J., Polkey, C. E., and Robbins, T. W. (1990). Planning and spatial working memory following frontal lobe lesions in man. *Neuropsychologia, 28,* 1021–1034.

Owen, A. M., Evans, A. C., and Petrides, M. (1996). Evidence for a two-stage model of spatial working memory processing within the lateral frontal cortex: A positron-emission tomography study. *Cerebral Cortex, 6,* 31–38.

Owen, A. M., Iddon, J. L., Hodges, J. R., Summers, B. A., and Robbins, T. W. (1997). Spatial and non-spatial working memory at different stages of Parkinson's disease. *Neuropsychologia, 35,* 519–532.

Owen, A. M., James, M., Leigh, P. N., Summers, B. A., Marsden, C. D., Quinn, N. P., Lange, K. W., and Robbins, T. W. (1992). Fronto-striatal cognitive deficits at different stages of Parkinson's disease. *Brain, 115,* 1727–1751.

Owen, A. M., Roberts, A. C., Hodges, J. R., Summers, B. A., Polkey, C. E., and Robbins, T. W. (1993). Contrasting mechanisms of impaired attentional set shifting in patients with frontal lobe damage or Parkinson's disease. *Brain, 116,* 1159–1175.

Paulesu, E., Frith, C. D., and Frackowiak, R. S. J. (1993). The neural correlates of the verbal component of working memory. *Nature, 362,* 342–345.

Petrides, M. (1982). Motor conditional associative-learning after selective prefrontal lesions in the monkey. *Behavioural Brain Research, 5,* 407–413.

Petrides, M. (1985). Deficits in non-spatial conditional associative learning after periarcuate lesions in the monkey. *Behavioural Brain Research, 16,* 95–101.

Petrides, M. (1991). Monitoring of selections of visual stimuli and the primate frontal cortex. *Proceedings of the Royal Society of London B246*, 293–298.

Petrides, M. (1994). Frontal lobes and working memory: Evidence from investigations of the effects of cortical excisions in nonhuman primates. In F. Boller and J. Grafman (Eds.), *Handbook of neuropsychology*, pp. 59–82. Amsterdam: Elsevier.

Pigott, S., and Milner, B. (1994). Capacity of visual short-term memory after unilateral frontal or anterior temportal lobe resection. *Neuropsychologia, 32*, 969–981.

Postle, B. R., Berger, J. S., and D'Esposito, M. (1999). Functional neuroanatomical double dissociation of mnemonic and executive control processes contributing to working memory performance. *Proceedings of the National Academy of Sciences (USA), 96*, 12959–12964.

Postle, B. R., and D'Esposito, M. (1998). Homologous cognitive mechanisms and neural substrates underlie dissociable components of set-shifting and task-switching phenomena. *Society for Neuroscience Abstracts, 24*, 507.

Postle, B. R., and D'Esposito, M. (1999). Dissociation of human caudate nucleus activity in spatial and nonspatial working memory. *Cognitive Brain Research, 8*, 107–115.

Postle, B. R., Jonides, J., Smith, E. E., Corkin, S., and Growdon, J. H. (1997). Spatial, but not object, delayed response is impaired in early Parkinson's disease. *Neuropsychology, 11*, 171–179.

Postle, B. R., Locascio, J. J., Corkin, S., and Growdon, J. H. (1997). The time course of spatial and object visual learning in early Parkinson's disease. *Neuropsychologia, 35*, 1413–1422.

Pribram, K. H., Ahumada, A., Hartog, J., and Roos, L. (1964). A progress report on the neurological processes disturbed by frontal lesions in primates. In J. M. Warren and K. Akert (Eds.), *The frontal granular cortex and behavior*, pp. 28–55. New York: McGraw-Hill.

Prisko, L.-H. (1963). Short-term memory in focal cerebral damage. Ph.D. diss., McGill University.

Ptito, A., Crane, J., Leonard, G., Amsel, R., and Caramanos, Z. (1995). Visual-spatial localization by patients with frontal-lobe lesions invading or sparing area 46. *NeuroReport, 6*, 1781–1784.

Rinne, J. O., Lonnberg, P., and Marjamaki, P. (1990). Age-dependent decline in human brain dopamine D1 and D2 receptors. *Brain Research, 508*, 349–352.

Roberts, R. J., Hager, L. D., and Heron, C. (1994). Prefrontal cognitive processes: Working memory and inhibition in the antisaccade task. *Journal of Experimental Psychology: General, 123*, 374–393.

Rogers, R. D., and Monsell, S. (1995). Costs of a predictable switch between simple cognitive tasks. *Journal of Experimental Psychology: General, 124*, 207–231.

Sarter, M., Bernston, G., and Cacioppo, J. (1996). Brain imaging and cognitive neuroscience: Toward strong inference in attributing function to structure. *American Psychologist, 51*, 13–21.

Sawaguchi, T., and Goldman-Rakic, P. S. (1991). D1 dopamine receptors in prefrontal cortex: Involvement in working memory. *Science, 251*, 947–950.

Schweickert, R., Guentert, L., and Hersberger, L. (1990). Phonological similarity, pronunciation rate, and memory span. *Psychological Science, 27*, 74–77.

Shallice, T. and Vallar, G. (1990). The impairment of auditory-verbal short-term storage. In G. Vallar and T. Shallice (Eds.), *Neuropsychological impairments of short-term memory*, pp. 11–53. New York: Cambridge University Press.

Smith, E. E., and Jonides, J. (1998). Neuroimaging analyses of human working memory. *Proceedings of the National Academy of Sciences, U.S.A., 95,* 12061–12068.

Stroop, J. R. (1935). Studies of interference in serial verbal reactions. *Journal of Experimental Psychology, 18,* 643–662.

Stuss, D. T., Alexander, M. P., Palumbo, C. L., Buckle, L., Sayer, L., and Pogue, J. (1994). Organizational strategies of patients with unilateral or bilateral frontal lobe injury in word list learning tasks. *Neuropsychology, 8,* 355–373.

Talairach, J., and Tournoux, P. (1988). *Co-planar stereotaxic atlas of the human brain.* New York: Thieme.

Taylor, A. E., Saint-Cyr, J. S., and Lang, A. E. (1986). Frontal lobe dysfunction in Parkinson's disease: The cortical focus of neostriatal outflow. *Brain, 109,* 279–292.

Thompson-Schill, S. L., D'Esposito, M., Aguirre, G. K., and Farah, M. J. (1997). Role of left inferior prefrontal cortex in retrieval of semantic knowledge: A reevaluation. *Proceedings of the National Academy of Sciences, U.S.A., 94,* 14792–14797.

Vallar, G., and Baddeley, A. D. (1984) Fractionation of working memory: Neuropsychological evidence for a phonological short-term store. *Journal of Verbal Learning and Verbal Behavior, 23,* 151–161.

Vallar, G., and Papagno, C. (1995). Neuropsychological impairments of short-term memory. In A. D. Baddeley, B. A. Wilson, and F. N. Watts (Eds.), *Handbook of memory disorders,* pp. 135–165. New York: Wiley.

Verin, M., Partiot, A., Pillon, B., Malapani, C., Agid, Y., and Dubois, B. (1993). Delayed response tasks and prefrontal lesions in man: Evidence for self-generated patterns of behavior with poor environmental modulation. *Neuropsychologia, 31,* 1379–1396.

Wechsler, D. (1981). *WAIS-R manual.* New York: Psychological Corporation.

Wechsler, D., and Stone, C. P. (1945). *Wechsler Memory Scale.* New York: The Psychological Corporation.

Wiegersma, S., van der Scheer, E., and Human, R. (1990). Subjective ordering, short-term memory, and the frontal lobes. *Neuropsychologia, 28,* 95–98.

Wong, D. F., Wagner, H. N., Dannals, R. F., Links, J. M., Frost, J. J., Ravert, H. T., Wilson, A. A., Rosenbaum, A. E., Gjedde, A., Douglass, K. H., Petronis, J. D., Folstein, M. F., Toung, J. K. T., Burns, H. D., and Kuhar, M. J. (1984). Effects of age on dopamine and serotonin receptors measured by positron tomography in the living human brain. *Science, 226,* 1393–1396.

27 Visual Affordances and Object Selection

M. Jane Riddoch, Glyn W. Humphreys, and
Martin G. Edwards

ABSTRACT Neuropsychological evidence indicates that actions may be evoked directly by visually presented objects. Such actions are affected by learned association with objects and by congruency between the parts of objects and (1) the goal state of the actor; and (2) the effectors used for action ("affordance"). Patients who are unable to conform to a task rule and who show aspects of frontal lobe utilization behavior can be shown to make inappropriate actions in response to objects, actions that are influenced both by object-action associations and by affordances, although such patients remain able to appropriately select objects for action. Thus the processes involved in selecting a visual object for action appear to precede and to be separated from those involved in selecting a given hand with which to respond (according to a predefined task rule). Further data suggest that once an object is selected for a manual reaching action, other objects in the trajectory of the reach compete for the action. This secondary stage of competition may be useful for navigating between objects in neurologically intact individuals, but can lead to difficulties for patients with problems in action selection.

Affordances are potential complementary relations between an organism and its environment, reflecting whether an object or an object's parts might be effective for goal-directed action (Gibson 1979; Shaw and Turvey 1981). Affordances based on the parts of objects may exist even when the object is unfamiliar. For instance, the affordance of a flint tool will be based on complementary relations between its parts (a graspable section, a sharp edge), the actions that may be conducted by means of these parts (grasping and then cutting with the edge), and the goal of the actor (to cut food). In addition, any one object may potentiate a number of different actions according to the goal state of the organism (the tool may afford cutting when the actor is hungry but throwing when angry). Goal states thus play a crucial role in determining which aspects of objects are relevant for behavior. This may impact on performance in several ways. For instance, when multiple objects are present, behavior would be most efficient if only stimuli relevant for actions were selected, in preference to stimuli that were irrelevant. Goal states should play a part in selecting objects for action. In addition, the actions selected in response to any objects should be consistent with the goals of a given task (having selected the flint tool, the actor should use it for cutting rather than throwing, provided the actor is hungry and the food close to hand).

Thus goal states should also determine the selection of action from objects. This chapter is concerned with the role of affordances—the complementary relations between objects (in particular, their structure) and at least some subset of behavioral goals—in both selecting the object for an action and in selecting the action appropriate to the object.

In everyday life, people occasionally either select the wrong object for an action (e.g., picking up their neighbor's rather than their own dinner roll) or the wrong action for an object (using a knife as if it were a spoon). Such "action errors" are usually (and fortunately) rare, but in each case they can be elicited when visual cues are partially consistent with the goal of the behavior, and when we fail to pay "full attention" to the task at hand (see Reason 1979, 1984). "Full attention" might correspond here to something like the appropriate setting of the goal structure for a task, so that only behaviorally relevant objects and actions are selected. In physiological terms, this may mean establishing appropriate templates for a task in the frontal lobes, which modulate both the selection of the target object from other objects present and the selection of any subsequent action (see Miller, chap. 22, this volume). Forcing subjects to respond under speeded-deadline conditions can increase the incidence of action errors, presumably because responses are then elicited based on partial activation of templates for actions (Rumiati and Humphreys 1998). In more complex everyday situations, errors may also reflect activation of some subset, but not all, of the task goals. Because action errors under deadline conditions tend to be related to the visual rather than other (e.g., semantic) properties of objects, and because they occur when objects—but not when words—are presented (Rumiati and Humphreys 1998), action templates may be directly activated by visual-structural properties of objects. Such activation may reflect the affordance of the objects for action.

Interestingly, abnormally large numbers of action errors have been reported in patients with frontal lobe lesions (Humphreys and Forde 1998; Humphreys, Forde, and Francis, chap. 18, this volume). Indeed, patients with frontal lobe damage are often described as "overresponsive" to environmental cues, and as lacking goal-based inhibition of actions activated by such cues. Luria, for instance, described a patient who "on seeing the button operating a bell, was involuntarily drawn to it and pressed it, and when the nurse came in response to the bell, he was unable to say why he had done so" (Luria 1973, 200). Such "utilization behaviors" (Lhermitte 1983) may occur when affordances within the environment are not modulated appropriately by goal-based structures (though there may be activation of some subset of goals, sufficient to produce the affordance). The precise factors that generate utilization behaviors remain poorly understood; detailed study of the conditions under which such behaviors are elicited can inform us about the role of affordances in selecting both the objects for action and the actions that are carried out.

The failure to "control" behavior according to the goals of the task is also observed in patients with "anarchic hand syndrome" (Della Sala, Marchetti, and Spinnler, 1991, 1994; Marchetti and Della Sala 1998), which may be defined as "the occurrence of movements of an upper limb that are unintended although clearly directed to some purpose. The "anarchic hand" seems to act autonomously, carrying out complex movements against the subject's verbally reported will, that can interfere with the development of an intentional action that the other hand has begun" (Della Sala, Marchetti, and Spinnler 1991, p. 1113). Anarchic hand syndrome is associated with anterior lesions of the corpus callosum and of the medial frontal cortex (Della Sala, Marchetti, and Spinnler 1991, 1994).

Recently we had the opportunity to study two patients whose actions in response to objects seem to be associated with and afforded by the objects, but who seem poor at inhibiting such actions when they are inappropriate to the task. This problem is manifested in action errors, where patients fail to select the appropriate effector required by the task when a competing action is activated for the other effector. We examine the behavior of such patients in an attempt to understand the relations between the selection of an action in response to an object (e.g., when we present a single object that can be used in several ways and by different limbs) and the selection of objects for action (e.g., when several objects are present).

27.1 CASE 1: USE OF THE WRONG EFFECTOR IN ANARCHIC HAND SYNDROME

Riddoch and collaborators (1998) attempted a systematic assessment of the factors underlying inappropriate hand responses in E.S., a female patient with anarchic hand syndrome. Over the preceding five-year period, E.S. had a history of increased clumsiness in both arms and increasing inability to perform activities of daily living such as dressing or managing a knife and fork. There was no known precipitating injury, and the symptoms had a gradual onset. She was 59 years old at the time of testing. On examination, her muscle strength was found to be normal, but tactile sensation and proprioception were bilaterally impaired. MRI suggested some changes to the posterior centrum-semiovale and the corpus callosum on the left due to small vessel disease (see Riddoch et al. 1998). Features of involuntary limb activity were apparent in both the dominant (right) and the nondominant limbs. E.S. reported that her right hand would sometimes undertake spontaneous, purposeful movements that interfered with the activities of her left hand. She was aware of these movements, but was unable to inhibit them, indicating that these movements may be described as those of an "anarchic" rather than "alien hand" (Della Sala, Marchetti, and Spinnler 1991, 1994). Intermanual conflict was also a feature of the left hand; on occasion, when E.S. was asked to perform a task with her right hand, the left hand would grip her right arm and would not let it go.

Our initial experiments focused on the factors determining the selection of an effector when E.S. was required to use only the left or the right hand to fulfil the goals of the task (Riddoch et al. 1998). A simple task rule applied to all the experiments: the left hand was to be used to respond to stimuli presented on the left of the patient, and the right hand to stimuli on the right side. E.S. was aware of the task rule, and responded verbally with the appropriate task rule when prompted. Targets were presented randomly to either left or right sides. These could be LEDs, cups, or cuplike nonobjects (plastic blocks stuck together to create a cylinder with a handle). Some items had an associated learned hand response (such as cups), some had a learned response but were placed in an unfamiliar orientation (e.g., upside-down cups), and some had no learned response but might elicit an affordance based on their parts or similarity to known objects (e.g., cuplike nonobjects). The task goals were either to point or to reach and pick up the stimuli. The stimuli were also positioned so that either the handle was compatible with the hand required by the task rule for the response (e.g., cup left, handle left) or it was incompatible with the hand required by the task rule, but compatible with the opposite hand (e.g., cup left, handle right). Irrespective of whether the object was familiar or unfamiliar, or placed in its normal or inverted orientation, the grasp response (when required) was similar: it involved a precision grip between the thumb and forefinger. The grasp response to stimuli when the handle was compatible with the effector was somewhat easier to make than when the handle was incompatible with the effector because the hand was turned away from the mass of the stimulus in the latter condition, but this held true whether or not the stimulus was a familiar object or in a familiar orientation.

We found that, presented with cups, E.S. made numerous errors in selecting the correct effector to use when the handle of the cup was incompatible with the hand demanded by the task rule (e.g., she would use her right hand to pick up the left-side cup with a right-side handle). Although these effector errors were made as often with her right as with her left hand, their frequency was affected by the task and by the familiarity of both the stimulus and its orientation. When pointing rather than grasping was required, left-hand effector errors still occurred (i.e., when using the left hand while pointing to the right cup; these also occurred while pointing to a right LED), whereas right-hand errors were eliminated. Thus changing the task goal had a moderating effect on response errors with the right hand. Effector errors were also reduced when we used cuplike nonobjects and when we used inverted rather than upright cups (Riddoch et al. 1998). These results were stable across repeated testing in the same conditions. Also, although kinematic data were not recorded, E.S. showed no signs of hesitation in her actions, and this held for all the conditions.

These effects of both the familiarity of the object and its orientation cannot be explained in terms of the difficulty in grasping stimuli in the incompatible condition because the grasping action was similar with upright cups, cuplike nonobjects, and inverted cups. Instead, the results suggest that there are effects of two factors on E.S.'s ability to select the correct effector for the task: (1) learned object-action associations; and (2) compatibility between object parts and a particular hand. The influence of both factors was modulated by the intended action (pick up versus point). Object-action associations are apparent in the strong effects with cups, although the remaining effects with nonobjects, and the effect of object orientation even with familiar objects, correspond much more to something like an affordance (cf. Gibson 1979). Performance is affected by the position of a graspable part relative to an effector, depending on the goal state of the actor (pick up versus point). The data provide evidence for the psychological reality of affordances, based on congruence between object structure and the task goal. Not all goal states are effective for E.S., however; otherwise, she would not make effector errors (which transgress the task rules). Performance appears to break down when the task rule is relatively novel, and when the stimulus corresponds to a subset, but not all, of the task goals (here the novel task requires both that grasp responses be made and that the hand used be specific to the location of the stimulus). Presumably, when pointing is required, grasping responses are not a subset of the goal states, and thus do not get activated.

How do the effector errors we have elicited relate to other forms of pathological behavior found in neuropsychological patients? Consider first anarchic hand syndrome. E.S. showed no awareness of making incorrect responses on our task (one of the critical defining features of anarchic hand syndrome). This suggests that the errors we elicited may arise from a source separate from that of her action errors in everyday life (which she showed awareness of). On the other hand, the consequences of some of the action errors that befell E.S. in everyday life could be severe, although in our experiments there were no adverse consequences for using the incorrect hand, and, as we have noted, a subset of the task goals were fulfilled even when this occurred: E.S. picked up the target object. Speculatively, we might suggest that awareness of inappropriate actions in anarchic hand syndrome actually reflects the *consequences* of actions rather than observations of the inappropriate actions per se. In this last case, the present errors may in fact be part of the anarchic hand syndrome in this patient. The "awareness" shown by patients diagnosed as having anarchic hand syndrome may apply only to consequential acts noted in the clinic.

How do the effector errors relate to deficits such as utilization behaviors in patients with frontal lobe lesions? In their most dramatic form, utilization behaviors seem to bear little relation to any task goal (see Lhermitte 1983). Because E.S. was influenced by task goals, it may be possible to distinguish effector from utilization errors. On the other hand, this may be a matter of degree. In some cases, patients may be unable to instantiate any task goals to override activation from familiar object-action associations; utilization behaviors then occur. In others, patients may instantiate a subset of task goals and responses, then depend on concordance between these goals and the stimuli. It may also be that the goals set by the patient fractionate. For example, E.S. seems to have an impaired ability to set novel goals specifying which effector to use in a task; there is then some deficit in selecting the effector for action. On the other hand, it may be that she is able to set novel goals that help her select which object should be used for action, when multiple objects are present. This was tested here.

Selection of Objects

Consider an everyday behavior such as reaching to pick up your cup of tea on a breakfast table holding many objects. Some of the other objects may be picked up by a handle (e.g., another cup, the teapot, the milk jug) others may be associated with a drinking action (other cups, glasses of orange juice). The parts of several objects may be spatially compatible with the required response. Do all of these objects evoke affordances or learned object-action responses? How can behavior in such circumstances be regulated? One possibility is that visually evoked actions (from affordances and learned associations alike) are regulated by an initial process of visual selection, which is functionally separate from the subsequent process of selecting an action to make in response to an object. Actions may only be made in response to a target object once it is selected from among the many objects that may be present. In a patient such as E.S., we witness a breakdown in goal-based control of the selection of action: she fails to select the appropriate hand for a task according to the rule. We assessed whether she might be able to select the object for action in the first place in experiments where E.S. was presented with two objects and had to select one for a response. Are effector errors evoked by distractors as well as targets, and what factors modulated any responses to the distractors?

General Methodology

As before, E.S. was instructed to use the left hand for all left-side targets, and the right to respond to right-side targets (targets were distinguished from distractors by color). Stimuli were positioned on a tabletop in left- or right-side or both positions in E.S.'s horizontal plane (40 cm away from E.S.), stimuli were placed 20 cm either to the left or the right of E.S.'s mid-sagittal plane. A large board obscured E.S.'s view while the stimuli were positioned; its removal triggered the onset of the trial. Trials consisted of unilateral or bilateral presentations, and each experiment consisted of a number of different experimental conditions. Both the conditions and uni- or bilateral presentations were randomized over trials. There were no time constraints. E.S. was asked to respond as accurately as possible.

Experiment 1: Selection by Color

A first experiment assessed whether E.S. could select a target object to make an action in response to when she was presented with a distractor as well as a target. Would any affordances evoke action only for the target object, or for both the target and the distractor? The selection cue was color (the target was green and the distractor was red). The stimuli were cups (identical, apart from their color) and the task was to pick up the green target. The use of either the left or right hand was again determined by the position of the target (to the left or right side of E.S.'s body).

Method E.S. was instructed to pick up the green cup when it appeared at either left or right or both locations but to ignore the red cup. When the green cup appeared in the left location, E.S. was to pick it up with the left hand; when it appeared in the right location, she was to pick it up with the right hand. There were 8 conditions with unilateral presentations where a single cup (either target or distractor) appeared in either the left or the right location. There were 8 bilateral presentation conditions where two targets or two distractor cups appeared on a given trial (when two targets appeared, E.S. was asked to pick up the left one with the left hand, and the right one with the right hand), and 8 bilateral presentation conditions where both a target and a distractor cup appeared on each trial (with all possible combinations of side of presentation and side of handle on the cup; see table 27.1). There were 10 trials per combination, creating a total of 80 single-object trials (40 with a target and 40 with a distractor), 80 trials with identical stimuli (40 two-target and 40 two-distractor trials), and 80 with one target and one distractor. On trials with only distractors, E.S. was required to make no response. On trials with two targets, she was required to respond simultaneously with her left and right hands. The conditions were presented randomly.

Results Table 27.1 displays the results of experiment 1. Collapsing over conditions, E.S. scored 172/240 (71.7%) correct. Performance was better in the unilateral than in the bilateral conditions: 82.5% and 66.3% correct,

Table 27.1A E.S. Picking Up a Target (Green) Cup and Ignoring a Distractor (Red) Cup: Unilateral Conditions

	Number correct	Error types	
		LH to RC	RH to LC
Condition 1: 	10/10		
Condition 2: 	10/10		
Condition 3: 	10/10		
Condition 4: 	10/10		
Condition 5: 	10/10		
Condition 6: 	3/10		7
Condition 7: 	3/10	7	
Condition 8: 	10/10		

Note:  = green cup;  = red cup; LH = left hand responds; RH = right hand responds; LC = left cup; RC = right cup.

respectively; chi-square (1) = 6.9, $p < 0.008$; although performance did not differ in the conditions where there were either two targets or two distractors relative to the conditions when both a target and a distractor were present: Chi-square (1) < 1.0 (see table 27.1B and 27.1C, respectively).

Summing over conditions, 75 errors were made. These were classified as either hand errors, where E.S. reached for the target (green) cup with the incorrect hand ($n = 60$); distractor errors, where E.S. reached for the distractor (red) cup ($n = 6$); or neglect errors, where E.S. failed to pick up the target cup when it was present ($n = 9$). The difference in the number of errors is significant: multinomial $p < 0.0001$.

There was no difference in the number of errors made with the left ($n = 34$) or the right hand ($n = 26$): binomial $p > 0.05$. These data support those reported in more detail by Riddoch et al. (1998).

Discussion E.S. made many hand errors, with both hands when the side of the target and the side of its handle were incompatible; in contrast, responses were rarely made to distractors. The results show that E.S. was much more likely to respond to the target than to a distractor, even

Table 27.1B E.S. Picking Up a Target (Green) Cup and Ignoring a Distractor (Red) Cup: Bilateral Conditions (Either 2 Green or 2 Red Cups)

	Number correct	Error types					
		Hand		Distractor		Neglect	
		LH to RC	RH to LC	LH	RH	LC	RC
Condition 9: 	10/10						
Condition 10: 	9/10			1[a]			1[a]
Condition 11: 	10/10						
Condition 12: 	10/10						
Condition 13: 	2/10	7					1
Condition 14: 	8/10	1					1
Condition 15: 	3/10	1	5[a]			1	1[a]
Condition 16: 	2/10	8[b]	7[b]				

Notes:

[a] On one trial in conditions 10 and 15, E.S. lifted the left cup with the right hand and neglected the right cup.

[b] On seven trials in condition 16, E.S. lifted the left cup with the right hand and the right cup with the left hand.

 = green cup;  = red cup; LH = left hand responds; RH = right hand responds; LC = left cup; RC = right cup.

though she often then failed to select the appropriate effector for the action (responding to the position, relative to her, of the handle rather than of the cup). The results demonstrate that E.S. is relatively successful at selecting between the target and the distractor, using the target's color as the selection criterion, although, having selected the target object, she remained prone to making the response afforded by the congruency between the cup handle and the effector. The large number of hand errors match prior data (Riddoch et al. 1998). In addition, experiment 1 replicates the pattern that right-hand errors are as likely to occur as left-hand errors in conditions where both the stimulus and the associated response are familiar.

Table 27.1C E.S. Picking Up a Target (Green) Cup and Ignoring a Distractor (Red) Cup: Bilateral Conditions (Both Red and Green Cups)

		Error types					
		Hand		Distractor		Neglect	
	Number correct	LH to RC	RH to LC	LH	RH	LC	RC
Condition 17:	6/10	4					
Condition 18:	7/10						3
Condition 19:	3/10	6					1
Condition 20:	8/10			1			1
Condition 21:	9/10				1		
Condition 22:	8/10				2		
Condition 23:	6/10		2		1	1	
Condition 24:	5/10		5				

Note: = green cup; = red cup; LH = left hand responds; RH = right hand responds; LC = left cup; RC = right cup.

Experiment 2: The Effects of Distractor Proximity

Visual selection of an object for action provides one means of controlling affordances from stimuli in the environment. Experiment 1 demonstrated that E.S. is generally able to select the target object for an action (at least by color), even though she is then impaired at selecting the appropriate effector for the target (according to the experimental instructions). This suggests that selection of the target precedes the selection of action, and may be dissociated from it. (In experiment 4, we show how the same instruction as used here cannot be implemented for action selection in another patient, even though it can be used in selecting the object for action.)

But what are the consequences of selecting an object for an action, such as reaching and grasping? Is information subsequently extracted only from the selected object (or from its associated location), or is information processed from other stimuli relevant to the action (e.g., other stimuli in the path of the action or close to the hand used for the action)? In experiments in which normal subjects are required to reach for and grasp objects, Tipper and colleagues (see Tipper, Howard, and Houghton, chap. 10, this volume; Tipper, Howard, and Jackson 1997) have shown effects of distractors according to their locations with respect to the hand of the actor (but see also Castiello 1996). For example, both the time to initiate and complete the movement, and the movement trajectories, are affected by distractors. Reaction times and movement times are slowed when distractors fall between the target and the hand for action, with movement times also slowed by distractors not in the movement path, provided these fall close to the hand used for action; and reach trajectories to far targets are displaced away from distractors near the hand (Tipper, Howard, and Jackson 1997). These data suggest that, in making a reaching and grasping action, items in addition to the target may be processed and influence performance, especially when such items are near the responding hand (see Pratt and Abrams 1997). We examined this possibility with E.S. in experiment 2. The task was to point to an LED with the left or right hand when the LED was on the left or the right side of her body, respectively. Riddoch et al. (1998) showed that E.S. makes many errors with the left hand under these conditions (pointing incorrectly to right-side target with her left hand). In experiment 2 we added a distractor cup to the displays. The cup fell either to the left or the right of each target LED. Suppose the right LED is lit. Normally, E.S. would be prone to make an error by pointing with her left hand to this light. But what if the distractor cup falls to the left of the LED (though both fall on the right side of space)? The distractor cup then falls closer to the interfering left hand than does the target, and it also falls close to the movement path to the target. If only the target is selected, the cup should not affect performance. If, however, the cup is also selected (being relatively close to the hand selected for the response, falling close to the movement path, or both), then it might also become linked to the left-hand response. Either of two events might follow. E.S. might point to the cup rather than the target LED. Or, because the cup does not correspond to the task goal for the target object ("point to the *light*"), she might reject it as not being the target. Linkage between the rejected cup and the potentiated response might then lead to inhibition of the left-hand response, enabling E.S. to make a right response to the target LED. Somewhat counterintuitively, the cup distractor may *improve* performance.

Method E.S. was presented with red LEDs that fell 20 cm to the left and right of her midline; she was 40 cm away from the virtual line connecting the LEDs. She had to point to the right light, when turned on, with her right hand and to the left light, when turned on, with her left hand. E.S. was also presented with a distractor cup, which fell either on the side of space close to the target or on the side of space opposite the target light. There were eight conditions when it fell on the opposite, and eight when it fell on the same side of space. When on the opposite side of space, the target could occupy the left or right locations; the cup could be left or right of the *other* light, and the cup could have its handle to the left or right (2 target positions × 2 cup positions × 2 handle positions). The same conditions were created when the cup fell on the same side of space as the target light (here the cup could fall on the left or right of the target but on the same side relative to the midline). There were 10 trials per condition, creating 160 trials in the study, these were presented in a randomized order. Before the onset of each trial, E.S. was verbally cued ("ready"). There were no time limits.

Results When the target light was on the left side of space relative to E.S.'s midline, she made only one error (scoring 79/80), namely, when the distractor cup was on the same side and to the left of the target, with the handle facing left (1/10 errors in this condition). She then responded with her right hand to the light. When the target light fell to the right of midline, many more effector errors occurred (E.S. responding with her left hand), as in Riddoch et al. 1998. When the distractor was on the opposite side of space (thus the target light appeared alone on the right side), she scored only 3/40 correct, with all the errors being made with the left hand. Neither the position of the distractor relative to the left light, nor the position of its handle, affected performance (she scored 1/10 correct in 3 of the 4 subconditions, and 0/10 when the cup was to the right of the left light and had its handle left). When, however, the distractor cup fell on the same side of space as the right target light, performance was affected by the distractor. When the cup fell to the right of the target (i.e., further from the target and out of the reach path), she scored 0/20. E.S. always pointed to the right target light with her left hand. When the cup fell to the left of the target (i.e., closer to the left effector and close to the path of its reach for the target), she scored 19/20, making only one error with her left hand. The position of the handle on the cup did not affect performance (the only error in the last condition was when the handle of the cup faced left). Performance on right-side targets was better when the cup fell closer to the left hand and thus close to the reach path to the target: Fisher's exact p value < 0.0001.

Discussion As in experiment 1, E.S. tended not to make selection errors by responding to the distractor rather than the target (indeed, no distractor errors occurred here). Nevertheless, the distractor did affect performance in one condition, improving performance by reducing errors made with the left hand when it fell to the immediate left of the right-side target light. This is consistent with the idea that rejection of the distractor in this condition inhibits the (inappropriate) response evoked by the target. For this to occur, the distractor would need to become linked to the left-hand response, which was typically activated to the right-side light (as shown in all the other conditions). Any subsequent rejection of the distractor as not conforming to the task goal specifying the target would then result in linked inhibition of the associated response. As a consequence, the other (right-hand) response "wins" any competition to be linked to the target. The result is that fewer left-hand errors occur. According to this proposal, one consequence of selecting a target for action is also to select objects close to the effector, or in the response pathway, or both, with these objects becoming linked to the response as well (see also Tipper, Howard, and Jackson 1997). In other studies with E.S. (Riddoch and Humphreys forthcoming), we have shown that errors resulting from a failure to comply with the task rule are not only blocked by distractors in a pointing task (as here) but also in reaching tasks (e.g., pick up a plastic block and ignore a cup). The result is not confined to the present procedure.

On the other hand, although these data are consistent with arguments about object selection, they do not necessarily demonstrate selection of an action path. It may be, for instance, that E.S. misunderstood the task instructions and thus responded to the relative rather than the absolute locations of the light. When the distractor cup fell to the left of the lights, E.S. may have made more right-hand responses because the lights then fell to the right of the cup. This "relative position" account still needs to explain why performance was only affected by the position of the light relative to the cup when the light was on the right of E.S.'s body. Nevertheless, the relative position account remains viable. We have gone on to test it, and the idea that objects close to the effector, or in the path of the action, or both are also selected and linked to the response to the target, in further work with F.K., a different patient with frontal lobe damage (see section 27.2). This also demonstrates the generality of the results because the effects are not confined to the single patient E.S. with relatively rare neuropsychological symptoms (anarchic hand syndrome).

27.2 CASE 2: VISUAL AFFORDANCE AND FRONTAL LOBE DAMAGE

As noted in the introduction, frontal lobe damage is associated with impulsive actions, poorly constrained by task goals. For instance, in the "action disorganization syndrome" (ADS), patients may pick up and use objects in the wrong sequence or when the task demands that other objects are used (Humphreys and Forde 1998; Schwartz and Buxbaum 1997; Schwartz et al. 1995). Similarly in frontal lobe "utilization behavior" goal-directed performance seems to be impaired (Lhermitte 1983; Shallice et al. 1989). Hence, in addition to patients with anarchic hand syndrome, patients with frontal lobe damage are good candidates to show responses that are inappropriately driven by affordances, learned stimulus-response relationships, or both in the selection of action. We can again ask whether any deficits in the selection of action (e.g., using the effector afforded by the stimulus rather than the effector consistent with the task rule) dissociate from the processes involved in visual selection.

To address this issue, we tested whether F.K., a patient with bilateral frontal lobe damage and symptoms of ADS (Humphreys and Forde 1998; Humphreys, Forde and Francis, chap. 18, this volume) would show evidence of hand errors when the currently inappropriate responses are "afforded" by the stimulus (as a function of the position of the object relative to the effector and also the goals of the task). In addition, we examined whether poor selection of action in such a patient may arise even if there is good selection of the object for action (as in E.S.), and we tested the consequences of object selection on performance.

F.K. was a right-handed male, 30 years old at the time of testing. He suffered carbon monoxide poisoning in 1989, which resulted in bilateral damage to the frontal and temporal cortices (Humphreys and Forde 1998). F.K. showed clear symptoms of frontal lobe damage, performing poorly on the Wisconsin Card-Sorting Test and the Stroop Test, and his errors on everyday tasks were consistent with a diagnosis of ADS (see Humphreys and Forde 1998 for a discussion of this and a full case report). We first tested F.K. under conditions similar to those used by Riddoch et al. (1998) to examine the selection of the appropriate effector according to the prescribed task rules.

Experiment 3: Selection of Hand and Selection of Objects

In a first study, we assessed F.K.'s ability to select (1) which of two objects he was required to make an action to; and (2) the appropriate effector (according to the task rules) to an object likely to elicit both a familiar and an afforded action (a cup). There were three conditions. In the first two, F.K. was asked to respond by picking up a cup presented on his left side with his left hand and a cup on his right side with his right hand. In condition 1, there was a single cup; in condition 2 there were always two cups and F.K. had to respond only to the red one and to ignore the other (green) one. In condition 3, a single cup was presented but F.K. was required to point at rather than pick up the cup. When pointing is required, there should be a reduction in the affordance of a grasp response (Riddoch et al. 1998) and in effector errors consistent with the affordance.

Method Conditions 1, 2, and 3 were conducted over separate weeks. Following condition 3, half the trials in condition 1 were repeated to ensure that the better performance observed in condition 3 was due to the task and not to a general improvement in performance. In conditions 1 and 3, a single red cup was presented, either to the left or right of F.K.'s midline and with the handle on the left or right of the cup. There were 20 trials for each possibility. In condition 2 (two objects), there were eight

Table 27.2A Number of Correct Trials Made by F.K. in Experiment 3: Conditions 1 and 3

	Condition 1 (Pick up)	Condition 3 (Point)	Condition 1 (Repeat; Pick up)
	17/20[a]	20/20	10/10
	8/20[b]	20/20	1/10[c]
	20/20	20/20	10/10
	20/20	20/20	10/10

Notes:

[a] Three right-hand errors.

[b] Twelve right-hand errors.

[c] Nine right-hand errors.

Table 27.2B Number of Correct Trials and Errors Made by F.K. in Experiment 3: Condition 2 (Pick Up the Red Cup)

		Number correct	Hand errors		Distractor errors	
			LH to RC	RH to LC	Left	Right
Target left		10/10				
		9/10		1		
		0/10		10		
		1/10		9		
Target right		8/10	1		1	
		9/10			1	
		10/10				
		9/10	1			

Note:  = green cup;  = red cup; LH = left hand responds; RH = right hand responds; LC = left cup; RC = right cup.

variations of presentation as a function of: target location (left or right), target handle position (left or right), and distractor handle position (left or right). There were 10 trials for each variation. The distances between stimulus items (from each other and from F.K.) were the same as those used for E.S. The cups were hidden from F.K.'s view before each trial. The trial was initiated by a verbal response ("ready"), and there were no time limitations.

Results The complete data for experiment 3 are shown in table 27.2. In condition 1 (pick up the single cup), F.K. made correct right-hand responses to all right-side cups, irrespective of the position of the handle (40/40 in total). With left-side cups, he performed reasonably well when the handle was left (responding correctly with the left hand on 17/20 trials) but made errors when the handle was right (reaching for the handle with the right hand on 12/20 of these trials). Performance to left-side cups was affected by the position of the cup handle: chi-square (1) = 8.64, $p < 0.003$. Similar data were obtained in the repeat of this condition. Like E.S., F.K. was unaware of his errors.

In condition 2 (red and green cups), F.K. made only 2 errors (in 80 trials) by picking up the green distractor, relative to 22 errors using the wrong hand to respond to the target (20 and 2 with right and left hands, respectively). Of the 20 errors made with the right hand, 19 were made in response to left-side targets whose handles faced right. Performance in response to left-side targets was better when the handle faced left than when it faced right: 19/20 versus 1/20 correct; chi-square (1) = 32.4, $p < 0.0001$. In condition 3, F.K. scored at ceiling. Like E.S., F.K. was able to report the rules of each task when asked after the completion of the trials in each condition.

Discussion The findings were similar to those obtained with E.S., the sole difference being that F.K. made errors predominantly with his right hand (E.S. made the same number of errors with each hand in experiment 1). These responses were again influenced by the position of the handle on the cup (being more likely when this faced right), suggesting that F.K. was sensitive to the affordance between the relevant part of the cup and the location of the effector. F.K. was poor at selecting the appropriate left-hand response when the cup afforded action with the right hand (when the handle of the cup faced to the right). Performance was also modulated by the task. Hand errors were eliminated when pointing was used rather than grasping. Finally, despite being impaired at selecting the appropriate effector for a target object in accordance with task rules, F.K. was able to select the object for action, based on its color. He made few errors by picking up the distractor when it was defined by a color different from that of the target. As with E.S., this demonstrates that the processes involved in selecting an object for action, indexed by color, can be dissociated from the processes involved in selecting an effector in accordance with task instructions. Moreover, F.K. seems able to set up at least some of the task goals that determine the selection of action because effector errors were eliminated in the pointing task. The deficit is revealed when action selection requires a relatively novel set of goals and stimuli activate a subset of goals linked to afforded responses ("grasp the cup").

Experiment 4: Consequences of Object Selection

In a final experiment, we examined the consequences of object selection on F.K.'s reaching and grasping performance. The task required F.K. to reach for a central cup, now using the hand indicated by the *handle* of the cup (left hand if the handle faced left, right hand if it faced right). With a single object, F.K. performed this task effortlessly. We then introduced a distractor, differing in color from the target. The task remained to respond to the central cup. Experiment 3 showed that F.K. can use color information to select the target. The contrast between this study and the earlier one, however, is that here distractors were placed directly in the path of either a left- or a right-hand response. If following the selection of the target, objects in the path of the response, close to the effector, or both are also selected for action, then distractors between F.K.'s hand and the target may affect performance (e.g., left-side distractors when the target's handle faces left; right-side distractors when the target's handle faces right). If F.K. was poor at grasping the target when a distractor cup fell in the reach trajectory, he might be unable to adjust his reach so that it bypassed the distractor. If problems in altering the reach trajectory alone are important, then grasp responses to target cups should also be affected by the block, which again provides an obstacle in reaching for the target. Contrasting performance when the distractor was another cup versus when it was a wooden block helped us to determine whether this was so. It also enabled us to test effects due to the relevance of the distractor to the task. The data from experiment 2, with E.S., suggest that actions to distractors close to the hand or path of a response may be blocked when the distractor is irrelevant to task goals. In experiment 3, we assessed what happens when the object is relevant to the task goals (comparing cups with wooden blocks). When distractors are of a relevant type with respect to task goals and in the path of the response (close to the effector or both) are responses assigned to them?

Method The task was to reach for a central red cup and to pick it up with the left hand when the handle was left and with the right hand when the handle was right. There were two conditions. In the first, a green distractor cup was added in a position halfway between the cup and either F.K.'s left or right hand. In the second condition, a wooden block (approximately the same size as the cup) was put in either of the same two locations. F.K.'s hands were positioned 6 cm to the right and left of his body midline, and the distractors were presented 3 cm either to the left or right of F.K.'s midsagittal plane, on a virtual horizontal line 20 cm away from F.K.. The target fell at F.K.'s midsagittal plane, 40 cm away. Before each condition, F.K. was given 40 trials in which he had to respond with the left or right hand (according to the position of the handle) to a target presented in isolation. He scored 40/40 on each occasion. There followed a block of 40 trials with either a cup or a wooden distractor, and two blocks per condition presented in an ABBA design. Within each block, the target faced left or right, and the distractor was presented in either the left or right location on 10 trials each. When the distractor was a cup, its handle faced either right or left on 5 trials per distractor location and position of target.

Table 27.3 F.K.'s Correct Responses to a Central Cup, As a Function of the Position and Type of Distractor, in Experiment 4

Target handle		Left		Right	
Distractor position		Left	Right	Left	Right
Distractor cup	Handle left	0/10	8/10	8/10	2/10
	Handle right	0/10	8/10	10/10	0/10
Distractor block		20/20	19/20	20/20	20/20

Results The number of correct responses made by F.K. on trials when the distractor was present are shown in table 27.3. F.K. reached for the distractor item on a large number of trials (unlike condition 2, experiment 3) but only when the distractor was a cup and only when it lay in the reach path to the target.

When the distractor was a wooden block, F.K. made virtually no errors. He was thus able to select the target cup for action when it differed in shape, color, and kind from the distractor. When, however, the distractor was a cup, performance was much worse. On 38 trials, he picked up the distractor rather than the target. All of these errors occurred when the distractor fell in the path of the response. Importantly, the hand of response was dictated by the position of the handle of the target cup and *not* by the position of the distractor's handle (i.e., errors occurred when the distractor was in the near-left location and the handle on the target faced left, or when the distractor was in the near-right location and the handle on the target faced right). When the target faced left, there were 10 distractor errors both when the distractor faced left and when it faced right; when the target faced right, there were 10 distractor errors when the target faced right, and 8 when it faced left. There were no distractor errors when the distractor fell on the opposite side of space to the effector linked to the target by the handle rule. The errors that did occur on these last trials (6/40) were all due to F.K. selecting the wrong hand to respond to the target (4 right-hand errors when the handle faced left). Performance was reliably better when the distractor fell on the opposite versus the same side of space for the effector indicated by the central target's handle: 34/40 versus 2/40 correct; chi-square (1) $= 51.7$, $p < 0.0001$.

Discussion The pattern of errors in this study tells us a great deal about the factors that determine F.K.'s performance. Consider first his performance with wooden distractor blocks. F.K. made no errors by misreaching for distractor blocks, even though he had to redirect his reaches to bypass such distractors to get to the target. He did make misreaches for distractor cups, but clearly this was not because he was unable to redirect his reaches. The absence of reaches for the wooden distractor blocks also indicates that F.K. was not merely responding to the relative positions of the stimuli (e.g., reach with the right hand for the rightmost of two objects).

Misreaches for distractor cups were affected by the position of the handle on the target rather than on the distractor. F.K. thus made effector errors in which he picked up the distractor with the hand that was incompatible with that object (but which was compatible with the handle of the target), and these errors occurred even though the response was then relatively difficult to effect (distractor cups were picked up with the hand facing away from the cups' center of mass). This indicates two points. First, the ease of the response was less important than the hand activated by the orientation of the target cup. The compatibility effects found with F.K. in experiment 3 are thus unlikely to be due to the difficulty of grasping incompatible cups. Second, the result confirms that F.K. was able to select the target cup (because the orientation of the target determined the hand of response). Having selected the target for action, however, the action was then transferred to a distractor falling in the reach trajectory, which suggests that, in addition to targets, objects in the reach trajectory to targets are selected for action. Distractors irrelevant to the goals of the task (wooden blocks) are rejected: F.K. redirected his actions to bypass such distractors. But distractors relevant to a subset of the task goals (e.g., other cups when the task specifies cup grasping) tended to become linked to the concurrently activated response, with the result that action errors were made by misreaching for distractors.

This transfer of action to the distractor can be attributed to F.K.'s deficit in instantiating relatively novel goals for the selection of actions in response to objects ("reach for the central red cup"). Distractors selected at the first stage of object selection (because they lie in the reach trajectory to targets) activate a subset of the goals for action selection (being cups). We suggest that F.K. is unable to override this activation because the full goal structure for action selection is not in place. It is interesting to note, however, that although the colors and locations of the targets and distractors were effectively ignored once distractors were visually selected because they lay in the path of the target, the same properties were used to visually select the target in the first place. Thus what is critical is not simply whether the task requires a novel rule but whether this rule is used for object selection or for the selection of action. F.K. is particularly poor at implementing novel rules for the selection of action.

27.3 GENERAL DISCUSSION

We have presented evidence from a patient with anarchic hand syndrome and a patient with symptoms of action disorganization syndrome, both of whom showed marked problems in selecting a task-specific response (e.g., use of the left versus the right hand) to an object. When the object either had a learned action other than the one required, or when the object afforded another action (when the parts of the object were congruent with the other effector and with some subset of goals for the task), the incorrect hand was used (i.e., the patients tended to use the hand congruent with the affordance of the object, rather than the one specified by the arbitary task rule). Changing the general goals of the task (e.g., from picking up a cup to pointing) decreased the number of errors for both F.K. and E.S. Thus some goal structures could be applied to enable actions to be selected in response to objects, but problems arose when the goals were relatively novel and the currently inappropriate affordance was congruent with a subset of these goals. The data support the argument that there can be direct activation of actions from visual representations of objects, but with this action modulated by (1) congruency between the effector and the object and (2) the task goals.

In contrast to their impaired selection of an effector for action (according to the task rule), both patients were able to select which of two objects to make an action in response to, when target objects were cued by their color (experiments 1 and 3). These results suggest that the selection of an effector can be functionally distinguished from the visual selection of an object for action, with only the former being impaired in these patients. Indeed, in experiment 4, F.K. was able to implement one rule for visual selection ("Select the central red cup") but then failed to apply the same rule for action, once distractors had become selected visually. Thus not only is the nature of the task goal important (e.g., whether it is novel), but whether the goal is used for visual selection or for selection of an effector for action. Distractors rejected in the first stage of object selection do not evoke an associated or afforded response (or at least not strongly enough to generate effector errors). It follows that object selection can provide a means by which behavior is controlled in complex environments containing multiple objects. Object-action associations, or visual affordances, are most potent when generated from selected objects.

There is an important proviso, however, namely, that following object selection, other objects in the path of the action to the object can also be selected. With patient F.K., we showed that actions were made in response to distractors that fell in the reach path and were nearer to the hand activated by targets. This is consistent with distractors being selected for action under this circumstance. Responses were not made to distractors that fell out of the reach path, and were nearer to the nonactivated effector. Whether the reach path is more important than being close to the effector is an issue awaiting further research (Tipper, Howard, and Jackson 1997). Whichever is the case, the results indicate that actions activated by targets can be transferred to other stimuli that are subsequently selected visually.

With E.S., the data suggest another consequence of selection on subsequent action. When asked to point to lights with the hand appropriate to the side of presentation, E.S. made mainly left-hand interference responses (pointing to the right light; see also Riddoch, et al. 1998). Positioning a cup to the left of the light (on the right side of space) reduced the frequency of these errors. We propose that the target light was selected visually and frequently activated an inappropriate (left-hand) response in E.S., but that when the cup fell close to this effector, its path of action, or both, the cup too was selected visually and linked to the concurrently activated response. If the distractor then failed to match any of the goals for action (being a cup rather than a light), it was rejected as the target for action. Rejection of the distractor had the consequence of linked rejection of the associated left-hand reach. This enabled the right hand to win the resultant competition for the pointing response to the right-side light, improving performance. We still need to explain why a similar effect was not found for F.K. in the condition with the distractor block in experiment 4. Our suspicion here is that the degree of inhibition produced on responses associated with stimuli may depend on how strong the stimuli are as competitors for action selection. In experiment 2 with E.S., the cups may be potent competitors for action with the lights, both of which may be pointed to. The strong inhibition of the cup results in linked inhibition of the associated pointing response. In experiment 4 with F.K., the blocks may not be potent competitors for an afforded grasp response to the cup, and thus inhibition of the response is less when the distractor block is rejected; the reach for the target cup continues. (For a model of selection in which the strength of inhibition is linked to the strength of competition between stimuli, see Tipper, Howard, and Houghton, chapter 10, this volume.)

The results with both patients indicate an account in which task goals can have dissociable effects on performance. Goals used for visual selection can be fractionated from those used for the selection of action (e.g., which effector is appropriate for action, as in experiment 3, or whether actions are made to particular items, even if they have been selected visually, as in experiment 4). Patients can have difficulties in implementing novel goals for the selection of action, while remaining able to set similar goals for visual selection of objects.

Neurological Considerations

F.K. had sustained damage to medial areas of the frontal lobes and bilaterally to the temporal lobes. We suggest that his impairment in action selection is linked to his frontal lobe damage. As noted previously, patients with frontal lobe damage are notoriously poor at making novel actions in response to objects and often produce inappropriate, but prepotent responses in their place (for evidence see Humphreys and Forde 1998). In work with the monkey, Passingham and colleagues (Passingham 1993) have demonstrated that medial frontal areas (particularly the supplementary motor area) are involved in volitional action, whereas more lateral frontal areas are involved in more automatic responses to visual stimuli. F.K.'s frontal lesions are consistent with the supplementary motor area being damaged, with a consequent reduced ability to make task-dependent volitional responses and an increased sensitivity to direct visual cues. In E.S.'s case, corticobasilar degeneration may have resulted in disconnection of the medial frontal lobe areas from visual sensory signals. The result is again a propensity to act in response to preexisting object-action associations and affordances rather than to relatively novel task instructions. The effect of object familiarity on E.S.'s right-hand responses further suggests that activation of frontal areas in the left hemisphere is based on object-action associations. In contrast, right-hemisphere activation seems linked to the spatial control of action and less by object-action associations. E.S.'s left-hand responses arise even when pointing responses are used and even with unfamiliar objects; Riddoch et al. (1998) showed these effects occurred only under conditions of spatial uncertainty.

The ability of these patients to visually select objects fits with the distinction made by Posner and Petersen (1990) between anterior and posterior attentional systems. In E.S. and F.K., anterior systems concerned with the selections of actions to objects are damaged, disconnected, or both. Nevertheless, posterior attentional subsystems concerned with visual selection of objects seem to operate relatively efficiently. These last systems likely depend on areas in the posterior parietal cortex as well as subcortical thalamic structures that may be primed top-down by more frontal regions (Miller, chap. 22, this volume). These areas are relatively spared in our patients. In this account, the posterior attentional system plays an initial role in selecting objects, with the selection of task-based action in response to these objects subsequently mediated by medial frontal regions.

NOTE

This work was supported by a grant from the Medical Research Council (U.K.).

REFERENCES

Castiello, U. (1996). Grasping a fruit: Selection for action. *Journal of Experimental Psychology: Human Perception and Performance, 22,* 582–603.

Della Sala, S., Marchetti, C., and Spinnler, H. (1991). Right-sided anarchic (alien) hand: A longitudinal study. *Neuropsychologia, 29,* 1113–1127.

Della Sala, S., Marchetti, C., and Spinnler, H. (1994). The anarchic hand: A fronto-mesial sign. In F. Boller and J. Grafman (Eds.), *Handbook of neuropsychology,* pp. 233–255. Amsterdam: Elsevier.

Gibson, J. J. (1979). *The ecological approach to visual perception.* Boston: Houghton Mifflin.

Humphreys, G. W., and Forde, E. M. E. (1998). Disordered frontal schema and frontal apraxia. *Cognitive Neuropsychology, 15,* 771–811

Lhermitte, F. (1983). "Utilisation" behaviour and its relations to lesions of the frontal lobes. *Brain, 106,* 237–255.

Luria, A. R. (1973). *The working brain.* London: Penguin.

Marchetti, C., and Della Sala, S. (1998). Disentangling alien and anarchic hand. *Cognitive Neuropsychiatry, 3,* 191–207.

Passingham, R. (1993). *The frontal lobes and voluntary action.* Oxford: Oxford University Press.

Posner, M. I., and Peterson, S. E. (1990). The attention system of the human brain. *Annual Review of Neuroscience, 13,* 25–42.

Pratt, J., and Abrams, R. A. (1997). Selective reaching to grasp: Evidence for distractor interference effects. *Visual Cognition, 4,* 1–38.

Reason, J. T. (1979). Actions not as planned: The price of automatisation. In G. Underwood and R. Stephens (Eds.), *Aspects of consciousness,* pp. 67–89. London: Academic Press.

Reason, J. T. (1984). Lapses of attention. In R. Parasuraman, R. Davies, and J. Beatty (Eds.), *Varieties of attention.* Orlando, FL: Academic Press.

Riddoch, M. J., Edwards, M. G., Humphreys, G. W., West, R., and Heafield, T. (1998). Visual affordances direct action: Neuropsychological evidence from manual interference. *Cognitive Neuropsychology, 15,* 645–683.

Riddoch, M. J., and Humphreys, G. W. (Forthcoming). Neuropsychological evidence distinguishing object selection from action (effects) selection. *Cognitive Neuropsychology.*

Rumiati, R. I., and Humphreys, G. W. (1998). Recognition by action: Dissociating visual and semantic routes to action in normal observors. *Journal of Experimental Psychology: Human Perception and Performance, 24,* 631–647.

Schwartz, M. F., and Buxbaum, L. J. (1997). Naturalistic action. In L. G. G. Rothi and K. M. Heilman (Eds.), *Management and treatment of limb apraxia,* pp. 269–289. Hove, U.K.: Psychology Press.

Schwartz, M. F., Montgomery, M. W., Fitzpatrick-DeSalme, E. J., Ochipa, C., Coslett, H. B., and Mayer, N. H. (1995). Analysis of a disorder of everyday action. *Cognitive Neuropsychology, 12,* 863–892.

Shallice, T., Burgess, P. A., Schon, F., and Baxter, D. M. (1989). The origins of utilisation behaviour. *Brain, 112,* 1587–1598.

Shaw, R., and Turvey, M. T. (1981). Coalitions as models for ecosystems: A realistic perspective on perceptual organization. In M. Kubovy & J. R. Pomerantz (Eds.), *Perceptual organization,* pp. 343–416. Hillsdale, NJ: Erlbaum.

Tipper, S. P., Howard, L. A., and Jackson, S. R. (1997). Selective reaching to grasp: Evidence for distractor interference effects. *Visual Cognition, 4,* 1–38.

28 Deficits of Task Set in Patients with Left Prefrontal Cortex Lesions

Steven W. Keele and Robert Rafal

ABSTRACT Subjects with lesions to left or right lateral prefrontal cortex were compared to control subjects in situations that did or did not require task set. When a single dimension (color or shape) was relevant for a block of trials and the irrelevant dimension was absent (a condition not requiring set), reaction time differed little between groups. When both dimensions were present and set was required to specify which was relevant, reaction time of the left frontal group increased markedly, not just when set was switched, but also when set was maintained for several trials, unlike the other groups. The three groups did not differ reliably in "local" shifting time as measured by the reaction time difference between switched and nonswitched sets.

One extensively tested left frontal subject exhibited little deficit in establishing the first set in a block of trials. The deficit greatly increased on subsequent sets within a block, only to abate between blocks. Thus set-shifting costs were not local, which would have indicated longer time to reconfigure set, but global, which may reflect difficulty in inhibiting prior sets.

The study reported in this chapter concerns the role of lateral prefrontal cortex in the executive process of switching task set. Generally speaking, executive function is invoked when the current stimulus and general task instructions do not provide sufficient information to determine an appropriate course of action (Norman and Shallice 1986). Additional information from specific instruction or immediately prior context is needed to codetermine the response. For example, when instructions specify order of responding to two concurrent stimuli, even the first response may be delayed, suggesting intervention of a control process to assure correct ordering (Umiltà et al. 1992). Even when the stimuli are not concurrent, order still may be dictated by a time-consuming executive process (Meyer and Kieras 1997). In a common situation requiring executive function, examined in the current experiments, different components of the present stimulus afford different, conflicting responses. The relevant component, such as color versus shape, is specified on a short-term basis.

Changing the basis for response adds additional processing time at the moment of change, reflecting an internal reconfiguration (e.g., Allport, Styles, and Hsieh 1994; Meiran 1996; Rogers and Monsell 1995). For present purposes, we call the reconfiguration time "local switch cost". In addition to local switching time, a former set may have residual effects that persist even after set is switched (e.g., Allport and Wylie, chap. 2, this volume). We call this "global shift cost". We examine both local and global cost, focusing on the former in experiment 1 and on the latter in experiment 2.

Our concern is with the role of prefrontal cortex in shifting processes. Patients with frontal lesions tend to perseverate on past action, such as mistakenly repeating a pen stroke or letter or word when writing, rather than progressing through an orderly series of actions (Shallice 1988). Such perseveration suggests difficulties in moving from one subsequence of activity to another, a process that we have argued requires set switching (Hayes et al. 1998). Frontal patients also exhibit increased difficulties on tasks modeled after the Wisconsin Card-Sorting Test (WCST), a task that requires shifting from one hypothesis to another until the correct sorting basis is obtained (e.g., Owen et al. 1993). Moreover, a recent fMRI study by Konishi et al. (1998) has shown both left and right Brodmann's areas 44 and 45 of prefrontal cortex are active when there is a shift in the basis of sorting.

While evidence suggests that the frontal lobes are involved in task setting and switching, a more precise specification of critical process is lacking. For example, in tasks related to the WCST, an error may induce a number of problem-solving and memory processes other than simple set switching. In the current study, we examine possible switching deficits in frontal patients more directly by observing the reaction times to stimuli when sets remain the same or are switched. We examine reaction times not only at the point of change (local switch costs) but also residual effects on reaction time long after the change (global switch cost).

Our first experiment compared three subject groups, one with lesions of the left lateral prefrontal cerebral cortex, another with similar lesions in the right cortex, and a control group. This experiment concentrated on local switching time, comparing reaction times on the first trial in which set was first switched with immediately succeeding trials in which the newly switched set was maintained. Such manipulation is similar to one recently conducted by Rogers et al. (1998), who report finding a local set-shifting deficit in left frontal patients. As will be seen, we do not find such a deficit, and we offer an alternative account for their results. The second of our experiments studied a single, left frontal patient over several sessions, allowing us to examine both global and local shifting effects.

28.1 EXPERIMENT 1

Two conditions differing in their set requirements were compared. One involved either two- or four-choice reaction times to unidimensional color or shape stimuli. The variation in choice difficulty allows an assessment of whether mere difficulty affects frontal patients more than controls. Because executive function is not required to specify the correct basis of response with unidimensional stimuli, which have no irrelevant dimension, there is no reason to suppose that increased choice would affect frontal and control patients differently. The second condition involved bidimensional stimuli, which varied both in color and shape. The word "color" or "shape" was given immediately before each stimulus, specifying the relevant dimension.

In the bidimensional condition, the same set (color or shape) was used for a series of 8 trials before switching to the alternate set. This procedure continued through blocks of 80 trials. Comparing the first trial of a set of 8 with subsequent trials allowed us to assess switching efficiency, or local switch cost. The paradigm also allowed us to compare the situation requiring set, regardless of whether it was switched, with the unidimensional situation requiring no set. To anticipate, this latter comparison turns out to be the most revealing.

Our patient pool was restricted to subjects with damage in lateral prefrontal regions of the left or right hemisphere. We focused on these areas because of the neuroscience literature on working memory. Typical working-memory paradigms contain elements of task set and set switching, where features of one stimulus must be held in memory until a comparison stimulus occurs, and then set switches for a second pair of stimuli. Studies of monkeys (e.g., Fuster 1985; Goldman-Rakic and Selemon 1990) and human neuroimaging work (Smith et al. 1995) have implicated lateral prefrontal cortex in such working memory, leading us to hypothesize involvement of these regions in setting processes per se. At the outset of these studies, we had little reason to suppose a difference between left and right lesions, but subsequent literature suggests that left frontal regions may be more critical than right in set shifting. We discuss this literature later.

Subjects

Eleven patients with chronic, unilateral lesions restricted to lateral prefrontal cortex participated in this study: 6 with left frontal lesions (mean age: 63 years) and 5 with right (mean age: 60 years). Patient details are provided in table 28.1. Three of the six left frontal patients exhibited signs of aphasia; three did not. In no case was aphasia severe enough to impair understanding of the nature of the task, as revealed by subjects' errors of response (detailed in "Results"). None of the right frontal patients showed signs of aphasia. Reconstructions of the anatomical locations of the lesions are shown in figure 28.1. Five subjects served as normal controls (mean age: 67 years; age range: 65–72 years), having no documented neurological damage.

Stimuli and Apparatus

Color or shape stimuli, or both, appeared on a computer monitor. Circular color patches of red or blue, subtending a visual angle of 0.76

Table 28.1 Clinical Information

Patient	Age/Sex	Lesion	Years since lesion onset	Lesion volume (cc)
Left-hemisphere lesions				
L.S.	67F	Tumor resection	16	28
R.T.	80M	Stroke	12	39
A.L.	66F	Stroke	16	51
O.A.	63M	Stroke	10	48
J.C.	70M	Stroke	9	10
A.A.	29F	Stroke	5	59
Right-hemisphere lesions				
W.T.	50M	Tumor resection	7	26
E.B.	78F	Stroke	12	17
M.G.	32M	AVM resection	11	25
S.R.	75F	Stroke	2	13
M.K.	63M	Stroke	16	200

Notes: AVM = arteriovenous malformation. Left frontal patients R.T., A.L., and J.C. exhibited clinical signs of aphasia, primarily anomia for R.T. and A.L. and anomia and some Broca's aphasia for J.C., who also exhibited signs of hemiplegia.

degree when viewed from 60 cm, were assigned to response keys 1 and 2. Shapes were a triangle or a square in black outline and were assigned to the same two keys as color. The shapes were 3.0 degrees of visual angle high and 3.0 (square) or 4.0 degrees (triangle) wide. Key 1 corresponded to the "0" key on the computer number pad and key 2 to the decimal key. Thin pieces of wood were attached to the keys to make them both larger and the same size.

Given the nature of the stimuli, where the outline shapes were larger than the color patches, the two dimensions could be manipulated independently, with either one or both present. When both were present, the shape surrounded the circular color patch.

Procedure

There were four conditions: two-choice, four-choice, four-choice cued, and "switch 8" (switching set every eight trials). The order of practice was fixed; all subjects starting with the two-choice condition and ending with the switch 8.

Two-Choice Unidimensional Subjects were shown a card explaining that red stimuli were to be responded to with key 1 and blue stimuli with key 2. They then received two blocks of 60 trials in this two-choice condition, each color occurring equally often. The stimulus stayed on until the correct key was pressed, with the next stimulus appearing 500 msec after onset of keypress. In these trial blocks, no shape surrounded the cir-

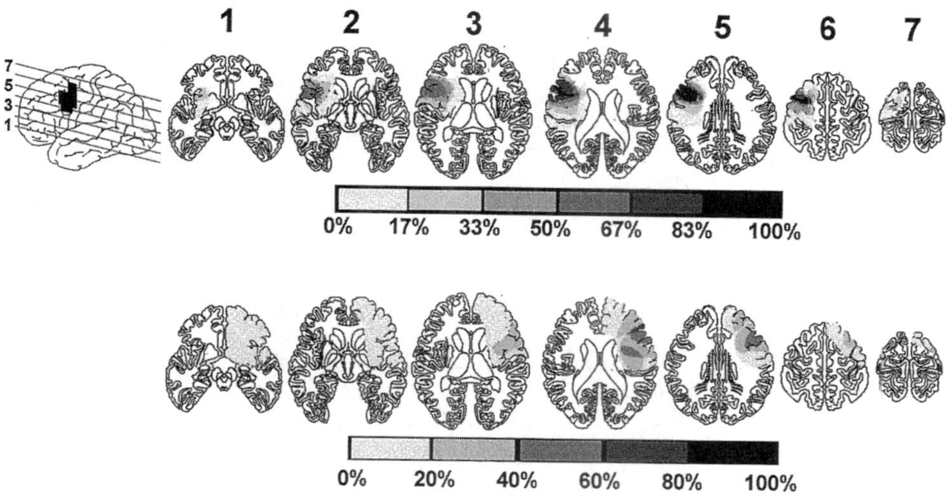

Figure 28.1 Neuroimage reconstructions of scans for patients with left-hemisphere lesions (top row) and right-hemiphere lesions (bottom row). The column numbers refer to the slices indicated on the lateral view. The scale indicates the percent of patients in the group that have a lesion in the region indicated. The lateral view shows the region of cortex involved in all the patients having left-hemisphere lesions (Brodmann's areas 44 and 6).

cular color patch. The procedure was repeated with shapes, triangle being assigned to key 1 and square to key 2, and the color patch being absent.

Four-Choice Unidimensional The procedure was the same, except that on each trial any of the four stimuli—red, blue, triangle, or square—could appear, each stimulus occurring equally often in each of the two blocks of 60 trials. When a color appeared, no surrounding shape was present; when shape was present, there was no color patch.

Four-Choice Cued The word "color" or "shape" appeared above the position at which the stimulus would appear indicating whether the following stimulus would be a color or a shape; 750 msec later, the word disappeared and the stimulus appeared, with the next cue appearing 500 msec after response. Again, two blocks of 60 trials were presented. Because the stimuli were unidimensional, subjects did not need a cue to determine the correct response, but in contrast to the four-choice unidimensional condition, the cue allowed subjects to anticipate the next dimension.

Switch 8 All stimuli were bidimensional—a color patch surrounded by a shape. On half the trials, the color and shape specified the same response (congruent); on the other half, they specified different responses (incongruent). An instructional word appeared 750 msec before each

Table 28.2 Reaction Times in Milliseconds, Error Rates in Percent for Experiment 1

	Two-choice	Four-choice	Four-choice cued	Switch 8 set switch	Switch 8 no set switch
Left frontals	580	638	653	1,058	996
	2.5	2.1	1.9	12.5	4.3
Right frontals	530	596	573	715	588
	1.9	2.4	2.7	9.0	2.6
Control subjects	439	538	492	707	502
	1.3	1.3	2.7	2.0	1.9

Note: "Switch 8 set switch" refers to trial 1 of sets of 8 trials with the same set; "Switch 8 no set switch" refers to the mean of trials 2–8 on those sets.

stimulus, indicating the relevant dimension (color or shape). The next instruction appeared 500 msec after response onset. Subjects were informed that the same instruction would be used for a series of 8 successive stimuli and then switched. There were 80 trials in each of 2 blocks of trials, and a brief rest was given after each 40 and between blocks.

Results

Summary reaction time results and error rates for experiment 1 are shown in table 28.2. The reaction time scores are based on correct responses only, ignoring the trial following an error. For the choice conditions, median reaction times were calculated for each block of trials for each subject. Means of these medians were then calculated across subjects for each condition. In the switch 8 condition, there were 8 trials in each subblock having the same set, and 10 such subblocks in 80 trials. Median reaction times were calculated from the 10 trials having the same position within a subblock (e.g., position of switch, trial first following a switch, etc.). Further analyses were based on the means of the medians. Medians were employed to eliminate the possibility that the pattern of results could be attributed to outlying reaction times that might appear differentially among the different groups. Table 28.2 shows reaction times of the switch 8 condition for the switch trials (i.e., the first in a run of 8 trials after changing set) and the average of the 7 trials of same set that follow. Figure 28.2 shows a more detailed breakdown of switch 8 reaction times, presenting times for each successive trial within a set.

Although the reaction times of the left frontal group are slightly longer than those of either the control or the right frontal group in the unidimensional conditions not requiring set (two-choice, four-choice, and four-choice cued), there is little differential effect of amount of choice. In the switch 8 condition, however, where set is required to specify the relevant dimension of bidimensional stimuli, reaction times increase sub-

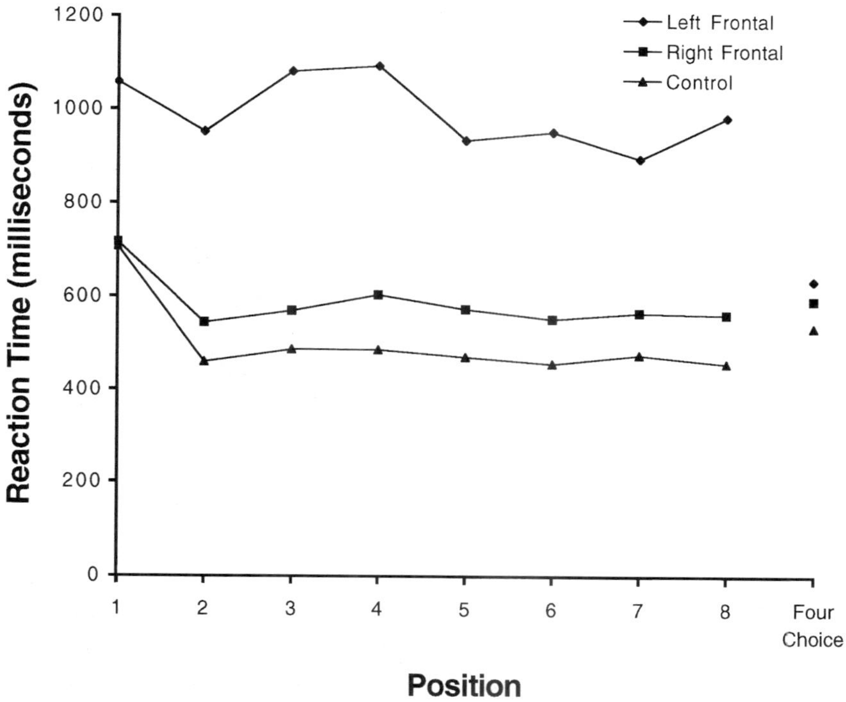

Figure 28.2 Effect on reaction time of position within subblocks of trials all with the same set in experiment 1. Position 1 refers to the initial trial with a new set, a switch trial, and the remaining positions involve retention of the same set. For comparison, the right margin of the figure portrays reaction times in the four-choice, unidimensional condition. Standard errors of the mean in the switch 8 condition are 35, 18, and 16 msec for the left frontal, right frontal, and control groups, respectively. Standard errors of the mean for the four-choice condition are 13, 28, and 9 msec for the respective groups.

stantially for the left frontal group compared to the other two groups. The greatly increased reaction time for the left frontal group when set is required is seen not only when set switches but throughout the last 7 of 8 trials where set remains the same (cf. figure 28.2), with subjects' reaction times nearly double their four-choice times. After a set switch, reaction times for the control group and the right frontal group drop immediately below the level of their four-choice times, indicating that these subjects are able to use the instructional cue to effectively filter the irrelevant dimension and to restrict their choice.

Despite their general difficulty in using set in the bidimensional case, left frontal patients are not particularly impaired at *local* switching time, as measured by the difference in reaction times on the trials of set switch versus the adjacent trials where set remains the same. (In experiment 2, we distinguish local cost from global shift cost, finding there may be a frontally based impairment in the latter.)

We turn now to a more thorough analysis.

Choice Reaction Time In the unidimensional conditions, the stimulus is sufficient to specify the correct response without a dimensional cue. Subjects in the different groups differed little in error rates, averaging 2.2, 2.3, and 1.8% in the unidimensional choice conditions for the left frontal, right frontal, and control subjects, respectively. An analysis of variance (ANOVA) was conducted on reaction times with factors of amount of choice and group. Reaction time was reliably affected by choice, increasing from the two- to the four-choice condition and falling between these two conditions when a cue specified which dimension would be presented on the next trial: $F(2, 26) = 24.01$, $p < 0.0001$. Although reaction time was in general longer for left frontal patients than for the right frontal patients and control subjects, a statistical analysis revealed group differences only to be marginally reliable: $F(2, 13) = 3.56$, $p < 0.06$. More important, the effect of amount of choice was no greater for the left frontal group than for either of the other two groups, indicating that choice per se is not affected by the left frontal lesions. If anything, the effect of amount of choice was greatest for the control group. Whereas a cue to specify the forthcoming dimension reduced reaction time below the four-choice level for the control and right frontal groups, the cue slightly increased reaction time for the left frontal group, although the ANOVA revealed the apparent interaction of group with choice to be not significant: $F(4, 26) = 1.89$, $p > 0.10$. These results suggest that the imposition of task set requirements, which will be seen to cause differential effects among groups, cannot be attributed to nonspecific increases in decision difficulty.

Switch 8 Reaction Times In the switch 8 condition, stimuli were bidimensional, with the irrelevant stimulus value being incongruent with the relevant stimulus value on half the trials. An instructional cue specified the correct dimension for runs of eight trials, and then changed to the other dimension.

Figure 28.2 shows mean reaction times as a function of the position within the eight trials of the same set. The first position is that at which set change occurs (Reaction times in the four-choice unidimensional condition are included for comparison.)

An ANOVA found reaction times to differ across position, primarily reflecting an increased reaction time when set was first switched, compared to later trials with the same set: $F(7, 91) = 5.77$, $p < 0.0001$. The three groups also differed reliably from each other, the left frontal group having a much longer reaction time: $F(2, 13) = 9.46$, $p < 0.003$. The interaction of group with position was not significant, suggesting the groups do not differ on the size of the switching effect: $F(14, 91) = 1.14$. More restricted analyses make these points more firmly.

An analysis of reaction times for positions 2 through 8, all after a set switch, found a significant effect of position: $F(6, 78) = 2.27$, $p < 0.05$; but again the interaction of group with position failed to be significant: $F(12, 78) = 1.14$. The position effect stems from the fact that reaction times improve on the trial after a set switch and then slow slightly for about two trials before set is more firmly established. The general trend, however, is similar for all groups.

Because there was no interaction of group with positions 2–8, a more powerful analysis compared reaction times of the initial switch position (position 1) with those of nonswitch positions averaged over positions 2 through 8. Switch versus nonswitch was significant, as was group: $F(1, 13) = 11.15$, $p < 0.005$; $F(2, 13) = 6.56$, $p < 0.01$, respectively. On the other hand, the interaction of group with switch was still not significant, despite this more powerful test: $F(2, 13) = 1.14$. If anything, the control group exhibited a larger switching effect (235 msec), though not significantly so, than either the left (71 msec) or right (145 msec) frontal groups. This pattern suggests that the lack of switching impairment in the patient groups is not a matter of a marginal effect failing to manifest itself.

Despite lack of a reliably different switching effect on reaction times among groups, the left frontal group is highly impaired in the situation requiring set, compared to the unidimensional choice conditions. That is, their problem is not a momentary one of switching per se but of using set even when it is unchanged over eight trials. This problem can best be appreciated by comparing reaction times on nonswitch trials with those on four-choice trials. The control and right frontal groups show shorter reaction times on the nonswitch than on the four-choice trials, approximating those in the four-choice cued condition, where a cue also restricts the possible choices. These two groups thus effectively employ set to filter the irrelevant dimension. In contrast, the left frontal group shows markedly longer reaction times than in the four-choice condition, even after set is switched. An ANOVA compared reaction time averaged over positions 2–8 with four-choice reaction time. The interaction of group with task was highly significant: $F(2, 13) = 11.5$; $p < 0.005$.

Switch 8 Errors As seen in table 28.2, the left frontal patients exhibit a larger error rate overall than either the right frontal or the control group, though even on switch trials the left frontal patients were correct on average 87.5% of the time, indicating general success in switching of set. The group difference was reliable: $F(2, 13) = 5.0$, $p < 0.03$. Moreover, the left frontal group exhibited a larger decrease in error rate from the initial switch position to the mean of the seven nonswitch positions, revealed in a significant interaction of group with position: $F(2, 13) = 4.2$, $p < 0.04$.

Although, on the surface, such results would suggest, contrary to the reaction time data, that left frontal patients show a larger switching effect than right frontal patients or control subjects, error data are dificult to interpret. One reason relates to scaling issues. Relatively large changes in error rate when the baseline established by the lower rate is itself high (from 12.5% error on the switch trial to 4.3% on nonswitch trials for left frontal patients) cannot easily be compared to smaller differences when baseline rates are lower (9% versus 2.6% for right frontal patients). Converting these to *z*-score differences, for example, reverses the order, producing a marginally larger switching effect for right frontal than for left frontal patients. While it might be argued that the 2.0 and 1.9% error rates in the switch and nonswitch conditions of the control subjects reflect no switching effect on errors (*z*-score difference of 0.01), both probability and *z*-score estimates are extremely unreliable for such small rates. When an ANOVA was performed using an arcsine transformation, a procedure commonly recommended with errors, a significant group difference remained: $F(2, 13) = 4.9$, $p < 0.03$. However, the interaction was no longer reliable: $F(2, 13) = 2.6$.

We also examined whether subjects showing a large switching effect on errors might show a reduced switching effect on reaction times. Based on all twenty-one subjects of experiment 1, the correlation between the two switching effects was near zero (−0.05). Based only on the six left frontal patients, the correlation remained near zero (0.26). Thus there appears to be no compensatory effects between reaction times and error rates on the size of the switching effect.

We must conclude that the evidence from both reaction times and error rates fails to support a deficiency in the local cost of switching set; what is more, the left frontal patients are markedly impaired in a situation requiring task set, though their difficulty appears not to be one of switching per se.

Switch 8 Congruency Analysis In the two- and four-choice conditions the stimuli were unidimensional, allowing no possible response conflict from competing values on two dimensions. In the condition requiring set, however, the stimuli were bidimensional, and on half the trials the values on the two dimensions were incongruent, specifying different responses, (on the other half, they were congruent). Because the sparsity of data does not allow a breakdown of the congruency effect by switch versus nonswitch trials, the congruency analyses ignored position within groups of trials all having the same set. Table 28.3 indicates that for all three subject groups, performance on incongruent trials was slower and more error prone than on congruent trials. The congruency effects were largest for the left frontal group. An ANOVA on reaction times revealed a significant effect of congruency: $F(1, 13) = 10.0$, $p < 0.007$, but a nonsignificant interaction of group with congruency at conventional levels for reaction time: $F(2, 13) = 2.7$, $p < 0.10$. Using an arcsine transformation of errors, an ANOVA revealed both a significant congruency effect and a marginally significant interaction of group with congruency: $F(1, 13) = 45.1$, $p < 0.0001$, and $F(1, 13) = 3.6$, $p < 0.055$, respectively.

Table 28.3 Reaction Times (RT) in Milliseconds, Error Rates (ER) in Percent for Congruent and Incongruent Trials in the Switch 8 Condition of Experiment 1

	Congruent		Incongruent	
	RT	ER	RT	ER
Left frontals	944	1.5	1,067	5.4
Right frontals	563	0.7	590	7.6
Control subjects	464	0.7	498	2.3

The two sources together, error rates and reaction times, suggest that the left frontal patients are less efficient in filtering the irrelevant dimension than are control subjects. This suggestion needs to be taken with caution, however, because the exact meaning of a larger congruency effect when superimposed on markedly larger baseline reaction times is uncertain and would depend on a more precise computational model.

Discussion

Subjects with lesions of frontal cortex in experiment 1 exhibited little decision time deficit when each stimulus uniquely specified the appropriate response, dependent only on general task instructions. No frontal deficit emerged with an increase in choice difficulty, even though the stimulus dimension (shape or color) was unpredictable. The increased reaction time in the choice case is contrary to earlier results of Spector and Biederman (1976), who found no increase in reaction time when dimensions alternated. The discrepancy is likely due to the unpredictability of our dimensional changes.

When set was required to specify which dimension of a *bidimensional* stimulus was relevant, compared to the unidimensional case, reaction times increased greatly for left frontal patients, but less for right frontal patients and controls. The large increase for the left frontal patients does not seem attributable to an increase in time to reconfigure set, after which processing is normal. Their reaction times remained deficient even on seven trials following set shift, where the same set was maintained. Indeed, the difference in reaction times between shift and nonshift trials (our measure of local switch cost, presumably reflecting reconfiguration time) was less for the left frontal group than for the two other groups, though not significantly so.

While the reaction time results suggest no increased reconfiguration time for left frontal patients, the results remain somewhat problematic because error rates exhibit an opposing trend. However, arcsine transformations on errors failed to reveal a significantly different switching effects among groups.

Overall, therefore, the results of experiment 1 provide no support for a local switching deficit defined by the difference in time between an initial trial with a new set and succeeding trials with the same set, a measure thought to reflect, at least in part, the time to reconfigure from one set to another (e.g., Rogers and Monsell 1995; Meiran 1996; Mayr and Keele 2000).

An alternative expectation might have been that in frontal patients a new set would improve over successive trials with the same set. This also was not the case. Set for all groups was more or less at its most efficient after one trial, a result also similar to one of Rogers and Monsell 1995. Although, on the surface, this result would suggest that frontal patients not only exhibit no local shift cost but also no global shift cost, experiment 2 will shed new light on the matter.

Left frontal subjects exhibit a significantly higher overall error rate than controls in the condition requiring set. They also show a larger congruency effect both on reaction times and errors, though only marginally reliable. These two features suggest that even an established set is less efficient for the left frontal patients than for control subjects in that the irrelevant dimension is less well filtered . Nonetheless, it is quite striking that left frontal patients, on average, are over 95% correct following a switch, and that their deficit is revealed primarily in long reaction times, whether set is switched or not. This suggests that mechanisms for resolving conflict between dimensions are largely retained in left frontal patients, though resolving such conflict takes more time.

Experiment 1 suggests that, consistent with Rogers et al. 1998, the set deficit arises from left rather than right prefrontal lesions, a result that does not seem due to differences in the location or size of the lesions, ignoring hemisphere. First, the lesions of the left and right frontal patients were all lateral prefrontal and were about the same size (see table 28.1). Second, there was considerable overlap in lesion location on homologous sides of the cortex for the two groups (see figure 28.1). Third, no subject in the right frontal group exhibited reaction times in the switch 8 condition as long as the average of the left frontal group; indeed, all members of the right frontal group exhibited shorter reaction times than any save one member of the left frontal group.

Within the left lateral prefrontal region, finer localization may be possible. Figure 28.1 reveals the area in common to all the left frontal patients is in Brodmann's area 44 and (part of) Brodmann's area 6. The left focus in lateral prefrontal cortex for a setting process corresponds to suggestions from other literature using paradigms similar to the current one. This holds both for patient analysis (Rubinstein, Evans, and Meyer n.d.; Rogers et al. 1998), and neuroimaging analysis (Meyer et al. 1997, 1998; Postle and D'Esposito 1998). These latter two studies have implicated left Brodmann's areas 9, 44, 45, and 46. Konishi et al. (1998), using fMRI, found activation of Brodmann's areas 44/45 to be associated with putative set switching on a task related to the Wisconsin Card-Sorting Test (WCST). In their case, however, right- as well as left-side frontal foci were activated. Although the right-side focus may reflect additional processes involved in the complex problem-solving activities of WCS-like tasks, to confirm such a conclusion would require additional investigation.

In summary, several studies suggest that left lateral prefrontal cortex centered in or near Brodmann's area 44 is critically involved in a situation where set is frequently switched. Nonetheless, a puzzle remains. While the left lateral prefrontal region is important where set frequently changes, experiment 1 provided no evidence that the left frontal deficit was confined to the local occurrence of a switch. Slowing, compared to the "no-set" unidimensional conditions, occurred not only at the point of the switch but also in relatively undiminished form over a series of as many as seven additional trials employing the same set. Experiment 2 provided insight into the nature of the left frontal problem, suggesting a global rather than a local set-shifting deficit.

28.2 EXPERIMENT 2

Experiment 2 examined performance of a single left frontal patient over four sessions. An original interest concerned whether difficulties in situations requiring frequent changes of set were reduced with practice, but this turned out not to be the case. Extended practice provided sufficient data, however, to compare performance on the initial set in a block of trials with performance on subsequent sets in the same block, which became the central feature of experiment 2. Because the patient's performance replicated comparable results of experiment 1, the patient can be fruitfully compared to control subjects of that experiment.

A version of the switch 8 condition of experiment 1, in this case, switching every 6 trials within a block of 48 trials, was employed, allowing more blocks and more observations of set switch points. The bulk of the experimental sessions involved shifting between colors (red and blue) and size, with the shape dimension and the colors yellow and green reserved to assess transfer of any learning. Because practice effects were negligible, making transfer issues moot, we simplify the details of the procedure and results, excluding further mention of trial blocks involving either shape or the colors yellow and green.

Subject

One of the left frontal patients who had participated in experiment 1 (patient LS of table 28.1) participated in four sessions of experiment 2. This patient had shown set deficits that were representative of the group as a whole. A reconstruction of lesion location, involving (most of) left Brodmann's area 44 and (parts of) 45 Brodmann's areas and 6, is shown in figure 28.3.

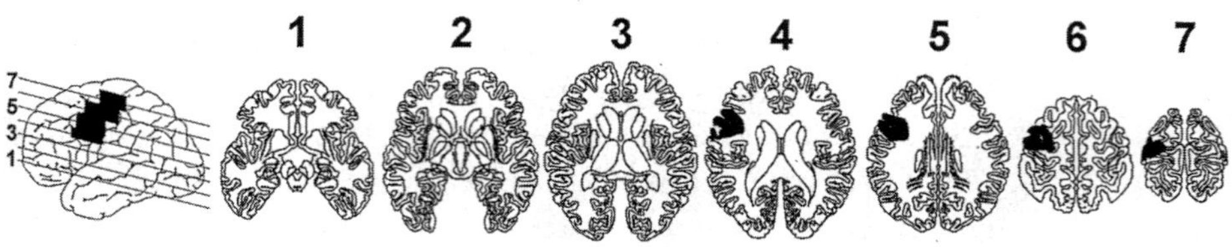

Figure 28.3 Neuroimage reconstruction of lesion location for left frontal patient L.S. of experiment 2.

Stimuli and Apparatus

The stimuli were an octagon in black outline surrounding a central, circular color patch. The octagon could be large in size (key 1), medium (no response), or small (key 2). The large octagon was approximately 6 cm in diameter; the medium, 3.5 cm; and the small, 1.5 cm. The color patch centered within the octagon was approximately 1 cm in diameter and was black (neutral and not assigned to a key), red (key 1) or blue (key 2). Responses were made on two keys approximately 1.5 cm square and separated by 10 cm.

Conditions

On each of four sessions spread over nine days, the subject participated both in a baseline ("unidimensional") condition where a single dimension, (color or size) was relevant for an entire block of trials (the irrelevant dimension taking the neutral value) and in a condition where nonneutral values occurred on both dimensions (the relevant dimension being specified by the instructional cue "size" or "color") and where task set was switched every six trials of a block.

Each session involved two unidimensional color blocks of 48 trials, size having a neutral value, followed by two similar blocks with size relevant. These were followed by blocks of 48 bidimensional trials in which the word "color" or "size" on the computer screen cued which dimension was relevant for each trial. There were three bidimensional blocks each on sessions 1 and 4, and five on sessions 2 and 3. In subsequent analyses, all blocks on a day (sessions 1–4) were averaged to yield a variable of session.

The 48 trials of each bidimensional block were divided into 8 subblocks of 6 trials each. The written set cue ("color" or "size") was presented 500 msec after one response and 500 msec prior to the next stimulus. The same set cue was used for 6 trials and then changed to the alternate cue. Altogether, 6 different block arrangements were used: 3 starting with a color set and 3 with a size set. Although not done systematically, the first set of a subsequent block was the same as (or different from) the last set of the prior block about half the time.

Results

Over the four sessions, error rates averaged 0.7% in the two-choice color and size unidimensional cases. In the bidimensional cases, where set shifted every six trials of a block, error rates averaged 8.6% on the first trial of a shift and 1.4% across the five nonshift positions. Given the sparsity of data for critical comparisons, no further error analyses were conducted.

Because, within a session, only 3 to 5 scores were available for each combination of position within 6 trials of the same set and the 8 subblocks of 6 trials each, too few for a reliable assessment of median, reaction times were analyzed based on means. Nonetheless, the subject of this study exhibited a pattern of results similar to that in similar conditions of experiment 1, where medians had been used to eliminate aberrant reaction times. In particular, the bidimensional reaction times at all positions within a set were considerably longer than the "unidimensional" reaction times, in contrast to the pattern exhibited by the right frontal patients and control subjects of experiment 1.

A preliminary analysis revealed no reliable difference in reaction time to the color and size dimensions. Subsequent analyses were therefore collapsed over dimension, leaving as factors "session" of the experiment (1–4), "subblock" (1–8), and "position within subblock" (1–6); each set, color or size, was applied for a subblock of 6 trials, yielding 6 positions within a set and 8 subblocks in a block of 48 trials).

An ANOVA on the bidimensional data revealed all three main effects of session, subblock, and position within subblock to be reliable: $F(3, 105) = 11.4$, $p < 0.001$; $F(7, 105) = 42.7$, $p < 0.001$; and $F(5, 105) = 3.7$, $p < 0.005$, respectively. None of the two-way interactions approached significance; the three-way interaction served as the error term for the analysis (using a higher-order interaction as the error term results in a conservative analysis because it includes not only error variance but potentially an additional variance component).

The effect of practice, as revealed by the session variable, is shown in table 28.4 both for the unidimensional two-choice case and for the bidimensional switch 6 case, which is subdivided into the first trials of subblocks of six where set was first switched and the average of the remaining five trials retaining the same set. Clearly, the left frontal patient of this study exhibits the same general phenomenon exhibited by the left frontal group as a whole in experiment 1. Reaction times in the bidimensional case are more than twice as long as those in the unidimensional case (control subjects of experiment 1 showed relatively similar reaction times in the two cases). There is no systematic improvement in reaction time in the two-choice case over sessions for the patient of this study. The significant effect of session in the bidimensional case is primarily due to a modest reduction in reaction time (~150 msec on average) from session

Table 28.4 Mean Reaction times in Milliseconds for Experiment 2

		Session			
		1	2	3	4
Two-choice		749	806	779	808
Switch 6	Switch	1,831	1,734	1,606	1,737
	Nonswitch	1,792	1,624	1,596	1,672
	Switch time	39	110	10	65

Table 28.5 Mean Reaction Times in Milliseconds within Subblocks of 6 Trials and for 8 Subblocks in Experiment 2

	Position in subblock						
	1	2	3	4	5	6	Mean
Subblock 1	1,268	1,057	1,012	1,122	1,076	1,160	1,116
Subblock 2	1,605	1,669	1,794	1,555	1,684	1,618	1,654
Subblock 3	1,940	1,670	1,907	1,690	1,674	1,627	1,751
Subblock 4	1,874	1,950	1,750	1,781	1,628	1,694	1,780
Subblock 5	1,891	2,041	1,944	1,890	1,662	1,629	1,843
Subblock 6	1,742	1,678	1,746	1,786	1,814	1,687	1,742
Subblock 7	1,722	1,845	1,759	1,765	1,637	1,671	1,733
Subblock 8	1,776	1,828	2,080	1,697	1,855	1,715	1,825
Mean	1,727	1,717	1,749	1,661	1,628	1,600	

1 to the subsequent sessions, and there is no systematic tendency for the switching effect to change with practice.

Table 28.5 breaks down the effect of position within a subblock of six trials by the eight subblocks. The overall effect of position is shown in the bottom margin of the table; the overall effect of subblock, on the right margin. Reaction time on the initial trial of a set (position 1) is not noticeably longer than on the second or even third trial. Reaction times improve slightly on positions 4, 5, and 6, indicating a gradually strengthening set. In contrast, experiment 1 showed longer reaction times on the initial trial where set was newly changed, and no systematic improvement beyond the second trial. Despite minor differences, the two experiments yield some common findings. For the subject in experment 2, there is no impairment in switching (based on comparisons with control group switching times of experiment 1), as would be indicated by a considerably longer reaction time for position 1 versus later positions with the same set. Rather, reaction times are uniformly long over all positions in the bidimensional condition. If switching time is indicated, not by the difference between initial trials of new sets and the trials immediately following, but by initial trials of new sets minus the last trials of the sets (i.e., trial 6 of a subblock), switching time is only 127 msec, still shorter than the average for control subjects of experiment 1. Error rates are larger on the initial trial of a new set (8.6%) than on the subsequent five trials with the same set (1.4%), suggesting an initial shift cost, but it is difficult to make comparison with the results of experiment 1.

The primary lesson to be drawn at this point is that the results for this single left frontal patient are similar to those of the left frontal group as a whole in experiment 1, justifying comparison of results from experiment 2 with those from the control subjects of experiment 1. Most critically, unlike the control subjects of experiment 1, the bidimensional reaction times of the patient in experiment 2 remain substantially longer than her unidimensional reaction times, even following set shift.

The most revealing result of experiment 2 concerns the reliable effect of subblock of trials, shown in overall form in the right margin of table 28.5. Recall that subblock refers to the 8 different sets within a block of 48 trials, set changing every 6 trials. Reaction times are strikingly shorter—on the order of 500 to 600 msec shorter—on the first subblock of trials than on any of the other subblocks, though still longer than reaction times in the unidimensional two-choice case (see table 28.4). Reaction time within the first subblock is relatively constant across all six trial positions, then increases on the first trial of the second subblock (trial 7 of a block), which is the first trial of a new set. Reaction time remains relatively constant across all remaining trials of the block regardless of whether the set changes once again or not.

In short, the reliable effect of subblock appears to reflect a global shifting deficit. The first set of a block of trials, while exhibiting some reaction time impairment, exhibits much less impairment than subsequent subblocks. Following shift from the first set of the block, reaction time increases markedly and remains high through out the remaining trial block. Such results indicate a residual effect of earlier sets on subsequent sets. This proactive influence abates to large degree in the substantial rest period between blocks of trials.

It is important to note that before the first trial of a block of 48 trials, the subject is unaware of whether color or size will be the attended dimension for the first subblock. It also is the case that on about half the occasions the set of the first subblock was changed from that of the last set of the preceding trial block. Despite the unpredictability of what the first set will be, its establishment is markedly easier than that of the subsequent set changes in the block. (Because these effects were unanticipated, the rest break was not rigidly controlled, varying from 20 or 30 seconds to a minute or so, and often involving conversation to keep the patient motivated over a difficult experiment of several sessions.)

Although there was no control group in experiment 2, given the similarity of paradigm and patient results in experiments 1 and 2, we can safely infer from experiment 1 that control subjects would have portrayed a different pattern (see table 28.2). In experiment 1, controls exhibited bidimensional reaction times on trials following set change that were both shorter than unidimensional four-choice reaction times and somewhat longer than two-choice times. Given this narrow bracketing, it would be unlikely that the first subblock of the bidimensional trials for control subjects could be much faster than the other subblocks. They would not be expected to fall below the unidimensional two-choice case. Thus, if there would be any tendency for control subjects to show reduced reaction times on the first subblock of bidimensional blocks compared to subsequent ones, the tendency must be markedly expanded for the left frontal patient in experiment 2.

Discussion

Experiment 2 provided sufficient data for a detailed look at performance at different points in the trial block. Patient L.S. had a greatly reduced problem of task set on the first set of each block of bidimensional trials, though some difficulty compared to unidimensional reaction time remained. On subsequent sets, difficulty was greatly magnified. Although there was a relatively small improvement in reaction time over successive trials with the same set, such times did not approach those of the first subblock. As it was for the left frontal patients of experiment 1, patient L.S.'s difficulty does not appear to be a local one of the time taken to reconfigure or switch set, followed by normal reaction times, but rather remains even after set shift.

The bulk of the setting difficulty first appears following change of an initial set of a trial block. Although this constitutes a switching cost, because it persists throughout a block and is not added to by additional set changes, with the possible exception of the first one or two, the cost is global: it represents *nondiminishing proactive interference* of initially established sets on later ones.

Such global shift cost appears quantitatively, perhaps even qualitatively, different from shift cost in young normal subjects. Allport, Styles, and Hsieh (1994; cf. also Allport and Wylie, chap. 2, this volume) have shown proactive effects of prior sets on the current one, but such effects rapidly diminished over a short set of trials. In our experiment 1, set difficulties of left frontal patients showed little evidence of abatement over eight trials, all with the same set; in our experiment 2, there was only modest abatement over six trials. What is more, left frontal patients' reaction times in the situation after several trials with the same set remained markedly above those in unidimensional conditions when set (beyond general instructions) was not required. In contrast, control subjects' reaction times in experiment 1 immediately after set shift returned to a point near reaction times where set was not required. Rogers and Monsell 1995 also showed no residual proactive effect of former set beyond a single trial in normal, young subjects. Thus the proactive pattern of prior set in patient L.S. in experiment 2 appears to differ from that in normal subjects.

Although the proactive effect of prior sets in experiment 2 is a form of perseveration, a commonly reported problem with frontal patients, two aspects are especially worthy of further emphasis. First, the perseveration does not manifest itself in an undue number of errors. Once set has been altered, error rates remain low (1.4%), whereas reaction times remain very long. Second, even when a new set is adopted, the proactive influence of the old set appears not to diminish to any degree until there is a substantial rest break between blocks of trials.

28.3 GENERAL DISCUSSION

The primary results from experiments 1 and 2 are as follows:

1. Patients with lesions of prefrontal cortex suffer little impairment in decisions where a unidimensional stimulus and general experimental instructions are sufficient to specify a response.

2. Patients with lesions to left (but not right) lateral prefrontal cortex suffer impairment when bidimensional stimuli require task set, at least set that changes from occasion to occasion. The critical lesion site appears centered in Brodmann's area 44 but may encompass parts of nearby areas.

3. The impairment manifests itself on trials following set shift as well as on the shift trials themselves. Even after several trials having the same set, reaction times of the left frontal patients for bidimensional stimuli fail to revert to the level for unidimensional stimuli, unlike reaction times of right frontal patients and control subjects. Thus the deficit is not a local one of immediate shift per se.

4. The analysis of the single left frontal patient in experiment 2 reveals a markedly reduced, though not absent, impairment on the first set of a block of trials where set frequently changes. The impairment increases substantially when the set changes in the same block of trials, and diminishes only marginally over several trials all with the same set. Thus a major component of the left frontal problem with task set appears to be one of a perseverative influence of prior set.

5. When set is required for bidimensional stimuli, there is increased susceptibility to conflict between the relevant and irrelevant dimensions in left frontal patients, compared to control subjects.

These phenomena speak to a number of issues regarding prefrontal cortex and the establishment and alteration of task set.

Do Prefrontal Patients Exhibit a Set-Switching Deficit?

Our results provide no evidence for an increase in reaction time cost in left or right frontal patients on shift versus nonshift trials. Such a result suggests that the left frontal deficit is not one of time to reconfigure set. In contrast, two indices suggest a global shift cost associated with left prefrontal cortex: (1) a greatly magnified cost in the situation where set is required regardless of whether a change has just occurred; and (2) a strong perseverative effect on later sets of an initial set.

How does this conclusion correspond to the existing literature? Consider first Rubinstein, Evans, and Meyer (n.d.) and Meyer et al. 1997, 1998. In the Rubinstein, Evans, and Meyer study, a written instruction (e.g., "shape") indicated which dimension of one stimulus subjects were to match to a series of other stimuli. Even when the same instruction was used for an entire block of trials, reaction times of the left frontal patients were substantially longer than for the control subjects. This contrasts with our results, where reaction times of patients and controls differed little for unidimensional stimuli. The discrepancy could be because our control involved unidimensional stimuli, whereas Rubinstein and colleagues' paradigm maintained neutral values on irrelevant dimensions. Their main finding, however, was that reaction time increased even more in left prefrontal patients than in control patients on trial blocks when set changed on each trial, consistent with our findings.

The studies of Meyer and colleagues employed neuroimaging analysis and showed regional cerebral blood flow to increase in left prefrontal cortical regions on blocks of trials that involved switching between color and shape sets as opposed to blocks using only one set or the other.

The studies of Rubinstein, Evans, and Meyer and of Meyer et al. are consistent with the current study in identifying left prefrontal cortex as relevant to task switching. Unlike the current study, however, they cannot distinguish between local and global shifting costs. They do not compare shifts versus nonshifts within the same trial block, a comparison central to our conclusion that shifting deficits from left frontal lesions are restricted to the global level.

Our conclusion is also consistent with analyses of problem-solving tasks akin to the Wisconsin Card-Sorting Test, such as Owen et al. 1993, which found that frontal patients take longer than control subjects to discover a new sorting basis after a switch of set. Owen and colleagues attributed the frontal patients' problem to difficulty in inhibiting a prior set. Were the switching difficulty simply one of a longer time to reconfigure set upon exposure to new conditions, as opposed to simply a longer time on an initial trial, there would be no reason to expect additional trials to solution. Their findings, like ours, suggested a prolonged perseverative effect of prior sets.

Most problematic for our conclusions is Rogers et al. 1998, which suggested a *local* set-switching deficit in left frontal patients. Subjects alternated every two trials between naming either the digit or the letter of letter-digit pairs. In contrast to our study, reaction times of left frontal patients increased more than those of control subjects on switch trials compared to immediately adjacent nonswitch trials.

Some perspective on the discrepancy is provided by Stablum et al. (1994), many of whose closed-head injury patients likely had frontal cortical damage. Stimuli were arrows pointing left or right or syllables. The two tasks alternated either every two trials, much as in Rogers et al. 1998, or every ten trials, closer our study, where set switched every six or eight trials. When set alternated every ten trials, Stablum et al.'s patients exhibited the same pattern as our left frontal patients: longer overall reaction times than those of controls, but no significant difference in local switching cost (154 msec for patients versus 137 msec for controls). In contrast, when set alternated every two trials, switching costs differed significantly (59 msec for patients versus 19 msec for controls), a pattern replicating the Rogers et al. 1998 result.

Why did Stablum et al.'s patients show an inflated switching cost when set switched every two, but not every ten, trials? When switching occurs every two trials, normal subjects likely anticipate when the switch will be needed, and they reconfigure set in the 500 msec interval between the response on the last trial and the next stimulus. In accordance with this view, switch cost is minuscule (19 msec). Patients appear not to take the same advantage of the predictable switching every two trials. When set alternates every ten trials, however, it is difficult to keep track of exactly when a switch will occur. In that case, we might presume that switching occurs only when the new stimulus appears, and reaction times reflect a full-blown, local switching cost. Notably, patients and controls do not differ in switching cost when set switches every ten trials. Stablum et al.'s results suggest that the impairment of their closed-head injury patients reflects lack of advance planning for a switch, that is, lack of *advance* reconfiguration, rather than lack of time per se to reconfigure set.

A similar interpretation may be made for Rogers et al. 1998, where set strictly alternated every two trials with a 1,000 msec interval between one response and the next trial. Left frontal patients exhibited switching costs larger than those of controls. Judging from Stablum et al.'s very similar paradigm, we might predict that the switching times of *control* subjects would increase in the Rogers et al. paradigm (where set switched every ten trials) and indeed that there would be no difference in switching times between frontal patients and controls—the precise result we found in experiment 1 (where set switched every eight trials). In this scenario, the frontal deficits apparent as local switch costs may reflect not so much the time to reconfigure set per se as the failure to use time between cue and imperative stimulus to effect the switch. Such a deficit might be called a "planning deficit". Although a planning deficit interpretation offers a possible resolution of the discrepancy in local switch costs between Rogers, et al.'s and our results, a resolution supported by Stablum et al.'s results, research to date has been inadequate to clearly differentiate planning from implementation deficits in frontal patients. To do so would require comparing predictable with unpredictable shifts and varying the time available between shift cue and imperative stimulus. Nevertheless, our data strongly indicate a global shifting deficit associated with left frontal damage; we consider the causality of such a deficit in the next subsection.

Possible Underlying Causes of a Global Switching Deficit

Although we suggest that left prefrontal damage around Brodmann's area 44 produces no local deficit in time to reconfigure set, experiment 2 suggests a global switching deficit. The single patient of that experiment exhibited a carryover effect of a preceding set that did not significantly abate over several trials of a new set and that only dissipated during rest breaks of a minute or so during which the patient engaged in different behavior. Such a global switching deficit could reflect either impaired excitatory or impaired inhibitory processes (cf. Kimberg and Farah, chap. 32, this volume).

By "impaired excitatory processes," we mean that sets are only weakly activated. In the face of any residual activation from prior sets, a weakly activated new set might require time-consuming competitive processes for correct action to dominate. Our data provide no direct evidence for or against this view except our finding that, even with extensive rest breaks between trial blocks, reaction time of the left frontal patient of experiment 2, though improving, still remained longer than normal where specific set was required.

Alternatively, left frontal patients may be deficient in inhibition of prior sets. Incomplete or weakened inhibition , like weakened excitation, may result in increased time on each trial to resolve conflict of the appropriate dimension. Our data regarding congruency provide one hint for impaired inhibitory processes. In incongruent trials, the irrelevant dimension specifies a response different from that of the relevant dimension; in congruent trials, the two dimensions correspond in response. The left frontal patients of experiment 1 exhibited a larger difference, marginally reliable, between these two conditions than did control subjects, suggesting the now irrelevant but previously relevant dimension was less inhibited in the patients.

Also consistent with the inhibition view is an observation of Rubinstein, Evans, and Meyer (n.d.), who found the shifting deficit of left frontal patients, which we have interpreted as a global switching deficit, to be reduced, though not eliminated, when the dimension of a prior set was removed upon switching to a new set. Thus at least a portion of the setting problem appeared attributable not to dealing with changes in set per se but to ignoring previously relevant dimensions. Similar results were found by Owen et al. (1993) from a paradigm related to the Wisconsin Card-Sorting Test. Problem-solving difficulty exhibited by frontal patients was reduced when a formerly relevant dimension was removed upon switching to a new dimension.

While these findings are consistent with a view of impaired inhibitory processes resulting from lesions to left prefrontal cortex, they are not compelling. As with the suggestion that left frontal patients are deficient in planning processes, to determine an inhibitory deficiency will require paradigms specifically targeted on such process (see, for example, Mayr and Keele 2000).

Might the global switching deficit be related to language deficits resulting from left frontal damage, especially given that the dimensional cue was verbal? Our experiments suggest not. Of the six left frontal patients of experiment 1, three exhibited signs of aphasia (table 28.1); three did not. A cardinal sign we have developed for a set deficiency is that reaction time following set shift fails to return to baseline levels established in the unidimensional task. Mean reaction time of the three aphasic subjects on the nonshift trials of the switch 8 condition was 1,102 msec; mean reaction time on the unidimensional four-choice trials was 723 msec. Comparable times for the nonaphasic left frontal patients were 872 and 554 msec. Thus, while the nonaphasic patients were faster over all, they still were much slower in the situation requiring set, even following a shift, than in unidimensional decisions, suggesting that set deficits are not tied to aphasia. Confirmation of this conclusion comes from the more extensive analysis of the nonaphasic patient in experiment 2, where a set deficit clearly remained.

Comparisons to Other Subject Populations

It is useful to compare the present results with those resulting from damage to the basal ganglia as a result of Parkinson's disease. In a paradigm with some similarities to our current one, we (Hayes et al. 1998) examined set shifting in Parkinson patients. Compared to control subjects, these patients are slower on switch than on nonswitch trials. Thus they appear impaired in local switch cost, presumably reflecting the time to reconfigure set. Unfortunately, Hayes et al. 1998 was not designed to assess global switch cost—that is, long-term carryover effects of prior sets.

Rubinstein, Evans, and Meyer (n.d.) found marked slowing on switch versus nonswitch trial blocks, not only for left frontal patients but also for left temporoparietal (posterior) patients. In contrast to the frontal patients, however, the deficit exhibited by the posterior patients was not ameliorated when the dimension of a prior set was removed upon changing to a new set. Such would suggest that whatever the cause of the posterior deficit, it does not lie in impaired inhibitory processes.

The current study has raised the need for more careful differentiation of the various processes related to task set. Although we did not directly examine "planning" in our study, the results of Stablum et al. 1994 lead us to predict that left frontal patients will be deficient in such a process, as reflected in relative failure to employ preparatory intervals to change set prior to stimulus onset. Our own results strongly suggest that left frontal patients also exhibit long-term carryover effects of prior sets, which we have called "global shift cost," and we hypothesize that these are due to impaired inhibitory processes. We found no evidence that patients with lesions of left prefrontal cortex, in contrast to Parkinson's patients, are deficient in time to reconfigure set, as reflected in local switching cost, leading us to hypothesize a dissociation between the two patient classes. We now have available procedures to test these hypotheses, and efforts are under way to do so.

NOTE

We very much appreciate the advice and help on experiments and analysis provided by Avishi Henik, Amy Hayes, Allen Day, Kristi Hiatt, Gregg DiGirolamo, Richard Ivry, and Ulrich Mayr, as well as the useful comments of three anonymous reviewers. This project was funded by National Institutes of Health grant NS17778 to Steven W. Keele, Robert Rafal, and Richard Ivry.

REFERENCES

Allport, D. A., Styles, E. A., and Hsieh, S. (1994). Shifting intentional set: Exploring the dynamic control of tasks. In C. Umiltà and M. Moscovitch (Eds.), *Attention and Performance XV*, pp. 421–452. Hillsdale, NJ: Erlbaum.

Fuster, J. M. (1985). The prefrontal cortex and temporal integration. In A. Peters and E. G. Jones (Eds.), *Cerebral Cortex*, vol. 4. pp. 151–177. New York: Plenum Press.

Goldman-Rakic, P. S., and Selemon, L. D. (1990). New frontiers in basal ganglia research. *Trends in Neurosciences, 13,* 241–243.

Hayes, A. E., Davidson, M. C., Keele, S. W., and Rafal, R. D. (1998). Toward a functional analysis of the basal ganglia. *Journal of Cognitive Neuroscience. 10,* 178–198.

Konishi, S., Kyoichi, N., Uchida, I., Kameyama, M., Nakahara, K., Sekihara, K., and Myashita, Y. (1998). Transient activation of inferior prefrontal cortex during cognitive set shifting. *Nature Neuroscience, 1,* 80–84.

Mayr, U., and Keele, S. W. (Forthcoming). Changing internal constraints on action: The role of backward inhibition. *Journal of Experimental Psychology: General, 129,* 4–26.

Meiran, N. (1996). Reconfiguration of processing mode prior to task performance. *Journal of Experimental Psychology: Learning, Memory, and Cognition, 22,* 1423–1442.

Meyer, D. E., Evans, J. E., Lauber, E. J., Rubinstein, J., Gmeindl, L., Junck, L., and Koeppe, R. A. (1997). Activation of brain mechanisms for executive mental processes in cognitive task switching. Poster presented at March Cognitive Neuroscience Society, Boston.

Meyer, D. E., Evans, J. E., Lauber, E. J., Rubinstein, J., Gmeindl, L., Junck, L., and Koeppe, R. A. (1998). The role of dorsolateral prefrontal cortex for executive cognitive processes in task switching. Poster presented at April Cognitive Neuroscience Society, San Francisco.

Meyer, D. E., and Kieras, D. E. (1997). A computational theory of executive control processes and multiple-task performance: 1. Basic mechanisms. *Psychological Review, 104,* 3–65.

Norman, D. A., and Shallice, T. (1986). Attention to action: Willed and automatic control of behavior. In R. J. Davidson, G. E. Schwartz, and D. Shapiro (Eds.), *Consciousness and self-regulation*, vol. 4, pp. 1–18. New York: Plenum Press.

Owen, A. M., Roberts, A. C., Hodges, J. R., Summers, B. A., Polkey, C. E., and Robbins, T. W. (1993). Contrasting mechanisms of impaired attentional shifting in patients with frontal lobe damage or Parkinson's disease. *Brain, 116,* 1159–1175.

Postle, B. R., and D'Esposito, M. (1998). Homologous cognitive mechanisms and neural substrates underlie dissociable components of set-shifting and task-switching phenomena. *Society for Neuroscience Abstracts, 24.*

Rogers, R. D., and Monsell, S. (1995). The cost of a predictable switch between simple cognitive tasks. *Journal of Experimental Psychology: General, 124,* 207–231.

Rogers, R. D., Sahakian, B. J., Hodges, J. R., Polkey, C. E., Kennard, C., and Robbins, T. W. (1998). Dissociating executive mechanisms of task control following frontal lobe damage and Parkinson's disease. *Brain, 121,* 815–842.

Rubinstein, J., Evans, J. E., and Meyer, D. E. (n.d.). The control of task switching in patients with prefrontal and parietal cortex damage.

Shallice, T. (1988). *From neuropsychology to mental structure.* Cambridge: Cambridge University Press.

Smith, E. E., Jonides, J., Koeppe, R. A., Awh, E., Schumacher, E. H., and Satoshi, M. (1995). Spatial versus object working memory: PET investigations. *Journal of Cognitive Neuroscience, 7,* 337–356.

Spector, A., and Biederman, I. (1976). Mental set and mental shift revisited. *American Journal of Psychology. 89,* 669–679.

Stablum, F., Leonardi, G., Mazzoldi, M., Umiltà, C., and Morra, S. (1994). Attention and control deficits following closed head injury. *Cortex, 30,* 603–618.

Umiltà, C. A., Nicoletti, R., Simion, F., Tagliabue, M. E., and Bagnara, S. (1992). The cost of a strategy. *European Journal of Cognitive Psychology. 4,* 21–40.

29 Executive Control Problems in Childhood Psychopathology: Stop Signal Studies of Attention Deficit Hyperactivity Disorder

Gordon D. Logan, Russell J. Schachar, and Rosemary Tannock

ABSTRACT Children with attention deficit disorder (ADHD) appear to have trouble controlling attention. The difficulties they have are readily apparent in their everyday behavior, particularly in the classroom, but it has been hard for researchers to pinpoint the source of the problem. Many aspects of attention appear to be intact in experimental studies of these children. Recently, the research focus has shifted from attention itself to the executive processes that control attention. This chapter reviews research on the stop signal paradigm, which requires subjects to inhibit an ongoing response. It has been particularly successful in distinguishing children with ADHD from children with related syndromes and from children with no psychiatric diagnoses. The deficit in response inhibition may be the key deficit underlying ADHD.

Disorders of executive control are interesting to basic and clinical researchers. From a basic research perspective, they reveal important properties of executive control processes, suggesting ways to distinguish executive processes from the subordinate processes they control and ways to distinguish among the executive processes themselves. From a clinical perspective, they provide new ways to understand disorders, in terms of the underlying processes that cause the characteristic behavioral disruptions that characterize the phenomenology of the disorder. Our purpose in this chapter is to characterize executive control problems in children with attention deficit hyperactivity disorder (ADHD), and to show what these problems reveal about executive control and the nature of the disorder.

29.1 ATTENTION DEFICIT HYPERACTIVITY DISORDER AND THE NATURE OF EXECUTIVE CONTROL

A fundamental distinction in the study of executive control is between *control processes* and subordinate processes. *Subordinate processes* do the basic computations involved in performing a task. They are part of the chain of processes that lead from stimulus to response, taking input from the stimulus or stimulus-driven processes, and giving output to response systems or the processes that drive them. *Executive processes*, are outside the chain, but they act on it: they control the subordinate processes, enabling them and directing them, turning them on and off (see, for example, Logan 1985; Meyer and Keiras 1997).

The cognitive deficits in children with ADHD are interesting because they provide support for the distinction between control and subordinate processes: the deficits seem to be specific to control processes, often, there is no apparent deficit in subordinate processes. From a clinical perspective, we might not expect to see deficits in subordinate processes (e.g., the stages described in Sternberg 1969) because they operate at a timescale in which tens of milliseconds are significant intervals. By contrast, the clinical presentation of inattentiveness and impulsivity unfold over seconds and minutes. This suggests the deficits may lie in the processes that organize and coordinate attention, which we identify as "executive control processes."

The hypothesis that ADHD is purely a disorder of executive control is intriguing. For basic researchers, a subject population with a pure executive deficit would provide a unique and powerful stimulus for theory and research, much as patients with amnesia have done for memory research and patients with neglect have done for attention research. For clinicians, evidence of a pure executive deficit would address questions about the etiology of ADHD that have remained unanswered since the disorder was first described. It might suggest new therapies specifically targeted at improving executive processes.

The hypothesis that the disorder in ADHD is purely executive is hard to evaluate decisively for two reasons. First, testing the hypothesis requires a method for distinguishing executive from subordinate processes, and there are no commonly accepted methods for distinguishing them. The distinction is hard to make in the abstract. It is more meaningful in the context of a theory that specifies the subordinate processes and the executive processes that control them. Unfortunately, no current theory that provides a complete account of subordinate and executive processes. Some theories, like that of Bundesen (1990), provide detailed descriptions of subordinate processes but say nothing about the executive processes that control them. Other theories, like that of Meyer and Keiras (1997), provide detailed descriptions of executive processes but say little about the subordinate processes they recruit and control. Thus, it is hard to be confident in identifying particular tasks or particular effects as instances of executive or subordinate processing.

Second, the effects of executive control processes can be quite subtle and thus, hard to detect. While some executive actions, like those that turn responses on and off, have dramatic, directly observable effects, many have less dramatic effects. For example, executive actions that adjust the parameters of subordinate processes (e.g., setting attention weights, biases, response criteria, etc.) are likely to have subtle effects. Most likely, parameter adjustments would modulate the effects of stimulus conditions on subordinate processes, producing interactions between stimulus conditions and parameter values. To measure those interactions, we would have to know when the parameter values changed (when the executive action occurred), and that may be hard to discern.

The effects of executive processes may also be subtle because one executive action may cascade through the whole system, affecting every process in one way or another. For example, changes in criteria for perceptual processes may increase or lessen the load on subsequent decision processes. Even drastic acts of control may have cascaded effects. One executive action may enable another, which in turn, enables another. Thus every effect we measure may contain a variety of cascaded executive effects, some closer to the source than others. In addition to separating executive processes from subordinate processes, we must separate executive processes from each other.

We deal with these problems by focusing on the ability to inhibit ongoing responses as it is manifest in the "stop signal paradigm" (Logan and Cowan 1984). We argue that response inhibition is an executive ability because the processes that underlie it operate directly on other processes, disabling them if they can. Moreover, the immediate effects of response inhibition are not subtle. The response to be inhibited either occurs or does not occur. This observation, together with a theory of the underlying processes, allows us to estimate the latency of the inhibitory act, which can be used to diagnose the effectiveness of the underlying control process. We also chose the stop signal paradigm because deficient response inhibition has been implicated as a causal factor in ADHD. The direct and cascaded effects of deficient inhibition manifest themselves in behavior typical of ADHD (Barkley 1997).

Our focus on the stop signal paradigm allows us to test some parts of the hypothesis that ADHD is purely an executive disorder, but not others. Finding a deficit on the stop signal task in children with ADHD would demonstrate they have an executive disorder, which is a necessary but not sufficient step in confirming the hypothesis. A deficit would not, however, tell us that subordinate processes were intact, which is a second necessary but not sufficient step in confirming the "purely executive" hypothesis. Moreover, a deficit in the stop signal task tells us little about other possible executive deficits, and it may not be sufficient in itself to account for the phenomenology of ADHD behavior (Barkley 1997).

A terminological note: The word "inhibition" has many senses. We use it in a behavioral sense; behaviors that are withheld are inhibited. This usage has a long tradition in behavioral research, dating perhaps to Pavlov, this usage is conventional in the literature on the stop signal task (responses that can be stopped are inhibited) and in the literature on ADHD, where it describes the phenomenology of the everyday behavior of children with ADHD. We do not mean to imply that the mechanisms underlying response inhibition are inhibitory in a neural sense or a computational sense. Whether the mechanisms are inhibitory in those senses is an empirical question that we do not address.

29.2 ATTENTION DEFICIT HYPERACTIVITY DISORDER

Clinical Features

The cardinal features of ADHD are developmentally excessive and impairing levels of activity, inattention, and impulsiveness. Children with ADHD have great difficulty remaining seated when required to in structured situations such as the classroom or at the dinner table. They are more active than their peers in unstructured situations (e.g., at the playground). They fail to pay attention to instructions in academic and social situations. They have serious difficulty withholding a response of any kind until the appropriate moment, interrupting an inappropriate course of action once initiated, and adjusting incorrect or maladaptive responses.

ADHD occurs in approximately 3–5% of school-age children (APA 1994). It is more prevalent in boys than in girls, the ratio varying from 2:1 to 4:1 (APA 1994). The incidence of ADHD symptoms varies from situation to situation. Roughly 73% of children with the diagnosis display symptoms in school settings but not at home; about 11% display them at home but not at school; and the remainder display symptoms pervasively, at home and at school (Szatmari, Offord, and Boyle 1989). The most common treatment is stimulant medication, such as methylphenidate (Ritalin), which paradoxically calms the children and allows them to focus more effectively.

Diagnosis

North Americans rely on diagnostic criteria described in the *Diagnostic and Statistical Manual* of the American Psychiatric Association, now in its fourth edition (DSM-IV; APA 1994). The diagnostic criteria for ADHD from DSM-IV are presented in table 29.1. Europeans rely on the World Health Organization's *International Classification of Diseases* for diagnosis, currently in its tenth edition (ICD-10; World Health Organization 1994). Earlier in this century, North American and European diagnoses diverged; North Americans diagnosed more liberally, whereas Europeans required evidence of neuropathology (e.g., closed-head injury; seizures) and pervasive symptoms. As a result, the incidence of diagnosis was much higher in North America than in Europe. The current versions of DSM-IV and ICD-10 list essentially the same criteria (i.e., those in table 29.1), except that ICD-10 considers "talks excessively" to be a symptom of impulsivity rather than hyperactivity, so the diagnosis is becoming more uniform. Despite the agreement on symptoms, the diagnostic algorithms are significantly different with DSM-IV and ICD-10. DSM-IV considers children to have inattentive-type ADHD if they manifest six of the nine symptoms of inattention; to have hyperactive/impulsive ADHD if

Table 29.1 North American (DSM-IV) Criteria for Diagnosis of Attention Deficit Disorder

Inattention
1a. Fails to give close attention to details, makes careless mistakes
1b. Difficulties in sustaining attention in tasks or in play activities
1c. Does not seem to listen
1d. Does not follow instructions or finish schoolwork, chores, or duties in workplace
1e. Difficulty organizing tasks and activities
1f. Avoids or dislikes tasks requiring mental effort
1g. Loses things
1h. Easily distracted by external stimuli
1i. Forgetful in daily activities
Hyperactivity
2a. Fidgets
2b. Leaves seat
2c. Runs or climbs excessively
2d. Difficulty playing or engaging in leisure activities quietly
2e. On the go, driven by a motor
2f. Talks excessively
Impulsivity
2g. Blurts out answers
2h. Difficulty in waiting for turn
2i. Frequently interrupts or intrudes

Source: APA 1994.

they manifest six of the nine symptoms of hyperactivity and impulsivity; and to have combined-type ADHD if they manifest six inattentive symptoms and six hyperactive/impulsive symptoms. By contrast, ICD-10 does not distinguish subtypes and requires children to have six inattentive symptoms, three hyperactive symptoms, and one impulsive symptom to meet diagnostic criteria. Thus ICD-10 focuses on the combined type, in which the problems are pervasive and more severe.

The diagnostic problem is made worse by a host of comorbid disorders (APA 1994). Children with ADHD often meet diagnostic criteria for learning disabilities, conduct disorder, oppositional disorder, and emotional or anxiety disorders, causing a "chicken or the egg" problem. A child with ADHD and conduct disorder, for example, may manifest ADHD symptoms because of the conduct disorder, or may manifest conduct disorder symptoms because of the ADHD. Some of the studies in the literature may be compromised by comorbidity. The solution to the comorbidity problem is to run several groups of subjects, including children who exhibit pure symptoms of the typical comorbid disorders and children who exhibit mixed symptoms and so get mixed diagnoses. The appropriate multigroup designs allow investigators to distinguish effects that are unique to children with ADHD from effects that are due to psychopathology in general.

History

The major features of ADHD—impulsivity, hyperactivity, and inattentiveness—were being discussed in the medical literature by the end of the nineteenth century (for a historical review, see Schachar 1986). Since then, although these symptoms have been interpreted in different ways, changing with the currents of developments in neurology and psychology, the essence of the disorder remains the same. Still (1902) and Tredgold (1908) are credited as the first to report detailed case histories of children with the disorder, describing the problem as a "defect in moral control" due to minimal brain dysfunction. Bradley (1937), intending to cure a headache, serendipitously discovered the beneficial effects of stimulant medication (benzedrine) on attention and behavior in children with ADHD.

In the 1940s and 1950s, the disorder was described as "minimal brain disorder" and attributed to some unknown and undetectable neural cause. By the 1960s, the emphasis in diagnosis shifted from unknown neurology to observable behavior, focusing on the excessive activity manifested by these children. The disorder became known as "hyperkinetic reaction of childhood" (APA 1968). Throughout the 1970s, the focus shifted once again from behavior to the cognitive processes that underlie it. Douglas (1972) and others suggested that an attention deficit rather than overactivity lay at the heart of the disorder. The prevailing opinion changed, especially in North America, and in 1980, the name changed to "attention deficit disorder with or without hyperactivity" (ADDH or ADD; APA 1980). In 1987, the name changed once more, to "attention deficit hyperactivity disorder," lumping all diagnostic criteria into one scale (APA 1987). Throughout the 1990s, the focus shifted once again from basic cognitive processes to the executive processes that control them. Self-regulation became an important issue (Barkley 1997; Pennington and Ozonoff 1996; Quay 1988). The name remains the same in DSM-IV (APA 1994), but the symptoms that underlie it are now divided into the two clusters seen in table 29.1: inattentiveness and hyperactivity/impulsivity.

Cognitive Psychopathology: The Search for an Attention Deficit
Cognitive research on ADHD began in the early 1970s with the adaptation of the "continuous performance task" (CPT), developed to assess brain damage in children and adults (Rosvold et al. 1956), to children with psychiatric disorders. The CPT is a kind of vigilance task. A series of letters are displayed on a screen, one at a time, and a target letter occurs on 10–15% of the trials. The child's task is to respond when the target appears and not when nontargets appear. Children with ADHD miss more targets than normal controls and they false alarm to nontargets more often (Sykes, Douglas, and Morgenstern 1973). Their poorer performance was interpreted as evidence of an attention deficit.

The experimental design in the early CPT tasks makes the results hard to interpret. Children with ADHD and controls were compared in a single condition of the CPT task, and the evidence for an attention deficit was a main effect of diagnostic group. No factors were manipulated that would allow insight into the processes underlying the main effect (cf. Sternberg 1969). Executive and subordinate, attentional and nonattentional processes could be responsible for the difference.

We conducted more analytic CPT experiments, manipulating factors such as warning events, exposure duration, and event rate that should affect attention and perhaps executive processing. We found no interactions between group and factors affecting preparation (Schachar et al. 1988), suggesting that preparatory attention may be spared. We found weak interactions between group and exposure duration and between group and event rate (Chee et al. 1989) such that ADDH children were disadvantaged by short exposure durations and slow event rates, but the interactions were quite small relative to the main effects. The attentional processes that are affected by exposure duration and event rate cannot account for all of the deficit we observed. Our CPT studies suggest that not all attentional or executive processes are deficient in ADHD.

Several studies found evidence suggesting that basic subordinate processes are spared in ADHD. Sykes et al. 1973 found no interactions between diagnostic group and the number of choices in a multiple-choice reaction time task, suggesting that response selection processes were intact in children with ADHD. Tannock, Schachar and Logan 1993 found no interaction between diagnostic group and the number of cues in a visual search task, suggesting that children with ADHD can focus attention as sharply as control children. In an influential set of studies, Sergeant, van der Meere and colleagues adapted Sternberg's "additive factors method" (1969) to locate the processes that are deficient in ADHD children. They manipulated factors that affected each of the four stages leading from stimulus to response in Sternberg-type visual and memory search tasks (i.e., encoding, comparison, decision, and response selection), looking for interactions between diagnostic group and factors that affected particular stages. They found no interactions between diagnostic group and any of the stage-defining factors, suggesting that the deficit was not specific to any stage between stimulus and response (see, for example, Sergeant and van der Meere 1990; van der Meere, van Baal, and Sergeant 1989).

Another strategy is to examine drug effects, to see whether stimulant medication (Ritalin) improves the ability to attend. We tested children with ADHD on a four-item visual search task, in which we cued 1–4 locations that might contain the target. We found a main effect of drug but no interaction with number of cues, as if the drug had no effect on ability to focus attention (Tannock, Schachar, and Logan 1993). Similarly, Reid and Borkowski (1984) found no interaction between Ritalin and number of alternatives in a multiple-choice task and no interaction between Ritalin and the level of match in a Posner letter-matching task: physical (e.g., AA) versus semantic (e.g., Aa). In a rare study that compared stimulant medication effects in children with ADHD to those in normal controls, Sostek, Buchsbaum, and Rapoport (1980) found that amphetamine improved CPT performance equally for both groups. The kinds of attention tapped in these tasks seem spared in ADHD as well.

Executive Control Deficits

The difficulty in finding deficits in elementary attentional processes and other subordinate processes shifted the scales in favor of the possibility that children with ADHD were deficient in executive control processes. The ability to inhibit responses became a popular candidate for the executive deficit. Quay (1988) applied Gray's theory of anxiety (1982) to ADHD, proposing that Gray's behavioral inhibition system was deficient in ADHD, rather than Gray's behavioral activation system. Pennington and Ozonoff (1996) proposed executive control deficits in several childhood disorders, including ADHD. Barkley's theory (1997) proposed inhibitory control as the core deficit in ADHD. He argued that a deficit in inhibition impaired the ability of ADHD children to engage various executive control strategies to optimize their behavior. The control strategies involve working memory, self-regulation, internal speech, and "reconstitution" (i.e., the ability to restructure behavior). All of these strategies require children to stop and think; a deficit in inhibitory control would allow them to act without thinking and therefore miss out on the benefits of these more carefully considered control strategies.

Of course, deficits in inhibition or, more broadly, self-regulation are not the only explanations of ADHD behavior in the current literature. Some interpret ADHD as a manifestation of a motivational deficit or an insensitivity to reinforcement (Barkley 1981; Glow and Glow 1979). Zentall (1985) suggested that children with ADHD were chronically underaroused and that their hyperactive behavior was intended to increase and optimize their level of arousal. Nevertheless, because the idea that a deficit in inhibition underlies ADHD is currently quite popular, it is important to test it.

29.3 STOP SIGNAL PARADIGM

Method

The stop signal paradigm involves two tasks: a go task and a stop task. The object of the go task is to respond to a stimulus as quickly as possible. Typically, the go task involves a choice among stimulus and response alternatives (e.g., discriminating an *X* from an *O*), but it need not (see e.g.,

Logan, Cowan, and Davis 1984; Ollman 1973). The object of the stop task is to inhibit the response to the go task. Subjects are presented with a "stop signal" (usually but not necessarily auditory) on occasion (typically, on 25% of the go trials), which instructs them to inhibit the response to the go task.

The most important dependent variable is the probability of successfully inhibiting the go response or its complement, the probability of responding given a stop signal. The mean and standard deviation of the go task and go task accuracy on trials without stop signals (*no-signal* trials) are also important dependent variables. Finally, go task reaction time for responses that escape inhibition when the stop signal occurs (*signal-respond* trials) is important as well.

The most important independent variable is the interval between the onset of the go signal and the onset of the stop signal (*stop signal delay*). If the stop signal is presented early enough—sometimes before the go signal appears—subjects will always inhibit. If it is presented late enough, subjects will always respond. The probability of responding, given a stop signal, increases monotonically as delay increases from early to late, forming an *inhibition function*. Other independent variables include the nature of the go task, the nature of the strategy subjects adopt to perform the go task, and the nature of the subject population (for reviews, see Logan 1994; Logan and Cowan 1984).

Race Model

Performance in the stop signal paradigm has been modeled as a race between the go task and the stop task: if the go task is faster than the stop task, the subject responds, whereas if the stop task is faster than the go task, the subject inhibits (Logan 1981; Logan and Cowan 1984; Osman, Kornblum, and Meyer, 1986, 1990; Ollman 1973). Stop signal delay handicaps the race in favor of one process or the other. Short delays are advantageous for the stop task, increasing the probability of inhibition; long delays are advantageous for the go task, increasing the probability of responding.

The race model is illustrated in figures 29.1–29.3. In these figures, reaction time to the go signal (*go reaction time* or go RT) is assumed to be a random variable (represented by the cumulative distribution function) and reaction time to the stop signal (*stop signal reaction time* or SSRT) is assumed to be constant. Logan and Cowan (1984) provide versions of the race model in which both go RT and SSRT are assumed to be random variables. The assumption that SSRT is constant makes it easier to explain the race model, and several analyses suggest that the assumption of constant SSRT does not compromise the main predictions of the race model (see Band 1997; De Jong et al. 1990; Logan and Cowan 1984).

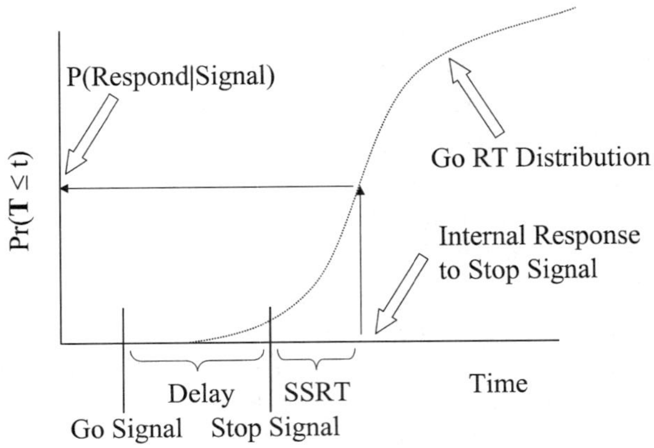

Figure 29.1 How the race between the go process, reflected in the go reaction time (RT) distribution, and the stop process, reflected in stop signal reaction time (SSRT), determine the probability of responding given a signal. The stop signal is presented Delay milliseconds after the go signal. SSRT milliseconds later, the internal response to the stop signal occurs. Go responses faster than that point in time are executed; go responses slower (later) than that point in time are inhibited. P(respond | signal) is estimated by determining the point in time at which the internal response to the stop signal occurs, relative to the go RT distribution (by extending a vertical line upward from the time axis until it intersects the distribution) and determining the probability the go response was faster than that point (by extending a vertical line from the go RT distribution to the y axis).

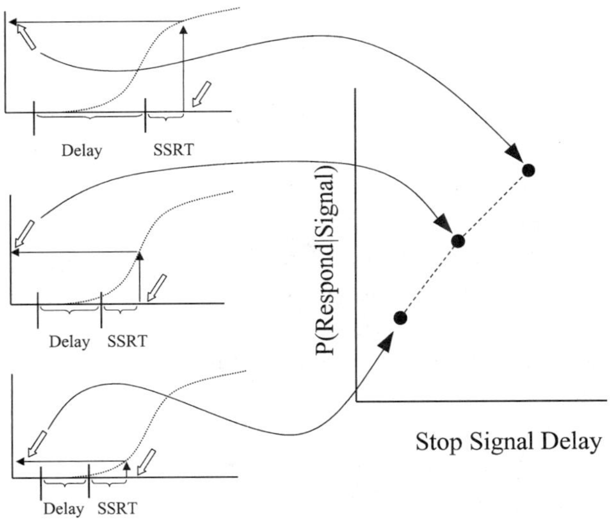

Figure 29.2 How the race model produces inhibition functions as stop signal delay is varied. In the top left panel, delay is long and the internal response to the stop signal occurs near the end of the go reaction time distribution, so P(respond | signal) is high. In the middle left panel, delay is intermediate and the internal response to the stop signal occurs near the middle of the go RT distribution, so P(respond | signal) is near 0.5. In the bottom left panel, delay is short and the internal response to the stop signal occurs early in the go RT distribution, thus P(respond | signal) is low.

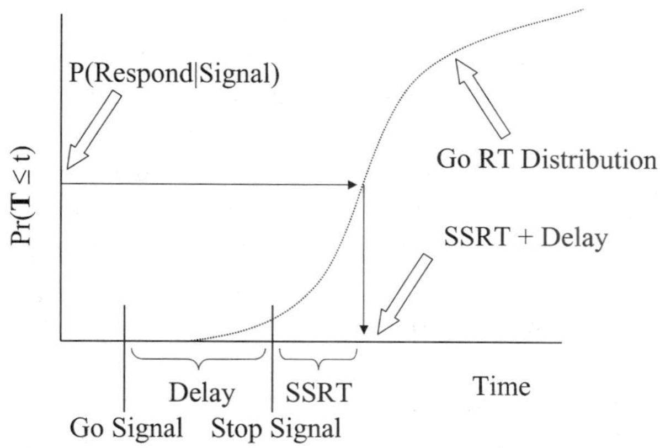

Figure 29.3 How the race model estimates reaction time to the stop signal (SSRT). The quantile on the (observed) cumulative distribution of go reaction times (RTs) corresponding to the (observed) probability of responding given a stop signal is determined. That quantile estimates the time at which the internal response to the stop signal occurred, relative to the onset of the go signal. To estimate the time of the internal response to the stop signal relative to the onset of the stop signal (to calculate SSRT), Delay must be subtracted from value of that quantile.

Figure 29.1 illustrates how the race between go processes and stop processes determines the probability of responding given a stop signal. The go signal occurs and generates a distribution of finishing times for the go processes, which is represented as a cumulative distribution. At some delay after the go signal, the stop signal is presented. The internal response to the stop signal occurs SSRT ms after the stop signal. According to the race model, go responses that occur before this point in time are executed and go responses that occur after this point in time are inhibited. The probability that a response is executed or inhibited can be determined by extending a line upwards from the point on the time axis at which the internal response to the stop signal occurred that intersects the cumulative distribution of go RTs. The point at which the line intersects the distribution can be extended horizontally to the y axis, which represents the probability that a go response occurs at or before the time at which the internal response to the stop signal occurred, or in other words, the probability of responding given a stop signal.

Figure 29.2 shows how variation in stop signal delay produces an inhibition function. In the top left panel, delay is long, thus the internal response to the stop signal occurs quite late, relative to the go RT distribution, and subjects are likely to respond. In the middle left panel, delay is intermediate, and subjects inhibit about half of the time. In the bottom panel, delay is short, and subjects inhibit most of the time. Probability of responding, given a stop signal, grows monotonically as delay increases.

Figures 29.1 and 29.2 suggest that variation in the inhibition functions depends on variation in the go RT distribution, and the data bear out that suggestion. Differences in inhibition functions between conditions, strategies, tasks, and subjects can be accounted for almost entirely by differences in parameters of the go RT distribution (Logan and Cowan 1984).

Most important, the race model provides ways to estimate SSRT (Colonius 1990; De Jong et al. 1990; Logan and Cowan 1984). Differences in SSRT contribute to differences in inhibition functions, particularly differences between subjects and between subject populations. Thus, differences in SSRT are a primary measure of the differences in executive ability we are interested in, so they are important to measure. Measurement of SSRT is difficult because only go RT is observable directly; SSRT must be inferred. If subjects inhibit successfully, there is no response whose latency can be measured. If subjects fail to inhibit, SSRT must have been slower than the observable go reaction time (*signal-respond reaction time*), but it is not clear how much slower it was. A formal model is necessary to estimate SSRT.

Figure 29.3 depicts one of three race model methods of estimating SSRT. In essence, the logic is the inverse of the logic used to explain how the race between stop and go processes produced an inhibition function. The race model explanation of the inhibition function worked from unobservables to observables. The race model estimation of SSRT works in the opposite direction, starting from the observed probability of responding, given a stop signal at some delay. According to the race model, the probability of responding given a signal represents the proportion of the go RT distribution that was faster than the internal response to the stop signal. In figure 29.3, SSRT is estimated by extending a horizontal line from the point on the y-axis representing the probability of responding, given a stop signal, until it intersects the go RT distribution, and then extending a line vertically from there to the time axis. The point at which the vertical line intersects the time axis represents the time at which the internal response to the stop signal occurred, relative to the onset of the go signal. To express it as SSRT, relative to the onset of the stop signal, the delay is simply subtracted.

The race model also predicts the speed of (signal-respond) responses that escape inhibition. In essence, the mean signal-respond RT is equal to the mean of the part of the distribution that occurs before the internal response to the stop signal (see figures 29.1–29.3). Tests of this prediction have shown excellent fits (De Jong et al. 1990; Jennings et al. 1992; Logan and Cowan 1984). More generally, the race model predicts that distributions of signal-respond RT will share a common minimum value and fan out as a function of delay as time increases (with longer upper tails associated with longer delays). This prediction has been confirmed many times as well (Lappin and Eriksen 1966; Osman, Kornblum, and Meyer 1986).

The race model accounts for the data very well and it seems to be accepted universally by researchers who study the stop signal paradigm.

This unusual state of affairs is advantageous. It allows us to use the race model to understand inhibitory control in various situations and subject populations. The model becomes background and the phenomena of inhibitory control become foreground.

Applications

The stop signal paradigm has been used with a variety of response modalities, including keypresses (e.g., Logan et al. 1984; Osman, Kornblum, and Meyer 1986, 1990), hand squeezes (De Jong et al. 1990, 1995), arm movements (McGarry and Franks 1997), eye movements (Hanes and Carpenter 1997; Logan and Irwin 2000), and typewriting (Logan 1982). It has been used with a variety of electrophysiological measures, including event-related brain potentials (ERPs; De Jong et al. 1990, 1995), heart rate (Jennings et al. 1992), electromyograph (De Jong et al. 1990; McGarry and Franks 1997), and single-cell activity (Hanes, Patterson, and Schall 1998).

Hanes, Patterson, and Schall 1998 study provides striking evidence for the validity of the race model. They recorded from cells in the frontal eye fields of macaque monkeys involved in making saccadic eye movements and in maintaining fixation. On trials with no stop signal, the firing rate of saccade cells increased monotonically after the go signal, reaching a maximum when the saccade began. On stop signal trials, the firing rate in saccade cells followed the no-stop-signal pattern up to a point, and then dropped precipitously. Estimates of SSRT derived from the monkeys' behavior predicted the point of divergence. The same thing was found with fixation cells. Their firing rate on no-signal trials decreased and reached a minimum during the saccade. On stop signal trials, the firing rate followed the same pattern up to the predicted SSRT, and then diverged, increasing as it would during a fixation.

The stop signal paradigm has been used with a variety of subject populations, including monkeys (Hanes and Schall 1995; Hanes, Patterson, and Schall 1998). It has been used to document an improvement in inhibitory ability across childhood (Schachar and Logan 1990a) and a decline in inhibitory ability in the elderly (Kramer et al. 1994; May and Hasher 1998). Recently, we completed a life span study of stopping, testing subjects from 6 to 81 years of age, confirming the improvement across childhood and the decline with old age (Williams, et al. 1999). Moreover, the paradigm has been used to study drug effects on inhibitory ability. Mulvihill, Skilling, and Vogel-Sprott (1997) found that low doses of alcohol impaired the ability to inhibit without affecting performance on the go task. Most relevant to our present focus on ADHD, however, Logan, Schachar, and Tannock (1997) found that subjects who rated high on the Eysenck impulsivity scale had longer SSRTs than subjects who rated low, suggesting that impulsive people have difficulty inhibiting impulsive

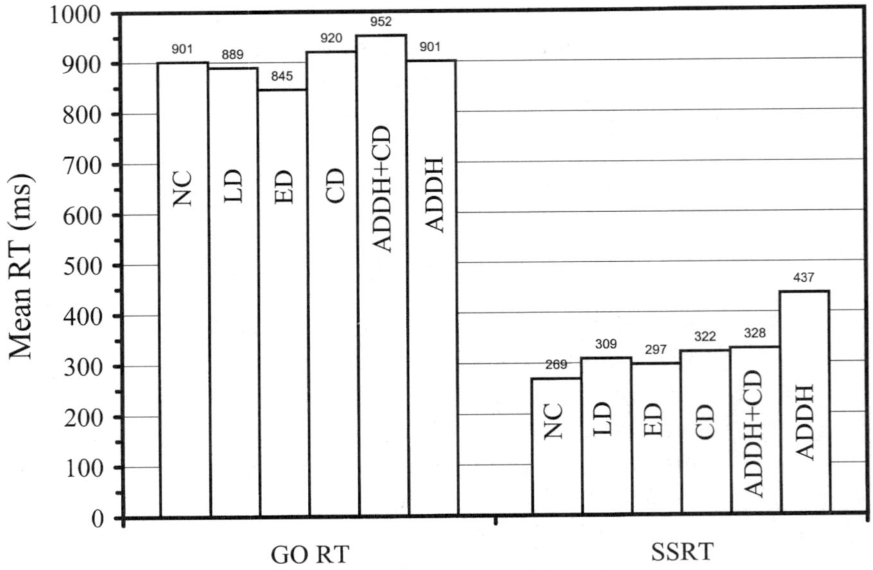

Figure 29.4 Mean go reaction time (RT) and stop signal reaction time (SSRT) in normal control children (NC), children with learning disabilities (LD), children with emotional disorder (ED), children with conduct disorder (CD), children with mixed conduct disorder and attention deficit disorder with hyperactivity (ADDH + CD), and children with attention deficit disorder with hyperactivity (ADDH). From Schachar and Logan 1990a.

behavior not because they go too quickly, but rather because they stop too slowly.

29.4 INHIBITORY CONTROL AND CHILDHOOD PSYCHOPATHOLOGY

Inhibitory Deficit in Attention Deficit Hyperactivity Disorder

It is clear that children with ADHD are deficient in their ability to inhibit responses in the stop signal paradigm (for a review, see Schachar, Tannock, and Logan 1993; for a meta-analysis, see Oosterlaan, Logan, and Sergeant 1998). When given the same opportunity, they inhibit less often than normal controls. Children with ADHD are less responsive to stop signal delay, producing inhibition functions that are flatter than those of controls (e.g., Schachar and Logan 1990a), though sometimes these differences can be accounted for in terms of differences in the mean and standard deviation of go RT. In the meta-analysis (Oosterlaan, Logan, and Sergeant 1998), there was no significant difference between the inhibition functions of ADHD and normal children, after the race model corrections had been applied. Thus the primary measure of inhibitory performance is SSRT rather than the inhibition function.

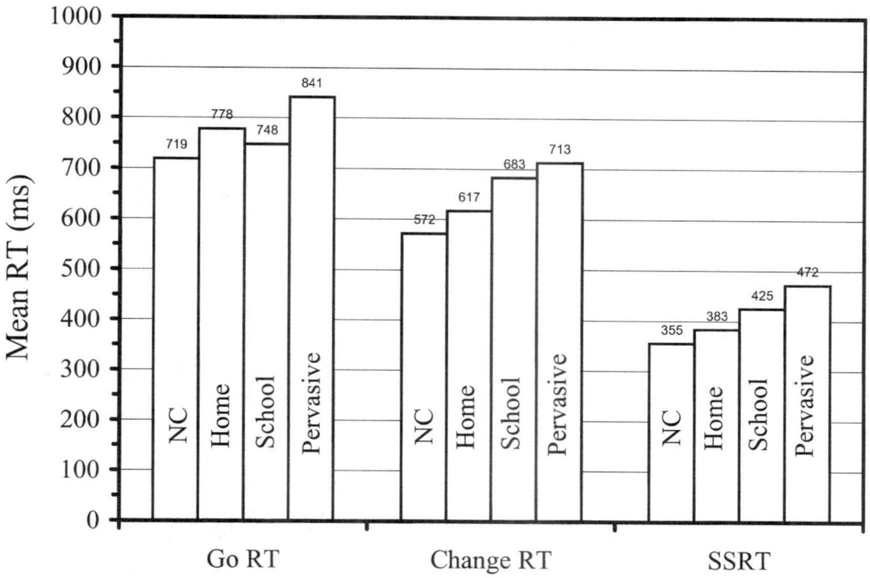

Figure 29.5 Mean go reaction time (RT), overt reaction time to the change signal, and stop signal reaction time (SSRT) in normal control children (NC), children with home-situational ADHD (home), children with school-situational ADHD (school), and children with pervasive ADHD (pervasive). From Schachar et al. 1995.

Children with ADHD consistently take more time than controls to respond to the stop signal, even when differences in go RT are taken into account. Figure 29.4 displays mean go RT and mean SSRT from Schachar and Logan 1990a which compared ADDH (DSM-III diagnosis) with normal controls and a number of psychiatric control groups. The key comparison is between normal controls (NC) and children with ADDH: there is no difference in mean go RT ($M = 901$ msec for both NC and ADDH) but a substantial difference in SSRT (168 msec). The meta-analysis (Oosterlaan, Logan, and Sergeant 1998) showed significant differences in SSRT between ADHD children and normal controls.

The inhibitory deficit in ADHD seems to be correlated with the severity of the disorder. Figure 29.5 presents data from Schachar et al. 1995, which compared normal control children and three groups of children with ADHD: *home-situational* children, who display ADHD symptoms only at home; *school-situational* children, who display ADHD symptoms only at school; and *pervasive* children, who display ADHD symptoms both at home and at school. These children were tested on a variation of the stop signal paradigm called the "change task" (Logan and Burkell 1986). Like the stop task, subjects see a go signal on each trial, and occasionally hear a "stop and change" signal. When the stop and change signal sounds, subjects have to do two things: (1) they have to stop their response to the go task, as in the stop signal paradigm; and (2)

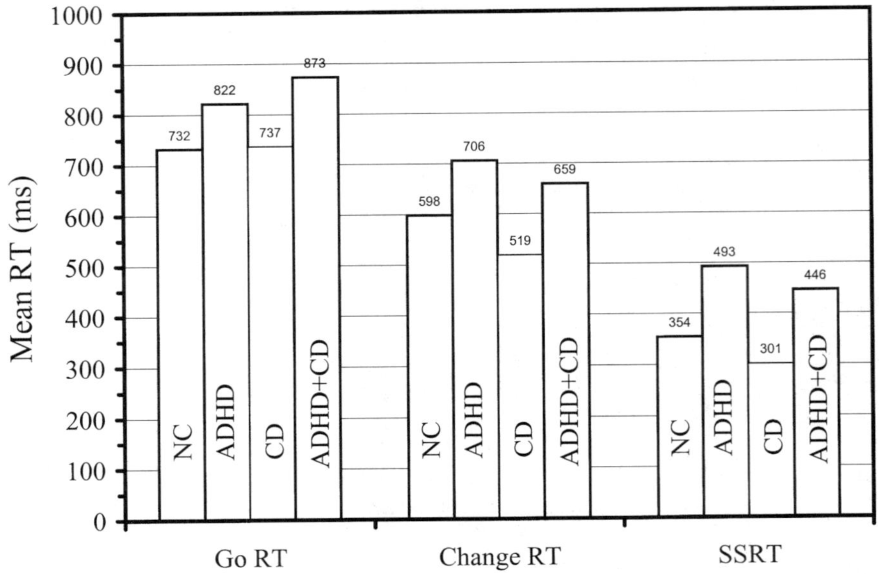

Figure 29.6 Mean go reaction time (RT), overt reaction time to the change signal, and stop signal reaction time (SSRT) in normal control children (NC), children with attention deficit hyperactivity disorder (ADHD), children with conduct disorder (CD), and children with attention deficit hyperactivity disorder and conduct disorder (ADHD + CD). From Schachar and Tannock 1995.

they have to make a separate, overt response to the change signal, as in a dual-task paradigm (see Logan and Burkell 1986).

The data, plotted in figure 29.5, show longer SSRTs for children with pervasive ADHD than for children with home- or school-situational ADHD, which in turn, are longer than SSRTs for normal controls. Only the difference between normal controls and pervasive ADHD was significant, however. Note that the differences between normal controls and pervasive ADHD children were about as large for SSRT (117 msec) as for go RT (122 msec) even though go RT was nearly twice as long as SSRT. Thus the difference in SSRT does not merely reflect overall differences in go RT.

Differences between Diagnostic Groups

Figure 29.4 also displays go RT and SSRT from four important control groups whose disorders are often comorbid with ADHD. The data show no clear differences in go RT but substantial differences between ADHD children and the clinical controls in SSRT. The clinical controls are slower than the normal controls, but not significantly so. Perhaps the most important contrasts in the data are between children with ADHD only on the one hand, and children with conduct disorder (CD) and children with

CD and ADHD on the other. The slower SSRT in the ADHD-only children suggests that the inhibitory deficit is specific to ADHD.

The conclusion that children with ADHD behave differently on the stop task than children with CD or ADHD + CD is controversial. Some studies replicate the difference and some do not. The meta-analysis (Oosterlaan, Logan, and Sergeant 1998) showed no significant difference in SSRT between ADHD and CD, though the numerical difference was more than 80 msec (349 msec in ADHD; 265 msec in CD). Our own research has produced inconsistent results. Displaying data from Schachar and Tannock 1995, which compared normal controls, children with ADHD only, children with CD only, and children with ADHD + CD, figure 29.6 shows that the SSRT difference between ADHD children and normal controls is robust, but in this sample, children with ADHD + CD were different from controls and not different from children with ADHD (cf. Schachar and Logan 1990a). Note that the difference between normal controls and children with ADHD is bigger for SSRT (139 msec) than for go RT (90 msec). Thus the effect cannot be attributed to differences in overall speed.

Several factors may underlie the difficulty in finding reliable differences across studies. Diagnostic procedures were not uniform, sample sizes were not always very large, and parameters of the stop task were not the same. The resolution of this controversy awaits further research. For the present, however, it is clear that children with ADHD have an inhibitory deficit that sets them apart from most other children.

Alternative Interpretations

One interpretation of the inhibitory deficiency in ADHD in terms of goal neglect: (Duncan et al. 1996; Duncan et al. 1997): children with ADHD may not "hear" the stop signal because they have trouble maintaining two goals simultaneously. Two lines of evidence allow us to rule out this possibility. First, if the program for the stop task is run as a dual-task paradigm, with subjects responding to the go stimulus on every trial and making a different overt response to the tone when it occurs, children with ADHD respond to as many tones as control children (i.e., almost all of them) and show refractory effects comparable to those of control children (Schachar and Logan 1990b). Thus children with ADHD are able to keep two goals in mind at once.

The second line of evidence against the idea that ADHD children do not hear the stop signal comes from the change paradigm, which requires subjects to respond overtly to the stop-and-change signal as well as to inhibit their responses to the go task. The change task requires children to keep three goals in mind: going to the go signal, stopping to the stop signal, and going to the change signal. In this task, children with ADHD

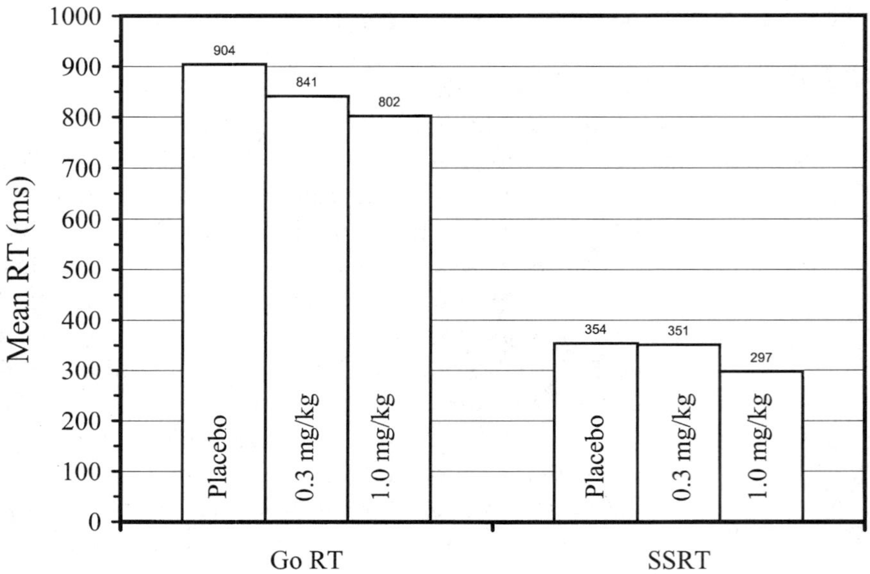

Figure 29.7 Mean go reaction time (RT) and reaction time to the stop signal (SSRT) in children with ADDH as a function of dosage of Ritalin (methylphenidate). From Tannock, Schachar, Carr, Chajczyk, and Logan 1989.

sometimes fail to inhibit go responses when the stop signal occurs, but on the same trial, they make the correct overt response to the tone (Schachar and Tannock 1995; Schachar et al. 1995). This suggests that they kept at least two goals in mind: going to the go signal and going to the change signal. Moreover, children with ADHD inhibit responses in the change task about as often as they do in the stop task (see Schachar, Tannock, and Logan 1993), and their SSRTs are not much different (compare SSRTs in figure 29.4 with those in figures 29.5 and 29.6). If goal neglect were the source of the deficit, subjects should inhibit less often and less rapidly with three goals in mind (change task) than with two (stop task).

It is possible that the nature of the goals is more important than the number of goals in producing goal neglect. The goals of the stop task are in direct opposition to the goals of the go task—so only one can be satisfied on a stop trial—goal neglect may be quite likely. In contrast, the goals in a dual-task paradigm merely compete: both of the goals can be accomplished; the competition determines which one finishes first and which one waits. Goal neglect may be less likely. The present data cannot rule out this version of the goal neglect hypothesis.

Stimulant Medication and Inhibitory Control

The stimulant medication methylphenidate (Ritalin) improves the behavioral symptoms and academic performance of children with ADHD (Tannock, Schachar, Carr, and Logan 1989). Tannock, Schachar, Carr,

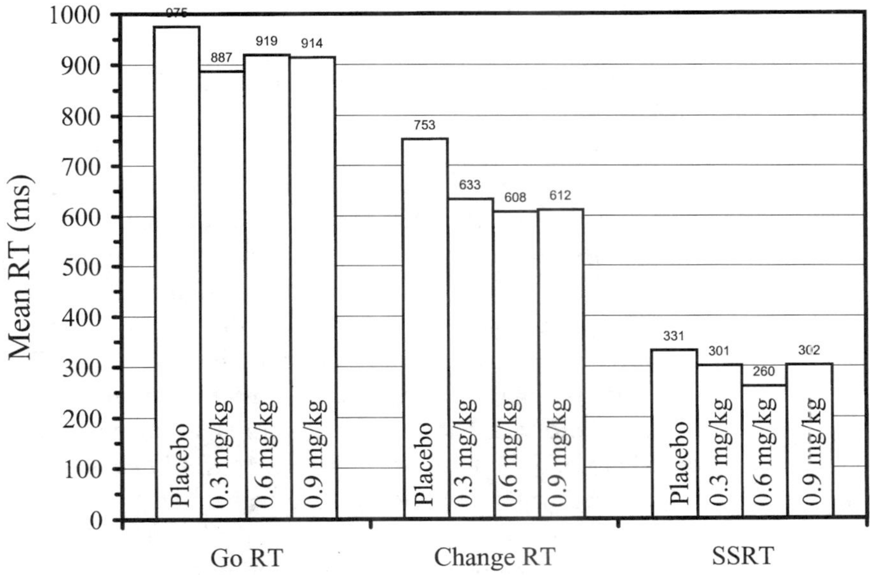

Figure 29.8 Mean go reaction time (RT), overt reaction time to the change signal, and stop signal reaction time (SSRT) in children with ADHD as a function of dosage of Ritalin (methylphenidate). From Tannock, Schachar, and Logan 1995.

Chajczyk, and Logan (1989) examined the effects of Ritalin on stop task performance in ADHD children, varying dose from 0 mg/kg (placebo) to 1.0 mg/kg. As dose increases from placebo to 1.0 mg/kg, ADHD children inhibit more often, producing inhibition functions that are steeper, like those of controls. Figure 29.7 shows the effects of Ritalin on go RT and SSRT. Ritalin had a strong effect on SSRT, moving it closer to the normal range.

A follow-up study (Tannock, Schachar, and Logan 1995) in the change paradigm produced results that were similar in some respects but tantalizingly different in others (go RT and SSRT from that study are plotted in figure 29.8). As in the stop task, SSRT was faster with methylphenidate than with placebo, but the dose-response function appeared curvilinear. SSRT decreased from low (0.3 mg/kg) to medium (0.6 mg/kg) dose but then increased as dose increased further (to 0.9 mg/kg). This result is very important because it suggests that cognitive performance (i.e., stopping) may have different dose-response function than behavior (e.g., fidgeting in class). High doses of methylphenidate may impair cognition even though they still improve overt behavior. The methylphenidate effects are important because they show that the same treatment that improves the clinical presentation of ADHD symptoms also improves inhibitory ability. This suggests that inhibitory ability may be the central deficit in ADHD (Barkley 1997; Pennington and Ozonoff 1996; Quay 1988).

29.5 DISCUSSION

Is Attention Deficit Hyperactivity Disorder Purely an Executive Deficit?

The research reviewed in this chapter provided some evidence in support of the hypothesis that ADHD is purely an executive deficit, and found nothing that disconfirmed it. On balance, however, the hypothesis is far from confirmed. The stop signal studies show that children with ADHD have one kind of executive deficit—the ability to inhibit ongoing responses. This does not imply that they have no other deficits, executive or otherwise. Barkley (1997) claimed that the direct and cascaded effects of an inhibitory deficit may account for all of the phenomena seen in children with ADHD. To test that claim, one would have to model the system through which the effects cascade to be able to separate the effects of deficient inhibition from the effects of other executive and subordinate processes. Such a theory is still beyond our grasp.

The studies we reviewed from other paradigms, such as the CPT, did not reveal deficits in subordinate processes. However, that review was nowhere near exhaustive and possibly not representative. Other studies may have revealed deficits in subordinate processes already. Only one such study is required to falsify the hypothesis.

Whatever the fate of the purely executive deficit hypothesis, the evidence of a distinct deficit in response inhibition is an important advance for research on ADHD. The stop signal paradigm provides a cognitive marker for the disorder that is closer to the underlying neurology than the overt symptoms are (see Hanes, Patterson, and Schall 1998). The paradigm may be useful in defining subtypes of ADHD and it may be a useful cognitive marker in future genetic and brain imaging studies (for a review of current studies, see Tannock 1998).

What is Deficient in Deficient Response Inhibition?

The deficit in response inhibition in children with ADHD manifests itself primarily as a slowing of SSRT that seems independent of the speed of go RT. What does this imply about the executive processes underlying performance on the stop task? The race model offers little help beyond estimating SSRT. It is abstract and general, addressing only the finishing times of the stop and go processes. It says nothing about the nature of the computation that produces the finishing times.

A minimal functional analysis suggests that the stop process involves at least three components: (1) maintaining a goal to inhibit the go response when a stop signal occurs; (2) detecting the stop signal; and (3) carrying out whatever it is that disables the go task. De Jong, Coles, and Logan (1995) distinguished two mechanisms for disabling the go task:

(3a) a central mechanism that interrupts central programming of movements; and (3b) a peripheral mechanism that inhibits the go "pulse" that drives the program. Earlier, we discussed the possibility that the incompatibility of the goals of stopping and going may slow go responses (e.g., if RT depends on goal activation) or occasionally deactivate the goal of stopping. That possibility remains viable. It is also possible that the processes that act on the go task are slower or fail occasionally. It seems unlikely, however, that the processes that detect the stop signal are deficient. For one thing, children with ADHD readily make overt responses to concurrent tones in dual-task (Schachar and Logan 1990b) and change paradigms (Schachar et al. 1995). For another, detecting a signal requires the kind of subordinate processes that may be spared in ADHD. Future research and more detailed models of the underlying processes will be necessary to distinguish among these possibilities.

29.6 CONCLUSIONS

Stop signal studies demonstrate a clear deficit in the executive processes underlying response inhibition that appears to be specific to children with ADHD, relative to age-matched psychiatric and normal controls. Moreover, the deficit is improved by the same stimulant medication that improves the behavioral symptoms of ADHD, which suggests that the deficit in inhibition plays a causal role in the clinical presentation of the disorder, supporting Barkley 1997.

NOTE

This research was supported by National Science Foundation (U.S.) grant 9808971, National Institutes of Health (U.S.) grant 1RO1HD31714-0142, Medical Research Council of Canada grant MT14366, and a grant from the National Health Research Development Program of Health Canada.

REFERENCES

APA (American Psychiatric Association). (1968). *Diagnostic and statistical manual of mental disorders.* 2d ed. Washington, DC.

APA. (1980). *Diagnostic and statistical manual of mental disorders.* 3d ed. Washington, DC.

APA. (1987). *Diagnostic and statistical manual of mental disorders.* 3d ed., rev. Washington, DC.

APA. (1994). *Diagnostic and statistical manual of mental disorders.* 4th ed. Washington, DC.

Band, G. (1997). Preparation, adjustment, and inhibition of responses. Ph.D. diss., University of Amsterdam.

Barkley, R. A. (1981). *Hyperactive children: A handbook for diagnosis and treatment.* New York: Guilford.

Barkley, R. A. (1997). Behavioral inhibition, sustained attention, and executive functions: Constructing a unifying theory of ADHD. *Psychological Bulletin, 121,* 65–94.

Bradley, C. (1937). The behavior of children receiving benzedrine. *American Journal of Psychiatry, 94,* 577–585.

Bundesen, C. (1990). A theory of visual attention. *Psychological Review, 97,* 523–547.

Chee, P., Logan, G. D., Schachar, R. J., Lindsay, P. H., and Wachsmuth, R. (1989). Effects of event rate and display time on sustained attention in hyperactive, normal, and control children. *Journal of Abnormal Child Psychology, 17,* 371–391.

Colonius, H. (1990). A note on the stop signal paradigm, or how to observe the unobservable. *Psychological Review, 97,* 309–312.

De Jong, R., Coles, M. G. H., and Logan, G. D. (1995). Strategies and mechanisms in nonselective and selective inhibitory motor control. *Journal of Experimental Psychology: Human Perception and Performance, 21,* 498–511.

De Jong, R., Coles, M. G. H., Logan, G. D., and Gratton, G. (1990). Searching for the point of no return: The control of response processes in speeded choice reaction performance. *Journal of Experimental Psychology: Human Perception and Performance, 16,* 164–182.

Douglas, V. I. (1972). Stop, look, and listen: The problem of sustained attention and impulse control in hyperactive and normal children. *Canadian Journal of Behavioral Science, 4,* 259–282.

Duncan, J., Emslie, H., Williams, P., Johnson, R., and Freer, C. (1996). Intelligence and the frontal lobe: The organization of goal-directed behavior. *Cognitive Psychology, 30,* 257–303.

Duncan, J., Johnson, R., Swales, M., and Freer, C. (1997). Frontal lobe deficits after head injury. *Cognitive Neuropsychology, 14,* 713–741.

Glow, P. H., and Glow, R. A. (1979). Hyperkinetic impulse disorder: A developmental defect of motivation. *Genetic Psychology Monographs, 100,* 159–231.

Gray, J. A. (1982). *The neuropsychology of anxiety.* New York: Oxford University Press.

Hanes, D. P., and Carpenter, R. H. S. (1997). Countermanding saccades in humans. *Society for Neuroscience Abstracts, 23,* 797.

Hanes, D. P., and Schall, J. D. (1995). Countermanding saccades in macaque. *Visual Neuroscience, 12,* 929–937.

Hanes, D. P., Patterson, W. F., and Schall, J. D. (1998). Role of frontal eye fields in countermanding saccades: Visual, movement and fixation activity. *Journal of Neurophysiology, 79,* 817–834.

Jennings, J. R., van der Molen, M. W., Brock, K., and Somsen, R. J. M. (1992). On the synchrony of stopping motor responses and delaying heartbeats. *Journal of Experimental Psychology: Human Perception and Performance, 18,* 422–436.

Kramer, A. F., Humphrey, D. G., Larish, J. F., Logan, G. D., and Strayer, D. L. (1994). Aging and inhibition: Beyond a unitary view of inhibitory processing in attention. *Psychology and Aging, 9,* 491–512.

Lappin, J. S., and Eriksen, C. W. (1966). Use of a delayed signal to stop a visual reaction-time response. *Journal of Experimental Psychology, 72,* 805–811.

Logan, G. D. (1981). Attention, automaticity, and the ability to stop a speeded choice response. In J. Long and A. D. Baddeley (Eds.), *Attention and Performance IX,* pp. 205–222. Hillsdale, NJ: Erlbaum.

Logan, G. D. (1982). On the ability to inhibit complex movements: A stop-signal study of typewriting. *Journal of Experimental Psychology: Human Perception and Performance, 8,* 778–792.

Logan, G. D. (1985). Executive control of thought and action. *Acta Psychologica, 60,* 193–210.

Logan, G. D. (1994). On the ability to inhibit thought and action: A users' guide to the stop signal paradigm. In D. Dagenbach and T. H. Carr (Eds.), *Inhibitory processes in attention, memory, and language,* pp. 189–239. San Diego, CA: Academic Press.

Logan, G. D., and Burkell, J. (1986). Dependence and independence in responding to double stimulation: A comparison of stop, change, and dual-task paradigms. *Journal of Experimental Psychology: Human Perception and Performance, 12,* 549–563.

Logan, G. D., and Cowan, W. B. (1984). On the ability to inhibit thought and action: A theory of an act of control. *Psychological Review, 91,* 295–327.

Logan, G. D., Cowan, W. B., and Davis, K. A. (1984). On the ability to inhibit responses in simple and choice reaction time tasks: A model and a method. *Journal of Experimental Psychology: Human Perception and Performance, 10,* 276–291.

Logan, G. D., and Irwin, D. E. (2000). Don't look! Don't touch! Inhibitory control of eye and hand movements. *Psychonomic Bulletin and Review, 7,* 107–112.

Logan, G. D., Schachar, R. J., and Tannock, R. (1997). Impulsivity and inhibitory control. *Psychological Science, 8,* 60–64.

May, C. P., and Hasher, L. (1998). Synchrony effects in inhibitory control over thought and action. *Journal of Experimental Psychology: Human Perception and Performance, 24,* 363–379.

McGarry, T., and Franks, I. M. (1997). A horse race between independent processes: Evidence for a phantom point of no return in the preparation of a speeded motor response. *Journal of Experimental Psychology: Human Perception and Performance, 23,* 1533–1542.

Meyer, D. E., and Keiras, D. E. (1997). A computational theory of executive cognitive processes and multiple-task performance: 1. Basic processes. *Psychological Review, 104,* 3–65.

Mulvihill, L. E., Skilling, T. A., and Vogel-Sprott, M. (1997). Alcohol and the ability to inhibit behavior in men and women. *Journal of Studies on Alcohol, 58,* 600–605.

Ollman, R. T. (1973). Simple reactions with random countermanding of the "go" signal. In S. Kornblum (Ed.), *Attention and Performance IV,* pp. 571–581. New York: Academic Press.

Oosterlaan, J., Logan, G. D., and Sergeant, J. A. (1998). Response inhibition in ADHD, CD, comorbid ADHD + CD, anxious and normal children: A meta-analysis of studies with the stop task. *Journal of Child Psychology and Psychiatry, 39,* 411–425.

Osman, A., Kornblum, S., and Meyer, D. E. (1986). The point of no return in choice reaction time: Controlled and ballistic stages of response preparation. *Journal of Experimental Psychology: Human Perception and Performance, 12,* 243–258.

Osman, A., Kornblum, S., and Meyer, D. E. (1990). Does response programming necessitate response execution? *Journal of Experimental Psychology: Human Perception and Performance, 16,* 183–198.

Pennington, B. F., and Ozonoff, S. (1996). Executive functions and developmental psychopathology. *Journal of Child Psychology and Psychiatry, 37,* 51–87.

Quay, H. C. (1988). The behavioral reward and inhibition systems in childhood behavior disorder. In L. M. Bloomingdale (Ed.), *Attention deficit disorder III: New research in treatment, psychopharmacology, and attention,* pp. 176–186. New York: Pergamon Press.

Reid, M. K., and Borkowski, J. G. (1984). Effects of methylphenidate (Ritalin) on information processing in hyperactive children. *Journal of Abnormal Child Psychology, 12,* 169–186.

Rosvold, H., Mirsky, A., Sarason, I., Bransome, E., and Beck, L. (1956). A continuous performance test of brain damage. *Journal of Consulting Psychology, 20,* 343–350.

Schachar, R. J. (1986). Hyperkinetic syndrome: Historical development of the concept. In E. Taylor (Ed.), *The overactive child,* pp. 19–40. Philadelphia: Lippincott.

Schachar, R. J., and Logan, G. D. (1990a). Impulsivity and inhibitory control in normal development and childhood psychopathology. *Developmental Psychology, 26,* 710–720.

Schachar, R. J., and Logan, G. D. (1990b). Are hyperactive children deficient in attentional capacity? *Journal of Abnormal Child Psychology, 18,* 493–513.

Schachar, R., Logan, G. D., Wachsmuth, R., and Chajczyk, D. (1988). Attaining and maintaining preparation: A comparison of attention in hyperactive, normal, and disturbed control children. *Journal of Abnormal Child Psychology, 16,* 361–378.

Schachar, R. J., and Tannock, R. (1995). Test of four hypotheses for the comorbidity of attention-deficit hyperactivity disorder and conduct disorder. *Journal of the American Academy of Child and Adolescent Psychiatry, 34,* 639–648.

Schachar, R. J., Tannock, R., and Logan, G. D. (1993). Inhibitory control, impulsiveness, and attention deficit hyperactivity disorder. *Clinical Psychology Review, 13,* 721–739.

Schachar, R., Tannock, R., Marriott, M., and Logan, G. (1995). Control of response processes in attention deficit hyperactivity disorder. *Journal of Abnormal Child Psychology, 23,* 411–437.

Sergeant, J. A., and van der Meere, J. (1990). Additive factor method applied to psychopathology with special reference to childhood hyperactivity. *Acta Psychologica, 74,* 277–295.

Sostek, A. J., Buchsbaum, M. S., and Rapoport, J. L. (1980). Effects of amphetamine on vigilance performance in normal and hyperactive children. *Journal of Abnormal Child Psychology, 8,* 491–500.

Sternberg, S. (1969). The discovery of processing stages: Extensions of Donders' method. In W. G. Koster (Ed.), *Attention and Performance II,* pp. 276–315. Amsterdam: Elsevier.

Still, G. F. (1902). The Coulstonian Lectures on some abnormal psychical conditions in children. *Lancet, 1,* 1008–1012, 1077–1082, 1633–1168.

Sykes, D. H., Douglas, V. I., and Morgenstern, G. (1973). Sustained attention in hyperactive children. *Journal of Child Psychology and Psychiatry, 14,* 213–220.

Szatmari, P., Offord, D. R., and Boyle, M. H. (1989). Ontario Child Health Study: Prevalence of attention deficit disorder with hyperactivity. *Journal of Child Psychology and Psychiatry, 30,* 219–230.

Tannock, R. (1998). Attention deficit hyperactivity disorder: Advances in cognitive, neurobiological, and genetic research. *Journal of Child Psychology and Psychiatry, 39,* 65–99.

Tannock, R., Schachar, R. J., Carr, R. P., Chajczyk, D., and Logan, G. D. (1989). Effects of methylphenidate on inhibitory control in hyperactive children. *Journal of Abnormal Child Psychology, 17,* 473–491.

Tannock, R., Schachar, R. J., Carr, R. P., and Logan, G. D. (1989). Dose response effects of methylphenidate on academic performance and overt behavior in hyperactive children. *Pediatrics, 84,* 648–657.

Tannock, R., Schachar, R. J., and Logan, G. D. (1993). Does methylphenidate induce overfocusing in hyperactive children? *Journal of Clinical Child Psychology, 22,* 28–41.

Tannock, R., Schachar, R., and Logan, G. D. (1995). Methylphenidate and cognitive flexibility: Dissociated dose effects on behavior and cognition in hyperactive children. *Journal of Abnormal Child Psychology, 23,* 235–266.

Tredgold, A. F. (1908). *Mental deficiency (Amentia).* New York: W. Wood.

van der Meere, J., van Baal, M., and Sergeant, J. (1989). The additive factor method: A differential diagnostic tool in hyperactivity and learning disability. *Journal of Abnormal Child Psychology, 17,* 409–422.

Williams, B. R., Ponesse, J. S., Schachar, R. J., Logan, G. D., and Tannock, R. (1999). Development of inhibitory control across the life span. *Developmental Psychology, 35,* 205–213.

World Health Organization. (1994). *ICD-10: Classification of mental and behavioral disorders.* London: Churchill Livingstone.

Zentall, S. (1985). A context for hyperactivity. In K. D. Gadow and I. Bailer (Eds.), *Advances in learning and behavioral disabilities,* vol. 4, pp. 273–343. Greenwich, CT: JAI Press.

VII Computational Modeling of Control

30 Modern Computational Perspectives on Executive Mental Processes and Cognitive Control: Where to from Here?

David E. Kieras, David E. Meyer, James A. Ballas, and Erick J. Lauber

ABSTRACT Formal concepts and algorithms from contemporary computer operating systems can facilitate efforts to precisely characterize the supervisory functions of executive mental processes. In particular, by helping to advance work with the "executive-process interactive control" (EPIC) architecture, a theoretical framework for computational modeling of human multitask performance, operating system fundamentals provide insights about how people schedule tasks, allocate perceptual-motor resources, and coordinate task processes under both laboratory and real-world conditions. Such insights may lead to discoveries about the acquisition of procedural task knowledge and efficient multitasking skills.

Following the cognitive revolution in scientific psychology (circa 1950), many experimental psychologists and cognitive scientists have assumed that human cognition shares fundamental similarities with symbolic information processing by electronic digital computers (Lachman, Lachman, and Butterfield 1979; Newell 1990). Although the operations of such computers are serial in some respects, they can emulate parallel processing of multiple information streams and implement algorithms for modeling the performance of perceptual-motor and cognitive tasks. As a result, the computer metaphor has inspired significant discoveries about perception, attention, learning, memory, language, and problem solving. Furthermore, as computational hardware and software continue to evolve, the computer metaphor may become increasingly apt.

Encouraged by this prospect, our work has focused on characterizing executive mental processes with a particular theoretical framework, the "executive-process interactive control" (EPIC) architecture. Using EPIC, we have formulated precise computational models of human multitask performance under both laboratory and real-world conditions (e.g., Kieras and Meyer 1997, forthcoming; Meyer and Kieras 1997a,b, 1999). EPIC models account well for quantitative data, predict new phenomena, and point toward promising directions for future research on cognitive control.

The functions of executive processes in EPIC correspond closely to ones provided by a computer operating system (OS) that supports parallel information processing for concurrent execution of multiple task programs (Stallings 1998). This correspondence suggests that studying the fundamentals of contemporary OSs may facilitate the development of EPIC. Such study may also advance the conceptualization of executive mental processes in other theoretical frameworks (e.g., Baddeley 1986; Braver and Cohen, chap. 31, this volume; Kimberg and Farah, chap. 32, this volume; Norman and Shallice 1986), thereby helping to banish the "homunculus" of cognitive control about which previous pundits have complained vociferously (e.g., Newell 1980; Neisser 1967).

In our opinion, the modern computer metaphor is relevant to answering several related questions: Do people have general executive processes that are used across many contexts? Exactly what functions do these processes serve? How might they influence the representation and acquisition of procedural task knowledge? Are there task-specific aspects of cognitive control for which general executive processes must be supplemented through special training? Which experimental procedures are especially suited for eliciting and analyzing particular control operations? Does the human brain really implement the types of function that an OS provides?

Toward answering these questions, section 30.1 introduces EPIC; section 30.2 describes results from applications of EPIC to modeling multitask performance and characterizing particular executive mental processes; section 30.3 presents additional relevant concepts from contemporary computer technology and OSs; section 30.4 discusses how these concepts may promote research with EPIC and guide theorizing about cognitive control; and section 30.5, in summarizing our conclusions, offers final thoughts on the directions of future research.

30.1 THE EXECUTIVE-PROCESS INTERACTIVE CONTROL ARCHITECTURE

Extending proposals by previous theorists (e.g., Anderson 1983; Card, Moran, and Newell 1983; Newell 1990), we have designed EPIC to integrate cognitive and perceptual-motor operations with procedural task analyses of skilled performance.

Components

EPIC has a central cognitive processor with a production-rule interpreter and a multipartition working memory (WM) surrounded by peripheral sensors, perceptual processors, motor processors, and effectors that all operate in parallel. These permanent interconnected components constitute EPIC's "hardware." Each perceptual and motor processor functions as a distinct limited-capacity channel of input or output. Task performance is modeled by programming the cognitive processor with production rules that make decisions and generate responses based on the contents of WM. The production rules, stimulus codes, and response codes may vary depending on specific task requirements.

Consistent with basic periodicities of human information processing (Kristofferson 1967), EPIC's cognitive processor operates in cycles that have stochastic durations whose mean is 50 msec. While doing so, the cognitive processor enables a high degree of parallelism in multitask performance. On each cycle, its production-rule interpreter tests the conditions of all rules in procedural memory, and executes the actions of all rules whose conditions match the current contents of WM. There is no set limit on how many rules can be applied simultaneously. Thus cognitive processes involving distinct sequences of rules may progress simultaneously, sharing system resources as time passes.

Basics of Control

The flow of information processing in EPIC is controlled with production rules like the following one, which selects and initiates a manual "poke" response to a red target stimulus during a tactical decision task (Kieras and Meyer 1997, forthcoming; Meyer and Kieras 1999):

```
IF
((GOAL DESIGNATE-TARGET-FOR-TACTICAL-TASK)
 (STRATEGY MAKE-POKE-RESPONSE-IMMEDIATELY)
 (STEP MAKE-POKE-RESPONSE)
 (TAG ?OBJECT IS STIMULUS)
 (VISUAL ?OBJECT COLOR RED)
 (NOT (VISUAL ??? SIZE LARGE))
 (STATUS TACTICAL-TASK-PROCESS-HAS-EYE)
 (MOTOR MANUAL PROCESSOR FREE))
THEN
((SEND-TO-MOTOR-MANUAL-PROCESSOR PERFORM-POKE-(LEFT INDEX)
  ?OBJECT)
 (ADDWM (GOAL WATCH-FOR-DESIGNATION-EFFECT))
 (DELWM (STEP MAKE-POKE-RESPONSE))
 (ADDWM (STEP WAIT-FOR-WATCHING-DONE)))
```

Sequential Rule Execution As illustrated here, EPIC production rules have conditions and actions that contain goal and step items. Adding and deleting step items in working memory enables the rules to be executed in particular sequences. For example, the preceding rule would be enabled by putting "STEP MAKE-POKE-RESPONSE" in WM with an add-to-WM (ADDWM) action. Taking this item out of WM with a delete-from-WM (DELWM) action would disable the rule, and then putting "STEP WAIT-FOR-WATCHING-DONE" in WM would enable another subsequent rule. Because information in WM is subject to loss or corruption, errors of sequencing may occur under EPIC, as they do under real-world circumstances.

Subroutine Calls Using the same goal item in a set of EPIC production rules lets them function like a computer program subroutine. The subroutine would be "called" by adding its shared goal item to working memory. After the call, a start-up rule in the subroutine would "fire" and add its first step item to WM. When the subroutine finishes, its termination rule would delete the routine's goal and last step items from WM, and signal that the subroutine has finished. For example, the preceding rule calls a subroutine for watching the visual effects of the manual poke response. This entails adding two items to WM: "GOAL WATCH-FOR-DESIGNATION-EFFECT," which is the goal item for the subroutine; and "STEP WAIT-FOR-WATCHING-DONE," which is used by another rule that waits for the subroutine to be completed.

Interrupts Thus EPIC implements capabilities analogous to computer interrupts. A production rule can have conditions such that it waits for a certain future event to occur regardless of other intervening activities. When these conditions are satisfied, the rule may start the execution of other rule sequences to deal with the interrupting event.

Task Processes

Procedural knowledge for performing tasks is represented by EPIC production rules that fire in particular sequences. Our models embody programming-style principles like those applied in computer software design. Each task and subtask has a set of rules with standard formatting of control items and input/output (I/O) information. Standard protocols are used for task start-up, completion, error detection, abort, and restart procedures.

Executive Processes

In modeling multitask performance, we formulate distinct sets of supervisory production rules that implement supraordinate executive processes, whose function is to add and delete working-memory items for controlling the execution of various task and subtask procedures. Under EPIC, an executive process may suspend a task process by deleting its goal item from working memory, and then resume the task process by adding its goal item to WM again. Similarly, an executive process may use strategy items to instruct a task process about which of several alternative paths to take. These control operations can be accomplished through rules whose conditions match status items that the task process adds to WM along the way.

30.2 APPLICATIONS OF EXECUTIVE-PROCESS INTERACTIVE CONTROL TO MULTITASK PERFORMANCE

To illustrate more fully how we characterize executive mental processes, this section describes four cases of several for which EPIC models of multitask performance have been developed: (1) discrete successive tasks; (2) discrete concurrent tasks; (3) elementary continuous tasks; and (4) compound continuous tasks. From them, it will become clearer how EPIC enables task coordination and scheduling to be described under a variety of conditions. Also, the stage will be set for examining cognitive control from the perspective of computer operating systems.

Discrete Successive Tasks

In the discrete successive-tasks procedure, also known as "task switching," participants either alternate between two different choice-reaction tasks or perform one task repeatedly during a series of discrete trials, with a *response-stimulus interval* (RSI) separating each response from onset of the next stimulus. Reaction time (RT) and accuracy are measured as a function of trial type, RSI, and other factors. Switching time costs (STCs) are calculated from differences between mean RTs on alternating-task and repeating-task trials (for a review, see Pashler, chap. 12, this volume).

According to some theorists, executive mental processes contribute substantially to STCs (e.g., Meiran 1996; Rogers and Monsell 1995; Rubinstein, Meyer, and Evans forthcoming; see also Goschke, chap. 14, De Jong, chap. 15, Meiran, chap. 16, and Keele and Rafal, chap. 28, this volume). Following their lead, we have formulated an EPIC model to account for some results from the successive-tasks procedure. The details of this formulation concern both the representation of procedural task knowledge and the cognitive control of task switching.

Lauber 1995 For now, our model deals with data from Lauber 1995 (exps. 4 and 5), which varied response-stimulus intervals, stimulus-response compatibility, and practice orthogonally. Additive and interactive effects of these factors strongly constrain the type of model that may account for them.

Twenty undergraduate students participated in Lauber's study. They were divided into two groups that performed basic choice-reaction tasks with different S-R mappings. Members of each group were tested individually during three 1-hour sessions. The stimuli for each task were printed digits. The responses were keypresses made with fingers of the right hand. Stimuli and responses were paired to form four alternative S-R mappings, each of which was used in one of four different tasks: compatible task A, compatible task B, incompatible task C, and incompatible task D. For task A, the digits 1, 2, 3, and 4 were mapped respectively to

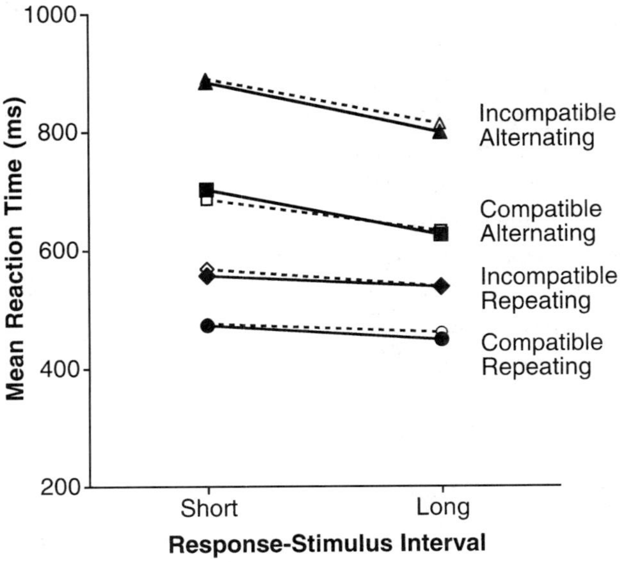

Figure 30.1 Second-task reaction times from the first session in Lauber 1995. The dark points connected by solid lines represent observed mean reaction times as a function of response-stimulus interval, task difficulty (compatible versus incompatible S-R mappings), and trial type (alternating-task versus repeating-task trials). The light points connected by dashed lines represent simulated mean RTs produced by the EPIC model in figure 30.2.

the index, middle, ring, and little fingers; for task B, this mapping was reversed. For task C, the digits 1, 2, 3, and 4 were mapped respectively to the middle, little, index, and ring fingers; for task D, this mapping was reversed.

During each test session, there were two types of trial block. One type contained a series of alternating-task trials, and the other contained a series of repeating-task trials. On each alternating-task trial, participants in group 1 performed task A followed by task B, or vice versa; on each repeating-task trial, they performed one of these tasks twice. A similar arrangement of tasks C and D was used for group 2. Before each trial block, subjects were told what their tasks would be. Each block included two RSIs, 50 and 750 msec, which varied randomly across trials. The intertrial intervals equaled 1 sec.

Empirical Results Figure 30.1 shows some results from the first session. Mean RTs of second-task (post-RSI) responses were reliably longer for alternating-task trials, incompatible S-R mappings, and short RSIs. Although some reliable two-way interactions occurred between these factor effects, S-R compatibility and RSI affected mean switching time costs almost additively. Furthermore, despite these effects, large switching time costs persisted after the longer RSI, as other investigators have found (e.g., Allport and Wylie, chap. 2, this volume; Allport, Styles, and Hsieh

1994; De Jong, chap. 15, this volume; Rogers and Monsell 1995).[1] It is this overall pattern for which EPIC accounts.

EPIC Models Of course, there are various ways that we could model task switching with EPIC. For example, one conceivable model would have two sets of task-specific, goal-sensitive production rules available simultaneously in procedural memory. In this case, the rules used to select responses for Lauber's incompatible tasks C and D might have the following forms:

```
IF
((GOAL PERFORM TASK C)
 (STEP MAKE PRESS-RESPONSE TO DIGIT 1)
 (VISUAL ?OBJECT DIGIT 1))
THEN
((SEND-TO-MOTOR MANUAL PERFORM PRESS (RIGHT MIDDLE))
 (DELWM (STEP MAKE PRESS-RESPONSE TO DIGIT 1))
 (ADDWM (STEP WAIT-FOR PRESS-DONE)))
IF
((GOAL PERFORM TASK D)
 (STEP MAKE PRESS-RESPONSE TO DIGIT 1)
 (VISUAL ?OBJECT DIGIT 1))
THEN
((SEND-TO-MOTOR MANUAL PERFORM PRESS (RIGHT RING))
 (DELWM (STEP MAKE PRESS-RESPONSE TO DIGIT 1))
 (ADDWM (STEP WAIT-FOR PRESS-DONE)))
```

Given the simultaneous availability of such rules, an executive process could switch tasks simply by changing the task goal items in working memory, disabling one task's rules and enabling the other's.

Yet this type of model would fail to account for persistent large switching time costs such as Lauber observed. Under EPIC, changing goal items takes only one cognitive-processor cycle, which should be completed within about 50 msec regardless of other prevailing factors. However, Lauber's STCs ranged from 200 to 300 msec, they endured after a relatively long (750 msec) RSI, and S-R incompatibility affected them reliably. Thus additional delays associated with other control operations besides changing goal items presumably contributed to task switching here. Perhaps these contributions occurred because the tasks had different S-R mappings but involved the same stimuli and responses. Such mapping conflicts might substantially increase the amount of practice needed to learn adequate task-specific, goal-sensitive production rules (Anderson 1983), requiring participants to rely initially on other types of procedural and declarative knowledge instead.

Thus our modeling of Lauber's results has taken an alternative direction. Consistent with some other theorists (e.g., Rubinstein, Meyer, and

Evans forthcoming), we assume that to reduce conflicts in switching between similar tasks, five constraints are imposed: (1) at each moment, symbolic S-R mapping information for performing just one task is kept in WM; (2) switching tasks involves removing currently irrelevant information from WM; (3) the irrelevant information is replaced with relevant information for the next task; (4) these "cleanup" and "setup" operations entail relatively slow interactions with long-term memory; and (5) setting up for the next task is triggered by its stimulus onset.

On the basis of these assumptions, we have formulated a model with a single set of generic production rules that perform both of Lauber's incompatible tasks. For each incompatible task, these rules select responses by using a particular list of S-R pairs in WM. This involves checking the stored S-R pairs serially to find one whose stimulus term matches the presented stimulus (cf. Theios 1973). When the match is found, its associated response term is sent to the manual motor processor. Given this protocol, task switching requires not only changing task goal items but also retrieving the next relevant S-R pairs from long-term memory.

For performing both of Lauber's compatible tasks, our model has another set of generic production rules. They assume that EPIC's visual perceptual processor directly recodes each presented stimulus into two response symbols appropriate for the alternative compatible S-R mappings (e.g., "1" → "index finger" and "1" → "little finger"). A task rule then chooses and sends one or the other of these response symbols to the manual motor processor. This choice is made by referring to a WM strategy item that indicates which S-R mapping is currently relevant. Given this protocol, task switching requires not only changing task goal items but also retrieving the relevant strategy item from long-term memory.

These operations are controlled by an executive process that takes different paths for alternating-task and repeating-task trials (figure 30.2). At the start of repeating-task trial blocks, the executive process calls a subroutine that sets up WM to perform a particular task, and then lets this task be performed twice during each trial. In contrast, at the start of each alternating-task trial, the executive process waits until the first-task stimulus has been recognized, next calls the subroutine that sets up WM for the first task, and then lets the first task be performed. After the first-task response has been made, the executive process calls another subroutine that cleans up WM, waits until the second-task stimulus has been recognized, calls the setup subroutine for the second task, lets the second task be performed, and finally cleans up WM again. Fitting our model to Lauber's data required adjusting the times taken by the WM setup and cleanup subroutines.

Simulated Results Figure 30.1 shows the mean second-task RTs produced by our model, which accounts well for the main effects of trial

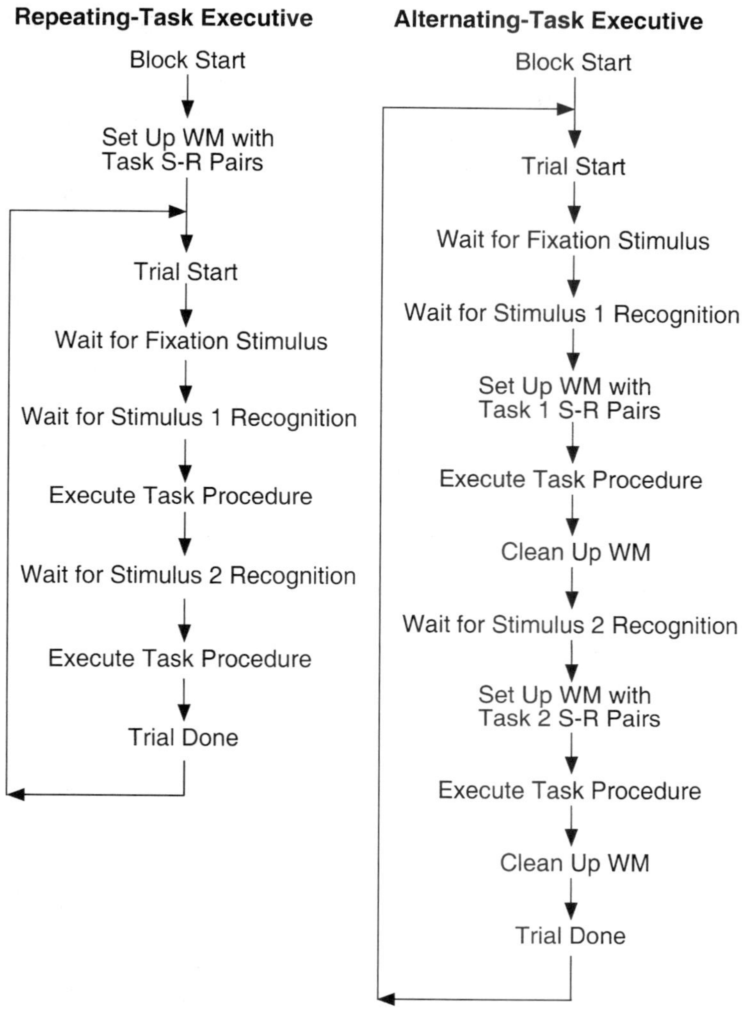

Figure 30.2 Flowchart of executive processes on repeating-task trials (left) and alternating-task trials (right) in the EPIC model for Lauber 1995. Mean reaction times produced by this model appear in figure 30.1.

type, RSI, and S-R mapping, as well as their additivities and interactions.[2] Our model succeeds much better than one that switches tasks simply by changing goal items in working memory.

Theoretical Implications The working-memory setup and cleanup operations that we needed to fit Lauber's data each took about 150 msec. Why so long? One possible answer is that in reality, these operations entail gradually activating relevant and inhibiting irrelevant symbolic long-term memory representations (cf. Allport, Styles, and Hsieh 1994; Anderson 1983; Goschke, chap. 14, this volume). This would explain why STCs persist at long RSIs and why WM setup is not started until the next task's stimulus has been recognized. Perhaps the executive process waits to start setting up WM because stimulus recognition helps amplify requisite memory activation. At present, EPIC does not implement such activation explicitly. Thus supplementing EPIC with appropriate activation mechanisms could prove worthwhile.

From our present perspective (figure 30.2), however, the executive processes for task switching seem relatively simple. Other than calling WM setup and cleanup subroutines, they contribute very little to STCs. This is consistent with claims of Allport, Styles, and Hsieh (1994), who questioned whether task-switching studies reveal much about executive mental processes per se. Nevertheless, such studies could have further benefits in other respects. For example, they may yield new insights about the representation of procedural task knowledge, extending what we have discovered already through EPIC modeling.

Discrete Concurrent Tasks

A second context in which EPIC has enabled us to learn more about executive mental processes is the "psychological refractory period" (PRP) procedure (Pashler 1994, chap. 12, this volume). In this procedure, subjects perform two concurrent choice-reaction tasks during series of discrete trials. Typically the tasks involve different stimuli and responses. On each trial, a first-task stimulus is followed by a second-task stimulus. Because the stimulus onset asynchrony (SOA) is relatively short, the second-task stimulus may precede the first-task response. However, subjects are instructed to give task 1 higher priority, and they may be encouraged to make the first-task response before the second-task response. RTs and response accuracy are measured as a function of the SOA and other task factors. The PRP procedure interests us because, despite its task prioritizing and stimulus sequencing, there is potentially ample opportunity for tasks 1 and 2 to be performed at least somewhat in parallel. By formulating EPIC models under these conditions, we can better understand how such cognitive control is achieved.

EPIC Model For example, figure 30.3 outlines the executive process of a model that has been tested extensively in our research concerning the PRP procedure (Meyer and Kieras 1997a,b). Here the executive process puts tasks 1 and 2 respectively in "immediate" and "deferred" modes at the start of each trial. This is done by adding strategy items (e.g., "STRATEGY TASK 1 IS IMMEDIATE") to WM. Putting task 1 in immediate mode lets its responses be selected and sent to their motor processor as quickly as possible for movement production. While task 2 is in deferred mode, its production rules can select symbolic identities of second-task responses and store them in working memory, but the selected second-

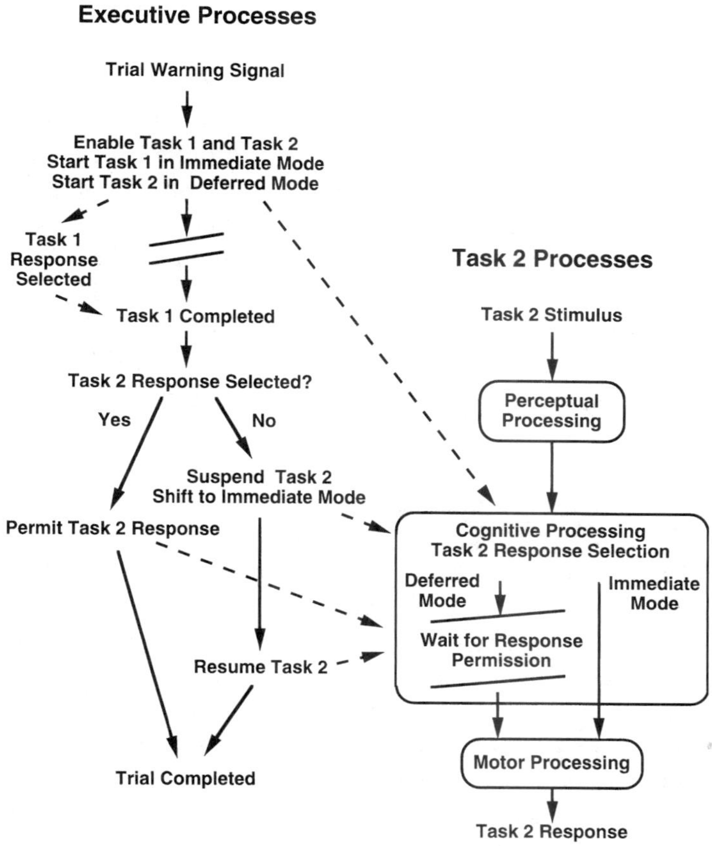

Figure 30.3 Flowchart of executive and secondary-task processes in the EPIC strategic response-deferment model for the psychological refractory period (PRP) procedure.

task response identities are not sent to a motor processor, and they are not produced as overt movements. When, however, a prespecified "unlocking event" occurs subsequently (e.g., the overt first-task response is initiated), the executive process shifts task 2 to immediate mode. Following this shift, previously selected second-task responses may be sent from WM to their motor processor for movement production. If response selection has not yet finished for task 2 before it is shifted to immediate mode, then subsequently the second-task production rules will both select and send the second-task responses directly to their motor processor.

Simulated versus Empirical Results Comparisons between simulated and empirical results from various studies with the PRP procedure have been encouraging. Our EPIC strategic response deferment model accounts accurately for differences between observed mean first- and second-task RTs as well as additive and interactive factor effects on them.

The model's goodness of fit is typically high ($R^2 > 0.95$) and involves only modest numbers of "free" parameters.

Theoretical Implications Our research has revealed that people schedule the tasks of the PRP procedure through a combination of various mechanisms. Symbolic response codes for tasks 1 and 2 may be selected concurrently under flexible strategic control, whereby physical movements are produced in proper serial order. Contrary to traditional response-selection bottleneck hypotheses (cf. Pashler, chap. 12, Jolicoeur, Dell'Acqua, and Crebolder, chap. 13, and Ivry and Hazeltine, chap. 17, this volume), we have found no evidence that skilled dual-task performance is constrained by immutable "hardware" decision or response-selection bottlenecks.

Elementary Continuous Tasks

The preceding conclusions based on the EPIC architecture have been strengthened by formulating computational models of executive mental processes for elementary continuous tasks (Kieras and Meyer 1997). Here the focus is on visuomanual tracking and choice-reaction tasks that must be performed without predictable pauses along the way. By fitting quantitative results obtained under such conditions, we further demonstrate the existence and generality of strategic cognitive control that judiciously overlaps stages of processing in human multitask performance.

Martin-Emerson and Wickens 1992 For this demonstration, our research has dealt especially with Martin-Emerson and Wickens 1992, in which subjects viewed upper and lower windows on a display screen. In the upper window were a circular target and crosshairs cursor. During 1-minute test intervals, the cursor's location was perturbed haphazardly by an accelerative forcing function. The subjects performed a compensatory tracking task, moving a right-hand joystick to keep the cursor on target. The tracking task was either hard or easy, requiring more or less frequent joystick movements. Meanwhile, in the lower window, horizontal arrows appeared intermittently. Depending on whether an arrow pointed right or left, subjects pressed a left-hand index or middle finger key. The centers of the task windows were separated by a visual angle that varied systematically across test intervals. As this angle increased, eye movements that traveled greater distances were required for the stimuli to be identified correctly. Both the tracking and arrow-discrimination tasks were supposed to receive high priority.

Empirical Results As shown in figure 30.4, mean RTs for the arrow discriminations increased reliably with the visual angle between display windows but were relatively unaffected by tracking difficulty. In contrast,

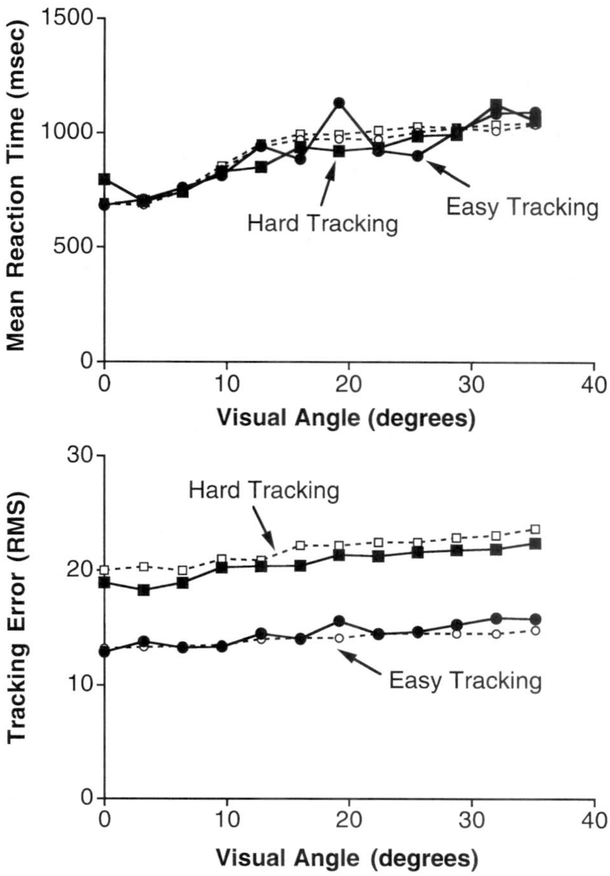

Figure 30.4 Results from Martin-Emerson and Wickens (1992). *Top*. Observed mean reaction times (dark points on solid lines) and simulated mean reaction times (light points on dashed lines) produced by the EPIC model in figure 30.5 for the arrow-discrimination task when it was performed concurrently with either an easy or hard visuomanual tracking task. *Bottom*. Observed and simulated root mean square (RMS) errors for the visuomanual tracking task when it was easy or hard.

root mean square (RMS) tracking errors were reliably greater for hard tracking, but the visual angle affected them relatively little. This occurred even though the tracking errors were measured during 2 sec intervals that started at the onsets of the stimuli for the arrow-discrimination task.

EPIC Models To account for these results, we first formulated an EPIC model that uses inefficient "lockout" scheduling, which let us test predictions based on the traditional response-selection bottleneck hypothesis (cf. Pashler, chap. 12, this volume). According to this model, whenever an arrow occurs, tracking is suspended as soon as possible, performance of the arrow-discrimination task proceeds until completion, and then tracking is resumed. Given realistic delays in EPIC's motor processors, such lockout scheduling yielded excessively large RMS tracking errors.

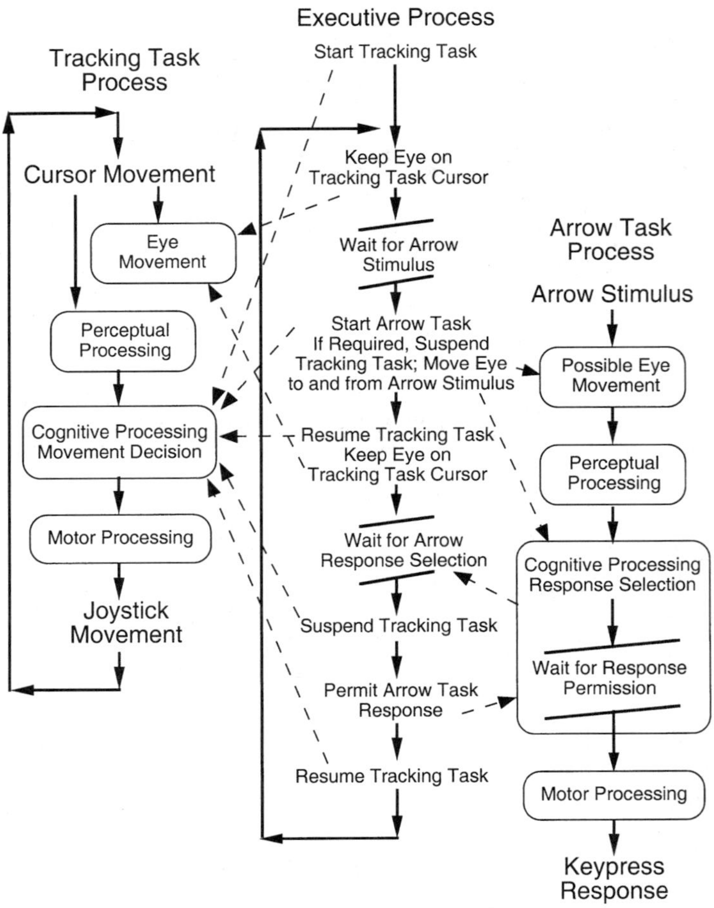

Figure 30.5 Flowchart of an EPIC model with a customized executive process that implements overlapped task scheduling for Martin-Emerson and Wickens 1992. Dashed diagonal arrows from the executive process to the concurrent tracking and arrow-discrimination task processes represent context-dependent supervisory control imposed under these conditions. Mean reaction times and root mean square (RMS) tracking errors produced by this model appear in figure 30.4.

These discrepancies led us to reject this first model and to formulate a second model, with more efficient overlapped task scheduling.

Figure 30.5 shows the task and executive processes of our second model. Here the executive process initially starts the tracking task and enables decisions about joystick movements to be made on the basis of perceived cursor movements. Next, the executive process enters an iterative loop in which it sends commands to the ocular motor processor for keeping the eyes on the tracking task cursor while waiting for an arrow to occur. During this wait, cursor movements may trigger the production rules of the tracking task, which send commands to the manual motor processor for producing joystick movements that keep the cursor on target. When the onset of an arrow is detected, the executive process starts the arrow-discrimination task and enables its production rules to select a keypress response in deferred mode. For an arrow in foveal or parafoveal vision, perceptual identification proceeds without further ado, and a keypress response is selected while tracking continues until the response's identity becomes available in working memory. Otherwise, for an arrow in peripheral vision, the executive process takes several additional steps: it suspends tracking, moves the eyes to look at the arrow so that its identification can proceed, returns the eyes to look at the cursor, and resumes tracking until a deferred-mode keypress response to the arrow has been selected. As far as possible, this lets tracking continue simultaneously with perceptual identification and response selection for the arrow. Furthermore, as soon as possible after a keypress response has been selected, the executive process also suspends tracking and permits the keypress's identity to be sent to the manual motor processor. Then the keypress response is produced, the arrow-discrimination task is terminated, and tracking is resumed again. Thus, this overlapped task-scheduling model is similar to our previous model for the PRP procedure (cf. figure 30.3).

Simulated Results Figure 30.4 shows simulated results from the present model, whose mean RTs and RMS tracking errors closely approximate those produced by actual participants. Unlike lockout scheduling, overlapped scheduling does not yield excessively large tracking errors.

Theoretical Implications The present model's success supports our claims about how executive mental processes may temporally overlap visual, response selection, ocular motor, and manual motor operations in multitask performance. Apparently, the types of control mechanisms and scheduling strategies we have proposed for discrete concurrent (e.g., PRP) tasks also contribute to efficient performance of elementary continuous tasks. These mechanisms seem to be used regardless of whether the tasks involve the same (e.g., visuomanual) or different (e.g., auditory-vocal and visuomanual) perceptual-motor modalities.

Compound Continuous Tasks

Our characterization of executive mental processes applies not only to elementary but also to compound continuous tasks that entail several distinct subtasks. For example, Ballas, Heitmeyer, and Perez 1992 studied concurrent visuomanual tracking and tactical decision making during simulated military aircraft operations. In tracking, subjects plied a joystick to superimpose a cursor over an evasive target plane. In tactical decision making, subjects pressed finger keys to designate the hostility of numbered icons that depicted jet fighters, bombers, and missile sites. Because there were various types of icon and designation criteria, this decision making constituted a compound task.

To account for performance under these quasi-realistic conditions, we have found that an EPIC model with a three-level hierarchy of executive and task processes fits empirical data well (Kieras and Meyer 1997, forthcoming; Meyer and Kieras, 1997b, 1999). As part of this model, a supraordinate dual-task executive process provides overall supervision for a tracking process, a display monitoring process, and a tactical executive process that coordinates three subprocesses—stimulus icon selection, hostility response selection, and track-number response selection—in tactical decision making. Through this hierarchical control, the relative priority of tactical decision making and the temporal overlap of its subprocesses are varied dynamically, contingent on the numerosity of potentially hostile icons in the display. The model, with its adaptive scheduling mechanisms, accounts well for observed sequences of tactical-decision RTs and RMS tracking errors.

Interim Status Quo

From the preceding illustrations, it should be clear that EPIC yields significant theoretical insights about executive mental processes. However, our progress thus far has been limited in some major respects.

Limitations of EPIC Models One limitation is that the executive processes of our models have been customized for particular task combinations. Although these processes may be somewhat similar across contexts, their formulation has incorporated considerable task-specific knowledge. For example, in modeling Martin-Emerson and Wickens 1992, we had the executive directly control eye movements from the stimulus arrows to the tracking cursor (figure 30.5). This enhances tracking performance, consistent with available data, but makes the executive context dependent and nonmodular. To be strengthened further, EPIC needs general executive processes that are context independent.

Previous theorists have also stressed the importance of general executive processes, as in proposals about the "central executive" (Baddeley 1986) and "supervisory attentional system" (Norman and Shallice 1986). Yet they have not provided explicit computational algorithms that achieve the required generality. Thus we must look elsewhere for ways to fulfill this need.

Accompanying EPIC's lack of general executive processes is a second, related deficiency. Competition among processes for access to limited "hardware" resources may cause miscommunication or deadlock, in which wrong information is transmitted or processes become perpetually stalled (Stallings 1998). EPIC does not yet solve these concurrency problems adequately. Without adequate solutions, veridical modeling of complex adaptive multitask performance will be impossible.

A third limitation is that EPIC does not yet deal with procedural learning in multitask performance. How do people learn to schedule and coordinate concurrent tasks efficiently? How are their multitasking skills transfered across situations? Deeper answers are needed for modeling skill acquisition and developing effective instructional techniques in practical applications (Gopher 1993).

Potential Contributions of Operating System Fundamentals Fortunately, contemporary computer operating systems may stimulate further theorizing. Fundamental principles that underlie their operation provide basic ways for implementing context-independent control and for solving problems of task concurrency (Stallings 1998). By considering these fundamentals, we may augment EPIC with needed general executive processes, concurrency solutions, and multitasking skill acquisition.

30.3 CONTEMPORARY OPERATING SYSTEMS AND COMPUTER TECHNOLOGY

Contemporary operating systems supervise information processing for task programs that are executed virtually or actually in parallel. However, limited capacities of computer hardware impose constraints on an OS trying to maximize process throughput. Consequently, we next consider aspects of both hardware design and OS functions that bear on these matters.

Hardware Design

Starting with early computers like ENIAC, hardware design has become increasingly sophisticated (Tucker 1997). As a result, modern computers typically have at least one central processing unit (CPU), at least one memory unit, and various input/output (I/O) peripherals. The CPU executes sequences of instructions for system and task programs. The memory unit stores programs and data, letting them be manipulated in similar ways. Thus generic information-processing capabilities are implemented by the hardware, whereas overall system control and task procedures are provided by the software.

Uniprocessor Architecture Many operating systems and task programs are used on computers with one CPU. Although this uniprocessor architecture executes instructions sequentially in some respects, its components enable extensive parallelism. For example, separate streams of data may be transmitted simultaneously to or from different I/O peripherals, and the CPU may perform multiple suboperations in parallel. Exploiting such capabilities, an OS can sustain concurrent threads of processing at least somewhat as if each program had its own CPU.

Multiprocessor Architectures Moreover, some operating systems and task programs have been implemented with multiple CPUs. These multiprocessor architectures enable true parallel processing and provide enormous, relatively inexpensive, computational power. Particularly relevant for us is the *shared-memory symmetric multiprocessor* (SMP), in which multiple CPUs function as equivalent "peers" that share one memory unit and I/O peripherals. This corresponds at least approximately to EPIC's organization. Although EPIC has one cognitive processor, it tests conditions and executes actions of multiple production rules in parallel. When programmed with two or more rule sets, the cognitive processor emulates a collection of peer CPUs; as in a SMP, these rule sets share WM and I/O peripherals.

Thus contemporary OS fundamentals should be applicable to EPIC. Indeed, computer scientists have discovered that OS fundamentals are extremely general, applying across many uniprocessor and multiprocessor architectures. This suggests that what OSs and EPIC teach us will likely hold as well for the human mind and brain, which also implement forms of multiprocessor parallelism. To appreciate OS fundamentals, more background about them is in order (see Stallings 1998; Tucker 1997).

Operating System History

Like computer hardware, operating systems have become increasingly sophisticated. For early computers (circa 1950), people loaded and started programs manually. Subsequently (circa 1960), primitive OS resident monitors were developed to automate these processes. Following this development, OS capabilities were gradually extended to enable overlapping CPU and I/O operations so that the CPU would not have to wait idly on slow mechanical devices. These advances led to multitasking, an overarching OS function (circa 1970).

In multitasking, an OS interleaves or overlaps execution of task programs requiring certain limited hardware resources. When an execution process has taken a set time or must wait for pending I/O, it is suspended, and the CPU is allocated to another process. After completion of I/O or other prerequisites, the suspended process is resumed. Consequently, multiple processes may advance efficiently without individual users' intervention. Software for multitasking on uniprocessors has been gracefully adapted for multitasking on multiprocessors.

Operating System Objectives

Systems programmers developed operating systems to keep CPU and memory hardware as busy as possible, increasing process throughput. OSs have also made it simpler and faster to formulate *noncooperating task programs,* which are executed asynchronously and compete for hardware resources. Given OS services, such a program can be formulated as if it were the only one executed and no intricate control of I/O were required. Furthermore, OSs have facilitated the formulation of *cooperating task programs,* which are executed synchronously and share their products interactively.

However, OSs are neither logically necessary nor maximally efficient in every respect. Nonhierarchical "flat" programs can be formulated to perform multiple tasks concurrently on "bare" computer hardware without OS support. Through this formulation, the computational overhead of hierarchical software can be eliminated, and even faster performance achieved. Nevertheless, such improvement has serious costs. Because it requires dealing directly with many levels of control, the time and effort needed to formulate flat programs can be exorbitant. Also, flat programs do not readily generalize beyond their original applications. In contrast, OSs provide a better compromise between speed of execution, on the one hand, and ease and generality of software development, on the other.

Operating System Functions

This compromise is enabled by operating system functions that solve a basic problem: detailed sequences of execution for independent task programs cannot be predicted accurately. An OS must ensure that execution proceeds correctly and rapidly despite unpredictable interruptions and resumptions. The solution entails judicious task scheduling, resource allocation, process coordination, and conflict resolution.

Task Scheduling In task scheduling, an operating system must make and implement decisions about when programs will be executed. Doing so requires prioritizing, preparing, initiating, suspending, preserving, resuming, and terminating each execution process at apt moments. OSs use various scheduling algorithms for this. Among them are "first come, first serve," "round robin," "shortest remaining time," "shortest process next," "highest response ratio," and "least-time-consumed scheduling," each of which may produce relatively high or low performance, depending on nuances of the prevailing context. Task scheduling by an OS must therefore be "tuned" adaptively to maximize overall throughput.

Resource Allocation An operating system must also allocate hardware resources judiciously to individual processes, depending on resource availability and process needs. For example, during execution, a process may request resources. If these are available, the OS may comply by allocating them immediately. Alternatively, if they have been committed to other processes already, then the OS may deny the current request temporarily, and perhaps suspend the requesting process until its needs can be satisfied. Exactly when processes request and release their resources, and how the OS handles them, contribute significantly to attained performance.

Process Coordination Among the processes being executed, some may need to share intermediate products of their computations. For this sharing to succeed, these cooperating processes must be coordinated, because interprocess communication involves writing to and reading from the same memory locations in proper serial order.

To facilitate interprocess communication, an operating system performs several coordinative functions, including mutual exclusion, process synchronization, and message passing. Relying on these functions, a receiving process may request that the OS suspend it until an expected message arrives from another sending process. When the sending process is ready to transmit this message, it may request that the message be passed to the receiving process. The OS may then pass the message and resume the receiving process.

Conflict Resolution Because concurrent processes impose high loads on hardware resources and may be noncooperative, serious conflicts can arise. An operating system has to avoid these conflicts as best it can, and resolve them gracefully when need be. This function is crucial for dealing with *deadlocks*, which entail closed chains of processes such that each process currently has exclusive ownership of some resource needed by the next process in the chain. Adaptive conflict resolution also helps deal with other undesirable situations such as *starvation*, in which some low-priority process is perpetually preempted by higher-priority processes.

30.4 COGNITIVE CONTROL AND OPERATING SYSTEM FUNDAMENTALS

Contemporary operating systems embody precise and comprehensive instantiations of executive processes. Such instantiations are scarce in current psychological theories. Thus, to promote further progress, we next discuss some stimulating theoretical concepts, multitasking models, and explanatory hypotheses inspired by these considerations.

Theoretical Concepts

The concepts that interest us here involve distinctions between various types of executive and task processes.[3]

Customized Executives One major distinction concerns customized versus general executives. By "customized executive" (CE), we mean a modular set of supraordinate mental processes that manage multitask performance based on unique context-dependent knowledge about the particular tasks and their temporal interrelations. A CE works for only one task combination and cannot be transferred readily across different situations.

Thus far, EPIC models have all used CEs. An instructive case is our model of performance in Martin-Emerson and Wickens 1992. The executive process of this model preallocates resources (i.e., ocular and manual motor processors) to tracking and arrow discrimination without these task processes requesting them explicitly (figure 30.5). The preallocation is possible here because the executive already "knows" the task processes' needs and satisfies them in proper sequence. Such use of context-dependent knowledge may be common after extensive practice under conditions in which high performance is desired.

Our theorizing need not be confined, however, to models with CEs. New EPIC models may be formulated on the basis of general executives that function at least partly like contemporary OSs. From testing them empirically, we learn more about the extent to which OS fundamentals characterize how human multitask performance is controlled.

General Executives A general executive (GE) is a modular set of supraordinate mental processes that manage multitask performance without using unique context-dependent knowledge about the tasks and their temporal interrelations. Given such generality, cognitive control can be achieved for different task combinations through standard functions like those of contemporary OSs. Implementing these functions in EPIC is straightforward because it resembles a shared-memory symmetric multiprocessor.

Nevertheless, determining whether a GE should be added to EPIC requires answering a fundamental question about cognitive control: Do people have GEs and use them for multitask performance? We might expect an affirmative answer, given the potential ease of preparing and efficiency of executing task programs based on GE functions. Yet the only way to be sure about this is to formulate and test EPIC models that rely on a GE. We take this course after introducing more distinctions that will facilitate our pursuits.

Managerial Styles Another relevant distinction concerns managerial styles of general executives. At one extreme, a *conservative* GE can have a strict regimented style of scheduling task processes and allocating limited resources to them. Under such regimentation, task processes may have to request resources before using them; processes may be suspended when their requested resources are unavailable; and processes not prone to make deferent resource requests may be kept from starting (i.e., locked out) until others have finished. Alternatively, a *liberal* GE can have a tolerant, laissez-faire managerial style, under which task processes may be allowed to proceed at least partially unabated while their requested resources are unavailable, and processes prone to use resources without requesting them may also be accommodated insofar as possible. In principle, a GE's managerial style is adaptable to particular situations. Such adaptability, contingent on the "manners" of task processes, will determine the attained level of multitask performance.

Process Manners and Etiquette Task processes can have various manners of interaction with a general executive. Proper etiquette for a task process entails requesting resources (e.g., motor mechanisms) immediately before they will be used, waiting for the GE's permission to use them, and then releasing the resources immediately after their use is complete. A "*polite* process" conforms to all of these rules. This establishes favorable circumstances for a laissez-faire managerial style through which relatively high multitask performance is attainable.

Theoretically, however, some task processes may be impolite. For example, a *presumptuous* process might use crucial resources without requesting them. An *impatient* process might request resources but not wait for permission to use them. A *greedy* process might request resources too early and release them too late. Such inconsiderate conduct will force a GE to be more conservative, curtailing the processes' temporal overlap and impeding their progress.

Cost-Benefit Assessment To assess the costs and benefits of alternative general executive managerial styles, various factors are relevant. One is *interaction overhead*, which includes scheduling, allocation, and abdication costs for supervising task processes. *Scheduling costs* are amounts of time consumed by adding and deleting goals in working memory to start, suspend, resume, and terminate processes selectively. *Allocation costs* are amounts of time consumed by making and fulfilling resource requests. *Abdication costs* are amounts of time consumed by releasing resources. Ideally, these costs should be paid in ways that decrease *resource possession times*, the amounts of time during which a task process possesses crucial resources. Also, as best they can, the payments should increase *process overlap intervals*, the intervals during which multiple processes are advancing simultaneously.

Taking these factors into account, impolite task processes may escape some interaction overhead, but they increase resource-possession times and force the GE to eliminate process-overlap intervals. In contrast, a liberal GE and polite task processes make an attractive compromise. Their process-overlap intervals and resource-possession times may be relatively long and short, respectively, thereby more than compensating for the GE's moderate interaction overhead.

Nevertheless, there are other ways to perform better on all scores. CEs (customized executives) tuned for particular task combinations can achieve even lower interaction overhead, shorter resource-possession times, and longer process-overlap intervals. As we shall see, this leads to interesting hypotheses about multitasking skill acquisition.

New Multitasking Models

To illustrate how these theoretical concepts help clarify the nature of cognitive control, we have implemented them in two new EPIC models for Martin-Emerson and Wickens 1992. Model 1 has a conservative general executive that supervises two impolite task processes. Model 2 has a more liberal GE that supervises two polite task processes. By comparing these models to our previous one that has a customized executive (figure 30.5), we examine the effects of managerial style and process manners on multitask performance.

Model 1: Conservative General Executive with Impolite Processes In model 1, tracking and arrow discrimination are assumed to be impolite processes. They do not request or release resources for producing eye and hand movements. Instead, each process tries to move the eyes and hands without regard for what is happening elsewhere in the system, creating prospects for "jams" in EPIC's motor processors.

To cope with this impoliteness, model 1 has a general executive that uses a first come, first serve (FCFS) algorithm for scheduling the tracking and arrow-discrimination task processes in strict lockout mode. Under it, these processes may be started optionally when their stimuli (arrows and suprathreshold tracking errors) are detected. However, the GE lets only one process proceed at a time. If stimuli for both processes occur simultaneously, then the lower-priority one (tracking) is postponed until the higher priority one (arrow discrimination) has responded to its current stimulus.

This protocol resembles the one of Norman and Shallice's "supervisory attentional system" (SAS; 1986), in which action schemata are activated by "trigger" stimuli and contend for limited response mechanisms. Precluding conflicts from this "contention scheduling," the SAS transmits top-down activation to the highest-priority schema, favoring it over lower-priority schemata. In our model 1, the lockout scheduling is like the selective prioritization imposed by the SAS. Thus we may test both model 1 and the SAS by comparing the performance of model 1 to real data.

Table 30.1 shows results of this comparison. When a small visual angle (<5 degrees) separates the displays of the tracking and arrow-discrimination tasks, simulated RTs from model 1 are considerably less than observed ones (mean difference = 103 msec), but at larger angles (>10 degrees), simulated RTs are considerably greater than observed ones (mean difference = 97 msec). Furthermore, the simulated RMS tracking errors of model 1 are much larger than the observed ones; when tracking is difficult, they differ by more than a factor of 2 at large visual

Table 30.1 Reaction Times and Tracking Errors Produced by Four Alternative EPIC Models to Account for Observed Data from Martin-Emerson and Wickens 1992

		Mean arrow-discrimination reaction times (msec)					Root mean square tracking error (pixels)				
Tracking Task	Visual angle	Model 1	Model 2	Model 3	Model 4	Data	Model 1	Model 2	Model 3	Model 4	Data
Easy	small	608	690	686	612	697	19.4	17.2	13.4	13.5	13.3
	medium	781	836	782	696	787	22.5	18.0	13.6	13.6	13.3
	large	1,070	1,106	978	886	997	28.6	20.1	14.3	14.5	15.0
Hard	small	632	688	686	608	749	28.7	25.2	19.9	19.7	18.6
	medium	802	842	787	700	786	34.8	27.6	20.6	20.9	19.6
	large	1,097	1,112	996	887	976	43.2	30.0	22.7	22.9	21.4

Note: Reaction times and tracking errors were measured as a function of the tracking-task difficulty (easy versus hard) and the magnitude of the visual angle ("small" < 5 degrees; "medium" 6 to 10 degrees; "large" > 10 degrees) between the display windows for the tracking and arrow discrimination tasks (cf. figure 30.4).

angles. Model 1 performed very poorly even though under it, tracking and arrow discrimination progress as fast as reasonably possible while they are under way, and there are no resource allocation or abdication costs of supervising them. Instead, the poor performance of model 1 stems from an absence of process overlap caused by its GE having to cope conservatively with the impoliteness of the task processes in their use of motor resources.

These results disconfirm both model 1 and the SAS with respect to Martin-Emerson and Wickens 1992. Contrary to these models, under at least some conditions, cognitive control for multitask performance is more efficient than a conservative GE and impolite task processes allow. We investigate the sources of this efficiency more fully by considering a second new model.

Model 2: Liberal General Executive with Polite Task Processes In model 2, tracking and arrow discrimination are assumed to be polite processes. Each task process requests motor resources immediately before it would use them, does not use them until the general executive grants permission, and releases them immediately after they have been used. Given this politeness, the GE lets these processes advance simultaneously insofar as possible, even after one of them has requested resources that the other is currently using. Such liberalism is feasible because the task processes make eye and hand movements in a considerate manner that avoids motor-processor "jams," thereby enabling more process overlap than model 1 allows.

Another virtue of model 2 is its straightforward flow of control. Compared to our original model for Martin-Emerson and Wickens 1992 (figure 30.5), model 2 has a relatively simple flowchart (figure 30.6). Consequently, during multitasking practice, the skill embodied in model 2 should be fairly easy to acquire.

Consistent with these points, table 30.1 shows that model 2 produces somewhat better performance than model 1 does. Especially when tracking is difficult, simulated RMS errors from model 2 are markedly smaller than those from model 1. Nevertheless, there remain significant discrepancies between the performance of model 2 and the observed data. Both the simulated tracking errors and simulated RTs are still excessively large, suggesting that actual participants achieved even more process overlap than model 2 allows.

Why and how might this be? An answer may come from reconsidering our original model for Martin-Emerson and Wickens 1992 which we now call "model 3."

Model 3: Customized Executive with Resource Preallocation and Enhanced Task Processes As depicted before (figure 30.5), model 3 uses a customized executive that exploits context-dependent knowledge

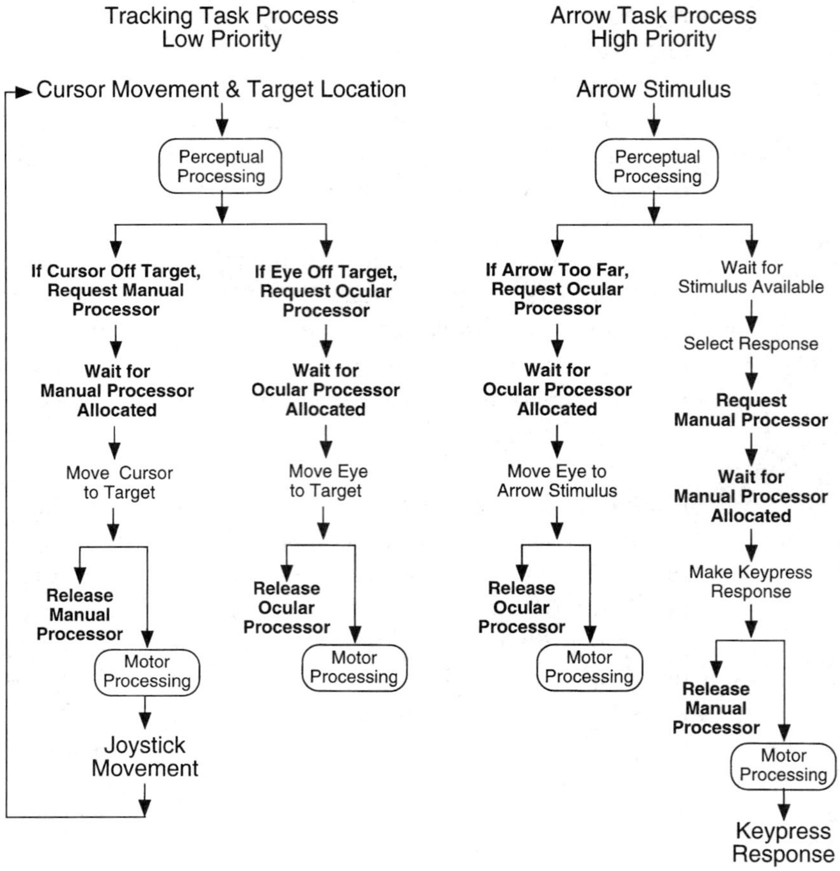

Figure 30.6 Flowchart of an EPIC model that performs the tracking and arrow-discrimination tasks of Martin-Emerson and Wickens 1992 with polite task processes and a general executive whose managerial style is liberal in task scheduling and resource allocation (cf. figure 30.5).

about the tasks and their temporal relationships. Based on this knowledge, the customized executive preallocates resources (i.e., ocular and manual motor processors) to tracking and arrow discrimination without being requested to do so. This enables the task processes to advance even more quickly than under model 2. Under model 3, the task processes also prepare eye movements beforehand. Together, these enhancements further facilitate performance so that the simulated RTs and tracking errors of model 3 are considerably less than those of model 2, closely approximating observed data (table 30.1).

The good fit of model 3 suggests that participants in Martin-Emerson and Wickens 1992 achieved excellent multitask performance through especially efficient cognitive control. Without this efficiency, limitations of perceptual-motor mechanisms would have precluded such performance. The customized executive of model 3 overcomes these limitations more so than a general executive can. Nevertheless, during the course of practice, participants may have relied on a GE to acquire their high level of multitasking skill. How this could happen is considered next.

Hypotheses about Skill Acquisition

Taken together, our results from models 1, 2, and 3 lead to hypotheses that explain various major aspects of multitask performance and skill acquisition.

Multitasking Skill-Acquisition Stages We hypothesize that multitasking skill acquisition progresses through five stages: preprocedural interpretative multitasking (stage 0); general hierarchical competitive multitasking (stage 1); general hierarchical cooperative multitasking (stage 2); customized hierarchical multitasking (stage 3); and customized heterarchical multitasking (stage 4). Each of these stages can be characterized with respect to its degree of efficiency, types of interaction between executive and task processes, and exploitation of context-dependent procedural knowledge.

Preprocedural interpretive multitasking is necessitated by a fundamental dependence between procedural and declarative task knowledge. We call this "stage 0" because it occurs at the start of practice before sets of production rules for the particular tasks have been created. During stage 0, people must use a generic interpretive process to execute propositional instructions about how the tasks should be performed. Here performance is presumably slow and error prone, placing heavy loads on working memory as people "think" their way verbally through each task. Nevertheless, it is from this explicit directed intentional activity that more efficient procedural knowledge for subsequent task performance emerges (Anderson 1983; Bovair and Kieras 1991; Kieras and Bovair 1986).

Once such knowledge becomes available, general hierarchical competitive multitasking may ensue. We call this "stage 1" because it is the first stage during which a general executive supervises task processes that are executed through individualized sets of production rules. Also during stage 1, task scheduling and coordination are managed as in our model 1 for Martin-Emerson and Wickens 1992. Here performance presumably entails a conservative GE with strict lockout scheduling of impolite task processes whose manners in using perceptual-motor resources are impulsive, presumptuous, and greedy. This impoliteness may be attributed to a need for more practice in order to acquire rules that conform with proper task etiquette.

As practice continues, general hierarchical cooperative multitasking may come next. During what we call "stage 2," task scheduling and coordination would be managed as in our model 2. Here performance presumably entails a liberal GE with temporal overlapping of task processes that request, use, and release system resources politely. This politeness enables the GE to be more permissive in letting these processes advance rapidly toward completion.

Customized hierarchical multitasking would involve an even higher skill level. During what we call "stage 3," task scheduling and coordination may be managed as in our model 3. Here unique context-dependent knowledge about the particular tasks and their temporal interrelations presumably is exploited to preallocate system resources without time-consuming requests for them, thereby further increasing temporal overlap among task processes. Also, as in model 3, these processes may be enhanced to prepare their motor responses anticipatorily.

Culminating this evolution is customized heterarchical multitasking. During what we call "stage 4," performance presumably is controlled without supraordinate executive processes. Instead, the task processes interact directly with each other, self-governing their resource usage as efficiently as possible. This interaction optimizes overall system throughput, completely eliminating scheduling, allocation, and abdication time costs that contribute to the transaction overhead of hierarchical cognitive control.

Table 30.1 shows some benefits of such optimization. Here we have included results from a fourth model ("model 4") that uses the customized heterarchical multitasking of stage 4 to simulate performance in Martin-Emerson and Wickens 1992. The RMS tracking errors of model 4 closely approximate the data, and its mean RTs are even shorter than observed ones. Although the subjects in Martin-Emerson and Wickens 1992 were highly skilled, they apparently had not yet reached this ultimate asymptotic performance level.

Executive Learning Mechanisms Operations within and transitions between the preceding five stages of multitasking skill acquisition may be mediated by various executive learning mechanisms (cf. Anderson 1983; Bovair and Kieras 1991; Chong and Laird 1997; Kieras and Bovair 1986). These mechanisms may entail several components: a *task interpreter*, which executes propositional instructions for performing single and multiple tasks during stage 0; a *task compiler*, which creates rudimentary sets of production rules for the initially impolite task processes of stage 1; a *task socializer*, which makes these processes more polite in stage 2; an *executive modulator*, which tailors the general executive's managerial style to be either conservative or liberal, depending how polite the task processes are; an *executive customizer*, which creates customized executives to enable even more efficient control in stage 3; and an *executive integrator*, which "flattens" the CEs, converting their flow of control from a hierarchical to heterarchical organization in stage 4.

We hypothesize that such mechanisms are sensitive to the evolving characteristics of performance. For example, during stage 1, simultaneous attempts by multiple impolite task processes to produce movements in the same response modality could generate motor-processor "jams." These jams might be detected by the executive modulator, leading it to have the GE be conservative during the period of time when the task socializer works toward making the task processes more polite. The task socializer and executive modulator also could operate partly on the basis of noticing that the task processes do not request and release resources properly. Later, after the task socializer achieves its objectives, the executive modulator perhaps would adjust the GE to be more liberal because motor-processor jamming has ceased. Accompanying the latter adjustment, the executive customizer might start creating a CE that later triggers hierarchical-to-heterarchical flattening by the executive integrator. Of course, future research will be needed to understand and model the details of such hypothetical learning mechanisms.

Multitasking Skill-Acquisition Phenomena By doing so, we may eventually explain and predict many empirical phenomena of multitasking skill acquisition. For example, Gopher (1993) has found that multitask performance is better after variable-priority rather than fixed-priority training. In his fixed-priority training condition, one group of participants gave equal priorities to visuomanual tracking and choice-reaction tasks throughout a series of practice sessions. In his variable-priority training condition, a second group of participants also gave the two tasks equal priorities on some occasions, but devoted higher priority to either tracking or choice reactions on other occasions. After variable-priority training, the second group performed better than the first group even when the two tasks received equal priorities. Similar results have been reported by Meyer et al. (1995). The benefits of variable-priority training could stem from the task socializer and executive modulator receiving a wider range of feedback, which guides them more quickly through successive stages of skill acquisition.

Our hypotheses likewise account for results obtained with some other laboratory paradigms. For example, RTs from the PRP procedure sometimes manifest a response-selection bottleneck (Pashler 1994, chap. 12, this volume). This seems to occur especially when participants receive relatively little practice at coordinating their primary and secondary tasks (Schumacher et al., 1999). A possible reason is that participants lack sufficient opportunity to socialize initially impolite task processes, so their GE has to deal with this impoliteness through strict lockout scheduling (cf. Meyer and Kieras 1997a,b).

30.5 CONCLUSIONS

Assimilating the fundamentals of contemporary computer operating systems into theories of cognitive control will make it possible to characterize a wider range of control functions more precisely, and to test more definitively for the existence of general as well as customized executive processes. These advances also will lead to more detailed and veridical analyses of multitasking skill acquisition. Computational modeling based on the EPIC architecture provides one vehicle whereby this progress can occur.

For the present prospects to be fully realized, future research must use a wide variety of empirical procedures to investigate multitask performance. This investigation should extend beyond basic laboratory paradigms like the task-switching and PRP procedures, which are helpful for isolating particular elementary control functions, but come nowhere near to engaging the whole host of executive mental processes that people presumably have. Rather, to explore these processes more completely, overlapping-task procedures with complex realistic tasks and unpredictable stimulus-response event sequences will be needed (e.g., Ballas, Heitmeyer, and Perez 1992).

Another major path for future research will involve identifying systematic relationships between underlying brain mechanisms and the executive mental processes revealed by taking operating system fundamentals into account. Because OS fundamentals apply quite generally to shared-memory symmetric multiprocessors, of which the brain is perhaps one type, it seems reasonable that the brain implements these fundamentals as well. If so, then insights from EPIC computational modeling, applied to results from studies of brain imaging and focal lesion analysis, could eventually yield fundamental solutions to the mind-body problem of cognitive control.

NOTES

Funding for this research was provided by U.S. Office of Naval Research grant N00014-92-J-1173 to the University of Michigan. We thank David Fencsik, Darren Gergle, Jennifer Glass, Leon Gmeindl, Cerita Jones, Shane Mueller, Eric Schumacher, Mollie Schweppe, and Travis Seymour of the Brain, Cognition, and Action Laboratory at the University of Michigan for their helpful assistance. Comments by Leon Gmeindl, Stephen Monsell, Travis Seymour, and two anonymous reviewers on drafts of this chapter are greatly appreciated.

1. First-task responses yielded a similar pattern of results. Although mean reaction times decreased across sessions, their pattern did not change qualitatively with practice. No significant asymmetries occurred in switching time costs. Error rates were moderately low (< 10%) on average and correlated positively with mean RTs, suggesting no systematic speed-accuracy trade-offs.

2. Our model also accounts well for mean first-task RTs and the factor effects on them.

3. Insofar as we know, the distinctions described here have not been made explicitly in operating system textbooks. They are introduced here to address issues about human cognitive control, which extend well beyond those associated with computer applications where experienced task programmers adhere consistently to common a priori conventions.

REFERENCES

Allport, A., Styles, E., and Hsieh, S. (1994). Shifting intentional set: Exploring the dynamic control of tasks. In C. Umiltà and M. Moscovitch (Eds.), *Attention and performance XV*, pp. 421–452. Cambridge, MA: MIT Press.

Anderson, J. R. (1983). *The architecture of cognition*. Cambridge, MA: Harvard University Press.

Baddeley, A. D. (1986). *Working memory*. Oxford: Oxford University Press.

Ballas, J. A., Heitmeyer, C. L., and Perez, M. A. (1992). *Direct manipulation and intermittent automation in advanced cockpits*. Technical Report NRL/FR/5534—92-9375. Washington, DC: Naval Research Laboratory.

Bovair, S., and Kieras, D. E. (1991). Toward a model of acquiring procedures from text. In R. Barr, M. L. Kamil, P. B. Mosenthal, and P. D. Pearson (Eds.), *Handbook of reading research*, vol. 2, pp. 206–229. New York: Longman.

Card, S. K., Moran, T. P., and Newell, A. (1983). *The psychology of human-computer interaction*. Hillsdale, NJ: Erlbaum.

Chong, R. S., and Laird, J. E. (1997). Towards learning dual-task executive process knowledge using EPIC-Soar. In *Proceedings of the Nineteenth Annual Conference of the Cognitive Science Society*, pp. 107–112. Hillsdale, NJ: Erlbaum.

Gopher, D. (1993). Attentional control: Acquisition and execution of attentional strategies. In D. E. Meyer and S. Kornblum (Eds.), *Attention and performance XIV: Synergies in experimental psychology, artificial intelligence, and cognitive neuroscience*, pp. 299–322. Cambridge, MA: MIT Press.

Kieras, D. E., and Bovair, S. (1986). The acquisition of procedures from text: A production-system analysis of transfer of training. *Journal of Memory and Language, 25*, 507–524.

Kieras, D. E., and Meyer, D. E. (1997). An overview of the EPIC architecture for cognition and performance with application to human-computer interaction. *Human-Computer Interaction, 12*, 391–438.

Kieras, D. E., and Meyer, D. E. (Forthcoming). The role of cognitive task analysis in the application of predictive models of human performance. In J. M. C. Schraagen, S. E. Chipman, and V. L. Shalin (Eds.), *Cognitive task analysis*. Mahwah, NJ: Erlbaum.

Kristofferson, A. B. (1967). Attention and psychophysical time. In A. F. Sanders (Ed.), *Attention and Performance*, pp. 93–100. Amsterdam: Elsevier.

Lachman, R., Lachman, J. L., and Butterfield, E. C. (1979). *Cognitive psychology and information processing: An introduction*. Hillsdale, NJ: Erlbaum.

Lauber, E. J. (1995). Executive control of task-switching operations. Ph.D. diss., University of Michigan, Ann Arbor.

Martin-Emerson, R., and Wickens, C. D. (1992). The vertical visual field and implications for the head-up display. In *Proceedings of the Thirty-sixth Annual Symposium of the Human Factors Society*. Santa Monica, CA: Human Factors Society.

Meiran, N. (1996). Reconfiguration of processing mode prior to task performance. *Journal of Experimental Psychology: Learning, Memory, and Cognition, 22*, 1423–1442.

Meyer, D. E., and Kieras, D. E. (1997a). EPIC—A computational theory of executive cognitive processes and multiple-task performance: Part 1. Basic mechanisms. *Psychological Review, 104*, 3–65.

Meyer, D. E., and Kieras, D. E. (1997b). EPIC—A computational theory of executive cognitive processes and multiple-task performance: Part 2. Accounts of psychological refractory-period phenomena. *Psychological Review, 104,* 749–791.

Meyer, D. E., and Kieras, D. E. (1999). Précis to a practical unified theory of cognition and action: Some lessons from computational modeling of human multiple-task performance. In D. Gopher and A. Koriat (Eds.), *Attention and Performance XVII,* pp. 15–88. Cambridge, MA: MIT Press.

Meyer, D. E., Kieras, D. E., Lauber, E., Schumacher, E., Glass, J., Zurbriggen, E., Gmeindl, L., and Apfelblat, D. (1995). Adaptive executive control: Flexible multiple-task performance without pervasive immutable response-selection bottlenecks. *Acta Psychologica, 90,* 163–190.

Neisser, U. (1967). *Cognitive psychology.* New York: Appleton Century Crofts.

Newell, A. (1980). Harpy, production systems, and human cognition. In R. A. Cole (Ed.), *Perception and production of fluent speech,* pp. 289–380. Hillsdale, NJ: Erlbaum.

Newell, A. (1990). *Unified theories of cognition.* Cambridge, MA: Harvard University Press.

Norman, D. A., and Shallice, T. (1986). Attention to action: Willed and automatic control of behavior. In R. J. Davidson, G. E. Schwartz, and D. Shapiro (Eds.), *Consciousness and self-regulation,* vol. 4, pp. 1–18. New York: Plenum Press.

Pashler, H. (1994). Dual-task interference in simple tasks: Data and theory. *Psychological Bulletin, 116,* 220–244.

Rogers, R., and Monsell, S. (1995). Costs of a predictable switch between simple cognitive tasks. *Journal of Experimental Psychology: General, 124,* 207–231.

Rubinstein, J. S., Meyer, D. E., and Evans, J. (Forthcoming). Executive control of cognitive processes in task switching. *Journal of Experimental Psychology: Human Perception and Performance*, accepted pending revision.

Schumacher, E. H., Lauber, E. J., Glass, J. M. B., Zurbriggen, E. L., Gmeindl, L., Kieras, D. E., and Meyer, D. E. (1999). Concurrent response-selection processes in dual-task performance: Evidence for adaptive executive control of task scheduling. *Journal of Experimental Psychology: Human Perception and Performance, 25,* 791–814.

Stallings, W. (1998). *Operating systems: Internals and design principles.* 3d ed. Upper Saddle River, NJ: Prentice Hall.

Theios, J. (1973). Reaction time measurements in the study of memory processes: Theory and data. In G. H. Bower (Ed.), *The psychology of learning and motivation,* vol. 7, pp. 43–85. New York: Academic Press.

Tucker, A., Ed. (1997). *The computer science and engineering handbook.* Boca Raton, FL: CRC.

31 On the Control of Control: The Role of Dopamine in Regulating Prefrontal Function and Working Memory

Todd S. Braver and Jonathan D. Cohen

ABSTRACT An important aspect of cognitive control is the ability to appropriately select, update, and maintain contextual information related to behavioral goals, and to use this information to coordinate processing over extended periods. In our novel, neurobiologically based, connectionist computational model, the selection, updating, and maintenance of context occur through interactions between the prefrontal cortex (PFC) and dopamine (DA) neurotransmitter system. Phasic DA activity serves two simultaneous and synergistic functions: (1) a gating function, which regulates the access of information to active memory mechanisms subserved by PFC; and (2) a learning function, which allows the system to discover what information is relevant for selection as context. We present a simulation that establishes the computational viability of these postulated neurobiological mechanisms for subserving control functions.

The need for a control mechanism in cognition has been long noted within psychology. Virtually all theorists agree that some mechanism is needed to guide, coordinate, and update behavior in a flexible fashion—particularly in novel or complex tasks (Norman and Shallice 1986). In particular, control over processing requires that information related both to current context and to behavioral goals be actively represented, such that these representations can bias behavior in favor of goal-directed activities over extended periods. Indeed, most computationally explicit theories of human behavior have included such a mechanism as a fundamental component. For example, in production system models, goal states represented in declarative memory are used to coordinate the sequence of production firings involved in complex behaviors (e.g., Anderson 1983). One critical feature of goal representations in production systems is that they must be actively represented and maintained throughout the course of a sequence of behaviors. Such formulations of a control (or "executive") mechanism closely parallel theorizing about the nature of frontal lobe function (Bianchi 1922; Damasio 1985; Luria 1969), and clinical observations of patients with frontal lesions who often exhibit impairments in tasks requiring control over behavior—the so-called dysexecutive syndrome. Shallice (Norman and Shallice 1986; Shallice, 1982, 1988) explicitly noted this relationship, using the production system framework to describe his theory of a "supervisory attentional system" (SAS) as a mechanism by which the frontal lobes coordinate complex cognitive processes and select nonroutine actions. While these efforts have provided insights into the types of processes that may be engaged by cognitive control, they do not map transparently onto underlying neural mechanisms. They have also not fully addressed several critical issues, such as how a control system can develop through learning.

A number of recently proposed connectionist models of prefrontal function incorporate some of the central features of control processes in production system models, such as the active maintenance of goal representations (Dehaene and Changeux 1992; Guigon et al. 1991; Levine and Prueitt 1989). Connectionist models have the advantage of both being mechanistically explicit and using a computational architecture that maps more naturally onto neural mechanisms than traditional production system models. In this chapter, we report on work that uses this framework to address a critical question about cognitive control: How can a system learn to choose and appropriately update representations in active memory that can be used to control behavior? This is an extension of our ongoing effort to specify the neural underpinnings of cognitive control (Braver et al. 1995a; Cohen, Braver, and O'Reilly 1996; Cohen and Servan-Schreiber 1992), reviewed briefly below as background.

A central hypothesis in our work is that a cardinal function of prefrontal cortex (PFC) is to actively maintain context information. We use the general term *context* to include not only goal representations, which have their influence on planning and overt behavior, but also representations that may have their effect earlier in the processing stream, on interpretive or attentional processes. We assume that a primary function of PFC is to maintain task-relevant context representations in an active state. These active context representations serve to mediate control by modulating the flow of information within task-specific pathways such that processing in the task-relevant pathway is favored over a (possibly stronger) competing pathway. This function of PFC can also be thought of as a component of *working memory* (WM), commonly defined as the collection of mechanisms responsible for the on-line maintenance and manipulation of information necessary to perform a cognitive task (Baddeley and Hitch 1994). From this perspective, context can be viewed as the subset of representations within WM that govern how other representations are used.

As noted above, there is long-standing recognition that control involves representation and maintenance of context information (e.g., goals). However, a more complete account of cognitive control has additional requirements. Here we focus on four. Context information must be (1) appropriately selected for maintenance; (2) held for arbitrary lengths of time; (3) protected against interference; and (4) updated at appropriate junctures. Inasmuch as we assume that context information is represented in PFC, our interest is in the mechanisms that regulate the selection and updating of representations in PFC. One type of system meeting these requirements uses a gating mechanism to regulate the flow of activity into PFC: when the gate is opened, activity can flow into the PFC and activate the appropriate context representations; when the gate is closed, the activated representations are protected from interference, and therefore can be maintained and exert control for extended periods. Such a system, however, must know when it is appropriate to deploy the gate. This additional requirement threatens to introduce a regress in the control of processing: If the gating mechanism controls the controller, "who" is controlling the gating mechanism? Moreover, how can this component of control be learned, and how can this be mediated in a neurobiologically plausible way?

In this chapter, we propose a computational and neurobiological solution to this dilemma that involves the dopamine (DA) neurotransmitter system. Specifically, we suggest that DA projections to PFC serve to gate access of context representations into active memory through simple neuromodulatory effects on processing units in the PFC. These effects serve both gating and learning functions, which enable the system to discover what information must be maintained for performing a given task, and to regulate when that information is updated. This avoids the "homunculus" that plagues many theories of executive control. Below, we review evidence for this hypothesis, including evidence that PFC supports active memory, computational analyses of simple and gated active memory systems, and evidence that the modulatory effects of DA can support both its gating and learning functions. Following this review, we present a simulation that establishes the model's computational viability.

31.1 A NEURALLY BASED ACCOUNT OF THE CONTROL OF ACTIVE MEMORY

Prefrontal Cortex and Control

Neurobiological Evidence The role of control mechanisms in PFC has long been suggested by neuropsychological evidence. Increased distractibility and perseveration are hallmarks of neurological damage to PFC (Damasio 1985; Engle, Kane, and Tuholski 1999; Milner 1963; Owen et al. 1991; Stuss and Benson 1986) and of psychiatric disorders known to involve PFC such as schizophrenia (Malmo 1974; Nuechterlein and Dawson 1984). Neurophysiological studies have begun to provide a more detailed characterization of PFC function. Miller (chap. 22, this volume) provides an excellent review of this literature, which demonstrates that units in PFC (1) selectively code information relevant to task performance and not distractor information; (2) can code multimodal, task-relevant contingencies (including sensory information from different modalities and sensorimotor mappings); (3) can maintain such information over extended delays, in the absence of sustained sensory input; and (4) exhibit a pattern of temporal dynamics that suggests they are the source of attentional bias in posterior systems directly responsible for sensory and motor processing. These findings are consistent with the control function that we have ascribed to PFC. Recent neuroimaging studies using event-related fMRI have begun to corroborate these neurophysiological findings in human subjects, demonstrating sustained activity of PFC during delay intervals in working memory tasks (Cohen et al. 1997; Courtney et al. 1997) and in tasks that engage the "executive" functions of working memory (D'Esposito and Postle, chap. 15, this volume; Frith, chap. 24, this volume)

Computational Analysis As noted above, we have hypothesized that PFC exerts control by biasing processing in the pathways responsible for task performance. This biasing function is illustrated by Cohen and colleagues' previous models of the Stroop task (Cohen, Dunbar, and McClelland 1990; Cohen and Huston 1994; Cohen and Servan-Schreiber 1992), in which activation of a context representation corresponding to the relevant task dimension (e.g., color) sends activity to all the hidden units in the processing pathway corresponding to that dimension. This favors the flow of activity along that pathway, allowing it to compete effectively with information flowing along an otherwise stronger but irrelevant pathway (word naming). Thus activation of the context representation biases processing in favor of the task-relevant dimension, establishing the sensorimotor mapping necessary to perform the task.

For context representations to bias processing, however, they must be actively maintained for the duration of the task. Although the previous models noted above did not include a mechanism for doing so; a number of mechanisms can support the short-term maintenance of information in connectionist models. The most commonly employed and best understood of these are fixed-point attractor networks (e.g., Hopfield 1982; Zipser 1991), which possess recurrent connections that "recirculate" activation among units, and are thus capable of supporting sustained activity. Such networks typically settle into stable states called "attractors," in which a particular pattern of activity is maintained, and which therefore can be used to store information actively. A number of computational models of simple maintenance tasks have demonstrated that both physiological and behavioral data regarding PFC function can be captured using attractor networks (Braver, Cohen, and Servan-Schreiber 1995a; Dehaene and Changeux 1989; Moody et al. 1998; Zipser et al. 1993).

On the other hand, simple attractor systems have limitations that pose problems in more realistic tasks. The state of an attractor system is determined by its inputs, so that presentation of any new input will drive the system into a new attractor state, overwriting previously stored information (Bengio, Frasconi, and Simard 1993; Mozer 1993), and making the system subject to interference from task-irrelevant inputs (i.e., distractors). Although attractor networks can be configured to display resistance to disruption from distractors (i.e., hysteresis), this impairs their ability to be easily updated. One way in which attractor networks can overcome these difficulties is through the addition of a gating mechanism. Gated networks respond to inputs, changing their attractor state only when the "gate" is opened. Compared to other types of recurrent networks, networks with a gating mechanism were found better able to learn and perform complex short-term memory tasks, especially when the tasks involved noisy environments, frequent updating, and relatively long periods of storage (Hochreiter and Schmidhuber 1997). These and other computational studies suggest that gated attractor systems can meet many of the requirements for active memory in a control system. Moreover, the physiological evidence reviewed above is consistent with the hypothesis that prefrontal cortex implements such a system. Zipser and colleagues (Moody et al. 1998; Zipser 1991; Zipser et al. 1993) have proposed gated attractor models of short-term memory, and have used these to simulate the patterns of delay period activity observed for PFC neurons, although these models have specified neither the source of the gating signal nor how its timing is learned.

Dopamine Modulation of Information Processing

Dopamine and Cognitive Control There has been a growing appreciation of the role of dopamine (DA) in higher cognitive function (see Robbins and Rogers, chap. 21, this volume). Several lines of evidence have shown a link between DA function and cognitive control. These include studies of cognitive deficits in patients suffering from brain disorders involving DA pathology, such as Parkinson's disease and schizophrenia (e.g., Cohen et al. 1999; Gold 1992; Robbins et al. 1994), pharmacological studies manipulating DA activity locally in the PFC of nonhuman primates (Brozoski et al. 1979; Sawaguchi and Goldman-Rakic 1991, 1994; Sawaguchi, Matsumura, and Kubota 1990), and systemic manipulation of DA in humans (Kimberg, D'Esposito, and Farah 1997; Luciana, Collins, and Depue 1995; Luciana et al. 1992; Servan-Schreiber et al. 1998). Based on these findings, several authors have proposed that DA activity serves to modulate the cognitive control functions mediated by PFC (Cohen and Servan-Schreiber 1992; Goldman-Rakic and Selemon 1997). Here, we extend this idea, by proposing more specifically that the DA system provides a mechanism for learning to predict reward and to update the contents of active memory correspondingly, so as to maximize the chance of receiving reward. We propose that this function is carried out by simple, but appropriately timed neuromodulatory effects on target neurons. We hypothesize that one effect of DA is to modulate the responsivity of PFC units to their input, allowing DA to gate inputs to PFC. Another effect of DA is to modulate the strength of the connection between these inputs and the DA neurons themselves, allowing the DA system to discover what information should trigger this gate, and thereby to update the contents of active memory in PFC appropriately. There is a substantial corpus of neurobiological data to support this view of DA function.

Modulatory Effects of Dopamine Like other catecholamines, dopamine is known to produce modulatory effects on target neurons (Chiodo and Berger 1986; Hernandez-Lopez et al. 1997; Penit-Soria, Audinat, and Crepel 1987). Our previous models, by implementing this neuromodulatory action as a change in the slope (or gain) of the activation function of processing units, have simulated a variety of the effects of DA, at both the physiological and behavioral levels (Braver, Cohen, and Servan-Schreiber 1995a; Cohen and Servan-Schreiber 1993; Servan-Schreiber et al. 1998; Servan-Schreiber, Printz, and Cohen 1990). A change in gain modulates the responsivity of units to their afferent input, and thus can function as a gate on the flow of activity into PFC. Detailed anatomic studies of PFC suggest that DA projections are well positioned to influence both excitatory inputs and local inhibitory interactions (Lewis et al. 1992; Sesack, Snyder, and Lewis 1995; Williams and Goldman-Rakic 1993), a pattern that is consistent with a role of DA in gating PFC (discussed below). Furthermore, although neuromodulatory influences are typically assumed to be slow acting and nonspecific in information content (Moore and Bloom 1978), recent findings have suggested that DA cells can exhibit fast and stimulus-specific responses, as required to serve a gating function (Grace 1991; Schultz, Apicella, and Ljungberg 1993).

Timing of Dopamine Responses Schultz and colleagues (Schultz 1992) have observed rapid, stimulus-locked and stimulus-specific activity in DA neurons ($\sim$100 msec in duration, occurring 80–150 msec after stimulus onset). For example, following training in a spatial delayed-response task requiring active maintenance (Schultz, Apicella, and Ljungberg 1993), DA cells came to respond to the cue to be maintained. The cue was the first stimulus in the sequence that itself was unpredictable, but that predicted subsequent reward (even when there were intervening distractors). This is precisely the timing that might be expected of a control mechanism responsible for updating context representations. When an unexpected cue indicates that a new desired state can be achieved, then this cue should elicit an updating of the context representation (e.g., goal) in active memory, replacing the current representation with one that will guide behavior toward the desired state.

Learning effects of Dopamine Findings from reward-conditioning paradigms suggest how the gating signal could be learned. DA has long been recognized to play a role in reward learning (Wise and Rompre 1989). In the Schulz and colleagues studies referred to above, DA responds initially only to the rewarding event, but with training this response "migrates" to predictive cues. Montague, Dayan, and Sejnowski (1996) have proposed a formal analysis of the role of DA in reward conditioning, in terms of a temporal difference (TD) learning algorithm (Sutton 1988; Sutton and Barto 1990). The TD algorithm provides a mechanism by which learning can chain backward in time, allowing the DA system to identify successively earlier predictors of reward, until the earliest possible predictor is found that cannot itself be predicted. In the Montague, Dayan, and Sejnowski model, DA responses are simulated as being proportional to the prediction error in the TD algorithm (i.e., the degree of mismatch between expected and received rewards), and DA release modulates the strength of synapses from units representing cues that predict reward to the DA units themselves.[1] In simulations as in empirical studies, the DA response decreases to events as they become more predictable (e.g., an expected reward), whereas it increases to events that predict reward but are themselves unpredicted. Intriguingly, the parameter used by Montague, Dayan, and Sejnowski to simulate the effects of DA on learning is analogous to the parameter we have used to simulate DA effects on unit responsivity. This raises the possibility that a single parameter can be used to account for both effects, which may occur simultaneously, in turn providing a means by which the gating signal can be learned.

A New Theory Although we have previously theorized that PFC is critical for the active maintenance of context information, and that DA activity serves to modulate the responsivity of PFC neurons to external input (Cohen and Servan-Schreiber 1992), the findings just discussed suggest a number of hypotheses revising and extending our original theory. These hypotheses provide an account of both the ability to update context representations and the means of learning how to do so:

- Context representations are actively maintained in a gated attractor system within PFC.
- Phasic changes in DA activity serve two functions:
  1. to gate information into active memory in PFC;
  2. to strengthen associations between stimuli that predict reward and the DA response.
- Both effects rely on a similar neuromodulatory mechanism.
- The gating effect occurs through the transient potentiation of both excitatory afferent and local inhibitory effects in PFC.
- The learning effect occurs through modulation of synaptic weights, driven by errors between predicted and received rewards (i.e., the TD learning algorithm).

• The coincidence of the gating and learning signals produces cortical associations between the information being gated and a triggering of the gating signal in the future.

In the studies presented below, we test the plausibility of these claims in a computer simulation of a model that implements our theory. Specifically, the simulation examines the hypothesis, suggested in the previous two subsections, that appropriate timing of gating signals can be acquired during task performance through reward-based learning mechanisms.

31.2 SIMULATION: REWARD-BASED LEARNING OF GATING SIGNALS

This study was conducted to establish the computational validity of the hypothesis that DA implements both gating and learning effects, and that such a system can learn to appropriately gate relevant context information into active memory. Although previous work has demonstrated that DA activity can be simulated accurately in a system governed by reinforcement learning (Montague, Dayan, and Sejnowski 1996), it has not been shown that the dynamics of DA activity can simultaneously be exploited as (and used to learn the timing of) a gating signal. Furthermore, this hypothesis poses the following dilemma. If gating the appropriate context representations into active memory is learned through a reward-based mechanism, but reward itself depends on gating the appropriate context representations, then how can the process get started? This is a classic "bootstrapping" problem, solutions for which are often best demonstrated by simulation. To do so, we constructed a model of a simple cognitive control task, where context information must be actively maintained across delay periods during which intervening distractor events may occur, and properly updated on a trial-to-trial basis.

Task

We used a variant of a delayed-response paradigm (the AX version of the continuous performance test, or AX-CPT; Nuechterlein and Dawson 1984; Rosvold et al. 1956) that we have used extensively to study the processing of context and its relationship to PFC and DA function in behavioral (Cohen et al. 1999; Cohen and Servan-Schreiber 1993; Servan-Schreiber, Cohen, and Steingard 1996), psychopharmacological (Braver 1997), and neuroimaging (Barch et al. 1997; Carter et al. 1998) studies. The AX-CPT paradigm has also been the subject of previous modeling work (Braver 1997; Braver, Cohen, and Servan-Schreiber 1995b; Cohen, Braver, and O'Reilly 1996). In this paradigm, a cue is presented at the beginning of each trial (e.g., the letter *A* or *B*), followed by a delay of variable length, and then a probe (e.g., the letter *X*) to which one of two responses must be made. The correct response to the probe is contingent on the identity

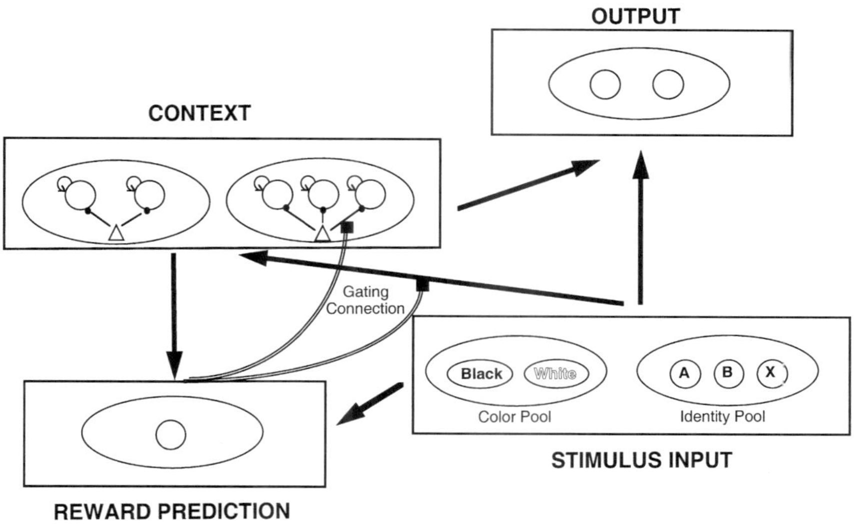

Figure 31.1 Learning/gating model used in simulation. Excitatory connections exist between layers (indicated by arrows), whereas lateral inhibitory connections exist within each layer (not shown). Input units make one-to-one connections with context layer units. Context units have self-excitatory connections, providing a mechanism for active maintenance. Low levels of baseline activity in the context layer are enforced by local inhibitory bias units (indicated by small triangles). The input and context layers are fully connected to the reward prediction/gating (RPG) unit. This unit, in turn, makes a gating connection with both afferent excitatory and local inhibitory input to the context layer. The RPG unit also modulates learning in all modifiable connections of the network.

of the cue. One reponse (e.g., press the left button) is required if the probe follows a specified cue (e.g., A-X, which we will refer to as "AX" trials), and the other response (e.g., right button) is required for all other cue-probe sequences (e.g., BX). Thus responding correctly to the probe requires maintenance of context information provided by the cue. Additionally, distractor stimuli are presented randomly, interspersed during the cue-probe delay and intertrial interval (ITI). Distractors are distinguished from the cue and probe stimuli by a particular feature (e.g., the color of the letters), but can have the same identity as the cue (e.g., A or B). Thus the AX-CPT paradigm engages cognitive control, insofar as correct performance requires the abilities to actively maintain context over a variable delay, ignore distractors, and update context selectively in response to cue stimuli but not distractors.

Architecture and Processing

Our model of this task is shown in figure 31.1. The network is composed of a stimulus layer (5 units), a context layer (5 units), a response layer (2 units), and a reward prediction/gating (RPG) unit. The stimulus and context layers are each separated into two pools, the first used to represent stimulus identity (*A*, *B*, *X*), and the second, stimulus color (black, white).

Units in the stimulus layer have one-to-one excitatory connections to corresponding units in the context layer. All units within the stimulus and context layers have excitatory connections to both units in the response (output) layer, which represent the two possible responses. Finally, there are lateral inhibitory connections among units within each layer. Thus between-layer excitatory connections mediate flow of information, while within-layer inhibitory connections mediate competition for representation, consistent with the computational framework proposed by McClelland (1993). The activation of each unit in the network is determined by the logistic of its time-averaged net input (with the exception of the RPG unit described below).[2] This allows units to integrate their inputs over time, and the model to simulate the temporal dynamics of processing.

In addition to the connectivity described above, units within the context layer have strong self-excitatory connections and an inhibitory input from a tonically active bias unit. This arrangement allows context units to assume a relatively low baseline of activity, yet self-sustain a higher level of activity following a sufficiently strong input, even after the input is removed.[3] We use this behavior to simulate active maintenance of context information in PFC. The weights of the one-to-one connections from the stimulus units to the context units, and among the context units, are fixed at values such that stimulus unit activity can activate context units when the entire context pool is at rest (i.e., no context units are active), but stimulus unit activity cannot alter an existing pattern of context unit activity.[4] Thus stimulus units are not able on their own to update the state of activated context units; this requires the "intervention" of the RPG unit (discussed below). The "hardwiring" of these connections reflects our assumption that the active maintenance properties of PFC, and its connections with task-specific processing pathways, arise by mechanisms different from the reward-based learning mechanisms described below, beyond the scope of current consideration.[5] The connection weights to and from the RPG unit and from the stimulus and context units to the output units are modifiable, and adjusted according to the learning rule described below.

The reward prediction/gating unit receives connections from all units in the stimulus and context layers. Its activity is computed as the weighted sum of the input received from the stimulus and context units on the current time step (current predicted reward) and the value of the actual reward for that trial ($+1$ for correct response and -1 for incorrect response) minus the stimulus and context input received on the previous time step (previously predicted reward), which is the temporal difference (TD) error.[6] The behavior of this unit serves as our simulation of phasic changes in dopamine activity, as in Montague, Dayan, and Sejnowski 1996. Accordingly, the activity of this unit (i.e., the value of the TD error) serves as a learning signal, used to adjust all modifiable weights in the network according to the TD learning algorithm.[7]

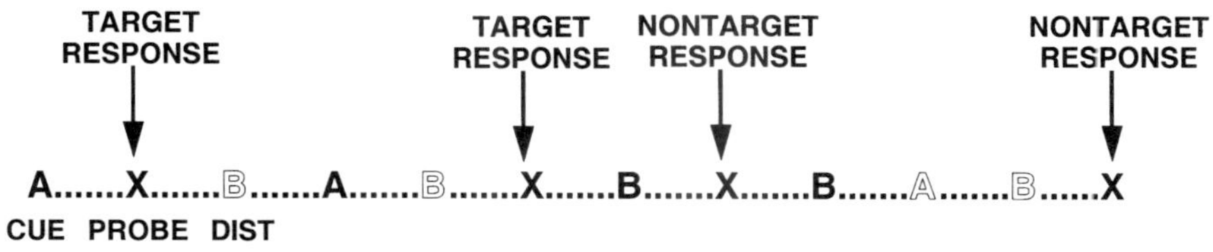

Figure 31.2 Typical sequence of training trials. Distractor stimuli are shown in outline type; task-relevant stimuli in solid type. Delay and intertrial interval (ITI) time steps are indicated by dashes. Distractors could occur during the delay period or ITI. Cue stimuli (*A* or *B*) need to be maintained over the delay in order to make the correct response to the probe (*X*)—one response when it follows an *A* cue, and a different response when it follows a *B* cue.

The RPG unit also exerts a gating effect on the context layer, allowing the current stimulus to change the state of (i.e., active representation in) the context layer. This occurs through potentiation of the strength of both afferent input (excitatory connections from stimulus to context) and local inhibition (inhibitory connections from the tonically active bias unit).[8] These potentiating effects have the following consequences. If a context unit, active when a gating signal occurs, does not receive excitation from any stimulus unit, but another context unit does, then the gating signal will favor activation of the competing context unit (due to potentiation of its excitatory input) and suppression of the current context unit (due to potentiation of inhibition from the competing context unit). Thus the gating influence of the RPG unit provides a mechanism for updating the state of activity in the context layer.

Training

We trained the network with a continuous sequence of task trials. (figure 31.2) Each trial consisted of the following events (simulated by activating the appropriate stimulus units): cue (*A* or *B*), delay interval, probe (*X*), and intertrial interval. Stimuli were presented for 3 time steps each; the minimum interval period was 7 time steps. Distractor events could be presented within both the delay and ITI, and each distractor increased the length of the interval by an additional 10 time steps (3 time steps for stimulus presentation +7 additional time steps for delay interval). The probability of a distractor appearing during any interval period was 0.50 for the first distractor in that period. The probability of an additional distractor appearing in that period decreased by half as the number of distractors increased (i.e., the probability of a second distractor appearing was 0.25, the probability of a third distractor appearing was 0.125, etc.). Each stimulus identity (*A* or *B*) was presented with equal frequency for both cues and distractors.

All modifiable weights were initialized to small random values prior to training (-0.25, 0.25). During training, weights were adjusted on every

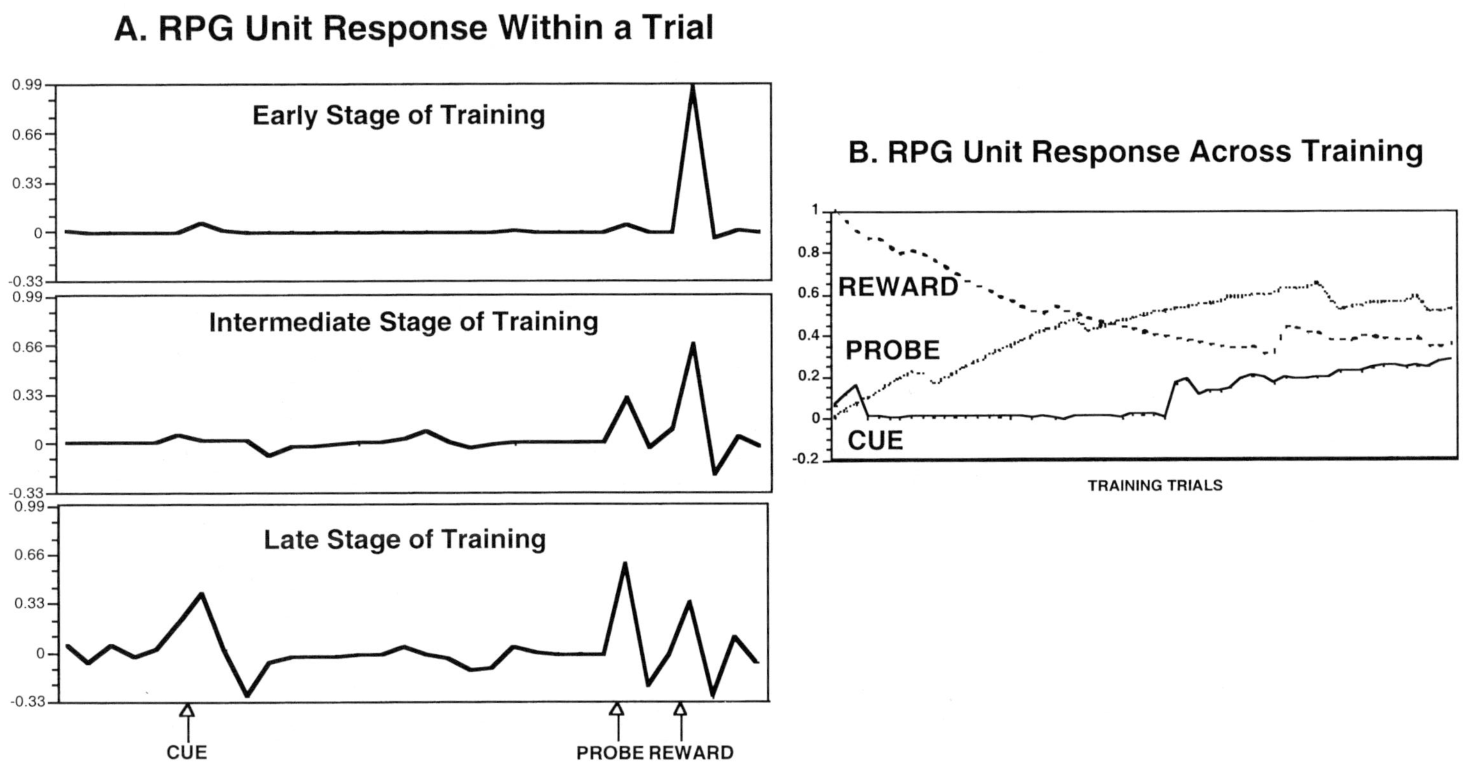

Figure 31.3 Reward prediction/gating (RPG) unit learning dynamics. A. RPG unit response (temporal difference error) within a trial in early, middle, and late stages of training. Note that in the early stage, the RPG unit only responds during reward delivery. In the middle stage, a response to the probe develops, whereas the reward response decreases. In the late stage, a response develops to the cue stimulus as well, whereas probe and reward responses continue to decrease. B. RPG unit response across training, where learning dynamics are shown continuously, as a function of the number of training trials.

time step in proportion to the activity of the RPG unit. Following presentation of the probe, the RPG unit received an input of +1 if the response was correct and −1 reinforcement if the response was incorrect, in addition to its usual input from the stimulus and context units. A response was considered correct if the activity of the left output unit was greater than 0.5 and greater than the right output unit for AX trials, and if the right output unit was greater than 0.5 and greater than the left output for BX trials. Thus, to perform the task correctly, the network had to learn to activate the context representation for the cue (*A* or *B*), maintain this over the delay, prevent distractor stimuli from disrupting this representation, and then use it to determine the correct response to the probe. During training, Gaussian noise was added to the net input of both context and output units, and was reduced in amplitude as error decreased (i.e., through a simple annealing schedule), consistent with the practice in other reinforcement learning simulations of having noise levels inversely related to the level of reward predicted (Gullapalli 1990).[9]

Results

Ten runs of the simulation were performed, each with randomly assigned initial weights for the modifiable connections in the network. The network converged to perfect performance on all ten runs. Learning followed a consistent pattern, comprising three stages (see figure 31.3). In the first stage, the connections from the stimulus and context units to the RPG unit remained weak, reflecting the lack of prediction or expectations of reward. Consequently, TD error (and the activity of the RPG unit) increased when reward was received because its delivery was unpredicted. In the intermediate stage, the stimulus unit for the probe (*X*) developed a positive connection with the RPG unit. Because reward (when it occurred) was delivered only following presentation of the probe, the network learned that the probe was a good predictor of reward. In reinforcement learning terms, the probe became a "secondary reinforcer," reducing the TD error (i.e., unexpectedness) at the time of reward delivery, and the response of the RPG unit to reward. Because the network had not yet learned to maintain the cue, however, the response to the probe was not always accurate, and reward was not delivered on every trial. Thus the probe was not a perfect predictor of reward, and a moderate level of TD error (and RPG unit activity) persisted for reward delivery. The third stage was reached when the TD algorithm allowed the network to learn the assocation between the cue stimuli and reward. Strong positive connections developed from the cue identity units (*A* and *B*) and the cue color unit (black) to the RPG unit, and a strong negative connection from the distractor color unit (white) to the RPG. As a consequence, activity in the RPG unit increased following presentation of a cue, but not following presentation of distractors. This increase in RPG unit activity produced a gating signal, which allowed the cue information to properly update the context representation, and be actively maintained over the delay. Moreover, because the cue information was being maintained as context, the context units began developing positive weights to the prediction unit, so that reward could be predicted based on the cue information. Once the cue information became a good predictor of reward, (because maintaining the cue increased the probability that reward was delivered), it became a "tertiary reinforcer," which further reduced the TD error both to the probe and reward delivery.

Note that noise in the context and output layers played a critical role in learning. In the output layer, noise encouraged response exploration, allowing the network to discover the correct response to the probe. Similarly, in the context layer, noise provided a way for the appropriate context unit to be active at the time of probe presentation (through random updating on some proportion of trials), before the network had learned to maintain the cue. This was critical for "bootstrapping" to take place. To summarize, the association between reward prediction and gating, coupled with noise, provided a mechanism for the network to discover how to regulate active memory so that cue information could selectively update the context representation.

Discussion

The results of this simulation provide preliminary support for the hypothesis that control over active maintenance of context representations can be achieved using a gating signal triggered by reward prediction errors. The pattern of RPG unit activity over the course of learning is very similar to that observed for DA neuronal activity over the course of learning in a delayed-response task (Schultz, Apicella, and Ljungberg 1993). In this respect, the results of our simulation replicate those of Montague, Dayan, and Sejnowski (1996), providing physiological support for the theory. However, our results go beyond those of Montague and colleagues, by demonstrating that the learning system can work synergistically with a gating signal to regulate control over active maintenance. By using the cue to predict reward, the network was also able to gate context information provided by the cue into active memory, where it could be used to bias subsequent responding. As a result, the probability of making the correct response was increased, and more rewards were achieved. Furthermore, because only cue stimuli elicited gating of the context layer, distractor stimuli were unable to disrupt the information maintained in the context layer. Thus the results also demonstrate that this type of control mechanism can protect context representations from the effects of interference. Moreover, the simulation makes clear how each of the two effects of the RPG unit are interdependent for learning the task properly. If RPG unit activity did not serve a gating function, the context representation would not be updated following cue presentation (or would be disrupted by every distractor). If the RPG unit activity did not modulate weight strengths based on reward prediction, the presentation of the cue input (*A* or *B* unit plus black unit) would never develop positive weights to the RPG unit, such that it could be activated by future cue presentations. Thus the simulations illustrate how both computational mechanisms associated with the RPG unit (gating and reward prediction learning) cooperate in the development of cognitive control over behavior in the task. The simulation also raises a number of more general conceptual issues regarding active maintenance, cognitive control, and reinforcement learning, which are discussed below.

Representation over Time A fundamental and unresolved issue in the application of reinforcement learning to classical and operant conditioning phenomena concerns the representation of perceptual information over time (Schultz, Dayan, and Montague 1997). For an organism to learn a relationship between a naturally reinforcing event (i.e., an unconditioned stimulus or US) and a predictive sensory cue (i.e., a conditioned stimulus or CS), the cue must still be represented when the reinforcement occurs. With very short delays, some perceptual trace of the cue may remain at the time of reinforcement. Although this is not likely at longer delays, when perceptual representations have presumably decayed.[10] To account for learning over such delays, some investigators (e.g., Sutton and Barto 1990) have proposed the mechanism of a decaying synaptic eligibility trace, which allows weights to be updated even when the cue is no longer actively represented. This does not solve an additional problem, however. Predictions of reward must continue at every time step from cue presentation until reward delivery for TD error to decrease and TD-based algorithms to function properly. Consequently, some form of active representation of the cue over an arbitrary period of time is required. Accordingly, most models of reinforcement learning represent each sensory cue as a vector, each element of which corresponds to the activity of that cue at a different point in time. In other words, the temporal dynamics of a cue are transformed into an explicit spatial representation (often referred to as a "complete serial compound" or CSC representation). Although it allows the system to learn an independent prediction of reward for every point in time (implemented as the connection strength from each element of the vector to the reward prediction unit), the CSC representation has a number of drawbacks, perhaps the most important of which is its neurobiological implausibility (Schultz, Dayan, and Montague 1997).

Our model implements a different solution to these problems. The context layer actively maintains representations that provide a continuous source of reward prediction necessary for TD learning to occur. Thus we propose that active maintenance within PFC may provide a mechanism for continuous reward prediction necessary for TD learning. As Hochreiter and Schmidhuber (1997) have observed, learning in difficult short-term memory tasks requires "constant error flow," which can be provided by computational units with activation that remains constant over time. One concern with such a solution, however, is that mechanisms for active maintenance must already be present for reward-based learning to occur. There are three principal ways that this could arise: (1) recurrent connectivity that develops as part of some intrinsic maturational process in PFC; (2) non-TD-based learning mechanisms that operate either prior to or interactively with reward-based learning (i.e., as another "bootstrapping" process); or (3) some other, innate mechanism (such as intrinsic bistability of neuronal activation states) that is preferentially expressed in PFC neurons. The available data do not adjudicate among these possibilities, although all three represent neurobiologically plausible mechanisms that are consistent with our model.

Alternative Control Mechanisms Another fundamental issue raised by the current study is whether gating is computationally required as a control mechanism for updating context representations. In principle, the answer is no. All that is required is a signal that differentiates task-relevant from task-irrelevant information and is derived in some form in the sensory input. This does not require a gating mechanism. For example, updating could occur through the proper conjunction of input features, previously maintained context representations, or both, coupled with the appropriate connection weights from input to context units (e.g., the conjunction of the *A* stimulus and the color black is sufficient to activate the *A* unit in the context layer and to overcome competition from other units in that layer). Thus, for any network that uses a gating signal to regulate access to active memory, an equivalent network can be constructed to perform the same functions without gating. There is a question, however, whether such a nongated network could be effectively learned through error-driven learning algorithms (either classical supervised or reward-based). The appropriate conjunction of weights required might be so precise as to be very difficult to learn using gradient descent procedures. We suspect that gated attractor networks coupled with TD learning provide a more powerful and robust computational mechanism for learning to perform tasks that require regulation of access to active memory. Although consistent with Hochreiter and Schmidhumber's analyses (1997) of simple recurrent networks and supervised learning algorithms, this conjecture remains to be tested for networks using TD learning to control the gating mechanism.

31.3 GENERAL DISCUSSION

In this chapter, we have presented a new model of the mechanisms underlying an important dimension of cognitive control: the ability to appropriately update context representations used to guide processing and the ability to learn how to do this. Furthermore, we have described simulation results that establish the computational plausibility of this model. On the other hand, the current model has important limitations, and significant challenges remain for a comprehensive theory of cognitive control. For example, we have not demonstrated that the mechanisms we propose can learn to gate into memory task-relevant information that itself is not directly predictive of reward. We have not provided an account of performance in more complex tasks, such as those which involve subgoaling. We have also not addressed the nature of context representations in the PFC—how these come about and how, without requiring infinite capacity, they can support the remarkable range and flexibility of behaviors of which humans are capable. These all remain challenges for further theoretical work.

Nevertheless, we believe that this model, even in its current limited form, has the potential to enrich our understanding of cognitive control. The model makes strong predictions about the engagement of PFC and DA in performance of simple control tasks, such as the AX-CPT, as well as the effects that disturbances of PFC and DA should have on task performance. We have begun to garner support for some of these predictions in related work using a wide variety of cognitive neuroscience methods. First, in behavioral studies, we have shown that patients with schizophrenia, who are thought to suffer from DA abnormalities in PFC, show a specific pattern of performance deficits in the AX-CPT consistent with a deficit in actively maintaining context (Braver, Barch, and Cohen 1999b; Cohen et al. 1999; Servan-Schreiber, Cohen, and Steingard 1996). Moreover, we have found a strikingly similar pattern of deficits in healthy subjects performing the AX-CPT under interference conditions (Braver, Barch, and Cohen 1999b). Second, in simulation studies we have found that the gating model can capture both of these patterns of deficits in terms of disturbances to the DA system (i.e., the reward prediction/gating unit). In particular, the model suggests that the deficits observed in schizophrenia might be due to increased noise in the RPG unit (Braver, Barch, and Cohen 1999a; Braver and Cohen 1999), while the deficits observed under interference can be captured by assuming that the distractor stimuli produce partial RPG unit activation (Braver, Cohen, and McClelland 1997). Third, preliminary results from a pharmacological study suggest that the interference-induced deficits in AX-CPT performance in healthy subjects may be ameliorated by low doses of amphetamine, a potent enhancer of DA transmission (Braver 1997). Finally, in functional neuroimaging studies, we have directly demonstrated the role of PFC in the active maintenance of context. During performance of the AX-CPT under conditions where the delay between cue and probe was manipulated, we observed greater PFC activity in long versus short delay trials, and further found that this activity was sustained throughout the delay period (Barch et al. 1997). In contrast, we observed that in the interference version of the AX-CPT, this activity is not sustained, but rather decays during the delay period, when distractors are presented (Braver, Barch, and Cohen 1999b).

Our model may also lead to new insights regarding cognitive control at the psychological level. For example, gated attractors may provide a useful theoretical framework within which to consider the effects of task switching that are addressed in detail in other contributions to this volume (e.g., Allport and Wylie, chap. 2, Jolicoeur, Dell'Acqua, and Crebolder, chap. 13, Goschke, chap. 14, De Jong, chap. 15, and Meiran, chap. 16, this volume). More generally, our model may help drive a reexamination of the relationship between motivational processes and cognitive control. The account of dopamine provided here suggests that it plays a unified role in motivation and cognition by configuring the system to optimize its predictions of reward and by regulating cognitive processes to increase the frequency with which rewards are obtained. This, in turn, offers an interesting perspective on prefrontal cortex function: the active maintenance of information in the service of maximizing rewards. From this perspective, one might imagine that PFC evolved at least in part to take control over the deployment of DA-mediated reinforcement by chaining together complex internal representations of reward prediction, and thus to support the construction of elaborate goal structures necessary for complex, temporally extended behaviors. This perspective suggests the intriguing possibility that the literature on the cognitive functions of PFC and DA can be linked with the growing, but heretofore separate, literatures on the affective and motivational functions of these brain systems (Bechara et al. 1996; Davidson and Sutton 1995; Willner and Scheel-Kruger 1991).

At the most general level, the model we have presented provides an illustration of how a system built of simple processing elements and general principles of learning can organize itself to regulate its own behavior in an adaptive fashion, without invoking the problem of a "homunculus." It also provides an example of how implementing a theory as an explicit computational model can lead to new and unexpected insights. Our hypotheses concerning the modulatory effects of dopamine (i.e., its role in gating) bear little surface resemblance to theories regarding the role of DA in reinforcement learning. It was only through a comparison of the formalisms of specific models that we were led to the observation that similar parameters were being used to implement these seemingly different DA effects, and to the idea that these effects may have synergistic effects. Our work also illustrates how efforts to understand the neural underpinnings of cognition can lead to insights at the psychological level. Our insights into the potential relationship between reward-based learning and gated attractors as a mechanism for the control of processing were driven in large measure by observations about the effects of a particular neurotransmitter and by efforts to account for its function. Thus, even in light of the limitations of our current model, we hope that our work may indicate how theories that draw simultaneously from, and bridge between, the neurobiological, psychological and computational domains can help advance our understanding of the mechanisms underlying cognitive control.

NOTES

1. The claim that dopamine modulates synaptic plasticity has received support in the neurophysiological literature (Calabresi et al. 1997; Law-Tho, Deuce, and Crepel 1995; Wickens, Begg, and Arbuthnott 1996).

2. The activation of unit a_i at time t is given as

$$a_i(t) = \frac{1}{1+ e^{-\gamma \mathrm{net}_i(t)}},$$

where γ is the gain on the activation function, while $\mathrm{net}_i(t)$ is given as

$$\mathrm{net}_i(t) = \tau \sum_j a_j(t)\, wi_{ij} + (1-\tau)\ \mathrm{net}_i(\mathrm{t}-1),$$

where τ is the time constant for averaging the net input (set at 0.5 for all simulations), and w_{ij} is the weight of the connection from each unit j that projects to unit i.

3. We should note that single, continuous-valued processing units in our model are used to simulate cell assemblies in the cortex (e.g., Amit 1989), and that recurrent self-connections simulate mutual excitatory synapses among cells belonging to a particular assembly.

4. Input-to-context module weights were set to $+3.0$; self-excitatory connections within the context module, to $+5.5$: lateral inhibitory connections within the context level, to -4.0; and local inhibitory input from the bias unit, to -2.5.

5. In work currently in progress, we have found preliminary evidence that both the active maintenance properties and connectivity pattern of context layer representations can be independently discovered through the application of learning algorithms, such as LEABRA (O'Reilly 1996), that combine correlational with error-driven learning. It remains a question for future research to determine whether this type of learning algorithm can be integrated with TD learning to provide more sophisticated models (i.e., ones that can address larger data sets and more complex cognitive tasks) and to reduce the number of parameters that need be fixed prior to learning.

6. More precisely, the TD error is computed according to the equation, derived from Sutton 1988:

$$\delta(t) = r(t) + \lambda P(t) - P(t-1),$$

where $r(t)$ is the reward input at time t, $P(t)$ is the total prediction input at time t, and λ is a discount factor, fixed at 0.95 for all simulations. This formulation suggests that an unexpected actual reward (for which predictions are zero) would lead to an increase in TD error (i.e., phasic activation of the RPG unit). Additionally, in the absence of actual reward (i.e., $r(t) = 0$), TD error increases when the current state is thought to be more predictive of reward than the previous state (i.e., $P(t) > P(t-1)$), such as when a salient cue appears in the environment.

7. Modifiable network weights are adjusted according to the learning rule:

$$\Delta w = \eta \delta(t) x_i(t-1),$$

where $x(t-1)$ is the activity of the sending unit at time $(t-1)$, η is the learning rate, and $\delta(t)$ is the TD error at time t (see note 6).

8. The modulatory effect of gating on afferent excitatory and local inhibitory input to context units was given as

$$w_{ij}' = \gamma(t) w_{ij},$$

where

$$\gamma(t) = 1 + \frac{k-1}{1+e^{-(S\delta(t)-0.5B)}} \quad k>1,$$

and where $\delta(t)$ is the TD error of the gating unit at time t, with k determining the maximum gain (γ) of the gating unit. The function $\gamma(t)$, a sigmoid in which gain monotonically increases with the level of TD error, is bounded such that the minimum gain is 1 and the maximum is k. S and B are additional parameters that determine the slope of the sigmoid and its baseline value (i.e., when $\delta(t)$ is zero). In the simulation, $k = 5$, $S = 20$, and $B = 5$. The results of the model were not found to depend critically on these parameter values, although it was important to choose a parameter that caused the slope of the function to be relatively steep, such that small increases in $\delta(t)$ had a nonnegligible effect. This allowed the RPG unit to exert a gating function early in the learning process, when activity is not very high.

9. The noise was drawn from a Gaussian distribution having zero mean. Its standard deviation was initialized to a value of 0.2. During training, this value was decreased by half whenever the TD error at the time of reinforcement delivery (averaged across a moving window of ten trials) also decreased by half. The noise parameter and annealing schedule primarily affected the speed of learning, and the results of the simulation were not found to depend upon the exact values used.

10. Although there is evidence that some presumably perceptual regions, such as posterior parietal and inferior temporal cortex, do show sustained active representations of stimuli over delay periods, these representations appear to be abolished by the presentation of new stimuli (Constantinidis and Steinmetz 1996; Miller, Erickson, and Desimone 1996). Thus they cannot serve as generally useful temporal representations of the sort desired for reward prediction learning.

REFERENCES

Amit, D. J. (1989). *Modeling brain function: the world of attractor neural networks.* Cambridge: Cambridge University Press.

Anderson, J. R. (1983). *The architecture of cognition.* Cambridge, MA: Harvard University Press.

Baddeley, A. D., and Hitch, G. J. (1994). Developments in the concept of working memory. *Neuropsychology, 8,* 485–493.

Barch, D. M., Braver, T. S., Nystrom, L., Forman, S. D., Noll, D. C., and Cohen, J. D. (1997). Dissociating working memory from task difficulty in human prefrontal cortex. *Neuropsychologia, 35,* 1373–1380.

Bechara, A., Tranel, D., Damasio, H., and Damasio, A. R. (1996). Failure to respond autonomically to anticipated future outcomes following damage to prefrontal cortex. *Cerebral Cortex, 6,* 215–225.

Bengio, Y., Frasconi, P., and Simard, P. (1993). The problem of learning long-term dependencies in recurrent networks. In *Proceedings of the IEEE International Conference on Neural Networks,* pp. 1183–1188.

Bianchi, L. (1922). *The mechanism of the brain and the function of the frontal lobes.* Edinburgh: Livingstone.

Braver, T. S. (1997). Mechanisms of cognitive control: A neurocomputational *model*. Ph.D. diss., Carnegie Mellon University, Pittsburgh.

Braver, T. S., Barch, D. M., and Cohen, J. D. (1999a). Cognition and control in schizophrenia: A computational model of dopamine and prefrontal function. *Biological Psychiatry, 46,* 312–328.

Braver, T. S., Barch, D. M., and Cohen, J. D. (1999b). Mechanisms of cognitive control: Active memory, inhibition and the prefrontal cortex. Technical Report PDP. CNS. 99.1. Pittsburgh: Carnegie Mellon University.

Braver, T. S., and Cohen, J. D. (1999). Dopamine, cognitive control, and schizophrenia: the gating model. *Progress in Brain Research, 121,* 327–349.

Braver, T. S., Cohen, J. D., and McClelland, J. L. (1997). An integrated computational model of dopamine function in reinforcement learning and working memory. *Society for Neuroscience Abstracts, 23,* 775.

Braver, T. S., Cohen, J. D., and Servan-Schreiber, D. (1995a). A computational model of prefrontal cortex function. In D. S. Touretzky, G. Tesauro, and T. K. Leen (Eds.), *Advances in neural information-processing systems,* vol. 7, pp. 141–148. Cambridge, MA: MIT Press.

Braver, T. S., Cohen, J. D., and Servan-Schreiber, D. (1995b). Neural network simulations of schizophrenic performance in a variant of the CPT-AX: A predicted double dissociation. *Schizophrenia Research, 15,* 110.

Brozoski, T. J., Brown, R. M., Rosvold, H. E., and Goldman, P. S. (1979). Cognitive deficit caused by regional depletion of dopamine in prefrontal cortex of rhesus monkey. *Science, 205,* 929–931.

Calabresi, P., Pisani, A., Centonze, D., and Bernardi, G. (1997). Synaptic plasticity and physiological interactions between dopamine and glutamate in the striatum. *Neuroscience and Biobehavioral Reviews, 21,* 519–523.

Carter, C. S., Braver, T. S., Barch, D. M., Botvinick, M. M., Noll, D., and Cohen, J. D. (1998). The role of the anterior cingulate cortex in error detection and the on line monitoring of performance. *Science, 280,* 747–749.

Chiodo, L., and Berger, T. (1986). Interactions between dopamine and amino-acid induced excitation and inhibition in the striatum. *Brain Research, 375,* 198–203.

Cohen, J. D., Barch, D. M., Carter, C. S., and Servan-Schreiber, D. (1999). Schizophrenic deficits in the processing of context: Converging evidence from three theoretically motivated cognitive tasks. *Journal of Abnormal Psychology, 108,* 120–133.

Cohen, J. D., Braver, T. S., and O'Reilly, R. (1996). A computational approach to prefrontal cortex, cognitive control, and schizophrenia: Recent developments and current challenges. *Philosophical Transactions of the Royal Society of London, B351,* 1515–1527.

Cohen, J. D., Dunbar, K., and McClelland, J. L. (1990). On the control of automatic processes: A parallel distributed processing account of the Stroop effect. *Psychological Review, 97,* 332–361.

Cohen, J. D., and Huston, T. A. (1994). Progress in the use of interactive models for understanding attention and performance. In C. Umiltà and M. Moscovitch (Eds.), *Attention and Performance XV,* pp. 1–19. Cambridge, MA: MIT Press.

Cohen, J. D., Perstein, W. M., Braver, T. S., Nystrom, L. E., Noll, D. C., Jonides, J., and Smith, E. E. (1997). Temporal dynamics of brain activation during a working memory task. *Nature, 386,* 604–608.

Cohen, J. D., and Servan-Schreiber, D. (1992). Context, cortex and dopamine: A connectionist approach to behavior and biology in schizophrenia. *Psychological Review, 99,* 45–77.

Cohen, J. D., and Servan-Schreiber, D. (1993). A theory of dopamine function and cognitive deficits in schizophrenia. *Schizophrenia Bulletin, 19,* 85–104.

Constantinidis, C., and Steinmetz, M. A. (1996). Neuronal activity in posterior parietal area 7a during the delay periods of a spatial memory task. *Journal of Neurophysiology, 76,* 1352–1355.

Courtney, S. M., Ungerleider, L. G., Keil, K., and Haxby, J. V. (1997). Transient and sustained activity in a distributed neural system for human working memory. *Nature, 386,* 608–612.

Damasio, A. R. (1985). The frontal lobes. In K. M. Heilman and E. Valenstein (Eds.), *Clinical neuropsychology,* pp. 339–375. New York: Oxford University Press.

Davidson, R. J., and Sutton, S. K. (1995). Affective neuroscience: The emergence of a discipline. *Current Opinion in Neurobiology, 5,* 215–225.

Dehaene, S., and Changeux, J. P. (1989). A simple model of prefrontal cortex function in delayed-response tasks. *Journal of Cognitive Neuroscience, 1,* 244–261.

Dehaene, S., and Changeux, J. P. (1992). The Wisconsin Card-Sorting Test: Theoretical analysis and modeling in a neuronal network. *Cerebral Cortex, 1,* 62–79.

Engle, R. W., Kane, M., and Tuholski, S. (1999). Individual differences in working memory capacity and what they tell us about controlled attention, general fluid intelligence and functions of the prefrontal cortex. In A. Miyake and P. Shah (Eds.), *Models of working memory: Mechanisms of active maintenance and executive control,* pp. 102–134. New York: Cambridge University Press.

Gold, J. (1992). Prefrontal function and schizophrenic symptoms. *Neuropsychiatry, Neuropsychology, and Behavioral Neurology, 5,* 253–261.

Goldman-Rakic, P. S., and Selemon, L. D. (1997). Functional and anatomical aspects of prefrontal pathology in schizophrenia. *Schizophrenia Bulletin, 23,* 437–458.

Grace, A. A. (1991). Phasic versus tonic dopamine release and the modulation of dopamine system responsivity: A hypothesis of the etiology of schizophrenia. *Neuroscience, 41,* 1–24.

Guigon, E., Dorizzi, B., Burnod, Y., and Schultz, w. (1991). Neural correlates of learning in the prefrontal cortex of the monkey: A predictive model. *Cerebral Cortex, 5,* 135–147.

Gullapalli, V. (1990). A stochastic reinforcement learning algorithm for learning real-valued functions. *Neural Networks, 3,* 671–692.

Hernandez-Lopez, S., Bargas, J., Surmeier, D. J., Reyes, A., and Galarraga, E. (1997). D1 receptor activation enhances evoked discharge in neostriatal medium spiny neurons by modulating an L-type Ca^{2+} conductance. *Journal of Neuroscience, 17,* 3334–3342.

Hochreiter, S., and Schmidhuber, J. (1997). Long short-term memory. *Neural Computation, 9,* 443–453.

Hopfield, J. J. (1982). Neural networks and physical systems with emergent collective computational abilities. *Proceedings of the National Academy of Sciences, U.S.A., 79,* 2554–2558.

Kimberg, D. Y., D'Esposito, M., and Farah, M. J. (1997). Effects of bromocriptine on human subjects depend on working memory capacity. *Neuroreport, 8,* 381–385.

Law-Tho, D., Deuce, J. M., and Crepel, F. (1995). Dopamine favours the emergence of long-term depression versus long-term potentiation in slices of rat prefrontal cortex. *Neuroscience Letters, 188,* 125–128.

Levine, D. S., and Prueitt, P. S. (1989). Modeling some effects of frontal lobe damage: Novelty and perseveration. *Neural Networks, 2,* 103–116.

Lewis, D. A., Hayes, T. L., Lund, J. S., and Oeth, K. M. (1992). Dopamine and the neural circuitry of primate prefrontal cortex: Implications for schizophrenia research. *Neuropsychopharmacology, 6,* 127–134.

Luciana, M., Collins, P. F., and Depue, R. A. (1995). DA and 5-HT influences on spatial working memory functions of prefrontal cortex. Paper presented at the Second Annual Meeting of the Cognitive Neuroscience Society, San Francisco, April.

Luciana, M., Depue, R. A., Arbisi, P., and Leon, A. (1992). Facilitation of working memory in humans by a D2 dopamine receptor agonist. *Journal of Cognitive Neuroscience, 4,* 58–68.

Luria, A. R. (1969). Frontal lobe syndromes. In P. J. Vinken and G. W. Bruyn (Eds.), *Handbook of clinical neurology,* vol. 2, pp. 725–757. New York: Elsevier.

Malmo, H. P. (1974). On frontal lobe functions: Psychiatric patient controls. *Cortex, 10,* 231–237.

McClelland, J. L. (1993). Toward a theory of information processing in graded, random, and interactive networks. In D. E. Meyer and S. Kornblum (Eds.), *Attention and Performance XIV: Synergies in experimental psychology, artificial intelligence, and cognitive neuroscience,* pp. 655–688. Cambridge, MA: MIT Press.

Miller, E. K., Erickson, C. A., and Desimone, R. (1996). Neural mechanisms of visual working memory in prefrontal cortex of the macaque. *Journal of Neuroscience, 16,* 5154–5167.

Milner, B. (1963). Effects of different brain lesions on card sorting. *Archives of Neurology, 9,* 90–100.

Montague, P. R., Dayan, P., and Sejnowski, T. J. (1996). A framework for mesencephalic dopamine systems based on predictive Hebbian learning. *Journal of Neuroscience, 16,* 1936–1947.

Montague, P. R., and Sejnowski, T. J. (1994). The predictive brain: Temporal coincidence and temporal order in synaptic learning mechanisms. *Learning and Memory, 1,* 1–33.

Moody, S. L., Wise, S. P., di Pellegrino, G., and Zipser, D. (1998). A model that accounts for activity in primate frontal cortex during a delayed matching-to-sample task. *Journal of Neuroscience, 18,* 399–410.

Moore, R. Y., and Bloom, F. E. (1978). Central catecholamine neuron systems: Anatomy and physiology of the dopamine systems. *Annual Review of Neuroscience, 1,* 129–169.

Mozer, M. C. (1993). Neural net architectures for temporal sequence processing. In A. Weigend and N. Gershenfeld (Eds.), *Predicting the future and understanding the past,* pp. 243–264. Redwood City, CA: Addison-Wesley.

Norman, D. A., and Shallice, T. (1986). Attention to action: Willed and automatic control of behavior. In T. Shallice and D. A. Norman (Eds.), *Consciousness and self-regulation,* pp. 1–18. New York: Plenum Press.

Nuechterlein, K. H., and Dawson, M. E. (1984). Information processing and attentional functioning in the developmental course of schizophrenia disorders. *Schizophrenic Bulletin, 10,* 160–203.

O'Reilly, R. C. (1996). The LEABRA model of neural interactions and learning in the neocortex. Ph.D. diss., Carnegie Mellon University, Pittsburgh.

Owen, A. M., Roberts, A. C., Polkey, C. E., Sahakian, B. J., and Robbins, T. W. (1991). Extra-dimensional versus intra-dimensional set shifting performance following frontal lobe excisions, temporal lobe excisions or amygdalo-hippocampectomy in man. *Neuropsychologia, 29,* 993–1006.

Penit-Soria, J., Audinat, E., and Crepel, F. (1987). Excitation of rat prefrontal cortical neurons by dopamine: An in vitro electrophysiological study. *Brain Research, 425,* 263–274.

Robbins, T., Roberts, A., Owen, A., Sahakian, B., Everitt, B., Wilkinson, L., Muir, J., DeSalvia, M., and Tovee, M. (1994). Monoaminergic-dependent cognitive functions of the prefrontal cortex in monkey and man. In A.-M. Thierry, J. Glowinski, P. Goldman-Rakic, and Y. Christen (Eds.), *Motor and cognitive functions of the prefrontal cortex,* pp. 93–111. Berlin: Springer.

Rosvold, H. E., Mirsky, A. F., Sarason, I., Bransome, E. D., and Beck, L. H. (1956). A continuous performance test of brain damage. *Journal of Consulting Psychology, 20,* 343–350.

Sawaguchi, T., and Goldman-Rakic, P. S. (1991). D1 dopamine receptors in prefrontal cortex: Involvement in working memory. *Science, 251,* 947–950.

Sawaguchi, T., and Goldman-Rakic, P. (1994). The role of D1 dopamine receptors in working memory: Local injections of dopamine antagonists into the prefrontal cortex of rhesus monkeys performing an oculomotor delayed-response task. *Journal of Neurophysiology, 63,* 1401–1412.

Sawaguchi, T., Matsumura, M., and Kubota, K. (1990). Effects of dopamine antagonists on neuronal activity related to a delayed response task in monkey prefrontal cortex. *Journal of Neurophysiology, 63,* 1401–1410.

Schultz, W. (1992). Activity of dopamine neurons in the behaving primate. *Seminars in Neurosciences, 4,* 129–138.

Schultz, W., Apicella, P., and Ljungberg, T. (1993). Responses of monkey dopamine neurons to reward and conditioned stimuli during successive steps of learning a delayed response task. *Journal of Neuroscience, 13,* 900–913.

Schultz, W., Dayan, P., and Montague, P. R. (1997). A neural substrate of prediction and reward. *Science, 275,* 1593–1599.

Servan-Schreiber, D., Bruno, R., Carter, C. S., and Cohen, J. D. (1998). Dopamine and the mechanisms of cognition: 1. A neural network model predicting dopamine effects on selective attention. *Biological Psychiatry, 43,* 713–722.

Servan-Schreiber, D., Cohen, J. D., and Steingard, S. (1996). Schizophrenic deficits in the processing of context: A test of a theoretical model. *Archives of General Psychiatry, 53,* 1105–1113.

Servan-Schreiber, D., Printz, H., and Cohen, J. D. (1990). A network model of catecholamine effects: Gain, signal-to-noise ratio, and behavior. *Science, 249,* 892–895.

Sesack, S. R., Snyder, C. L., and Lewis, D. A. (1995). Axon terminals immunolabeled for dopamine or tyrosine hydroxylase synapse on GABA-immunoreactive dendrites in rat and monkey cortex. *Journal of Comparative Neurology, 363,* 1735–1780.

Shallice, T. (1982). Specific impairments of planning. *Philosophical Transactions of the Royal Society of London, 298,* 199–209.

Shallice, T. (1988). *From neuropsychology to mental structure.* Cambridge: Cambridge University Press.

Stuss, D. T., and Benson, D. F. (1986). *The frontal lobes.* New York: Raven Press.

Sutton, R. (1988). Learning to predict by the method of temporal difference. *Machine Learning, 3,* 9–44.

Sutton, R. S., and Barto, A. G. (1990). Time-derivative models of Pavlovian reinforcement. In M. Gabriel and J. Moore (Eds.), *Learning and computational neuroscience: Foundations of adaptive networks*, pp. 497–537. Cambridge, MA: MIT Press.

Wickens, J. R., Begg, A. G., and Arbuthnott, G. W. (1996). Dopamine reverses the depression of rat corticostriatal synapses which normally follows high frequency stimulation of cortex in vitro. *Neuroscience, 70,* 1–5.

Williams, M. S., and Goldman-Rakic, P. S. (1993). Characterization of the dopaminergic innervation of the primate frontal cortex using a dopamine-specific antibody. *Cerebral Cortex, 3,* 199–222.

Willner, P., and Scheel-Kruger, J. (1991). *The mesolimbic dopamine system: From motivation to action*. New York: Wiley.

Wise, R. A., and Rompre, P.-P. (1989). Brain dopamine and reward. *Annual Review of Psychology, 40,* 191–225.

Zipser, D. (1991). Recurrent network model of the neural mechanism of short-term active memory. *Neural Computation, 3,* 179–193.

Zipser, D., Kehue, B., Littlewort, G., and Fuster, J. (1993). A spiking network model of short-term active memory. *Journal of Neuroscience, 13,* 3406–3420.

32 Is There an Inhibitory Module in the Prefrontal Cortex? Working Memory and the Mechanisms Underlying Cognitive Control

Daniel Y. Kimberg and Martha J. Farah

ABSTRACT Studies of development, aging, and the cognitive deficits of patients with prefrontal cortical damage have led theorists to postulate an inhibitory function of the prefrontal cortex that is separable from other hypothesized prefrontal functions, specifically working memory. Particularly suggestive data come from two tasks: the A-not-B task, interpreted as an index of prefrontal development, and the antisaccade task, a test of volitional control over reflexive eye movements. We provide an alternative account for evidence from these tasks, one that is consistent with earlier working-memory views, but that does not require a distinct, dedicated inhibitory component. We conclude that the case for inhibition as a primary function of the prefrontal cortex has yet to be made.

Damage to prefrontal cortex often leads to behavior that can be described as "disinhibited." In social contexts, frontal patients may say whatever embarrassing thing crosses their minds, or grab for objects or people they find desirable regardless of the consequences (Stuss and Benson 1983). When performing cognitive tests, these patients tend to respond impulsively, before they have fully considered the available information, and are unduly influenced by the most salient features of the stimuli and the readiest or most habitual options for response (Kimberg, D'Esposito, and Farah 1997). In infancy, when prefrontal cortex is immature, behavior is similarly bound by perceptual salience and habit (Diamond 1990). These observations have led many theorists to conclude that a fundamental psychological function of prefrontal cortex is behavioral inhibition.

In this chapter, we offer a different view of the role of prefrontal cortex in guiding behavior in general and inhibiting inappropriate responses in particular, based on the concept of working memory. We will argue that inhibition depends on prefrontal cortex only in the same weak sense that constructional ability might be said to depend on parietal cortex. Just as parietal cortex houses certain spatial perceptual abilities that are heavily taxed in tasks requiring construction, but does not house a "constructional faculty," so prefrontal cortex houses certain basic psychological functions that are heavily taxed in tasks requiring inhibition, but does not house "inhibition" as a fundamental psychological process. We will argue that the contribution of prefrontal cortex to the performance of tasks requiring inhibition is working memory, and that the weakening of working memory leads to disinhibited behavior.

We introduce our argument with a brief review of the empirical literature linking inhibition to prefrontal cortex, and a summary of contemporary theoretical perspectives on inhibition and prefrontal function. We then describe how a limitation of working memory could affect tasks apparently requiring inhibition, focusing on the two tasks most frequently cited in support of a inhibitory function of the prefrontal cortex: the antisaccade task and the A-not-B delayed-response task.

32.1 EMPIRICAL RELATION BETWEEN INHIBITION AND PREFRONTAL CORTEX

The conclusion that prefrontal cortex plays a role in behavioral inhibition comes from a variety of clinical and laboratory tasks, as well as the more impressionistic observations of patients with prefrontal damage mentioned at the outset. Tasks that are relevant to the study of inhibition are those in which some response other than the correct one is highly prepotent and must therefore be inhibited. Failures to withhold such responses are often reported in subjects with prefrontal damage, infant subjects, and normal adults under conditions of distraction or cognitive load. Roberts and Pennington (1996) identify five tasks that are particularly relevant to the study of inhibition and prefrontal cortex:

1. *Stroop task*. Subjects are shown color names printed in colored ink (e.g., the word "red" in blue ink), and asked to name either the color or the word. Normal subjects have particular difficulty naming colors when the word name is in conflict because they must inhibit the prepotent action of reading (most individuals have read many more words than they have named colors). Perret (1974) has shown that left frontal damage renders subjects particularly susceptible to errors in this condition.

2. *Go/no-go task*. Subjects are typically given training that associates stimuli with particular responses and then instructed to respond differently to the same stimuli. As a simple example, subjects may be asked to mimic the experimenter, who will tap the table either once or twice with a finger. After training at this task, subjects may then be asked to tap twice in response to a single tap, and not at all to a double tap. Frontal patients are often impaired at a variety of tasks like this (e.g., Drewe 1975). In particular, they have difficulty inhibiting the previously correct response.

3. *Wisconsin Card-Sorting Test* (WCST). Subjects must sort a series of cards into piles according to a changing criterion. The cards vary according to the shape on the card (square, circle, cross, or star), the color of the shapes (red, blue, green, or yellow), and the number of shapes present (from one to four). Four reference cards are placed before the subject such that each value of each attribute is represented on exactly one card (i.e., only one of the cards is red). Subjects are told only that they are to place each card with one of the four reference cards. After each card is placed,

subjects are told only "right" or "wrong," with no other feedback. The experimenter begins by giving positive feedback only if subjects sort according to color. After ten correct responses, the experimenter switches to shape. This pattern continues until subjects achieve six categories, or until the experimenter runs out of cards (128 are used).

Frontal patients typically encounter difficulty with the WCST when the category changes. They will persist in sorting cards according to the previous category, sometimes even stating that they are wrong or stating the correct category while they perseverate in sorting according to the old category. This behavior has been interpreted as disinhibited in that the patients' verbal behavior indicates that they "know better" even as they make the perseverative errors (Walsh 1987).

4. *Antisaccade task.* Subjects are required to respond to a visually presented cue by looking to a location away from the cue. The cue is presented to one side of a fixation point, and subjects are required to look to the other side. Because of the reflexive nature of saccades to the cue under such conditions, the antisaccade task appears to require inhibition of the normal reflex. Guitton, Buchtel, and Douglas (1985) reported that frontal patients have trouble with this task. While normal at making prosaccades (eye movements toward the cue), these subjects are impaired at making antisaccades.

5. *A-not-B task.* Infants are shown a toy, which is then hidden in one of two wells. After a brief delay, the infants are allowed to reach for the toy. After several repetitions of this with the toy hidden in the same well each time, the toy is then hidden in the other well. Infants between 7 1/2 and 9 months typically continue to reach for the previously correct well, even though they saw the new hiding location (Diamond 1990). Monkeys with prefrontal lesions are also impaired at this task, as are children treated early and continuously for phenylketonuria (PKU), in whom brain damage is relatively restricted to dopaminergic projections to the prefrontal cortex (Diamond et al. 1997). This task has been taken to be a test of both behavioral inhibition and memory, and indeed much of Diamond's theory (Diamond et al. 1997) of inhibition and prefrontal cortex is based on research with this task.

32.2 THEORETICAL-PERSPECTIVES ON INHIBITION

The behaviors reviewed above can be described as failures of inhibition in the descriptive, theoretically neutral sense that prepotent but incorrect responses tend to occur and have therefore plainly not been inhibited. Many theorists believe, however, that a failure of inhibition is more than just descriptive of the behavior of frontal patients. They suggest that the failure to inhibit is caused by damage to an inhibition mechanism, that is, a component of the cognitive architecture dedicated to response inhibition.

Diamond (e.g., 1990; Diamond, Cruttenden, and Neiderman 1994) has proposed that prefrontal cortex houses both working memory and inhibition. She argues that memory deficit alone cannot explain the behavior of subjects with immature or damaged prefrontal cortices, saying "dorsolateral prefrontal cortex is required whenever any information at all must be remembered within a trial, as long as the task also demands inhibition of a prepotent response as well.... The pattern of error ... cannot be accounted for by forgetting alone.... [The error pattern] follows what would be predicted on the basis of inhibiting the predominant response" (Diamond 1990, 293). Similarly, Roberts, Hager, and Heron (1994, p. 374) state that there appear to be "two principal prefrontal functions: ... working memory, and the inhibition of prepotent but inappropriate responses.... Little is known about whether and how such processes interact in the generation of behavior."

The fact that prefrontal cortex is essential for behavioral inhibition in the descriptive sense does not, of course, imply that response inhibition as a mechanism is a basic element of the cognitive architecture, any more than the dependence of constructional ability on parietal cortex implies that constructional ability is such an element. In computer simulations (Kimberg and Farah 1993), we have shown how disinhibited behavior in the WCST and the Stroop task follow from damage to working memory in a system that has no separate inhibition mechanism. Cohen and O'Reilly (1996, p. 272) have made a similar point, writing that "memory and inhibition reflect the operation of the context processing mechanism under different task conditions." In both cases, the context mechanism is still performing the same basic function: supporting representations necessary to perform the task.

In section 32.3, we will summarize previous simulation results on inhibition and working memory in the WCST and the Stroop task, as well as two other tasks from the literature on prefrontal dysfunction. We will then turn to the two tasks that have been the primary focus of researchers investigating inhibition and its relation to working memory and prefrontal cortex: the antisaccade task and the A-not-B task. We will show that disinhibited behavior in these tasks as well can be explained by an impairment of working memory without a separate inhibitory mechanism.

32.3 A WORKING-MEMORY ACCOUNT OF BEHAVIOR AFTER PREFRONTAL DAMAGE

Our previous computer simulation used the ACT-R production system architecture to implement a simple response selection model of behavior in four different tasks: WCST, Stroop, motor sequencing, and context memory (Kimberg and Farah 1993). We chose these four tasks because they are sensitive to prefrontal function, and have traditionally been viewed as measures of executive function. Our goal was to demonstrate how a disparate set of behaviors that seem, on the surface, to result from the loss of a "central executive" can be understood more simply in terms of the weakened influence of working memory.

The essence of the model can be summarized in two main points. First, response selection is determined by the levels of activation of production rules representing the competing responses, with the most activated response being selected. Second, four distinct sources of activation jointly determine the activation level of each candidate response:

1. *Working-memory activation.* We follow Anderson 1993 in viewing working memory as the subset of long-term declarative memory that is currently activated, rather than as a separate buffer into which certain memory contents are transferred. Anderson's framework allows for degrees of activation of working-memory elements, rather than requiring an element to be either in or out of working memory. Working-memory activation in our model consists of activation added directly to an element of working memory, as when the representation of a lever is activated by presentation of a lever to the subject, and activation that spreads among associated elements, as when the representation of the lever activates the representation of the associated gesture of pulling. All other things being equal, the response to pull a lever will be most strongly activated when there is a lever present, subjects are thinking about pulling, and there is a strong connection between the working-memory representations of levers and pulling so that these two active elements further activate each other.

2. *Priming activation.* Here we have a fast-decaying form of activation whereby a recently executed response is made temporarily more available than usual. For example, if one has just pushed a button, the button-pushing response is temporarily primed.

3. *Baseline activation.* An enduring level of activation associated with each response, baseline activation reflects the differences in availability of different responses that result from long-term differences in their frequency of use. For example, on the assumption that we have pushed more buttons in our lives than we have pulled levers, button pushing would be a more available response. This is reflected in the model by a higher baseline activation level for the representation of button pushing than for that of level pulling.

4. *Noise activation.* A random source of activation contributing to response selection, noise activation reflects the imperfect information processing of the cognitive system.

The original simulations included four tasks, chosen to be different from one another and yet similar in their sensitivity to prefrontal damage. In a simple motor sequencing task, the model was required to make the correct manual response to each of a series of devices (e.g., lever, button, knob). This task is similar to Kimura's manual sequence box paradigm (1977). Within the theoretical framework of the model, strong learned associations between each device and its corresponding response (e.g., between levers and pulling) enable the correct response to be made when confronted with a given device. When these associations are weakened by damage to working memory, the difference in spreading activation to response representations in working memory becomes smaller, and noise activation can more often overwhelm these differences, resulting in sequencing errors. Priming activation biases these errors toward perseverations.

The context memory task was based on tasks used to demonstrate impaired context memory in frontal patients (e.g., Janowsky, Shimamura, and Squire 1989). The model is endowed with a memory of multiple items in different contexts, represented by associations between items and context features; presented with an item, it is required to indicate the item's original context. When associations are weakened, the discriminability of the different contexts is reduced, causing increased errors in context memory judgments.

In the WCST simulation, information about which sorting category is the correct one is maintained through a connection between eligible categories and their corresponding attributes. Although the system normally prefers eligible categories over ineligible ones, when these associations are weakened, this preference becomes much smaller. The presence of noise makes it possible to select an ineligible category. Priming makes it more likely that recently used sorting strategies will be preferred over others.

In the Stroop task, the model is required to "name" the colors in which word stimuli are printed. The words themselves are color names that conflict with the ink colors, a conflict that ordinarily causes interference. While word naming is a more routine task than color naming, and therefore has a higher baseline activation in the model, activation of the color-naming task representation and working-memory associations between the task representation and the stimulus normally allows the system to bias itself toward naming colors. Weakening of these associations increases the relative contribution of the baseline activation in determining the response, resulting in word-reading intrusions.

Although we did not originally undertake these simulations for this purpose, the results are nevertheless informative with respect to inhibition. In the WCST and the Stroop simulations, disinhibited behavior results not from disabling a mechanism whose normal function is specifically to inhibit prepotent responses, but rather from a weakening of working-memory associations; thus our model contains no dedicated inhibitory mechanism.

32.4 EXTENDING THE MODEL TO ACCOUNT FOR OTHER INHIBITORY FAILURES

Although the WCST and the Stroop task have been considered paradigmatic tests of the ability to inhibit a prepotent response, two other tasks have figured more prominently in recent research on behavioral inhibition. They are the antisaccade task and the A-not-B task, described earlier. To support our claim that disinhibited behavior does not require a separate inhibitory mechanism, we need to show how a working-memory account can explain disinhibited behavior in these tasks. Although we have verified our account with computer simulations, in the interest of brevity and focus, we will not report their details here. We instead provide a conceptual explanation of why the weakening of working-memory associations will manifest itself as disinhibition in these two tasks. (Interested readers may contact the author Kimberg for a full report of the computational work.)

Antisaccade Task

In this task, two potential responses compete for activation: looking toward the stimulus, and looking away from it to the other stimulus. Looking toward the stimulus is of course the prepotent response, which is reflected in its higher baseline activation. The working-memory representation of the antisaccade instructions provides activation to the response of looking away, to boost it above its normally very low baseline activation.

Any weakening of working memory has the effect of reducing the impact of instructions by reducing the influence of working-memory activation on response selection. In the prosaccade task, this makes little difference because the baseline activation of looking toward is so high. This strong bias makes the discriminability, when looking toward is correct, very high, regardless of whether the instructions are properly represented in working memory. In the antisaccade task, however, weakening the contribution of working memory will work against performing the task correctly because it is only the contribution from working memory that allows the system to override the strong bias in favor of looking toward stimuli. Not surprisingly, this result is obtained in simulations using a wide variety of parameter settings for the baseline activations of the two responses, the contribution of working memory, the level of noise, and the degree of damage.

Roberts, Hager, and Heron (1994) note that the proportion of reflexive saccade errors increases if the prosaccade task is performed first. This result is consistent with the response-priming assumption: doing the prosaccade task should increase the strength of the looking-toward response, at least briefly.

A-Not-B Task

The two competing responses in this task are reaching toward the two wells, A (where the object is initially hidden) and B (where it is later hidden). Working memory includes representations of wells A and B and of the object. The selection of a reaching response depends on two main features of the model. First, the two responses receive activation from their corresponding wells. In other words, activation of the working-memory representation of well A favors the selection of the response of reaching toward well A. Second, hiding the object in a well creates an association between those two representations in working memory. It follows that when the stimulus is hidden in well A, because of spreading activation from the object representation, the representation of well A receives more activation than that of well B, and is still more active when subjects are allowed to reach. This will normally lead subjects to reach to the correct location.

When associations within working memory are weakened, the correct well's representation receives less activation from the hiding of the stimulus, and other sources of activation will weigh relatively more heavily in determining action. Although initially the baseline activations of the two reaching responses are similar, repeated use of one reach raises its baseline activation. With weak working-memory associations, the response specified by association with the current location of the object is more weakly favored, and differences in baseline activation are more likely to cause the previous response to be the most active.

The model also accommodates some finer-grained features of infant performance with the A-not-B task. For example, the classic A-not-B error is more likely after a delay between the hiding of the object and the reach. This follows naturally from our account on the reasonable assumption that working-memory activation decays over time.

The model can also accommodate the finding that infants sometimes make the A-not-B error even when the well covers are transparent, eliminating the apparent memory load (e.g., Diamond 1985). Although the transparent well covers will certainly aid working memory for the hiding location, they do not guarantee that the working-memory representation will remain at the same strength for the entire delay period. Indeed, if the working-memory representation does not decay during the delay with transparent covers, then inhibition accounts are also unable to explain the error because it is unclear why inhibitory processes would themselves be stronger in the shorter delay conditions.

32.5 GENERAL DISCUSSION

We have argued that the disinhibited behavior of patients with prefrontal damage, and of infants with immature prefrontal cortices, does not imply the existence of a specific inhibitory mechanism in prefrontal cortex. In support of this, we accounted for performance in two tasks frequently used to elicit disinhibited behavior, the antisaccade task and the A-not-B task, using a simple computational model that lacks specific inhibitory mechanisms. Damage to working memory produces the patterns of behavior that had previously been interpreted in terms of damage to inhibitory mechanisms.

This type of account, which we have previously used to explain disinhibitory patterns in the WCST and the Stroop task, works because inhibition and working memory play the same computational role in these paradigms. In fact, because the proposed inhibitory role of the prefrontal cortex is directed toward prepotent responses, these two factors will be difficult to unconfound—any information that supersedes a prepotent response would be expected to be held in working memory.

In the remainder of this chapter, we will address some of the broader issues to which these results relate.

From Behavioral Deficit to Cognitive Architecture

The present findings can viewed as an instance of a more general principle in neuropsychology concerning the relation between the behavior of brain-damaged patients and the architecture of the normal cognitive system. Although an apparently selective impairment in one cognitive ability (here, inhibitory control) might seem to imply the loss of a component of the normal cognitive architecture dedicated to this ability (here, the loss of an inhibitory mechanism), such direct inferences are not necessarily correct. They are based on the implicit assumption that the components of the cognitive architecture operate autonomously, with little or no interaction. The components of such a system are, in Fodor's terms (1983), "informationally encapsulated"—and, by that criterion, also "modular."

It is true that for a modular system, behavior after damage can be understood in terms of the normal operation of the undamaged components, with the contribution of the damaged component attenuated or eliminated. On the other hand, to the extent that a system is not modular in this sense, but instead involves some degree of interaction among its components, the chain of inference from circumscribed behavioral deficit to dedicated cognitive module will be vulnerable to error. This is because the behavior of such systems after damage results not only from the attenuation or loss of the damaged component, but also from the altered functioning of the remaining, undamaged components.

Farah (1994) summarized a series of neuropsychological inferences that could be reinterpreted in a more parsimonious way within the framework of an interactive cognitive architecture. Although, in each of these cases, the reinterpretation was supported by a distributed connectionist computer simulation, where interactivity is assumed to derive from the way information processing is implemented in the brain, one need not subscribe to the theoretical framework of connectionism to appreciate the role of interactivity in cognition and its implications for neuropsychological inference. Our model of prefrontal function (Kimberg and Farah 1993) was implemented in a production system architecture, and the simulations of the present project were implemented in a simplified version of the same. Yet in these cases, too, the interaction among different sources of information guiding a response and changes in the weighting of these sources of information after working-memory damage played a crucial role in explaining behavior after damage.

In Kimberg and Farah 1993, we argued that central executive models postulated a level of complexity beyond the simple components needed to perform each task, a level unnecessary to explain the pattern of deficits observed in frontal patients. In this chapter, we have shown that apparent inhibitory functions can also emerge from a simple response competition mechanism where activation from working memory contributes to the selection of responses. When the working-memory activation is reduced, the interaction between these components results in performance that can be characterized as a "loss of inhibition."

Physiological Evidence of Inhibition

Our preference for the view that prefrontal cortex subserves working memory, and does not implement inhibition as a distinct functional element of the cognitive architecture, is based on parsimony: an inhibitory mechanism is not needed to account for "disinhibited" behavior. Can we find some grounds other than parsimony to select between the two accounts? Prospects for a behavioral test seem dim because whenever behavioral inhibition is called for, it seems likely that the information needed to guide an appropriate response will reside in working memory. What about physiological evidence?

Although some individual neurons inhibit other neurons, and this might appear to be prima facie evidence for the kind of inhibitory process we argue against, the claim we have criticized has to do with the psychological function of a neural system, not the microstructure of its physiological implementation. Consider an unrelated example, which may make the point more clearly. Suppose that a content-addressable memory module is implemented by linking individual units so as to represent patterns of covariation between features in the memory representation, and that many of these links are inhibitory. Even though the function of individual units within this module would be to inhibit other units, one would not characterize the function of the module as a whole as "inhibitory."

A more relevant physiological observation is that some whole systems of neurons serve to inhibit the function of other systems. Indeed, systems for eye movement control involved in the antisaccade task operate by competing excitatory and inhibitory influences acting on the superior colliculus. Does this mean (1) that performance in the antisaccade task requires inhibition after all, and (2) that inhibitory control is what has been damaged in frontal patients who fail this task?

Neurophysiologists have answered the first question with a clear yes. The inhibition of reflexive saccades is a function of the substantia nigra, a subcortical nucleus to which prefrontal cortex projects. Most critical to the issues discussed in this chapter, however, the answer to the second question is no. The prefrontal cortex exerts control over the substantia nigra, but the control is not in itself inhibitory. The ultimate ability of the prefrontal cortex to inhibit reflexive saccades is indirect, mediated by the substantia nigra. That prefrontal cortex provides information to other brain centers whose function is to inhibit particular responses does not mean that prefrontal function should be characterized as "inhibitory." Indeed, the *indirect* nature of this control suggests a more representational role for the prefrontal cortex in the antisaccade task, consistent with the working-memory account.

Conclusions

The idea that prefrontal cortex houses an inhibitory mechanism, above and beyond its working-memory functions, seems to follow from the disinhibited behavior of patients after prefrontal damage. We have questioned this conclusion, pointing out that damage to working memory alone can account for the disinhibited behavior that follows prefrontal damage. This does not imply that other parts of the brain do not contribute to behavioral control by inhibition. Although our claim is more circumscribed, concerning only the function of prefrontal cortex, it nevertheless contradicts a number of recent and influential hypotheses about prefrontal cortex, offering in their place the simpler hypothesis that prefrontal cortex implements working memory, and a framework for understanding the crucial role working memory plays in performing tasks that tax behavioral inhibition.

NOTE

This research was supported by National Institutes of Health grants RO1NS34030, RO1AG14082, and KO2AG00756 to Martha J. Farah. We would like to thank Adele Diamond, Jon Driver, Stephen Monsell, George Houghton, Gordon Logan, and two anonymous reviewers for their helpful comments on an earleir draft of this chapter.

REFERENCES

Anderson, J. R. (1993). *Rules of the mind*. Hillsdale, NJ: Erlbaum.

Cohen, J. D., and O'Reilly, R. C. (1996). A preliminary theory of the interactions between prefrontal cortex and hippocampus that contribute to planning and prospective memory. In M. Brandimonte, G. O. Epstein, and M. A. McDaniel (Eds.), *Prospective memory: Theory and applications*, pp. 267–295. Hillsdale, NJ: Erlbaum.

Diamond, A. (1985). Development of the ability to use recall to guide action, as indicated by infants' performance on AB. *Child Development, 56*, 868–883.

Diamond, A. (1990). The development and neural bases of memory functions as indexed by the AB and delayed response tasks in human infants and infant monkeys. In A. Diamond (Ed.), *Annals of the New York Academy of Sciences, v. 608, The development and neural bases of higher cognitive functions*, pp. 267–317. New York: New York Academy of Sciences.

Diamond, A., Cruttenden, L., and Neiderman, D. (1994). AB with multiple wells: 1. Why are multiple wells sometimes easier than two wells? 2. Memory or memory + inhibition? *Developmental Psychology, 30*, 192–205.

Diamond, A., and Goldman-Rakic, P. S. (1989). Comparison of human infants and rhesus monkeys on Piaget's AB task: Evidence for dependence on dorsolateral prefrontal cortex. *Experimental Brain Research, 74*, 24–40.

Diamond, A., Prevor, M. B., Callender, G., and Druin, D. P. (1997). Prefrontal cortex cognitive deficits in children treated early and continuously for PKU. *Monographs of the Society for Research in Child Development, 62*, i–v.

Drewe, E. A. (1975). Go/no-go learning after frontal lobe lesions in humans. *Cortex, 11*, 8–16.

Farah, M. J. (1994). Neuropsychological inference with an interactive brain: A critique of the "locality" assumption. *Behavioral and Brain Sciences, 17*, 43–104.

Fodor, J. A. (1983). *The modularity of mind*. Boston: MIT Press.

Guitton, D., Buchtel, H. A., and Douglas, R. M. (1985). Frontal lobe lesions in man cause difficulties in suppressing reflexive glances and in generating goal-directed saccades. *Experimental Brain Research, 58*, 455–472.

Heron, C. M., and Roberts, R. J. (1996). The role of processing speed in the antisaccade task: Implications for frontal lobe function. Paper presented at the Third Annual Meeting of the Cognitive Neuroscience Society, March.

Janowsky, J. S., Shimamura, A. P., and Squire, L. R. (1989). Source memory impairment in patients with frontal lobe lesions. *Neuropsychologia, 27*, 1043–1056.

Kimberg, D. Y., D'Esposito, M., and Farah, M. J. (1997). Effects of bromocriptine on human subjects depend on working memory capacity. *NeuroReport, 8*, 3581–3585.

Kimberg, D. Y., and Farah, M. J. (1993). A unified account of cognitive impairments following frontal lobe damage: The role of working memory in complex, organized behavior. *Journal of Experimental Psychology: General, 122*, 411–428.

Kimura, D. (1977). Acquisition of a motor skill after left-hemisphere damage. *Brain, 100*, 527–542.

Lhermitte, F. (1983). "Utilisation behaviour" and its relation to lesions of the frontal lobes. *Brain, 106*, 237–255.

Perret, E. (1974). The left frontal lobe of man and the suppression of habitual responses in verbal categorical behavior. *Neuropsychologia, 12*, 323–330.

Roberts, R., J., and Pennington, B. F. (1996). An interactive framework for examining prefrontal cognitive processes. *Developmental Neuropsychology, 12,* 105–126.

Roberts, R. J., Hager, L. D., and Heron, C. (1994). Prefrontal cognitive processes: Working memory and inhibition in the antisaccade task. *Journal of Experimental Psychology: General, 123,* 374–393.

Stuss, D. T., and Benson, D. F. (1983). Frontal lobe lesions and behavior. In A. Kertesz (Ed.), *Localization in neuropsychology,* pp. 429–454. New York: Academic Press.

Walsh, K. W. (1987). *Neuropsychology: a clinical approach.* New York: Churchill Livingstone.

Author Index

Subject Index